Tunnel Design Methods

Tunnel Design Methods covers analytical, numerical, and empirical methods for the design of tunnels in soil and in rock. The material is intended for design engineers looking for detailed methods, for graduate students who are interested in tunneling, and for researchers working on various aspects of ground–support interaction under static and seismic loading.

The book is divided into seven chapters, covering fundamental concepts on ground and support behavior and on ground-excavation-support interaction and provides detailed information on analytical and numerical methods used for the design of tunnels, with applications, and on the latest developments on empirical methods. The principles and formulations included are used, throughout the book, to provide insight into the response of tunnels under both simple and complex loading conditions, thus providing the reader with fundamental understanding of tunnel behavior.

Both authors have experience in tunneling and have worked extensively in practice, designing tunnels both in the United States and abroad, and in research.

Tunnel Design Methods

Antonio Bobet
Herbert H. Einstein

CRC Press is an imprint of the
Taylor & Francis Group, an **informa** business

Front cover: Design by Violeta M. Ivanova. Photo by Herbert H. Einstein.

First edition published 2024
by CRC Press
6000 Broken Sound Parkway NW, Suite 300, Boca Raton, FL 33487-2742

and by CRC Press
4 Park Square, Milton Park, Abingdon, Oxon, OX14 4RN

CRC Press is an imprint of Taylor & Francis Group, LLC

Library of Congress Cataloging-in-Publication Data
Names: Bobet, Antonio, author. | Einstein, Herbert H., author.
Title: Tunnel design methods / Antonio Bobet and Herbert H. Einstein.
Description: First edition. | Boca Raton : CRC Press, 2023. | Includes
bibliographical references and index.
Identifiers: LCCN 2022045917 | ISBN 9781032358444 (hbk) | ISBN
9781032358451 (pbk) | ISBN 9781003328940 (ebk)
Subjects: LCSH: Tunnels--Design and construction.
Classification: LCC TA805 .B63 2023 | DDC 624.1/93--dc23/eng/20221202
LC record available at https://lccn.loc.gov/2022045917

ISBN: 978-1-032-35844-4 (hbk)
ISBN: 978-1-032-35845-1 (pbk)
ISBN: 978-1-003-32894-0 (ebk)

DOI: 10.1201/9781003328940

Typeset in Sabon
by SPi Technologies India Pvt Ltd (Straive)

Contents

Preface

The idea for the book started as a learning aid for the courses that the two authors teach at their universities. The scope then grew from a textbook, to complement class material, to an introductory volume on tunneling. The goal of the book is to present fundamental concepts of tunneling with particular emphasis on ground–structure interaction and the role of the construction process. Clearly, it is not possible to cover all aspects of tunneling in a single volume, and so we decided, perhaps somewhat arbitrarily, to cover fundamental aspects and design tools in the first five chapters and some special aspects in Chapters 6 and 7. The fundamentals of ground–structure interaction are introduced in Chapter 1, followed by Chapters 2 and 3 in which, respectively, aspects that are specific to rock and soil are covered. Analytical and numerical tools are then described in Chapters 4 and 5, all with example applications. The text also covers aspects of tunneling that are of particular interest to the authors. These are found in Chapter 6, with overviews of seismic design and response of tunnels, pressure tunnels, rockburst effects, and tunnels in swelling ground. Given that the book emphasizes design, we are, in Chapter 7, providing details on the design of structural components used for tunnel support. The chapter goes into some depth for rockbolts, because they are widely used in tunneling, and also because of the particular interest of the authors. We would like to emphasize that we extensively referenced the text. In this, we tried to be as up-to-date as possible, but we encourage our readers to also look for more recent material.

The authors hope that the book will be useful to graduate students interested in tunneling, to educators as a learning aid, and to engineers and other tunneling professionals as a reference. In this sense we made a particular effort to create many figures to facilitate the understanding of the material presented and to help with explaining concepts. Finally, we would like to say that we are grateful to colleagues who allowed us to use material from their publications and to Taylor and Francis for all the help received through the publication process. Most importantly we would like to thank our families, students and colleagues for their support, which ranged from strong encouragement to help with many details.

About the authors

Antonio Bobet is the Edgar B. and Hedwig M. Olson Professor in Civil Engineering at Purdue University, West Lafayette, Indiana, USA. He worked in industry before joining Purdue University, as senior geotechnical engineer for Euroestudios, S.A. and later as construction manager for Ferrovial, S.A. in Spain. His areas of expertise are rock mechanics and tunneling, and he has been involved in the seismic design of several tunnels around the world. His research has focused on the interaction, during tunnel construction, between the surrounding ground, groundwater, and tunnel support, and on the seismic performance and design of underground structures. He has investigated the failures of the Daikai station during the 1995 Kobe earthquake and the Longxi tunnel during the 2008 Wenchuan earthquake, among others.

Herbert H. Einstein is a professor of Civil and Environmental Engineering at the Massachusetts Institute of Technology in Cambridge, Massachusetts, USA. He has been working in research and practice in tunneling since 1966. His research involves the mechanical behavior of rock and soil interacting with the tunnel, ranging from experiments in the laboratory to field studies to analytical models. The latter includes, among others, the widely used Schwartz–Einstein method. In addition, he has been developing the Decision Aids for Tunnelling with which tunnel cost, time, and resources considering the effect of uncertainties can be determined. This tool has been widely applied, including on most new transalpine railroad tunnels. Newer activities include the use of AI and ML in tunneling. He was also involved in the design of several tunnels in the United States and Switzerland.

Chapter 1

Principles of ground–structure interaction

1.1 INTRODUCTION

Ground–structure interaction characterizes most geotechnical problems but is particularly important in tunneling, where the ground contributes both to the load on the structure and is loaded by the structure. Additional issues arise from the fact that not only loading but also unloading occurs. Depending on the geometry and depth of the tunnel, it may influence the ground surface and other structures, including neighboring tunnels (see Figure 1.1 for schematic examples).

Further complexities are associated with the fact that the tunnel construction is a three-dimensional process. The conditions near the tunnel face are quite different from those further back in the tunnel. Also, as this is analogous to other geotechnical problems, there is time dependence, usually, but not always, associated with the effect of groundwater when the tunnel construction or operation changes the groundwater regime, e.g. the tunnel acts as a drainage conduit (Figure 1.2).

The construction process itself has a major effect on ground–structure interaction. Creating an opening in the ground often requires applying an initial support, which is later strengthened. Such a support may be installed at different locations related to the excavation and also at different times. Also, the support can be purely internal, i.e. at the perimeter of the tunnel, or it can be in the ground by strengthening it or combinations thereof (Figure 1.3).

Analysis, be this with closed form methods (see Chapter 4) or with numerical methods (see Chapter 5), has to consider ground–structure interaction. This is complicated by the above-mentioned fact that there is not a simple final design but that design and construction occur in integrated and incremental stages, particularly when using the observational method (see Chapters 2 and 3).

This chapter is meant to present the basic ground–structure interaction principles considering as many of the influencing factors as possible. As such, it serves as a basis for the following chapters on empirical, analytical (closed form), and numerical methods. The chapter is organized into four sections. In the first one, numerical examples are used to describe the interplay between ground, construction, and support and to identify the most important parameters. The second section defines the characteristic curves method as a conceptual tool to qualitatively, and to some extent quantitatively, evaluate ground–structure interaction. The third section brings together design and construction and introduces the observational method. The last section describes the arching phenomenon, as it applies to tunneling.

DOI: 10.1201/9781003328940-1

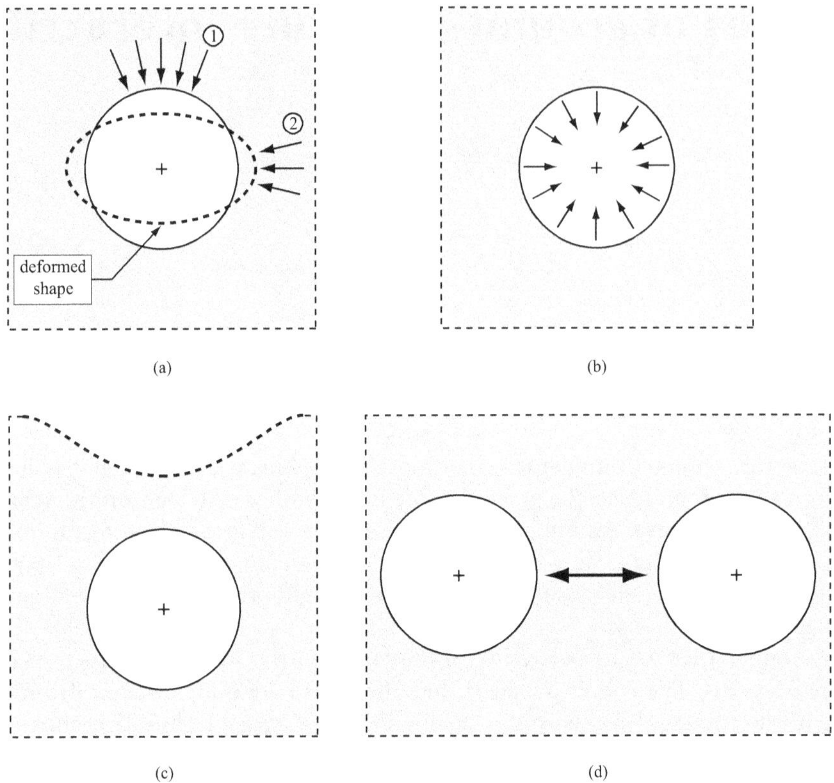

Figure 1.1 Various effects of tunnel on ground and vice versa. (a) Load of ground on tunnel (1); load of tunnel on ground (2). (b) Unloading by excavation. (c) Surface settlement caused by tunnel. (d) Tunnels affecting each other.

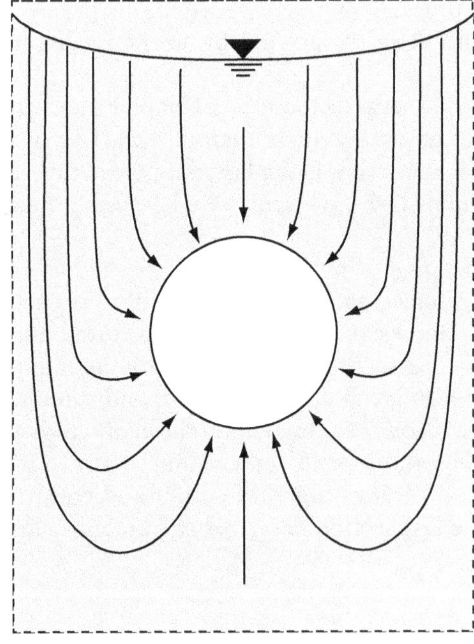

Figure 1.2 Groundwater flow into tunnel.

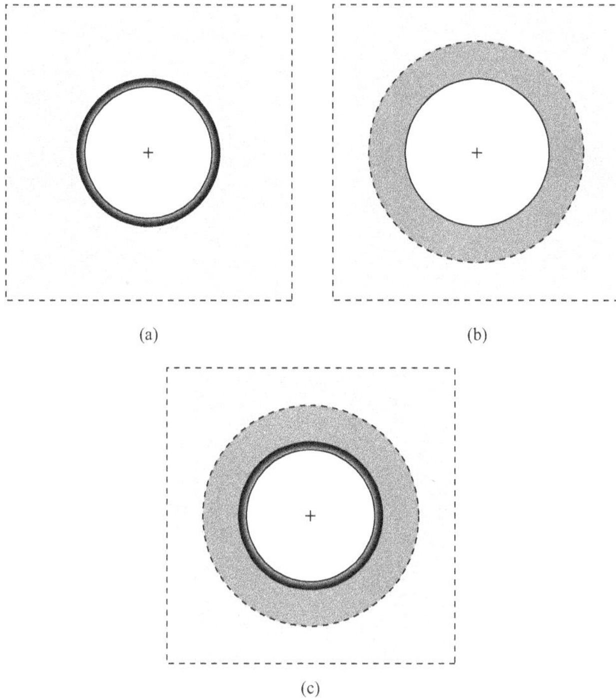

Figure 1.3 Tunnel support. (a) Internal. (b) External (e.g. Grouting of Ground). (c) Combination of Internal and External.

1.2 GROUND–STRUCTURE INTERACTION

The construction of a tunnel changes the stress field of the ground surrounding the structure. The changes affect not only the material around the opening but also the soil or rock ahead of the excavation. This is illustrated in the following by observing the behavior of an unsupported tunnel and then of a tunnel with support.

Figure 1.4 shows the construction process of an unsupported tunnel. In this case, the tunnel is continuously excavated without providing any support to the ground. Figure 1.4a shows an arbitrary starting point in the excavation process and Figure 1.4b the next excavation round. The figures also illustrate the convention used in the book to denote locations with respect to the face. Thus, "ahead of face" refers to points in the unexcavated ground, while "behind the face" is used for points in the excavated tunnel.

A simple numerical study has been done to analyze the stresses in the ground as a result of the excavation of a tunnel. A full-face excavation is assumed, but the fundamental behavior that will be discussed applies to other types of excavation processes. It is assumed that a deep horizontal tunnel with radius r_o is excavated in an elastic isotropic medium with far-field uniform stresses. That is, $\sigma_{vertical} = \sigma_{horizontal} = \sigma_{axial} = 1$ MPa (see Figure 1.4). The effects of different principal stresses, i.e. the effect of $K_o \neq 1$, are discussed in Chapter 4 (K_o is the coefficient of earth pressure at rest and is defined as the ratio between the horizontal and vertical effective stresses, i.e. $K_o = \sigma'_h/\sigma'_v$). The example is not intended to duplicate actual conditions, but, because of its simplicity, is well suited to illustrate fundamental principles. In the analysis, the following properties and geometry are assumed: Young's modulus E = 500MPa, $\nu = 0.25$, and $r_o = 2$ m. The Finite Element Method (FEM) code ABAQUS is used for the analysis. A discussion of models and modeling conditions is provided in Chapters 4 and 5. The results from the analysis are shown in Figure 1.5 and illustrate the effects of the excavation on the

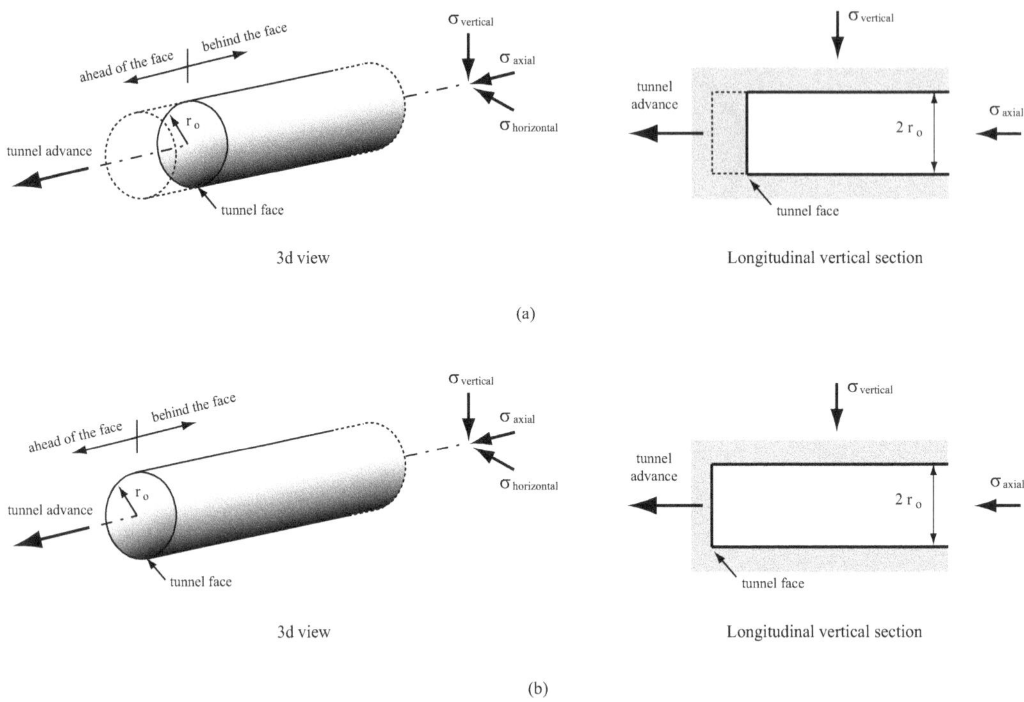

Figure 1.4 Excavation of unsupported tunnel. (a) Initial position. (b) Excavation.

ground surrounding the tunnel (Figure 1.5a for the axial stresses, 1.5b for the radial or vertical stresses, 1.5c for the tangential or horizontal stresses, and 1.5d for displacements). What is shown in the figure is stress contours on any section that contains the axis of the tunnel (for discussion purposes, it can be thought as a vertical section). The plots show that, as it should, the unsupported excavation carries zero normal and shear stresses at the face of the tunnel and on the perimeter of the tunnel. This results, in terms of the axial stresses (Figure 1.5a), in unloading ahead of the face (in the example over a length of one tunnel diameter) and a slight loading behind the face. The radial (vertical) stresses are zero at the perimeter of the tunnel, behind the face, and increase as the distance from the tunnel increases (Figure 1.5b) until they reach far-field conditions (1 MPa) at about 3–4 times the radius of the tunnel, measured from the axis of the tunnel. They also increase ahead of the face, up to a distance of about one tunnel radius. In contrast, the tangential (horizontal) stresses are the largest at the perimeter of the excavation behind the face and decrease as the distance from the tunnel decreases. Similar to the radial/vertical stresses, the far-field conditions are recovered at about 3–4 radii from the axis of the tunnel. Ahead of the tunnel face, the tangential/horizontal stresses are not much affected and show a slight unloading. It is interesting to note that beyond a distance of about 4–5 radii behind the face, the stresses do not change much. This is where plane-strain conditions apply and the solution is independent of the presence of the face of the tunnel.

The magnitude and direction of the ground displacements are given by the length and orientation of the arrows in Figure 1.5d. The figure shows that the displacements are the largest close to the tunnel. At the face, ground movements are mostly parallel to the tunnel, while behind the face they are mostly radial (vertical). The figure also shows the area of influence of the excavation. At about 2–3 tunnel radii ahead of the face, the displacements are small. Regarding the radial displacements, they are very small or negligible at about 5–6 times the

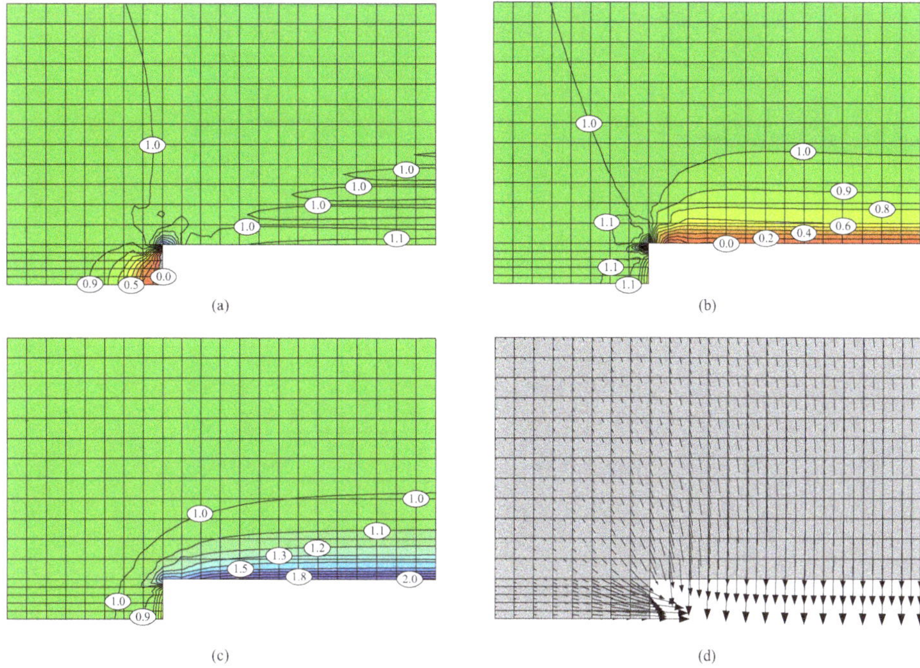

Figure 1.5 Stresses and displacements for an unsupported tunnel. Axisymmetric analysis with r_o = 2 m, σ_v = 1.0 MPa, K_o = 1.0. (a) σ_{axial}. (b) $\sigma_{radial/vertical}$. (c) $\sigma_{tangential/horizontal}$. (d) Displacements.

radius, from the axis. There is a clear rotation of the displacements, from mostly axial ahead of the face, to radial behind the face.

In essence, the excavation can be thought of as a void inside the ground that cannot transfer any stresses. Figure 1.6 is a conceptual illustration of the stress redistribution around the tunnel. The figure plots the stress trajectories from the far field for the axial stresses in Figure 1.6a, and for the radial/vertical stresses in Figures 1.6b and c; the same concept applies for the tangential/horizontal stresses. The axial stresses are deflected by the face of the tunnel toward the perimeter of the tunnel, as shown in Figure 1.6a. This requires that the stress lines separate close to the face, which leads to a reduction of stresses. This is consistent with the axial reduction of stresses shown in Figure 1.5a. The trajectories of the axial stresses are not much disturbed behind the face of the tunnel, except near the intersection between the tunnel face and the tunnel perimeter where some stress concentration occurs due to the sharp corner at this location. Figures 1.6b and c show the trajectories of the radial/vertical stresses deflected by the tunnel along a vertical longitudinal cross section and along a cross section perpendicular to the tunnel axis, respectively. Because the stresses cannot be transmitted across the tunnel, the stress trajectories have to deflect toward the unexcavated portions of the ground. As shown in Figure 1.6b, the stress transfer results in additional loading right in front of the face; see also Figure 1.5b. Perpendicular to the tunnel axis, Figure 1.6c, the stress trajectories separate at the crown of the tunnel (springline for horizontal stresses) and close up at the springline (crown for horizontal stresses). This corresponds to unloading at the crown and loading at the springline, as depicted in Figure 1.5b. For the tangential/horizontal stresses, this mechanism produces loading at the crown.

Figure 1.7 is a plot of the radial/vertical displacement along the crown of the tunnel. The displacements are normalized by the factor E/σ_v, and the axial distance from the face by the tunnel radius, r_o. Negative displacements denote inward displacements; also, the axial

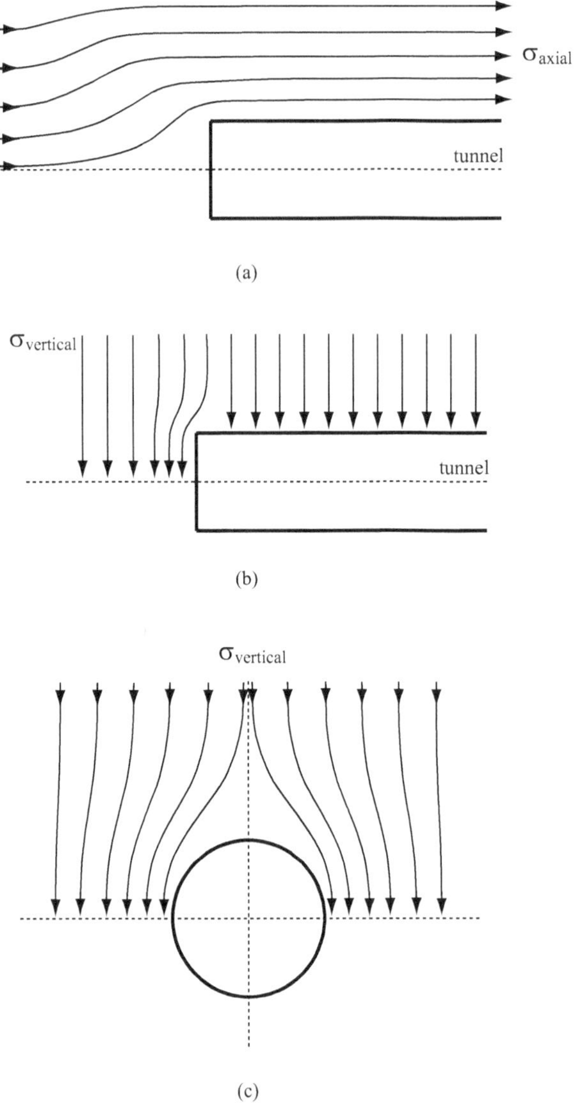

Figure 1.6 Stress redistribution. (a) Axial stress. Longitudinal vertical section. (b) Vertical stress. Longitudinal vertical section. (c) Vertical stress. Vertical cross section.

distance from the face is taken as negative ahead of the face (inside the ground) and positive behind the face (inside the tunnel). Ahead of the face of the tunnel (see Figure 1.4), beyond about 4 radii from the face, the settlements are small. The deformations increase significantly very close to the face and reach a maximum magnitude at about four to five radii behind the face; at this point plane strain conditions can be assumed.

Analogous effects are found in a tunnel where a liner is installed to provide support. The construction sequence is idealized in Figure 1.8. Construction is a cyclic process where, starting from an initial position, Figure 1.8a, the liner is placed, Figure 1.8b, and a new excavation round is performed, Figure 1.8c. Thus, the construction of the tunnel progresses in discrete steps of magnitude d (Figure 1.8), which is the length of the excavation and support rounds. As with the unsupported tunnel case, the stresses induced

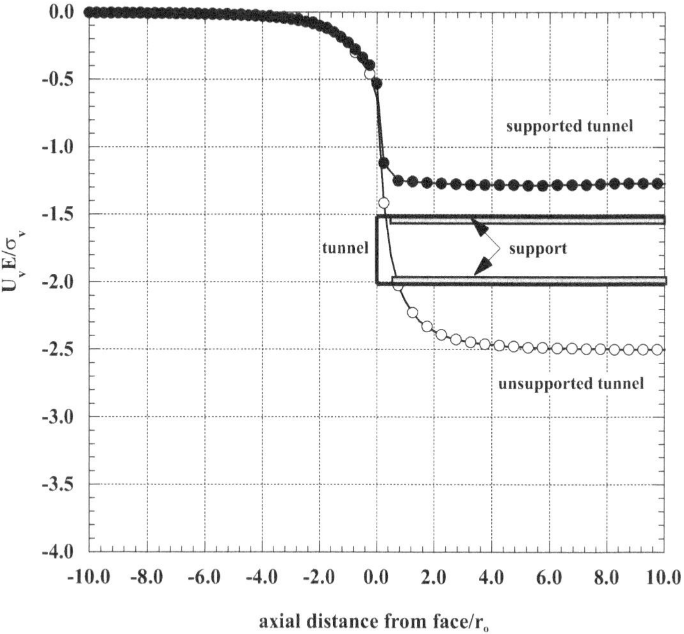

Figure 1.7 Vertical displacements at the crown of supported and unsupported tunnels, r_o = 2 m, K_o = 1.0.

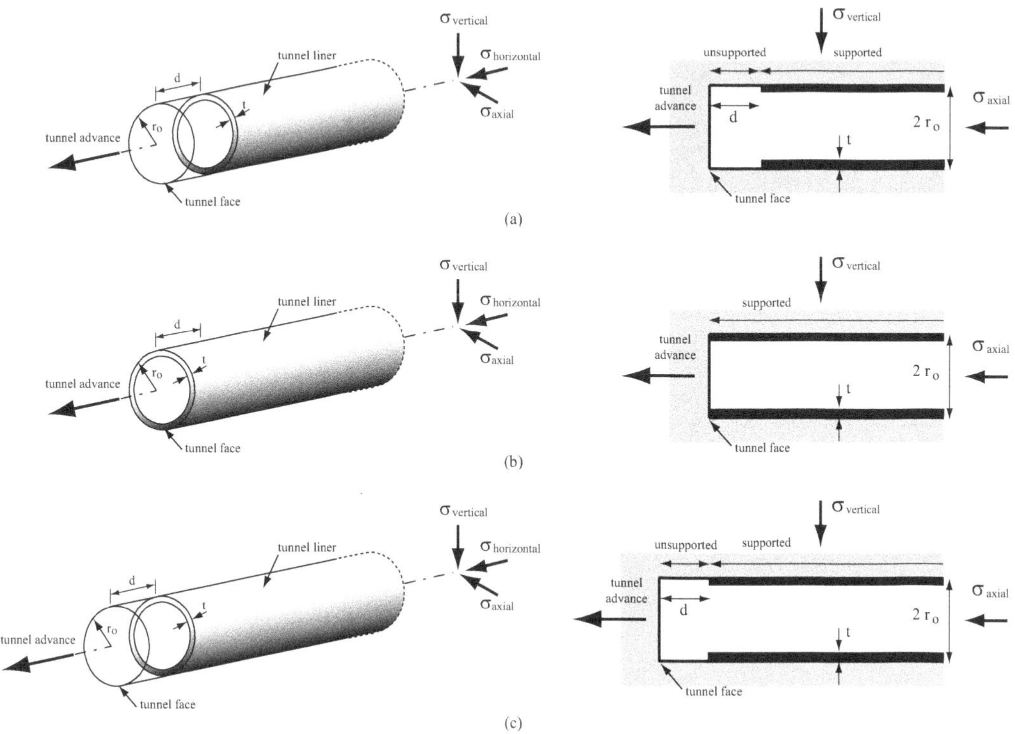

Figure 1.8 Excavation and support of a lined tunnel. (a) Initial position. (b) Liner installation. (c) Excavation.

in the ground are analyzed using the FEM code ABAQUS, with the condition that both the ground and the liner are elastic. The same tunnel dimensions, ground properties, and far-field stresses are used (e.g. axisymmetric analysis). For the liner, the following properties are employed: Young's modulus, E = 24,000 MPa; Poisson's ratio, ν = 0.25, and liner thickness, t = 0.2 m. An unsupported length d = 1.0 m is assumed. Figure 1.9 shows the results. Figure 1.9a is a contour plot of the axial stresses, Figure 1.9b of the radial/vertical stresses, Figure 1.9c of the tangential/horizontal stresses, and Figure 1.9d of the displacements. The lined tunnel induces a significant reduction of axial stresses ahead of the tunnel face with a magnitude similar to that observed for the unlined tunnel. The mechanism is the same as shown in Figure 1.6a. The radial/vertical stress contours show unloading, as with the unlined tunnel, and as described in Figures 1.6b and c. However, the magnitudes of the stresses are smaller with the lined tunnel than with the unlined tunnel. This is because the liner of the supported tunnel can take some of the stresses from the ground (as explained in Chapter 4, the magnitude of the stresses taken by the liner depends on the relative stiffness between the liner and the ground). The reduced loading of the ground occurs at the expense of the liner, which now takes load. What is interesting to note, and quite different from the case of the unsupported tunnel, is that the ground close to the tunnel experiences a stress path reversal, first of unloading due to the excavation, and then of loading as the liner is installed and takes load. This phenomenon also occurs if the ground yields (Cantieni and Anagnostou, 2009). Similar to what is shown in Figure 1.5, the stresses in the ground experience a rotation of principal stresses near the tunnel during excavation and support operations.

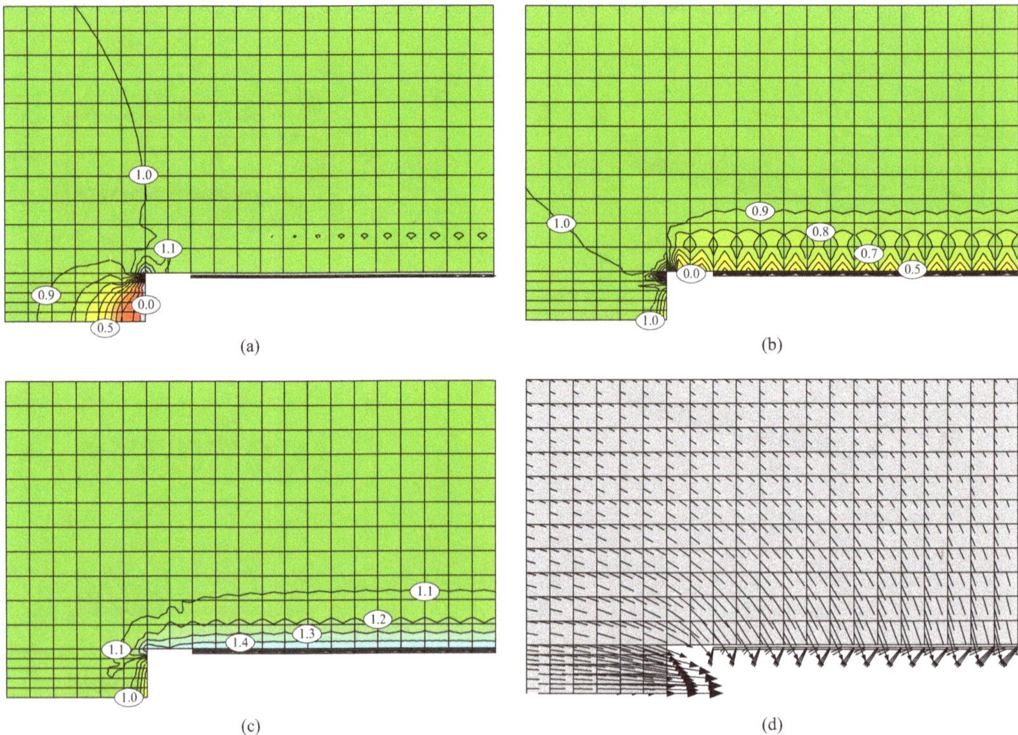

Figure 1.9 Stresses and displacements for a supported tunnel. Axisymmetric analysis with r_o = 2 m, σ_v = 1.0 MPa, K_o = 1.0. (a) σ_{axial}. (b) σ_{radial}. (c) $\sigma_{tangential}$. (d) Displacements.

The displacements induced in the ground are plotted in Figure 1.9d. As with the unsupported tunnel, significant ground displacements occur ahead of the tunnel and have a strong component along the direction of the excavation. With increasing distance behind the face of the tunnel, the deformations rotate and tend to be mostly radial. The displacements decrease as the distance from the tunnel increases. A comparison of displacements for the unsupported and supported tunnels is provided in Figure 1.7. Similar to the unsupported tunnel, the displacements at distances larger than about two diameters ahead of the tunnel face are negligible. As the tunnel approaches, the displacements increase and are significant immediately after passage of the tunnel face. At about two to three diameters behind the face, the displacements do not increase anymore and plane strain conditions are reached. The most substantial effect of the liner is the large decrease of displacements of the ground; for this particular problem, by a factor of about two.

Hence, the installation of a liner reduces the change of stresses in the ground compared with the unsupported tunnel, and so the ground deformations are also smaller. This occurs at the expense of the load of the liner. The interplay between the ground and the support depends on the stress field, on the properties of the ground, of the support, dimensions of the tunnel, and on the construction process; it is this interplay that has to be considered in tunnel design and analysis. This is, in essence, the underlying topic of this book and will now be looked at in the characteristic curves section.

1.3 THE CHARACTERISTIC CURVES METHOD

How ground and support interact, i.e. what stresses and displacements result, is the underlying topic of this book. The empirical, analytical, and numerical instruments discussed in this book will provide the tools to do so. It is, however, very important to conceptually understand the factors that influence ground–structure interaction, and in particular the concept of relative stiffness/flexibility of support and ground, and this will be done in the following. The tools used to explain this are known under different names; most common are the terms "characteristic curves method" (used in this book) and "convergence confinement method." Characteristic curves have been introduced by several authors (Daemen and Fairhurst, 1972; Lombardi, 1973; Peck, 1969b), but the principle has been used earlier, although not with these terms (e.g. Terzaghi, 1936). The convergence confinement method has been discussed, e.g. by Panet and Guellec (1974). For a discussion of the development of these methods, see e.g. Schwartz and Einstein (1980b).

As the name indicates, the characteristic curves method describes the characteristic behavior of ground and support and of their interaction in two dimensions. They relate displacements to the state of stress in a cross section perpendicular to the tunnel axis. The underlying reasoning behind this is that, in simple terms, one can look at the construction of a tunnel as a change from the primary to the secondary to the tertiary stress state as indicated in Figure 1.10. In the original, primary state of stress, the far-field stresses act everywhere. When the opening is created, stresses change producing displacement at the interface and in the ground (secondary state of stress). Finally, when the support is installed, further stress changes occur. These are caused by the fact that the ground will usually deform further even after the structure is installed. This produces stresses in the support and creates "counter-stresses" in the ground – the tertiary stress state.

The relative deformation of ground and support has a significant effect on the support loads and thus on the resulting stresses. This can be best illustrated with Peck's (1969b)

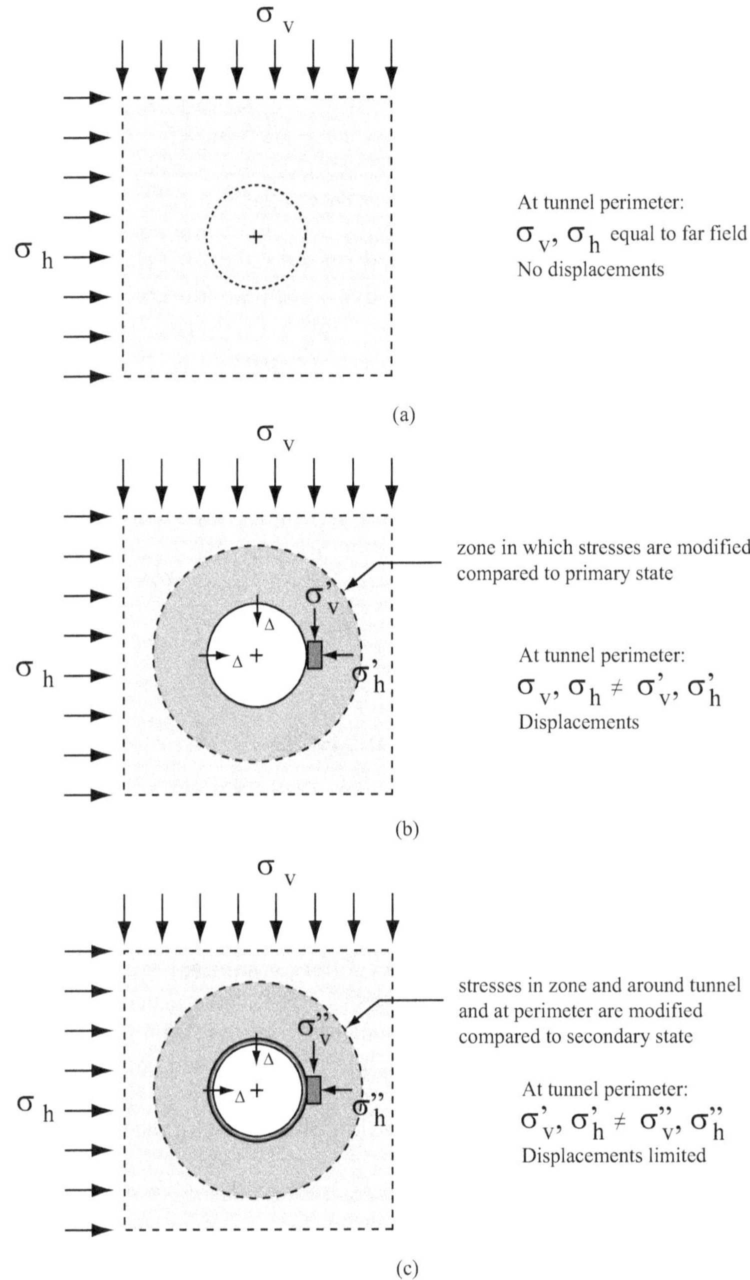

Figure 1.10 Different stress states caused by tunneling. (a) Primary state of stress. No opening. (b) Secondary state of stress. Opening. (c) Tertiary state of stress. Support installed.

comparison of supports that are stiff or flexible in bending (Figure 1.11). Assuming a far-field stress state of σ_v and $\sigma_h = K_o\,\sigma_v$ ($K_o < 1$; see Figure 1.11a), and simultaneous excavation and support installation, a flexible circular support will deform to adapt to the non-uniform stresses until the pressures acting on the support are equal (Figure 1.11b). If the support shape is elliptical with an axis ratio corresponding to $1/K_o$, i.e. $a/b = \sigma_v/\sigma_h = 1/K_o$, no deformation but a non-uniform applied stress will result (Figure 1.11c). For supports that are stiff

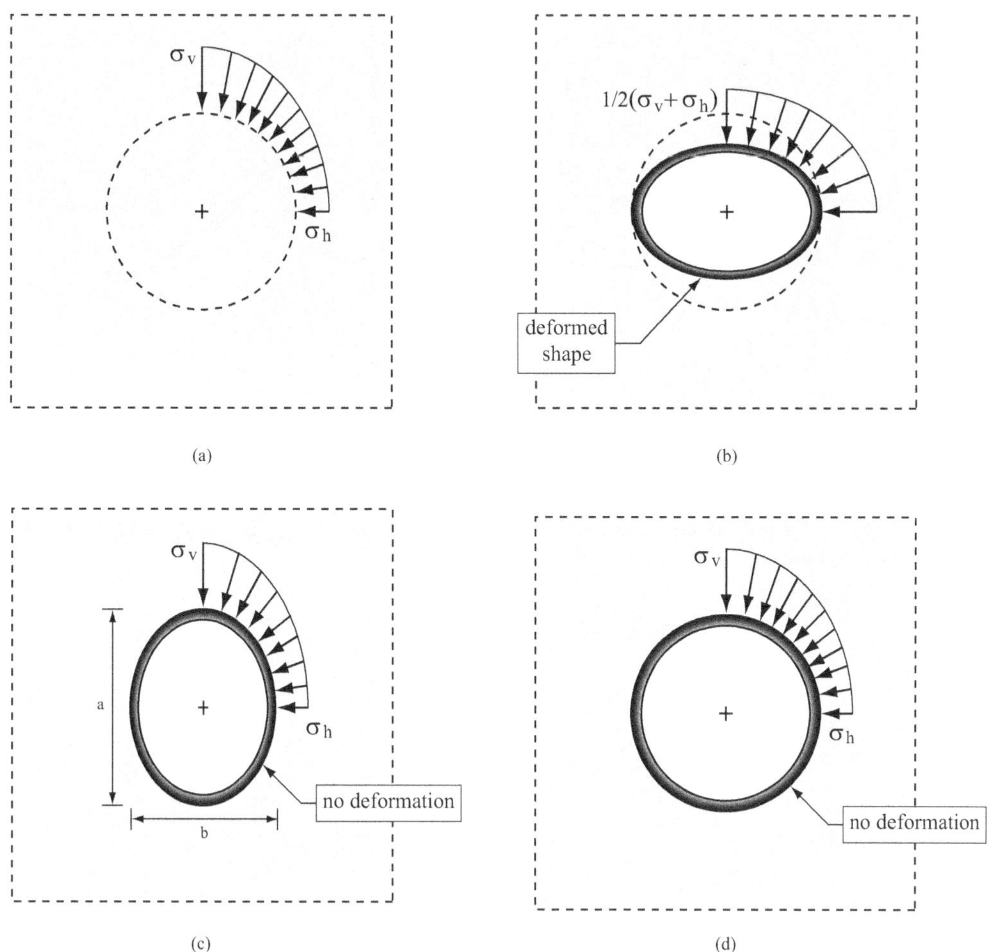

Figure 1.11 Stresses and deformation of tunnel support, depending on far-field stresses, shape and rigidity (after Peck 1969b). (a) Non-uniform stresses at perimeter of future tunnel. (b) Perfectly flexible support. (c) Elliptical Support with a/b – σ_v/σ_h. (d) Stiff support.

in bending, case 1.11d is identical to 1.11c. However, for circular openings, the non-uniform applied pressure will cause bending. Note that in all cases a "thrust," i.e. axial force will exist in the support. These conclusions reached by Peck through a thought process are proven analytically in Section 4.2.1.2.

The characteristic curve method conceptually includes the response of the ground as it is excavated, the response of the support given its stiffness/flexibility, and the resulting soil–structure interaction in terms of compatibility of stresses and deformations at the ground–support interface. This will be illustrated by first discussing ground characteristic curves, then support characteristic curves and, finally, by combining them.

The ground characteristic curve can be conceptualized as follows (Figure 1.12): The ground is subjected to a far-field stress, σ_o. An opening is conceptualized (created) in the ground. Simultaneously with creating the opening, a counterstress σ_o is applied at the opening perimeter; since the stresses applied are the same as the far field, no displacements occur. Then the counterstress σ_o is reduced and, as this happens, displacements Δ occur. In essence, this represents the change from the primary state to the secondary state

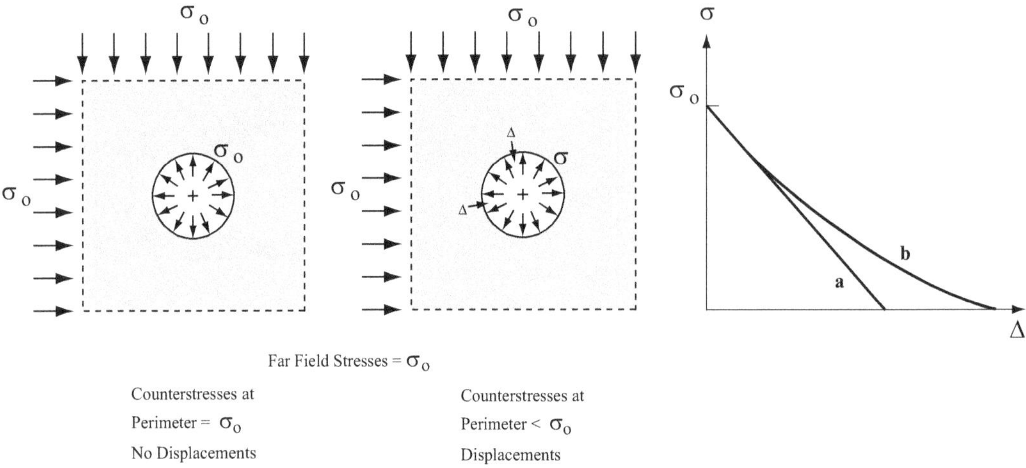

Figure 1.12 Concept for ground characteristic curves. Curve a: linear elastic; Curve b: nonlinear elastic or elasto-plastic.

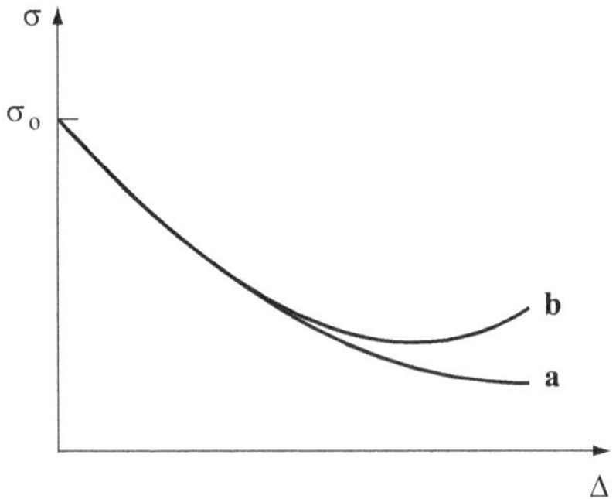

Figure 1.13 Unstable ground characteristic curves. Curve a: Unstable elasto-plastic; b: Unstable with stress increase.

(Figure 1.10b). If the ground behaves linearly elastic, the ground characteristic curve "a" results; for non-linear behavior, it is curve "b." Essentially, curve 'b' represents nonlinear elastic or elasto-plastic behavior. Both curves "a" and "b" in Figure 1.12 represent stable behavior in that for zero internal stresses, the displacements are finite. In Figure 1.13, two unstable ground characteristic curves are shown. Curve "a" in Figure 1.13 represents behavior with displacement approaching infinity for no internal stresses. Curve b represents behavior that is often called "loosening." A few comments are useful in the context of explaining elasto-plastic behavior (curve b in Figure 1.12 and curves a, b in Figure 1.13). What occurs is that the ground zone near the opening yields. Depending on the assumptions as to what happens in the yielded zone, different curves are obtained. Schwartz and Einstein (1980b) conducted some investigations on the effect of different assumptions, and this is shown in Figure 1.14.

In all the cases shown in Figure 1.14, one assumes either ideally plastic or strain hardening behavior, i.e. as the displacements increase, the internal stresses decrease. For stress softening behavior, however, the mobilized resistance of the ground decreases with increased displacements. This in turn leads to an internal (at the opening boundary) stress increase as the

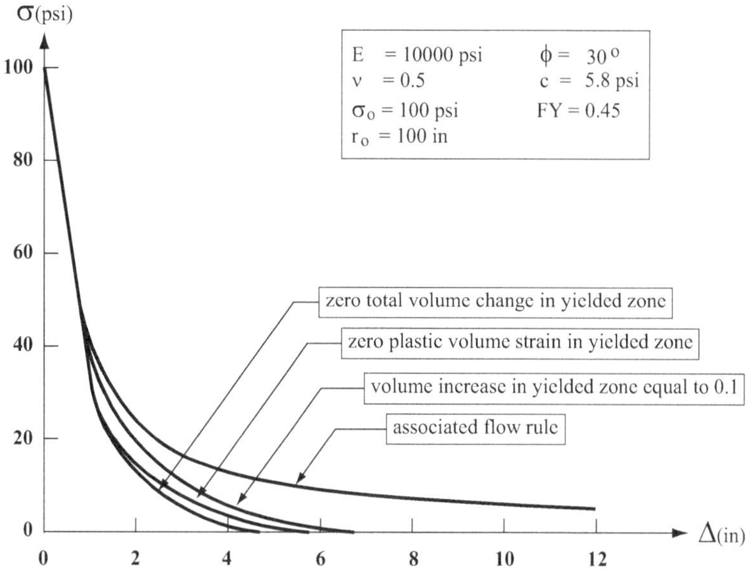

Figure 1.14 Ground characteristic curves for different conditions in the yielded (plastic) zone. FY = internal pressure at which yield occurs/applied internal pressure (after Schwartz and Einstein, 1980b).

displacements increase. One can best associate this with an enlargement of the yielded zone (loosening, as it is often used in this context, can be associated with strain softening but this is not generally recognized). As will be seen in the discussion on arching, this stress (load) increase can actually be observed, for instance in trapdoor tests, and can be associated with an increase of the yielded zone.

So far, the ground characteristics curves have been used to represent instantaneous behavior, i.e. the stress-displacement behavior as the imagined internal stress is placed inside the imagined opening and reduced to zero. The concept can, in principle, be extended to time-dependent behavior where the displacements at the perimeter of the underground opening increase with time, e.g. due to creep or swelling. This is schematically shown in Figure 1.15.

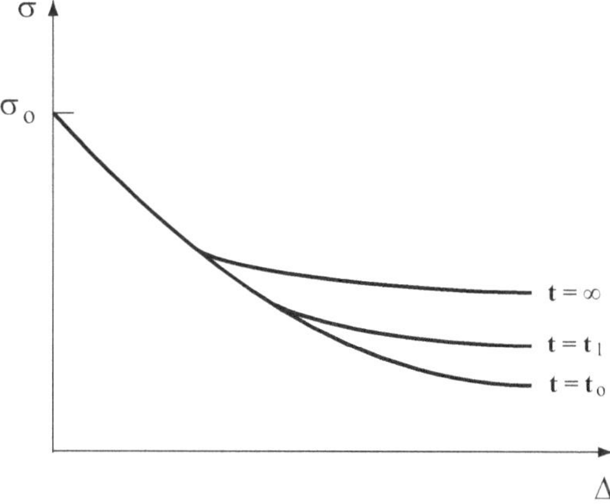

Figure 1.15 Typical characteristic curves for time-dependent behavior.

Analogous to the ground characteristic curves, one can construct support characteristic curves. Figure 1.16 shows a circular support loaded by a uniform stress σ and the corresponding linear elastic and elastic-perfectly plastic support characteristic curves. Figure 1.17 shows how, in principle, different supports can be characterized.

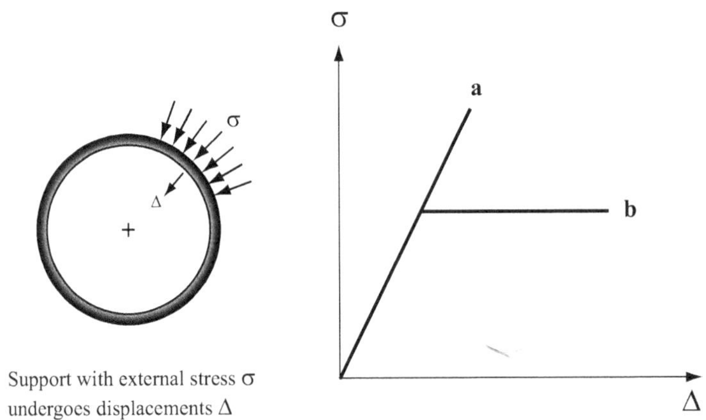

Figure 1.16 Concept of support characteristic curves. Curve a: Linear elastic; curve b: Elastic-perfectly plastic.

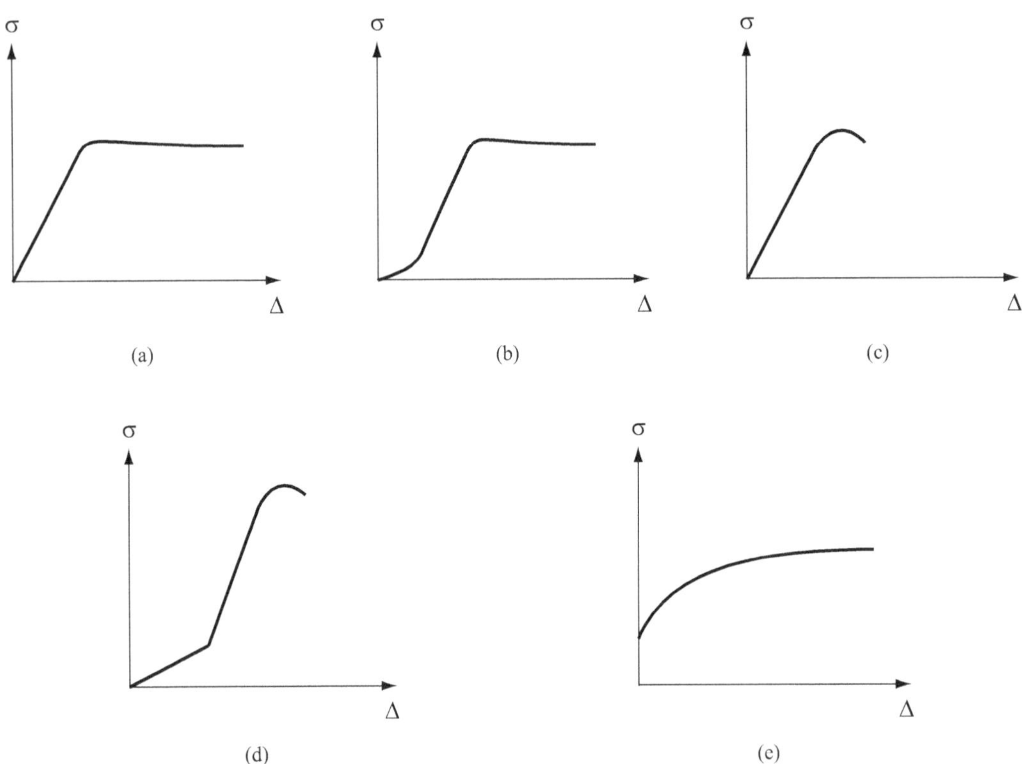

Figure 1.17 Typical support characteristic curves for a variety of supports (not to same scale). (a) Steel set. (b) Steel set with timber blocks. (c) Concrete support. (d) Initial layer of shotcrete followed by final concrete liner. (e) Bolts.

Ground–support interaction can then be presented by combining the two types of curves, as shown in Figure 1.18 for the linear elastic (a) and a non-linear (b) case. One has to visualize the mechanisms shown in Figure 1.18 by imagining the opening with the counterstresses acting in the ground while the support is simultaneously fitted into this opening. The counterstress in the ground is then released (reduced). This produces ground displacements, which load the support. The point of equilibrium is reached where the two stresses are equal. A reduction in support load can be accomplished with a softer support (Figure 1.19a) or installing the support with some "delay" (Figure 1.19b). Of course, a reduced load on the support carries larger displacements and stresses in the ground.

The concept of "support delay" is illustrated in Figure 1.8 that shows that the support is installed after excavation of the round. This delay can be longer, i.e. support is installed at a great distance from the face, or it can be shorter or even "negative" with shield supported excavation or with the umbrella method (one can imagine a ground that cannot support itself at all, i.e. no unsupported length is admissible. This requires pre-support of the tunnel, e.g. through a shield or the umbrella method, as schematically shown in Figures 1.20a, b.

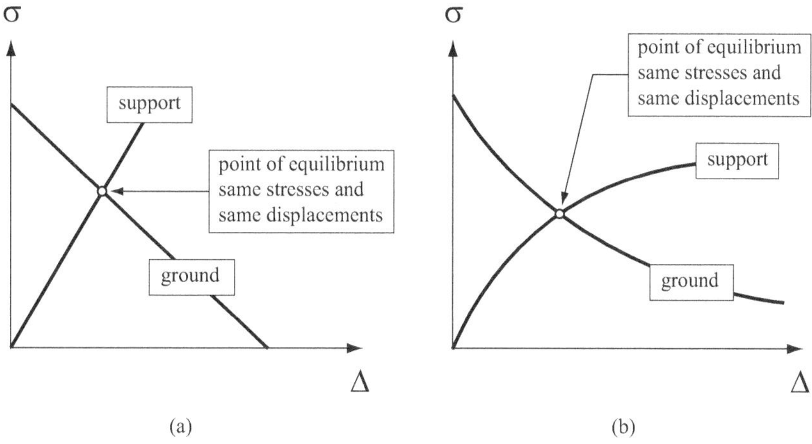

Figure 1.18 Ground support interaction. (a) Linear elastic. (b) Non-linear.

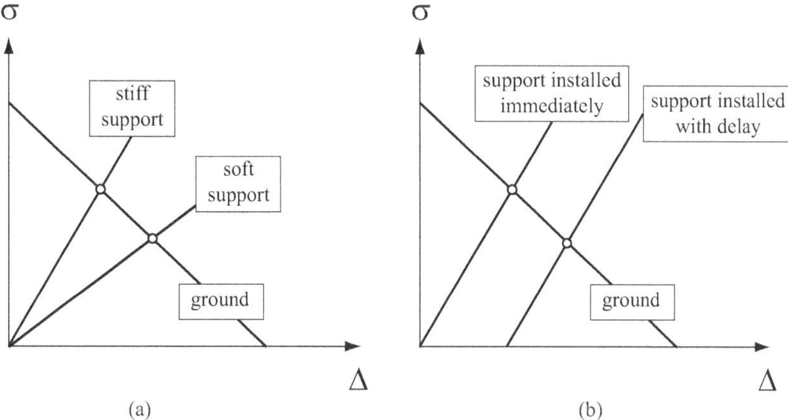

Figure 1.19 Reduced support load through. (a) Reduced support stiffness. (b) Support delay.

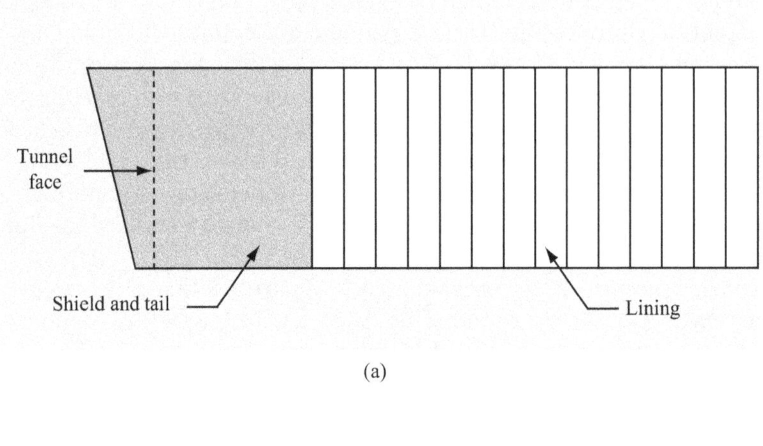

(a)

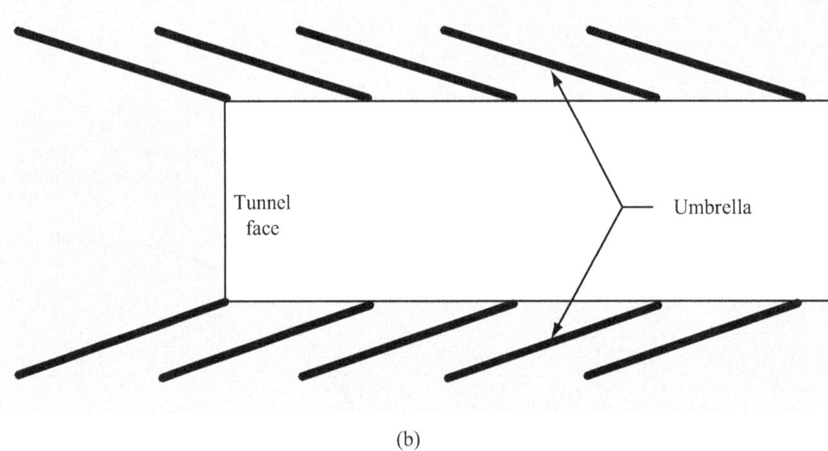

(b)

Figure 1.20 Examples of "Presupport" ahead of the face. (a) Shield method. (b) Umbrella method.

Installing a support will reduce displacements as was already shown in Figure 1.7. A comparison between Figures 1.18 and 1.19 and Figures 1.7 to 1.9 provides information on the extent to which a two-dimensional representation can capture three-dimensional effects. It is evident that the installation of the support at a distance from the face – a three-dimensional effect – can be conceptually captured in two-dimensions with a support delay. However, when looking in detail at Figures 1.7 or 1.9 one sees that ground displacements also take place ahead of the face. Lombardi (1973) extended the characteristic curve approach to include this effect. Observations in the field and numerical calculations have shown that for unsupported tunnels, about 50% of the total displacement takes place ahead of the face, while for supported tunnels, this can be as low as 30% of the total displacement. Figure 1.7 shows an example calculation. The support delay represented by the two-dimensional characteristic curves therefore has to represent the total displacement before the support is installed. The principle is still the same, as shown in Figure 1.19b, but one has to be aware of what "delay" means.

The displacements shown in Figures 1.5, 1.7, and 1.9 are "instantaneous." It is possible that the ground displacement and stresses ahead of the face, in the unsupported part of the tunnel and in the supported part of the tunnel, change with time. Similarly, the support

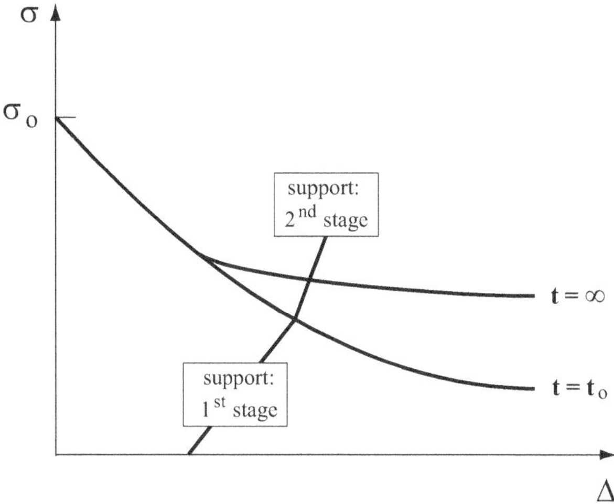

Figure 1.21 Support of different stiffness staged to handle time-dependent ground behavior.

behavior can change with time. The principles can be illustrated as was done in Figure 1.5. One strategy is the stage of support installation, as shown in Figure 1.21, e.g. by installing a flexible support first and then increasing its stiffness.

At this point, it is also appropriate to mention that the spatial and temporal relationships in excavating and supporting a tunnel are interrelated. Both excavation and support installation take time, and the distance from the face at which the support is installed is associated with the duration of the preceding excavation and additional time spent till the support is installed and interacts with the ground. Clearly one wants to avoid excessive displacements, particularly if these lead to a stress increase. Figure 1.22 shows an unstable ground characteristic curve involving very large displacements. To prevent excessive displacements and/or high stresses in the support, it may be necessary to install the support with as small a delay

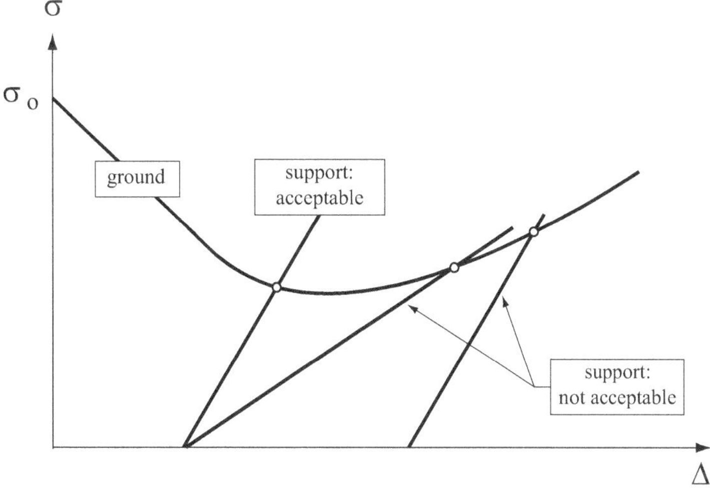

Figure 1.22 Support that is not stiff enough or installed too late can lead to load/stress increase.

as possible, i.e. close to or at the face (Figure 1.19b). This is done by reducing the round (cycle) lengths and thus reducing the unsupported length of tunnel. Different possibilities are schematically shown in Figure 1.23. Reducing the unsupported span is not only important in the longitudinal direction, i.e. the direction of advance. The size of the tunnel opening in the transverse direction can also be reduced as shown in Figure 1.24. The schematic examples in Figures 1.24b, c, d are simply examples, and there are many other possibilities.

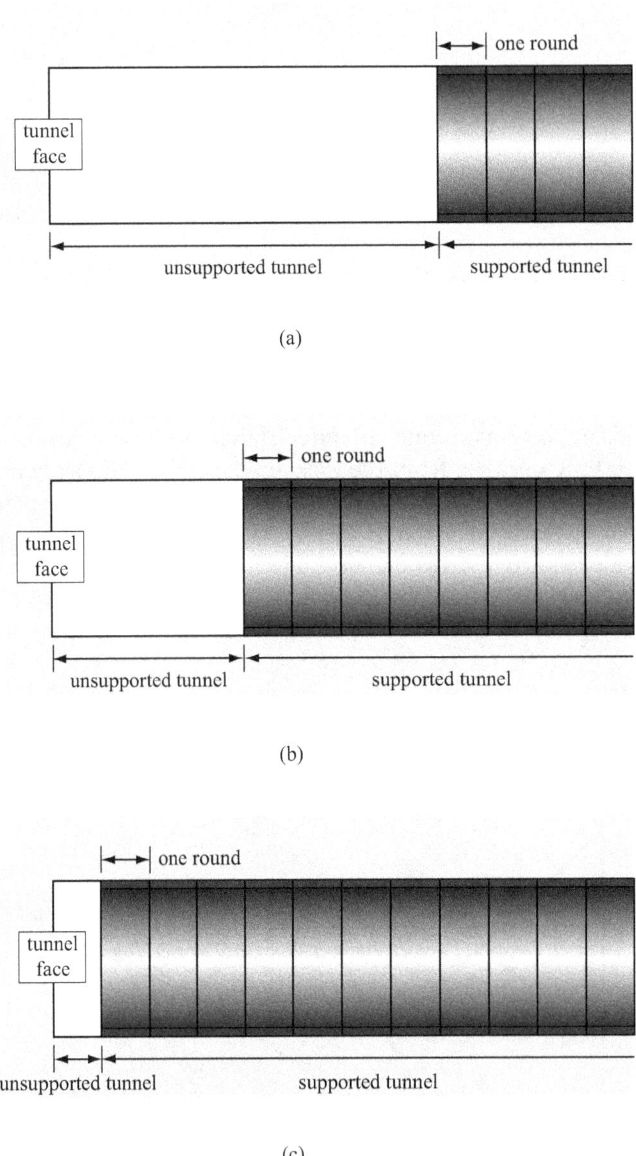

Figure 1.23 Distance of support installation from face. (a) "Good Ground" - Support installed independent of round (Cycle). (b) "Normal Ground" - Support installed one round (Cycle) from face. (c) "Difficult Ground" - Support installed with reduced round (Cycle) from face.

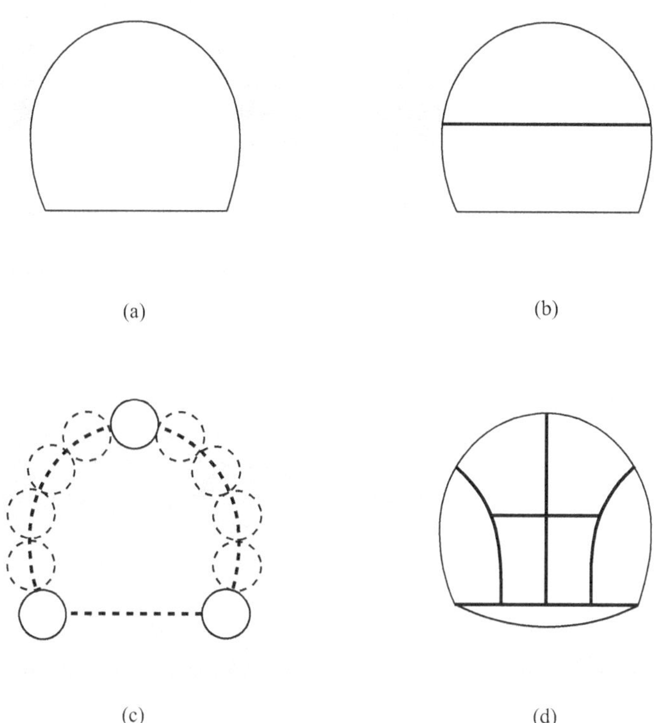

Figure 1.24 Reducing the unsupported span (size) of a tunnel in transverse section. In cases (b), (c), (d), each "heading/drift" is supported separately. (a) Full face. (b) Heading and bench. (c) and (d) Multiple drift.

1.4 PUTTING EVERYTHING TOGETHER – INTEGRATION OF DESIGN AND CONSTRUCTION, AND THE OBSERVATIONAL METHOD

1.4.1 Integration of design and construction introductory comments

When looking more closely at "ground–support interaction" one realizes that "support" is tunnel construction involving excavation and installation of initial and final supports, as shown in Figures 1.18, 1.19, 1.22, and 1.23. Tunnel design, as described and supported with empirical, analytical, and numerical approaches in the following chapters, then uses these construction related concepts. Tunneling, similar to but more strongly than other geotechnical and civil engineering projects in general, is truly an integration of design and construction. In other words, design not only addresses the performance of the final structure/tunnel but how the tunnel is built.

There is, however, more to integration of design and construction: How a tunnel behaves/performs during construction can be compared to design predictions for construction stages and, based on this, predict the final performance (Figure 1.22) and thus assess the adequacy of the design. This possibility of assessing and then possibly adapting the design based on observations during construction leads to the Observational Method.

1.4.2 History and fundamentals of the observational method

When designing, one does not know with certainty what ground conditions will be encountered during construction. This is the case in most geotechnical projects but is particularly so in tunneling where the ground is not only providing support but also produces the loads on the structure. The uncertainty associated with the ground conditions has led Terzaghi in "Past and Future of Applied Soil Mechanics" (1961) to state:

> The case records presented under the heading 'Foundation Design' showed that many problems of earthwork engineering can be solved without a detailed and accurate forecast of performance. Satisfactory solutions of such problems can be obtained on the basis of our knowledge of the fundamental principles of soil mechanics supplemented by a moderate amount of boring and testing. However, there are others in which the geological conditions preclude the possibility of securing in advance of construction all the essential information required for adequate design. If this condition prevails, sound engineering calls for design on the basis of the most unfavorable assumptions compatible with the results of the subsoil explorations. This rather uneconomical procedure can be avoided only on the condition that the project permits modifications in the design during or after construction in accordance with the results of significant observational data which are secured after construction is started. This can be called an *'observational procedure.'*
>
> (Emphasis by the authors of this book)

Terzaghi then illustrates the application of the observational procedure and from this it becomes clear that this is NOT "design as you go" but the thinking through of the entire design-construction process with performance predictions for intermediate stages, and the implementation of prepared contingency designs if the observed performance does not correspond to the predicted one. This fundamental formulation of the observational effort is represented in the flowchart of Figure 1.25.

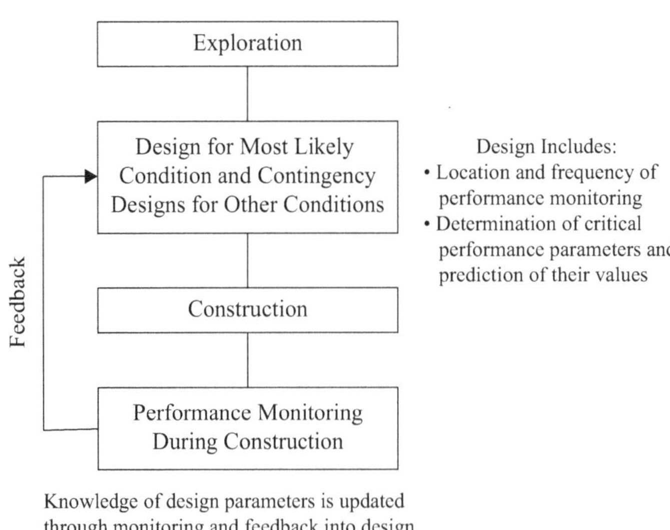

Figure 1.25 Principle of observational methods.

The necessity to have carefully prepared contingency designs is emphasized by Peck (1969a) in his Rankine Lecture "Advantages and Limitations of the Observational Method in Applied Soil Mechanics." In the synopsis of the 1969 paper Peck states:

> The observational method, used so successfully by Terzaghi in applied soil mechanics, often permits maximum economy and assurance of safety, provided the design can be modified as construction progresses. The essential features of the method are set forth and examples are given of its application.
>
> The method is not without its pitfalls and limitations. It should not be used unless the designer has in mind a plan of action for every unfavorable situation that might be disclosed by the observations. The observations must be reliable, must reveal the significant phenomena, and must be so reported as to encourage prompt action. The possibility of progressive failure may introduce a serious element of uncertainty.
>
> In spite of the limitations, the potential for savings of time and money without sacrifice of safety is so great that every engineer who deals with applied soil mechanics needs to be informed of its principal features.

Both Terzaghi and Peck describe examples of applications of the observational method, and, in Peck's contribution, also cases where the method incurred problems – the problems could be associated with the failure to anticipate unfavorable conditions, choosing the correct (relevant) observations and progressive failure.

It is both interesting and relevant that the history of geotechnical cases played such an important role in what Terzaghi (1961) and Peck (1969a) discuss. As a matter of fact, Terzaghi (1961) describes what in his opinion is a first application of the observational procedure in large-scale earthwork engineering, namely its application in slope stability problems in sensitive clays by the Swedish Railways ("Statens Järnvägars," 1922). In the historical context one can also find earlier cases such as the design and construction of the Eiffel Tower. As Harriss (1975) writes, it was absolutely essential that the first platform (see Figure 1.26) was absolutely horizontal to prevent any distortion of the subsequent parts of the tower. This was achieved by a combination of jacks and wedges to lift, and boxes filled with sand that could flow out to lower the tower legs with a precision of one tenth of a millimeter. Note also

Figure 1.26 The Eiffel Tower.

that this last example illustrates not only the observational method but also the integration of design and construction.

Reviewing the principles and history of the observational method, one can say that one prepares several designs and adopts the optimal design based on observations during construction. However, one has to be aware of the fact that optimization in civil works takes place on many levels (Figure 1.27) from the planning stage where e.g. several major alternatives of subway networks or transmountain routes are compared, down to the selection of a proper mix of aggregate and cement in concrete.

The principles illustrated in Figure 1.27 will now be used to discuss the use of the observational method in tunneling.

1.4.3 The observational method in tunneling

The optimization levels of Figure 1.27 are illustrated with a few examples at the corresponding levels of the observational methods in tunneling:

On the *Detailed Construction Level*, one observes geologic/geotechnical conditions and tunnel performance at a location and, based on this, either continues with the same design/construction option or selects a better suited one for the next cycle.

For example, in the application of the New Austrian Tunneling Method as shown in Figure 2.20a (discussed later in Chapter 2), different support designs are available. The performance is measured by deformation measurements at 10m intervals and compared to predefined acceptable values. Depending on this comparison, the same design option is used for the next tunnel cycle or one with more or less support is chosen. Importantly, the design/construction options also include round/cycle length and full face/partial face (see Figure 1.24) options.

Another more widely known example from geotechnical engineering is the selection of the number of tiebacks in a retaining structure as e.g. used in cut-and-cover tunneling. Figure 1.28 illustrates schematically this, after excavation of a stage and installation of one tieback level wall. Displacements exceeding acceptable values are observed and lead to the selection of the predesigned option with additional tiebacks in the following stage of excavation.

On the *Project Level*, one again relies on observations of the geologic conditions and of the performance of opening and supports, however on a larger scale, namely applying to what is observed and interpreted in the entire tunnel or large sections of a tunnel. The obvious examples are test sections and pilot tunnels, which are widely used. The associated performance observations and interpretation can be used to provide the basis for different design options that can then be used at the detailed design level. It is absolutely necessary to combine the

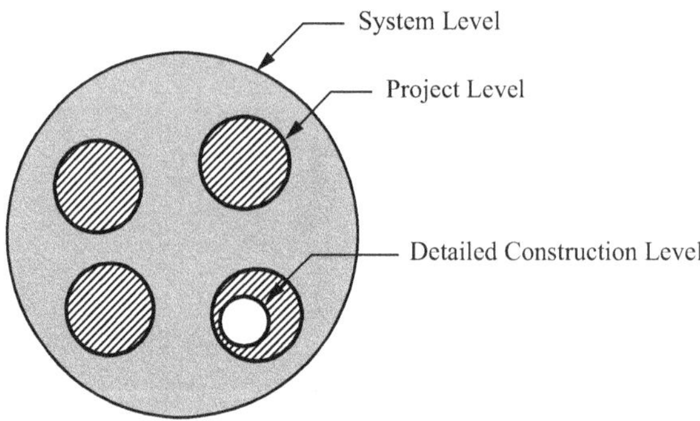

Figure 1.27 Levels of optimization.

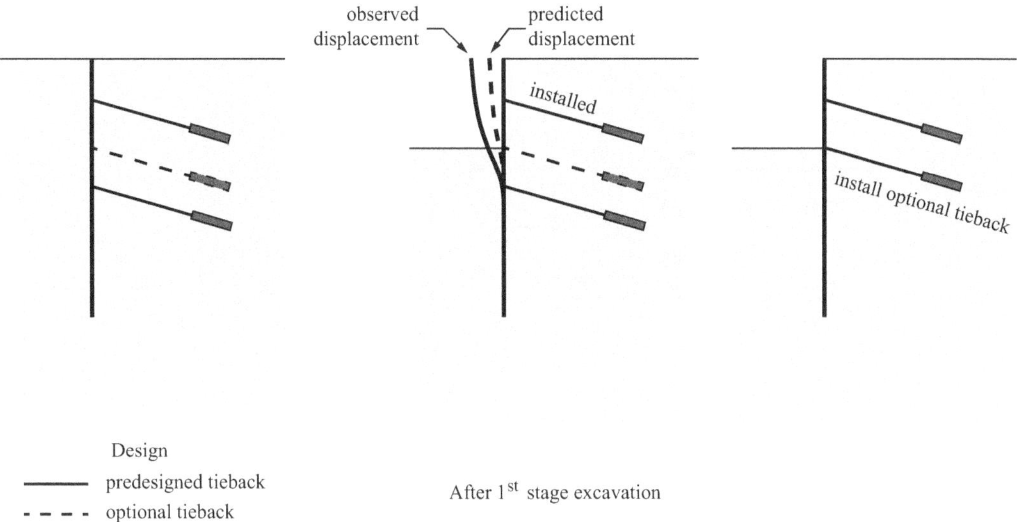

Figure 1.28 Installation of predesigned optional tiebacks if observed displacements exceed predicted displacements.

project level with the detailed level since e.g. the local geologic conditions or the size of pilot and final tunnels may be significantly different.

An interesting alternative to specific test sections is what has been done in the NATM application in the Arlbergtunnel. Observations in the initial part (500 m) of the tunnel were used to modify the design/construction alternatives as evident from the comparison of Figure 2.20a and b, where the as-built versions deviate from the original ones. Finally, a small example of the project level observational method is the use of test panels to optimize shotcrete mixes.

It is also possible to apply the observational method on the *Systems Level*. Examples in which this has been done are subway systems or long tunnels. In both, one gains experience at the beginning that can be used to optimize what follows. It is also possible to implement new technical developments.

As implied in Figure 1.27 and illustrated with the comments on the project and systems level, optimization on the lower level affects optimization on the higher level and vice versa. In other words, integration of design and construction can and should be done on all levels, and the observational method lends itself to do so.

1.5 ARCHING

1.5.1 Basic phenomenon: observations in the field and in the laboratory

Arching can be defined as the phenomenon in which the earth pressure on a moving (deformable) part of the support is less or more than on the neighboring non-movable part (Figure 1.29). Even though Figure 1.29 depicts arching on a horizontal surface, the phenomenon can occur on surfaces of any inclination. In the remainder of this section, arching will be mostly discussed in the context of pressure acting on horizontal surfaces corresponding more or less to tunnel crowns.

The reduced pressure on moveable supports has been observed in reality, and the reader is referred to Chapter 2 where, for example, the empirical method by Terzaghi (1946) is based

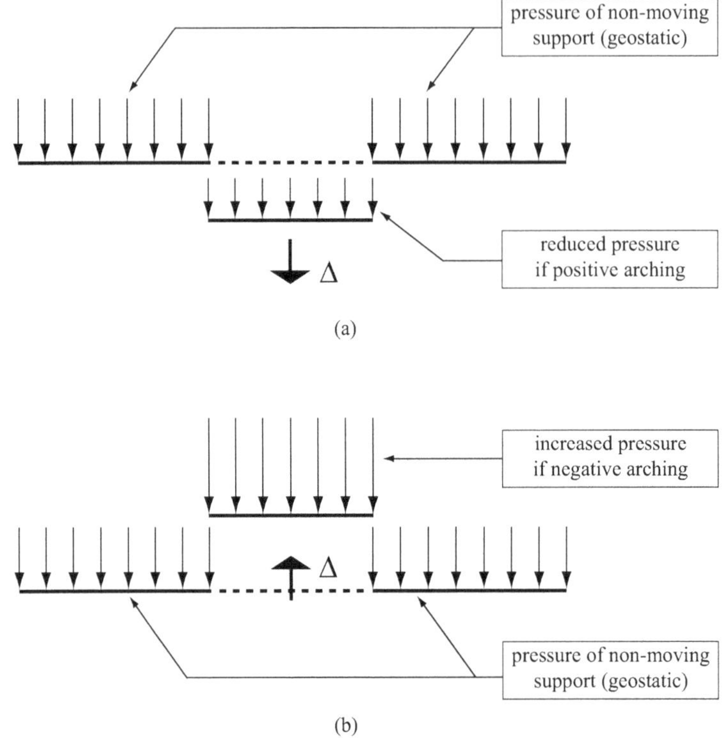

Figure 1.29 The arching phenomenon. (a) Positive arching. (b) Negative arching.

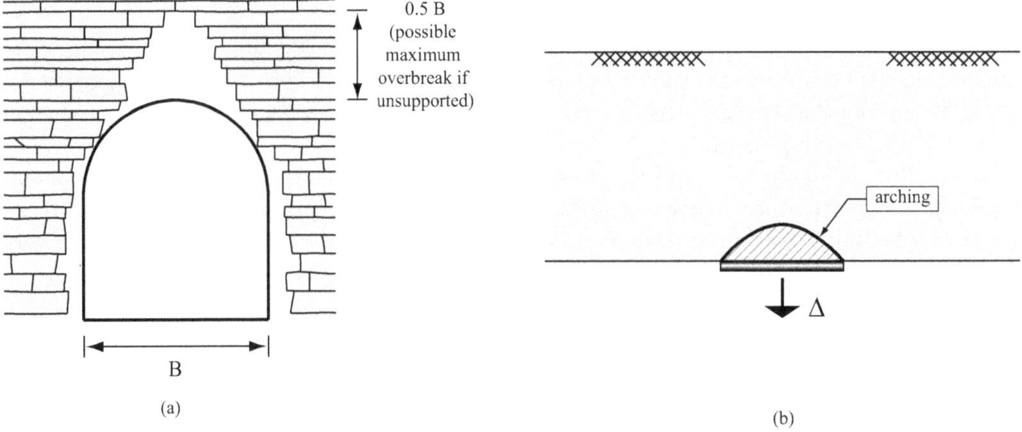

Figure 1.30 Arching. (a) Physical arch formed in stratified rock (Terzaghi, 1946). (b) Equivalent arch assumed by Engesser, 1882.

on the fact that the support pressure on the tunnel crown is usually less than the overburden pressure. In some instances, for example for stratified or blocky rock, Terzaghi (1946) assumed that the rock mass forms a physical arch such that only the rock below this arch has to be carried by the support (Figure 1.30a). Many others have made similar observations and this not only in rock, e.g. Marston and Anderson (1913). Similarly, and quite a bit earlier,

Engesser (1882) assumed that the ground, whatever its composition, forms a parabolic arch above the moveable support or tunnel opening for that matter (Figure 1.30b).

The fact that deformation of a support leads to pressure reduction is reflected in the typical ground characteristic curve that was described in Section 1.3; see e.g. Figure 1.12. Recall that the pressure reduction can be associated with the mobilization of shearing resistance in the ground, a phenomenon for which much additional evidence will be produced in the context of this discussion on arching.

Other evidence on arching from reality is, for instance, the fact that the flow of granular material from silos is often inhibited by the formation of an arch between the vertical silo walls (Janssen, 1895). (The stress transfer to the silo walls may also cause failure of these walls.) Arching, leading to a reduction of pressure on a vertical support, is made use of in soldier pile and lagging walls, where the arch is supported by the soldier piles while the lagging moves (the effect is often increased by placing a soft material, e.g. straw, behind the lagging).

While most of the arching phenomena described, except for the silo case, have advantageous effects, namely the reduction of support pressure, it should be kept in mind that this is only so for positive arching. If the support is stiffer than the neighboring ground such as when relatively stiff pipes or culverts are placed in an open cut and the ground around them is backfilled, negative arching occurs (Figure 1.31). This leads to a load increase on the structure compared to the simple overburden load. Some of this can be prevented by staged refilling/compacting and shaping of the structure.

Given all these field observations, it is not surprising that many experiments have been conducted in the laboratory to investigate the arching phenomena, e.g. Engesser (1882), Terzaghi (1936), Atkinson and Potts (1977), Stone (1988), Stone and Newsom (2002), Vardoulakis et al. (1981), Adachi et al. (2003), Tanaka and Sakai (1993), Sterpi and Cividini (2004), Ladanyi and Hoyaux (1969), Meguid et al. (2008). The best-known experiments in geotechnical engineering are those by Terzaghi (1936) in which he lowered a trapdoor below a soil mass (Figure 1.32a) and observed the load on the trapdoor as a function of the trapdoor displacement (Figure 1.32b).

As can be seen in Figure 1.32b, this has been done for different densities of the soil (sand in this case), which reach different minimum loads but the same ultimate load. This fact will be further discussed below. The same behavior was also observed in centrifuge trapdoor tests by Iglesia (1991). Iglesia (1991) also conducted tests with different overburden heights, which show particularly interesting results (Figure 1.33). For all the different depths, the load on the trapdoor decreased with increasing displacement of the trapdoor, essentially producing the ground characteristic curve shown in Figure 1.32b. For small depths (overburden), the final

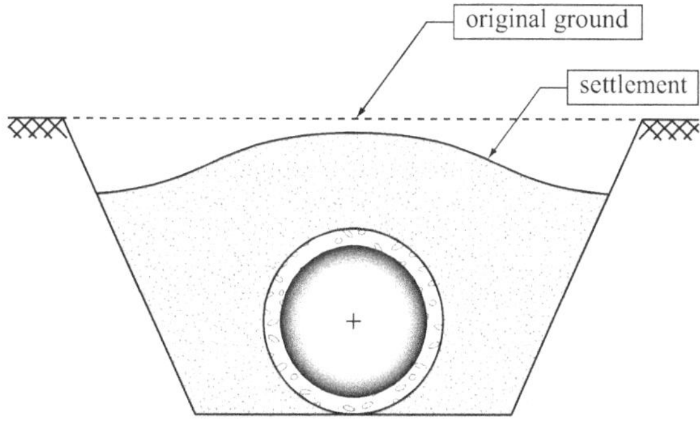

Figure 1.31 Negative arching on a stiff structure in deformable backfill.

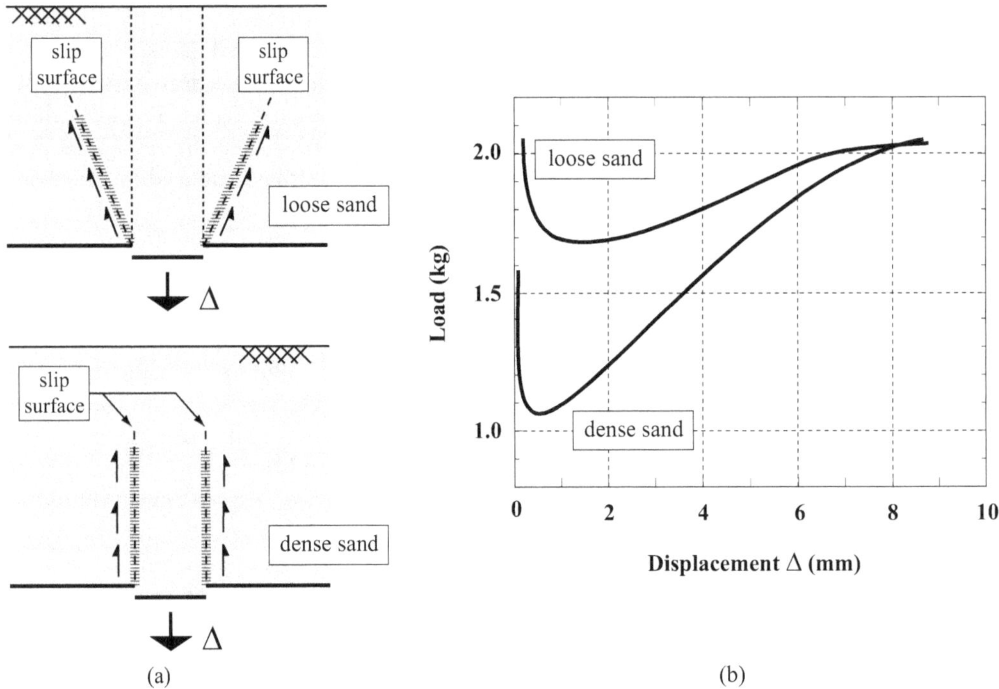

Figure 1.32 Trapdoor experiments by Terzaghi (1936). (a) Schematic of experiments. (b) Load-displacement curves.

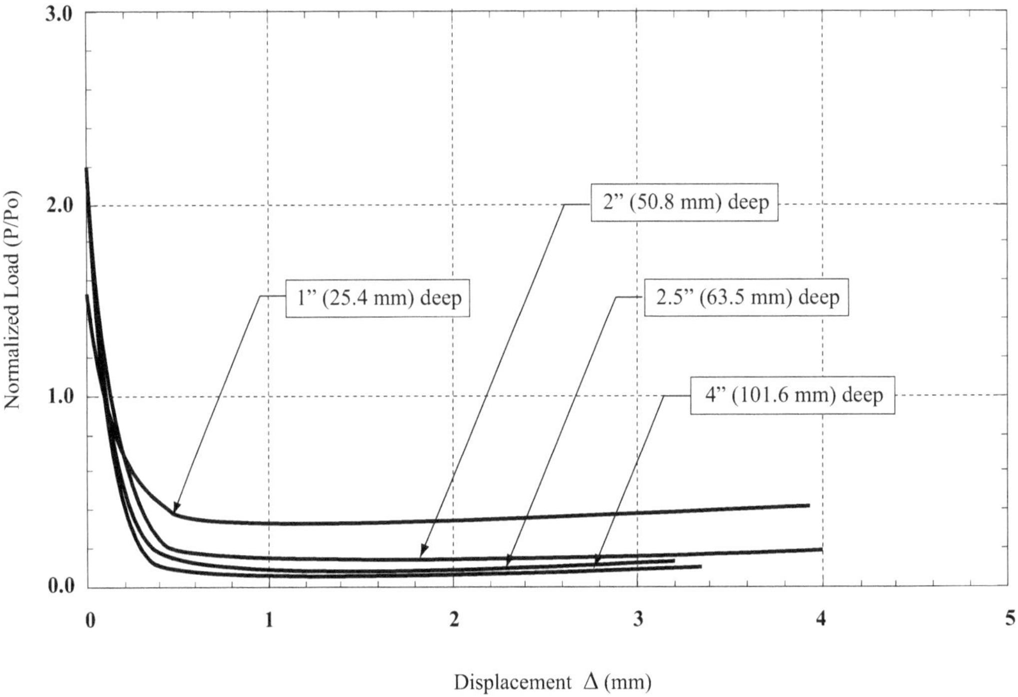

Figure 1.33 Load displacement curve on trapdoor in centrifuge experiment by Iglesia (1991). Effect of different depths.

load increased with increasing ground depth. For greater ground depths, however, the final load (at large displacements) was the same, independent of ground depth.

Both Terzaghi (1936) and Iglesia (1991) and many others, e.g. Stone (1988), also observed that the load-displacement (ground characteristic) curve after reaching a minimum load increased again to reach a somewhat higher load (ultimate load according to Terzaghi, 1936). This behavior, which has already been discussed in Section 1.3 (Figure 1.13) where it was associated with "loosening," can be related to what is physically happening in the ground mass. Both Iglesia (1991) and Stone (1988) in their trapdoor centrifuge tests observed the movement of particles over the trapdoor (Figure 1.34) and saw the behavior schematically illustrated in Figure 1.35.

As can be seen in Figures 1.34 and 1.35, the shape of the "arch" in the moving zone changes from a "gothic" (curved) arch to a triangular one, to end up with vertical boundaries as displacements increase. The latter occurs when the ultimate load is reached, while the minimum load seems to be related to "arches" of curved/triangular shapes. All this is mentioned in some detail because it explains the load development with displacement and also because it provides boundary conditions for the different theories discussed later in Chapter 2.

To conclude the comments on experiments, a few more results on centrifuge testing shall be made. Centrifuge testing has the advantage that real stress magnitudes can be applied while maintaining their geostatic distribution (see e.g. Joseph and Einstein, 1987 and Joseph et al., 1987) and also benefiting from the simple scalability of stresses and strains. (This is also a limitation of centrifuge tests in cases where displacements and not strains govern the behavior such as in jointed rock.) It is, therefore, not surprising that many centrifuge tests have been run for the tunnel problems (not only to investigate arching). As a matter of fact, a large number of centrifuge tests have been run to get a better idea of tunnel behavior. Meguid et al. (2008) reviewed many of these tests. Well-known examples are studies by Pokrovsky (see Pokrovsky and Fedorov, 1936), Panek (1952), Mair (1979), and Atkinson and Potts (1977). Interestingly, some of the earliest centrifuge tests were run by Bucky (1931) and Bucky and Fentress (1934) to investigate mine roof stability.

1.5.2 Arching theories

The plural in the title of this subsection is intentional. Some theories such as those by Janssen (1895), Marston and Anderson (1913), and Terzaghi (1943) are based on the ultimate state (Figures 1.32b and 1.34c). Others assume that a stable physical (equivalent) arch is formed in the ground (e.g. Engesser, 1882; Bierbaumer, 1913; Evans, 1983). The "ultimate" and "equivalent" arching theories will be summarized below, the latter including extensions based on work at MIT (Evans, 1983; Iglesia, 1991; Iglesia et al., 2011).

1.5.2.1 Ultimate arching theories

Terzaghi's (1943) theory is shown here as an example of the ultimate theory where a block of ground above the moving support slides down and mobilizes shearing resistance along its vertical boundaries. This theory has its origin in the silo theory (see Janssen, 1895). Terzaghi proposed the sliding and slip surfaces shown in Figure 1.36 above a moving trapdoor. In essence, the movement of the trapdoor reduces the vertical stress on top of it and increases the vertical stress on the fixed surface next to it. This difference causes sliding along the slip surfaces. As can be seen from the slip lines in Figure 1.36a and b, this leads to an inverse bearing capacity behavior. The sliding surface

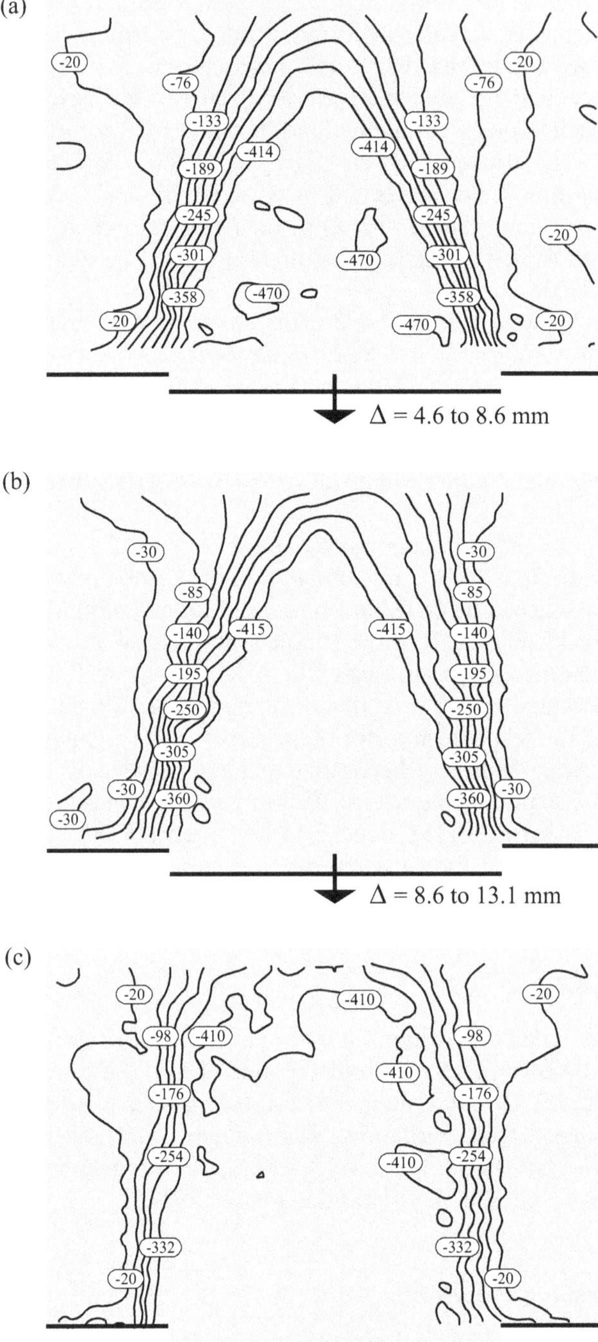

Figure 1.34 Particle movement in trapdoor centrifuge tests as observed by Stone (1988); Iglesia (1991) observed the same behavior. The tests were conducted on sand 14/25. As the trapdoor displacement Δ increases, the shape of the moving part changes from parabolic, to triangular to rectangular.

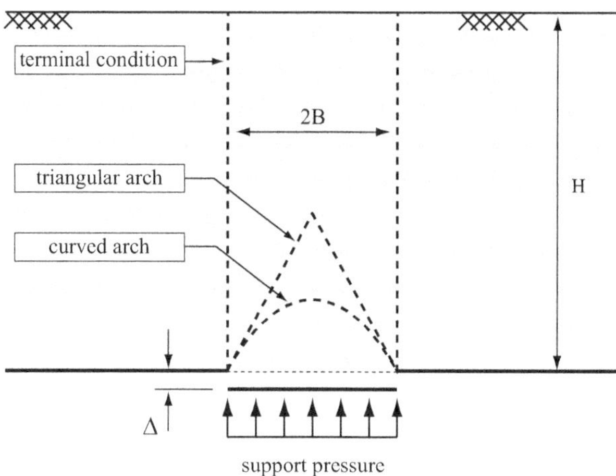

Figure 1.35 Schematic illustration of arching evolution mechanism (from Iglesia et al., 2011).

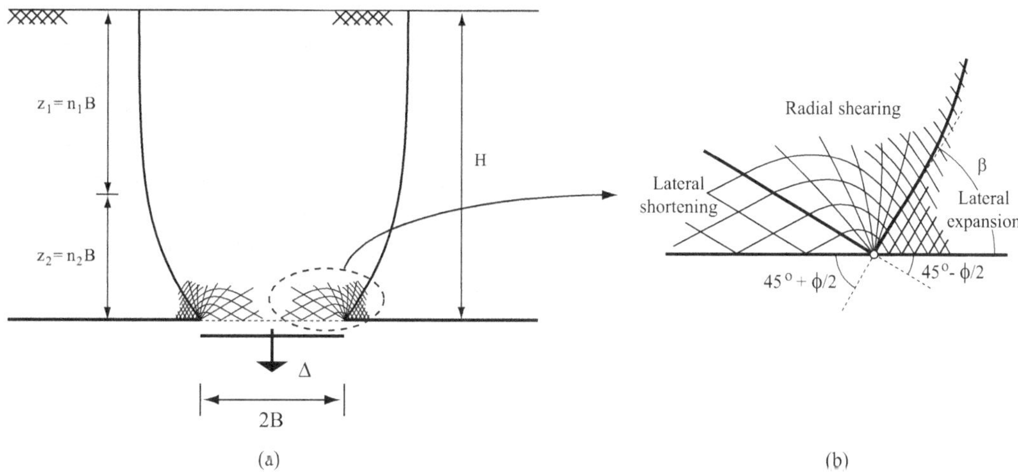

Figure 1.36 Slip lines above a moving trapdoor (from Terzaghi, 1943). (a) Sliplines above trapdor. (b) Detail.

shown in Figure 1.36a is one of the slip lines. Note, however, that the angle β is not well known; it appears to range from β = 90° for shallow depths to β = 45° + ϕ/2 for greater depths. When formulating the arching theory, Terzaghi simplified the geometry to the one shown in Figure 1.37.

Assuming that the lateral stress σ_h and vertical stress are related as

$$\sigma_h = K\sigma_v \tag{1.1}$$

The equilibrium equation for the slice shown in Figure 1.37 (according to Terzaghi, 1943) is:

$$2B\gamma dz = 2B(\sigma_v + d\sigma_v) - 2B\sigma_v + 2cdz + 2K\sigma_v dz\tan\phi \tag{1.2}$$

where γ is the unit weight and all other terms are as defined in Figure 1.37; also $\sigma_v = q$ for $z = 0$. Note that K is not K_o (coefficient of earth pressure at rest). Equation (1.2) can be solved for σ_v

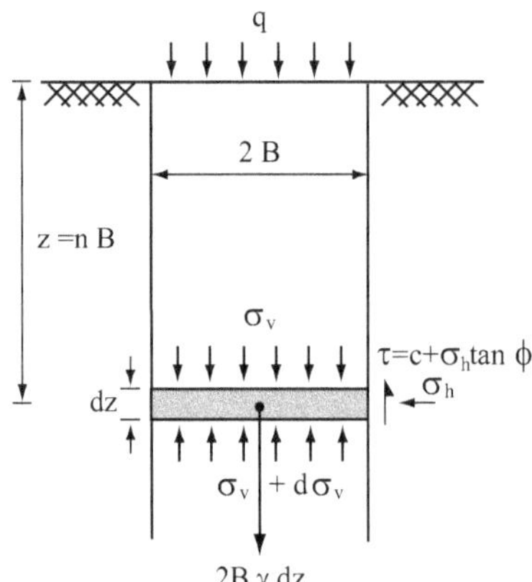

Figure 1.37 Assumption for Terzaghi's arching theory (1943). Pressure in sand acting on an element (slice) between two vertical sliding surfaces. Note that the shearing resistance is assumed to be $\tau = c + \sigma \tan \phi$.

$$\sigma_v = \frac{B\left(\gamma - \frac{c}{B}\right)}{K \tan\phi}\left(1 - e^{-K\frac{z}{B}\tan\phi}\right) + qe^{-K\frac{z}{B}\tan\phi} \tag{1.3}$$

This can be simplified for a number of special cases:

$$c > 0\ q = 0\ \text{(no surcharge)}: \sigma_v = \frac{B\left(\gamma - \frac{c}{B}\right)}{K \tan\phi}\left(1 - e^{-K\frac{z}{B}\tan\phi}\right) \tag{1.4}$$

$$c = 0\ q > 0: \sigma_v = \frac{B\gamma}{K \tan\phi}\left(1 - e^{-K\frac{z}{B}\tan\phi}\right) + qe^{-K\frac{z}{B}\tan\phi} \tag{1.5}$$

$$c = 0\ q = 0: \sigma_v = \frac{B\gamma}{K \tan\phi}\left(1 - e^{-K\frac{z}{B}\tan\phi}\right) \tag{1.6}$$

All equations lead to a limit value at great depth i.e. $z = \infty$, e.g. Equation (1.5) or (1.6) will produce

$$\sigma_v = \sigma_{v\infty} = \frac{B\gamma}{K \tan\phi} \tag{1.7}$$

This corresponds to what the experiments show in Figure 1.32, i.e. a final constant stress value is reached independent of depth.

As mentioned earlier, K relates the horizontal with the vertical stresses; see Equation (1.1). Terzaghi suggested, based on his trapdoor experiments, to use K = 1. Figure 1.38, from these experiments, shows that K is about 1 just above the trapdoor and increases further above the trapdoor to a level of 1.5 at 2B above the trapdoor. At about 5B above the trapdoor, the ground is not affected by the movement and the original horizontal and vertical stresses in the ground remain. According to Terzaghi, one can, therefore, consider that the upper part of

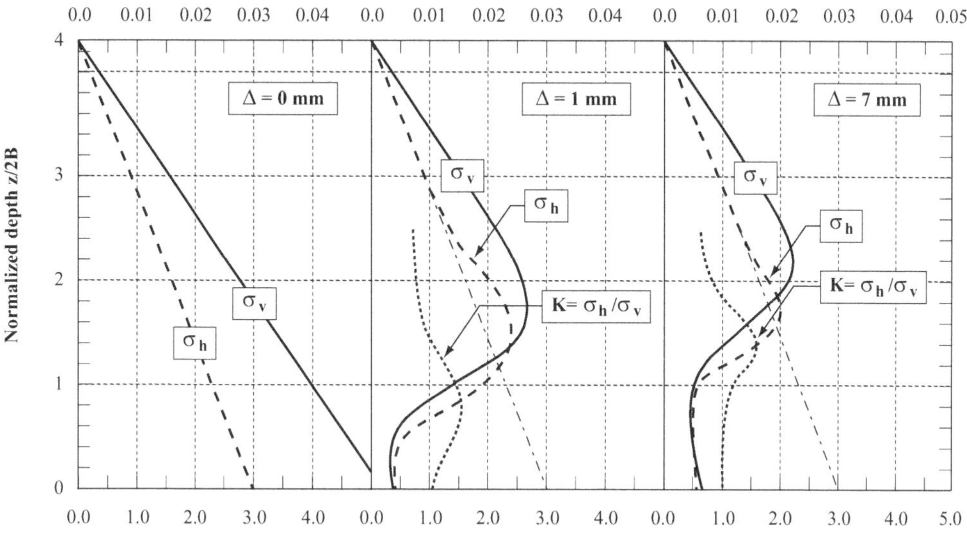

Figure 1.38 Ratio of vertical to horizontal stresses in Terzaghi's 1936 experiments on sand. Trapdoor dimensions: 7.3 cm (= 2B) × 46.3 cm.

the soil column, where no shear stresses along vertical sliding surfaces are mobilized, acts as a load (surcharge) on the lower moving part, which leads to the following equation for c = 0:

$$\sigma_v = \frac{\gamma B}{K \tan\phi}(1 - e^{-Kn_2 \tan\phi}) + \gamma B n_1 e^{-Kn_2 \tan\phi} \tag{1.8}$$

where $z_2 = n_2$ B, for the lower part, and $z_1 = n_1$ B for the upper part (acting as a surcharge; see also Figure 1.36). For $n_2 \to \infty$

$$\sigma_{v\infty} = \frac{\gamma B}{K \tan\phi} \tag{1.9}$$

i.e. the upper part has no effect and σ_v reaches a constant value as before, Equation (1.7).

Figure 1.39 is a plot of Equation (1.8) and of its components for K = 1 and $\phi = 40°$. (It corresponds to the example in Terzaghi, 1943). The curve with $n_1 = 0$ is the theoretical curve for the case of very shallow overburden. One sees that the load is different for different overburden (e.g. at z = 0.5B, $\sigma_v = 0.5\ \gamma B$; at z = 1B, $\sigma_v = 1.75\ \gamma B$). The other curve, with $n_1 = 4$, can be viewed as the theoretical curve for greater depths; it consists of two parts. The upper part is the non-moving part; if one sets $n_2 = 0$ in Equation (1.8), $\sigma_v = \gamma B n_1 = \gamma z_1$, i.e. the geostatic stress. The mobilization of the shear stress in the lower part leads to a reduction of the vertical stress and is represented by the lower part of the curve. The final value of σ_v is as predicted by Equation (1.7) or (1.9). Clearly the shape of the curve depends on the assumed n_1 ($n_1 = 4$ in this case), but in principle it represents well what happens above a moving trapdoor. Figure 1.40 recapitulates the vertical stress distribution above a shallow and deep tunnel (trapdoor), i.e. with small or large overburden.

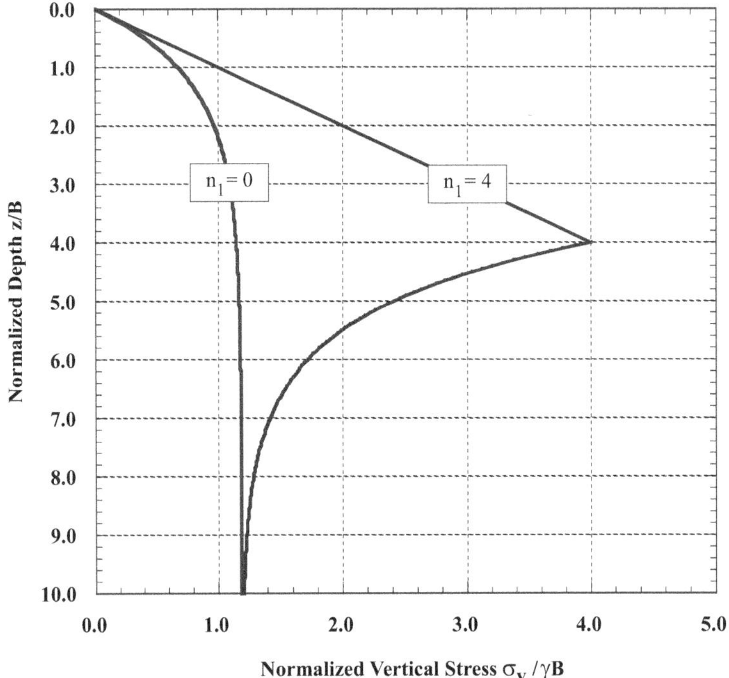

Figure 1.39 Stress distribution in ground above a moving trapdoor based on Equation (1.8).

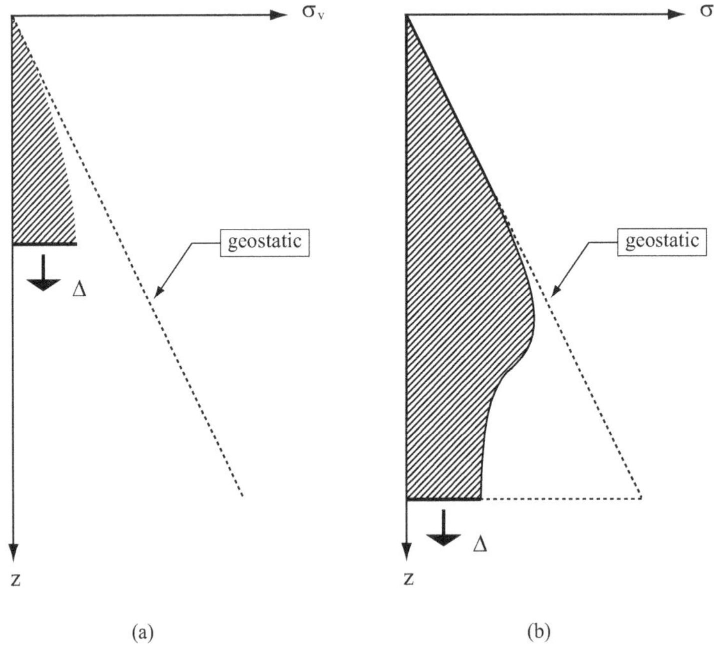

Figure 1.40 Distribution of vertical stress σ_v in ground above a moving trapdoor tunnel. (a) shallow tunnel. (b) deep tunnel.

Terzaghi (1943) also puts arching in the context of analyses for tunnels and recommends, for instance, to consider the fact that sidewall movement should be included by increasing the width B in Equation (1.8) to B_1 (see Figure 1.41).

1.5.2.2 Equivalent arching theory

This subsection essentially follows the discussion by Iglesia et al. (2011), which is based on the original arching theory by Engesser (1882) and extensions by Bierbaumer (1913) and Evans (1983). Engesser, for instance, assumes a stable physical arch such as the one shown in Figure 1.42. This physical arch has a parabolic shape. The angle θ is the angle between the horizontal and the tangent to the parabola at its end points. Engesser (1882) assumes that θ equals the friction angle ϕ. Further development of this concept by Evans (1983), albeit

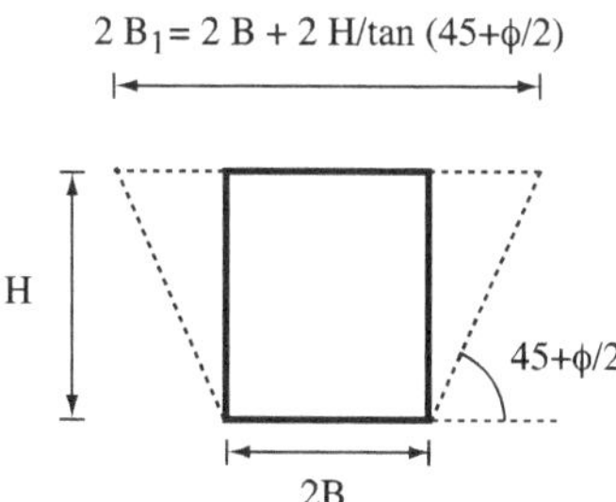

Figure 1.41 Increased width B_1 to consider sidewall movement.

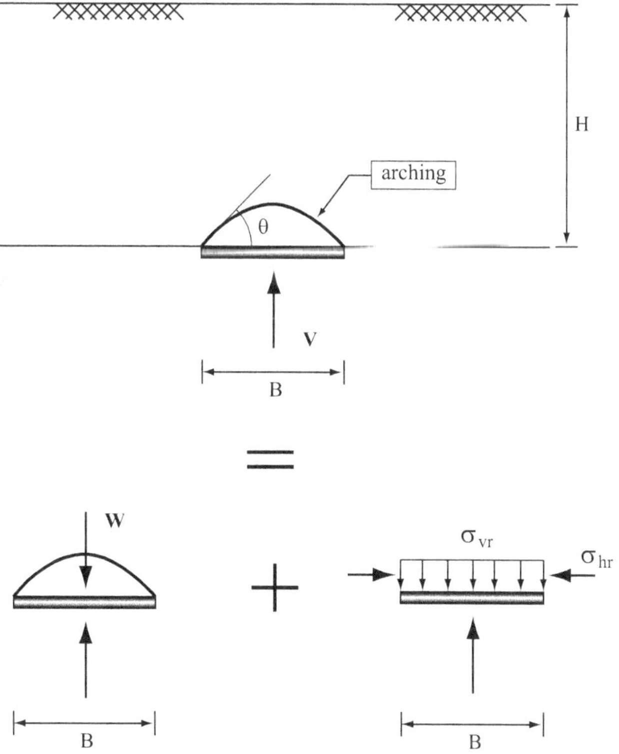

Figure 1.42 Engesser's (1882) equivalent arch approach.

for a triangular arch, shows that the boundary of this arch is a slip line at a dilation angle ν relative to the vertical direction of movement. Assuming an associated flow rule, this angle ν approaches ϕ and thus $\theta = 90 - \phi$. This is what Bierbaumer (1913), also for a triangular arch, assumed and the resulting tall arch is what is actually observed by experiments (e.g. Iglesia, 1991; Stone, 1988; see also Figure 1.35). Nevertheless, Engesser's concept can still be applied, and this stipulates that the material below the arch gets separated from it while the load on top of the arch gets transferred to the sides, increasing the horizontal stresses at this location. As shown in Figure 1.42, the resulting load on the trapdoor is then the weight of the soil (material) under the arch and the vertical stress σ_{vr} induced by the horizontal stress increment σ_{hr}. These assumptions can then be used to compute the stresses on the trapdoor based on Iglesia (1991):

For a parabolic arch with inclination angle θ and with $\theta = 90 - \phi$:

$$W = \frac{\gamma B^2 \tan\theta}{6} = \frac{\gamma B^2 \cot\phi}{6} \tag{1.10}$$

The additional vertical stress can be computed based on the assumption of a structural arch of thickness t and loaded uniformly by q, as shown in Figure 1.43. Engesser (1882) derived q as the overburden load γH (see Figure 1.42) minus the vertical normal stress σ_{vr} acting on the trapdoor (tunnel crown). This leads to:

$$q = t\left(\gamma - \frac{\sigma_{vr}}{H}\right) \tag{1.11}$$

The lateral thrust F_h (Figure 1.39), due to a uniform load, is (e.g. based on Leontovich, 1959):

$$dF_h = \frac{qB^2}{8f} = \frac{qB}{2\tan\theta} = \frac{qB}{2\cot\phi} \tag{1.12}$$

where $f = B \tan \theta/4$ and f is the rise of the arch (Figure 1.43). This allows one to compute σ_{hr} using Equations (1.11) and (1.12).

$$\sigma_{hr} = \frac{dF_h}{dh} = \frac{B}{2\cot\phi}\left(\gamma - \frac{\sigma_{vr}}{H}\right) \tag{1.13}$$

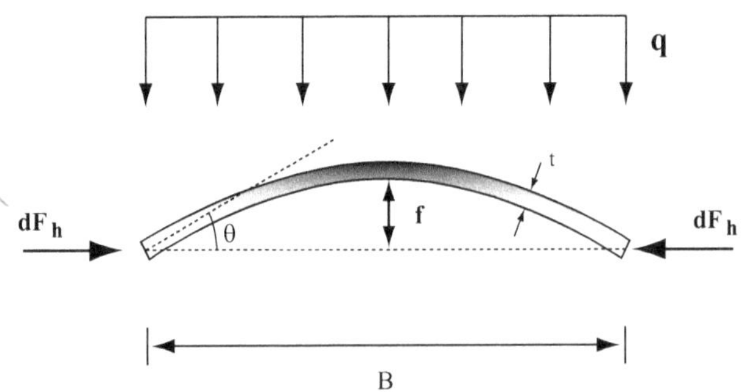

Figure 1.43 Loading of equivalent structural arch in Engesser's (1882) analysis.

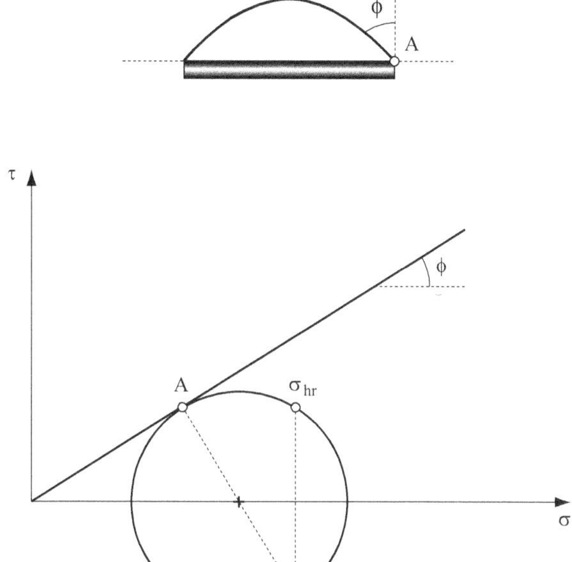

Figure 1.44 Arch and failure stress state at edge of trapdoor. Mohr diagram assuming failure stress state at edge of trapdoor.

The stresses σ_{hr} and σ_{vr} at failure can be related to each other using the Mohr diagram (see Figure 1.44).

$$\sigma_{vr} = \frac{\cos^2\phi}{1+\sin^2\phi}\sigma_{hr} \tag{1.14}$$

Recall that, based on Evans (1983) and Iglesia (1991), the arch, at least at the edge of the trapdoor, is a slip surface justifying the failure assumption just made. Equation (1.14) is often expressed as:

$$\sigma_{vr} = K_E\sigma_{hr}$$
$$K_E = \frac{\cos^2\phi}{1+\sin^2\phi} \tag{1.15}$$

Combining Equations (1.13) and (1.15) leads to:

$$\sigma_{vr} = \frac{\gamma BHK_E}{2H\cot\phi + BK_E} \tag{1.16}$$

The vertical load on the trapdoor is the sum of $\sigma_{vr} \times B$ and the weight W:

$$\text{Vertical Load: } \gamma B^2\left[\frac{HK_E}{2H\cot\phi + BK_E} + \frac{\cot\phi}{6}\right] \tag{1.17}$$

$$\text{Vertical stress on trapdoor} = \gamma B\left[\frac{HK_E}{2H\cot\phi + BK_E} + \frac{\cot\phi}{6}\right] \tag{1.18}$$

The expression leads to values that are very similar to those obtained by Evans (1983) for a triangular arch. However, if one compares this to results obtained with Terzaghi's arching theory, one obtains, e.g. for an infinite depth and a friction angle of $\phi = 40°$ (the case described in Figure 1.39):

Terzaghi:	$1.19\gamma B$
Engesser/Evans/Iglesia:	$0.48\gamma B$

This is not surprising if one compares the "curved/triangular" arch on the one hand and the "rectangular" arch on the other hand, as shown in Figures 1.34 and 1.35. As to be expected, the difference between the two types of theories diminishes with decreasing friction angle (the sides of the parabolic/triangular arch become steeper). Also, comparing the experimental load-displacement curves (ground characteristic curves), particularly those by Terzaghi (Figure 1.32b) a doubling of the load between the minimum and the ultimate can be expected. Hence using the ultimate approach, as compared to the equivalent approach, is more conservative. As shown in Chapter 2 (empirical methods), Terzaghi (1946) uses his arching theory for one of the rock load classes and clearly distinguishes the rock loads based on the displacement that the rock mass above a tunnel undergoes. One can thus conclude that the two types of arching theories provide boundaries and that the actual load to be chosen for design depends on the displacement of the ground.

One final point needs to be made. Both types of arching theories produce values of σ_v that are independent of depth at great depths. This does not usually correspond to reality. It can be explained by the action of a fixed surface next to the moving trapdoor or tunnel crown. This is an admissible assumption if the ground adjacent to the tunnel remains in elastic conditions, i.e. undergoing small displacements relative to those of the crown (possibly including the sidewall effect, see Figure 1.41). If the ground yields over greater distances from the tunnel, the "fixed support" is correspondingly further away. Although it is, in principle, possible to extend the arching theory to such conditions, it is better to use analytical or numerical elasto-plastic approaches as will be described in Chapters 4 and 5. Nevertheless, the concept of ground arching around the opening is very useful and will be used when discussing the elasto-plastic theories and analyses.

Chapter 2

Empirical methods and classifications for rock tunnels

2.1 INTRODUCTION

Empirical methods describe the rock mass qualitatively or quantitatively and relate these descriptions to tunnel support requirements, occasionally also considering the construction process. They are empirical since the relations between rock mass characteristics, so-called rock mass classifications or rock classifications, and tunnel support are based on a collection of cases. As will be seen when discussing the specific methods, some of the empirical relations have been obtained using formal (frequentist) statistical approaches such as regression analysis, while most rely at least to some extent on subjective evaluations. Most of the rock mass classifications discussed here have been specifically developed for use with empirical relations to tunnel support, but some have been extended to be also used in empirical relations regarding slope stability. Also, some of the classifications are used with empirical relations to "classic" mechanical parameters, specifically shearing resistance (friction angle and cohesion) and/or deformability (modulus of elasticity). The shearing resistance and deformability can then be used in analytical and numerical methods for designing tunnels or other structures in or on rock.

Clearly, the question arises why empirical methods are used so much in tunneling and particularly rock tunneling. Beginning in the late 19th century, when engineering entered the phase of formal modeling, this dichotomy between empirical and formal modeling could be seen to emerge even in the work of individuals. Culmann (1866), Ritter (1879), Kommerell (1940) were simultaneously developing analytical approaches to bridge and building design and empirical approaches to tunnel design. Of particular interest in this respect is Ritter's (1879) work. He derived a theoretical solution using the arching concept (see Chapter 1). However, being aware of the limitations, he explicitly called for an empirical approach to be used in conjunction with it. The dichotomy between tunneling and other branches of civil engineering remains. Empirical or simple analytical methods are used at most in the preliminary design of bridges and buildings, while they are often used in all phases of tunnel design and construction, particularly in rock tunnels. Tunnels are built through a largely unknown environment. This environment is composed of different materials that vary over a wide range and, especially in rock, adequate formal mechanical models are sometimes difficult to construct. In contrast, other disciplines of civil engineering, such as bridges or buildings, use materials that are fairly well understood, as are other influencing factors such as loads.

In tunneling, the principles that might serve as a starting point for the theoretical deduction still remain, to some extent, unknown, and the empirical facts are often treated in a manner isolated from the underlying principles. Although this is not entirely satisfactory from a theoretical point of view, the consideration of empirical facts, the establishment and testing of empirical relations promises to bring us closer to the discovery of the underlying principles. In addition, and this is the reason for the treatment in this chapter, empirical methods

DOI: 10.1201/9781003328940-2

are practically very attractive in view of the complexity of ground structure interaction in tunneling. In fact, the past 30 years have seen a resurgence of interest in empirical methods, and, with it, discussion and controversy. As will be seen, the empiricism has other advantages and disadvantages beyond those discussed so far. So the goal of this chapter is to provide the reader not only with a practically useful description of empirical methods but also with a critique allowing the user to judge when and where these methods should be applied with caution.

This chapter will, in the following Section 2.2, structure the empirical methods. This is necessary since the approaches are quite different regarding the characterization of the rock mass and the empirical relationships. Section 2.3 is the central part of the chapter and describes the most frequently used empirical methods in detail. To put everything into context, that section will start with a brief historical review of empirical methods before concentrating on those that are most used at present. For each of these methods, positive and problematic aspects (in the view of the authors of this book) will be mentioned. Finally, Section 2.4 provides an overall assessment of rock classification and empirical tunnel designs and methods. It is important to mention that this chapter relies on the report "Improved Design of Tunnel Supports - Volume 5 Empirical Methods in Rock Tunnelling Review and Recommendations" (Steiner and Einstein, 1980) and the corresponding PhD thesis by Steiner (1979). The reader is, therefore, often referred to these sources. Clearly, as much happened since that report was written, the material had to be completely updated.

2.2 STRUCTURE OF EMPIRICAL METHODS AND CRITERIA FOR THEIR USE

2.2.1 Introductory comments

Both empirical and analytical approaches to tunnel design attempt to relate rock mass conditions to support requirements and construction procedures. These relations must be specified using parameters, which represent the physical systems. Generally, empirical approaches are developed without an explicit behavioral model, while analytical approaches require one. Empirical approaches derive from a collection of prototype observations ("cases" as mentioned in Section 2.1). Analytical approaches derive from "first principles." Nevertheless, empirical approaches are also based on a behavioral model, even if that model is sometimes vague or only implicit.

Empirical and analytical methods serve the same purpose: to determine dimensions and quantities of tunnel support and possibly construction procedures. Empirical methods are often used where there is insufficient information to establish an explicit model, i.e. when the parameter states of such a model cannot be estimated, or when time and cost limitations prevent either. This means that empirical methods are primarily found in two applications:

1. Before construction ("limited" geotechnical information): design of initial support, determination of construction procedure, preliminary design of final support.
2. During construction (limited time): determination of (details of) initial support or adaptation of initial support, determination of construction procedure.

Since the various empirical methods and the associated rock characterization ("classification") differ quite a bit, it is necessary to systematize them by identifying typical underlying structures (Section 2.2.2). Also, to provide the user with some ideas on the applicability and limitations of empirical methods in general and specifically, a set of criteria to judge them will be given and discussed (Section 2.2.3).

2.2.2 Structure of empirical methods

Structuring empirical methods in tunneling was done first by Steiner (1979). In that work (see also Steiner and Einstein, 1980 and Einstein et al., 1977) a taxonomy (Figure 2.1) of empirical methods was proposed, and this has been expanded and somewhat modified for use in this book. As can be seen, the primary distinction is whether geology, i.e. the rock mass, is characterized quantitatively or qualitatively. This distinction is based on the predominant character of the method, since both quantitative and qualitative elements are present in every method. In a truly quantitative description, scaling can be on a physical scale (dimensions of force, length, time). In a qualitative characterization, a comparison is only possible on an ordinal scale, i.e. one can say that condition I is better or worse than condition II or that it is the same for conditions II and III. However, it cannot be said to what degree I is better or worse than II or III.

The structure in Figure 2.1 will be used to group the detailed descriptions of empirical methods in Section 2.3. To provide the reader with some initial explanation of what the taxonomy and structure practically mean, a few explanations are given here.

2.2.2.1 *Ground conditions and other parameters*

Ground or rock mass conditions are characterized qualitatively or quantitatively (or both) as indicated above and is shown in detail in Section 2.3. In a number of methods, tunnel size, depth, and construction procedures are used as modifiers of the rock classifications. Some methods limit their applicability to specific size ranges, while others include size as a factor analogous to the geologic characteristics. Since construction procedures can change the rock mass characteristics (the most obvious difference is between use of tunnel boring machines and drilling and blasting), they are considered together with the characterization of ground conditions. It is important to note that construction procedures are not only an "input"

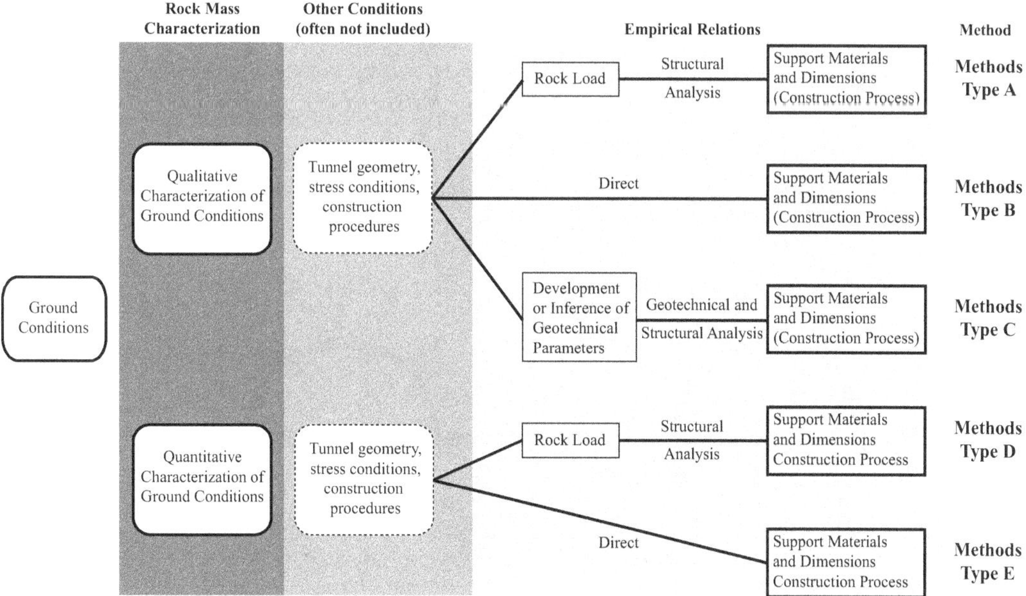

Figure 2.1 Taxonomy of rock classifications and empirical methods.

to the empirical relations but can also be a result, e.g. round length differs with ground characteristics.

2.2.2.2 *Methods A*

A largely qualitative description of the characteristics of the rock mass is transformed into a "rock load" exerting external forces on the support structure. Such loads usually consist of vertical loads only but can also include vertical and horizontal components depending on the quality of the rock mass. With these loads, a standard structural engineering analysis (closed form or numerical) can be used to determine the support dimensions. Such analyses can be done with or without consideration of ground support interaction (Figure 2.2). The Terzaghi (Proctor and White, 1946) method is the best-known Type A method.

2.2.2.3 *Methods B*

A verbal, mostly qualitative, description of the ground conditions is used to describe anywhere from three to five rock (ground) classes. This is what is usually done in application of the New Austrian Tunneling Method (see e.g., John, 1978a, 1978b). A ground class may thus be described as, for instance: "Completely crushed and broken ground, chemically altered. The overburden stress is large compared to the strength of the ground. Thus, the ground will squeeze at the springline, and invert heave will occur. Water reduces stability of the ground." The classes are then directly related to the initial support and to construction procedures (heading and bench, round length, sequence of initial support application).

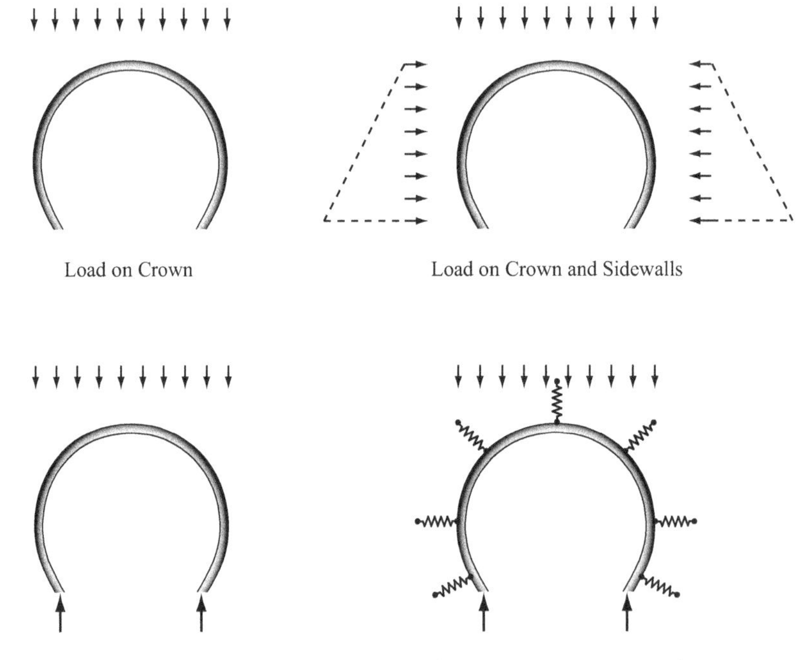

Figure 2.2 Principle of load application and structural analysis.

As the reader will already realize by looking at this procedure, rock classes, support and construction procedures are site specific, i.e. the number of classes and particularly the class description and support and construction details will differ from site to site.

This concept has been expanded to make it more generally applicable in the new guidelines of the "Richtlinie" (2001), in which the ground is described by so-called ground behavior types. These are then related together with opening size, construction parameters, and support types to so-called system behavior types. In this manner, the classification process itself is not site-dependent.

2.2.2.4 *Methods C*

Recognizing the difficulty of developing complete analytical (first principle) models for rock masses, attempts have been made to use simplified analytical models and to obtain the parameters of these models using qualitative or quantitative rock mass descriptions. Specifically, parameters such as the modulus of elasticity, the friction angle ϕ and cohesion c in the Coulomb model, or the Mohr curve can be obtained through relations with qualitative or quantitative descriptions of the rock mass. Examples are the Deere/Hendron (RQD based, e.g. Deere et al., 1967) and the Hoek/Brown (GSI based, e.g. Hoek, 1994b) approaches.

2.2.2.5 *Methods D*

Here a quantitative description of the rock, usually a description of joint geometry, is used to associate the rock mass with different classes (Deere et al., 1969). These classes are in turn related to rock loads which are used with structural analyses to design tunnel supports analogous to what is done in Methods A.

2.2.2.6 *Methods E*

These methods, to which most of the empirical methods in present use belong, are analogous to methods B. This means that the rock mass characteristics are directly related to support requirements; in some methods, also to the construction process. What is different from methods B is that the rock mass characteristics, i.e. the classifications are expressed in form of a number. Depending on the method, the rock mass is quantitatively described by a single number or a two-step approach is used, in which the different rock mass characteristics are described quantitatively and these numbers are then combined into a single one. The geologic characteristics represented by these methods are predominantly the joints (discontinuities) but also water and in some cases stress conditions and tunnel size. An interesting aspect is that some of these numbers can be measured/observed in the field while others are associated with verbal descriptions. The best-known methods belonging to category E are the Deere (e.g. Deere et al., 1969) and the Lauffer (1958) methods, which use a single numerical description, and the Q (e.g. Barton et al., 1974a, 1974b), RMR (e.g. Bieniawski, 1989) and RSR (Wickham et al., 1974a, 1974b) methods, which use the two-step approach.

2.2.3 Criteria for empirical methods

Empirical and, for that matter, also analytical methods should fulfill two sets of criteria: one which might be called user-requirements relating to practicality, economy, and safety; the other which might be called theoretical-requirements relating to correctness of derivation and appropriateness of methodology. The criteria described in this section were originally developed with the help of G. B. Baecher when writing the report Steiner and Einstein (1980). Also, the text and figures below are often taken from that report.

2.2.3.1 Criteria 1 – user requirements

The user needs a generally applicable method, requiring easily measurable parameters, that results in an economic, safe opening. Specifically, these criteria mean:

i. Economy and Safety.
Supports and construction procedures should not be overly conservative nor should they fail. Ideally, the degree of safety should be known. For initial supports, this means a factor of safety near one. For final supports, this means a design with a pre-determinable factor of safety or probability of failure (reliability).
Safety, particularly in the case of initial supports, does not only concern structural failure but also excessive deformations. In squeezing ground for instance, the support may fail without the tunnel collapsing but with large deformations, which may be entirely acceptable. These deformations should, however, not exceed some limit to prevent re-excavation, i.e. unfavorable economic consequences.
ii. General Applicability and Robustness
The method should be applicable to a wide range of ground conditions, opening sizes and shapes, and to different construction procedures and support types. If this is not the case, the range of applicability should be explicitly described.
Although some experience in tunnel design and construction may be a prerequisite, empirical methods should not require such skills, when applying them, that the users could just as easily have developed their own method. This particularly concerns subjective aspects of the method. After a few applications, the user should be able to easily and confidently make the required judgmental decisions. In particular, the method should be relatively insensitive to vagaries in judgment either by the same user or by different users, i.e. the method should be robust and repeatable.
iii. Readily Determinable Parameters
Parameters are determined from boring logs, outcrops, maps, general knowledge of the area, and from observations in the tunnels. Some limited physical testing may also be used. Only a limited number of parameters can be determined from boreholes, outcrops and maps (particularly concerning conditions at tunnel grade). These explorations are usually made prior to construction, and time is thus usually not a limiting problem. In contrast, observations during construction in the tunnel can detect many details, but time is often limited. Parameters that can be easily obtained from outcrops and boreholes or quickly observed (or measured) in the tunnel are desirable.

2.2.3.2 Criteria 2 – requirements regarding the methodology and derivation of empirical methods

While the previously discussed user requirements are concerned with applicability, theoretical requirements involve "basic correctness." As discussed, empirical methods do not rely on explicit formal models of ground conditions, of supports or of construction. Nevertheless, empirical relations are almost always based on some implicit model. The completeness and adequacy of underlying models are of paramount importance.

i. Model Accuracy
Figure 2.3 is a schematic characterization of ground–support interaction where ground is summarized by several influencing factors. The proportions and states (properties) of

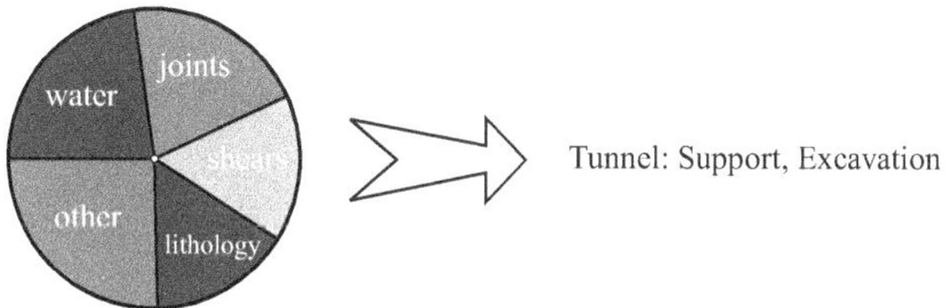

Figure 2.3 Characterization of ground support interaction.

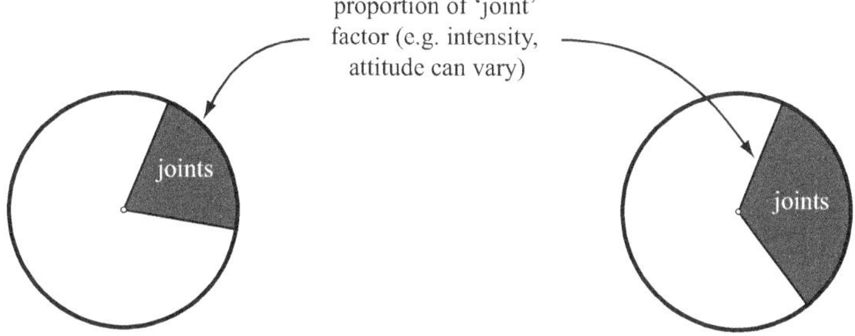

Figure 2.4 Variation of properties and proportions of influencing factors.

these factors can vary (Figure 2.4). The jointing factor, which is shown as an example, represents the influence of joints, i.e. the influence of the number, orientation, spacing, persistence, resistance of joints on the behavior of the ground around a tunnel. These detailed jointing characteristics, "the properties of the jointing factor," can vary and, with them, the effect of the jointing factor on the tunnel. Also, in some instances, jointing may have a smaller effect relative to other factors, e.g. in low-strength rock the intact strength may become relatively more important. The jointing factor may also depend on the size of the tunnel.

Ideally, one would like to know which factors affect the tunnel, their relative importance, and their state (properties). Empirical as well as analytical methods should represent as exactly as possible these relative influences and states. Of course, this is usually not the case. Instead, the ground is characterized by parameters that are not exactly congruent with the true factors. For example, RQD (Rock Quality Designation, which is a modified core recovery defined as the percentage of the length of intact core that is longer than 10 cm (4 inches); see Deere, 1963, Deere et al., 1967) represents a part of the "jointing factor" and a part of the "intact rock factor." Further, the parameters may or may not be exhaustive or mutually exclusive (Figure 2.5). For example, "RQD" and "spacing" do not fully describe the "jointing factor," and to some extent they express the same property of the jointing factor. Also, even if parameters are mutually exclusive, they may still be correlated. For example, joint spacing and water inflow rate.

The best representation of ground for tunneling would be by parameters that are congruent with the influencing factors. This is prevented by limitations that affect all geotechnical design approaches, namely, different types of uncertainty.

- The ground and other in situ conditions are spatially and temporally variable
- Techniques for determining geotechnical parameters introduce sampling and measurement errors: (sample disturbances, random errors, biases, statistical fluctuation) often called parameter uncertainty

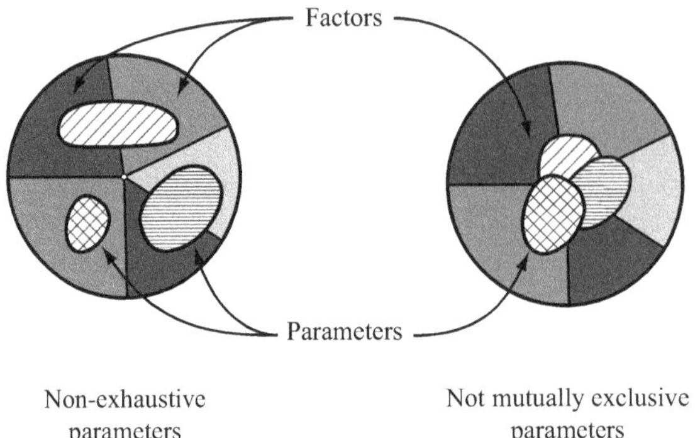

Figure 2.5 Parameters representing influencing factors.

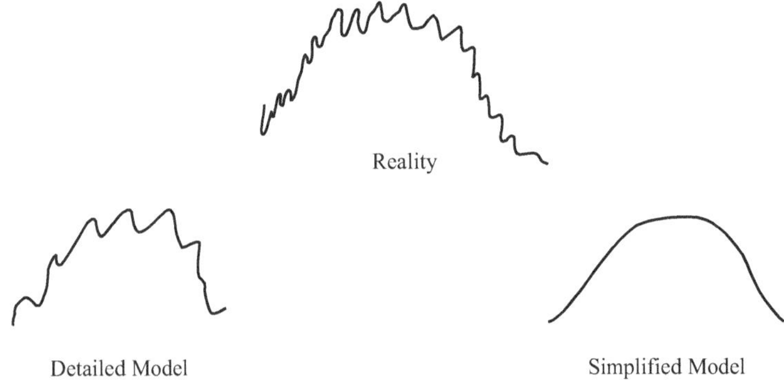

Figure 2.6 Models of different degrees of detail representing reality.

- The underlying model(s) only partially represent(s) reality, i.e. there is model uncertainty.

These uncertainties have been discussed extensively by Einstein and Baecher (1982). A few comments on model and parameter uncertainty are necessary in the context of this book: Empirical (and also analytical) methods relate tunnel features and processes to observed ground parameters, and thus usually involve three underlying models. One model represents geologic and geometric conditions of the ground, one represents interaction of the ground and tunnel, and one represents the connection of observations to actual geologic or geometric conditions of the ground. Again, these models are usually implicit.

It would be desirable to have models that represent the inherent variability of the ground and construction processes as closely as possible (Figure 2.6). However, there are trade-offs to be made in adopting a model. A detailed model is more accurate but requires more parameters than a simplified one. Since the measurement and inference of parameters involve statistical uncertainty, the more interdependent parameters there are, the more uncertainty there is in inferences of each.

There is thus a trade-off between model uncertainty and parameter uncertainty; an increase of the number of parameters does not guarantee a better representation. This trade-off can be seen in the classic Coulomb (c, ϕ) description of granular materials. This model is a gross simplification of intricate particulate mechanics; however, the relative ease with which c, ϕ can be determined makes testing feasible and thus reduces the amount of experimental work necessary to reduce parameter uncertainty. On the other hand, a model considering the details of particle interaction would be more accurate but would require more parameters that are difficult to determine, thus increasing parameter uncertainty.

ii. Subjective Character of Empirical Methods
Because models are abstractions of reality and because parameter estimates involve uncertainties, no procedure for establishing support requirements can be "objective"; they are all substantially subjective. This may not be obvious to the user. On one extreme is an empirical method using a subjective identification of ground parameters, which are subjectively related to support requirements. On the other extreme is an empirical method using a "measurement" of ground conditions, which is related by regression analysis to support requirements. The latter may be more consistent or repeatable and more explicit than the former, but both are fundamentally subjective. Nature is variable and the conditions of test cases must be summarized subjectively by the developer of an empirical method. Furthermore, a large number of implicit assumptions are required in formulating hypotheses for data analysis. Underlying models of an empirical method have a subjective character reflecting the developer's ideas. If the user does not recognize this, errors may result. For instance, rock loads in Methods Type A and D are usually not directly measured but have been subjectively derived.

In going from one extreme to the other, subjectivity (or discretion) by the user is simply replaced by that of the developer. One advantage of the more quantified methods, however, is that repeatability of factor estimates makes an intercomparison of results from one project to another possible. Nevertheless, the extension of regression analysis results to new situations requires a large number of unprovable assumptions taken on faith. The most important of these is that the cases to be predicted are homogeneous in all important ways with the calibrating cases (base cases).

iii. Representative Modeling and Completeness
If an empirical method is based on cases, in which a particular factor was important, generalization may be erroneous. For instance, the significance of geologic structure is recognized in all methods, but some limit the effect of attitude of the geologic structure relative to the opening. These cannot be generalized if the calibration cases were systematically biased with respect to attitude. Some methods have been developed in tunnels in a low strain environment. Such methods cannot be generalized to tunnels in squeezing or swelling rock. Thus, a method should be able to distinguish between different types of mechanisms that influence support requirements. Most importantly, many methods do not take construction effects into account. Unless at least TBM versus Drill and Blast is specified (or preferably details like partial or full face excavation, round lengths and support installation distance and sequence) generalization is difficult.

To summarize, empirical methods should attempt to satisfy the following, in some ways, incompatible, objectives:

1. They should promote economical, yet safe, designs.
2. They should be generally applicable and robust. If they are not generally applicable, the methods should clearly indicate their limits of applicability. The methods must be insensitive to vagaries of use.
3. The required parameters should be readily determinable without restrictions due to time, equipment, or accessibility.
4. The subjective character of empirical methods has to be recognized. Models are abstractions of reality and thus fundamentally subjective. A method that may be considered

"objective" by a user since it is based on measurable quantities is still fundamentally subjective.
5. The model underlying the method whether explicit or implicit should be correct. It should thus consider all relevant factors and differentiate between various types of ground–structure behavior.

2.3 DESCRIPTION OF EMPIRICAL METHODS

2.3.1 Introductory comments

The descriptions in these sections follow Figure 2.1, i.e. the empirical methods will be grouped into the categories shown in that figure and briefly described in Section 2.2. Where possible, the tables and figures in the original papers describing the particular methods will be used. In order to avoid overloading the reader, only the most frequently used methods will be described in detail and, if applicable, only the most recent version will be presented. This will allow the reader to use this book to apply the methods. In addition, some less frequently used methods and methods of historical interest will also be described but in a summary mode. What is important to note, however, is that the relevant information on the development and underlying principles of all of the methods will be discussed. This is very important since only this allows one to properly use each of the methods. As will be seen, some of the comments on the underlying principles are critiquing the method. Again, note that more detailed descriptions can be found in Steiner and Einstein (1980) and, particularly, in the original references describing the methods.

2.3.2 Methods Type A – qualitative indirect (rock load) methods

The principle of these methods, which is contained in Figure 2.1, is shown in detail in Figure 2.7. The methods described here are Bierbaumer (1913), Stini (1950), and Terzaghi (1946).

The first two, especially the Bierbaumer reference, are of significant historical interest. The Terzaghi method is still used in practice, particularly for steel rib/set supports, for which it was developed. Before going into detail, it is necessary to introduce briefly what rock load means and how it is expressed (Figure 2.8). Essentially the volume of rock V times its specific weight is the rock load, usually expressed as $B \times H_p \times \gamma$ per unit length of tunnel; alternatively, the vertical stress is given as $H_p \times \gamma$. The methods either list H_p for a particular tunnel width, as function of width B, i.e. $H_p = f(B)$ or as actual load. As mentioned in Section 2, this load is then used in a structural analysis of the support. Depending on rock type and method, lateral loads are also considered (see details below).

2.3.2.1 Bierbaumer (1913) method

Bierbaumer, an engineer for the Imperial Royal Austrian Railway, was probably one of the first engineers to formally use an observational approach. Specifically, he used observations on initial (temporary!) timber support to design the final masonry liner. This observational aspect will be discussed later in this subsection. Table 2.1 represents Bierbaumer's ground descriptions; namely, the rock load recommendations for the initial (temporary) timber support and the recommendations for the final support dimensions and material. The part on initial support of Table 2.1 follows the "classic" principle of a qualitative description related to rock loads. Here, this is done by grouping the ground into five classes. Interesting is the fact that the long-term loads are higher than the short-term loads, which can be associated with

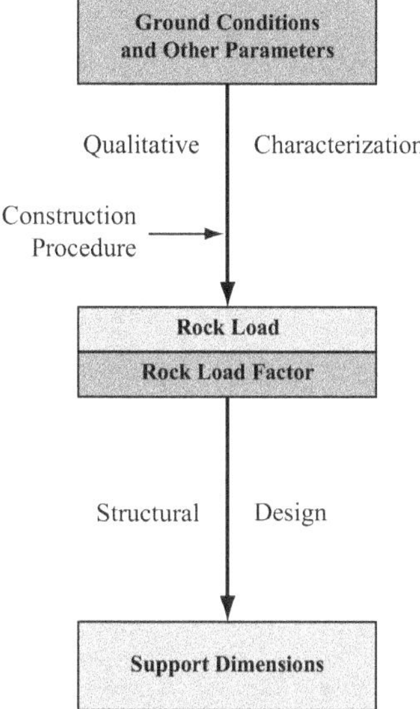

Figure 2.7 Qualitative indirect or rock load methods.

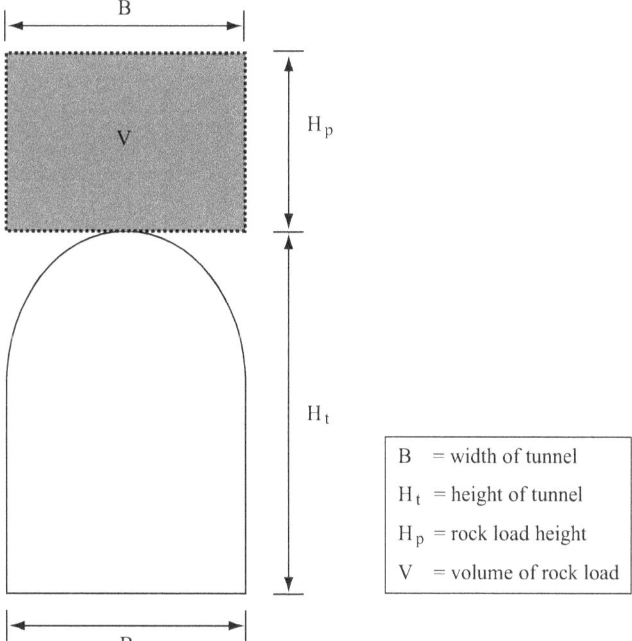

Figure 2.8 Terminology for rock load.

the loosening mechanism described in the context of ground structure interaction (Chapter 1) and which will also be seen in Terzaghi's (1946) approach. The ground–rock load relations are based on a literature review and mostly on Bierbaumer's observation of the deformation of the timber supports. Specifically, the deformation of the timber crossbeams (blocks) where they were intersected by the timber posts was observed. Bierbaumer conducted tests

Table 2.1 Bierbaumer's (1913) rock load and support recommendations

	Rockload (t/m²)		*Initial support: Temporary timber support*				*Final liner: Thickness of liner (cm)*				
							For single-track tunnel		*For double-track tunnel*		
Ground description	*During excavations*	*Long-term*	*Type*[a]	*Observed 'straining'*[b]	*Remarks for rock load*	*Type of cross section*	*Crown*	*Abutment*	*Crown*	*Abutment*	*Remarks for final liner*
Rock more or less afterbreaking	0	8–12	Wide and light	None or minor	Little "Loosening" Pressure	Liner in quarry stone	50–60	70–80	50–80	80–100	No Invert
Cohesive gravel, very after-breaking rock "Mild" rock mass with small overburden	10	35	Wide and strong	Little	Larger "Loosening" pressures that are experienced during mining	For lightly squeezing ground Abutment: Quarry stone Crown: Strong arch in quarry stones	70–80	100–120	90–100	130–150	Invert in quarry stone for cohesionless ground
Cohesionless gravel, strongly afterbreaking rock (crown failures)	20–25	35	Dense and strong	Not excessive	Large "Loosening pressures that are experienced during mining. Probably difficult to stabilize.	For lightly squeezing ground Abutment: Quarry stone Crown: light arch in quarry blocks	60	90	80	120	Invert arch in quarry stone for cohesionless ground
"Mild" ground, larger overburden, squeezing	35	50	Very dense and strong	Important		For heavily squeezing ground Crown: Rough rock blocks Abutment: Quarry stones	80	120	120	180	Invert arch in quarry stone
"Mild" ground very large overburden, heavily squeezing	50	120	Very dense and as strong as possible (hard timber)	Until failure		For very heavily squeezing ground Crown: "trimmed" rock blocks Abutment: Rough rock blocks	80–100	120–150	120–150	180–200	Invert arch in rough rock blocks

[a] Literally translated: wide refers to spacing, light and strong refers to strength of timber support
[b] This is a translation of "Inanspruchnahme" which could also be translated as state of loading. Since timber is quite compressible, a typical deformation is associated with each stress level.

on different types of timber (soft, hard) and different shapes of blocks as shown in Figure 2.9, in which he measured the applied load and observed the deformation, the latter subdivided into four stages ranging from "initial contact but no biting" to "general destruction" of the block (e.g. a stress of 2.5 MPa corresponds to stage 1 in soft timber and 10 MPa to stage 4). These types of deformations observed in the field were then used to backfigure the load. In other words, the crossbeams/blocks were used as load cells.

These observed loads were then not only used to calculate and modify the temporary support but to design the final supports. This is specifically prescribed by Bierbaumer, who

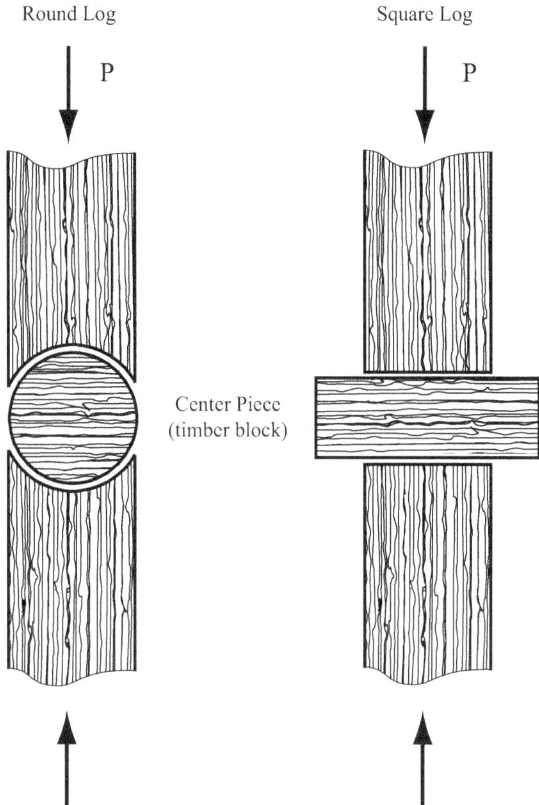

Figure 2.9 Load tests on timber blocks (after Bierbaumer, 1913).

also made a number of additional design recommendations: prevent loosening with careful execution and tight backpacking; in swelling ground the initial load may be deceivingly low. In all this there can be no doubt that there is a formal application of the observational approach.

2.3.2.2 *Stini (1950) method*

Stini used nine (9) ground classes to describe ground conditions and relate them to rock loads (Table 2.2). The relations were developed based on a literature review, which included Bierbaumer's (1913) as well as Kommerell's (1940) and Stini's (an engineering geologist) own experience. The rock load values in Table 2.2 are for tunnel widths between 4 and 5 m. Stini recommends increasing (decreasing) the rock load values by one step for every meter of increase (decrease) in tunnel width.

2.3.2.3 *Terzaghi (1946) method*

Terzaghi's approach is a part of the book (monograph) *Rock Tunnelling with Steel Supports* by Proctor and White (1946), specifically Section I of that publication. That section consists of five chapters (Chapter 1, Principal types of rocks; Chapter 2, Mechanical defects of rocks; Chapter 3, Chemical defects of rocks; Chapter 4, Load on tunnel supports; and Chapter 5,

Table 2.2 Stini's rock load recommendations (after Stini, 1950)

Height of rock pressure for tunnel with 4 to 5 m span	*Ground description*	*Details of ground behavior*	*Ground types*
0–0.5 m	Stable to very stable ground	Very little loosening along the circumference due to excavation work.	
0.5–1.0 m	Ground with satisfactory stability	Afterbreaks only caused by more or less unavoidable loosening caused by the excavation, becomes only important with time.	Mica Schist (rich in Mica) Schistose Gneisses.
1–2 m	Lightly afterbreaking ground	Small loads on support, only little loosening during excavation becomes more "lively" after some time.	Heavily jointed quartzitic phyllites, chlorite schists, laminated calcitic mica schists rich in mica.
2–4 m	Medium afterbreaking ground	Ground becomes afterbreaking after initial stability.	Heavily jointed dolomites in shear zones (Störungsstreifen).
4–10 m	Afterbreaking ground	The ground is quite stable after excavation, however, afterbreaks follow rapidly and strongly. Timbering is lightly loaded.	Clayshale, some thinly laminated brittle sandstones. "Squeeze" dolomites in shear zones.
10–15 m	Strongly afterbreaking ground	Loosening immediately after excavation, local rooffalls.	Thinly laminated marly/clayey sandstones, phyllites rich in mica, some hard marl, laminated calcitic schists, lateral moraine (in the Zwenberg Tunnel up to 15m).
15–25 m	Lightly squeezing ground	The dense and strong timbering is intensely loaded.	Black schists, lightly laminated schists, quartzitic "silk" schists rich in mica, hard rock with narrowly spaced clay filled discontinuities, rock in medium shear zones, many marlshales, humid clay ("dry" clay in Ratkonya Tunnel 14 to 19m), humid till (in the Dössener Tunnel: 20 to 25 m).
25–40 m	Medium squeezing ground	The very dense and strong timbering is heavily loaded.	Brittle (mürbe) thinly laminated "silk" schists, laminated schists, phyllites, soft marls, graphitic schists (black schist), wet clay (in the Ratkonya Tunnel: 27m).
40–60 m	Heavily squeezing ground	The timbering crushes even if it is very strong and dense.	Clay Shales, marls, large shear zones. Quetschgesteine.

Forecast of pressure and water conditions in rock tunnels). As can be gathered from this, Terzaghi goes into quite a bit more detail than what is usually referred to as the Terzaghi rock classification; namely, Table 3 of Chapter 5 of Proctor and White, i.e. Table 2.3 in this book. An attempt is made here to go beyond this table and use Terzaghi's comments related to each of the rock classes. Nevertheless, it is recommended that users of the Terzaghi rock classification consult the original references.

Table 2.3 Terzaghi's rock load recommendations (from Terzaghi, 1946). Rock load H_p in feet of rock on roof of support in tunnel with width B (ft) and height H_t (ft) at depth of more than 1.5 (B + H_t)[a]

	Rock condition	*Rock load H_p in feet*	*Remarks*
1.	Hard and intact	Zero	Light lining, required only if spalling or popping occurs
2.	Hard stratified or schistose[b]	0 to 0.5 B	Light support*
3.	Massive, moderately jointed	0 to 0.25 B	Load may change erratically from point to point
4.	Moderately blocky and seamy	0.25 B to 0.35 (B + H_t)	No side pressure*
5.	Very blocky and seamy	(0.35 to 1.10) (B + H_t)	Little or no side pressure*
6.	Completely crushed but chemically intact	1.10 (B + H_t)	Considerable side pressure. Softening effect of seepage toward bottom of tunnel requires either continuous support for lower ends of ribs or circular ribs*
7.	Squeezing rock, moderate depth	(1.10 to 2.10) (B + H_t)	Heavy side pressure, invert struts required. Circular ribs are recommended.
8.	Squeezing rock, great depth	(2.10 to 4.50) (B + H_t)	
9.	Swelling rock	Up to 250 ft. irrespective of value of (B + H_t)	Circular ribs required. In extreme cases use yielding support

Footnotes by Terzaghi:

[a] The roof of the tunnel is assumed to be located below the water table. If it is located permanently above the water table, the values given for types 4 to 6 can be reduced by 50%;

[b] Some of the most common rock formations contain layers of shale. In an unweathered state, real shales are no worse than other stratified rocks. However, the term shale is often applied to firmly compacted clay sediments which have not yet acquired the properties of rock. Such so-called shale may behave in the tunnel like squeezing or even swelling rock. If a rock formation consists of a sequence of horizontal layers of sandstone or limestone and of immature shale, the excavation of the tunnel is commonly associated with a gradual compression of the rock on both sides of the tunnel, involving a downward movement of the roof. Furthermore, the relatively low resistance against slippage at the boundaries between the so-called shale and rock is likely to reduce very considerably the capacity of the rock located above the roof to bridge. Hence, in such rock formations, the roof pressure may be as heavy as in a very blocky and seamy rock.

* Footnote by the authors of this book: In the original table there is a reference to photos of typical supports and rock mass conditions.

The basis for Terzaghi's rock classification can be inferred as:

- Potential overbreak for stratified, schistose, massive or moderately jointed rock
- Field observations, mostly Bierbaumer's
- Arching tests in Sand (Terzaghi, 1936)

Most importantly, Terzaghi used this information, together with his experience, to provide detailed discussions on each of the rock load classes. This is done in Chapter 4 of Proctor and White (1946). In the following, the relevant comments by Terzaghi on each of the classes or groups of classes will be summarized when discussing the classes in Table 2.3. Terzaghi describes the rock types as follows:

> Intact rock contains neither joints nor hair cracks. Hence, if it breaks, it breaks across sound rock. On account of the injury to the rock due to blasting, spalls may drop off the roof several hours or days after blasting. This is known as spalling condition. Hard intact

rocks may also be encountered in the popping condition involving the spontaneous and violent detachment of rock slabs from sides or roof.

Stratified rock consists of individual strata with little or no resistance against separation along the boundaries between strata. The strata may or may not be weakened by transverse joints. In such rock, the spalling condition is quite common.

Moderately jointed rock contains joints and hair cracks but the blocks between joints are locally grown together or so intimately interlocked that vertical walls do not require lateral support. In rocks of this type both the spalling and the popping condition may be encountered.

Blocky and seamy rock consists of chemically intact or almost intact rock fragments, which are entirely separated from each other and imperfectly interlocked. In such rock, vertical walls may require support.

Crushed but chemically intact rock has the character of a crusher run. If most or all of the fragments are as small as fine sand grains and no recementation has taken place, crushed rock below the water table exhibits the properties of a water-bearing sand.

Squeezing rock slowly advances into the tunnel without perceptible volume increase. Prerequisite for squeeze is a high percentage of microscopic and sub-microscopic particles of micaceous minerals or of clay minerals with a low swelling capacity.

Swelling rock advances into the tunnel chiefly on account of expansion. The capacity to swell seems to be limited to those rocks which contain clay minerals such as montmorillonite with a high swelling capacity.

In practice, there are no sharp boundaries between these rock categories and the properties of the rocks indicated by each one of these terms can vary between wide limits!!!

(Exclamation marks added by the authors.)

Even a very conscientious and expertly conducted geologic survey of the site of a proposed tunnel cannot accomplish more than a very crude estimate of the length of the tunnel sections in which each of the principal types of rock conditions will be encountered. Further differentiation cannot be expected. Hence, even if methods for accurately computing the rock load under given rock conditions were available, they would have very little practical value on account of the inevitable uncertainties associated with predicting the rock conditions. Approximate values for the rock loads to be anticipated under the principal rock conditions are all that tunnel practice requires. The geologic factors which determine the rock load are discussed in Chapter 4 in Terzaghi, (1946). Table 3 (Chapter 5 in Terzaghi, 1946), i.e. Table 2.3 here, contains a summary of the conclusions. Since there are no well-defined boundaries between the different conditions, the rock load corresponding to each rock condition is represented in the table not by a single value but by a range.

Comments on each of the rock conditions (excerpted from Terzaghi, 1946):

Class1: Hard and Intact

"Popping" rock, i.e. superficial rock bursting, is mentioned as a potential problem, i.e. slabs detaching suddenly from the tunnel perimeter. The support required is in essence for protection of the workers. Terzaghi mentions that liners resisting a load of 400 lbs/sq.ft. are sufficient unless the phenomenon progresses to greater depths.

Class 2: Hard Stratified or Schistose. Rock Load 0 to 0.5 B.

The rock load in this class is based on potential overbreak. A distinction is made between horizontally, vertically and inclined stratified rock, which leads to the different rock loads in the range 0 to 0.5 B.

Horizontally and vertically stratified rock.

Terzaghi lists the following factors as most important in influencing overbreak:

1. Spacing between joints,
2. Shattering effect of blasting of the rock located beyond the payline.
3. Distance between the working face and the roof support.
4. Length of time which elapses between the removal of the natural support of the roof and the installation of the artificial support.

For widely spaced transverse joints with respect to the opening (see Figure 2.10) a rectangular cross section will be stable. For more closely spaced joints, a dome is formed in the crown. If the required cross section of the tunnel does not follow the natural arch, the rock that tends to drop out of the crown has to be supported, i.e. overbreak tends to occur.

The effect of distance from the face to the support is illustrated in Figure 2.11. The closer the support is placed to the face, the smaller the overbreak due to a three-dimensional dome action (Figure 2.11). The maximum possible height of overbreak in horizontally stratified rock occurs for weak highly stratified rock and is at most 0.5 B (Figure 2.12). The potential overbreak shown in Figures 2.10 to 2.12 indicates what the maximum rock load might be, i.e. the rock volume detaching below the stable geometries shown in these figures. However, if the support is installed quickly after excavation, the loosening will be minimized and smaller loads will occur. The recommended rock load thus ranges from 0 to 0.5 B.

The derivation for vertically stratified rock striking parallel to the tunnel is similar to that for horizontally stratified rock (Figure 2.13). The rock load in this class is also based on potential overbreak. In this case, the maximum roof load assessed by Terzaghi is 0.25 B, leading to a range of rock load from 0 to 0.25 B. Terzaghi does not separately consider rock strata striking perpendicularly to the tunnel; for this case the values will, however, not exceed those for strata striking parallel to the tunnel.

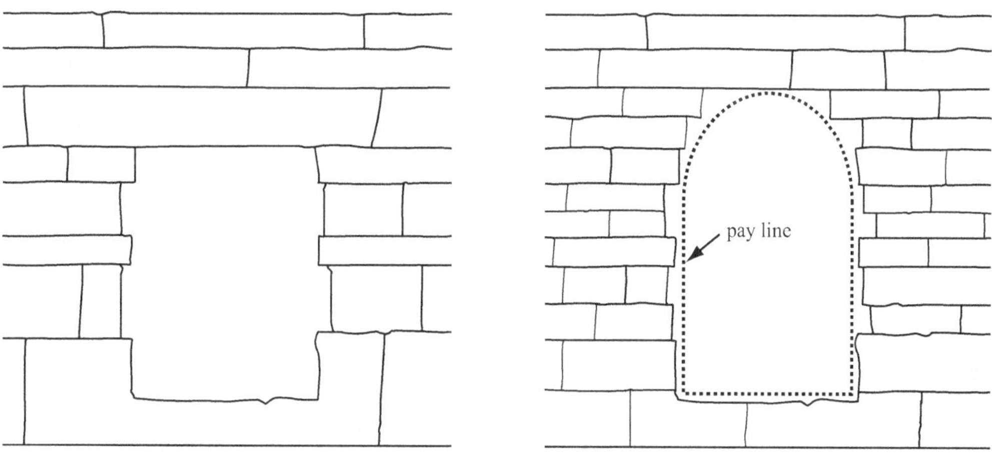

Figure 2.10 Effect of transverse jointing on tunnel stability (from Terzaghi, 1946).

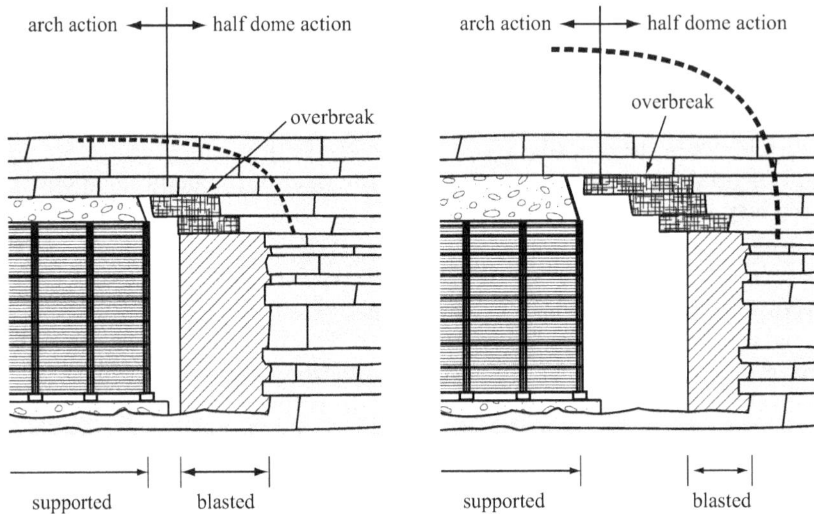

Figure 2.11 Effect of unsupported length on overbreak (from Terzaghi, 1946).

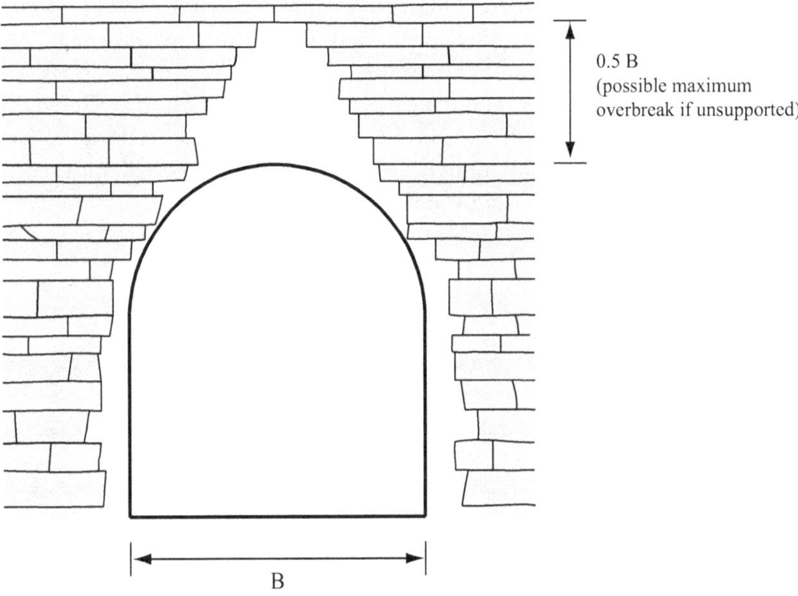

Figure 2.12 Maximum probable overbreak in horizontally stratified rock (from Terzaghi, 1946).

Inclined Strata:

For this case, Terzaghi states that the overbreak tends to produce a peaked roof, the maximum value of the roof load being in the order of 0.25 B for steep strata and 0.5 B for gently inclined strata. More important, however, is the consideration of the possibility of sliding wedges at the springlines (Figure 2.14). For this case Terzaghi proposes to perform a graphic analysis; as shown in Figure 2.14. The shear resistance is assumed to be caused by friction only but needs to also consider pore pressure if applicable.

Class 3. Massive and moderately jointed rock. Rock Load 0 to 0.25 B

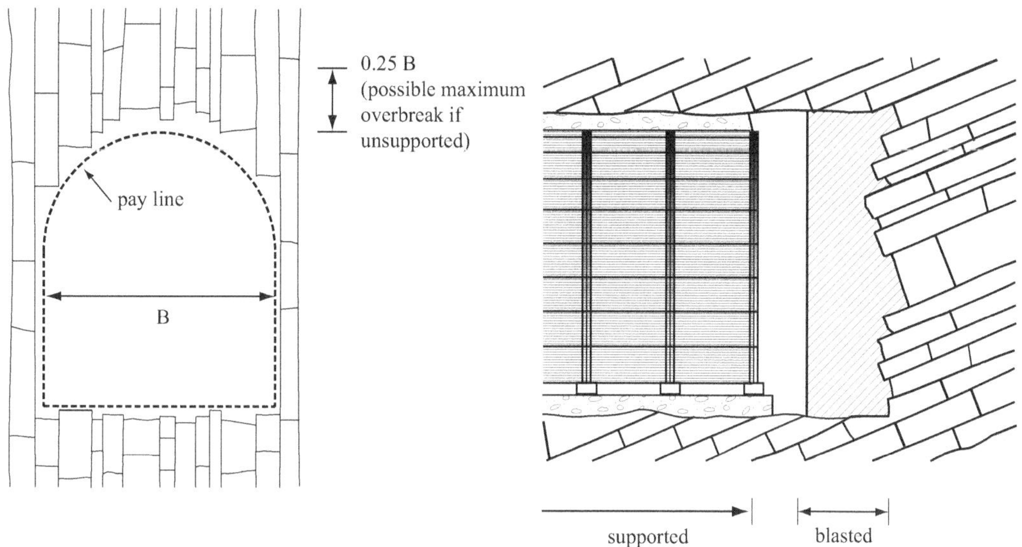

Figure 2.13 Tunnel in vertically stratified rock (from Terzaghi, 1946).

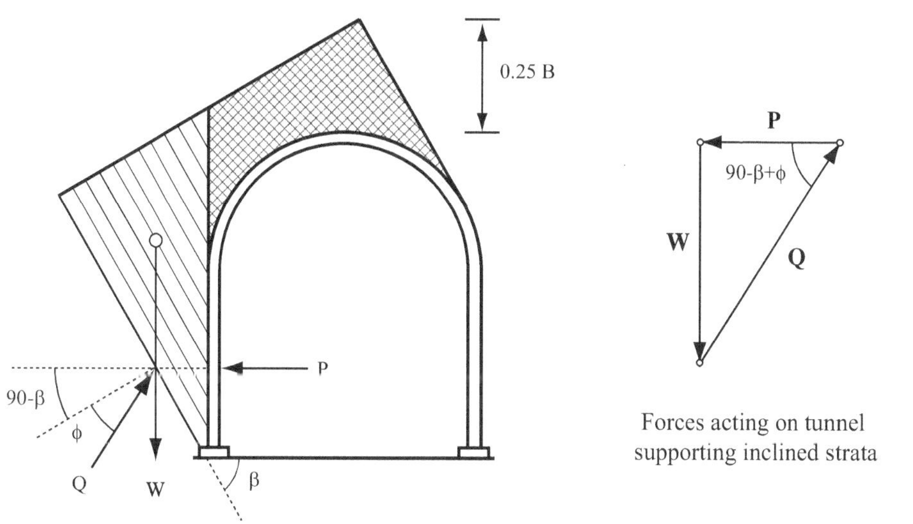

Figure 2.14 Unstable wedges at springlines (from Terzaghi, 1946).

As can be taken from Figure 2.15, the joint surfaces are irregular causing a tight interlock, and consequently there is only shallow overbreak. Depending on the timing and quality (tight wedging of support against rock) no or very small rock loads result, specifically 0 to 0.25 B.

The following rock load recommendations for classes 4 to 6 are based on field measurements and, for class 6, also on Terzaghi's (1936) arching experiments. The field data used are those reported by Bierbaumer (1913) (measurements in timber supported tunnels) with some interpretation by Terzaghi.

Class 4: Moderately blocky and seamy, Rock Load 0.25 B to 0.35 (B + H_t*); and*
Class 5: Very blocky and seamy, Rock Load (0.35 to 1.10) (B + H_t*)*

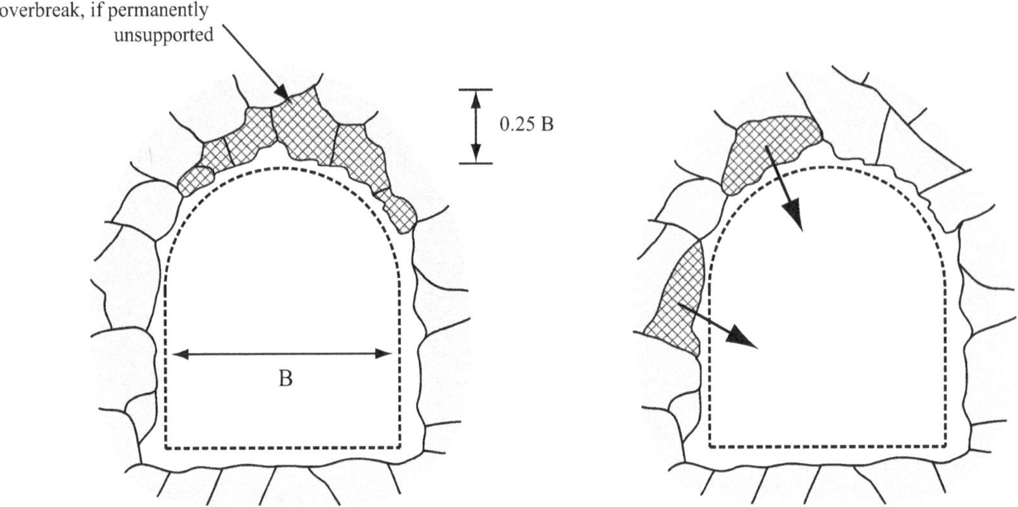

Figure 2.15 Overbreak and popping in moderately jointed rock (from Terzaghi, 1946).

Note that while these are the terms used in Table 2.3 and also used in this manner in Terzaghi's (1946) corresponding table, the accompanying text in Terzaghi (1946) uses the terms "moderately blocky" for Class 4 and "very blocky and shattered" for Class 5. These later terms are a bit more obvious (see also the explanation later in this book).

The question arises as to where the rock load values come from. The actual values seem to come from "observed rock loads in railroad tunnels" as mentioned in a footnote in Table 2 in Terzaghi 1946. Also, Terzaghi (1946) says:

> Our knowledge of the intensity of rock loads is derived chiefly from the results of tests which were done in various railroad tunnels in the eastern Alps. In these tests, wooden blocks with known strength were inserted between the individual members of timber sets and the load H_p on the timbers was estimated from the visible modification of the progressive failure of the blocks.

It appears that these are Bierbaumer's observations but this is not said so. In addition, Terzaghi draws conclusions from the generic behavior of support loads as tunnel excavation proceeds and also compares the rock loads of classes 4 and 5 to those he obtained from sand based on arching tests. These two latter arguments will be discussed in more detail below.

An important assumption is that, in contrast to Classes 2 and 3, there is only minimal interlocking between rock blocks and the rock mass behaves thus in principle similar to sand as discussed with Class 6. Specifically, this means that the rock load is exerted over a width greater than the tunnel width B but this width is a function of tunnel height H_t (see Figure 2.18), that is: Rock Load $H_p = f\ (B + H_t)$.

Another set of basic assumptions for these classes relates to what happens near the face as shown in Figures 2.16a, b. In the unsupported part between the face and the first initial support, a "half dome" is formed which implies greater stability/less overbreak than for a comparable two-dimensional arch, as shown in Figure 2.16a. This transition "half dome-arch" will happen independently of how well supported (backpacked) the supported tunnel is. However, and, as will be discussed in relation to Class 6, more movement in badly backpacked cases will produce higher ultimate loads. The half dome-arch transition needs to be considered in conjunction with what Terzaghi calls "bridge action." This is a combined

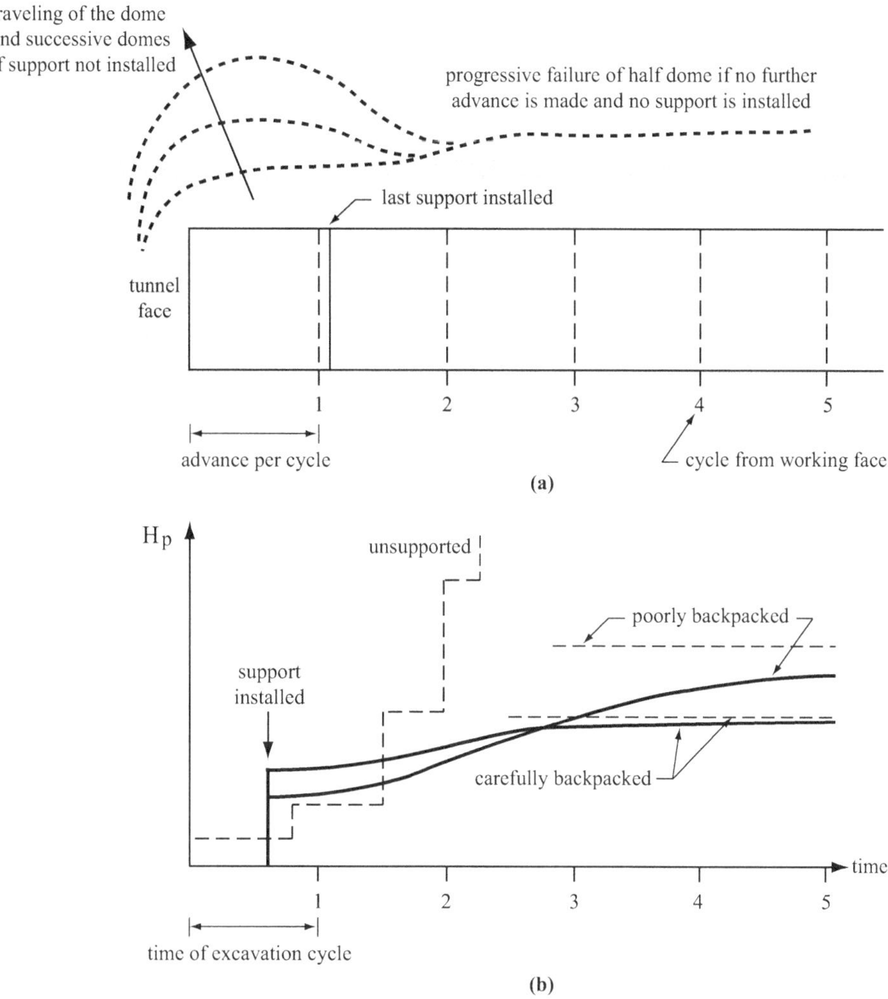

Figure 2.16 Relation between time, overbreak and rock load in blocky and seamy rock (from Terzaghi, 1946). (a) Dome action. (b) Rock load.

time-distance function, i.e. it depends on how long the space between face and first initial support is left unsupported and on the distance between the face and the first initial support. The effect of this "support delay" is discussed in Chapters 1 and 4 of this book. Depending on this bridge action, the rock mass may deform significantly or literally fall out. Large deformations lead to a larger half dome (Figure 2.16a) and very high rock loads (e.g. "unsupported") in Figure 2.16b.

Based on these assumptions, and mainly on the previously mentioned observed rock loads, Terzaghi assumes the values for rock loads in blocky and seamy rock, in "wet" tunnels (under the ground water table), given in Table 2.4. Two important comments are made by Terzaghi: (1) no definite boundary can be drawn between moderately jointed (i.e. Class 3) and very blocky and shattered rock (i.e. Class 5); and (2) as mentioned above, the values in Table 2.4 are for "wet tunnels" which might be interpreted as being under the ground water table. By analogy with arching tests in sand (see discussion of Class 6), one can assume that for dry tunnels the rock loads (see Table 2.3) can be reduced to ½ of the values given there. Terzaghi also says that unless dry conditions are guaranteed, one should work with the values in Table 2.4.

Table 2.4 Rock loads for closed areas ("Wet" tunnels)

	Initial value	*Ultimate value*
Moderately blocky rock (Class 4)	Hp = 0	$H_p = G\ H_{p\ ult} = 0.25\ B$ to $0.35\ (B + H_t)$
Very blocky rock (Class 5)	$H_p = 0$ to $0.6\ (B + H_t)$	$H_{p\ ult} = 0.35\ (H_t + B)$ to $1.10\ (B + H_t)$

From all the above and naturally from a detailed study of Terzaghi's original text, one can conclude that although using measured values as a base, there is a large degree of uncertainty. Terzaghi attempts to bracket this uncertainty on the basis of engineering judgment and by comparison with the extreme boundary case "sand" (see below).

Class 6: Completely crushed but chemically intact rocks, Rock Load 1.10 (B + H_t)
Terzaghi describes the physical behavior of crushed rock as being analogous to that of sand. Specifically, this relates to the fact that arch action develops if sand moves above an opening which leads to a significant reduction of the load compared to the actual overburden. The basis for all this is Terzaghi's trapdoor experiments (Terzaghi, 1936) which show this load reduction. Details of arching action are discussed in Chapter 1 of this book. Nevertheless, the principle of arch action providing the basis for Class 6 is (again) shown in Figure 2.17. Essentially the crosshatched zone moves toward the tunnel and mobilizes friction along its boundaries (ac, bd). The thickness of the crosshatched zone is D >1.5 B; the overburden above the crosshatched zone does not contribute to the load on the tunnel. As a matter of fact, a very small downward movement of this zone mobilizes friction to such an extent that only a small portion of the weight of the crosshatched zone acts as load on the tunnel. As the relative displacement increases and dilation is overcome (see Chapter 1)

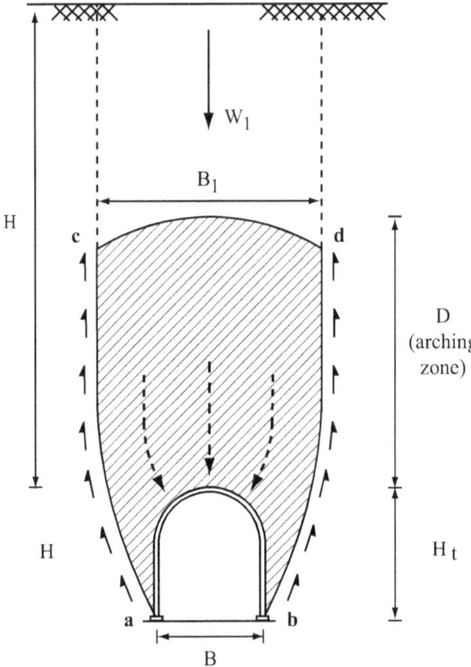

Figure 2.17 Ground arching (after Terzaghi, 1946).

the rock load increases. Since dilation has a significant effect on mobilizing friction and thus arching action, the initial small displacement "rock load" values are different for loose and dense sand while they are identical for large displacements. All this is shown in Table 2.5, where numbers are those that Terzaghi has taken from (his) model experiments.

What is also evident from Figure 2.17 is that not only the roof (crown) of the tunnel moves but also the sides of the tunnel. Hence, as shown in Figure 2.18, the rock load is computed as a function of $B_1 = B + H_t$, i.e. the inclined sides of the two lateral wedges are assumed to be inclined at the slope of 2V:1H. Two very important additional comments by Terzaghi need to be taken into account: (1) as indicated in Table 2.5, small displacements lead to the small rock loads. The tunnel support should, therefore, be tightly and quickly backpacked (recall that the Terzaghi method is based on steel supports); and (2) Terzaghi

Table 2.5 Rock loads for class 6

Dense Sand	$H_{p\ min} = 0.27\ (B + H_t)$ for yield of $0.01\ (B + H_t)$
	$H_{p\ max} = 0.60\ (B + H_t)$ for yield of $0.15\ (B + H_t)$
Loose Sand	$H_{p\ min} = 0.47\ (B + H_t)$ for yield of $0.02\ (B + H_t)$
	$H_{p\ max} = 0.60\ (B + H_t)$ for yield of $0.15\ (B + H_t)$

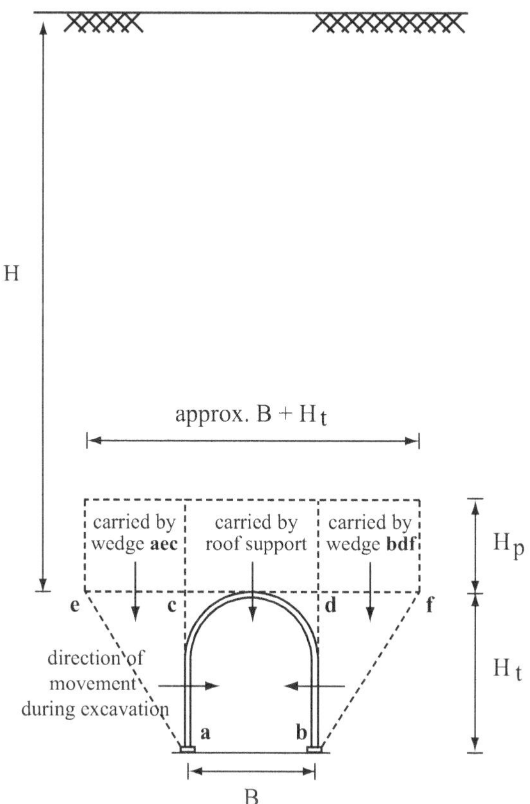

Figure 2.18 Loading of tunnel support in sand.

also states that regardless of the initial state and yield and after backpacking, the loads will eventually increase by 15%.

Class 7: Squeezing rock, moderate depth, Rock Load = (1.10 to 2.10) (B + H_t); and Class 8: Squeezing rock, great depth, Rock Load = (2.10 to 4.50) (B + H_t)

To quote Terzaghi:

> Squeezing rock is merely rock which contains a considerable amount of clay. The clay may have been present originally, as in some shales, or it may be an alteration product. The rock may be mechanically intact, jointed or crushed. The clay fraction of the rock may be dominated by the inoffensive member of the kaolinite group or it may have the vicious properties of montmorillonite. Therefore, the properties of squeezing rock may vary within as wide a range as those of clay.

As can be concluded from Chapter 1 (Ground Structure Interaction), associating squeezing with clay content may be too restrictive.

The rock load for Classes 7 and 8 appear also to be based on Bierbaumer's (1913) work. In Proctor and White (1946), Terzaghi has not referenced Bierbaumer; however, he uses Bierbaumer's table in the chapter on tunnel geology in Redlich et al. (1929). Bierbaumer has given his rock load recommendations as absolute load values in timber supported tunnels. Terzaghi modified them and related them to the dimensions of the tunnel.

It is important to note that the pressure is not only exerted in the crown but also in the sidewalls and invert for which Terzaghi recommends:

Side wall pressure = 1/3 roof load pressure
Invert pressure = ½ roof load pressure

Class 9: Swelling rock. Rock Load up to 250 ft irrespective of the value (B + H_t)
In his comments, Terzaghi mentions that, in deep tunnels, high swelling rock pressures up to 10 tons/sq.ft. and even 20 tons/sq. ft. have been observed, although it is not clear where these observations were made. The latter value corresponds roughly to an overburden of 270 ft., explaining the number 250 above. What is, however, interesting and related to the comments in Chapter 6 on design of tunnels in swelling rock are observations and recommendations made by Terzaghi in this context. He mentions cases where the initial timber support was crushed but the crushed support was then perfectly capable of maintaining the load (after the ground has undergone this deformation). Alternatively, final liners built with a space between the extrados and the rock have not sustained damage. All this is related to the fact that swelling pressure might be reduced if the swelling rock is allowed to deform. As discussed in Chapter 6, this may or may not be so, depending on the stress path. While Terzaghi is quite evidently aware of the problematic aspect of allowing too much deformability, saying that in most cases, including slightly swelling rock, "tight backpacking" should be applied and even in strongly swelling rock a support allowing initial allowable deformation with a more rigid limiting behavior (e.g. compressible elements) is suggested.

Further comments about the rock loads made by Terzaghi (1946):

1. The side pressure in swelling rock can be estimated at $p_H = 0.3\ \gamma\ (0.5\ H_t + H_p)$ where γ is the unit weight per cu. ft.

2. The rock load in the tunnel increases with time even if the support is tightly backpacked. Hence the rock load and side pressure increase by about 15%, i.e. $H_{ult} = 1.15\ H_p$.
3. The values in Table 2.5 are for dry sand (above the groundwater table). Terzaghi's experiments show that seepage toward the tunnel does not disturb the arching action but doubles the values in Table 2.5. The two last comments explain the value of 1.10 $(B + H_t)$ in Table 2.3 (maximum load for loose sand = $0.47 \times 1.15 \times 2 \cong 1.10$). Using this value may thus be quite conservative!
4. Finally, a comment is made that upward directed groundwater flow at the tunnel invert might reduce the bearing capacity of the steel rib foundation.

2.3.3 Methods Type B – qualitative direct methods

The principles of these methods are shown in Figure 2.1 and are repeated in the schematic of Figure 2.19. In this approach, ground (rock mass) conditions are described largely verbally although where necessary also including numbers such as joint spacing, overburden and water inflow. These conditions are then related to detailed descriptions of the initial support characteristics, i.e. dimensions and materials as well as to the construction procedures; very often the design characteristics of the final liner are also included. In essence, this is the original approach for designing tunnels before the rock load and similar methods (Type A) were initiated. Hence, many older tunnels were designed in this way but one finds this approach also in more recent tunnels, e.g. the Eisenhower tunnel in Colorado (see Steiner, 1979, for details).

It is clear from the above that the rock mass description and the relations between rock mass characteristics and support/construction procedure can differ from application to application. This is the reason why the term "rock mass classification" is not used in the context of Methods Type B.

Two closely related methods are described in this context, namely, the New Austrian Tunnel Method (NATM) and the "Rock Mass Behavior Type" (Classification). The reason is that the NATM is widely used and, most importantly, the rock mass description and relation to support/construction have been systematized in the "Rock Mass Behavior Type" classification. The discussion below deals exclusively with this ground characterization and relation to support/construction. It does not deal in detail with the underlying principles of the

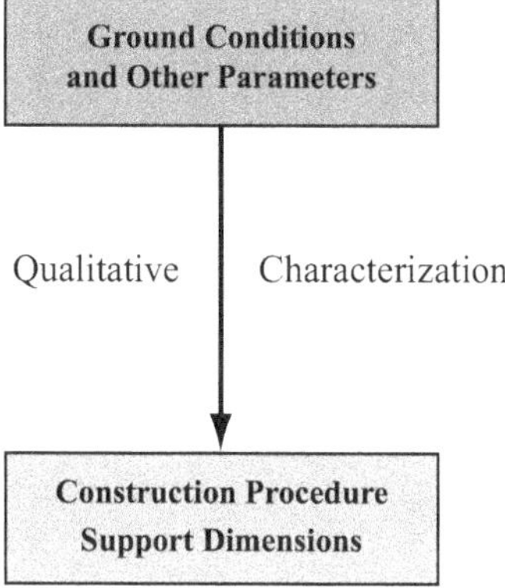

Figure 2.19 Qualitative direct methods.

NATM (observational method, control deformability, sequential excavation). Nevertheless, the underlying principles are reflected in the support types and construction procedures as will be shown below. In the following, a brief description of the history of the NATM rock mass characterization is given followed by some early examples and then emphasizing the most recent systematic approach from the Austrian standards (Richtlinie, 2001). Again, it has to be emphasized that the descriptions and relations differ from case to case, particularly in the original NATM approach, but that it is, nevertheless, possible to provide a systematic framework. Also, very important is the fact that since the NATM is an observational approach, the typical ground characteristics and relations to support/construction allow one to choose the appropriate previously designed support/construction combination on site. As will also be shown, it is also possible that the design and construction characteristics can be modified during construction if the observed performance so requires.

Historically (Steiner et al., 1979), it appears that one can trace back the NATM ground characterization to Lauffer's classes in the Prutz-Imst tunnel, which formed the basis both for Lauffler's standup classification (Type E Method, Section 2.3.6) and for its use in the NATM. The latter started with Rabcewicz' (1957, 1963) in the Schwaikheim tunnel with further development in the Tauern (see e.g. Pöchhacker, 1974) and Katschberg tunnels. The early and mid-1970s can be considered as the major development phase of the NATM in rock tunnels (see also Pacher et al., 1974). The practical experience and the increasing number of cases led, as one expects of empirical methods, to a corresponding development (change) of the characterization and the relation to support or construction procedures. Table 2.6, after John (1978a, 1978b), represents the status of rock mass description and design and construction consequences at that stage of the NATM. Particularly noteworthy are the geomechanical indicators and the great detail with which the construction (excavation, support) is described. Usually the NATM descriptions also involve detailed figures. An example is shown in Figure 2.20 for the Arlberg tunnel where "ground classes" IIIb, IV, and V are shown both prior to construction and as modified based on observed performance during construction.

The newest development regarding Methods Type B (direct relation between a mostly "qualitative description" of rock mass behavior and support/construction consequences) is documented in "Richtlinie" (Guideline for the Geomechanical Design of Underground Construction with a "Cyclic Advance," B 2201-1, and Guidelines for Geomechanical Design of Underground Construction with continuous excavation, B 2201-2) by the Austrian Society for Geomechanics (2001). These guidelines, as the titles indicate, are actually going beyond the use in the narrow context of the NATM and have been applied in tunnels with and without the principles of the NATM. Also, they can be applied both to rock and soil.

The methodology used in these guidelines is shown in Figure 2.21. As can be seen, it encompasses the entire process from describing the rock mass to determination of tunneling methods (classes) as needed for the bid documents. The process starts with the determination of ground types for which Table 2.7 (Table 1 in "Richtlinie" – Guideline) serves as an aid. As can be seen there, this includes both rock masses and soil deposits. Appendix A in the "Guidelines" contains detailed references to procedures and standards, which provide descriptions on how to determine the "relevant parameters." The next step leads to the Ground Behavior Types. (The translated version of the guidelines in Schubert et al., 2003, and Tenschert et al., 2003, shown in Figure 2.21, refers to "Behavior Type.") The ground behavior types are obtained by combining the ground types with aspects related to the underground space. Specifically, for tunnels (~ one dimension much larger than the others), an infinitely long unsupported opening is assumed, while for more equi-dimensional openings

Table 2.6 Ground classification for the Dalaas tunnel (after John, 1978)

			Excavation					*Support procedure*				
Class	*Ground behavior*	*Geomechanical indicators*	*Section*	*Round length*	*Method*	*Stand-up time (guidelines)*	*Construction procedure*	*Principle*	*Crown*	*Springlines*	*Invert*	*Face*
I	Intact rock (free standing, = standfest)	The stresses around the opening are less than the rock mass strength; thus the ground is standing. Due to blasting separations along discontinuities are possible. For high overburden danger of popping rock.	Full face possible	No limit	Smooth blasting	Crown: weeks springlines: unlimited	Check Crown for loose rock. When Popping rock is present placement of support after each round	Support against dropping rock blocks	Shotcrete: 0.5 cm Bolts: Cap = 15 t Length = 2 to 4 m locally as needed	Bolts: Cap = 15 t Length= 2 to 4 m locally	NO	NO
II	Lightly afterbreaking (nachbrüchig)	Tensile stresses in the crown or unfavorably oriented discontinuities together with blasting effects lead to separations.	Full face possible	3 to 5m	Smooth blasting	Crown: days springlines: weeks	The crown has to be supported after each round.	Shotcrete support in crown Bolted arch in crown	Shotcrete: 5–10 cm with wirefabric (3.12 kg/m²) Bolts: Cap = 15 t Length = 2 to 4 m one per 4–6 m²	Shotcrete: 0–5cm Bolts: Length = 2 to 4m locally	Bolts L= 3.5m if necessary	
III (formerly IIIa)	Afterbreaking to overbreaking	Tensile stresses in the crown lead to rooffalls that are favored by unfavorably oriented discontinuities. The stresses at the springlines do not exceed the mass strength. However, afterbreaking may occur along discontinuities (due to blasting).	Full face with short round lengths	Full face: 2 to 4m	Smooth blasting	Crown and springlines: several hours	Shotcrete after each round other support can be placed in stages.	Combined shotcrete -bolted arch in crown and at springlines	Shotcrete: 5 to 15 cm with wirefabric (3.12 kg/m²) Bolts: Cap = 15–25 t Length = 3 to 5 m one per 3 to 5 m²	Shotcrete: 5–10 cm Bolts: 15 to 25 t Length: 3 to 5 m 1 per 3 to 5 m²	Adapt Invert Support to local conditions	Adapt Face Support to local conditions
IV (formerly IIIb)	Afterbreaking to lightly squeezing	(1) The rock mass strength is substantially reduced due to discontinuities thus resulting in many afterbreaks; or (2) The rock mass strength is exceeded leading to light squeezing.	Heading and benching (heading max 45 m²)	Full face: 2 to 3m heading: 2 to 4m	Smooth blasting and local trimming with jackhammer	Crown and springlines: a few hours	Shotcrete after each round. The bolts in the heading have to be placed at least after each second round.	Combined shotcrete-bolted arch in crown and springlines, if necessary closed invert	Shotcrete: 10 to 15 cm with wirefabric (3.12kg/m²) Bolts: fully grouted Cap = 25 tons Length = 4–6 m one per 2–4 m²	Same as crown	Slab: 20 to 30 cm[1]	
V	Heavily afterbreaking to squeezing	Due to low rock mass strength squeezing ground conditions that are substantially influenced by the orientation of the discontinuities.	Heading and benching (heading max. 40 m²)	Heading: 1 to 3m benches: 2 to 4m	Smooth blasting or scraping or hydraulic excavator	Crown and springline: very short free stand-up time	All opened sections have to be supported immediately after opening. All support placed after each round.	Support ring of shotcrete with bolted arch and steel sets	Locally linerplates Shotcrete: 15–20 cm with wirefabric (3.12 kg/m²) Steelsets: TH21 Spaced: 0.8–2.0m Bolts: fully grouted Cap = 25 t Length: 5 to 7 m one per 1 to 3 m²	Same as crown but no linerplates necessary.	Invert arch >40 cm or Bolts L = 5 to 7m if necessary	Shotcrete 10 cm in heading (if necessary) 3 to 7 cm in bench
VI	Heavily squeezing	After opening the tunnel squeezing ground is observed on all free surfaces, the discontinuities are of minor importance.	Heading and benching (heading max 25 m²)	Heading 0.5 to 1.5m benching 1 to 3m	Scraping or hydraulic excavator	Very limited stand-up time	as Class V	Support ring of shotcrete with steelsets Incl. invert arch and densely bolted arch.	Linerplates where necessary, shotcrete: 20–25 cm with wirefabric, steelsets TH21 0.5–1.5m Bolts: Cap: 25 t L = 6–9 m, 1 per 0.5–2.5 m²	Same as crown	Invert: >50cm Bolts: 6–9 m long if necessary	Shotcrete 10 cm and additional face breasting
VII	Flowing		Requires special techniques e. g., chemical grouting, freezing, electro-osmosis									

the actual lengths (again unsupported) have to be considered. The ground behavior type is determined by: Ground type; original stress state; shape, size, and location of openings as well as excavation methods; orientation of opening relating to discontinuity patterns; and

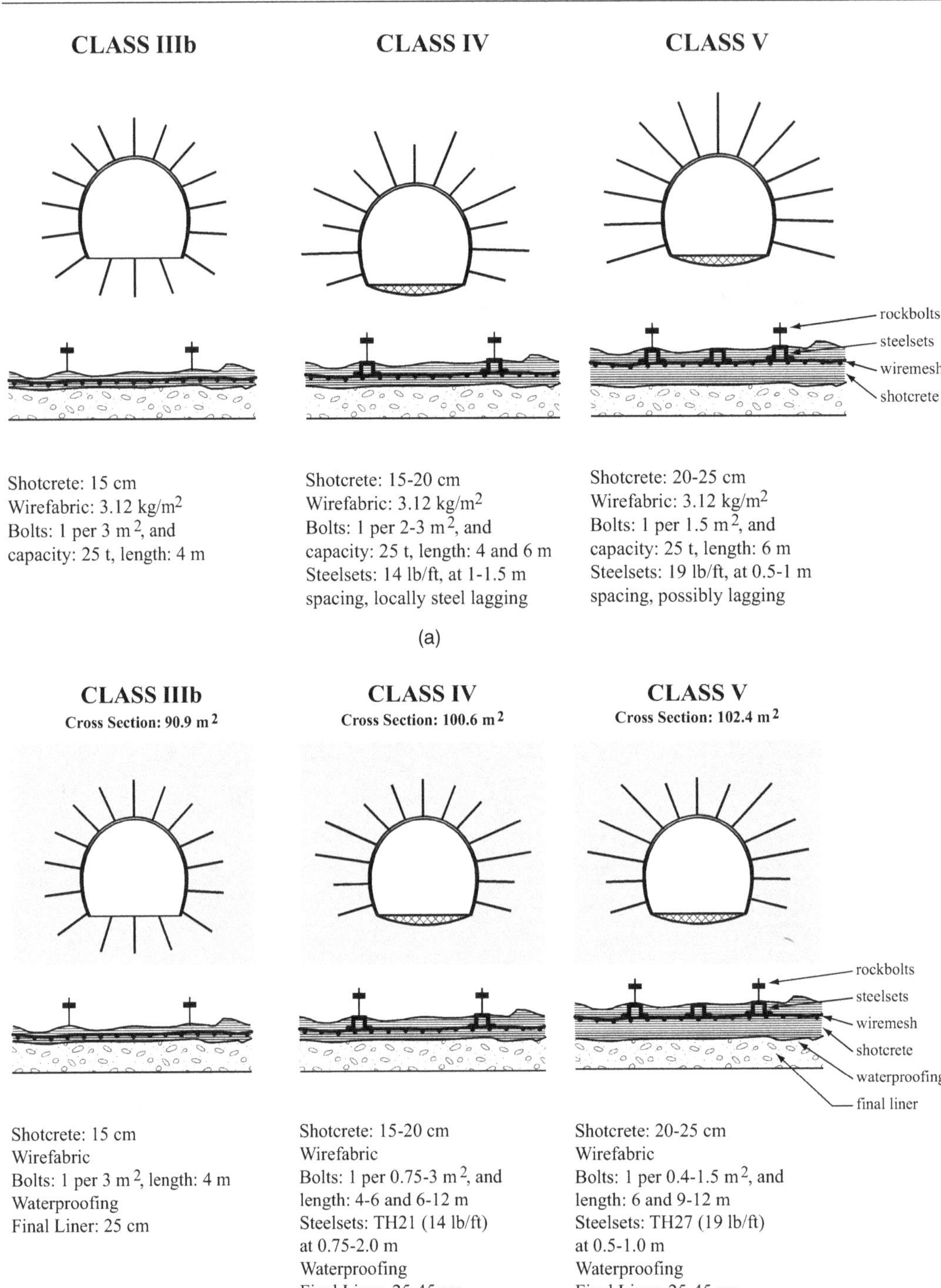

Figure 2.20 Groundclasses for Arlberg tunnel, (a) as designed (bid documents, after John, 1976), (b) as built (after Lässer-Feizlmayr, 1978).

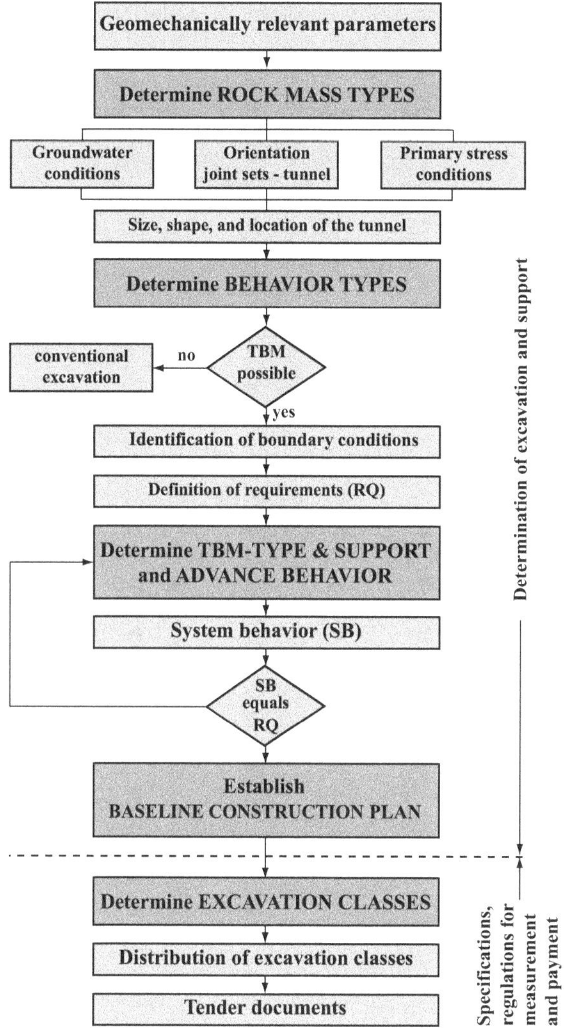

Figure 2.21 Flowchart for the geomechanical design for continuous excavation ("Richtlinie" Austrian Society for Geomechanics, 2001).

water (existence, hydrostatic and seepage pressure). Eleven ground behavior types are distinguished. They are described in Table 2.8 (Table 3 in "Richtlinie").

Each behavior type needs to be described at least with:

- Sketch of structure and failure mechanisms
- Ground type
- Orientation of discontinuities
- Description of effects of opening on ground
- Water and its effect on behavior
- Ground behavior (behavior during excavation at perimeter and face, deformation and failure mechanisms, long-term behavior)
- Displacement of perimeter

Table 2.7 Example of selected key parameters for different general rock types (after Austrian Society of Geomechanics, 2010)

		Relevant parameters																			
		Intact material properties													*Discontinuity properties*						
	Ground (Intact Material Type)	*Mineral content*	*Clay minerals, qualitative*	*Clay minerals, quantitative*	*Cementation*	*Grain size*	*Texture*	*Relation matrix/components*	*Porosity*	*Alteration/weathering*	*Dissolution features*	*Swelling properties*	*Strength*	*Anisotropy*	*Orientation of the dominant set*	*Number and orientation of sets*	*Block size*	*Persistence*	*Aperture*	*Shearing resistance/roughness*	*Filling*
Rock	Plutonic	■				■	■			□			■			■	■	□	■	□	□
	Volcanic massive	□					□		■	■			■			■	■	□	■	□	■
	Volcanic clastic	□	□		□	□		■	■	■		□	□								
	Coarse grained clastic, massive	□		□	■	■	□	■	□	□			■			□	□	□	□		
	Fine grained clastic, massive		■	■	■	■				□		■	□			□	□				
	Coarse grained clastic, layered	□		□	■	■		■	□	□			■	■	■					■	
	Fine grained clastic, layered		■	■	■	■				□		■		■	■					■	□
	Carbonates massive	■									■		■			□	■	□	■		□
	Carbonates layered	■									■		■		■	■				□	□
	Sulfatic rocks	■									■	■	□								
	Metamorphic massive	■				■	■			□			■			■	■	□	■		
	Metamorphic layered	■				■	■			□			■	■	■	■	■	□		■	■
	Fault zone materials	□	■	■	■			■		□		■	■								
Soil	Coarse grained, gravel					■		■	□				■								
	Coarse grained, sand					■		□	□				■								
	Mixed-grained	□		■		■		■	□			□	■								
	Fine grained, silt					■			□				■								
	Fine grained, clay	□		■		■			□			■	■								

Legend: ■ Property of great significance; □ Property of small significance

Note that these are more or less literal translations; several of these descriptions overlap. The type sketches (see below) are more clear in this regard.

It should be noted that the categories of Table 2.8 can be further subdivided if they differ regarding water, disturbance, etc. Figures 2.22a, b show typical ground behavior sketches from the guidelines. Other examples, which are taken from the "thesis" by Goricki (2003), show that photos and results of numerical methods can be used in the ground behavior type descriptions (Figures 2.23a–k).

The third step according to the framework (Figure 2.21) involves the deformation of excavation and support leading to the so-called system behavior. The system behavior depends

Table 2.8 General categories of ground behavior (after Austrian Society of Geomechanics, 2010)

Ground behavior type	*Description of behavior without support*
1. Stable	Potential of gravity based falling/sliding of small blocks
2. Structurally caused instabilities	Deeper reaching instabilities, partial exceedance of discontinuity shearing resistance
3. Instability near opening	Destabilization including plastification and rockfalls near opening
4. Deep instabilities	Deep destabilization/plastification with large deformations
5. Rockburst	Sudden detachments in brittle rock
6. Buckling	Buckling of plates
7. Shear-instabilities at low stress level	Potentially large instabilities with progressive shear failure
8. Rolling ground	Movement of cohesionless material
9. Flowing ground	Flowing with high water content
10. Swelling ground	Time-dependent volume increase
11. Ground with rapidly changing properties	Large variation of stresses and deformations

on: (1) ground behavior type; (2) shape and size of the opening; (3) three-dimensional development of construction processes; (4) time-dependent properties of ground and support; and (5) support types, placement, and time. Again there is some overlap in these factors, e.g. three-dimensional development and support phases.

The "baseline construction plan" (framework plan) in the fourth step (Figure 2.21), in principle, corresponds to a geotechnical report, i.e. specifying the distribution of the different

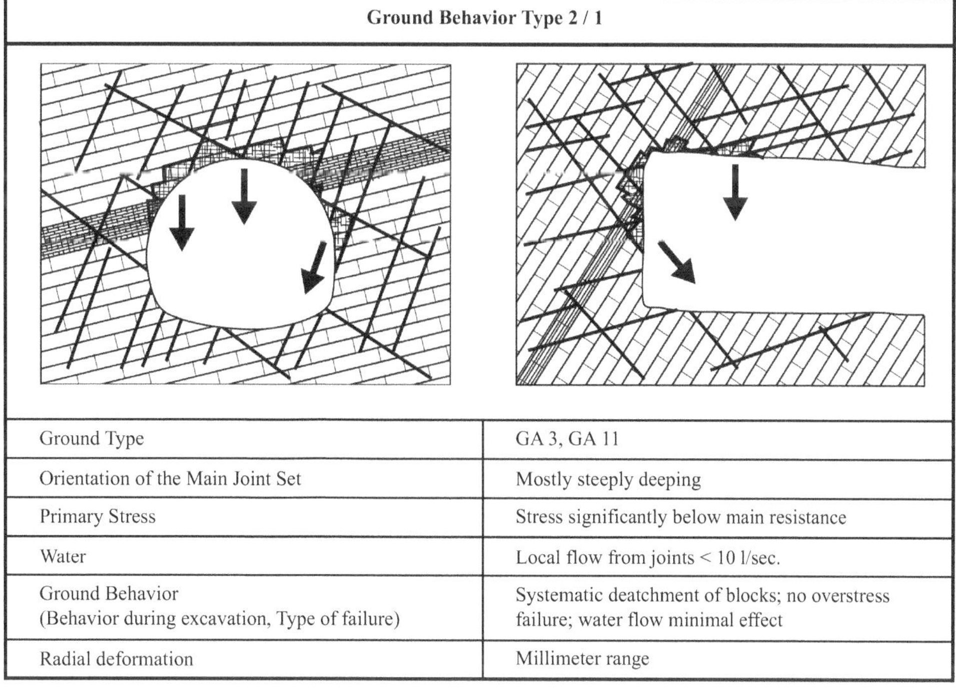

Ground Behavior Type 2 / 1

Ground Type	GA 3, GA 11
Orientation of the Main Joint Set	Mostly steeply deeping
Primary Stress	Stress significantly below main resistance
Water	Local flow from joints < 10 l/sec.
Ground Behavior (Behavior during excavation, Type of failure)	Systematic deatchment of blocks; no overstress failure; water flow minimal effect
Radial deformation	Millimeter range

Figure 2.22a Ground behavior type 2/1 (from "Richtlinie" Austrian Society for Geomechanics, 2001).

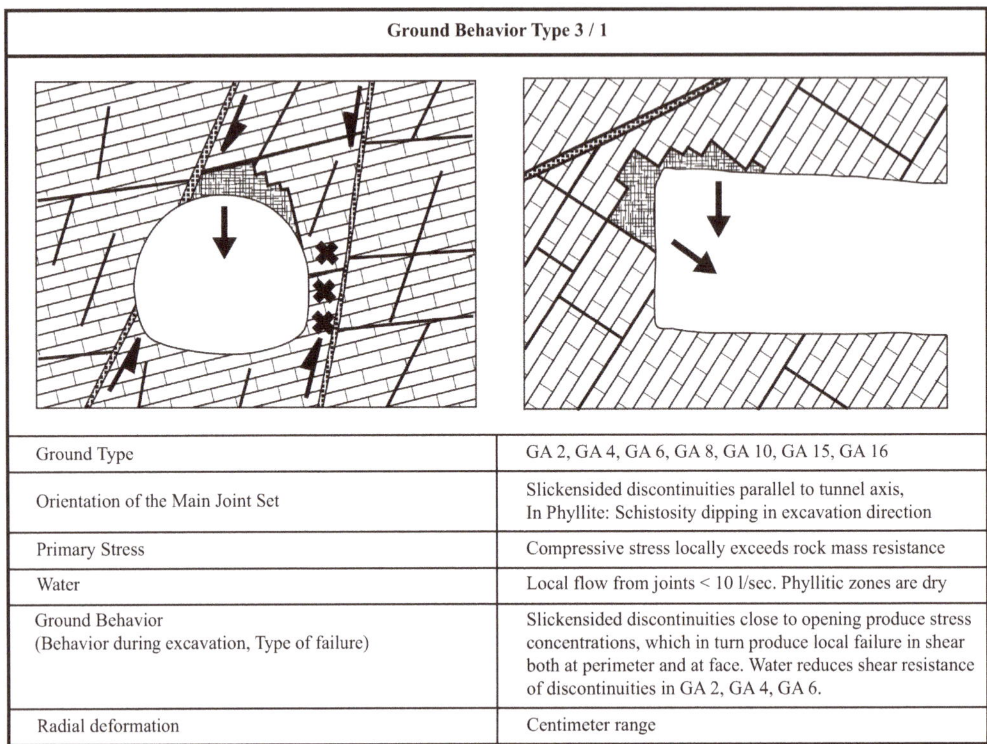

Ground Behavior Type 3 / 1	
Ground Type	GA 2, GA 4, GA 6, GA 8, GA 10, GA 15, GA 16
Orientation of the Main Joint Set	Slickensided discontinuities parallel to tunnel axis, In Phyllite: Schistosity dipping in excavation direction
Primary Stress	Compressive stress locally exceeds rock mass resistance
Water	Local flow from joints < 10 l/sec. Phyllitic zones are dry
Ground Behavior (Behavior during excavation, Type of failure)	Slickensided discontinuities close to opening produce stress concentrations, which in turn produce local failure in shear both at perimeter and at face. Water reduces shear resistance of discontinuities in GA 2, GA 4, GA 6.
Radial deformation	Centimeter range

Figure 2.22b Ground behavior type 3/1 (from "Richtlinie" Austrian Society for Geomechanics, 2001).

Behavior Type	Stable	
N°	1	
Description	Stable rock mass with the potential of small local gravity induced falling or sliding blocks	
Failure of excavation	• generally no failure • local potential for falling, sliding, rotating of blocks due to utilization of tensile strength or shear strength along discontinuities	
Causes	• local unfavorable orientations of discontinuities • primarily gravitational forces, the induced stresses influence the equilibrium	
Effects	• generally stable conditions • local falling/sliding of blocks into the excavation	
Example of a rock mass condition and an effect		

Figure 2.23a Behavior Type 1 – stable (from Goricki, 2003). (Reproduced with permission of Dr. A Goricki.)

ground and behavior types and assumptions and criteria regarding construction and support. Importantly, the criteria regarding safety, e.g. critical deformation values should be included.

The next (fifth) step (Figure 2.21) is the definition of construction classes which form the basis for bid calculations. It is important to note that the construction classes do not have

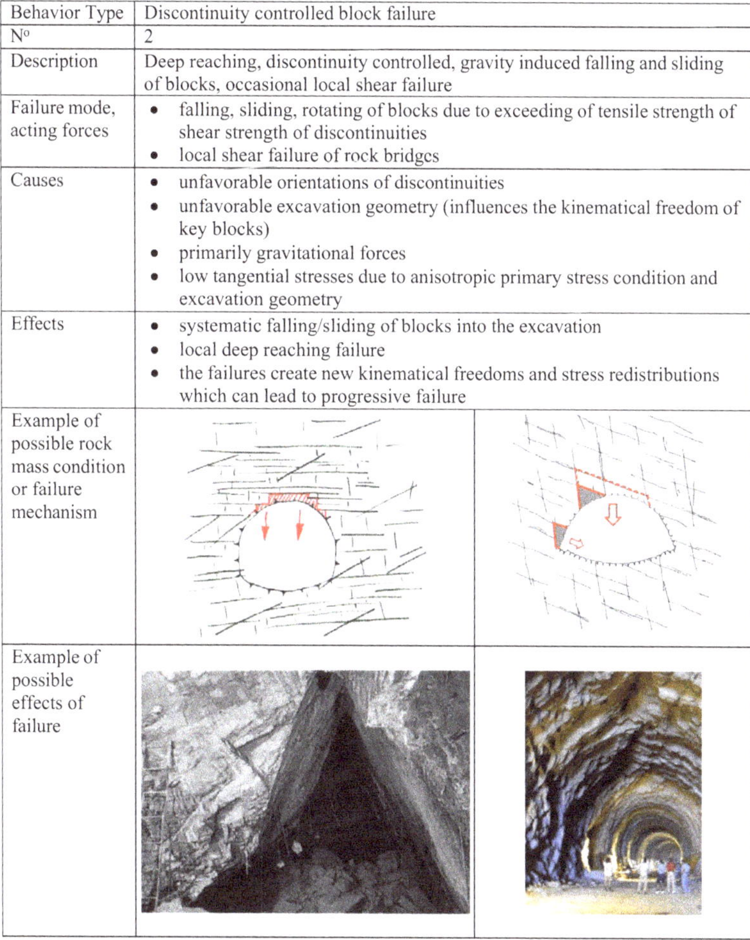

Behavior Type	Discontinuity controlled block failure	
N°	2	
Description	Deep reaching, discontinuity controlled, gravity induced falling and sliding of blocks, occasional local shear failure	
Failure mode, acting forces	• falling, sliding, rotating of blocks due to exceeding of tensile strength of shear strength of discontinuities • local shear failure of rock bridges	
Causes	• unfavorable orientations of discontinuities • unfavorable excavation geometry (influences the kinematical freedom of key blocks) • primarily gravitational forces • low tangential stresses due to anisotropic primary stress condition and excavation geometry	
Effects	• systematic falling/sliding of blocks into the excavation • local deep reaching failure • the failures create new kinematical freedoms and stress redistributions which can lead to progressive failure	
Example of possible rock mass condition or failure mechanism		
Example of possible effects of failure		

Figure 2.23b Behavior Type 2 – Discontinuity Controlled Block Failure (from Goricki, 2003). (Reproduced with permission of Dr. A Goricki.)

to have a one-to-one relation with the ground behavior types since the same construction procedures might cover several behavior types or vice versa.

All the preceding is obviously done prior to construction. The principles of the framework in Figure 2.21 are then also used during construction. This reflects the philosophy of the NATM as an adaptable, observational approach but is actually more generally valid in that a comparison of predicted and encountered conditions and deciding on the best construction approach should always be done. Figure 2.24, analogously to Figure 2.21, represents what is done during construction. What is different is that steps 1 and 2 are now the determination of the actual ground type and ground behavior type (rather than predictions) and that step 3 is the actual (in situ) specification of excavation and support procedures. Specifically, this should be done (according to the guideline) as follows:

i. Ground Type:
   Face (and perimeter) mapping together with measurements (tests), e.g. of unconfined compressive strength, joint spacing etc., are the basis for filling in a form sheet as shown in Figure 2.25.

Behavior Type	Shallow stress induced failure
N°	3
Description	Shallow stress induced brittle and shear failures in combination with discontinuity and gravity controlled failure of the rock mass
Failure mode, acting forces	• brittle failure of intact rock – spalling: propagation of new cracks in brittle rock mass • yielding of rock mass (shallow squeezing failure) • shear failure of intact rock • shear failure along existing discontinuities and zones of weakness in combination with new cracks along rock bridges • discontinuity controlled block failure as described in Behavior Type 2
Causes	• changing stress condition due to the excavation (stress transfer and redistribution) • anisotropic primary stress condition (influence of tectonic and rock mass structure) • stress concentrations at the tunnel walls due to inhomogeneities like slickensides, fault zones, or unit boundaries close to the excavation • discontinuity controlled block failure as described in Behavior Type 2
Effects	• loosening of rock mass at shallow depth coupled with reduction of rock mass strength • detaching rocks at the wall due to new created kinematical freedom • displacement with relatively small magnitudes • discontinuity controlled block failure as described in Behavior Type 2
Example of possible rock mass condition or failure mechanism	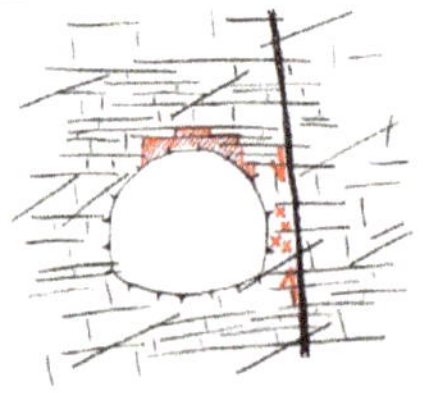 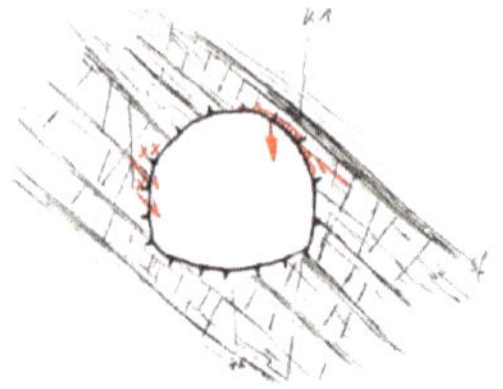
Example of possible effects of failure	 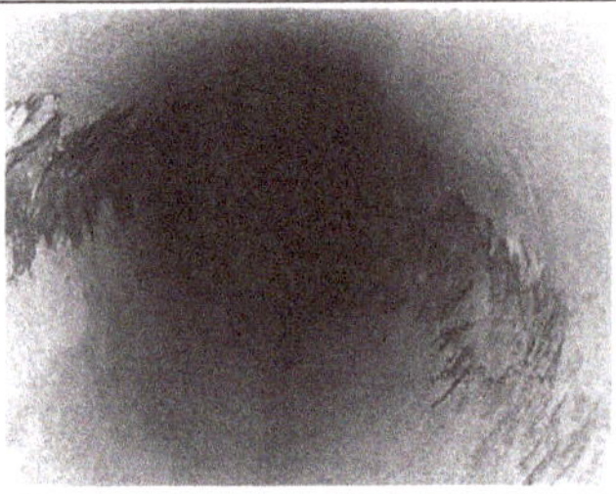

Figure 2.23c Behavior Type 3 – Shallow Stress Induced Failure (from Goricki, 2003). (Reproduced with permission of Dr. A Goricki.)

ii. Ground Mechanics Type:
This is an extension of the ground type characterization through inclusion of structural (discontinuity) aspects, stress, water conditions and, very importantly, observed displacements and their behavior with time. Sketches such as those in Figure 2.26 are used for documentation.

The observed ground behavior type will then be used together with the corresponding planned support and excavation based on the "Baseline Construction Plan" (framework plan). In case of discrepancies between predicted and encountered conditions, the framework has to be adapted. Also, more detailed parameter determinations and, particularly, the displacement measurements make it possible to refine the predictions. Updated predictions should be made for the following 10–20 meters. It is possible that, in addition to behavior types different from what was anticipated, the relation behavior type-support/excavation does not materialize. This can be both in the more favorable or the less favorable direction

Behavior Type	Deep seated stress induced failure
N°	4
Description	Deep seated stress induced brittle and shear failures in combination with large displacements
Failure modes	• brittle failure of intact rock – spalling: propagation of new cracks in brittle rock mass • yielding of rock mass (squeezing) • squeezing of blocks due to high confining pressure (cherry pit effect) • shear failure of intact rock and rock mass • shear failure along existing discontinuities and zones of weakness in combination with new cracks along rock bridges
Causes	• changing stress condition due to the excavation (stress transfer and redistribution) • anisotropic primary stress condition (influence of tectonic and rock mass structure) • stress concentrations at the tunnel walls due to inhomogeneities like slickensides, fault zones, or unit boundaries close to the excavation
Effects	• loosening of rock mass at shallow depth coupled with reduction of rock mass strength • detaching rocks at the wall surface due to new created kinematical freedom • displacement with relatively small magnitudes
Example of possible rock mass condition or failure mechanism 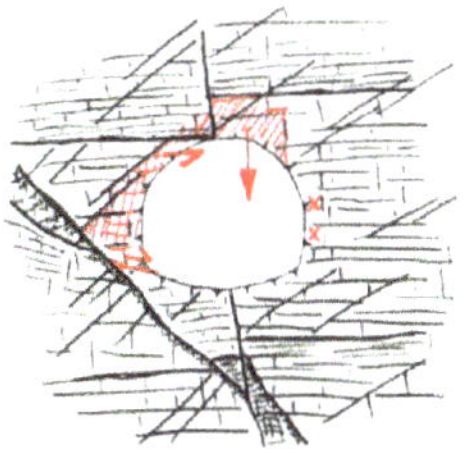 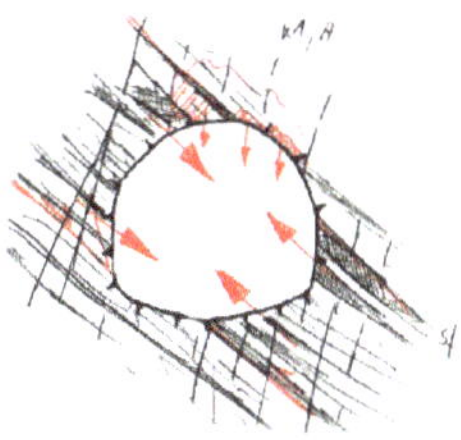	
Example of possible effects of failure  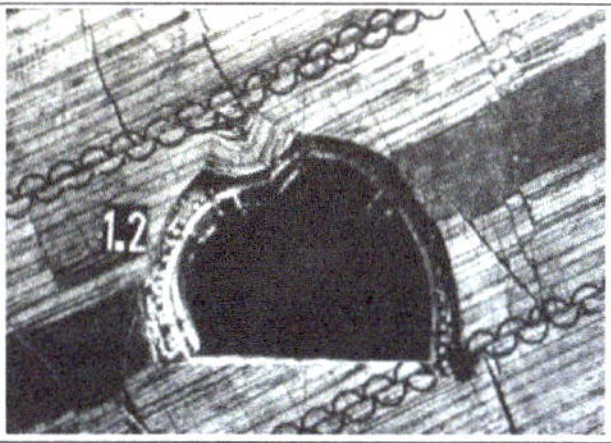	

Figure 2.23d Behavior Type 4 – Deep Seated Stress Induced Failure (from Goricki, 2003). (Reproduced with permission of Dr. A Goricki.)

and can be caused by differences in the parameters and/or the models relating predicted and encountered conditions. Corresponding changes in the predictions and, if necessary, in the framework plan should be made. Clearly, in the case of "worse than anticipated" performance, correcting measures to maintain safety have to be implemented. Deviations from what was anticipated naturally have cost/time consequences. The feedback regarding predicted and observed systems behavior is clearly shown in the lower part of Figure 2.24.

To conclude, this description of the Type B empirical methods and specifically the ground behavior-support/excavation relations as proposed in the "Richtlinie" (guideline), one can state:

i. The process proceeds logically and systematically from ground type to behavior type to support/excavation recommendations and associated construction classes.
ii. The characterization and relations are based on a mix of qualitative descriptions, analysis, and relations based on experience.

Behavior Type	Rock burst
N°	5
Description	Sudden and violent failure of the rock mass, caused by highly stressed brittle rocks and the rapid release of accumulated strain energy
Failure modes	• sudden expansion of the rock mass, the stored strain energy does not dissipate in a gradual manner • brittle failure associated with a sudden loss of rock mass strength • bulking of the rock due to fracturing • superficial spalling • buckling of rock columns and slabs • rock falls due to seismic shakings
Causes	• the stress in the rock mass exceed the rock mass strength • decrease of rock mass strength with time or due to loss of confinement • unfavorable stiffness of rock and excavation
Effects	• violent ejection of rock fragments • outward explosion of slabs • rock falls • seismic events
Example of possible rock mass condition or failure mechanism	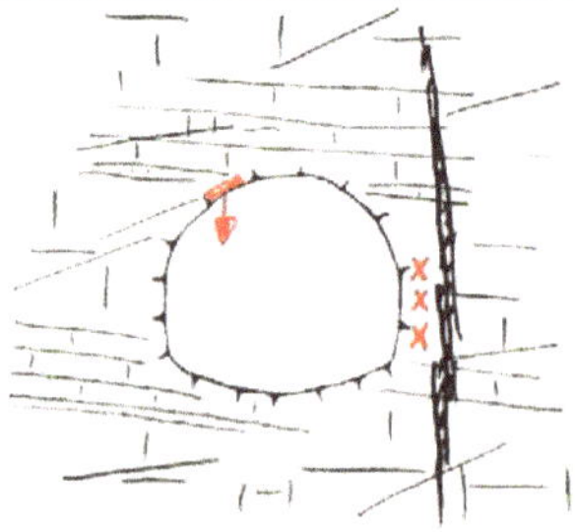
Example of possible effects of failure	

Figure 2.23e Behavior Type 5 – Rock Burst (from Goricki, 2003). (Reproduced with permission of Dr. A Goricki.)

iii. The required documentation ensures that the process is transparent, which then makes the necessary modifications of the predictions and their consequences equally transparent.
iv. The procedure leads not only to support material and dimensions but also to detailed descriptions of the excavation/support sequence, i.e. the construction process. This is an aspect which goes beyond that of most other empirical approaches.
v. The procedure does not stop with the pre-construction predictions but includes observations during construction, which again have to be documented in detail.
vi. The construction observations have to be used to compare predicted with actual behavior and, if necessary, lead to corrected (updated) predictions. Other empirical methods can also be used with such updating, but they do not specify such an integrated observational process.

Behavior Type	Buckling failure
N°	6
Description	Buckling of rocks with a narrowly spaced discontinuity set, frequently associated with shear failure
Failure mode	• rotational failure of blocks (three or four hinge buckling) or rock columns due to bending • shear and tension failure
Causes	• thin layered rock mass • high stress parallel to the rock mass layers • large displacements (cause temporary shear and tensile failures)
Effects	• kinking of layers into the excavation • deep reaching fracturing of the rock mass
Example of possible rock mass condition or failure mechanism	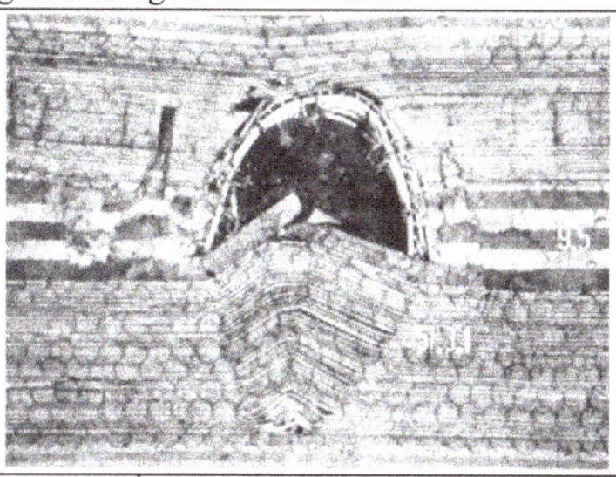
Example of possible effects of failure	 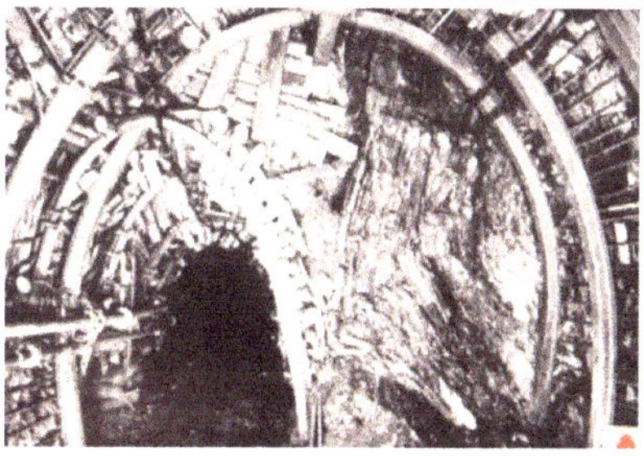

Figure 2.23f Behavior Type 6 – Buckling Failure (from Goricki, 2003). (Reproduced with permission of Dr. A Goricki.)

2.3.4 Methods Type C – relations between qualitative or quantitative rock mass descriptions and analytical model parameters

The principle of this method is indicated in Figure 2.1 and is repeated in Figure 2.27. The methods described here are, on the one hand, relations to obtain the deformability of rock masses (such as the one by Deere et al., 1967) and, on the other hand, rock mass strength and deformability (Hoek and Brown, 1980a, 1980b). As Figure 2.27 shows, the ground can be described qualitatively or quantitatively and Methods Type C, therefore, represent a transition from methods A, B on the one hand to D, E on the other hand. What distinguishes them from most other methods is that these descriptions are then related to "classic" analytical parameters describing the deformability and/or resistance of the rock mass. Note that the Q and RMR methods discussed in the context of Type E methods in Section 2.3.6 also include such relations, but not as the major underlying principle.

Behavior Type	Shear failure under low confining pressure
N°	7
Description	Potential for excessive overbreak and progressive shear failure with the development of chimney type failure, caused mainly by a deficiency of side pressure in combination with low shear strength of rock mass
Failure mode	• shear failure of highly fractured rock mass • shear along existing discontinuities • combined tensile and shear failure in case of excessive overbreak
Cause	• gravity and secondary stress conditions
Effect	• movement of ground into the excavation along new developed shear planes
Example of possible rock mass condition or failure mechanism 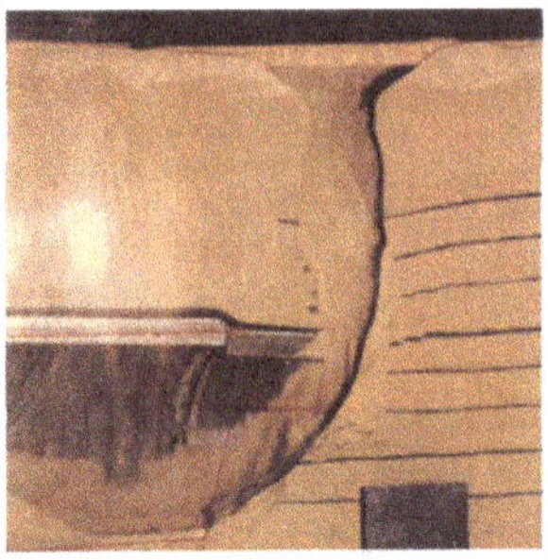 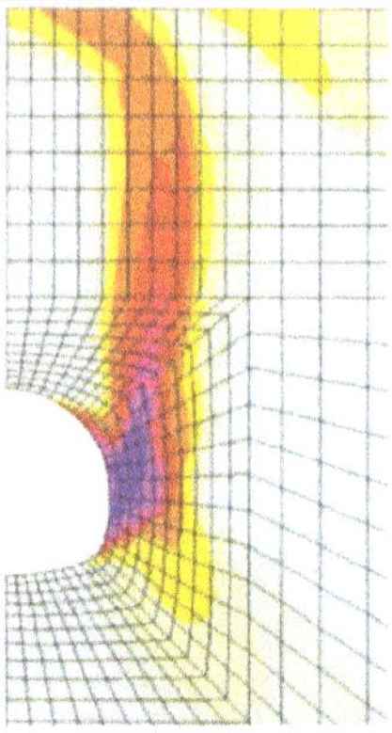	
Example of possible effects of failure	

Figure 2.23g Behavior Type 7 – Shear Failure under Low Confining Stress (from Goricki, 2003). (Reproduced with permission of Dr. A Goricki.)

Behavior Type	Raveling ground
N°	8
Description	Flow of cohesionless dry or moist, intensely fractured rock mass
Failure mode	Disintegration of the structure, falling, sliding and rotating of singular particles (grains, block) or coherent parts of rock mass
Cause	Gravitational forces
Effect	Flow of ground into the excavation

Figure 2.23h Behavior Type 8 – Flowing Ground (from Goricki, 2003). (Reproduced with permission of Dr. A Goricki.)

Behavior Type	Flowing ground
N°	9
Description	Flow of intensely fractured rock or soil with high water content and or pressure
Failure mode	• tensile failure of the particle to particle bonding, loss of cohesion • disintegration of the structure, floating of singular particles (corn, block) or coherent parts of rock mass
Causes	• gravitational forces • water pressure (hydrostatic and flow pressure)
Effect	Flow of ground into the excavation like a fluid
Example of possible effects of failure	

Figure 2.23i Behavior Type 9 – Flowing Ground (from Goricki, 2003). (Reproduced with permission of Dr. A Goricki.)

Behavior Type	Swelling
N°	10
Description	Time dependent volume increase of the rock mass caused by physical-chemical reaction of rock and water in combination with stress relief, leading to inward movement of the tunnel perimeter
Failure mode	• volume increase • disintegration of rock mass
Causes	• change of stress condition • contact with water
Effect	Invert heave, additional stresses in the rock mass which can lead to various secondary failures
Example of possible rock mass condition or failure mechanism	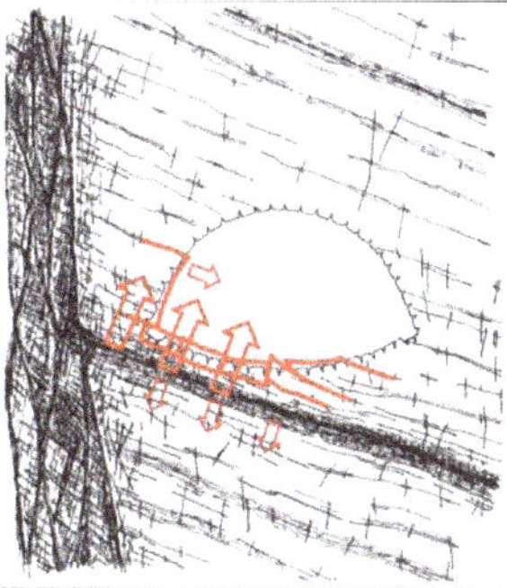

Figure 2.23j Behavior Type 10 – Swelling (from Goricki, 2003). (Reproduced with permission of Dr. A Goricki.)

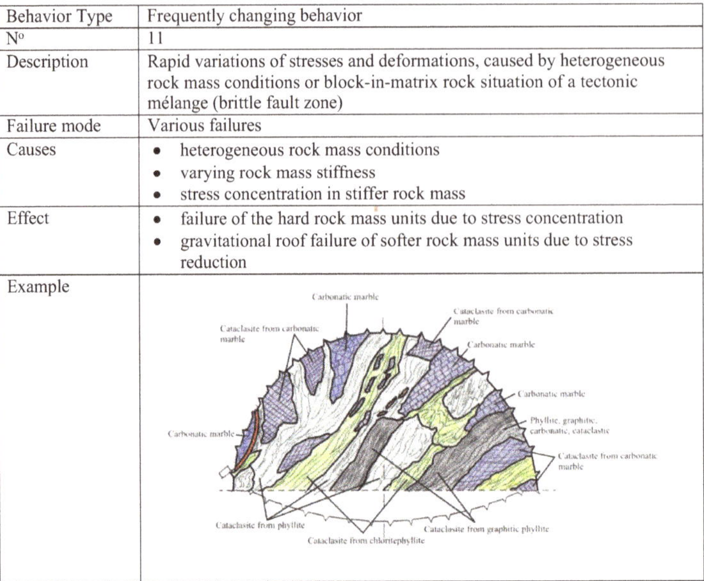

Behavior Type	Frequently changing behavior
N°	11
Description	Rapid variations of stresses and deformations, caused by heterogeneous rock mass conditions or block-in-matrix rock situation of a tectonic mélange (brittle fault zone)
Failure mode	Various failures
Causes	• heterogeneous rock mass conditions • varying rock mass stiffness • stress concentration in stiffer rock mass
Effect	• failure of the hard rock mass units due to stress concentration • gravitational roof failure of softer rock mass units due to stress reduction
Example	

Figure 2.23k Behavior Type 11 – Frequently Changing Behavior (from Goricki, 2003). (Reproduced with permission of Dr. A Goricki.)

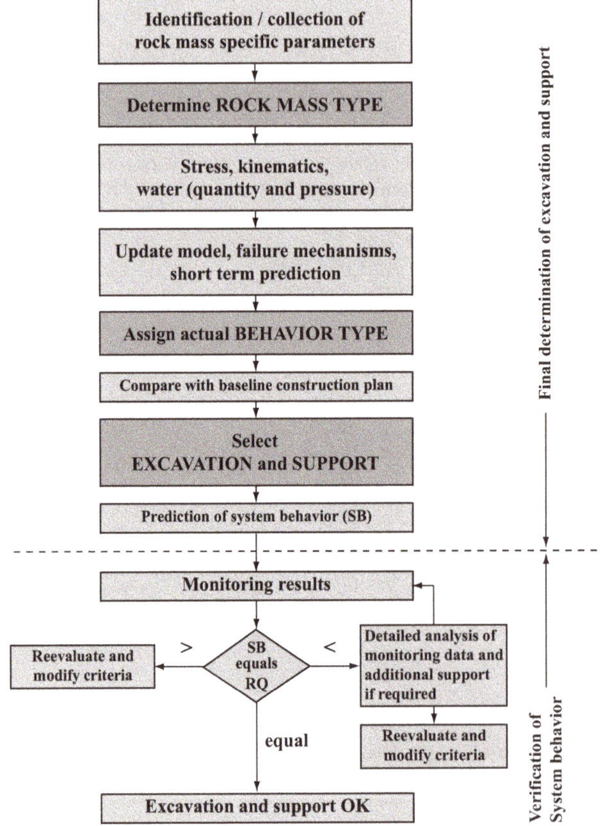

Figure 2.24 Flowchart of basic procedure for excavation and support selection and verification of the system behavior during construction (from "Richtlinie," Austrian Society for Geomechanics, 2001).

	Soils	GA 1	GA 2	GA 3	GA 4	GA 5
Lithology	Talus and Alluvial Debris	Calcareous Dolomitic Marble	Same	Same	Same	Same
Schistosity Anisotropy	NA	> 60 cm	> 60 cm	60 - 20 cm	60 - 20 cm	20 - 6 cm
Blocksize	NA	> 20 cm	> 20 cm	20 - 5 cm	20 - 5 cm	< 5 cm
Discontinuity Conditions	NA	Foliation and joints: rough	Foliation: Rough Joints: rough, Silty clayey filling	Foliation and joints: rough	Foliation: Rough Joints: rough, Silty clayey filling	Rough
Discontinuity-Exposed Extent	NA	Mostly low	Mostly low	Mostly low	Mostly low	High
Discontinuity Aperture	NA	Mostly closed	Open	Mostly closed	Open	Mostly closed

Figure 2.25 Ground type as determined during construction, examples for GA (Ground Type). ("Richtlinie" Austrian Society for Geomechanics, 2001).

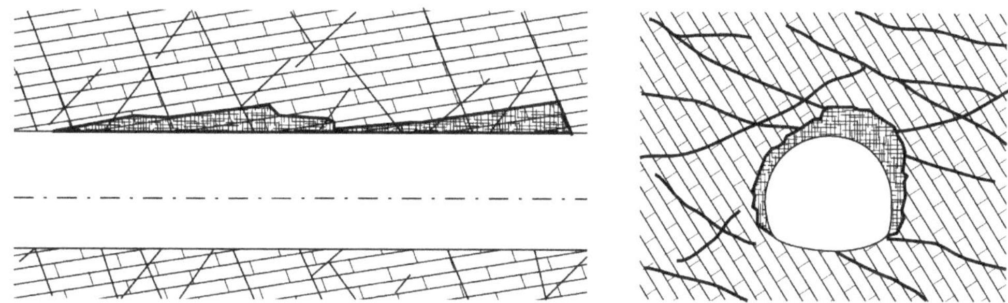

Figure 2.26 Sketch of in situ construction (from "Richtlinie" Austrian Society for Geomechanics, 2001).

TYPE C METHODS

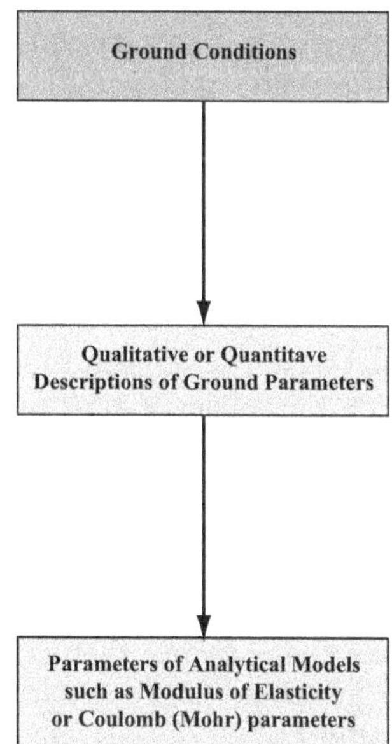

Figure 2.27 Principle of Type C methods.

2.3.4.1 Empirical determination of rock mass deformability

Rock mass deformability plays an important role in the performance of foundations, notably pile and dam foundations and tunnels where either high stresses or large influence zones or both are involved. Consequently, quite a few approaches exist, in which the modulus of elasticity of the rock mass is related to other parameters describing the rock mass, notably lithology and fracturing. Most of these relations are based on in situ measurements of the modulus of elasticity and qualitative or quantitative descriptions of other parameters. The in situ measurements can be performed statically (pressure chamber, plate loading, borehole deformation, borehole jack, back calculated from displacements of the actual structure) or dynamically (seismic velocity related analytically to modulus of elasticity). There are a wide variety of measurement methods, which differ substantially in the way the load is applied, the stress level, and the influence zone. Given the substantial differences between the static and dynamic field tests, these are usually looked at separately, i.e. relations between field modulus of deformability and rock mass characteristics are obtained from either the statically or the dynamically obtained moduli but not combinations (with exceptions). Several researchers have also found that relating rock mass classification to the modulus ratio E_{field}/E_{lab} produces tighter relations (see e.g. Deere et al., 1967). The laboratory tests are usually conducted on intact rocks and E_{lab} is, therefore, also called E_{intact}. This is certainly correct but introduces another, often unknown, factor, i.e. the definition of E_{lab}. In many cases this is E_{50} (see Deere et al., 1967), but it is often measured at the last straight portion of the loading curve or the first portion of the unloading curve. So, unless the different influencing factors are specifically mentioned, it is very difficult to produce unambiguous relationships between moduli (deformability) and rock mass characteristics. Ideally, such relationships should be established only for the same type of field (and lab) test and the same definition of Young's modulus. Finally, and this is important for tunneling, most of the reported field and lab moduli have been obtained in loading while in usual tunnel applications one is actually concerned with unloading. While the modulus of intact rock is typically greater in unloading than loading, field tests can produce unloading moduli that are either greater or smaller than the loading moduli. It is, therefore, even difficult to say if the empirical deformability–rock mass classification relationships are conservative or not.

The following provides a historical perspective of the advances made in relating rock mass deformability with a qualitative or quantitative description of the rock mass. It is not intended to be inclusive, given the large literature on the subject, but to provide a perspective of the different thinking and approaches taken over time. With any historical sequence, there is the tendency of assuming that the most recent contributions are the most relevant. While there may be some truth in this, as recent proposals take into account or include previous data, the interpretation of the data may have been different. As in any empirical correlation, one has to be careful when extrapolating results to cases that fall outside the domain of data used to establish the correlation.

It was perhaps Deere et al. (1967) who first formally presented a relation between a quantitative description of the rock mass, in the form of RQD, and rock mass deformability. Deere used the concept of rock quality to explain why previous results (Link, 1964) showed that seismically obtained moduli were larger than when obtained statically. This was consistent with Lane's (1964) conclusions, that the reduction of field versus laboratory modulus values can be related to jointing. A formal consideration of rock mass quality based on RQD and rock modulus of elasticity was done based on plate load tests at Dvorshak Dam (Deere et al., 1967). The RQD values were corrected for the fact that fractures near the surface (where the load was applied) were subject to greater stresses than those at depth. The resulting relations between RQD and the modulus ratio $E_{static\ field}/E_{static\ lab}$ are plotted in Figure 2.28 (Figure 19 in Deere et al., 1967). The $E_{static\ field}$ is the secant modulus at a load of 1000 psi, and for the

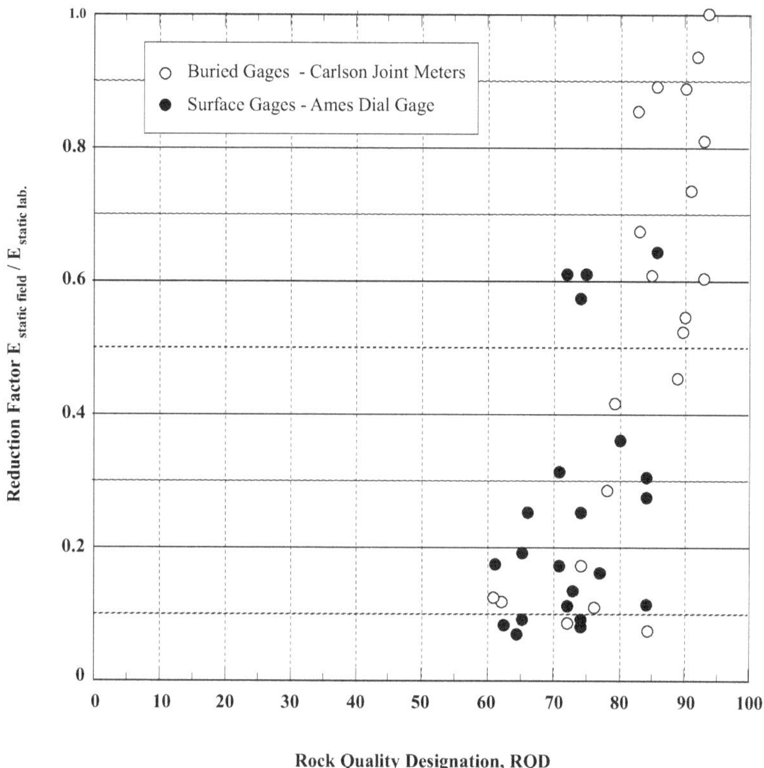

Figure 2.28 Variation of reduction factor with rock quality from plate jacking tests, Dvorshak Dam (Deere et al., 1967).

lab, the modulus E_{50} was used (the latter did not vary much, evidently). The concept of a modulus ratio (reduction factor) was then generalized by including data from the literature (Figure 2.29, which is Figure 16 in Deere et al., 1967). However, several comments are necessary in interpreting Figure 2.29. As can be seen, the rock mass quality is expressed either by RQD or the ratio of seismic field versus lab sonic velocities V_F/V_L. For quite a few of the data points in Figure 2.29, only RQD or V_F/V_L was available but not both. While Deere et al. (1967) in their research obtained a reasonably clear relation between RQD and V_F/V_L, this is not always the case as Deere et al. (1967) shows in Figure 8 of their paper. Also, the generalized relation becomes very flat below a reduction factor of 0.2 or RQD below 60%, respectively. Since this represents quite a large range of rock mass conditions, further work was recommended by Deere et al. (1967) and actually done by several researchers/practitioners, which will be discussed later.

Deere's database was expanded by Coon and Merritt (1970), who added results from three dam sites (Two Forks, on gneiss; Yellowtail, on limestone; and "other sites," on sandstone), with the adjustment of RQD with depth, similar to what was done by Deere et al. (1967). The Modulus Ratio versus RQD results are very similar to those in Figure 2.28. It is not quite clear why some of the data for the same tests (Dvorshak Dam) are different than in Deere et al. (1967); this, however, has practically no effect on the relationship between modulus ratio and RQD in the range of RQD>60%. However, when plotting the velocity ratio versus the modulus ratio, a much wider scatter was observed. This is important since it indicates that in situ seismic tests affect a much larger volume than what one can conclude from RQD. As a matter of fact, Coon and Merritt (1970) suggest using the relationship only in this

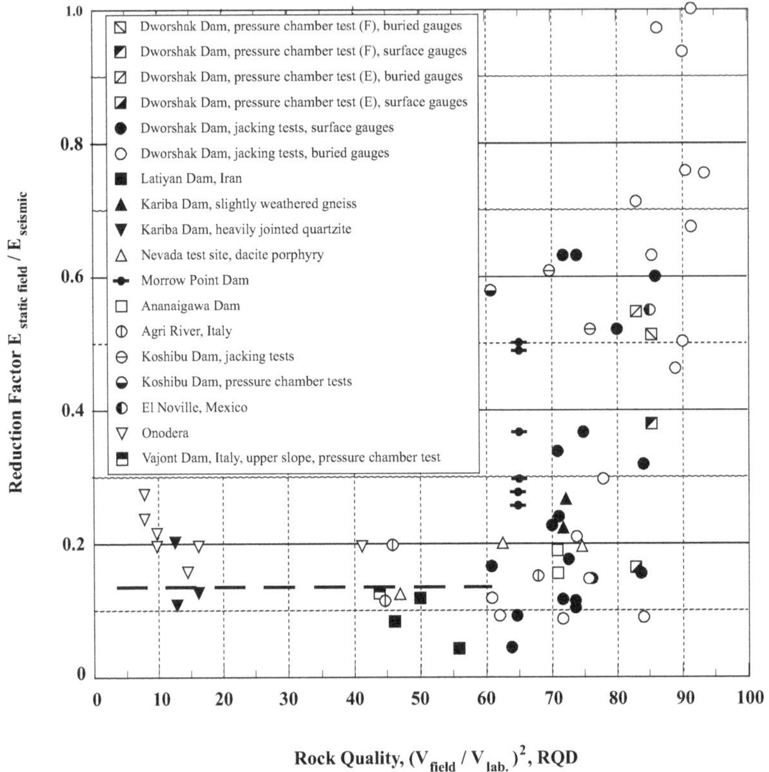

Figure 2.29 Variation of reduction factor with rock quality (Deere et al., 1967).

(>60%) range of RQD. In fact, their data come from a limited set of sites mostly involving rock masses in the fair to excellent range (RQD > 40%).

Clearly, the RQD, as a single value description of the rock mass, provides a very limited representation of the rock mass. Other parameters such as joint orientation, intensity, joint surface characteristics, weathering, and state of stress are not captured. Dershowitz (1979) showed, analytically, the shortcomings of the RQD when used to estimate stiffness for different orientations of the rock joints. They found that the deterministic relation between modulus ratio and RQD is very similar to the empirical relation reported by Deere et al. (1967) and Coon and Merritt (1970). They also found that steep joints increase the modulus ratio (not unexpected since this means that it is mostly the intact rock and not the joints affecting the rock mass modulus) and that dispersion (introduced in their analysis through the Fisher's distribution constant) decreases the modulus ratio, particularly for steep joints since this means that there are actually many more flatly inclined joints. The conclusion one can draw from this, which is important for the empirical relations, is that for low orientation dispersion, the RQD–modulus ratio relations are reasonably good while for large orientation dispersion, they are not (see also Figure 2.33).

Other methods, notably the RMR, Q, and GSI (discussed later in this chapter), include multiple value descriptors of the rock mass, which arguably can do a better job than the single-value RQD in qualitatively and/or quantitatively describing the rock mass. These rock mass classification indices have been used to estimate the rock mass stiffness.

Bieniawski (1978) was the first to initiate the transition from the relations based on Deere and his students to relations used later. Bieniawski used three cases in South Africa, which involved dam foundations, tunnels, and caverns and most importantly, considered a large number of laboratory and in situ tests, including in situ jacking tests and pressure chamber, petit sismique, and borehole jack tests. The rock mass was characterized by RQD and Bieniawski's RMR (using the 1976 classification). The description was, however, not necessarily obtained from boreholes directly below the plate load tests. An interesting observation not directly related to these comments on moduli relations is the observation by Bieniawski that the in situ moduli for the two directions (top-bottom, left-right) for the plate jacking conducted in tunnels differ by factors up to 4, confirming similar observations by Rocha and Da Silva. Initially, Bieniawski added his data, i.e. the modulus reduction factor to Deere's data, which increased the scatter. A plot of RMR versus the modulus ratio did not much better regarding the resulting scatter. However, when the in situ modulus of deformation (not the modulus ratio) was plotted against RMR, a linear relation resulted (Figure 2.30). The least square fit for the data in Figure 2.30 is:

$$E_m = 1.76\,RMR - 84.3 \tag{2.1}$$

The straight line in Figure 2.30 is rounded off to produce an easily useable relation:

$$E_m = 2\,RMR - 100 \tag{2.2}$$

Serafim and Pereira (1983) were not satisfied with the fact that Bieniawski's relation did not work for RMR < 50 and extended the relation with their own data. It is important to note that they used the 1979 Bieniawski classification for their data, which does not quite correspond to the 1976 relation. Also, in contrast to Bieniawski, they do not report how they obtained the in situ moduli. Figure 2.31 is a plot of the data using a logarithmic scale for the modulus, which resulted in a straight line relationship, with the formula:

$$E_m = 10^{\frac{RMR-10}{40}} \tag{2.3}$$

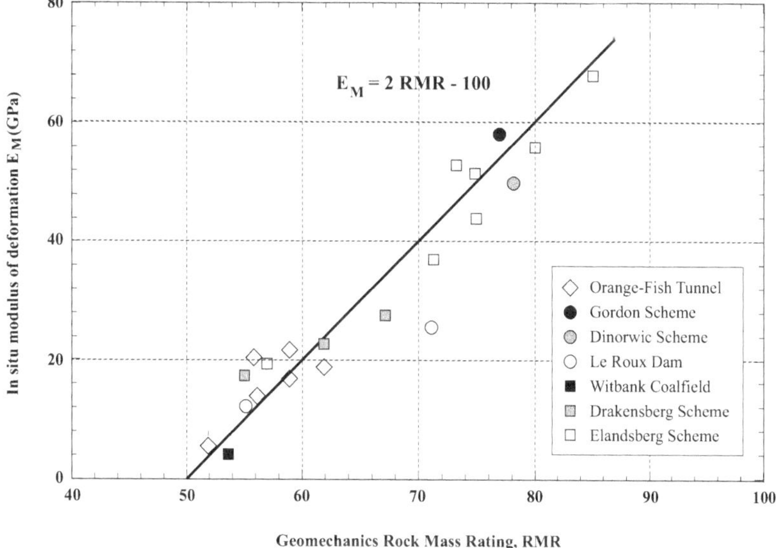

Figure 2.30 Relationship between in situ modulus and Rock Mass Rating (Bieniawski, 1978).

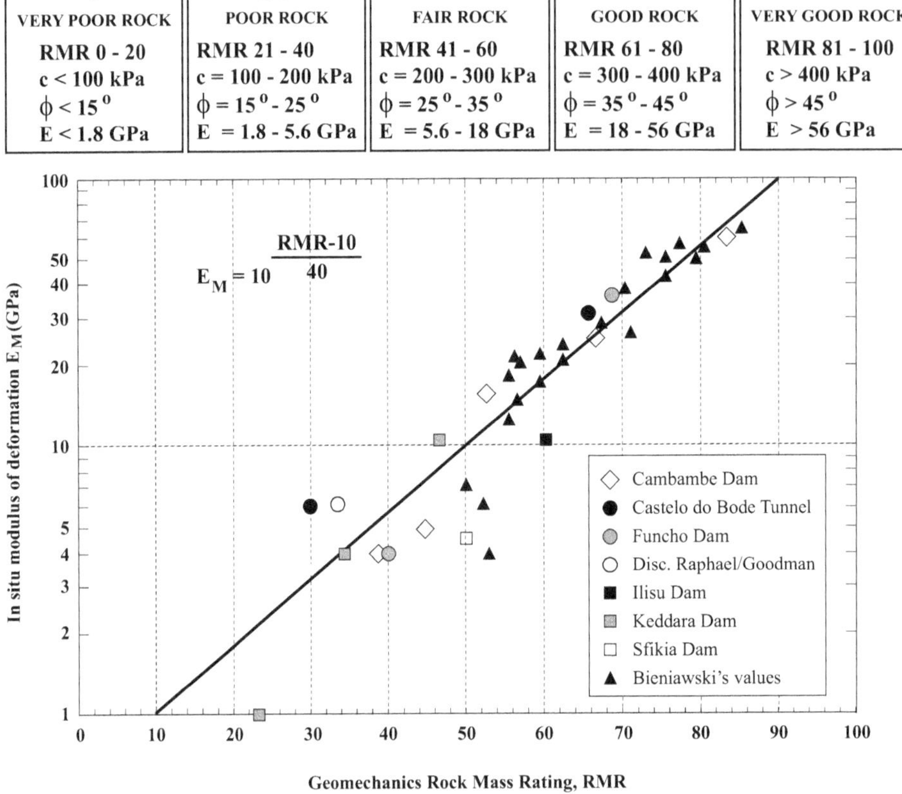

Figure 2.31 Relation RMR to in situ modulus E_M – semi-logarithmic scale (Serafim and Pereira, 1983).

Barton et al. (1980) used the original Bieniawski relation (see above) and extended it with field tests to examine the dispersion of measured and interpreted (from Q) moduli. This led to a range of possible values:

$$
\begin{aligned}
E_{max} &= 40 \log Q \\
E_{min} &= 10 \log Q \\
E_{mean} &= 25 \log Q
\end{aligned}
\tag{2.4}
$$

Barton concluded that the dispersion due to rock mass variability may be of similar magnitude as the dispersion due to the data interpretation!! Later, Barton (2002) modified the relation such that it covered most rocks (see Section 2.3.6.5 for comments on Q_c):

$$
\begin{aligned}
E_m &= 10 Q_c^{1/3} \\
Q_c &= Q \frac{\sigma_c}{100}
\end{aligned}
\tag{2.5}
$$

Among the many other attempts at deriving empirical relations for moduli, those by Ebisu et al. (1992) and Onodera and Asoka-Kumara (1980) are of some interest.

Probably, the most recent discussion of modulus-rock mass property relations is the one by Hoek and Diedrichs (2006). They address the issue that relations between rock mass modulus and rock mass characteristics (classifications) (Q, GSI, RMR, and RQD) do not have the expected asymptote at high values of the rock mass characteristics. What they say is that for a massive (sparsely jointed) rock the modulus should gradually approach the intact modulus. An analogous issue arises when plotting the ratio E_m/E_i (ratio between the modulus of rock mass and that of intact rock) versus rock mass characteristics. This, in principle, should reach a value of 1 for the best characteristics. But as Figure 2.32 shows, there is a wide scatter. The analytical investigations by Dershowitz (1979) may shed some light on the underlying causes. Introduction of the variability of joint geometry (stochastic model) in Figure 2.33 (Figure 9 in Dershowitz 1979) shows two interesting effects. Steep (~ 90°) joints increase the modulus ratio Ê/E (Ê modulus of jointed rock, E modulus of intact rock), which is not unexpected since this means that it is mostly the intact rock and not the joints affecting the rock mass modulus. Dispersion (κ = 1) decreases the modulus ratio, particularly for the steep joints since this means that there are actually many more flatly inclined joints. The conclusion one can draw from this, which is important for the empirical relations, is that for low orientation dispersion, RQD–modulus ratio relations are reasonably good relations, while for large orientation dispersion, they are not.

Hoek and Diedrichs (2006) relate rock mass characteristics to the in situ deformation modulus. Their conclusions are based on an extensive set of test data obtained by Chern in Taiwan and China. Most of the data come from flat jack and plate loading tests but some come from back-calculations. Hoek and Diederichs mention that clearly only back-calculations would provide information on large-scale moduli but that this information is limited. They then propose a simplified and a complete expression to represent the sigmoid (asymptotic) properties for the GSI–modulus relation (see the next section):

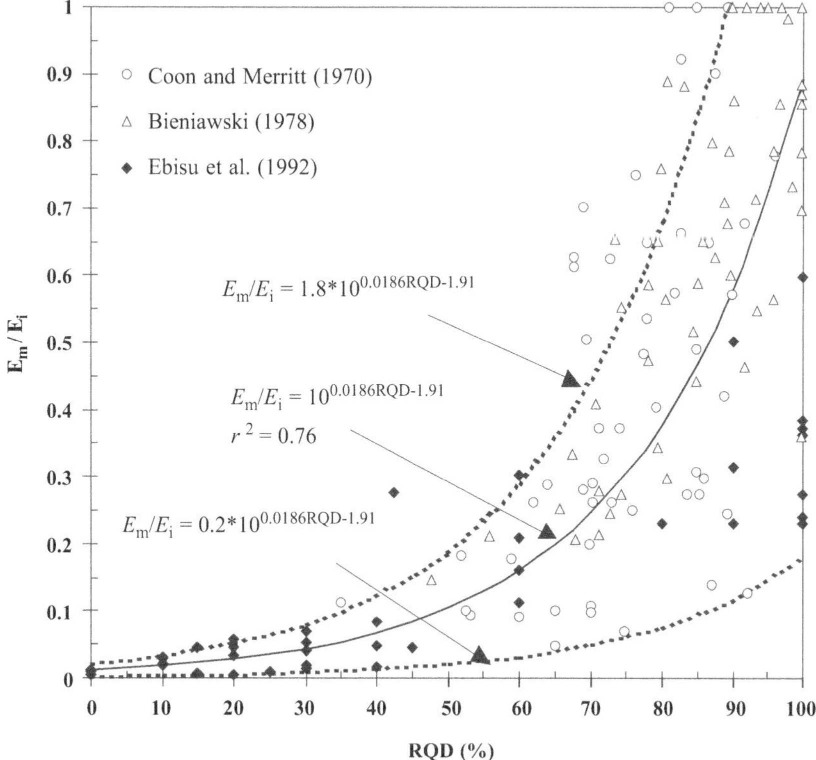

Figure 2.32 Recommended relations between RQD and E_m/E_i (Zhang and Einstein, 2004).

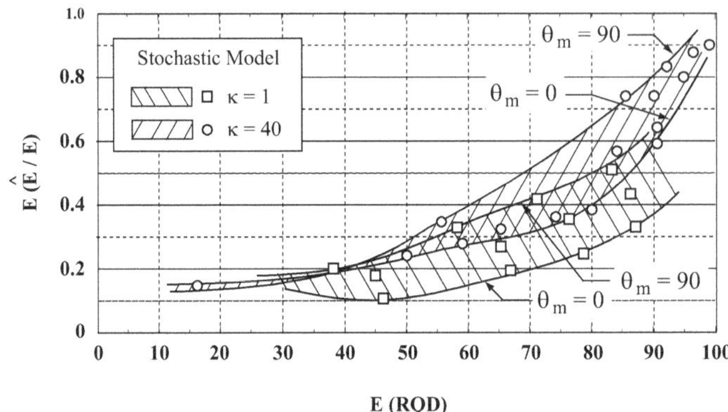

Figure 2.33 Relationship between the mean of the modulus ratio E[Ê/E] and the mean E[RQD] for different mean joint orientations θ_m and two different Fisher distributions: κ = 40, narrow distribution and κ = 1, wide distribution (After Dershowitz, 1979).

$$\text{Simplified:}\quad E_m(\text{MPa}) = 100{,}000\left(\frac{1-D/2}{1+e^{(75+25D-\text{GSI})/11}}\right) \tag{2.6a}$$

$$\text{Complete:}\quad E_m(\text{MPa}) = E_i\left(0.02 - \frac{1-D/2}{1+e^{(60+15D-\text{GSI})/11}}\right) \tag{2.6b}$$

where E_m is the in situ modulus and D the disturbance factor. Both curves (functions) were obtained by curve fitting to the Chinese/Taiwanese data. Note that in the second equation the intact modulus is not the modulus measured in the laboratory but backfigured with E_i = MR σ_{ci} where σ_{ci} is the intact unconfined compressive strength and MR is the "modulus ratio," as defined by Deere et al. (1968). If one plots σ_{ci} and E_i, both in logarithmic scale, of a variety of different rocks, they plot on a diagonal that is constant E_i/σ_{ci}= MR. The MR from Deere et al. (1968) was evidently modified by Hoek and Diederichs. Typical cases for E_m – GSI, are shown in Figures 2.34a to c.

2.3.4.2 Rock mass strength and deformability – Hoek and Brown criteria and GSI

The original Hoek and Brown criterion for (intact) rock was published in Hoek and Brown (1980a, 1980b) and had the form:

$$\sigma_1 = \sigma_3 + \sqrt{m\sigma_c\sigma_3 + s\sigma_c^2} \tag{2.7}$$

where σ_1 is the major principal stress at failure, σ_3 is the minor principal stress at failure, σ_c is the uniaxial compressive strength of the intact material, and m and s are constants that depend on the properties of the rock and to the extent to which it has been broken before being subjected to the stresses σ_1 and σ_3. As the explanations of the terms indicate, the expression is not for intact rock but more general. The relationship (2.7) can be obtained from triaxial tests when plotting σ_1 versus σ_3 or τ/σ as shown in Figure 2.35. By substituting $\sigma_3 = 0$ in expression (2.7), Hoek and Brown obtain the unconfined compressive strength of the rock as:

$$\sigma_{cs} = \sqrt{s}\,\sigma_c \tag{2.8}$$

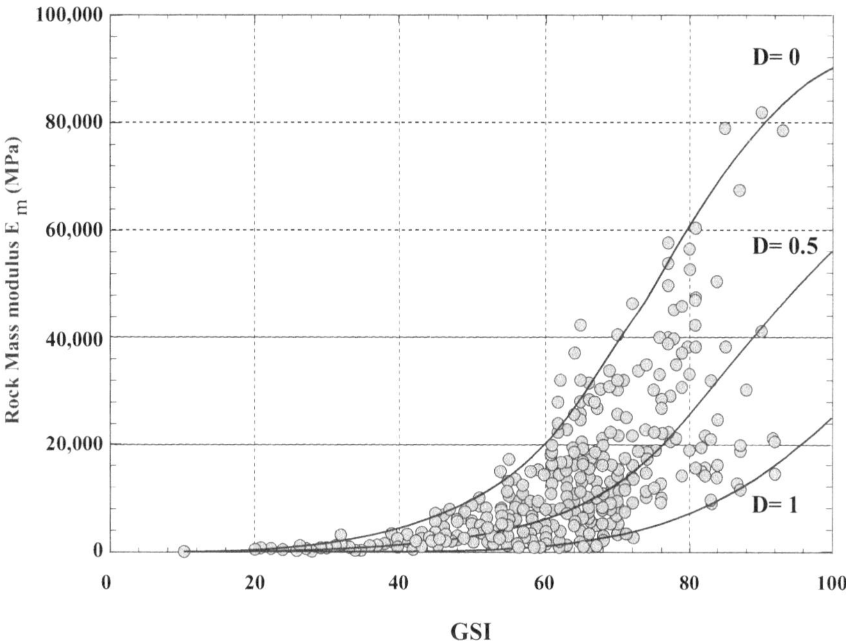

Figure 2.34a Plot of in situ modulus of deformation E_m versus GSI for Chinese and Taiwanese data and curves for simplified Hoek and Diedrichs (2006) Equation (2.6a).

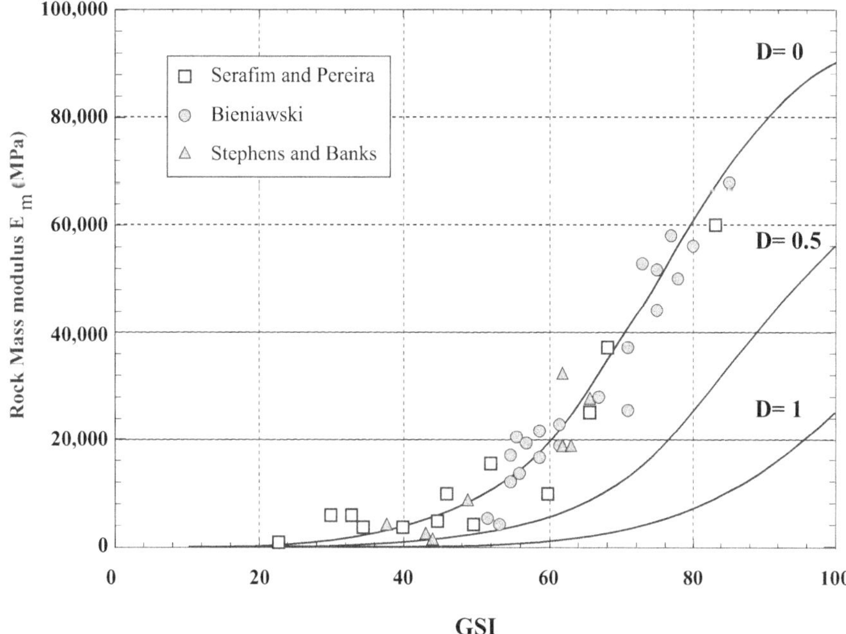

Figure 2.34b Plot of in situ rock mass deformation modulus data from Serafim and Pereira, (1983); Bieniawski (1978) and Stephens and Banks (1989) versus GSI and Simplified Hoek and Diedrichs Equation (2006).

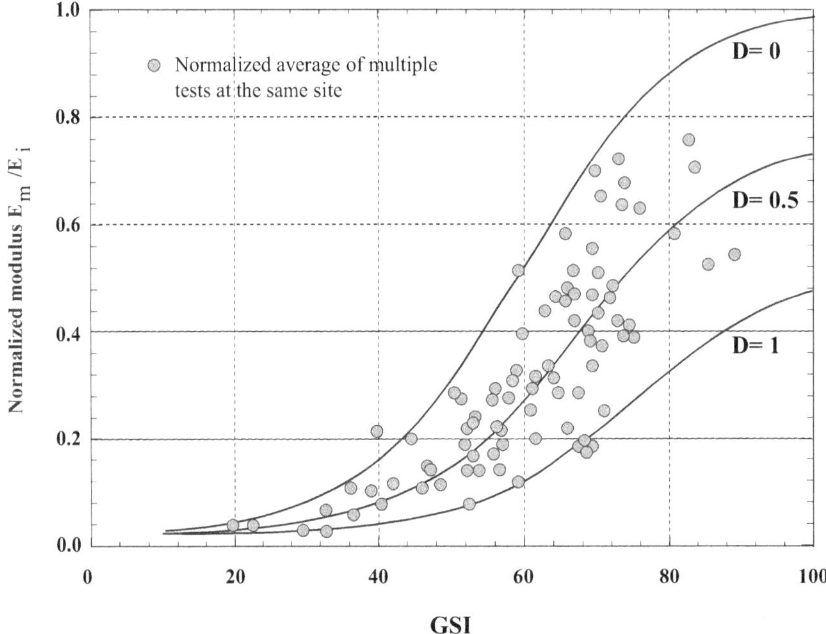

Figure 2.34c Plot of normalized in situ rock mass deformation modulus from China and Taiwan versus GSI and Hoek and Diedrichs Equation (2006). Each data point represents the average of multiple tests at the same site in the same rock mass.

And thus, for intact rock, since $\sigma_{cs} = \sigma_c$, the constant s = 1. This assumption, s = 1, is then used by Hoek and Brown when interpreting a large number of triaxial tests cited in the literature. Hoek (1994a, 1994b) makes it very clear that the expression (2.7), i.e. a parabola, was used by analogy to the Griffith stress criterion for brittle materials, which can be represented by a parabolic form. He also mentioned that the particular expression (2.7) was found by trial and error and that, therefore, there is no further mechanical meaning to the parameters in the expression. Equation (2.7) and the corresponding curves in the $\sigma_1 - \sigma_3$ or $\tau - \sigma_n$ diagram, therefore, require only two material parameters, namely σ_c and m. Hoek and Brown present their interpretation results in table form (Table 9 in Hoek and Brown). In making this interpretation, Hoek and Brown consider only the range of $\sigma_t < \sigma_3 < \frac{\sigma_1}{3.4}$ where the upper limit comes from Mogi, who found in his tests that the brittle–ductile transition occurs for $\sigma_1 = 3.4\ \sigma_3$. Implicit is, therefore, that the original Hoek and Brown criterion is applicable for brittle behavior and that the values listed in their original publication are for intact rock. Note that Hoek and Brown state that they looked at the original publications to make certain that the tests were run under consistent boundary conditions. Given the reputation of these two authors, there can be no doubt about the care they took. However, many of the tests were run in quite different apparati and this may have an effect on the results. This comment only relates to the specific numbers in the table but not to the criterion per se.

At the end of the original publication on the Hoek-Brown criterion, there is already an extension to applying it to the rock mass. Specifically they look at a rock from New Guinea, Panguna Andente, for which triaxial tests on intact rock and on specimens of various degrees of disturbance (weathering) were conducted. The data were then used to relate m/m_i (ratio of rock mass m versus intact rock m_i) and s to the Q and Geomechanics-RMR (1974) version ratings. Finally, a table listing different numerical expressions using the original criterion,

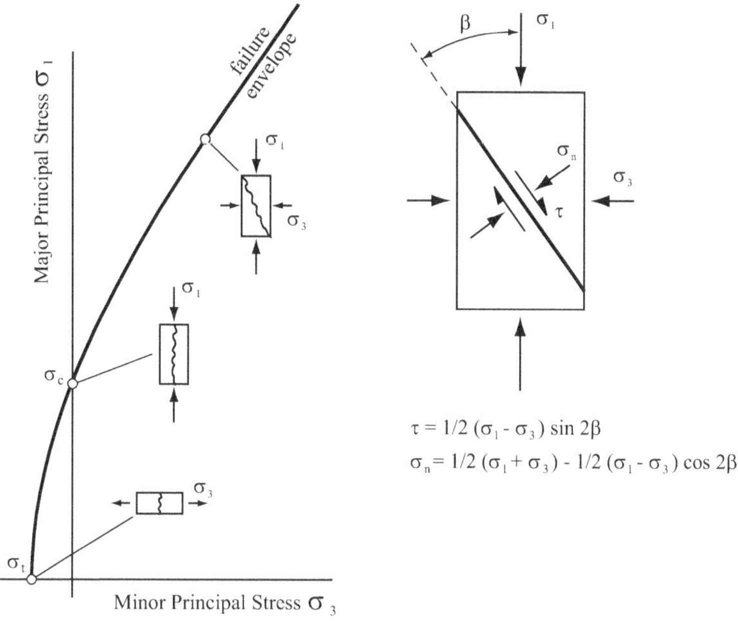

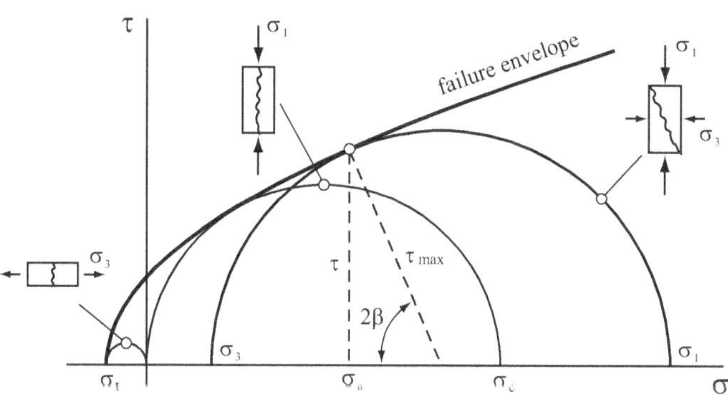

Figure 2.35 Graphical representation of stress conditions for failure of intact rock (Hoek and Brown, 1980a, 1980b).

Equation (2.7), for different rock types and rock mass variables was developed (Table 2 in Hoek and Brown). Looking at this table one already sees the GSI concept, which will be introduced later.

In 1988 (Hoek and Brown, 1988), 1992 (Hoek et al., 1992), and 1994 (Hoek, 1994a, 1994b) the Hoek and Brown criterion was extended to jointed rock masses, since these materials should have no strength when not confined ($\sigma_3 = 0$), i.e. also not have any tensile strength since the Mohr envelope would start at "0" (Hoek et al., 1992). This was somewhat modified such that well-interlocked jointed rock masses could have strengths under no confinement (Hoek, 1994a, 1994b), leading to:

$$\sigma_1' = \sigma_3' + \sigma_c \left(m_b \frac{\sigma_3'}{\sigma_c} + s \right)^a \tag{2.9}$$

where m_b is the value of the constant m for the rock mass; s and a are constants that depend on the characteristics of the rock mass; σ_c is the unconfined compressive strength of the intact rock; and σ_1', σ_3' are the major and minor principal effective stresses, respectively. Effective stresses are used from this point on by Hoek and Brown. Hoek and Brown (1997) make some general comments that depending on the situation, pore pressure should be considered or that the conditions may be fully drained. (It is important to emphasize that when using the Hoek and Brown criterion as an expression of rock mass behavior, it is necessary to consider the actual pore pressure regime – note by the authors of this book.)

For heavily disturbed rock, s in expression (2.9) will equal '0' and

$$\sigma_1' = \sigma_3' + \sigma_c \left(m_b \frac{\sigma_1'}{\sigma_c} \right)^a \tag{2.10}$$

In order to use these expressions σ_c, s (if applicable), a and m_b need to be estimated or backfigured from the Mohr ($\tau - \sigma$) curves from triaxial tests on the rock mass in question. Since conducting such tests is often impossible, Hoek (1994a, 1994b) provides information on how to estimate the values of σ_c and m_i. Note, and this is very important, that similar tables have been produced by Hoek and co-authors prior and after the 1994 tables. The values in these tables differ, reflecting the additional experience and information gained through applications of the method. Hence, only the most recent tables, which are reproduced here as Tables 2.9a and 2.9b from Marinos and Hoek (2000), should be used, and, if newer information becomes available, this latest information should be used. The other parameters m_b, s, and a need to be extracted from information on the rock mass characteristics. Originally (Hoek and Brown, 1988), such relations were based on the Geomechanics RMR characterization (Bieniawski, 1974 and 1976), which takes the form:

$$\begin{aligned} \frac{m_b}{m_i} &= \exp\left(\frac{RMR - 100}{14}\right) \\ s &= \exp\left(\frac{RMR - 100}{6}\right) \end{aligned} \tag{2.11}$$

For undisturbed, interlocking rock masses:

$$\begin{aligned} \frac{m_b}{m_i} &= \exp\left(\frac{RMR - 100}{28}\right) \\ s &= \exp\left(\frac{RMR - 100}{9}\right) \end{aligned} \tag{2.12}$$

These equations have been used to construct a table (Table 1 in Hoek and Brown, 1988) in which m_b and s for different undisturbed/disturbed rock masses and lithologies are related to the Hoek-Brown constants, as well as to RMR and Q, the latter using Bieniawski's (1976) relation: RMR = 9 ln Q + 44. In the 1988 paper, Hoek and Brown also made suggestions on how to avoid double counting effects when deriving the Hoek and Brown constants from RMR or Q:

> RMR: Use a rating of 10 (completely dry) for groundwater and consider the effect of water in determining the stresses.

Table 2.9a Field estimates of uniaxial compressive strengths of intact rock (Marinos and Hoek, 2000)

Grade[a]	*Term*	*Uniaxial comp. strength (MPa)*	*Point load index (MPa)*	*Field estimate of strength*	*Examples*
R6	Extremely strong	>250	>10	Specimen can only be chipped with a geological hammer	Fresh basalt, chert, diabase, gneiss, granite, quartzite
R5	Very strong	100–250	4–10	Specimen requires many blows of a geological hammer to fracture it	Amphibolite, sandstone, basalt, gabbro, gneiss, granodiorite, peridotite, rhyolite, tuff
R4	Strong	50–100	2–4	Specimen requires more than one blow of a geological hammer to fracture it	Limestone, marble, sandstone, schist
R3	Medium strong	25–50	1–2	Cannot be scraped or peeled with a pocket knife, specimen can be fractured with a single blow from a geological hammer	Concrete, phyllite, schist, siltstone
R2	Weak	5–25	[b]	Can be peeled with a pocket knife with difficulty, shallow indentation made by firm blow with point of geological hammer	Chalk, claystone, potash, marl, siltstone, shale, rocksalt
R1	Very weak	1–5	[b]	Crumbles under firm blows with point of a geological hammer, can be peeled by a pocket knife	Highly weathered or altered rock, shale
R0	Extremely weak	0.25–1	[b]	Indented by thumbnail	Stiff fault gouge

[a] Grade according to Brown (1981).

[b] Point load tests on rocks with a uniaxial compressive strength below 25 MPa are likely to yield highly ambiguous results.

Use "0" (very favorable) for joint orientation and decide on the applicability of the Hoek and Brown criterion based on the fact if the joint pattern produces a continuum behavior (Hoek and Brown applicable) or failure of discrete bodies (Hoek and Brown not applicable).

Q: Use a value of 1 for both the joint water reduction factor (J_w) and the stress reduction factor (SRF) in a dry rock mass subject to medium stresses, and consider the stress and water effects in the stress determination.

Further comments on the use RMR were made later. Specifically, RMR is not entirely satisfactory for low values (RMR < 25). This led to the introduction of GSI, the Geologic Strength Index. Originally, i.e. in Hoek (1994a, 1994b), GSI was derived from either RMR 1976 (Bieniawski, 1976) or RMR, 1989 (Bieniawski, 1989) and Q (Barton et al., 1974a, 1974b) with detailed comments on how to avoid double counting, i.e. considering essentially dry rock for both RMR and Q and not including the stress reduction factor in Q (see above). The reason for this is that when Equation (2.10) is used to represent the rock mass strength in an analytical or numerical model, it will be expressed in terms of effective stresses with the water pressure as additional boundary conditions, and included as such. Rock mass strength must, therefore, not also include these boundary conditions. Hoek (1994a, 1994b) provides detailed expressions on how to derive GSI from RMR and Q. They will not be mentioned

Table 2.9b Values of the constant m_i, for intact rock group. Note that values in parenthesis are estimates

			Texture			
Rock type	*Class*	*Group*	*Coarse*	*Medium*	*Fine*	*Very fine*
Sedimentary	Clastic		Conglomerates[a] Breccias[a]	Sandstones 17 ± 4	Siltstones 7 ± 2 Greywackes (18 ± 3)	Claystones 4 ± 2 Shales (6 ± 2) Marls (7 ± 2)
	Non-Clastic	Carbonates	Crystalline Limestone (12 ± 3)	Sparitic Limestones (10 ± 2)	Micritic Limestones (9 ± 2)	Dolomites (9 ± 3)
		Evaporites		Gypsum 8 ± 2	Anhydrite 12 ± 2	
		Organic				Chalk 7 ± 2
Metamorphic	Non-Foliated		Marble 9 ± 3	Hornfels (19 ± 4) Metasandstone (19 ± 3)	Quartzites 20 ± 3	
	Slightly foliated		Migmatite (29 ± 3)	Amphibolites 26 ± 6	Gneiss 28 ± 5	
	Foliated[b]			Schists 12 ± 3	Phyllites (7 ± 3)	Slates 7 ± 4
Igneous	Plutonic	Light	Granite 32 ± 3	Diorite 25 ± 5		
			Granodiorite (29 ± 3)			
		Dark	Gabbro 27 ± 3	Dolerite (16 ± 5)		
			Norite 20 ± 5			
	Hypabyssal		Porphyries (20 ± 5)		Diabase (15 ± 5)	Peridotite (25 ± 5)
	Volcanic	Lava		Rhyolite (25 ± 5) Andesite 25 ± 5	Dacite (25 ± 3) Basalt (25 ± 5)	
		Pyroclastic	Agglomerate (19 ± 3)	Breccia (19 ± 5)	Tuff (13 ± 5)	

Note: The range of values quoted for each material depends upon the granularity and interlocking of the crystal structure – the higher values being associated with tightly interlocked and more frictional characteristics. (Marinos and Hoek, 2000)

[a] Conglomerates and breccias may present a wide range of m_i values depending on the nature of the cementing material and the degree of cementation, so they may range from values similar to sandstone, to values used for fine-grained sediments (even under 10).

[b] These values are for intact rock specimens tested normal to bedding or foliation. The value of m_i will be significantly different if failure occurs along a weakness plane.

Note that this table contains several changes from previously published versions. These changes have been made to reflect data that have been accumulated from laboratory tests and the experience gained from discussions with geologists and engineering geologists.

here since this approach has been superseded by directly estimating GSI from information on the rock mass. GSI, be it derived directly or from Q, RMR, is used to estimate the parameter m_b, s and a.

$$\frac{m_b}{m_i} = \exp\left(\frac{GSI - 100}{28}\right) \tag{2.13}$$

For GSI > 25 (undisturbed rock masses)

$$s = \exp\left(\frac{GSI - 100}{9}\right)$$
$$a = 0.5 \tag{2.14}$$

For GSI < 25 (undisturbed rock masses; see note below)

$$s = 0.5 \quad \text{or} \quad 0$$
$$a = 0.62 - \frac{GSI}{200} \tag{2.15}$$

Note: In Hoek (1994a, 1994b), the term undisturbed is used for both GSI > 25 and s = 0.5. In later publications, e.g., Hoek and Brown (1997), the terms used are as follows:

> GSI > 25: "rock masses of good or reasonable quality";
> GSI < 25: "rock masses of very poor quality";
> GSI < 25: s = 0.

The newest development in 2002 (Hoek et al., 2002) replaces the expressions with a continuous one, i.e. not having a limit at GSI = 25. The expressions are:

$$\frac{m_b}{m_i} = \exp\left(\frac{GSI - 100}{28 - 14D}\right)$$
$$s = \exp\left(\frac{GSI - 100}{9 - 3D}\right) \tag{2.16}$$
$$a = \frac{1}{2} + \frac{1}{6}\left(e^{-GSI/15} - e^{-20/3}\right)$$

D is a disturbance factor expressing the effect of blasting damage or stress relaxation and recommendations (guidelines) are given by Hoek et al. (2002) and reproduced below as Table 2.10. Professor Hoek added, per personal communication on August 4, 2022:

> The Blast Damage factor D should only be applied to the rock mass surrounding the excavation and not to the entire rock mass. I found that some readers were making this mistake, with disastrous consequences. I tend to limit the extent of blast damage to about one tunnel diameter of the rock surrounding a tunnel.

As can be seen, for no disturbance, D = 0 and the expressions above are equal to those for GSI > 25, while for D = 1 and low GSI the expressions above approach those for GSI < 25. The disturbance factor was also included in relations between modulus ratio E_m/E_i or E_m,

Table 2.10 Guidelines for estimating disturbance factor D (Hoek et al., 2002)

Appearance of rock mass	*Description of rock mass*	*Suggested value of D*
	Excellent quality controlled blasting or excavation by Tunnel Boring Machine results in minimal disturbance to the confined rock mass surrounding a tunnel.	**D = 0**
	Mechanical or hand excavation in poor quality rock masses (no blasting) results in minimal disturbance to the surrounding rock mass. Where squeezing problems result in significant floor heave, disturbance can be severe unless a temporary invert, as shown in the photograph, is placed.	**D = 0** **D = 0.5** No invert
	Very poor quality blasting in a hard rock tunnel results in severe local damage, extending 2 or 3 m, in the surrounding rock mass.	**D = 0.8**
	Small-scale blasting in civil engineering slopes results in modest rock mass damage, particularly if controlled blasting is used as shown on the left-hand side of the photograph. However, stress relief results in some disturbance.	**D = 0.7** Good blasting **D = 1.0** Poor blasting
	Very large open pit mine slopes suffer significant disturbance due to heavy production blasting and also due to stress relief from overburden removal. In some softer rocks excavation can be carried out by ripping and dozing and the degree of damage to the slopes is less.	**D = 1.0** Production blasting **D = 0.7** Mechanical excavation

Source: Reproduced by permission of Prof. E. Hoek.

Note: Note that the blast damage factor D should only be applied to the rock mass surrounding the excavation and not to the entire rock mass.

MR, σ_{ci} and GSI mentioned earlier using extensive data from China and Taiwan (Hoek and Diedrichs, 2006; note that GSI was not directly determined in that paper but that it was derived from RMR). In Hoek et al. (2002), the authors provide equations for the modulus of elasticity considering the disturbance:

For σ_{ci} < 100 MPa

$$E_m(GPa) = \left(1 - \frac{D}{2}\right)\sqrt{\frac{\sigma_{ci}}{100}}\,10^{(GSI-10)/40} \tag{2.17}$$

and for σ_{ci} > 100 MPa

$$E_m(GPa) = \left(1 - \frac{D}{2}\right)10^{(GSI-10)/40} \tag{2.18}$$

For the Coulomb parameters c, ϕ obtained from a linearization of the Hoek and Brown criterion,

$$\begin{aligned} \phi &= \sin^{-1}\left[\frac{6am_b(s + m_b\sigma_{3n})^{a-1}}{2(1+a)(2+a) + 6am_b(s + m_b\sigma_{3n})^{a-1}}\right] \\ c &= \frac{\sigma_{ci}[(1+2a)s + (1-a)m_b\sigma_{3n}](s + m_b\sigma_{3n})^{a-1}}{(1+a)(2+a)\sqrt{1 + \frac{6am_b(s + m_b\sigma_{3n})^{a-1}}{(1+a)(2+a)}}} \end{aligned} \tag{2.19}$$

where $\sigma_{3n} = \sigma_{3max}/\sigma_{ci}$

In parallel to the work on the Hoek-Brown criterion, and as evident from the discussion above, GSI has evolved over time (see Hoek, 1994a, 1994b for the original version). A reasonably final version (see also comments below on the 2018 version) is shown in Tables 2.11, from Marinos and Hoek (2000). As mentioned earlier, estimates of uniaxial compressive strength and intact material parameter m_i, which have to be used together with GSI in the generalized Hoek and Brown criterion, have also undergone some changes. Tables 2.9a and b for σ_c and m_i are also from the Marinos and Hoek (2000) paper. In contrast to earlier papers by Hoek, this only refers to the tables without mentioning that it actually would be preferable to determine σ_c and m_i from triaxial tests.

Marinos and Hoek (2000) also include graphs that allow the user to estimate cohesion (cohesive strength), friction angle, rock mass strength (uniaxial strength of the rock mass), and modulus of elasticity for tunnels at depths greater than 30 m (computation with Hoek and Brown criterion, essentially a linearization of the τ–σ diagram using confining pressure $0 < \sigma_c < 0.25\,\sigma_{ci}$; see Hoek and Brown, 1997). Note that the range of confining pressure has also changed throughout the development of the Hoek and Brown criterion. These graphs are shown as Figures 2.36–2.38.

Most recently, Hoek and Brown (2019) published the paper "The Hoek-Brown Failure Criterion – 2018 Edition." This provides a summary of the Hoek and Brown–GSI process as summarized in the flow chart of Figure 2.39. Note that what is shown as "back analysis" corresponds to the observational method discussed early in Section 2.3. Importantly, the "2018 edition" also includes some comments on limitations as well as additions. The limitations concern in situ stress conditions, in which rock mass behaves in a ductile manner as well as aspects of GSI. Important additions are the possibilities to include a tension cutoff in the failure envelopes and the distribution of unconfined compressive strength. There are also

Table 2.11a Most common GSI ranges for typical sandstones[a] (Marinos and Hoek, 2000)

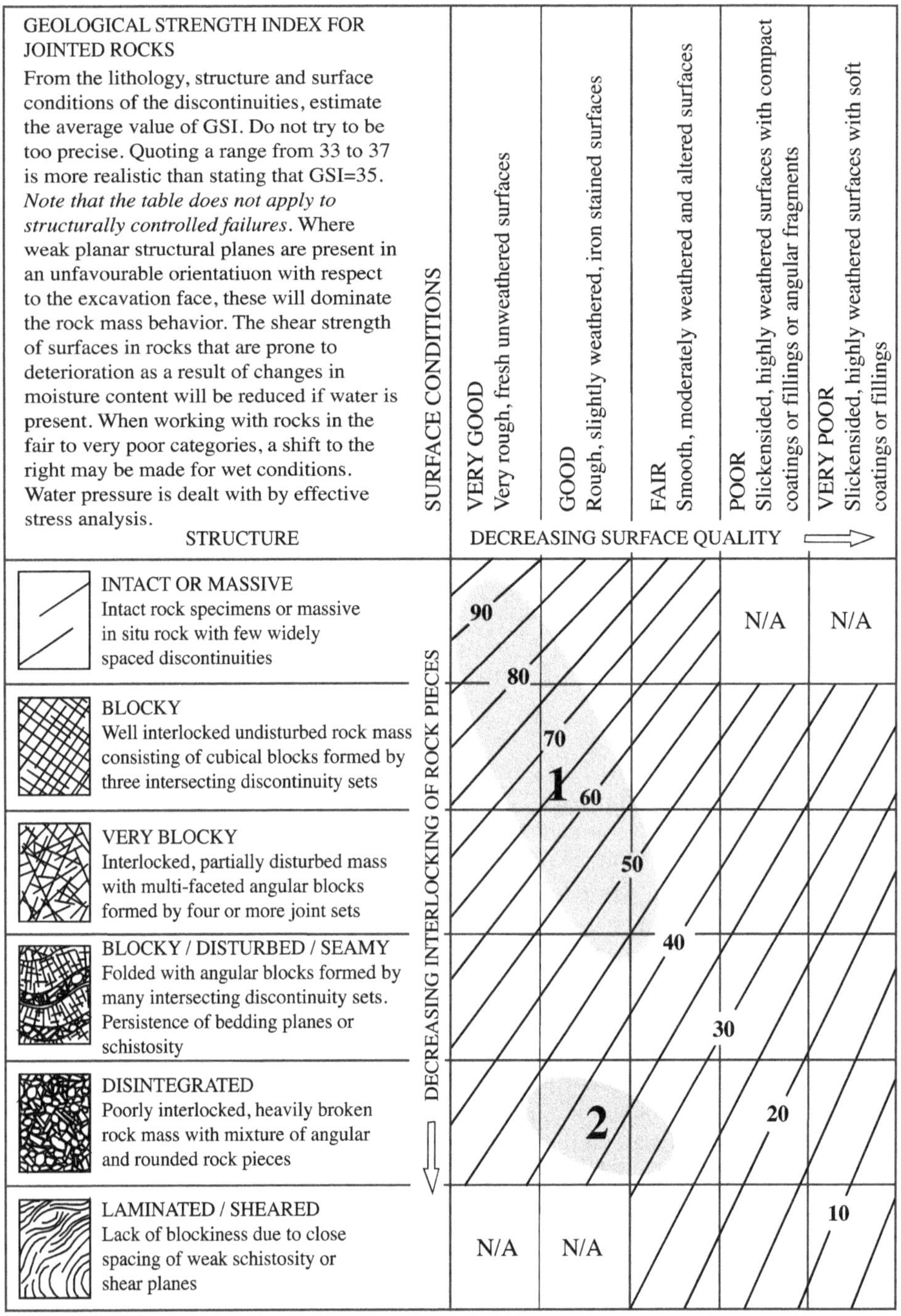

GEOLOGICAL STRENGTH INDEX FOR JOINTED ROCKS From the lithology, structure and surface conditions of the discontinuities, estimate the average value of GSI. Do not try to be too precise. Quoting a range from 33 to 37 is more realistic than stating that GSI=35. *Note that the table does not apply to structurally controlled failures*. Where weak planar structural planes are present in an unfavourable orientatiuon with respect to the excavation face, these will dominate the rock mass behavior. The shear strength of surfaces in rocks that are prone to deterioration as a result of changes in moisture content will be reduced if water is present. When working with rocks in the fair to very poor categories, a shift to the right may be made for wet conditions. Water pressure is dealt with by effective stress analysis. STRUCTURE	SURFACE CONDITIONS	VERY GOOD Very rough, fresh unweathered surfaces	GOOD Rough, slightly weathered, iron stained surfaces	FAIR Smooth, moderately weathered and altered surfaces	POOR Slickensided, highly weathered surfaces with compact coatings or fillings or angular fragments	VERY POOR Slickensided, highly weathered surfaces with soft coatings or fillings
		DECREASING SURFACE QUALITY ⇨				
INTACT OR MASSIVE Intact rock specimens or massive in situ rock with few widely spaced discontinuities	DECREASING INTERLOCKING OF ROCK PIECES ⇩	90			N/A	N/A
BLOCKY Well interlocked undisturbed rock mass consisting of cubical blocks formed by three intersecting discontinuity sets		80	70			
VERY BLOCKY Interlocked, partially disturbed mass with multi-faceted angular blocks formed by four or more joint sets			1 60	50		
BLOCKY / DISTURBED / SEAMY Folded with angular blocks formed by many intersecting discontinuity sets. Persistence of bedding planes or schistosity				40	30	
DISINTEGRATED Poorly interlocked, heavily broken rock mass with mixture of angular and rounded rock pieces			2		20	
LAMINATED / SHEARED Lack of blockiness due to close spacing of weak schistosity or shear planes		N/A	N/A			10

Source: Reproduced by permission of Taylor and Francis Group, LLC, a division of Informa plc.

[a] *WARNING*

The shaded areas are indicative and may not be appropriate for site specific design purposes. Mean values are not suggested for indicative characterization; the use of ranges is recommended

1. Massive or bedded (no clayey cement present)
2. Brecciated (no clayey cement present)

Table 2.11b Most common GSI ranges for typical siltstones, claystones and clay shales[a] (Marinos and Hoek, 2000)

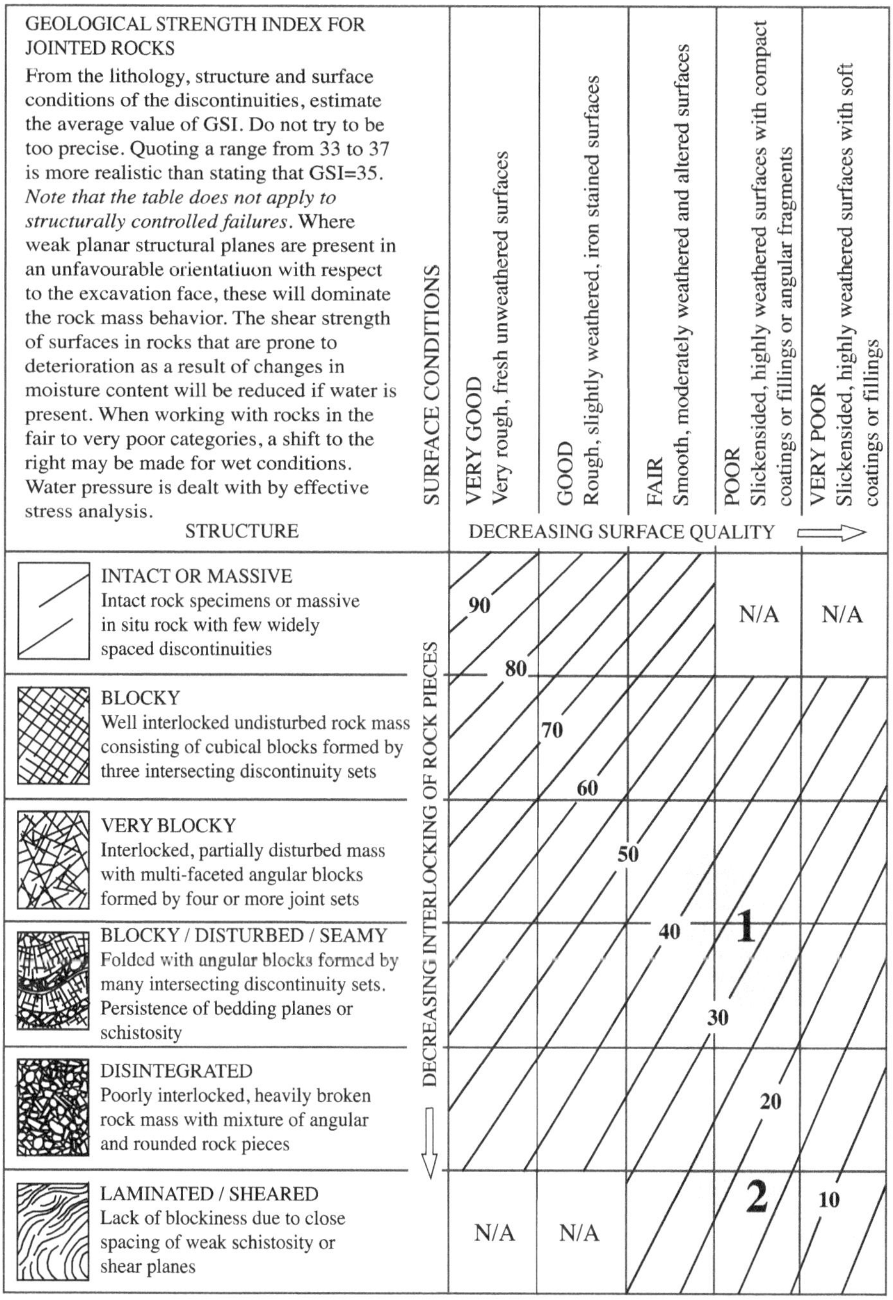

GEOLOGICAL STRENGTH INDEX FOR JOINTED ROCKS From the lithology, structure and surface conditions of the discontinuities, estimate the average value of GSI. Do not try to be too precise. Quoting a range from 33 to 37 is more realistic than stating that GSI=35. *Note that the table does not apply to structurally controlled failures.* Where weak planar structural planes are present in an unfavourable orientatiuon with respect to the excavation face, these will dominate the rock mass behavior. The shear strength of surfaces in rocks that are prone to deterioration as a result of changes in moisture content will be reduced if water is present. When working with rocks in the fair to very poor categories, a shift to the right may be made for wet conditions. Water pressure is dealt with by effective stress analysis. STRUCTURE	SURFACE CONDITIONS	VERY GOOD Very rough, fresh unweathered surfaces	GOOD Rough, slightly weathered, iron stained surfaces	FAIR Smooth, moderately weathered and altered surfaces	POOR Slickensided, highly weathered surfaces with compact coatings or fillings or angular fragments	VERY POOR Slickensided, highly weathered surfaces with soft coatings or fillings
		DECREASING SURFACE QUALITY ⇒				
INTACT OR MASSIVE Intact rock specimens or massive in situ rock with few widely spaced discontinuities	DECREASING INTERLOCKING OF ROCK PIECES ⇓	90			N/A	N/A
BLOCKY Well interlocked undisturbed rock mass consisting of cubical blocks formed by three intersecting discontinuity sets		80	70			
VERY BLOCKY Interlocked, partially disturbed mass with multi-faceted angular blocks formed by four or more joint sets			60	50		
BLOCKY / DISTURBED / SEAMY Folded with angular blocks formed by many intersecting discontinuity sets. Persistence of bedding planes or schistosity				40	1 30	
DISINTEGRATED Poorly interlocked, heavily broken rock mass with mixture of angular and rounded rock pieces					20	
LAMINATED / SHEARED Lack of blockiness due to close spacing of weak schistosity or shear planes		N/A	N/A		2	10

Source: Reproduced by permission of Taylor and Francis Group, LLC, a division of Informa plc.

[a] *WARNING*

The shaded areas are indicative and may not be appropriate for site specific design purposes.

Mean values are not suggested for indicative characterization; the use of ranges is recommended

1. Bedded, foliated, fractured
2. Sheared, brecciated

These soft rocks are classified by GSI if their mass is disturbed as associated with tectonic processes. Otherwise, GSI is not recommended. The same is true for typical marls.

Table 2.11c Most common GSI range of typical limestone.[a] (Marinos and Hoek, 2000)

GEOLOGICAL STRENGTH INDEX FOR JOINTED ROCKS From the lithology, structure and surface conditions of the discontinuities, estimate the average value of GSI. Do not try to be too precise. Quoting a range from 33 to 37 is more realistic than stating that GSI=35. *Note that the table does not apply to structurally controlled failures.* Where weak planar structural planes are present in an unfavourable orientatiuon with respect to the excavation face, these will dominate the rock mass behavior. The shear strength of surfaces in rocks that are prone to deterioration as a result of changes in moisture content will be reduced if water is present. When working with rocks in the fair to very poor categories, a shift to the right may be made for wet conditions. Water pressure is dealt with by effective stress analysis. STRUCTURE	SURFACE CONDITIONS VERY GOOD Very rough, fresh unweathered surfaces	GOOD Rough, slightly weathered, iron stained surfaces	FAIR Smooth, moderately weathered and altered surfaces	POOR Slickensided, highly weathered surfaces with compact coatings or fillings or angular fragments	VERY POOR Slickensided, highly weathered surfaces with soft coatings or fillings
	DECREASING SURFACE QUALITY ⇨				
DECREASING INTERLOCKING OF ROCK PIECES ⇩					
INTACT OR MASSIVE Intact rock specimens or massive in situ rock with few widely spaced discontinuities	90 80			N/A	N/A
BLOCKY Well interlocked undisturbed rock mass consisting of cubical blocks formed by three intersecting discontinuity sets		70 1 60			
VERY BLOCKY Interlocked, partially disturbed mass with multi-faceted angular blocks formed by four or more joint sets			50 40		
BLOCKY / DISTURBED / SEAMY Folded with angular blocks formed by many intersecting discontinuity sets. Persistence of bedding planes or schistosity		2		30	
DISINTEGRATED Poorly interlocked, heavily broken rock mass with mixture of angular and rounded rock pieces		3		20	
LAMINATED / SHEARED Lack of blockiness due to close spacing of weak schistosity or shear planes	N/A	N/A			10

[a] *WARNING*

The shaded areas are indicative and may not be appropriate for site specific design purposes. Mean values are not suggested for indicative characterization; the use of ranges is recommended

1. Massive
2. Thin bedded
3. Brecciated

Table 2.11d Most common GSI range for typical granite[a] (Marinos and Hoek, 2000)

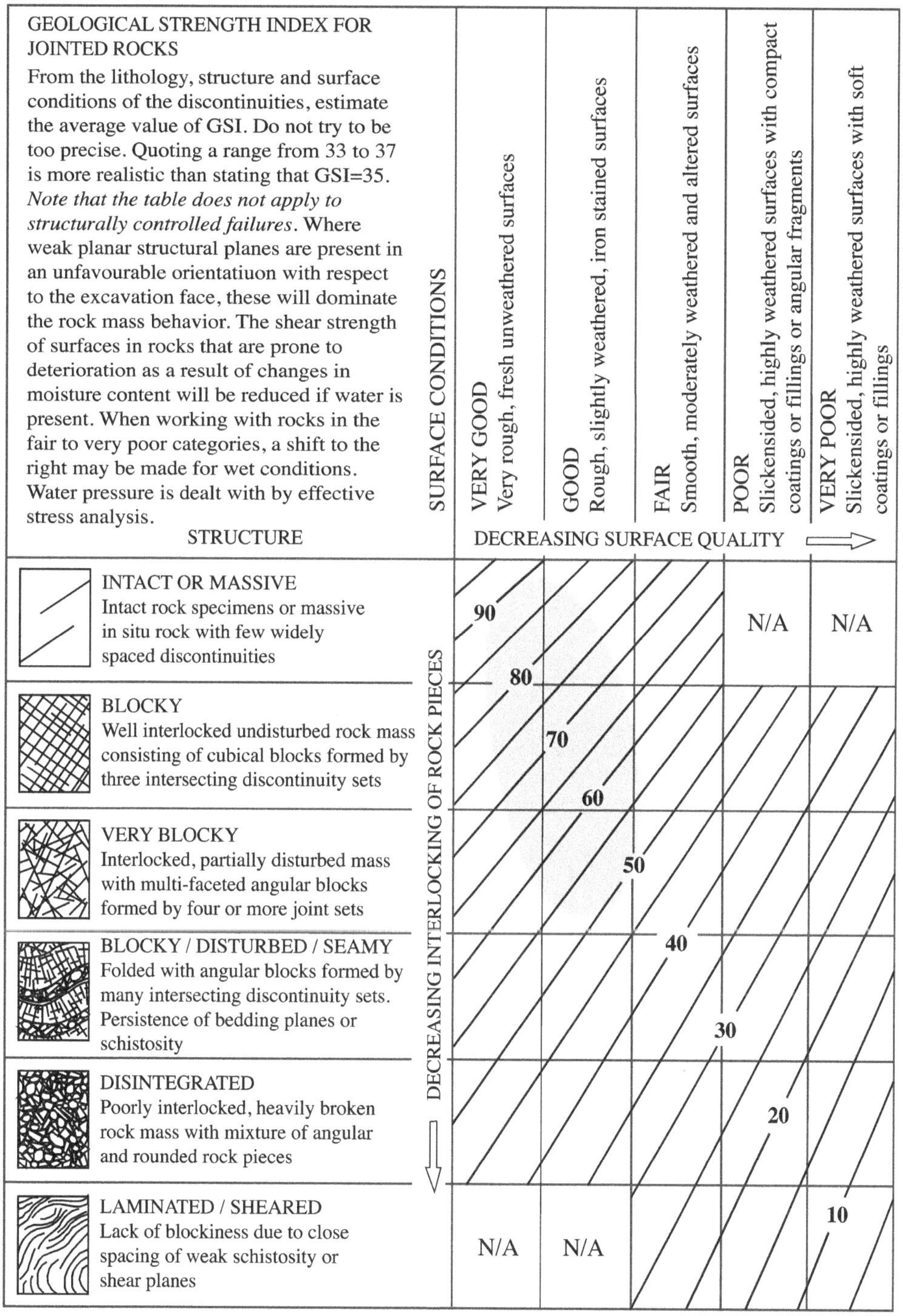

GEOLOGICAL STRENGTH INDEX FOR JOINTED ROCKS From the lithology, structure and surface conditions of the discontinuities, estimate the average value of GSI. Do not try to be too precise. Quoting a range from 33 to 37 is more realistic than stating that GSI=35. *Note that the table does not apply to structurally controlled failures*. Where weak planar structural planes are present in an unfavourable orientatiuon with respect to the excavation face, these will dominate the rock mass behavior. The shear strength of surfaces in rocks that are prone to deterioration as a result of changes in moisture content will be reduced if water is present. When working with rocks in the fair to very poor categories, a shift to the right may be made for wet conditions. Water pressure is dealt with by effective stress analysis. STRUCTURE	SURFACE CONDITIONS	VERY GOOD Very rough, fresh unweathered surfaces	GOOD Rough, slightly weathered, iron stained surfaces	FAIR Smooth, moderately weathered and altered surfaces	POOR Slickensided, highly weathered surfaces with compact coatings or fillings or angular fragments	VERY POOR Slickensided, highly weathered surfaces with soft coatings or fillings
		DECREASING SURFACE QUALITY ⇒				
INTACT OR MASSIVE Intact rock specimens or massive in situ rock with few widely spaced discontinuities	DECREASING INTERLOCKING OF ROCK PIECES ⇓	90			N/A	N/A
BLOCKY Well interlocked undisturbed rock mass consisting of cubical blocks formed by three intersecting discontinuity sets		80	70			
VERY BLOCKY Interlocked, partially disturbed mass with multi-faceted angular blocks formed by four or more joint sets			60	50		
BLOCKY / DISTURBED / SEAMY Folded with angular blocks formed by many intersecting discontinuity sets. Persistence of bedding planes or schistosity				40	30	
DISINTEGRATED Poorly interlocked, heavily broken rock mass with mixture of angular and rounded rock pieces					20	
LAMINATED / SHEARED Lack of blockiness due to close spacing of weak schistosity or shear planes		N/A	N/A			10

Source: Reproduced by permission of Taylor and Francis Group, LLC, a division of Informa plc.

[a] *WARNING*

The shaded areas are indicative and may not be appropriate for site specific design purposes. Mean values are not suggested for indicative characterization; the use of ranges is recommended

Only fresh rock masses are shown. Weathered granite may be irregularly illustrated on the GSI chart, since it can be assigned greatly varying GSI values or even behave as an engineering soil.

Table 2.11e Most common GSI range for typical ophiolites (Ultrabasic Rocks)[a] (Marinos and Hoek, 2000)

GEOLOGICAL STRENGTH INDEX FOR JOINTED ROCKS

From the lithology, structure and surface conditions of the discontinuities, estimate the average value of GSI. Do not try to be too precise. Quoting a range from 33 to 37 is more realistic than stating that GSI=35. *Note that the table does not apply to structurally controlled failures*. Where weak planar structural planes are present in an unfavourable orientatiuon with respect to the excavation face, these will dominate the rock mass behavior. The shear strength of surfaces in rocks that are prone to deterioration as a result of changes in moisture content will be reduced if water is present. When working with rocks in the fair to very poor categories, a shift to the right may be made for wet conditions. Water pressure is dealt with by effective stress analysis.

STRUCTURE / SURFACE CONDITIONS (DECREASING SURFACE QUALITY ⇒)	VERY GOOD Very rough, fresh unweathered surfaces	GOOD Rough, slightly weathered, iron stained surfaces	FAIR Smooth, moderately weathered and altered surfaces	POOR Slickensided, highly weathered surfaces with compact coatings or fillings or angular fragments	VERY POOR Slickensided, highly weathered surfaces with soft coatings or fillings
INTACT OR MASSIVE Intact rock specimens or massive in situ rock with few widely spaced discontinuities	90			N/A	N/A
BLOCKY Well interlocked undisturbed rock mass consisting of cubical blocks formed by three intersecting discontinuity sets	80	70	1		
VERY BLOCKY Interlocked, partially disturbed mass with multi-faceted angular blocks formed by four or more joint sets		60	50		
BLOCKY / DISTURBED / SEAMY Folded with angular blocks formed by many intersecting discontinuity sets. Persistence of bedding planes or schistosity			40	30	
DISINTEGRATED Poorly interlocked, heavily broken rock mass with mixture of angular and rounded rock pieces				20	2
LAMINATED / SHEARED Lack of blockiness due to close spacing of weak schistosity or shear planes	N/A	N/A			10

(DECREASING INTERLOCKING OF ROCK PIECES ⇓)

Source: Reproduced by permission of Taylor and Francis Group, LLC, a division of Informa plc.

[a] *WARNING*

The shaded areas are indicative and may not be appropriate for site specific design purposes. Mean values are not suggested for indicative characterization; the use of ranges is recommended

1. Fresh
2. Serpentinised with brecciation and shears

Table 2.11f Common GSI range for typical sound Gneiss[a] (Marinos and Hoek, 2000)

GEOLOGICAL STRENGTH INDEX FOR JOINTED ROCKS From the lithology, structure and surface conditions of the discontinuities, estimate the average value of GSI. Do not try to be too precise. Quoting a range from 33 to 37 is more realistic than stating that GSI=35. *Note that the table does not apply to structurally controlled failures.* Where weak planar structural planes are present in an unfavourable orientatiuon with respect to the excavation face, these will dominate the rock mass behavior. The shear strength of surfaces in rocks that are prone to deterioration as a result of changes in moisture content will be reduced if water is present. When working with rocks in the fair to very poor categories, a shift to the right may be made for wet conditions. Water pressure is dealt with by effective stress analysis. STRUCTURE	SURFACE CONDITIONS	VERY GOOD Very rough, fresh unweathered surfaces	GOOD Rough, slightly weathered, iron stained surfaces	FAIR Smooth, moderately weathered and altered surfaces	POOR Slickensided, highly weathered surfaces with compact coatings or fillings or angular fragments	VERY POOR Slickensided, highly weathered surfaces with soft coatings or fillings
		DECREASING SURFACE QUALITY ⇨				
INTACT OR MASSIVE Intact rock specimens or massive in situ rock with few widely spaced discontinuities	DECREASING INTERLOCKING OF ROCK PIECES ⇩	90 80			N/A	N/A
BLOCKY Well interlocked undisturbed rock mass consisting of cubical blocks formed by three intersecting discontinuity sets			70 60			
VERY BLOCKY Interlocked, partially disturbed mass with multi-faceted angular blocks formed by four or more joint sets				50		
BLOCKY / DISTURBED / SEAMY Folded with angular blocks formed by many intersecting discontinuity sets. Persistence of bedding planes or schistosity				40	30	
DISINTEGRATED Poorly interlocked, heavily broken rock mass with mixture of angular and rounded rock pieces					20	
LAMINATED / SHEARED Lack of blockiness due to close spacing of weak schistosity or shear planes		N/A	N/A			10

Source: Reproduced by permission of Taylor and Francis Group, LLC, a division of Informa plc.

[a] *WARNING*

The shaded areas are indicative and may not be appropriate for site specific design purposes. Mean values are not suggested for indicative characterization; the use of ranges is recommended

Sound gneiss. Shaded area does not cover weathered rockmasses.

Table 2.11g Common GSI range for typical schist[a] (Marinos and Hoek, 2000)

GEOLOGICAL STRENGTH INDEX FOR JOINTED ROCKS From the lithology, structure and surface conditions of the discontinuities, estimate the average value of GSI. Do not try to be too precise. Quoting a range from 33 to 37 is more realistic than stating that GSI=35. *Note that the table does not apply to structurally controlled failures.* Where weak planar structural planes are present in an unfavourable orientatiuon with respect to the excavation face, these will dominate the rock mass behavior. The shear strength of surfaces in rocks that are prone to deterioration as a result of changes in moisture content will be reduced if water is present. When working with rocks in the fair to very poor categories, a shift to the right may be made for wet conditions. Water pressure is dealt with by effective stress analysis. STRUCTURE	SURFACE CONDITIONS VERY GOOD Very rough, fresh unweathered surfaces	GOOD Rough, slightly weathered, iron stained surfaces	FAIR Smooth, moderately weathered and altered surfaces	POOR Slickensided, highly weathered surfaces with compact coatings or fillings or angular fragments	VERY POOR Slickensided, highly weathered surfaces with soft coatings or fillings
DECREASING INTERLOCKING OF ROCK PIECES ⇩	DECREASING SURFACE QUALITY ⇨				
INTACT OR MASSIVE Intact rock specimens or massive in situ rock with few widely spaced discontinuities	90			N/A	N/A
BLOCKY Well interlocked undisturbed rock mass consisting of cubical blocks formed by three intersecting discontinuity sets	80	70 60			
VERY BLOCKY Interlocked, partially disturbed mass with multi-faceted angular blocks formed by four or more joint sets			50		
BLOCKY / DISTURBED / SEAMY Folded with angular blocks formed by many intersecting discontinuity sets. Persistence of bedding planes or schistosity		1	40	30 2	
DISINTEGRATED Poorly interlocked, heavily broken rock mass with mixture of angular and rounded rock pieces				20 3	
LAMINATED / SHEARED Lack of blockiness due to close spacing of weak schistosity or shear planes	N/A	N/A			10

Source: Reproduced by permission of Taylor and Francis Group, LLC, a division of Informa plc.

[a] *WARNING*

The shaded areas are indicative and may not be appropriate for site specific design purposes. Mean values are not suggested for indicative characterization; the use of ranges is recommended

1. Strong (e.g. micaschists, calcitic schists)
2. Weak (e.g. chloritic schists, phyllites)
3. Sheared schist

Table 2.11h GSI estimates for heterogeneous rock masses such as flysch (Marinos and Hoek, 2000)

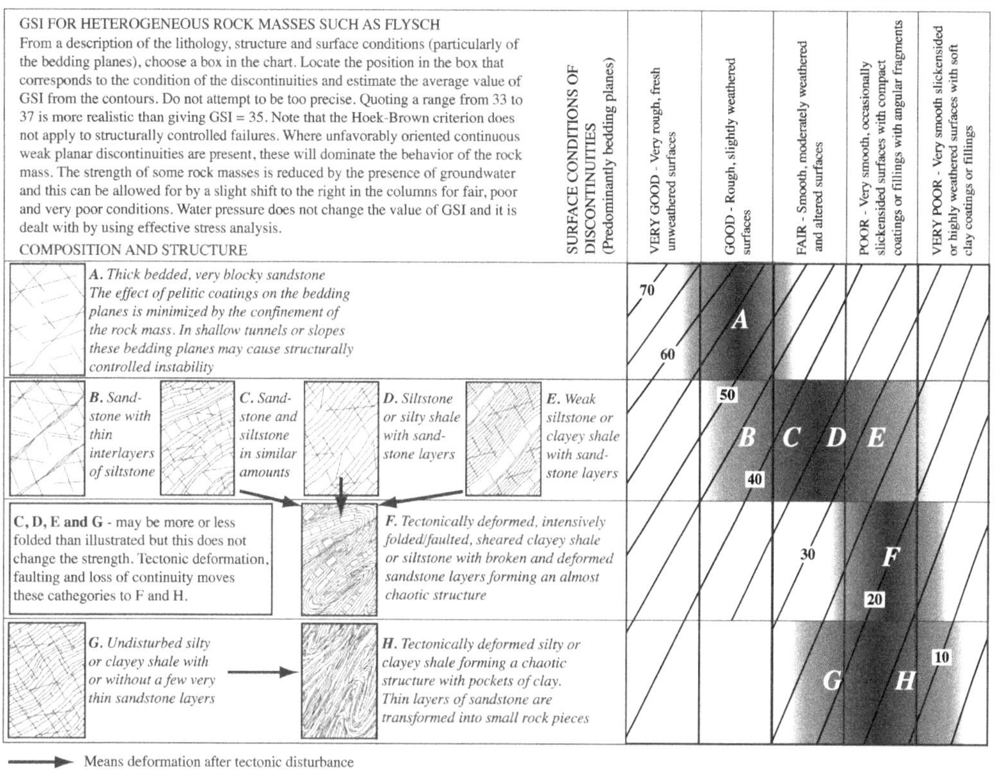

Source: Reproduced by permission of Taylor and Francis Group, LLC, a division of Informa plc.

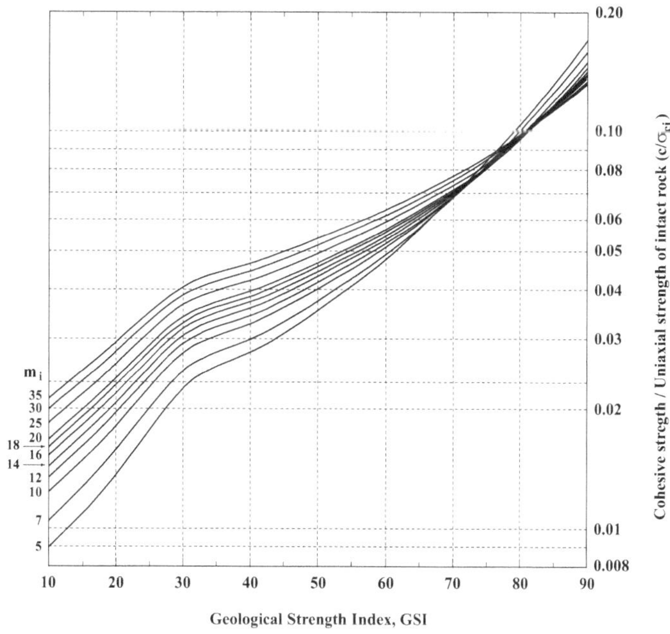

Figure 2.36a Relationship between ratio of cohesive strength to uniaxial compressive strength of intact rock c/σ_{ci} and GSI for different m_i values, for depths of more than 30m (Marinos and Hoek, 2000). (Reproduced by permission of Taylor and Francis Group, LLC, a division of Informa plc.)

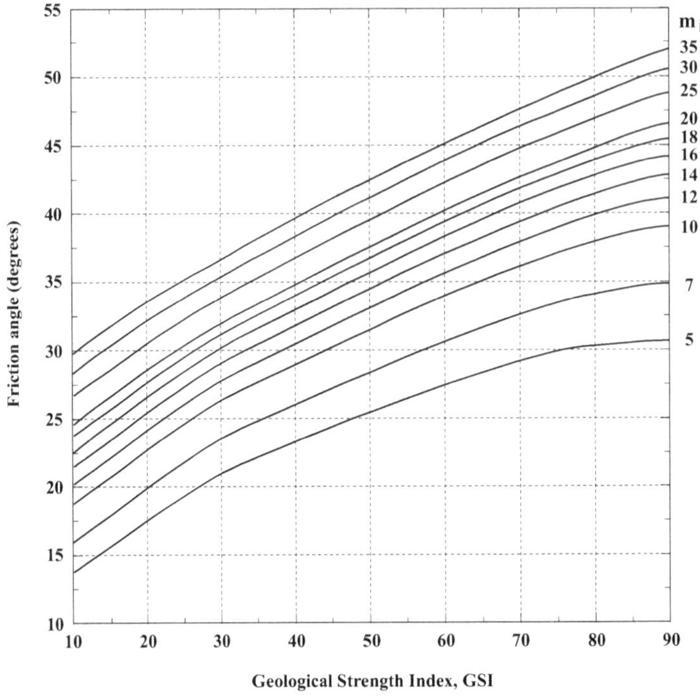

Figure 2.36b Friction angle φ for different GSI and m_i values, for depths more than 30m (Marinos and Hoek, 2000). (Reproduced by permission of Taylor and Francis Group, LLC, a division of Informa plc.)

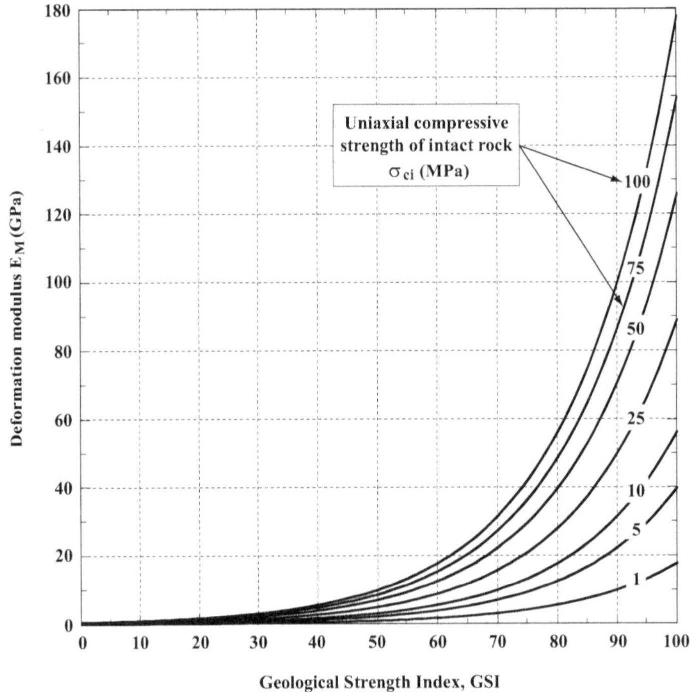

Figure 2.37 Rock mass deformation modulus E versus Geological Strength Index GSI (Marinos and Hoek, 2000). (Reproduced by permission of Taylor and Francis Group, LLC, a division of Informa plc.)

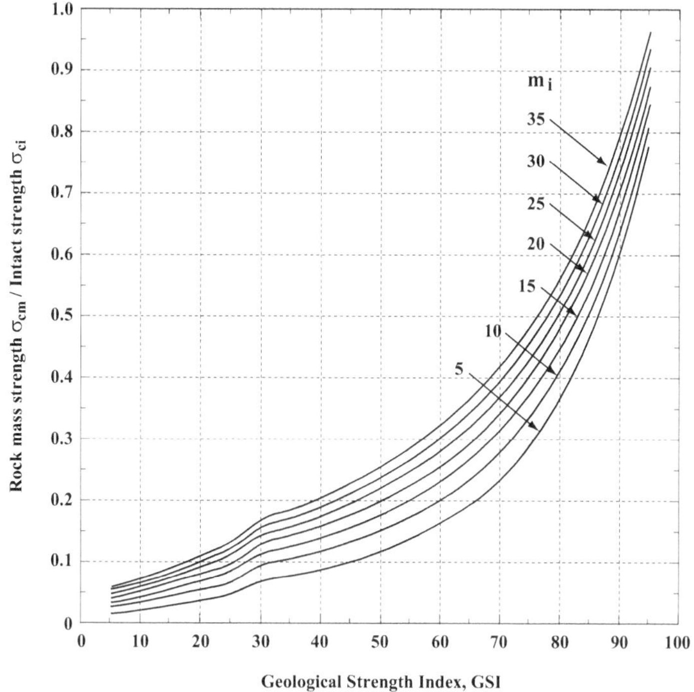

Figure 2.38 Relationship between rock mass strength σ_{cm}, intact rock strength σ_{ci}, constant m_i and the Geological Strength Index GSI for depths of more than 30 m (Marinos and Hoek, 2000). (Reproduced by permission of Taylor and Francis Group, LLC, a division of Informa plc.)

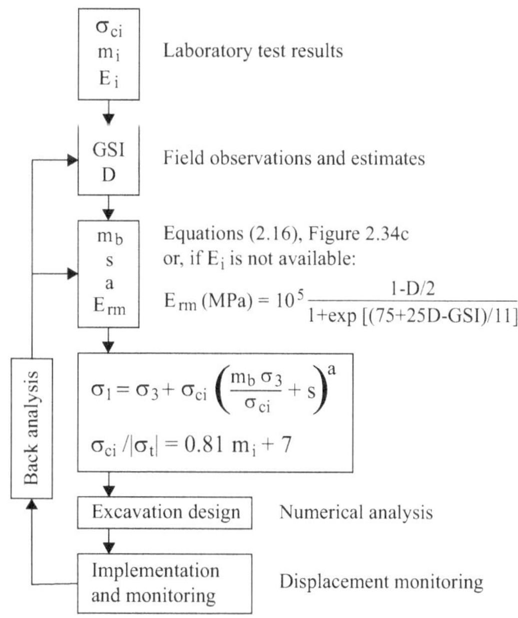

Figure 2.39 Flow chart for the application of the Hoek-Brown criterion and GSI system to an excavation design (after Hoek and Brown, 2019).

references to extensions of GSI classes to other rocks such as Marinos and Hoek (2001) and Marinos and Carter (2018) where the latter includes weathered rock masses.

2.3.5 Methods Type D – quantitative indirect (rock load) relations

Ground conditions are described quantitatively (Figure 2.40) but then related to a rock load or rock load factor. The support is designed by using the rock load for structural design as in Methods A.

Empirical relations of this type are the RQD–rock load factor developed by Deere et al. (1969) and Monsees (1970). Also in this category fall the wedge type analysis developed by Deere and his group (Deere et al., 1970; Cording and Deere, 1972).

2.3.5.1 RQD–rock load relation

Please note that these are different from the direct relations between RQD and support, a Type E Method, which will be discussed later.

Figure 2.41 shows the rock load factor – RQD relation. Curve "a" is the relation between RQD and the Rock Load factor for the average of Terzaghi's recommendations (this will be discussed below). Curve "b" is the recommended load curve for a drill and blast tunnel based

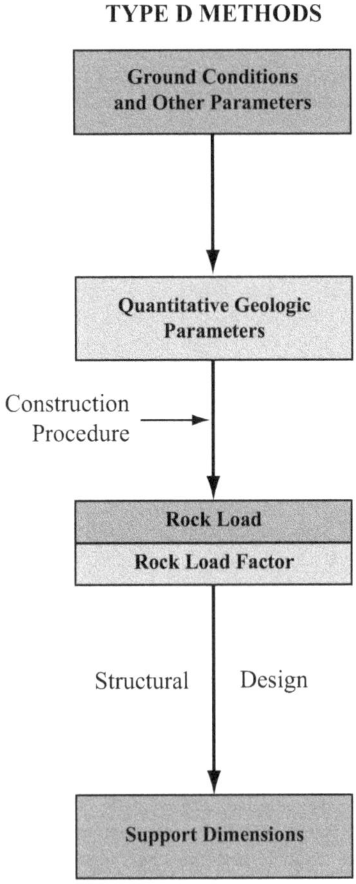

Figure 2.40 Quantitative indirect (rock load) methods.

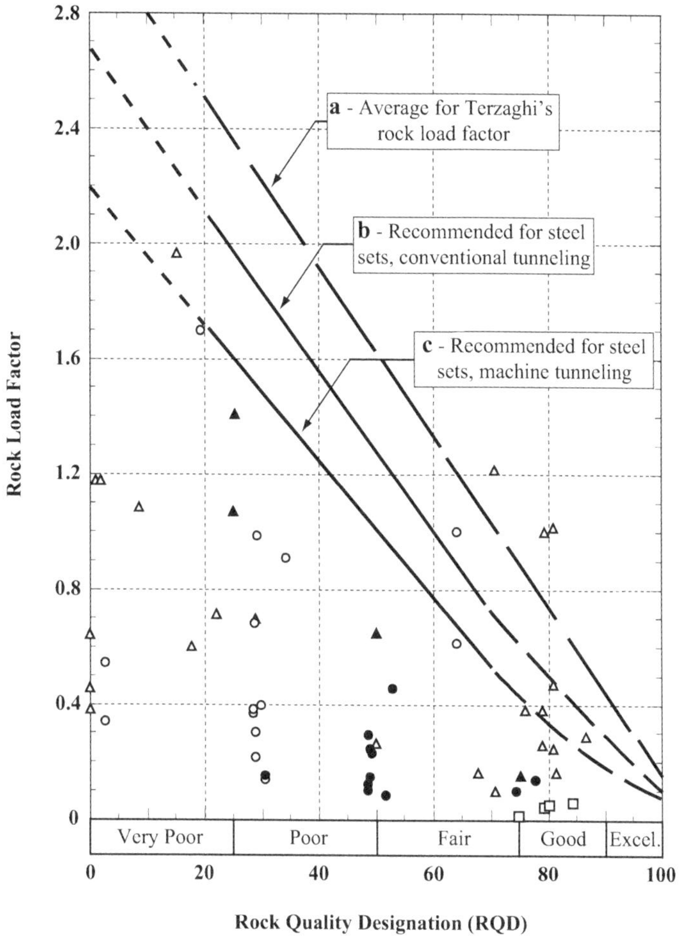

Figure 2.41 RQD rock-load factor relation (from Deere et al., 1969).

on field measurements. Curve "c," which is 25% below curve "b," is the recommended rock load curve for machine bored tunnels indicating that TBM excavation will cause less disturbance than drilling and blasting. These three curves should be used for steel set–supported tunnels.

The development of the RQD–Rock load correlation has been described in detail by Monsees (1970). The "estimated" correlation between RQD and Terzaghi's rock condition (from Monsees, 1970) is shown in Figure 2.42a and Table 2.12. For each ground class, expressed by a range of Terzaghi's Road Load Factors (RLF), a range of RQD has been assigned. Recall that Terzaghi defines his rock load either as a function of the span B or of the sum of span (B) and height (H_t) of the tunnel (see Section 2.3.2.3). Monsees assumed that H_t = B, thus the rock load can be expressed solely as a function of the width B. Each of the Terzaghi's classes is represented as a box in Figure 2.42a. Curve "a" in Figure 2.42a links the centroids of each "box" of Terzaghi's rock load recommendations and thus produces the curve relating RQD and Terzaghi's rock load, as mentioned above (this is the same as curve a in Figure 2.41). Curve 'b' in Figure 2.42a is the proposed envelope for the design of steel supports based on field measurements collected by Monsees. It represents a 90% envelope, i.e. 90% of the measurement points fall below this line. It should be noted that while the Terzaghi's rock load factor ranges are contiguous or overlap (vertical axis in Figure 2.42a) the RQD ranges do not for low values of RQD.

Table 2.12 Estimated correlation between RQD and Terzaghi's Rock Load Factor (RLF) (from Monsees, 1970)

Rock condition	*RQD*	*RLF*
Hard and intact	95–100	0
Hard, stratified or schistose	90–100	0–0.50
Massive, moderately Jointed	85–95	0.25–0.50
Moderately blocky and seamy	75–85	0.25–0.70
Very blocky and Seamy	40–75	0.70–2.20
Completely crushed but chemically intact	0–25	2.20–3.00

Note: The ranges of RLF correspond to the "rock classes" in Figure 2.42a.

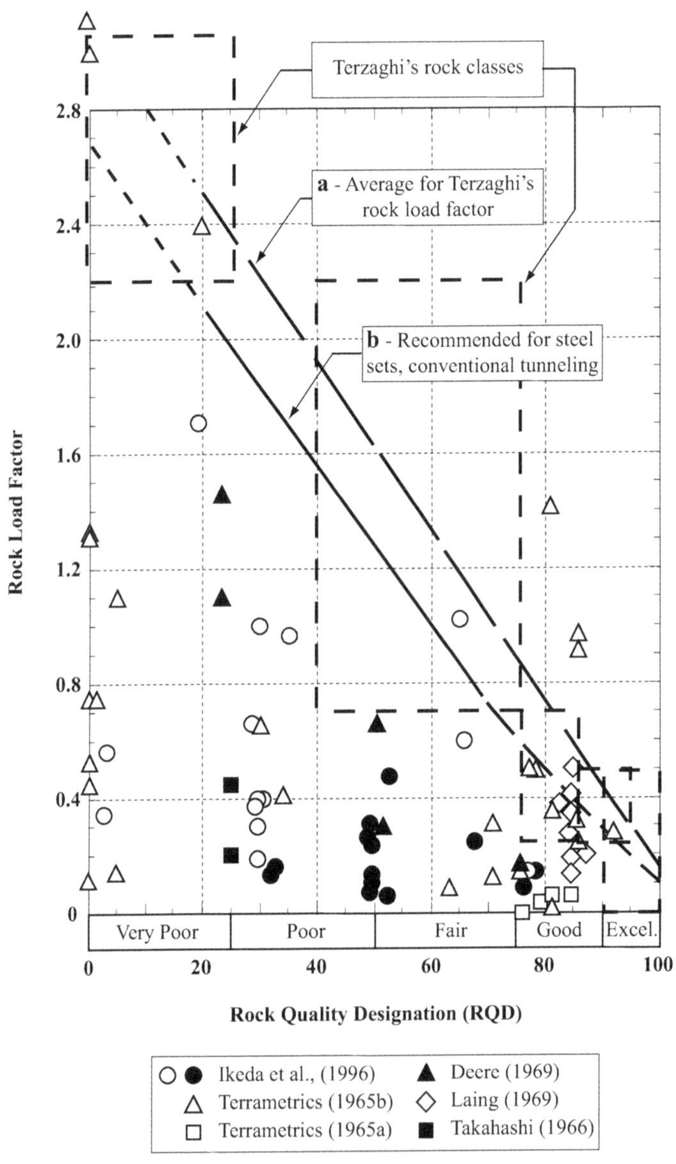

Figure 2.42a Relation of RQD to Terzaghi's rock load recommendations and field measurements (from Monsees, 1970).

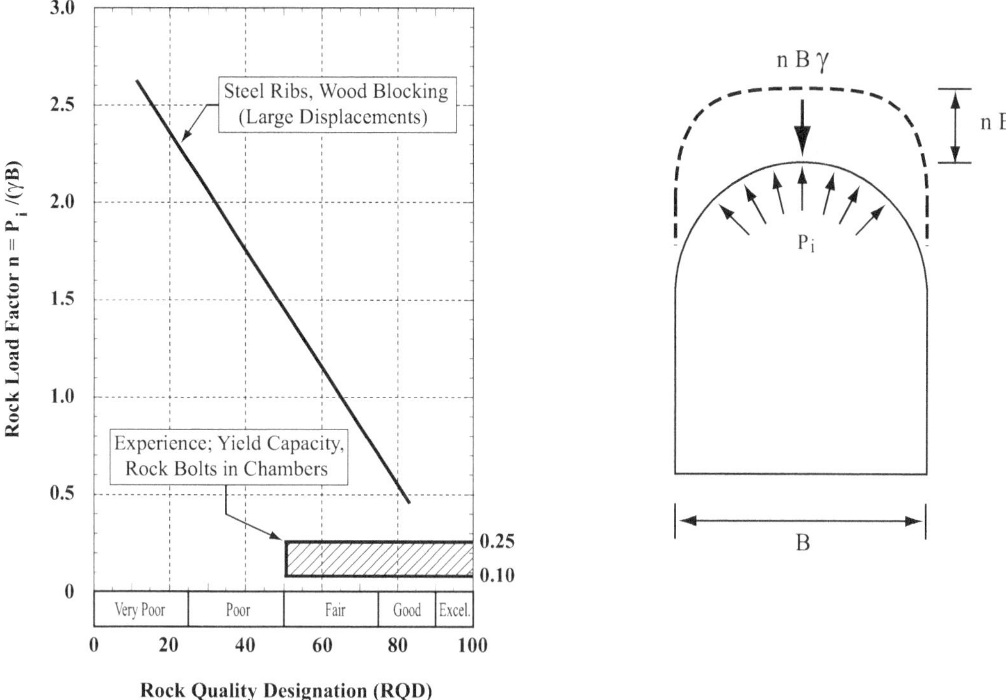

Figure 2.42b Comparison of support loads in steel-supported tunnels and rock-bolted caverns (from Cording and Deere, 1972).

For large rock chambers, inconsistent trends were noted. It was found that the "equivalent support pressure," i.e. the capacity of bolts at yield expressed as a rock load factor was lower than rock loads in steel set–supported tunnels (Figure 2.42b). Typical bolt support gave a rock load factor in the order of 0.1 to 0.25. This result possibly suggests a different support mechanism for rock bolts (rock reinforcement effect).

2.3.5.2 Rock loads due to unstable wedges

For larger chambers or tunnels in ground with widely spaced shear zones or other predominant discontinuities, a design based on RQD may not be conservative. Thus, a wedge type approach was developed by Cording and Deere (1972), which was further refined by Cording and Mahar (1978). This approach (Figure 2.43) considers the actual shape of wedges and the shear strength along the discontinuities. The support design uses the weight of the wedges as input parameter (see also Cording and Deere, 1972). In this approach, a wedge corresponding to the maximum tunnel width is considered (Figure 2.44). The support pressure, P_i, required to keep the symmetrical wedge in place is given by

$$P_i = P_N\left(1 - \frac{\tan\phi}{\tan\theta}\right) + \frac{\gamma B}{4\tan\theta} \tag{2.20}$$

where P_N is the stress acting on the sides of the symmetrical wedge (see Figure 2.44); ϕ is the friction angle on the sides of the wedge; θ is half the angle of the wedge; and B is the width of the tunnel. Hence, the required support pressure is a function of stresses on the wedge and

α Dip Angle	θ Half Angle	n B Height of Equivalent Rock Load	Minimum Condition for Failure	Sketch of Failure
0°- 30°	90°- 60°	(0 - 0.15) B	Both planes wavy, offset	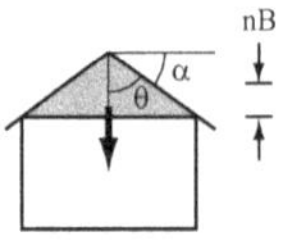
30°- 45°	60°- 45°	(0.15 - 0.25) B	One plane wavy or offset One plane smooth to slightly wavy	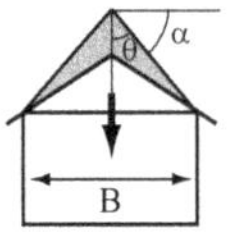
45°- 60°	45°- 30°	(0.25 - 0.45) B	One plane sheared, continuous and planar One plane slightly wavy	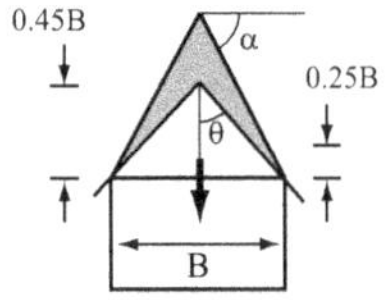
60°- 75°	30°- 15°	(0.45 - 1.0) B	Both planes sheared, continuous and planar	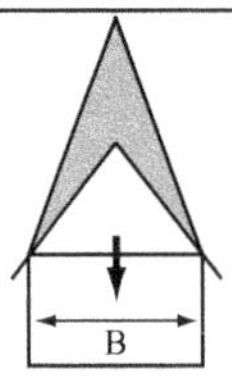
75°- 90°	15°- 0°	> 1.0 B	Low lateral stresses in arch Surfaces planar, smooth, possibly open or progressive failure aided by separation along low angle joint	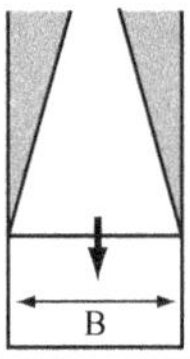

Figure 2.43 Rock loads due to dropping wedges in tunnel crown (from Cording and Mahar, 1978).

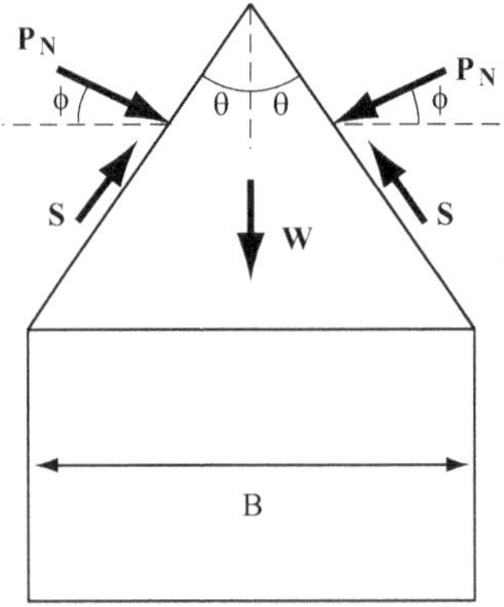

Figure 2.44 Wedge stability in crown (from Cording and Deere, 1972).

the weight of the wedge. To analyze the stability of the wedge, the normal stress on the wedge should be predicted along with ϕ, c, and B. For opening half angles θ that are larger than the friction angle, the wedge tends to drop out and the support has to carry the entire wedge. Note that, once the wedge tends to drop, the normal stress on the wedge, P_N, changes and ultimately reduces to zero; thus, the required support capacity equals the weight of the wedge (provided no additional loosening takes place and leads to an increase in support loads, i.e. more wedges falling out).

Cording and Deere (1972) make a further assumption: the wedge that can drop out is the largest one, i.e. $\theta = \phi$. With this assumption, the required support pressure for a dropping wedge is:

$$P_i = \frac{\gamma B}{4\tan\theta} \tag{2.21}$$

Thus, the rock load factor is

$$\frac{P_i}{\gamma B} = \frac{1}{4\tan\theta} \tag{2.22}$$

This relation, which considers a maximum wedge that drops out due gravity, is shown in Figure 2.45 together with experience in rock-bolted chambers; the actually placed support in large chambers is less than predicted with Equation (2.20). Equation (2.20) yields the

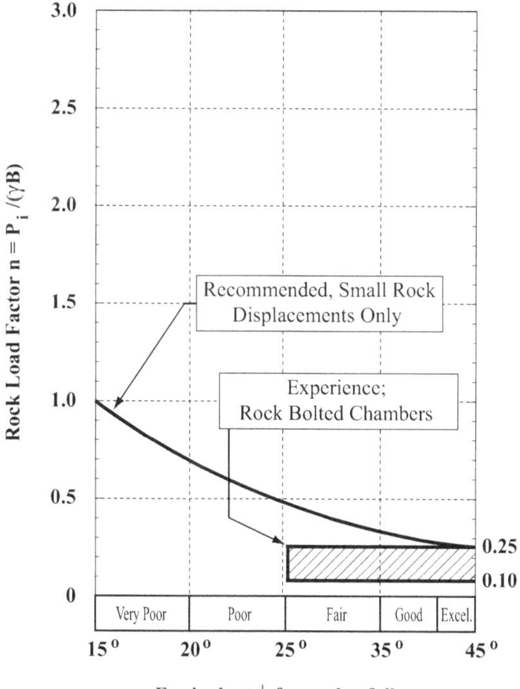

Figure 2.45 Relation of rock load factor to friction angle along discontinuities (from Cording and Deere, 1972).

maximum rock load for wedge type failure; if the actual discontinuity attitude and resistance were known, one could use the less conservative Equation (2.22).

Both the RQD–rock load relations and the wedge stability approach represent extensions of what Terzaghi proposed based on the extensive additional experience of Deere and his students. The RQD–rock load relations have the advantage, compared to Terzaghi, of a more quantitative assessment of the geology, which some engineers prefer. On the other hand, the judgment, which is specifically needed for the Terzaghi approach, is not explicit in the RQD–rock load relations. The wedge approach is a logical quantitative extension of the approach Terzaghi developed for the better rock mass categories. It is entirely logical but simplified in that the joint surface geometry and the in situ stress state are not considered. Nevertheless, a wedge stability computation is often a first step in tunnel stability analyses.

2.3.6 Methods Type E

Ground conditions are described quantitatively either by a single parameter (Type E1) or by multiple parameters (Type E2). The parameters are then directly related to support requirements and excavation procedures (Figure 2.46).

Single parameter methods are:

i. Lauffer's stand-up time–span relation
ii. RQD–support relations by Deere and his group

Multiparameter methods are:

i. Rock Structure Rating – RSR – Method by Wickham and coworkers
ii. Rock Mass Rating – RMR – Method by Bieniawski

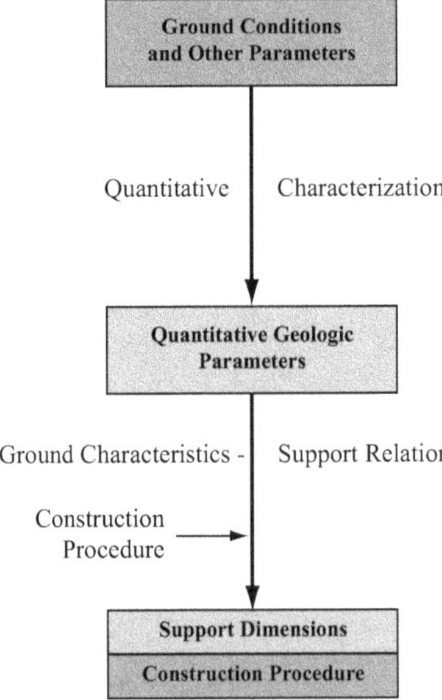

Figure 2.46 Quantitative direct methods.

iii. The Q-System – developed at NGI by Barton and coworkers
iv. Louis' classification (not described here, see Steiner, 1979, for details)
v. Franklin's classification (not described here, see Steiner 1979, for details).

Methods Type E, particularly the RQD, RMR, and Q methods, are by far the most frequently used empirical methods today. The Lauffer and RSR methods are listed and briefly described mostly for historical reasons (Lauffer) and because they contain interesting concepts (RSR).

2.3.6.1 Lauffer's stand-up time-span-support relation

The central aspect of the Lauffer method (Lauffer, 1958, 1960, 1988) is the so-called stand-up time, which is the time period during which an underground opening can remain unsupported without collapse or other serious disturbance. Ground conditions and the free (unsupported) span affect stand-up time and, based on them, stand-up time has to be estimated, since an actual collapse should not occur. Detzlhofer (1974) described how the stand-up time can be estimated:

> An estimate of the order of magnitude of the expected stand-up time, which is equal to the time until the support has to be placed, is possible. It will become easier to classify the ground after a time of adaptation to the actually encountered ground conditions, even though the actual stand-up time is unknown. Occasionally the acquired skill in classification may be verified with the occurrence of a so-called 'inevitable' collapse.

In Detzlhofer (1974), no information is provided as indicators that would allow one to predict a collapse.

Stand-up time or stand-up behavior, by the way, has been used by others (Terzaghi, Stini, Rabcewicz) and Lauffer credits them for this. As indicated, stability of an opening or stand-up time is affected not only by ground conditions but also by the unsupported span, ℓ, which is defined as the smallest value of the three following distances (Figure 2.47): the tunnel span (diameter); the distance from the last placed support member to the face; and for breasted face, the width of the opening at the face that can be left unsupported.

Unsupported span or "effective span" and stand-up time define a ground class, as shown in Figure 2.48, which is in turn related to support requirements. The ground classes are defined in Table 2.13. For each ground class, Lauffer has given alternate support types, i.e. timber support, steel sets in combination with shotcrete, rock bolts in combination with wire mesh and shotcrete, and shotcrete alone (Table 2.14). It should be noted that Figure 2.48, based on Lauffer (1958), shows a band (range) of most frequent applications, which is affected by the tunneling procedure – the upper boundary reflects the time required to install the support but needs to be shorter than the stand-up time while the lower boundary reflects the required time to keep the unsupported length as large as possible to permit simultaneous occupation of the space near the face by different types of equipment. A modified version of the Lauffer classification, which was widely used and quoted, has been published by Linder (1963) (Figure 2.49): in this case, shotcrete thickness is shown directly in the stand-up time span diagram. The dimensions of the shotcrete are those given by Lauffer (1960); only the format of presentation is different.

The stand-up time span relation can only be tested in the tunnel and a prediction of the stand-up time is difficult. According to Spaun (1974), stand-up time or ground class are predicted based on comparisons of the ground conditions in the "new" tunnel with experience gained in already built tunnels.

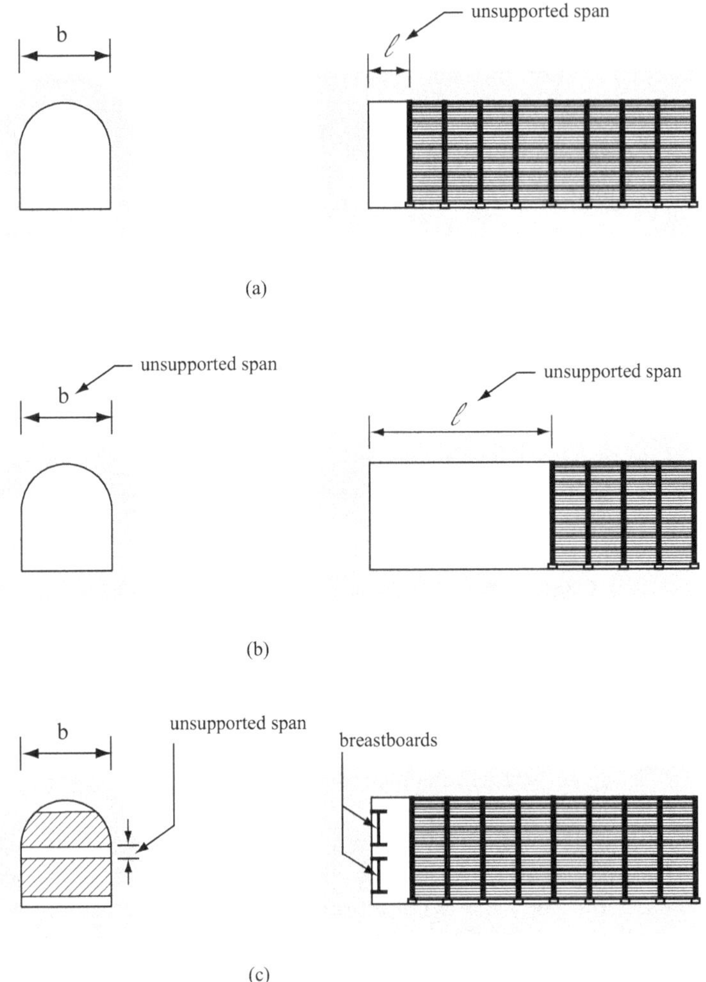

Figure 2.47 Definition of unsupported span (after Lauffer, 1958).

As mentioned earlier, the Lauffer method is mostly of historical interest, since it is one of the first methods to directly relate quantitative ground descriptions (ground classes) to support (see also Steiner, 1979). Also, it played an important role in the development of the NATM (see Section 2.3.3). Specifically, the Lauffer stand-up time relation was developed during the construction of the Prutz-Imst Hydropower tunnel in Austria. Initially, Lauffer (1958) only published the chart relating the ground classes to stand-up time and unsupported span (Figure 2.48), without support quantities. The latter were only included later (Lauffer, 1960) reflecting the experience gained during the construction of the Prutz-Imst Hydropower tunnel.

In the context of the evolution of empirical methods, it is also interesting to mention how Lauffer (1958) considered factors that have an influence on stand-up time: (1) orientation of geologic structure (Figure 2.50a); (2) shape of tunnel cross section (Figure 2.50b); (3) type of excavation (Figure 2.50c); and (4) type of support procedure (Figure 2.50d). The influence of the orientation, cross section, and method of excavation factors can be easily related to the mechanisms they represent, e.g. different disturbance by the excavation method, interaction with the geologic structure and/or stress field for the orientation and cross-section

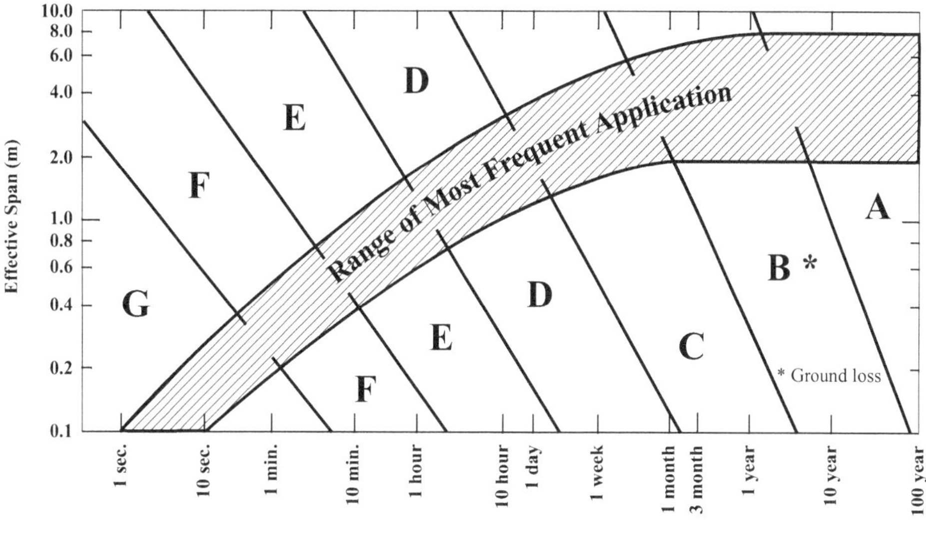

Figure 2.48 Stand-up time versus unsupported span classification (after Lauffer, 1958).

Table 2.13 Typical examples for stand-up time and span and equations for the boundary lines (ℓ^* "effective span" as in Figure 2.47 in m) (after Lauffer, 1958)

Ground class	*Examples for stand-up time and span*	*Equations of the assumed boundary lines (t in hours, ℓ^* in m)*
A stable	20 years....4.0m	
		$t \times \ell^{1.0} = 1.0\ 10^5$
B afterbreaking	6 months...4.0m	
		$t \times \ell^{1.2} = 2.5\ 10^3$
C very afterbreaking	1 week...3.0m	
		$t \times \ell^{1.4} = 6.3\ 10^1$
D breaking	5 hours...1.5m	
		$t \times \ell^{1.6} = 1.6\ 10^0$
E very breaking	20 mins...0.8m	
		$t \times \ell^{1.8} = 4.0\ 10^{-2}$
F squeezing	2 mins...0.4m	
		$t \times \ell^{2.0} = 1.0\ 10^{-3}$
G very squeezing	10 secs...0.15m	

factors; the support factor is somewhat more complex in that it incorporates the fact that some support types may take longer to install or, if quickly installed, may actually affect the ground behavior. Very important in all this is the fact that Lauffer fully acknowledges that the method is subjective, that it is actually mostly a record of what was done in the Prutz-Imst tunnel (Detzlhofer, 1960, 1974), and that it "might" be applied to other tunnels. It is particularly interesting in this regard to look at comments by Detzlhofer (1974), who compares the actual ground classes of the final 5.3 m diameter (~22 m^2 cross section) tunnel and a pilot

Table 2.14 Support recommendations (after Lauffer, 1960)

Ground class	*Description*	*Timber support*	*Stand-up time for unsupported span*	*Shotcrete*	*Rock bolt with wedges (not grouted)*	*Steel support remaining in tunnel*
A	Standing	None	20 years for 4.0 m	No	No	Not necessary
B	Afterbreaking	Head Protection	6 months for 4.0 m	2 to 3 cm in crown only	Bolts spaced 1.5 to 2.0 m in crown with wirefabric	Application not economical
C	Highly afterbreaking	Crown Support	1 week for 3.0 m	3 to 5 cm in crown only	Bolts spaced 1.0 to 1.5 m in crown with wirefabric or 2 cm shotcrete afterward	Application not economical
D	Breaking	Light Timbering	5 h for 1.5 m	5 to 7 cm primarily in crown with wirefabric	Bolts spaced 0.7 to 1.0 m with wirefabric or 3 cm shotcrete after placements of bolts	If necessary like class E
E	Very breaking	Heavy Timbering	20 min. for 0.8 m	7 to 15 cm with wirefabric	Only when bolt heads can be seated. Bolts spaced 0.5 to 1.2 m with immediate shotcrete	Steel or concrete lagging on steel sets
F	Squeezing	Forepole Timbering Without face support	2 min. for 0.4 m	15 to 20 cm with wirefabric and steel sets, if necessary face support with shotcrete	Cannot be bolted[a]	Lagging on braced steel sets with additional shotcrete support afterward
G	Very squeezing	Forepole Timbering with face support	10 sec. for 0.15 m	cannot be executed	Cannot be bolted[a]	Lagging on braced steel sets with immediate shotcrete application

[a] In classes F and G standard rock bolts (with wedges) cannot be used because they cannot be anchored (sliding, bearing capacity of wedges). Newer types of bolts (fully grouted bolts) may work, but were not available in the 1950s.

tunnel in the middle section of this tunnel (Figure 2.51). In the 10 m^2 pilot tunnel, a much larger percentage was considered to be in the best conditions (a, b classes) while no such classes were encountered in the final tunnel.

So while the Lauffer classification appears to explicitly include tunnel size, it is evidently problematic to extrapolate the classification to another tunnel even in the same geology. This relates to the "transferability problem" affecting empirical methods, which will be discussed further in Section 2.4. Nevertheless, the Lauffer method is important because it directly or indirectly led to a number of other empirical methods. For additional comments on Lauffer's method and possible expansions, see Steiner (1979).

2.3.6.2 *Deere's empirical RQD relations*

In the context of the type E1 methods, emphasis will be placed on the RQD–support relations (Deere et al., 1969, and Merritt, 1972) but with some comments on the descriptive classification by Deere et al. (1974). Note that the RQD–rock load relations, the large cavern support – aspects and the wedge stability – methods developed by Deere and co-workers were discussed in Section 2.3.5.

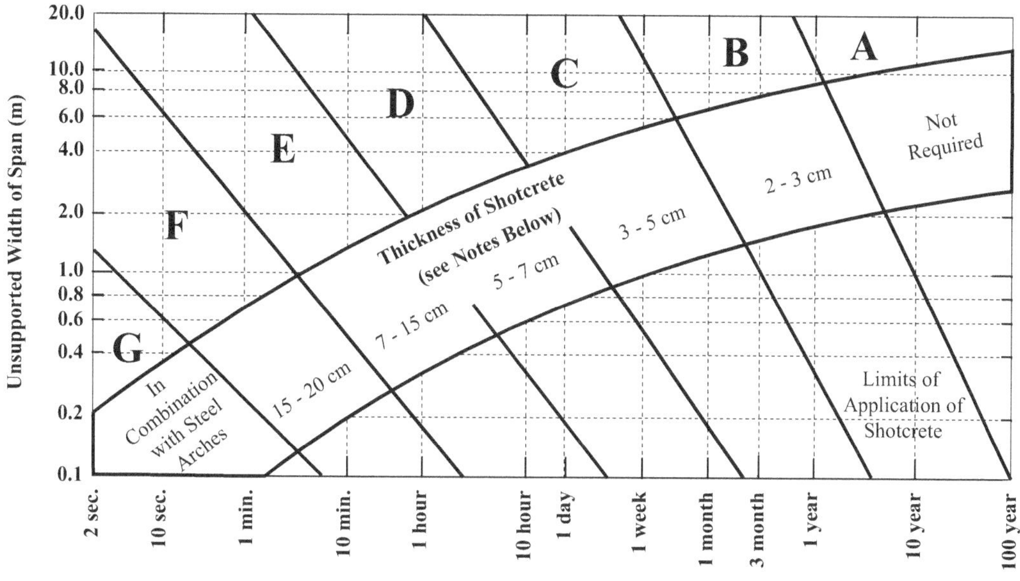

Stand-up Time Without Support

Notes:

(B) Alternatively rock bolts on 1.5-2 m spacing with wire net, occasionally reinforcement needed only in arch
(C) Alternatively rock bolts on 1-1.5 m spacing with wire net, occasionally reinforcement needed only in arch
(D) Shotcrete with wire mesh; alternatively rock bolts on 0.7-1 m spacing with wire net and 3 cm shotcrete
(E) Shotcrete with wire net; rock bolts on 0.5-1.2 m spacing with 3-5 cm shotcrete sometimes suitable; alternatively steel arches with lagging
(F) Shotcrete with wire net and steel arches; alternatively strutted steel arches with lagging and subsequent shotcrete
(G) Shotcrete and strutted steel arches with lagging

Figure 2.49 Modified Lauffer classification, after Linder (1963); (from Deere et al., 1969).

In the RQD–support relations, the ground is described by RQD. Based on RQD, the support can be estimated with Table 2.15. Different support quantities are given for tunnels excavated by drilling and blasting and for machine-bored tunnels. Support quantities distinguish between steel sets, rock bolts, or shotcrete (sometimes a combination of these three support systems). The limitations and simplifying assumptions of these RQD–support relations are clearly expressed in the footnotes of Table 2.15. In particular, these support recommendations should be only used for tunnels of 20 to 40 ft. diameter. Deere et al. (1969) clearly state that their recommendations reflect 1969 US tunneling technology. Also, during actual construction, the performance should be monitored and the ground support relations should be updated accordingly.

The support recommendations listed in Table 2.15 are based on several simplifying assumptions, the most important of which are the following:

1. The RQD adequately describes the quality of the rock.
2. The support systems are installed as close to the face as possible; for steel sets and for rock bolts this would be about 2 to 4 ft., and for shotcrete, essentially zero. Furthermore, it is assumed that the support systems are properly installed, i.e. lagging and blocking is tightly placed behind steel sets and rock bolts are properly tensioned.
3. The tunnel has a cross section (either horse-shoe or circular) with the height approximately equal to the width.

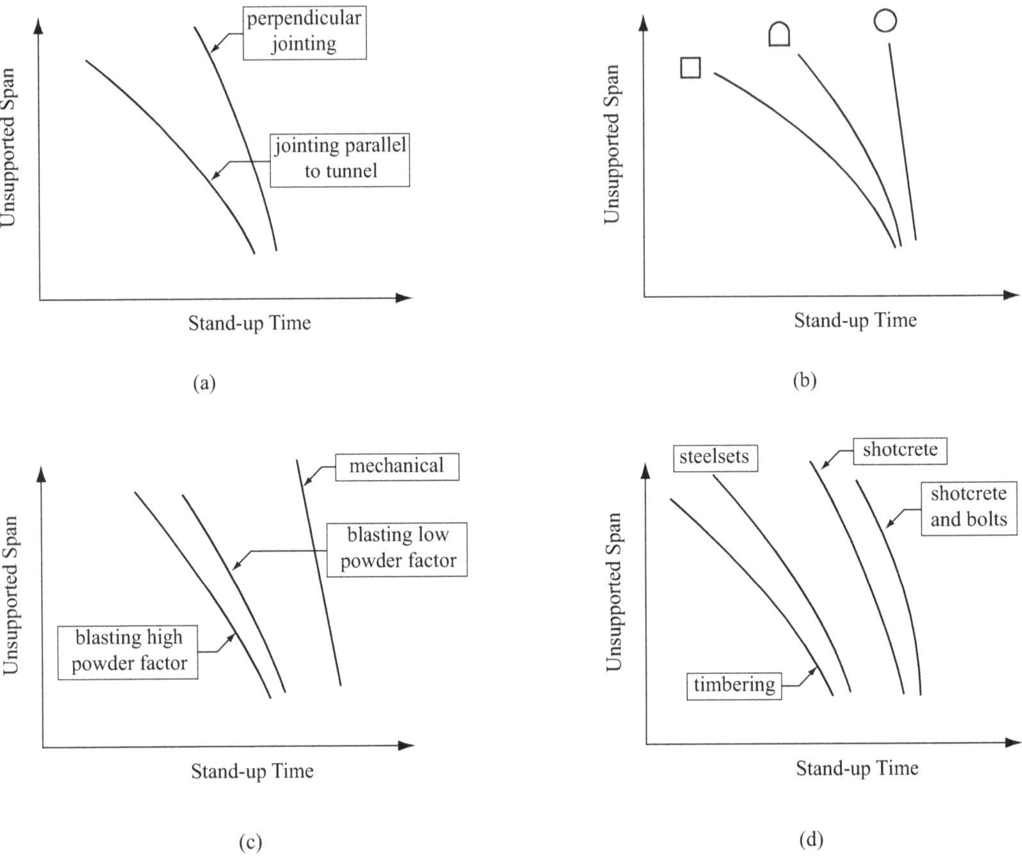

Figure 2.50 Qualitative effects on stand-up time span relations (after Lauffer, 1958). (a) Influence of orientation. (b) Cross section. (c) Method of excavation. (d) Type of support.

4. The tunnel is approximately 20 to 40 ft. in width.
5. The natural stresses in the ground are low enough such that stress concentrations around the periphery generally do not exceed the compressive strength of the rock.

It is interesting to briefly review the development of the RQD–support relations (Coon, 1968; Deere et al., 1968, 1969; and some further developments, Cecil, 1970, 1975; Merritt, 1972), all of which were performed by Deere and/or his students.

The original correlations were based on a limited amount of data, Coon (1968) and Deere et al., (1968). The database included a total of 14 tunnels; four of these tunnel studies were described in some detail (for a comprehensive summary of these cases, see Steiner, 1979; otherwise refer to the original publications). In three of the cases (Interstate 40 tunnels in North Carolina and Tehachapi No.3 tunnel) actually measured RQD values were available. In the fourth case (Straight Creek pilot tunnel), RQD was estimated from joint spacing data corrected for alteration of the rock. Interesting comments in the original publications show that these cases were systematically studied to determine if a RQD support relation was at all possible. The data for the tunnels are listed in Table 2.16. From these data, a plot of RQD versus opening widths (Figure 2.52) was obtained. The support quantities have been divided into three ranges (minimum, intermediate, maximum). This

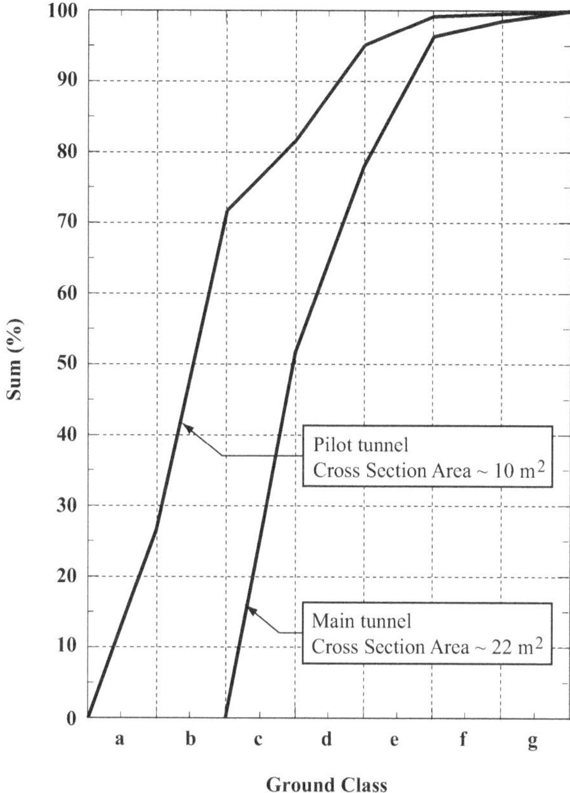

Figure 2.51 Size effect on ground classes observed in section Wenns of the Prutz-Imst tunnel (after Detzlhofer, 1974).

figure seems to have served as a basis for the development of the support recommendations (Table 2.15). Note that in Table 2.15 rock load factors are also listed, which correspond to those empirically derived for the RQD rock load correlation (see Section 2.3.5). The 14 case studies did not provide information on all the categories in Table 2.15. Rock loads were then used to develop support recommendations. Also, the 14 cases did not include any TBM tunnels. The TBM values in Table 2.15 were arrived at by reducing the rock load factor for TBM driven tunnels by 25% compared to those for drill and blast tunnels.

The RQD–support relations were then expanded by Cecil (1970, 1975) and Merritt (1972). Cecil studied a total of 97 tunnel sections mostly in Scandinavia in detail and attempted to correlate RQD, tunnel width and support requirements. Cecil also noticed that in some cases it was not possible to use RQD to predict support, e.g. with closely spaced tight joints a low RQD results, but no support is needed; conversely thin coatings or softening clay may lead to instability even with RQD > 75%. Merritt (1972) considered, in addition, 60 cases and plotted them on tunnel width – RQD charts (Figure 2.53). Figure 2.53 includes Merritt's results and the preceding work (Deere, Coon, Cecil). The figure shows the continued development of the RQD–support relations and a corresponding increase in the range of applicability, in particular the use of rock bolts for a wider range of RQD. Not surprisingly, Deere and his colleagues explicitly recommend updating ground support relations by monitoring performance

Table 2.15 Deere et al. RQD–support relations (from Deere et al., 1969). Guidelines for selection of primary support for 20-ft to 40-ft tunnels in rock

			Alternative support systems						
			Steel sets		*Rock bolts*[a] *(Conditional use in poor and very poor rock)*		*Shotcrete*[b] *(Conditional use in poor and very poor rock)*		
							Total Thickness		
Rock quality	*Construction method*	*Rock Load (B = Tunnel Width)*	*Weight of Sets*	*Spacing*[c]	*Spacing of Pattern Bolts*	*Additional Requirements and Anchorage Limitations*[a]	*Crown*	*Sides*	*Additional Support*[b]
Excellent[d] RQD > 90	Boring machine	(0.0 to 0.2)B	Light	None to occasional	None to occasional	Rare	None to occasional Local application	None	None
	Drilling and blasting	(0.0 to 0.3)B	Light	None to occasional	None to occasional	Rare	None to occasional Local application 2 to 3 in.	None	None
Good[d] RQD = 75 To 90	Boring machine	(0.0 to 0.4)B	Light	Occasional to 5 to 6 ft	Occasional to 5 to 6 ft	Occasional mesh and straps	Local application 2 to 3 in.	None	None
	Drilling and blasting	(0.3 to 0.6)B	Light	5 to 6 ft	5 to 6 ft	Occasional mesh or straps	Local application 2 to 3 in.	None	None
Fair RQD = 50 to 75	Boring machine	(0.4 to 1.0)B	Light to medium	5 to 6 ft	4 to 6 ft	Mesh and straps as required	2 to 4 in.	None	Provide for rock bolts
	Drilling and blasting	(0.6 to 1.3)B	Light to medium	4 to 5 ft	3 to 5 ft	Mesh and straps as required	4 in. or more	4 in. or more	Provide for rock bolts
Poor RQD = 25 to 50	Boring machine	(1.0 to 1.6)B	Medium circular	3 to 4 ft	3 to 5 ft	Anchorage may be hard to obtain. Considerable mesh and straps required	4 to 6 in.	4 to 6 in.	Rock bolts as required (~4–6 ft cc)
	Drilling and blasting	(1.3 to 2.0)B	Medium to heavy circular	2 to 4 ft	2 to 4 ft	Anchorage may be hard to obtain. Considerable mesh and straps required	6 in. or more	6 in. or more	Rock bolts as required (~4–6 ft cc)
Very poor RQD < 25 (excluding squeezing and swelling ground)	Boring machine	(1.6 to 2.2)B	Medium to heavy circular	2 ft	2 to 4 ft	Anchorage may be impossible. 100% mesh and straps required	6 in. or more on	whole section	Medium sets as required
	Drilling and blasting	(2.0 to 2.8)B	Heavy circular	2 ft	3 ft	Anchorage may be impossible. 100% mesh and straps required	6 in. or more on	whole section	Medium to heavy sets as required
Very poor, squeezing or swelling ground	Both methods	up to 250 ft	Very heavy circular	2 ft	2 to 3 ft	Anchorage may be impossible. 100% mesh and straps required	6 in. or more on	whole section	Heavy sets as required

Note: Table reflects 1969 technology in the United States. Groundwater conditions and the details of jointing and weathering should be considered in conjunction with these guidelines particularly in the poorer quality rock.

[a] Bolt diameter = 1 in. length = 1/3 to 1/4 tunnel width. It may be difficult or impossible to obtain anchorage with mechanically anchored rock bolts in poor and very poor rock. Grouted anchors may also be unsatisfactory in very wet tunnels.

[b] Because shotcrete experience is limited, only general guidelines are given for support in the poorer quality rock.

[c] Lagging requirements for steel sets will usually be minimal in excellent rock and will range from up to 25% in good rock to 100% in very poor rock.

[d] In good and excellent quality rock, the support requirement will in general be minimal but will be dependent on joint geometry, tunnel diameter, and relative orientations of joints and tunnel.

Table 2.16 Summary of initial case studies for RQD support relation (from Deere et al., 1968)

	Project	*Width of opening (ft)*	*Support*	*RQD or velocity index*
1	Pigeon River No. 1	36	Unsupported	87
2	Pigeon River No. 2	36	8 in. WF 10 to 4 ft	29
3	Tehachapi Site 3	21		See Fig. 3.5.8[a]
4	Tehachapi Site 1	28	8 × 8 in. 6 ft o.c. and bolts 5 ft o.c.	54–80
5	Straight Creek	13	Heavy support to unsupported	See Fig. 3.5.9[a]
6	Cavity I, NTS	Hemisphere with radius of 60 ft	Top 32 ft – bolts, 3 ft o.c. Mid 24 ft – bolts, 3 ft o.c. Bot 16 ft – bolts, 6 ft o.c.	72 90
7	Cavity, II, NTS	Hemisphere with radius of 60 ft	Top 32 ft – bolts, 3 ft o.c. Mid 24 ft – bolts, 3 ft o.c. Bot 16 ft – bolts, 6 ft o.c.	69
8	Cavity III, NTS	Hemisphere with radius of 35 ft	Top 24 ft – bolts, 3 ft o.c. Mid 16 ft – bolts, 3 ft o.c. Bot 8–16 ft – bolts, 6 ft o.c.	75
9	Adit at Two Forks	5		78
10	Adit at Yellowtail Dam	8		
11	Adits at Dvorshak Dam	5		81
12	Diversion Tunnel at Dvorshak dam granite gneiss 960 ft	30	47% ribs, 53% bolts	Estimated 85
	Schistose gneiss 760 ft	11	77% ribs, 23% bolts, ribs 5 ft o.c.	Estimated 70
13	Shaft, East Coast	20	Temporary timber and bolts – concrete lining w/i 20 ft of face	44–93
14	Tunnel, NTS	10	Unsupported	75

[a] For Figures cited, refer to Deere et al. (1968)

and incorporating such experience. It is, therefore, appropriate to quote Deere et al. (1969) and (repeated verbatim in Monsees, 1970):

> Coon (1968) has shown that a qualitative relationship exists between the RQD and the support required for tunnels in rock. For steel sets or precast-concrete segments, it should be possible to design the supports to resist a load that is a function of the rock quality, the size of opening, and the construction technique employed. The tentatively recommended Rock Loads are given in Table 2.15 (this book). By relating the load to the RQD, rather than to Terzaghi's qualitative description of rock quality, the proposed system is less sensitive to variations in personal observations. For other tunnel support methods (such as rock bolts and shotcrete) it is necessary to consider the possible mechanisms of instability and the manner in which each of the support methods acts to maintain stability. This consideration, combined with an evaluation of the types and amounts of support that have proven to be successful, leads to guidelines (see Table 2.15) on which to base the design. The use of the recommendations in Table 2.15 will produce designs that are equal to or more economical than designs obtained by methods currently in use. It is desirable to make observations of the tunnel behavior during and following the construction of the tunnel. Such observations will indicate tendencies toward instability, which might occur in areas of particularly bad rock. Thus, potential problems can be detected early enough to take corrective action before a failure occurs.

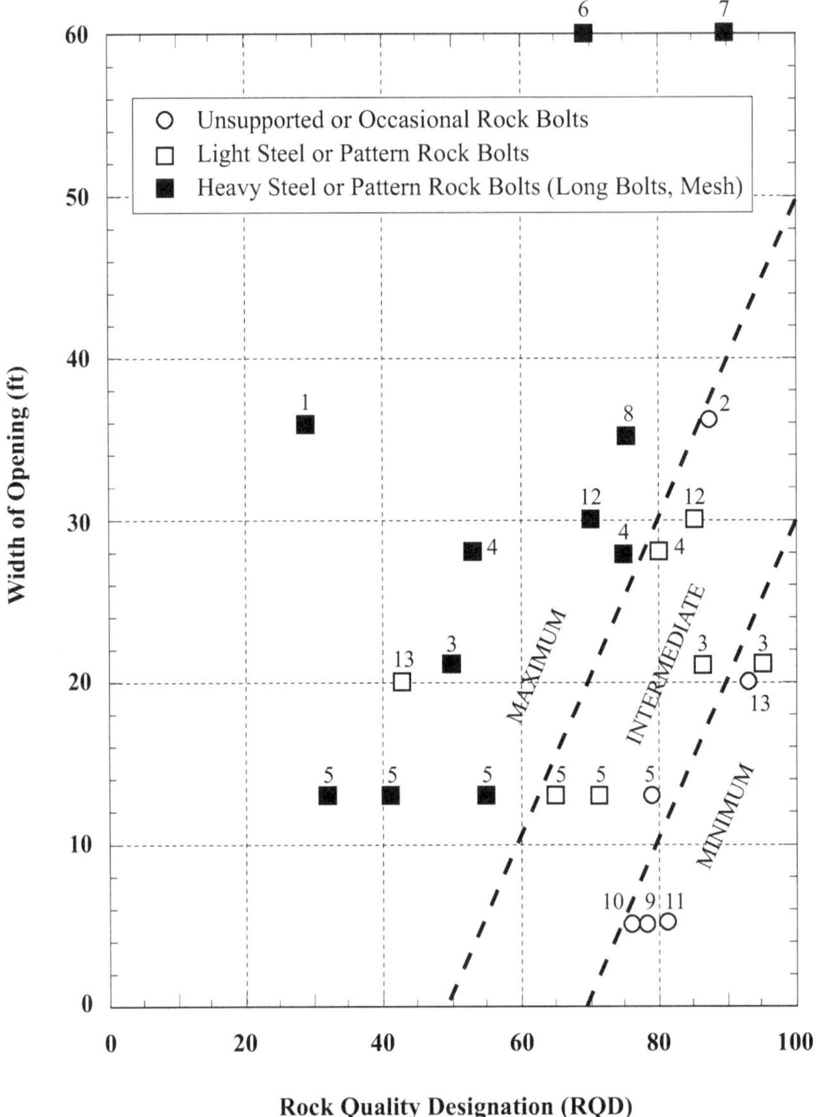

Figure 2.52 Rock Quality Designation (RQD) versus tunnel width plot (from Deere et al., 1968). The numbers in the graph refer to Table 2.16.

Not surprisingly, given this statement, Deere et al. (1974) developed also a "descriptive" classification, as an outgrowth of their experience in tunnel design and construction. They saw the need for an overview type classification that could be applied initially in a project before proceeding with a more detailed empirical design method. This descriptive classification considers three principal categories of ground: I – Good Tunneling Ground; II – Average to Difficult Tunneling Ground; and III – Very Difficult to Hazardous Tunneling Ground. For each of these three categories, the ground and hydraulic conditions leading to this category are discussed as are the consequences on construction. Specifically, orientation, shear strength of the discontinuities, the influence of water conditions, and construction procedure are considered, as are geologic features that lead to a particular ground condition and general construction procedures. The method is, however, not intended to provide detailed support

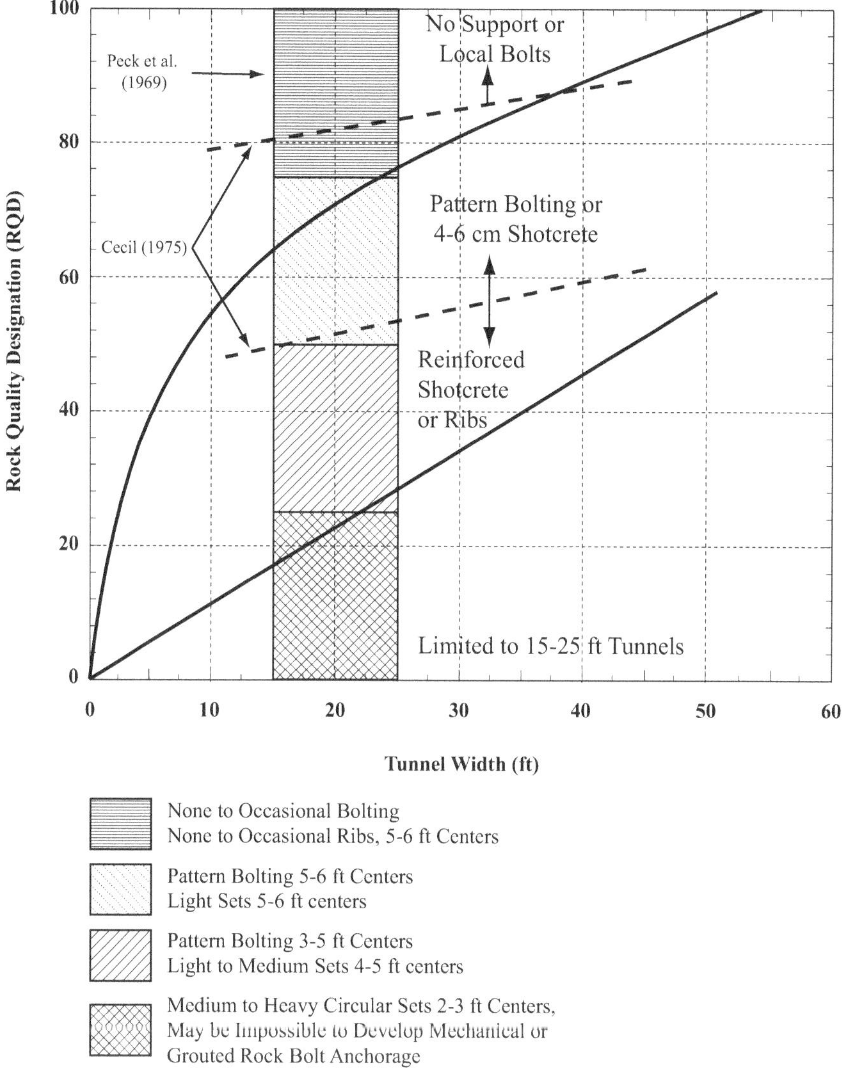

Figure 2.53 Comparison of RQD–support relation (from Merritt, 1972).

recommendations for which the RQD relations (or similar detailed methods) have to be used. The underlying philosophy and intended application are made clear by Deere's et al. (1974) statement:

> The three new categories of tunneling ground proposed herein (Category I – Good; Category II – Average to Difficult; and Category III – Very difficult to Hazardous) provide a broad and simple grouping which would appear to have merit as general terms. Each category requires its own methods of construction and support. It is believed that we will see the concurrent use of two systems: one a general system (Categories I, II, and III), and two, a more specific system such as the RSR or RQD modified to take into account the structural attitude and the surface characteristics of the joint surfaces.

Discussion

It is the authors' opinion that RQD support relations were a major step in relating information available from rock cores to anticipated tunneling characteristics. The use of the method increased rapidly. Support predictions based on RQD are easy to perform and they give a first indication of the support requirements based on information gathered from borings. However, RQD alone does not include all the factors that influence the stability and support requirements, and this has been fully acknowledged by Deere and co-authors.

2.3.6.3 The Rock Structure Rating – RSR method by Wickham et al. (1972, 1974a, 1974b); Wickham and Tiedemann (1972)

This is the first of three multiparameter methods (E2-methods) described here. Since it is not much used anymore, it is mostly described for historical interest and because it is based on a solid systematic approach when characterizing geology on the one hand, but also showing limitations typical of empirical approaches. Also, some detailed comments, namely the consideration of joint orientation, have been included in the RMR method by Bieniawski (1979a, 1979b).

The Rock Structure Rating (RSR) Method uses three parameters A, B, C to assess ground conditions. Parameter A (Table 2.17) assesses the general area geology and is a function of rock type (i.e. lithology), degree of decomposition and geologic structure. Parameter B (Table 2.18) rates the joint pattern (joint spacing) and includes also the orientation of joints relative to the tunnel axis. Parameter C (Table 2.19) considers ground water and joint conditions, essentially conditions affecting the shearing resistance. The sum of the parameters A+B+C is the RSR rating for drill and blast tunnels. As can be seen from Tables 2.17 to 2.19 the maximum possible RSR is 100. For tunnels driven with TBM, the Rock Structure Rating is increased by the TBM-factor (Figure 2.54). RSR is used to determine the rib ratio RR, a support scale, i.e. a numerical value that corresponds to certain support dimensions and quantities. The relation between Rock Structure Rating and rib ratio is based on the following equation:

$$RR = \frac{8800}{RSR + 30} - 80 \qquad (2.23)$$

Table 2.17 Rock structure rating: Parameter "A" (from Wickham et al., 1974)

ROCK STRUCTURE RATING
PARAMETER "A"
GENERAL AREA GEOLOGY

MAX. VALUE 30

Basic rock type					Geological structure			
	Hard	Med.	Soft	Decomp.				
Igneous	1	2	3	4				
Metamorphic	1	2	3	4				
Sedimentary	2	3	4	4	Massive	Slightly faulted or folded	Moderately faulted or folded	Intensely faulted or folded
	TYPE 1				30	22	15	9
	TYPE 2				27	20	13	8
	TYPE 3				24	18	12	7
	TYPE 4				19	15	10	6

Table 2.18 Rock structure rating: Parameter "B" (from Wickham et al., 1974)

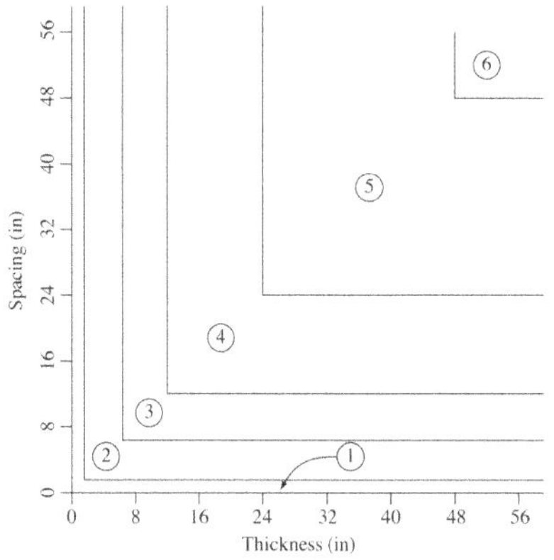

	Rock structure rating Parameter "B" Joint pattern Direction of drive					*Max. Value 45*		
	Strike perpendicular to axis					*Strike parallel to axis*		
	Direction of drive					*Direction of drive*		
	Both	*With dip*		*Against dip*		*Both*		
	Dip of prominent joints					*Dip of prominent joints*		
	Flat	*Dipping*	*Vertical*	*Dipping*	*Vertical*	*Flat*	*Dipping*	*Vertical*
1. Very closely jointed	9	11	13	10	12	9	9	7
2. Closely jointed	13	16	19	15	17	14	14	11
3. Moderately jointed	23	24	28	19	22	23	23	19
4. Moderate to blocky	30	32	36	25	28	30	28	24
5. Blocky to massive	36	38	40	33	35	36	34	28
6. Massive	40	43	45	37	40	40	38	34

Note: Flat 0–20°; Dipping 20°–50°; Vertical 50°–90°

Table 2.19 Rock structure rating: Parameter "C" (from Wickham et al., 1974)

	Rock Structure Rating Parameter "C" Ground Water Joint Condition			*Max. Value 25*		
	Sum Of Parameters A + B					
	13–44			*45–75*		
	Joint Condition					
Anticipated Water Inflow (Gpm/1000')	*Good*	*Fair*	*Poor*	*Good*	*Fair*	*Poor*
None	22	18	12	25	22	18
Slight (<200 Gpm)	19	15	9	23	19	14
Moderate (200–1000 Gpm)	15	11	7	21	16	12
Heavy (>1000 Gpm)	10	8	6	18	14	10

Joint Condition: Good = Tight or Cemented; Fair = Slightly Weathered or Altered; Poor = Severely Weathered, Altered, or Open

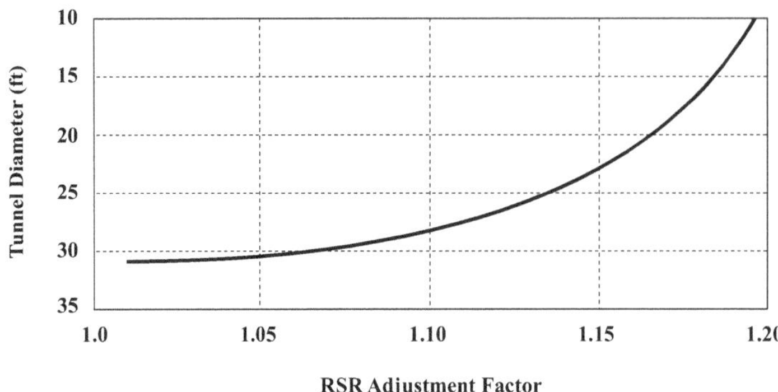

Figure 2.54 RSR adjustment factor for machine-bored tunnels (from Wickham et al., 1972).

A rib ratio of 100 corresponds to a steel support which is designed for the long-term load of loose sand below the water table according to Terzaghi (1946); see Section 2.3.2. This is the highest possible rock load considered and, according to Terzaghi (1946), it is:

$$P = 1.38(B + H)\gamma \tag{2.24}$$

where B is the width of the tunnel, H the height of the tunnel, and γ the rock unit weight.

This maximum rock load is then used as a basis to determine the type and spacing of steel rib support. For this, the tunnel is assumed to be of equal width and height (B = H) and that the sand has a unit weight of 120 pcf. With the rock load P from Equation (2.24), a load per foot of tunnel length is computed:

$$P(\text{per foot of tunnel}) = 331\,B^2 \text{in lbs./ft.} \tag{2.25}$$

The rock load from Equation (2.25) and tables in Proctor and White (1946) are then used to calibrate the theoretical ("datum") spacing of steel sets of various sizes for different tunnel dimensions (Table 2.20). As mentioned earlier, these spacings are for the worst conditions, i.e. the closest spacings for the particular "rib size." The actual spacing of the steel sets is then determined by:

$$\text{Spacing} = \text{Datum Spacing} \times \frac{100}{RR} \tag{2.26}$$

Wickham et al. (1972) and Wickham and Tiedemann (1972) also derive an equation that directly relates RSR and rock load per unit area:

$$W_r(\text{ksf}) = \frac{D(\text{ft})}{302}\left[\frac{8800}{RSR + 30} - 80\right] \tag{2.27}$$

where D is the diameter of the tunnel. Equation (2.27) is based on the fact that a rib ratio of 100 corresponds to a rock load of 1.38 γ(B + H), given in Equation (2.24) (for a complete derivation, see Wickham and Tiedemann, 1972). The rock load W_r can be used to estimate shotcrete and rockbolt supports. The design equation for bolt spacing in a square pattern is:

$$s = \sqrt{\frac{\text{Bolt Capacity}}{FWr}} \tag{2.28}$$

Table 2.20 Reference ("Datum") spacing for a Rib Ratio = 100% (from Wickham et al., 1974)

	Tunnel diameter										
Rib size	*10'*	*12'*	*14'*	*16'*	*18'*	*20'*	*22'*	*24'*	*26'*	*28'*	*30'*
417.7	1.16										
4H13.0	2.01	1.51	1.16	0.92							
6H15.5	3.19	2.37	1.81	1.42	1.14						
6H20		3.02	2.32	1.82	1.46	1.20					
6H25			2.86	2.25	1.81	1.48	1.23	1.04			
8WF31				3.24	2.61	2.14	1.78	1.51	1.29	1.11	
8WF40					3.37	2.76	2.30	1.95	1.67	1.44	1.25
8WF48						3.34	2.78	2.35	2.01	1.74	1.51
10WF49								2.59	2.22	1.91	1.67
12WF53										2.19	1.91
12WF65											2.35

where F is the Factor of Safety and W_r the Rock Load. For shotcrete, the empirical relation between rock load W_r (in ksf) and shotcrete thickness, t (in inches), proposed is

$$t(\text{inches}) = 1 + \frac{W_r(\text{ksf})}{1.25} \tag{2.29}$$

Discussion

The RSR method resulted from research by Jacobs Associates of San Francisco under a contract to the US Bureau of Mines. A first report was published in 1972 (Wickham and Tiedemann, 1972). The research was extended and resulted in a modified version (Wickham et al., 1974a, 1974b). For the first study, data from 33 tunnels were available. For the second study, additional tunnel data were included bringing the total number of tunnels to 53. Wickham and Tiedemann (1972) subdivided the tunnels into zones with similar ground conditions, thus leading to 134 sample sections for the 1972 study and 187 sections for the 1974 report. The tunnels include a large range of sizes from 70 sq. ft. (approx. 8.5 × 8.5 ft., i.e. 2.6 × 2.6 m = 6.8 m^2) to approximately 1050 sq. ft. (36 ft. or 10.9 m diameter, 93 m^2). For detailed tables describing these cases, see Wickham and Tiedemann (1972). The development of the method required the determination of: (1) Rib Ratio (RR) and (2) Rock Structure Rating (RSR), with its three individual parameters A, B, C. RR was determined for each individual case by relating the actual spacing to the "datum" spacing (RR = 100%) for the tunnel and steel set type. The determination of A, B, C, and thus RSR was more involved in that several methods were used for the rating of the ground conditions and finally one method was selected. Wickham and Tiedemann (1972) comment (in the 1972 study, Wickham, et al. presented RSR method #1 and #2. Method #1 presents the initial classification attempts and included more parameters than the three that were subsequently used in Method #2):

> As the study progressed, the original formats and assigned values were revised to more nearly reflect the data and findings of the research effort. RSR values as determined by several methods were compared and subsequently correlated with actual ground support used in the respective tunnels. These comparisons and evaluation of results, in conjunction with other information obtained from case studies, were used in finalizing the RSR method #2 which is proposed in this study.

Specifically, the development involved the assignment of a data pair (RR, RSR) for each of the tunnel sections and performing a regression analysis resulting in the design Equation (2.23). The regression curve of the 1974 version is shown in Figure 2.55. This curve and the regression Equation (2.23) resulting for the 1974 study are somewhat different from the earlier one (Wickham and Tiedemann, 1972). The differences between the two studies are the addition of some tunnel section and also a change in the RSR ratio. Many points of the original data were excluded from the regression analysis, which were said to be outside the scope of the model. The criteria for exclusion are not stated in detail for each individual case. Some of the excluded data points are "oversupported" sections, while other omitted data points represent squeezing ground conditions.

While the RSR method appears to be mathematically constructed and relatively straightforward to use, there are some issues that have to be kept in mind:

1. The method has been tested by applying it to several case studies. Prebid information was used to predict the support, which was then compared with the actually placed support. For details on this comparison, the reader is referred to the report by Wickham et al. (1974a).
2. Determination of the Parameter A is unproblematic. Parameters B and C can be measured. However, difficulties arise if there is more than one joint set; in this case, the user has to select the appropriate (average?) value for the parameters B and C (considering the fact that there are multiple joint sets).
3. Most important is the fact that the base cases are primarily steel-supported tunnels and appeared to be conservatively designed using the upper levels of Terzaghi's

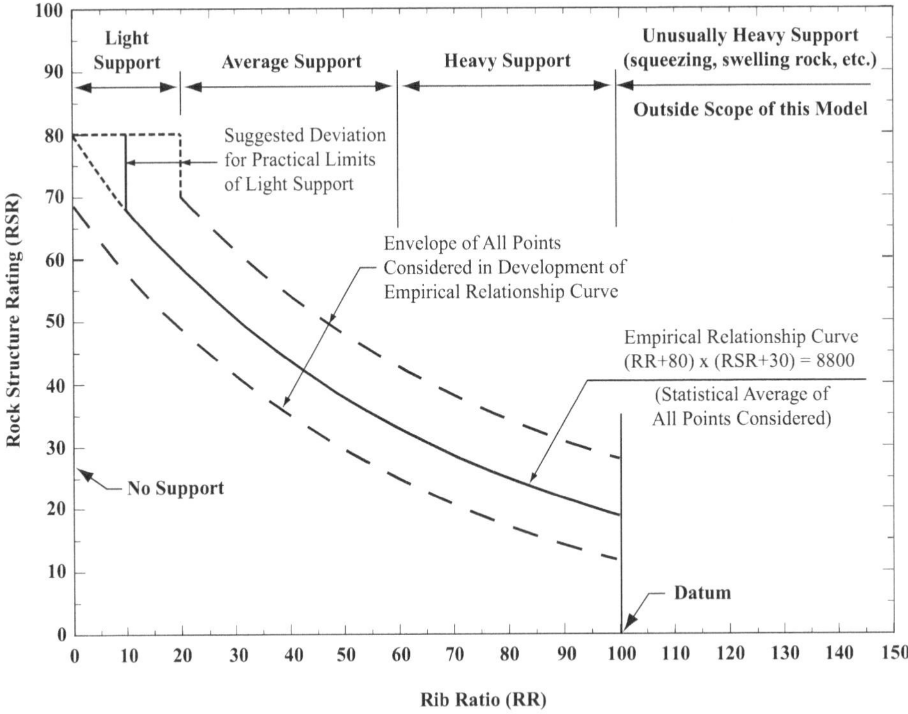

Figure 2.55 Correlation of Rock Structure Rating and Rib Ratio for 1974 database (from Wickham et al., 1974a, 1974b) (Note that the original diagram in Wickham et al., 1974a, 1974b contains all the data points).

recommendations. This has to be kept in mind when using the method for steel-supported tunnels. The application to other supports involves the intermediate step of calculating the equivalent rock load and, hence, adds another unknown factor.

2.3.6.4 *Geomechanics or Rock Mass Rating (RMR) classification*

This classification system was developed by Bieniawski starting in 1972, Bieniawski (1972), with further developments by Bieniawski (1973, 1974, 1976, 1979a, 1979b, 1979c) and others (Laubscher, 1977; Kendorski et al., 1983). The version presented here is the one published in 1989 by Bieniawski (1989). It should be stated that Bieniawski used (extended) his method also to foundations, slope and dam design.

The discussion will concentrate on the RMR classification for tunneling with some comments on its application to mining. At the very end of this section, a new development to apply RMR to predict excavation performance (Bieniawski and Celada, 2008) will be briefly discussed. Bieniawski (1989) emphasizes that although the classification method evolved over the years, it is essentially the same as originally developed.

2.3.6.4.1 The Rock Mass Rating (RMR) classification

The classification is based on six (Table 2.21) parameters: (1) strength of intact rock; (2) the Rock Quality Designation (RQD); (3) spacing of joints; (4) condition of joints; (5) groundwater inflow; and (6) orientation of joints. Table 2.21 is grouped into four parts (A–D). In

Table 2.21 Rock Mass Rating (from Bieniawski, 1989)

A. Classification Parameters and Their Ratings									
Parameter			*Ranges of values*						
1	Strength of intact rock material	Point-load strength index (MPa)	>10	4–10	2–4	1–2	For this low range, uniaxial compressive test is preferred		
		Uniaxial compressive strength (MPa)	>250	100–250	50–100	25–50	5–25	1–5	<1
	Rating		15	12	7	4	2	1	0
2	Drill core quality RQD (%)		90–100	75–90	50–75	25–50	<25		
	Rating		20	17	13	8	3		
3	Spacing of discontinuities		>2 m	0.6–2 m	200–600 mm	60–200 mm	<60 mm		
	Rating		20	15	10	8	5		
4	Condition of discontinuities		Very rough surfaces Not continuous No separation Unweathered wall rock	Slightly rough surfaces Separation < 1 mm Slightly weathered walls	Slightly rough surfaces Separation < 1 mm Highly weathered wall	Slickensided surfaces or Gouge < 5 mm thick or Separation 1–5 mm Continuous	Soft gouge > 5 mm thick or Separation > 5 mm Continuous		
	Rating		30	25	20	10	0		
5	Groundwater	Inflow per 10 m tunnel length (L/min)	None or	<10 or	10–25 or	25–125 or	>125 or		
		Ratio of: Joint water pressure/Major principal stress	0 or	<0.1 or	0.1–0.2 or	0.2–0.5 or	>0.5 or		
		General conditions	Completely dry	Damp	Wet	Dripping	Flowing		
	Rating		15	10	7	4	0		

(Continued)

Table 2.21 (Continued)

B. Rating Adjustment for Discontinuity Orientations						
Strike and dip orientations of discontinuities		*Very favorable*	*Favorable*	*Fair*	*Unfavorable*	*Very unfavorable*
Ratings	Tunnels and mines	0	−2	−5	−10	−12
	Foundations	0	−2	−7	−15	−25
	Slopes	0	−5	−25	−50	−60

C. Rock Mass Classes Determined from Total Ratings					
Rating	100←81	80←61	60←41	40←21	<20
Class no.	I	II	III	IV	V
Description	Very good rock	Good rock	Fair rock	Poor rock	Very poor rock

D. Meaning of Rock Mass Classes					
Class no.	*I*	*II*	*III*	*IV*	*V*
Average stand-up time	20 yr for 15-m span	1 yr for 10-m span	1 wk for 5-m span	10 h for 2.5-m span	30 min for 1-m span
Cohesion of the rock mass (kPa)	>400	300–400	200–300	100–200	<100
Friction angle of the rock mass (deg)	>45	35–45	25–35	15–25	<15

part A, the first five of the material parameters are used to obtain a rating for each parameter. This part is independent of the engineering applications of the method. In part B, the influence of orientation, which differs for different engineering applications (Tunnels, Slopes, Foundations), is rated. The ratings in parts A and B are then summed to obtain RMR, and RMR is then related to rock mass classes (Part C). Each of these classes is associated with engineering parameters (standup time, cohesion, friction angle) in Part D. The rock mass classes are then linked with excavation and support procedures (Table 2.22).

Some detailed comments on the ratings and some comments on the classification will now be made largely based on Bieniawski (1989):

Uniaxial strength: Either the uniaxial strength or the point load index are determined, and rated on a scale from 0 to 15.

RQD: RQD is assessed on a scale from 3 (RQD < 25%) to 20 (RQD = 90 to 100%).

Spacing of joints: The third step involves the assessment of the joint spacing. The maximum rating is 30 for joints spaced more than 300 mm, and the minimum is 5 for joints spaced less than 50 mm. It is important to note that the rating was developed for three joint sets. In cases where there are fewer than three sets, the rating is conservative and, according to Bieniawski (1989), the rating may be increased by 30%. In such cases, it may be advisable to use the chart in Figure 2.56, originally developed by Laubscher and Taylor (1976), to determine the rating for spacings in "multi-joint systems." This chart can be used for up to three joint sets in the manner indicated on the chart; for more than three joint sets, the three closest spaced sets are used.

To facilitate the practical use of the RMR rating for the three parameters (uniaxial compressive strength, RQD, joint spacing), Bieniawski (1989) provides several charts

Table 2.22 Guidelines for excavation and support of rock tunnels in accordance with the Rock Mass Rating systems[a] (from Bieniawski, 1989)

Rock mass class	*Excavation*	*Support*		
		Rock bolts (20-mm Diam. Fully Grouted)	*Shotcrete*	*Steel Sets*
Very good rock I RMR: 81–100	Full face 3-m advance	Generally, no support required except for occasional spot bolting		
Good rock II RMR: 61–80	Full face 1.0–1.5-m advance Complete support 20 m from face	Locally, bolts in crown 3 m long, spaced 2.5 m, with occasional wire mesh	50 mm in crown where required	None
Fair rock III RMR: 41–60	Top heading and bench 1.5–3 m advance in top heading Commence support after each blast Complete support 10 m from face	Systematic bolts 4 m long, spaced 1.5–2 m in crown and walls with wire mesh in crown	50–100 mm in crown and 30 mm in sides	None
Poor rock IV RMR: 21–40	Top heading and bench 1.0–1.5-m advance in top heading. Install support concurrently with excavation 10 m from face	Systematic bolts 4–5 m long, spaced 1–1.5 m in crown and wall with wire mesh	100–150 mm in crown and 100 mm in sides	Light to medium ribs spaced 1.5 m where required
Very poor rock V RMR: <20	Multiple drifts 0.5–1.5-m advance in top heading. Install support concurrently with excavation. Shotcrete as soon as possible after blasting	Systematic bolts 5–6 m long, spaced 1–1.5 m in crown and walls with wire mesh. Bolt Invert	150–200 mm in crown, 150 mm in sides, and 50 mm on face	Medium to heavy ribs spaced 0.75 m with steel lagging and fore-poling if required. Close invert

[a] Shape: Horeshoe: width: 10m; vertical stress <25 MPa; construction: drilling and blasting

(Figures 2.57 to 2.59) to allow the user to determine the specific rating, essentially providing an aid to interpretation. Bieniawski (1989) also shows a chart (Figure 2.60) in which RQD is related to discontinuity spacing based on the Priest and Hudson (1976) relation. Users should be warned, however, that the Priest and Hudson relation between spacing and RQD does not consider "non-existing core" that is not considered when determining RQD by measuring core lengths. The chart (Figure 2.60) is provided here for completeness only and is not recommended for use.

Condition of Joints: This rating, which ranges from 0 to 30, includes a variety of factors associated with the individual discontinuities, namely, roughness, continuity (persistence), aperture (separation), filler (gouge), and weaknesses of the rock surface. Table 2.23 (this is Chart E from Bieniawski, 1989) provides guidelines on how to consider the different factors.

Water Conditions: These conditions, which rate from 0 to 15, can be assessed on the basis of flow, water pressure (relative to the major principal stress) or general conditions.

Joint Orientation: As mentioned above, this parameter is rated separately since it takes on different values depending on the engineering application. In part B of Table 2.21, orientation is verbally described (very favorable to very unfavorable). These descriptions

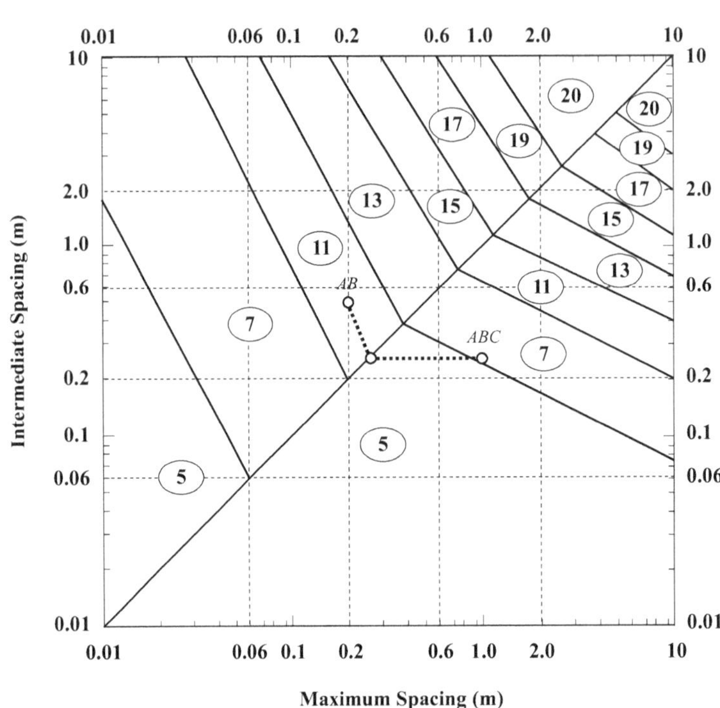

Figure 2.56 Geomechanics classification for hard-rock mining applications: ratings for multi-joint systems (from Laubscher and Taylor, 1976).

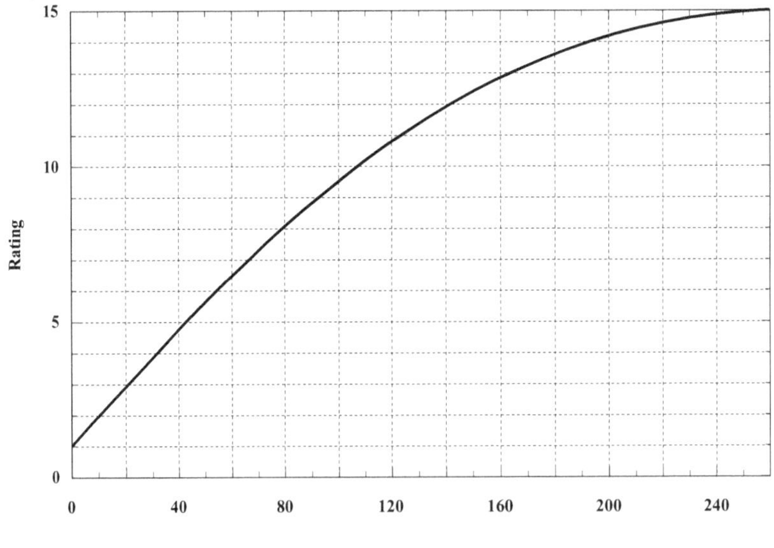

Figure 2.57 Chart A. Ratings for strength of intact rock (Bieniawski, 1989).

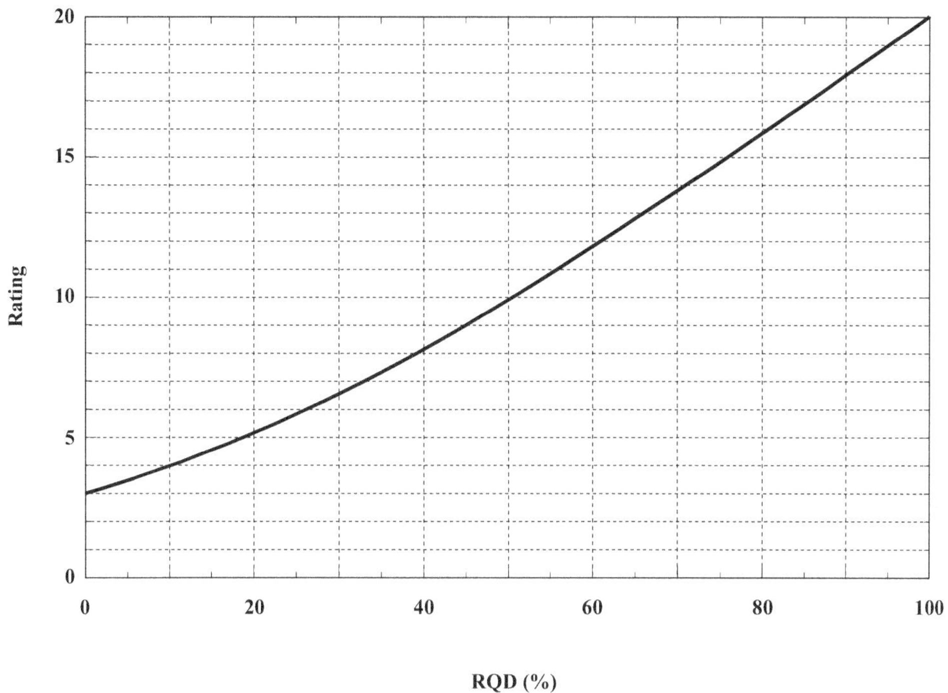

Figure 2.58 Chart B. Ratings for RQD (Bieniawski, 1989).

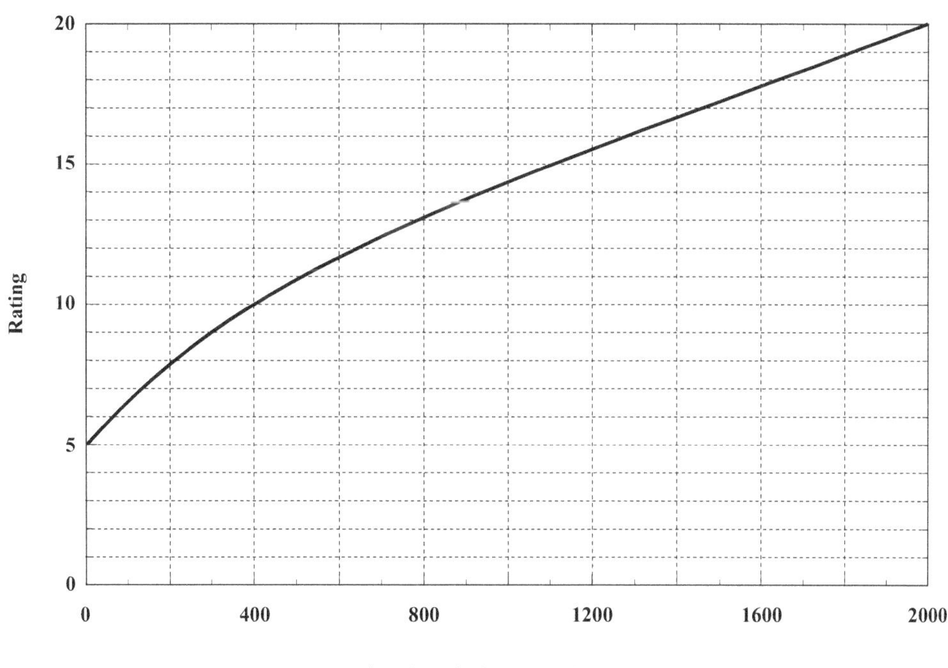

Figure 2.59 Chart C. Ratings for discontinuity spacing (Bieniawski, 1989).

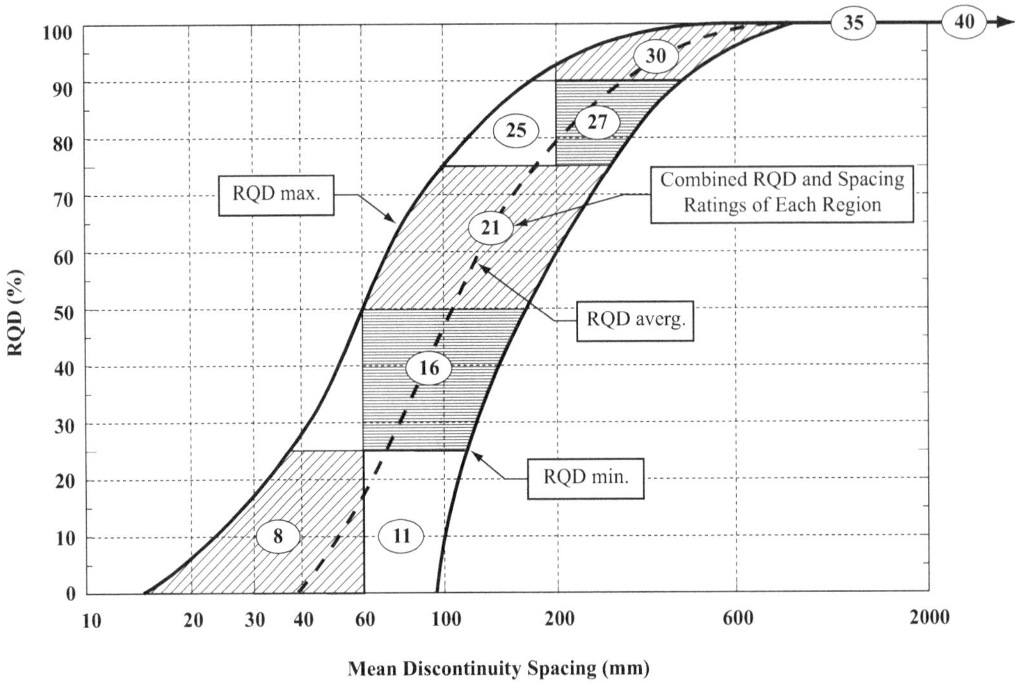

Figure 2.60 Chart D. Chart for correlation between RQD and discontinuity spacing (Bieniawski, 1989).

Table 2.23 Guidelines for classification of discontinuity conditions (Bieniawski, 1989)

Parameter	*Ratings*				
Discontinuity length (persistence/ continuity)	<1 m 6	1–3 m 4	3–10 m 2	10–20 m 1	>20 m 0
Separation (aperture)	None 6	<0.1 mm 5	0.1–1.0 mm 4	1–5 mm 1	>5 mm 0
Roughness	Very rough 6	Rough 5	Slightly rough 3	Smooth 1	Slickensided 0
		Hard filling		Soft filling	
Infilling (gouge)	None 6	<5 mm 4	>5 mm 2	<5 mm 2	>5 mm 0
Weathering	Unweathered 6	Slightly weathered 5	Moderately weathered 3	Highly weathered 1	Decomposed 0

Note: Some conditions are mutually exclusive. For example, if infilling is present, it is irrelevant what the roughness may be, since its effect will be overshadowed by the influence of the gouge. In such cases, use Table 2.21 directly.

can be related to specific values of the discontinuity orientations on the one hand and the tunnel axis on the other hand, as shown in Table 2.24. As indicated, this table, which was developed on the basis of work done by Wickham et al. (1972) in the context of RSR, is only to be used for tunneling. For slopes and foundations, the reader is referred to papers by Romana (1985) and Bieniawski and Orr (1976).

Table 2.24 Effect of discontinuity strike and dip orientations in tunneling (Bieniawski, 1989)[a]

Strike perpendicular to tunnel axis			
Drive with dip		*Drive against dip*	
Dip 45–90	Dip 20–45	Dip 45–90	Dip 20–45
Very favorable	Favorable	Fair	Unfavorable
Strike Parallel to Tunnel Axis			*Irrespective of Strike*
Dip 20–45	Dip 45–90		Dip 0–20
Fair	Very unfavorable		Fair

[a] Modified after Wickham et al. (1972)

Finally, the six parameter ratings are then summed and associated with a classification. This is followed by using the classification to determine tunnel excavation and support based on Table 2.22. Bieniawski (1989) specifically comments that these are *permanent* and not initial supports and for drill and blast excavation. Note that this statement somewhat contradicts the statement in Bieniawski and Celada (2008): "Rock mass classification on their own should only be used for preliminary planning programs and not for final rock reinforcement." Further comments about the use of the RMR classification for initial or final (permanent) supports will be made later. The classification Part D of Table 2.21 lists standup time, a concept Bieiniawski took from Lauffer (1958). Accordingly, the Lauffer chart was also used by Bieniawski and the latest versions (Bieniawski, 1989) are shown in Figure 2.61a and b. Figure 2.61a is for tunnels/mines using drill and blast excavation and Figure 2.61b is a version

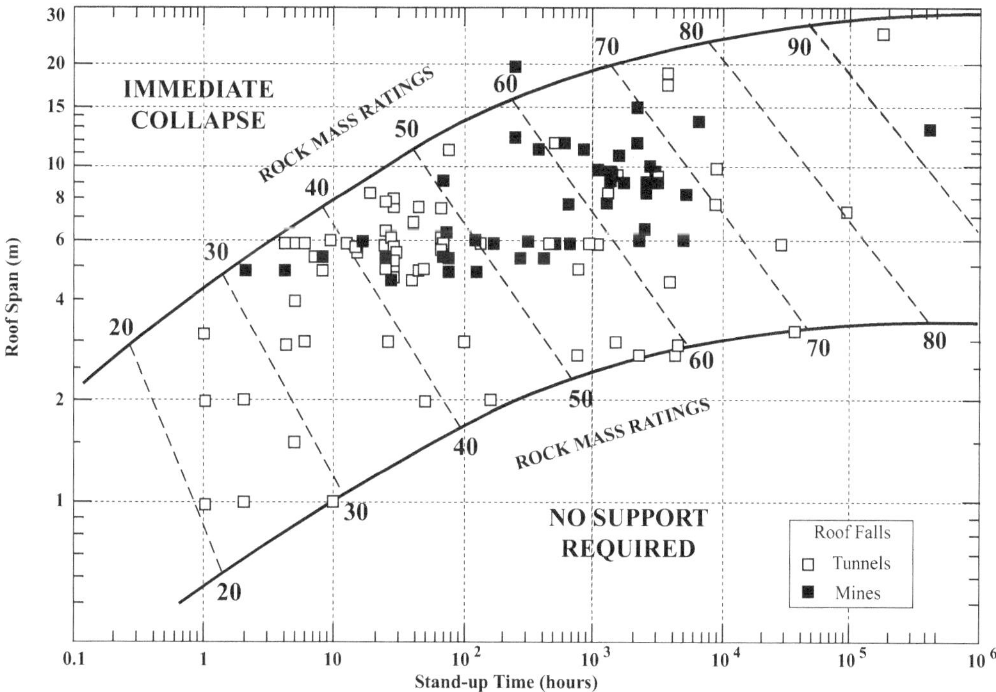

Figure 2.61a Relationship between the stand-up time and span for various rock mass classes, according to the Geomechanics classification. Note that the data points are for roof falls. The contour lines are limits of applicability (Bieniawski, 1989).

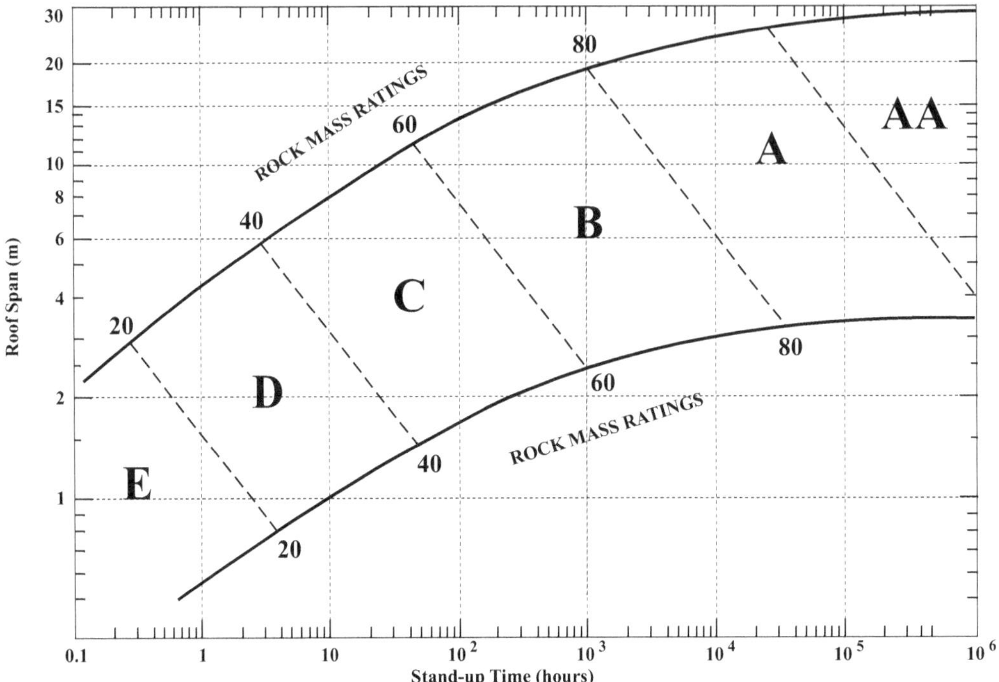

Figure 2.61b Modified 1988 Lauffer diagram depicting boundaries of rock mass classes for TBM applications (Bieniawski, 1989 after Lauffer, 1988).

modified by Lauffer (1988) to include TBM excavation. When comparing Figures 2.61a and b, one notices a shift of the class boundaries reflecting the fact that TBM excavation causes less disturbance. One could basically use these two figures to adapt the support dimensions in Table 2.22 to TBM tunnels (this is a comment by the authors of this book and not by Bieniawski).

It is important to note that Bieniawski (e.g. 1979a, 1979b, 1979c, 1989) recommends that the Geomechanics classification should be used in conjunction with monitoring of tunnel performance. In Bieniawski (1989) this comment is made together with a remark that the support characteristics obtained with the RMR method are rather conservative.

Several authors (e.g. Laubscher, 1977; Kendorski et al., 1983; Cummings et al., 1982; Unal, 1983; Venkateswarteu, 1986) extended the RMR system to hard rock-and coal mines. Bieniawski (1989) shows how the RMR rating can be adjusted for mining (Figure 2.62). More details on how the RMR system can be applied in mining are given in Chapter 8 of Bieniawski (1989).

Although empirical methods type E essentially provide quantitative relations between rock mass characteristics and excavation/support characteristics of tunnels, some are also used to provide mechanical properties of the rock mass. This is also the case for the RMR method. Friction angle and cohesion are related to the rock mass classes (see Table 2.21, part D). In addition, the modulus of elasticity can be obtained from RMR based on (Bieniawski, 1978):

$$E_m = 2RMR - 100 \tag{2.30}$$

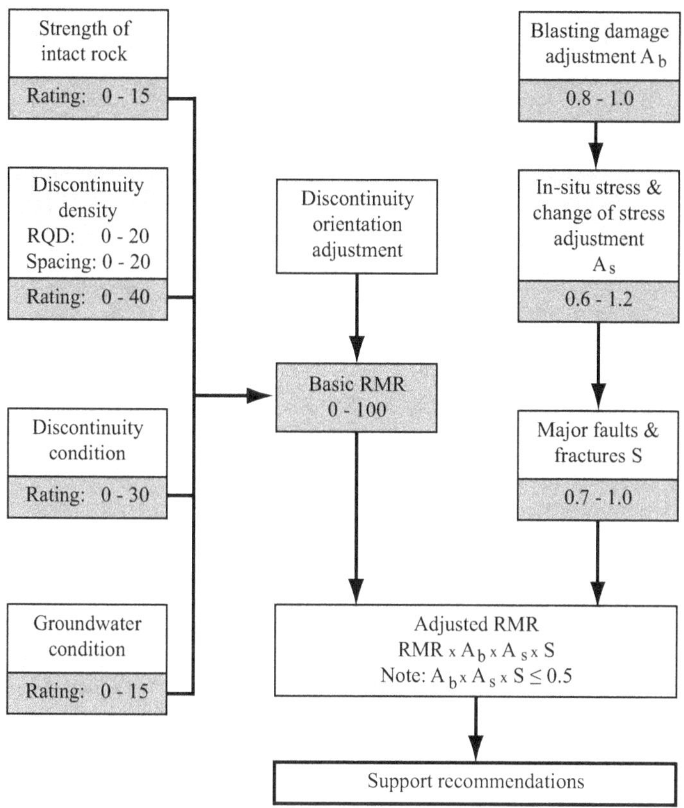

Figure 2.62 Adjustments to the Rock Mass Rating system for mining applications (Bieniawski, 1989).

and extended by Serafim and Pereira (1983) by including results for low RMR ranges:

$$E_m = 10^{(RMR-10)/40} \tag{2.31}$$

These relations are shown in Figure 2.63, and all this has also been discussed earlier in Section 2.3.4. Other issues regarding the RMR system in the context of the Hoek and Brown relation and GSI were also commented upon in Section 2.3.4.

Finally, it is also possible to use Unal's (1983) relation to determine rock load with RMR:

$$P = \frac{100 - RMR}{100}\gamma B \tag{2.32}$$

where P is the support load in kN; B is the tunnel width in m; and γ is the rock unit weight in kN/m^3

Discussion

As Bieniawski states in a number of his publications (e.g. Bieniawski, 1989) the method was systematically developed starting with a data base of 49 case histories collected and interpreted by Bieniawski and co-authors. Many others have used the method and some have

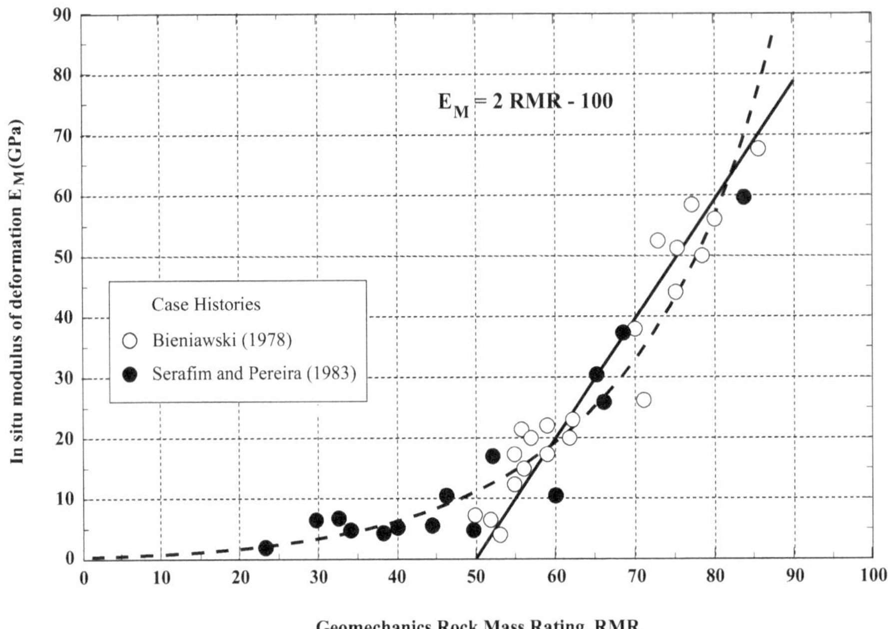

Figure 2.63 Correlation between the in situ modulus of deformation and RMR, (Bieniawski, 1989 after Serafim and Pereira, 1983).

published their cases. It is clear from Bieniawski's comments that his experience and that of others was used to modify and refine the method. The only problematic aspect in the view of the authors of this book is the change for the support recommendations from those for initial supports up to 1976 to final supports thereafter without a significant change in actual support dimensions. As mentioned earlier, Bieniawski justifies this in the 1979 paper by stating that in modern tunnel construction, the initial support is designed to maintain the lifelong stability of the tunnel. He then goes on in 1984, 1989 to state that the support recommendations are for the final support. This may also alleviate the fact that the support recommendations are (somewhat) conservative. The authors of this book do not agree, however, with the statement made by Bieniawski (1979a, 1979b, 1979c) that the initial tunnel support is entirely sufficient for the tunnel lifetime. Bieniawski (1997) later modified this somewhat (see Preposition 2, below) by saying that many tunnels are designed in such a way that the final support plays a role, e.g. support of water pressure, other stress effects. The user has to decide what to use the RMR method for, keeping these comments in mind. Clearly, and as recognized by Bieniawski, the RMR method is an empirical method limited by the underlying cases. It is, therefore, appropriate to quote what Bieniawski (1989) states:

> Finally, the RMR classification – as any other – is not to be taken as a substitute for engineering design. This classification is only a part of the empirical design approach, one of the three main design approaches in rock engineering (empirical, observational, and analytical). It should be applied intelligently and used in conjunction with observational and analytical methods to formulate an overall design rationale compatible with the design objectives and site geology.

Another interesting set of comments were made by Bieniawski in 1997 some of which are partially quoted here:

Preposition 1:

Rock mass classifications, either rating methods (RMR or Q) or descriptive methods (NATM) are most effective if not used on their own but used within the overall engineering design process. This leads to alternative design and comparison of the different design methods.

Preposition 2:

Rock mass classifications on their own should only be used for preliminary planning purposes and not for final tunnel support. This is my personal conviction and this does not mean diminishing the importance of rock mass classifications; rather the emphasis is shifted. For preliminary design and planning purposes, rock mass classifications are excellent; they quantify rock mass conditions, enable an estimate of rock mass properties, and provide the reference basis for expected tunneling conditions. But I would not recommend using the RMR system alone for final support design.

Preposition 3 (and the main idea):

The two main rating methods of rock mass classifications – the RMR and the Q-systems – are absolutely essential for monitoring rock conditions during construction, to enable effective comparison of predicted conditions from site investigations with those encountered during construction.

2.3.6.4.2 The Rock Mass Excavability (RME) Index

Bieniawski and Celada (2008), (see also Bieniawski, 2004 and Bieniawski et al., 2006) developed this index initially on the basis of three tunnels in Spain excavated by double shield TBM's (total length of 22.9 km and analyzed in 387 sections). This was recently extended by including the analysis of experience gained in the Guadarrama tunnels in Spain (Bieniawski and Celada, 2008). A strong comment is made by Bieniawski and Celada that this classification is separate from RMR. It specifically addresses excavation performance. As will be seen below, this statement is not entirely correct since the standup time derived from RMR is used in determining RME. Table 2.25 shows the parameters and ratings and, as can be seen there, the maximum possible RME is 100. The parameters UCS and groundwater flow are self-explanatory. The others require some comments:

Discontinuities: This consists of three "subparameters" (homogeneity, number of joints per meter, orientation) which have to be summed.
Drillability Index: This is based on the drilling rate index as proposed by Bruland (1998).
Standup Time: This is where the relation to RMR occurs. The standup time can be obtained from the chart shown in Figure 2.61a (Figure 41 in Bieniawski, 1989). Since this chart is for drill and blast tunnels, the RMR for TBM needs to be adapted as follows:

$$RMR_{TBM} = 0.8RMR_{D+B} + 20 \tag{2.33}$$

Table 2.25 Input ratings for Rock Mass Excavability (RME) Index (Bieniawski and Celada, 2008)

	UCS of intact rock (0–25 points)									
σ_c (MPa)	<5	5–30	30–90	90–180	>180					
Average rating	4	14	25	14	0					
	Drillability (0–15 points)									
Drilling Rate Index	>80	80–65	65–50	50–40	<40					
Average rating	15	10	7	3	0					
	Discontinuities at tunnel face (0–30 points)									
Homogeneity			Number of joints per meter					Orientation with respect to tunnel axis		
	Homogeneous	Mixed	0–4	4–8	8–15	15–30	>30	Perpendicular	Oblique	Parallel
Avg. rating	10	0	2	7	15	10	0	5	3	0
	Standup time (0–25 points)									
Hours	<5	5–24	24–96	96–192	>192					
Average rating	0	2	10	15	25					
	Groundwater inflow (0–5 points)									
L/sec	>100	70–100	30–70	10–30	<10					
Average rating	0	1	2	[1]4	5					

[1] Zero for argillaceous rocks

According to Bieniawski and Celada (2008), this adjustment is based on Alber (1996). RME is then used to determine the average advance rate (ARA) for single/double shielded TBM's on the basis of Figure 2.64. Since the base cases are tunnels with diameters close to 10 m, a correction for other diameters needs to be applied:

$$k_D = -0.007D^3 + 0.1637D^2 - 1.2859D - 4.5158 \tag{2.34}$$

where k_D is the correction factor with which the numbers obtained from Figure 2.64 need to be multiplied. Bieniawski and Celada (2008) also introduce other correction factors to express the "learning effect" and "crew effectiveness." These correction factors are not provided here since they are only based on the information from one tunnel.

This leads to a final evaluation by the authors of this book: While the RME includes factors that, in principle, affect excavation performance, the data base is too limited for general application. It is entirely possible that further developments will make the approach more widely applicable. Readers should, therefore, consider future publications by Bieniawski and co-authors.

2.3.6.5 The Q system

Barton et al. (1974b) published the Q-System based on their work at NGI (NGI Report, in Barton et al., 1974a). The original development was based on 212 cases, mostly from the literature but also many that the authors were involved in. Interestingly, the literature included the 97 cases studied by Cecil (Cecil, 1970, 1975); see also the comments in the discussion of the RQD relations. Since Cecil's cases were mostly Scandinavian, this means that, together with Barton's et al. (1974a, 1974b) cases, a majority of cases were in Scandinavian rock conditions and excavated by drilling and blasting. Since then, the Q-system was used in many more ("thousands" see Barton and Grimstad, 2014). All this led to some, but not major changes in the system, which will be discussed later.

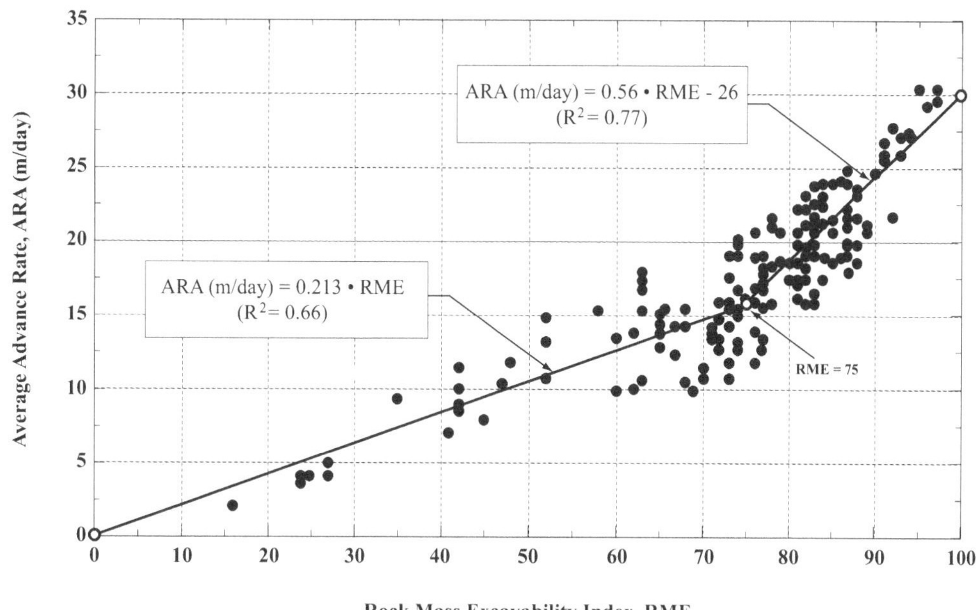

Figure 2.64 Correlation between the RME index and the average rate of advance (m/day) for single- and double-shield TBM's.

2.3.6.5.1 Description of the Q system

Q is defined as follows:

$$Q = \frac{RQD}{J_n}\frac{J_r}{J_a}\frac{J_w}{SRF} \tag{2.35}$$

where

RQD is the Rock Quality Designation (Deere, 1963)
J_n is the Joint Set Number
J_r the Joint Roughness Number
J_a the Joint Alteration Number
J_w the Joint Water Reduction Factor, and
SRF the Stress Reduction Factor.

Each of these parameters is rated using Tables 2.26a–f (A1 – A6 from Barton, 2002). When using these tables, special attention should be given to the footnotes. After calculating Q, this is used together with the equivalent dimensions (Excavation Support Ratio) in the chart, Figure 2.65, to obtain ground classes and the supports shown in this chart. "Note that Figure 2.65 is a reproduction of corresponding figures in Grimstad (2007a, 2007b), Barton and Grimstad (2014) and Barton (2017), reflecting the newest developments in the application of the Q-method" (personal communication by E. Grimstad, August 2022). The support types and dimensions in Figure 2.65 are for final (permanent) support (comments on the usage of the system for initial supports will be made later).

The Q system has remained largely the same as far as the ratings (Tables 2.26a–f) are concerned (some change in SRF). However, the relations to specific supports, as shown in Figure 2.65, are different from the original ones, and were derived by a combination of

Table 2.26a Parameters of the Q-system

Rock quality designation		*RQD (%)*
A	Very poor	0–25
B	Poor	25–50
C	Fair	50–75
D	Good	75–90
E	Excellent	90–100

Notes: (i) Where RQD is reported or measured as <10 (including 0), A nominal value of 10 is used to evaluate Q, (ii) RQD intervals of 5, i.e. 100, 95, 90, etc., are sufficiently accurate.

Table 2.26b

Joint set number		J_n
A	Massive, no or few joints	0.5–1
B	One joint set	2
C	One joint set plus random joints	3
D	Two joint sets	4
E	Two joint sets plus random joints	6
F	Three joint sets	9
G	Three joint sets plus random joints	12
H	Four or more joint sets, random, heavily jointed, 'sugar-cube," etc.	15
J	Crushed rock, earthlike	20

Notes: (i) For tunnel intersections, use (3.0 × J_n) and (ii) for portals use (2.0 × J_n)

Table 2.26c

Joint roughness number		J_r
(a) Rock-wall contact and (b) rock-wall contact before 10 cm shear		
A	Discontinuous joints	4
B	Rough or irregular, undulating	3
C	Smooth, undulating	2
D	Slickensided, undulating	1.5
E	Rough or irregular, planar	1.5
F	Smooth, planar	1.0
G	Slickensided, planar	0.5
(b) No rock-wall contact when sheared		
H	Zone containing clay minerals thick enough to prevent rock-wall contact	1.0
J	Sandy, gravely or crushed zone thick enough to prevent rock-wall contact	1.0

Notes: i. Descriptions refer to small-scale features and intermediate-scale features, in that order;
ii. Add 1.0 if the mean spacing of the relevant joint set is greater than 3 m;
iii. J_r = 0.5 can be used for planar, slickensided joints having lineations, provided the lineations are oriented for minimum strength;
iv. J_r and J_n classification is applied to the joint set or discontinuity that is least favorable for stability both from the point of view of orientation and shear resistance, τ (where $\tau \approx \sigma_n \tan^{-1}(J_r/J_a)$).

Table 2.26d

Joint alteration number		*Φ_r approx. (deg)*	*J_a*
(a) Rock-wall contact (no mineral fillings, only coatings)			
A	Tightly healed, hard, non-softening impermeable filling i.e. quartz or epidote	-	0.75
B	Unaltered joint walls, surface straining only	25–35	1.0
C	Slightly altered joint walls, non-softening mineral coatings, sandy particles, clay-free disintegrated rock, etc.	25–30	2.0
D	Silty- or sandy-clay coatings, small clay fraction (non-softening)	20–25	3.0
E	Softening or low friction clay mineral coatings, i.e. kaolinite or mica. Also chlorite, talc, gypsum, graphite, etc. and small quantities of swelling clays	8–16	4.0
(b) Rock-wall contact before 10 cm shear (thin mineral fillings)			
F	Sandy particles, clay-free disintegrated rock, etc.	25–30	4.0
G	Strongly over-consolidated non-softening clay mineral fillings (continuous, but < 5 mm thickness)	16–24	6.0
H	Medium or low over-consolidation, softening, clay mineral fillings (continuous but <5 mm thickness)	12–16	8.0
J	Swelling-clay fillings, i.e. montmorillonite (continuous, but < 5 mm thickness) Value of J_a depends on percent of swelling clay-size particles, and access to water, etc.	6–12	8–12
(c) No rock-wall contact when sheared (thick mineral fillings)			
KLM	Zones or bands of disintegrated or crushed rock and clay (see G, H, J for description of clay condition)	6–24	6, 8, or 8–12
N	Zones or bands of silty- or sandy-clay, small clay fraction (non-softening)	-	5.0
OPR	Thick, continuous zones or bands of clay (see G, H, J for description of clay condition)	6–24	10, 13, or 13–20

Table 2.26e

Joint water reduction factor		*Approx. water pres. (kg/cm²)*	*J_w*
A	Dry excavations or minor inflow, i.e. < 5 l/min locally	<1	1.0
B	Medium inflow or pressure, occasional outwash of joint fillings	1–2.5	0.66
C	Large inflow or high pressure in competent rock with unfilled joints	2.5–10	0.5
D	Large inflow or high pressure, considerable outwash of joint fillings	2.5–10	0.33
E	Exceptionally high inflow or water pressure at blasting, decaying with time	> 10	0.2–0.1
F	Exceptionally high inflow or water pressure continuing without noticeable decay	> 10	0.1–0.05

Notes: i. Factors C to F are crude estimates. Increase J_w if drainage measures are installed;

ii. Special problems caused by ice formations are not considered;

iii. For general characterization of rock masses distant from excavation influences, the use of J_w = 1.0, 0.66, 0.5, 0.33, etc. as depth increases from say 0–5, 5–25, 25–250 to >250 m is recommended, assuming that RQD/J_n is low enough (e.g. 0.5–25) for good hydraulic connectivity. This will help to adjust Q for some of the effective stress and water softening effects, in combination with appropriate characterization values of SRF. Correlations with depth-dependent static deformation modulus and seismic velocity will then follow the practice used when these were developed.

Table 2.26f

Stress reduction factor	*SRF*		
(a) Weakness zones intersecting excavation, which may cause loosening of rock mass when tunnel is excavated			
A. Multiple occurrences of weakness zones containing clay or chemically disintegrated rock, very loose surrounding rock (any depth)	10		
B. Single weakness zones containing clay or chemically disintegrated rock (depth of excavation ≤ 50m)	5		
C. Single weakness zones containing clay or chemically disintegrated rock (depth of excavation > 50m)	2.5		
D. Multiple shear zones in competent rock (clay-free), loose surrounding rock (any depth)	7.5		
E. Single shear zones in competent rock (clay-free), (depth of excavation ≤ 50m)	5.0		
F. Single shear zones in competent rock (clay-free), (depth of excavation > 50m)	2.5		
G. Loose, open joints, heavily jointed or "sugar cube," etc. (any depth)	5.0		
	σ_c/σ_1	σ_θ/σ_c	SRF
(b) Competent rock, rock stress problems			
H. Low stress, near surface, open joints	>200	<0.01	2.5
J. Medium stress, favorable stress condition	200–10	0.01–0.3	1
K. High stress, very tight structure. Usually favorable to stability, may be unfavorable for wall stability	10-5	0.3–0.4	0.5-2
L. Moderate slabbing after > 1 h in massive rock	5-3	0.5–0.65	5–50
M. Slabbing and rockburst after a few minutes in massive rock	3-2	0.65–1	50–200
N. Heavy rockburst (strain-burst) and immediate dynamic deformations in massive rock	<2	>1	200–400
	σ_θ/σ_c	SRF	
(c) Squeezing rock: plastic flow of incompetent rock under the influence of high rock pressure			
O. Mild squeezing rock pressure	1–5	5–10	
P. Heavy squeezing rock pressure	>5	10–20	
	SRF		
(d) Swelling rock: chemical swelling activity depending on presence of water			
R. Mild swelling rock pressure	5–10		
S. Heavy swelling rock pressure	10–15		

Notes:
i. Reduce these values of SRF by 25–50% if the relevant shear zones only influence but do not intersect the excavation. This will also be relevant for characterization;
ii. For strongly anisotropic virgin stress field (if measured): When $5 \le \sigma_1/\sigma_3 \le 10$, reduce σ_c to $0.75\ \sigma_c$. When $\sigma_1/\sigma_3 > 10$, reduce σ_c to $0.5\sigma_c$, where σ_c is the unconfined compression strength, σ_1 and σ_3 are the major and minor principal stresses, and σ_θ the maximum tangential stress (estimated from elastic theory);
iii. Few case records available where depth of crown below surface is less than span width, suggest an SRF increase from 2.5 to 5 for such cases (see H);
iv. cases L, M and N are usually most relevant for support design of deep tunnel excavations in hard massive rock masses, with RQD/J_n ratios from about 50–200;
v. For general characterization of rock masses distant from excavation influences, the use of SRF = 5, 2.5, 1.0 and 0.5 is recommended as depth increases from say 0–5, 5–25, 25–250 to > 250m. This will help to adjust Q for some of the effective stress effects, in combination with appropriate characterization values of J_w. Correlations with depth-dependent static deformation modulus and seismic velocity will then follow the practice used when these were developed;
vi. Cases of squeezing rock may occur for depth $H > 350Q^{1/3}$. Rock mass compression strength can be estimated from $SIGMA_{cm} \approx 5\gamma Q_c^{1/3}$ (MPa) where γ is the rock density in t/m^3, and $Q_c = Q \times \sigma_c/100$, Barton (2000).

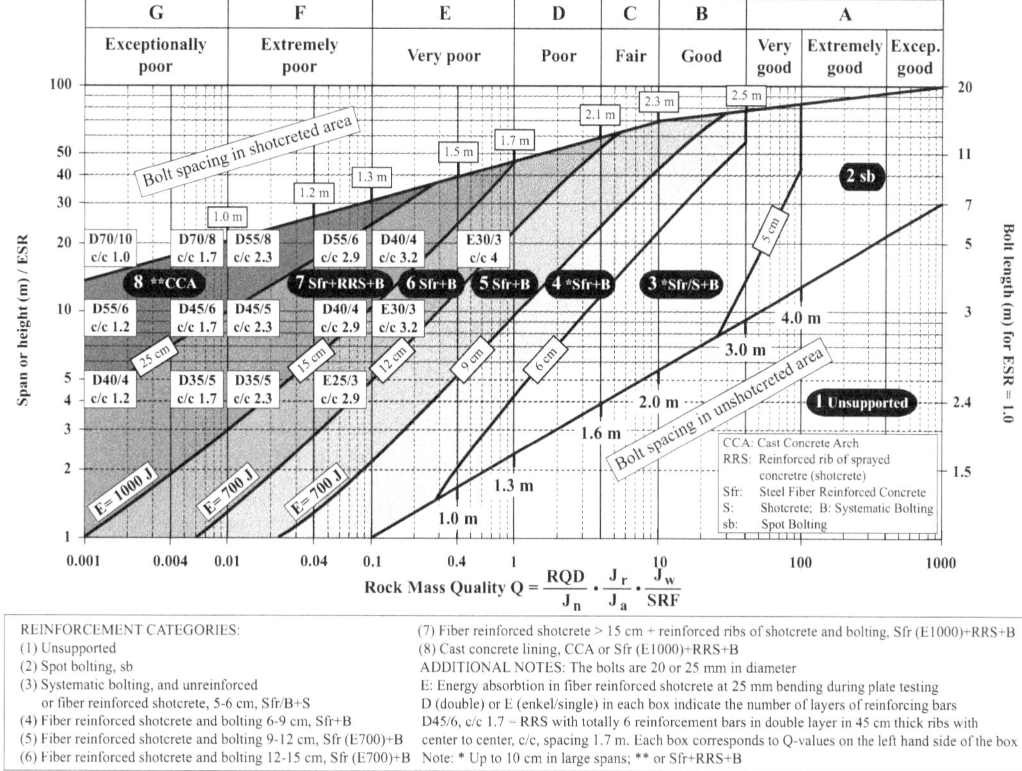

Figure 2.65 Updated Q-support chart (Grimstad, 2007a, 2007b). Also published in Barton and Grimstad (2014) and Barton (2017). (Reproduced by permission of E. Grimstad.)

empiricism and numerical modeling by a small team at NGI under the direction of Grimstad. The reason for the change of shotcrete thickness in the lower thickness range, compared to the Q-support chart in Grimstad and Barton 1993 (see Figure 2.69), is that 4 cm shotcrete is no longer acceptable. Also, some additional applications such as the relation to seismic velocity involved changes. Recently, and analogous to RME, a Q_{TBM} for evaluating TBM penetration/advance rate was developed. These changes will be discussed in more detail later. Before doing so, some comments on the individual parameters will be given. This is based on Barton's (e.g. Barton et al., 1974a, 1974b; Barton et al., 1975) explanations.

RQD, Rock Quality Designation: This is input as is, i.e. in percentage numbers (in contrast to what is done in the RMR system where it is transformed into a rating) and assumed to be derived from bore cores (not from scan lines on the rock surface!). Barton et al., (1974a, 1974b) also suggest that, if no bore core is available and in case of clay-free rock masses, Palmström's (1974, 2005) relation between joints per volume can be used to derive RQD:

$$RQD = 115 - 3.3J_v \text{(approx)} \tag{2.36}$$

where J_v is the total number of joints per m^3; this leads to RQD = 100 for $J_v < 4.5$

J_n, Joint Set Number: Here the comment is made that strongly developed schistosity and foliation surfaces should be counted as joint set(s) and if not strongly developed, they should be counted as random joints (see Table 2.26b).

J_r, Joint Roughness Number, and J_a, Joint Alteration Number: These parameters represent shear strength and should be based on the weakest "significant joint set" or "clay filled discontinuity" (Barton et al., 1974a, 1974b, 1975). It is also stated that if the joint set with the minimum J_r/J_a is favorably oriented with regard to the rock mass stability another less favorably oriented joint set may be more relevant. Barton et al. (1975) then provide detailed comments why in the Q-system favorable/unfavorable orientation (relative to the excavation) is not considered as a separate parameter, in contrast to the RMR and RSR systems. The main argument is that the J_n, J_r, J_a (and SRF) parameters indirectly consider also orientation since it is not only orientation but also the shearing resistance that plays a role if a joint/joint set is unfavorable. Specifically, it is stated (Barton et al., 1975) that "observing the presence of unfavorable joints/joint sets is related to the fact that they are visible." In the opinion of the authors of this book, this argument may be affected by the particular geologic conditions in Scandinavia from which the initial base cases in the Q system were taken.

J_w, Joint Water Reduction Factor: No specific comments were made by Barton et al., (1974a, 1974b, 1975) for J_w by itself but for the value J_w/SRF which will be given below.

SRF, Stress Reduction Factor: This is the most difficult and most controversial factor. It is also the one for which the ratings were changed between the original (1974) and (2002) publications. The names of the categories, a – d (see Table 2.26f), however, remained the same. As can be easily seen in Table 2.26f, the four categories describe quite different conditions. In particular, categories "a" and "b" are in a way extremes at the opposite end of possible conditions. For category "a," mechanisms may lead to "loosening loads" (Barton et al., 1974a, 1974b) while in category "b" the strength of intact rock may be its "weakest link" (ratio rock stress/rock strength). As a matter of fact, it is category "b" where the present parameter ranges and, particularly, the limits were changed between 1974 and 2002. Specifically, the maximum possible SRF in this category is now "400" while it was "20" in the 1974 classification. This change occurred in 1993 (Grimstad and Barton, 1993) and is explained in that publication in that "tunnels in massive rock under high stress require far more support than recommended by the corresponding Q values." Other smaller changes relate to the notes regarding category "b" where some of the ratios were changed and σ_θ (the tangential stress for a circular opening) is now used instead of the tensile strength. In addition, the notes make some comments about squeezing rock where, based on Singh's et al. (1992) investigation, squeezing is related to overburden, i.e. squeezing occurs when

$$H > 350Q^{1/3}(\mathrm{m}) \tag{2.37}$$

Where H is the height of overburden

And the rock mass compressive strength, σ_{cm}, can be estimated as:

$$\sigma_{cm} \sim 5\gamma Q_{co}{}^{1/3}(\mathrm{MPa}) \tag{2.38}$$

where γ is the rock density (unit weight: note from the authors) in t/m^3, and where:

$$Q_{co} = Q_o \sigma_c / 100 \tag{2.39}$$

where σ_c is the unconfined compressive strength of intact rock and Q_o is determined with RQD_o, i.e. in the tunnel direction (note by the authors: we use here Q_{co} rather Q_c, as in Barton, 2002, to differentiate when Q_o or Q is included in the definition; Q_c is defined later. This distinction is not made in Barton, 2002).

Comments on these values will be made in the discussion of the development of the Q method, later in this section. Before doing so, some comments made by Barton et al. (1974a, 1974b and later) regarding rating ratios (RQD/J_n, J_r/J_a, J_w/SRF) need to be made:

The ratio RQD/J_n can be considered a representation of the relative block size, which is useful for distinguishing massive rock (see note IV in Table 2.26f).

J_r/J_a is the relative frictional strength of the least favorable joint set or filled discontinuity. In the original publication (Barton et al., 1974a, 1974b), as well as in Barton (1995, 2002), a set of tables is provided which list $\tan^{-1}(J_r/J_a)^\circ$, i.e. "friction angles" (quotes also in original publication, Barton et al., 1974a, 1974b). These "friction angles" express both cohesion

Table 2.27a The Excavation Support Ratio (ESR) appropriate to a variety of underground excavations

Type of excavation	*ESR*	*No. of cases*
A Temporary mine openings, etc.	ca. 3–5?	(2)
B Vertical shafts: (i) circular section	ca. 2–5?	(0)
(ii) rectangular/square section	ca. 2.0?	(0)
C Permanent mine openings, water tunnels for hydro power (exclude high pressure penstocks), pilot tunnels, drifts and headings for large excavations, etc.	1.6	(83)
D Storage rooms, water treatment plants, minor road and railway tunnels, surge chambers, access tunnels, etc. (cylindrical caverns?)	1.3	(25)
E Power stations, major road and railway tunnels, civil defense chambers, portals, intersections, etc.	1.0	(79)
F Underground nuclear power stations, railway stations, sports and public facilities, factories, etc.	ca. 0.8?	(2)

Note: Original table from Barton et al. (1974b)

Table 2.27b ESR table in Barton and Grimstad (2014)

Type of excavation	*ESR – 1993*	*ESR – 2014*
A Temporary mine openings, etc.	ca. 2–5	ca. 2–5*
B Permanent mine openings, water tunnels for hydro power (exclude high pressure penstocks), pilot tunnels, drifts, and headings for large openings, surge chambers	1.6–2.0	1.6–2.0*
C Storage caverns, water treatment plants, minor road and railway tunnels, access tunnels	1.2–1.3*	0.9–1.1 Storage caverns: 1.2–1.3*
D Power stations, major road and railway tunnels, civil defense chambers, portals, intersections	0.9–1.1	Major road and railway tunnels: 0.5–0.8
E Underground nuclear power stations, railway stations, sports and public facilities, factories, major gas pipeline tunnels	0.5–0.8	0.5–0.8*

Note: Left ESR: values recommended in 1993; Right ESR: updated in 2014 reflecting "increased conservatism in some Sections of civil engineering when applying single shell NMT"
*Unchanged values from 1993 recommendations

Table 2.28 (a–c) Inter-block frictional behaviour. $\tan^{-1}$ (Jv/Ja) shows apparently dilatant (ϕ + i) friction angles for many joints, and apparently contractile (ϕ − i) friction angles for many mineral filled discontinuities. (Figure 12 in Barton, 2002)

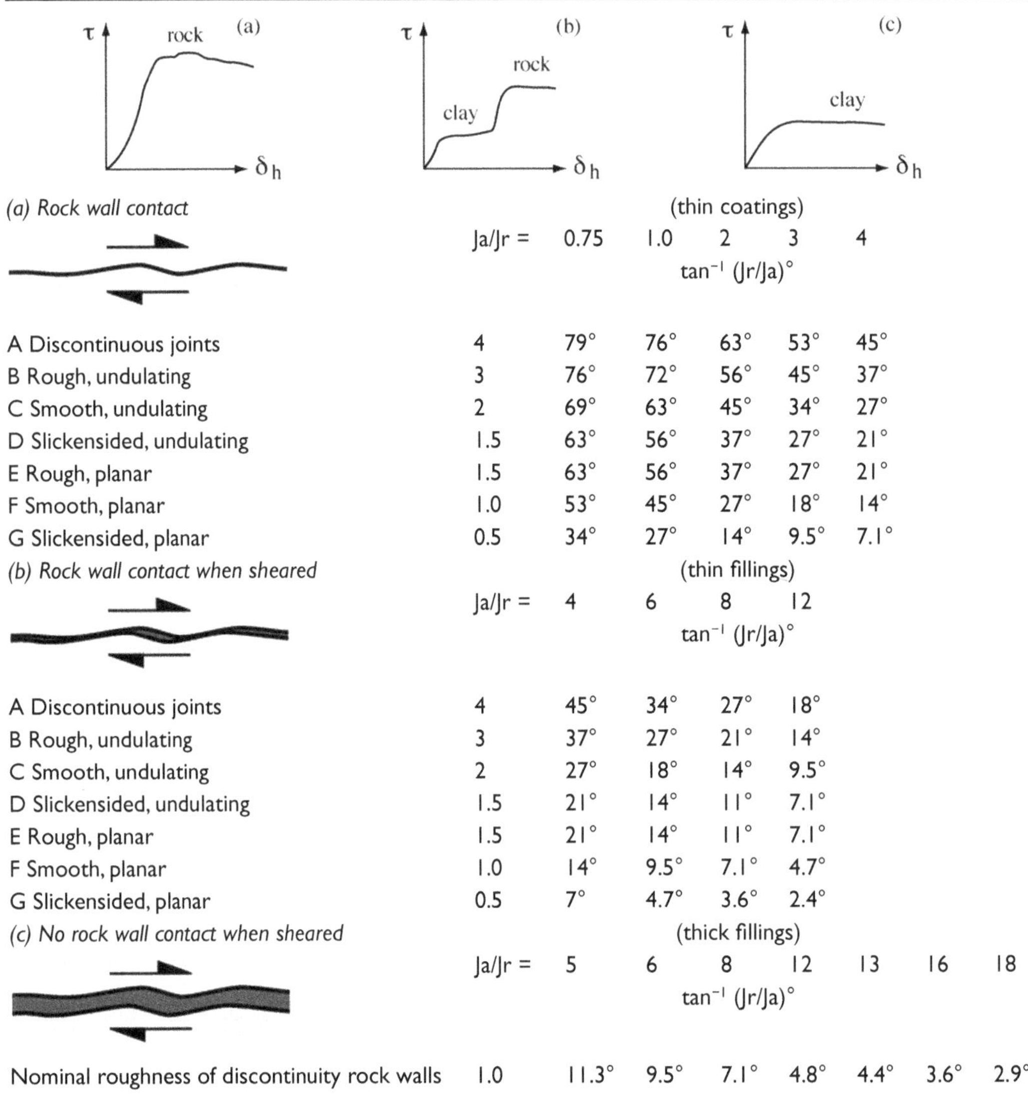

(a) Rock wall contact (thin coatings)

	Ja/Jr =	0.75	1.0	2	3	4
		$\tan^{-1}$ (Jr/Ja)°				
A Discontinuous joints	4	79°	76°	63°	53°	45°
B Rough, undulating	3	76°	72°	56°	45°	37°
C Smooth, undulating	2	69°	63°	45°	34°	27°
D Slickensided, undulating	1.5	63°	56°	37°	27°	21°
E Rough, planar	1.5	63°	56°	37°	27°	21°
F Smooth, planar	1.0	53°	45°	27°	18°	14°
G Slickensided, planar	0.5	34°	27°	14°	9.5°	7.1°

(b) Rock wall contact when sheared (thin fillings)

	Ja/Jr =	4	6	8	12
		$\tan^{-1}$ (Jr/Ja)°			
A Discontinuous joints	4	45°	34°	27°	18°
B Rough, undulating	3	37°	27°	21°	14°
C Smooth, undulating	2	27°	18°	14°	9.5°
D Slickensided, undulating	1.5	21°	14°	11°	7.1°
E Rough, planar	1.5	21°	14°	11°	7.1°
F Smooth, planar	1.0	14°	9.5°	7.1°	4.7°
G Slickensided, planar	0.5	7°	4.7°	3.6°	2.4°

(c) No rock wall contact when sheared (thick fillings)

	Ja/Jr =	5	6	8	12	13	16	18
		$\tan^{-1}$ (Jr/Ja)°						
Nominal roughness of discontinuity rock walls	1.0	11.3°	9.5°	7.1°	4.8°	4.4°	3.6°	2.9°

and friction, i.e. they are equal to $\tan^{-1}(\tau/\sigma)°$. The table from Barton (2002) is reproduced as Table 2.28 in this book. It is slightly different from the preceding one regarding some of the entries and facilitates its use through the inclusion of representative geometries and shear stress/displacement diagrams. An important note made by Barton (Barton et al., 1974a, 1974b and Barton, 2002) is that this possible use of J_r/J_a was discovered after the original development of the ratings. This (i.e. the concept of a frictional component was further generalized in Barton, 2002) will be discussed in the context of the development of the Q system (see later).

Finally, J_w/SRF, which is called the active stress term, attempts to combine two stress related factors. However, particularly given the ambiguity in the definition of SRF, it is difficult to relate J_w/SRF to rock mass behavior in a general sense.

As mentioned above, the structure of the Q-system and the individual ratings remained the same since the 1974 publication (the paper Barton et al., 1974b and the underlying NGI report, Barton et al., 1974a) with the exception of SRF as just mentioned. What has changed, however, is the relation to support design:

Barton et al. (1974a, 1974b) used some of the case records, in which support pressures were measured or where the design support pressure was known, to develop an empirical relation between support pressure and Q:

$$P_r = \frac{2.0}{J_r Q^{\frac{1}{3}}} \text{ in kg/cm}^2 \tag{2.40}$$

In Barton (2002) the relation for support pressure is given as $P_r = \frac{J_r}{20Q^{\frac{1}{3}}} \text{ in MPa}$. However, the authors of this book think that this is incorrect since lower J_r should lead to higher support pressures. The support pressure corresponds to the rock load as used by Terzaghi but in terms of stresses (rather than load) per linear dimension of the tunnel. In Barton et al. (1974a, 1974b), a somewhat extended support pressure relation is offered:

$$P_r = \frac{2\sqrt{J_n}}{30 J_r Q^{\frac{1}{3}}} \text{ in MPa} \tag{2.41}$$

and, in the same publication, it is suggested to compute wall support pressure, in MPa, based on the following modified Q values: 5 Q for Q > 10; 2.5 Q for 0.1 < Q < 10; and Q for Q < 0.1.

These support pressure relations are based on plotting case records against Q, as shown in Figure 2.66. However, only cases for which support pressures were measured or the design

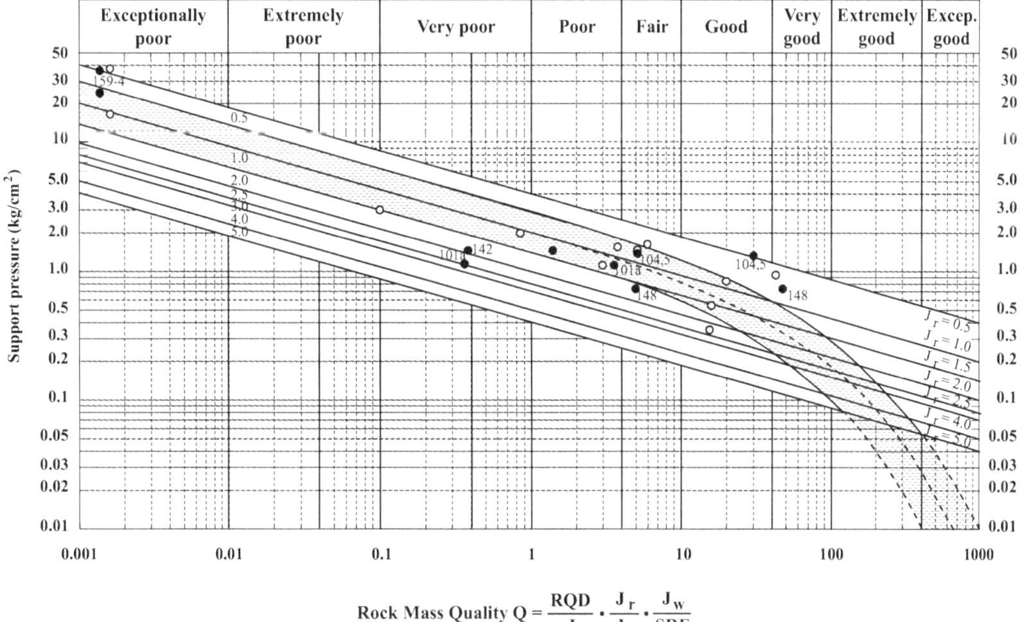

Figure 2.66 Q-support pressure relation from Barton et al. (1974a, 1974b).

pressures were available were considered. In parallel, all case records were recorded in detail, which led to a chart in form of Figure 2.67, where the numbered blocks represent 38 categories, for which the recommended supports were listed in the examples in Tables 2.29a and b (a for good and b for worst conditions). It is important to note that several cases represented by Barton et al. (1974a, 1974b) plot below the lower boundary since they are unsupported. Two other comments from the original publication Barton et al. (1974a, 1974b) are also important: First, it is clearly stated that the majority of the cases are from Europe and, in particular, 90 cases from Scandinavia, which explain the preponderance of shotcrete and bolt

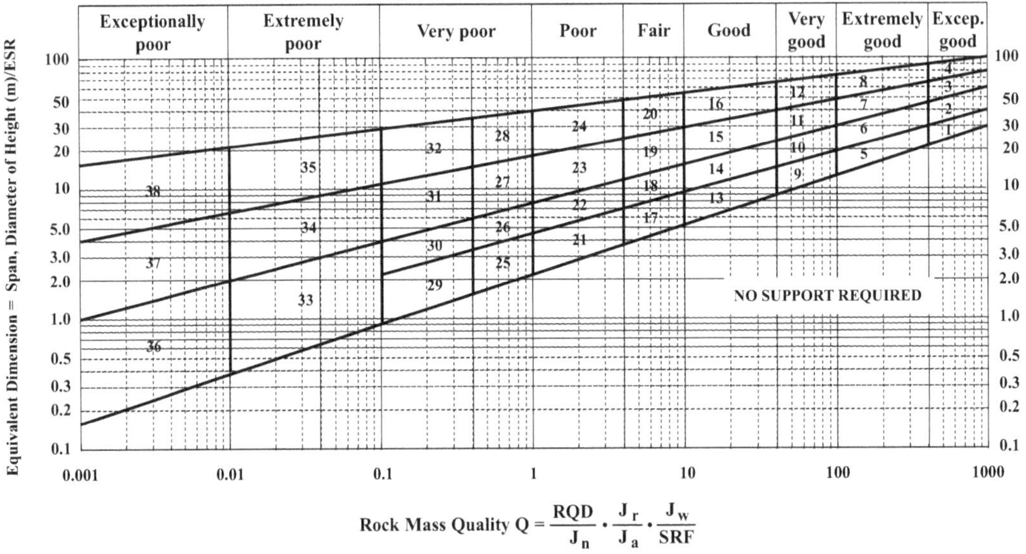

Figure 2.67a Support categories in relation to Q and equivalent dimension (Barton et al., 1974a, 1974b).

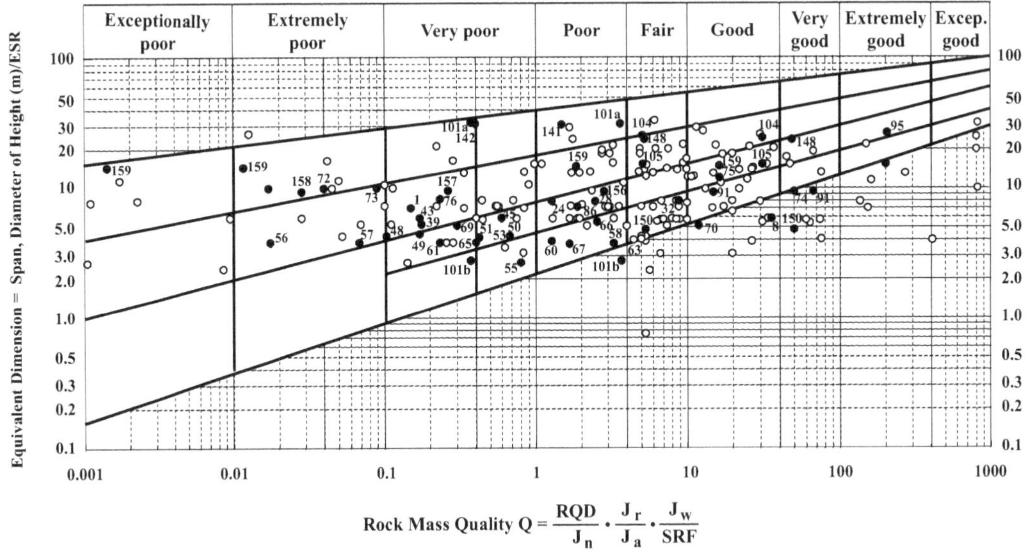

Figure 2.67b Relation of Q and equivalent support dimension showing the original cases (Barton et al., 1974a, 1974b).

Table 2.29a Example support category descriptions for "Good Conditions", (Barton et al., 1974). For "Notes", see Table 2.29c

Support category	*Q*	*Conditional factors* *RQD/J_n*	*J_r/J_n*	*SPAN/ESR (m)*	*P kg/cm²* *(approx.)*	*SPAN/ESR (m)*	*Type of support*	*Note See Table 2.29c*
1*	1000–400	—	—	—	<0.01	20–40	sb (utg)	—
2*	1000–400	—	—	—	<0.01	30–60	sb (utg)	—
3*	1000–400	—	—	—	<0.01	46–80	sb (utg)	—
4*	1000–400	—	—	—	<0.01	65–100	sb (utg)	—
5*	400–100	—	—	—	0.05	12–30	sb (utg)	—
6*	400–100	—	—	—	0.05	19–45	sb (utg)	—
7*	400–100	—	—	—	0.05	30–65	sb (utg)	—
8*	400–100	—	—	—	0.05	48–88	sb (utg)	—
9	100–40	≥20	—	—	0.25	8.5–19	sb (utg)	—
		<20	—	—			B (utg) 2.5–3 m	—
10	100–40	≥30	—	—	0.25	14–30	B (utg) 2–3 m	—
		<30	—	—			B (utg) 1.5–2 m + clm	—
11*	100–40	≥30	—	—	0.25	23–48	B (tg) 2–3 m	—
		<30	—	—			B (tg) 1.5–2 m + clm	—
12*	100–40	≥30	—	—	0.25	40–72	B (tg) 2–3 m	—
		<30	—	—			B (tg) 1.5–2 m + clm	—
13	40–10	≥10	≥1.5	—	0.5	5–14	sb (utg)	I
		≥10	<1.5	—			B (utg) 1.5–2 m	I
		<10	≥1.5	—			B (utg) 1.5–2 m	I
		<10	<1.5	—			B (utg) 1.5–2 m + S 2–3 cm	I
14	40–10	≥10	—	≥15	0.5	9–23	B (tg) 1.5–2 m + clm	I, II
		<10	—	≥15			B (tg) 1.5–2 m +S (mr) 5–10 cm	I, II
		—	—	<15			B (utg) 1.5–2 m + clm	I, III
15	40–10	>10	—	—	0.5	15–40	B (tg) 1.5–2 m + clm	I, II, IV
		≤10	—	—			B (tg) 1.5–2 m + S (mr) 5–10 cm	I,II, IV
16* See Note XII	40–10	>15	—	—	0.5	30–65	B (tg) 1.5–2 m + clm	I,V,VI
		≤15	—	—			B (tg) 1.5–2 m + S (mr) 10–15 cm	I,V,VI

*Authors' estimates of support. Insufficient case records available for reliable estimation of support requirements.
The type of support to be used in categories 1 to 8 will depend on the blasting technique. Smooth wall blasting and thorough barring-down may remove the need for support. Rough-wall blasting may result in the need for single applications of shotcrete, especially where the excavation height is > 25 m. Future case records should differentiate categories 1 to 8.

Key to Support Tables:
sb = spot bolting
B = systematic bolting
(utg) = untensioned, grouted

Table 2.29b Example support category descriptions for "Worst Conditions", (Barton et al., 1974). For "Notes", see Table 2.29c

Support category	*Q*	*Conditional factors* RQD/J_n	J_r/J_n	*SPAN/ESR (m)*	*P kg/cm² (approx.)*	*SPAN/ESR (m)*	*Type of support*	*Note See Table 2.29c*
33*	0.1–0.01	≥2	—	—	6	1.0–3.9	B (tg) 1 m + S (mr) 2.5–5 cm	IX
		<2	—	—			S (mr) 5–10 cm	IX
		—	—	—			S(mr) 7.5–15 cm	VIII, X
34	0.1–0.01	≥2	≥0.25	—	6	2.0–11	B (tg) 1 m +S (mr) 5–7.5 cm	IX
		<2	≥0.25	—			S (mr) 7.5–15 cm	IX
		—	<0.25	—			S (mr) 15–25 cm	IX
		—	—	—			CCA (sr) 20–60 cm + B (tg) 1 m	VIII, X XI
35 See note XII	0.1–0.01	—	—	≥15 m	6	6.5–28	B (tg) 1 m + S (mr) 30–100 cm	II, IX, XI
		—	—	≥15 m			CCA (sr) 60–200 cm + B (tg) 1 m	VIII, X, XI, II
		—	—	<15 m			B (tg) 1m + S (mr) 20–75 cm	IX, XI, III
		—	—	<15 m			CCA (sr) 40–150 cm + B (tg) 1 m	VIII, X, XI, III
36*	0.01–0.001	—	—	—	12	1.0–2.0	S (mr) 10–20 cm	IX
		—	—	—			S (mr) 10–20 cm + B (tg) 0.5–1.0 m	VIII, X, XI
37	0.01–0.001	—	—	—	12	1.0–6.5	S (mr) 20–60 cm	IX
		—	—	—			S (mr) 20–60 cm + B (tg) 0.5–1.0 m	VIII, X, XI
38 See note XIII	0.01–0.001	—	—	≥10 m	12	4.0–20	CCA (sr) 100–300 cm	IX
		—	—	≥10 m			CCA (sr) 100–300 cm + B (tg) 1 m	VIII, X, II, XI
		—	—	<10 m			S (mr) 70–200 cm	IX
		—	—	<10 m			S (mr) 70–200 cm + B (tg) 1 m	VIII, X, III, XI

*Authors' estimates of support. Insufficient case records available for confident prediction of support requirements.

Key to Support Tables:
(tg) = tensioned, (expanding shell type for competent rock masses, grouted post-tensioned in very poor quality rock masses; see Note XI)
S = shotcrete
(mr) = mesh reinforced
clm = chain link mesh
CCA = cast concrete arch
(sr) = steel reinforced

supports. Second, Barton et al. (1974a, 1974b) emphasize that the main difference between their method and Terzaghi's is the fact that the Q value does not depend on the size of the openings, while Terzaghi's (1946) classes do. The effect of opening size enters the system through charts such as those in Figure 2.67 (originally) and 2.65 (now).

In Barton et al. (1975), six base cases that actually failed were investigated by comparing the design "Q_D" with the actually encountered "Q"; the ratios of Q_D/Q varied between

Table 2.29c Supplementary notes to Tables 2.29a, b (Barton et al., 1974)

I. For cases of heavy rock bursting or "popping", tensioned bolts with enlarged bearing plates often used, with spacing of about 1 m (occasionally down to 0.8 m). Final support when "popping" activity ceases.
II. Several bolt lengths often used in same excavation, i.e. 3, 5, and 7 m.
III. Several bolt lengths often used in same excavation, i.e. 2, 3, and 4 m.
IV. Tensioned cable anchors often used to supplement bolt support pressures. Typical spacing 2–4 m.
V. Several bolt lengths often used in some excavations, i.e. 6, 8 and 10 m.
VI. Tensioned cable anchors often used to supplement bolt support pressures. Typical spacing 4–6 m.
VII. Several of the older generation power stations in this category employ systematic or spot bolting with areas of chain link mesh, and a free span concrete arch roof (25–40 cm) as permanent support.
VIII. Cases involving swelling, for instance montmorillonite clay (with access of water). Room for expansion behind the support is used in cases of heavy swelling. Drainage measures are used where possible.
IX. Cases not involving swelling clay or squeezing rock.
X. Cases involving squeezing rock. Heavy rigid support is generally used as permanent support.
XI. According to the authors' experience, in cases of swelling or squeezing, the temporary support required before concrete (or shotcrete) arches are formed may consist of bolting (tensioned shell-expansion type) if the value of RQD/Jn is sufficiently high (i.e. > 1.5), possibly combined with shotcrete. If the rock mass is very heavily jointed or crushed (i.e. RQD/Jn < 1.5, for example a "sugar cube" shear zone in quartzite), then the temporary support may consist of up to several applications of shotcrete. Systematic bolting (tensioned) may be added after casting the concrete (or shotcrete) arch to reduce the uneven loading on the concrete, but it may not be effective when RQD/Jn < 1.5, or when a lot of clay is present, unless the bolts are grouted before tensioning. A sufficient length of anchored bolt might also be obtained using quick setting resin anchors in these extremely poor quality rock masses. Serious occurrences of swelling and/or squeezing rock may require that the concrete arches are taken right up to the face, possibly using a shield as temporary shuttering. Temporary support of the working face may also be required in these cases.
XII. For reasons of safety, the multiple drift method will often be needed during excavation and supporting of roof arch. Categories 16, 20, 24, 28, 32, 35 (SPAN/ESR > 15 m only).
XIII. Multiple drift method usually needed during excavation and support of arch, walls and floor in cases of heavy squeezing. Category 38 (SPAN/ESR > 10 m only).

13 and more than 80 (note that Barton et al., 1980, use Q_o which is replaced here by Q_D to avoid confusion with an earlier Q_o). Since there are only six cases, of which four did not even include design support pressures, the relevance of this otherwise interesting investigation is questionable. More important in that the paper is a brief discussion of the use of the Q system for temporary (initial?) support; it is recommended to increase ESR to 1.5 ESR and analogously both the roof and wall – Q's by 5.

With Grimstad and Barton (1988) (preceded by Grimstad et al., 1986 in Norwegian) through Barton (1991) – World Tunneling – and then in Grimstad and Barton (1993) several major changes occurred. Specifically, the support classification was modified from the support category chart in Figure 2.67 to a chart (Figure 2.68) that essentially was similar to the present one shown in Figure 2.65. The chart in Figure 2.68 was slightly modified in Barton (1991) and then led to the chart in Figure 2.69 (Grimstad and Barton, 1993), which is similar to but with a few differences in the category boundaries compared to Figure 2.65. The new format simplifies the support selection. In addition, reflected in the comments in Figures 2.69 and 2.65, widespread use of fiber reinforced shotcrete led to its inclusion in the charts. One support type listed in Figures 2.65 and 2.69 "reinforced rib of shotcrete" needs some explanation, which is given through Figure 2.70 (from Grimstad and Barton, 1993). Related to the use of fiber reinforced shotcrete is the plot shown in Figure 2.71.

Also, a format for systematic recording of the Q system parameters was introduced in Barton (1991); see Figure 2.72. Another interesting piece of information in this process is a plot relating Q and "driving rate" considering different support types as shown in

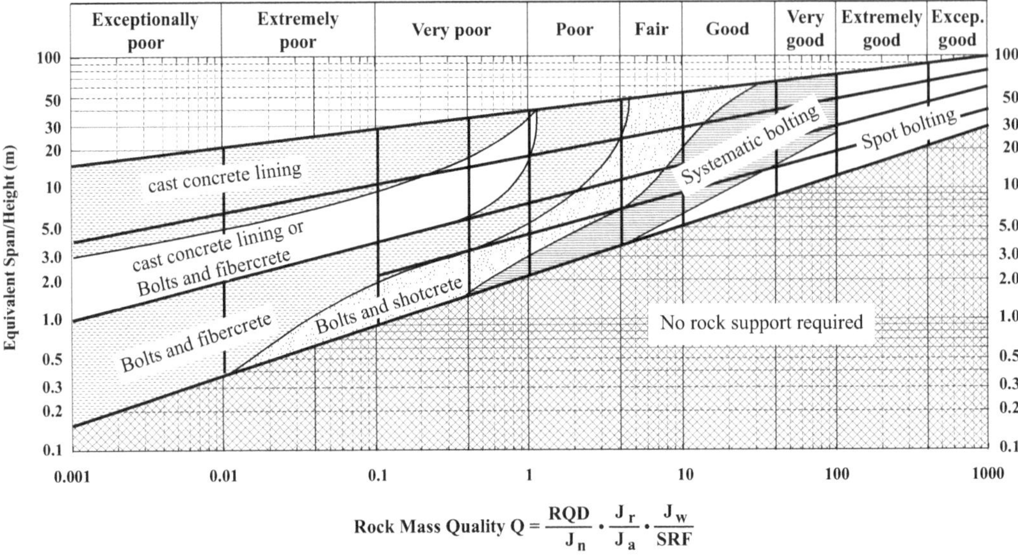

Figure 2.68 Simplified diagram for design of rock support based on the Q-system (Grimstad and Barton, 1988).

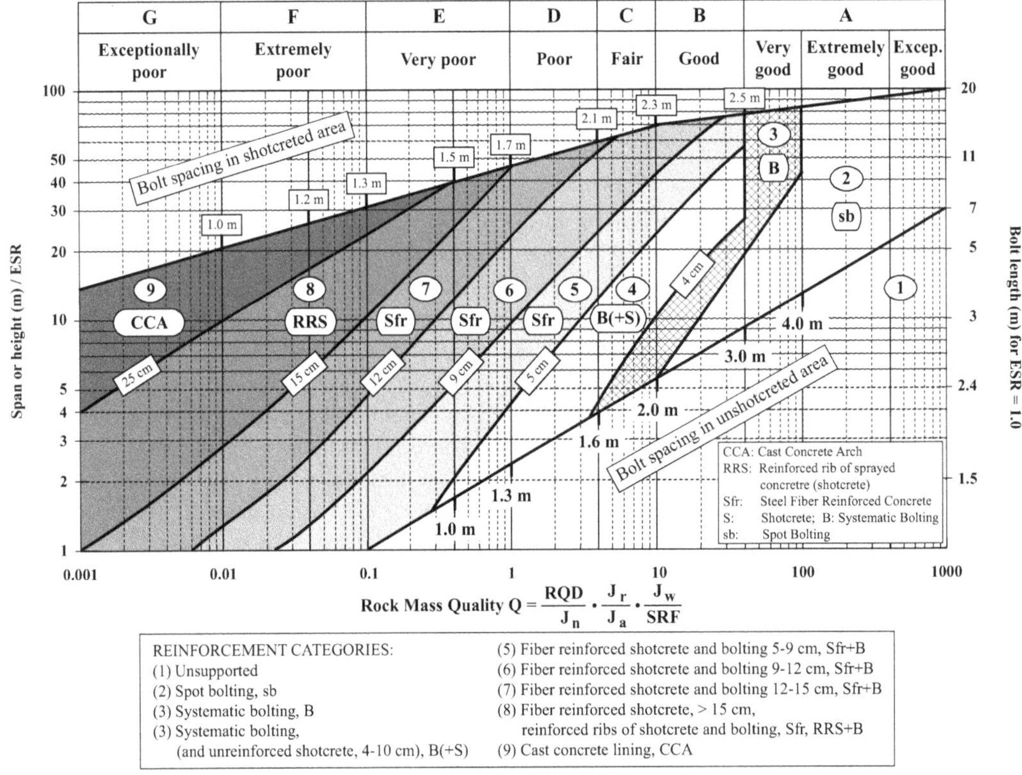

Figure 2.69 Rock mass classification – permanent support recommendation based on Q and NMT (Norwegian Method of Tunneling) after Figure 1 in Grimstad and Barton (1993). Note the differences in shotcrete thicknesses and support categories between this figure and Figure 2.65. (Reproduced by permission of Elsevier.)

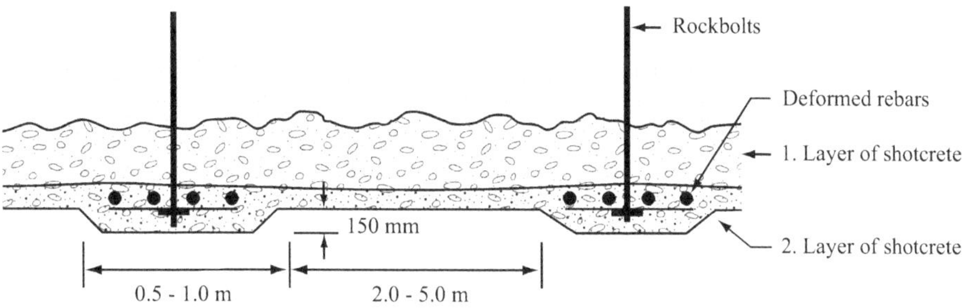

Figure 2.70 Reinforced Ribs of Sprayed Concrete (RRS) (Grimstad and Barton, 1993).

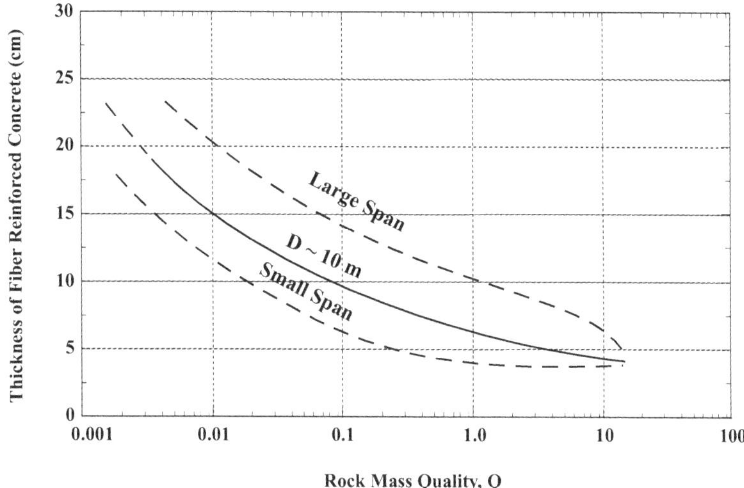

Figure 2.71 Typical thickness of fiber reinforced shotcrete used at Norwegian tunnel sites in the late 1980s (Grimstad and Barton, 1993).

Figure 2.73. This was modified (generalized in Grimstad and Barton, 1993) and is shown in Figure 2.74.

Simultaneously with the above-mentioned development of the Q-system per se, an extension of its use took place by relating Q to seismic velocity V_p. The earliest work on this was done in Barton (1991), which was then somewhat modified and led to the present relation:

$$V_p = 3.5 + \log Q_c (\text{in km/sec}) \tag{2.42}$$

Q_c was introduced to consider also weaker rocks which were not part of the evaluation of the original (1991) relation. It is defined as $Q_c = Q\ \sigma_c/100$. As a matter of fact, it is best to show Figure 2.75 from Barton (1995) and Barton (2002), which relates V_p not only to Q but also to the depth and inherent porosity. It also shows the corresponding static modulus of deformation E_{mass}. The latter was obtained using relations developed by Bieniawski and Serafim and Pereira (see earlier) and the relation between RMR and Q:

$$\text{RMR} = 15 \log Q + 50 \tag{2.43}$$

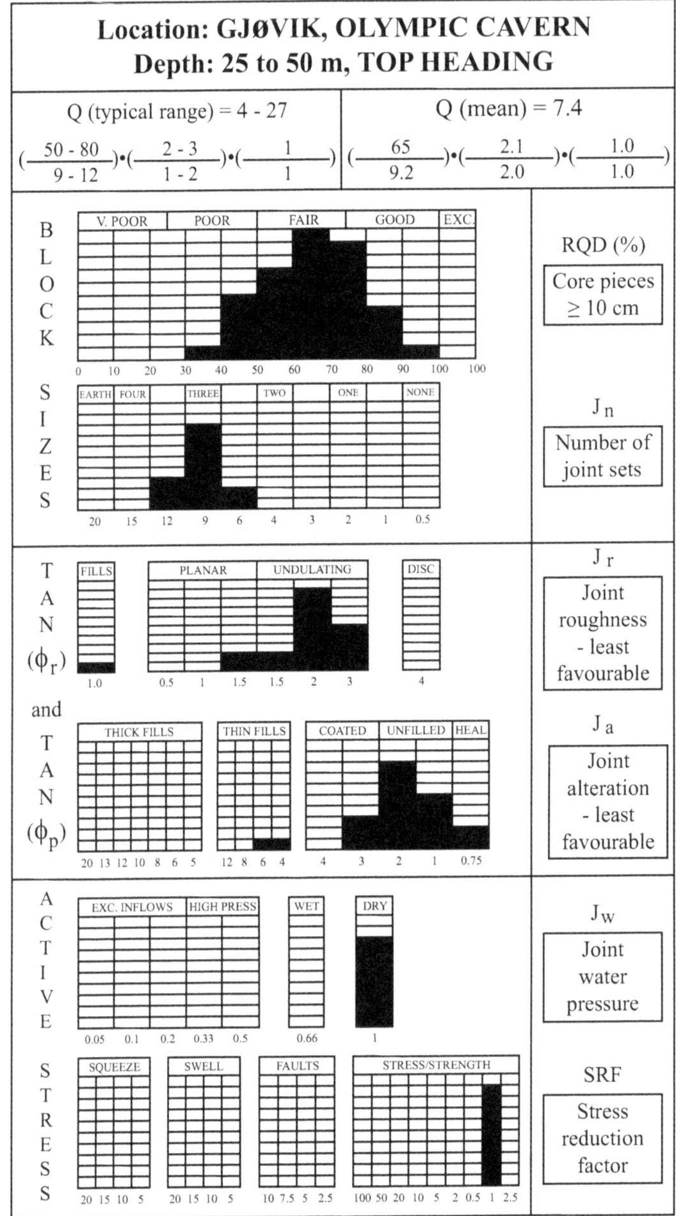

Figure 2.72 Systematic recording of Q-system data in the pilot heading of the 60 m span Olympic Ice-Hockey cavern (Løset and Bhasin, 1991 (from Barton, 1991).

Which then leads to

$$E_{mass} = 10Q_c^{1/3} \tag{2.44}$$

Barton (2002) discusses this more extensively also with other relations but mentions that the relation (2.44) is conservative.

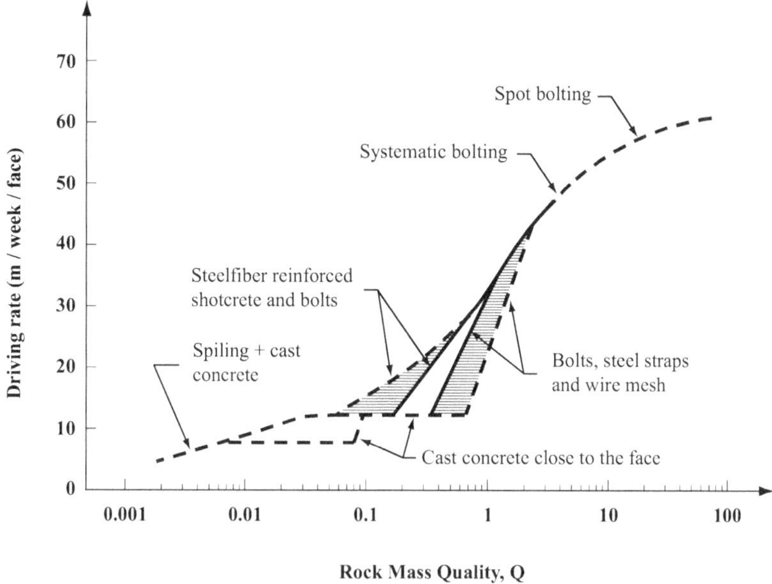

Figure 2.73 Typical driving rate including completed temporary support for 70 m^2 and 90 m^2 (10 m to 14 m span) Road tunnels, based on Grimstad, 1981 (from Barton, 1991).

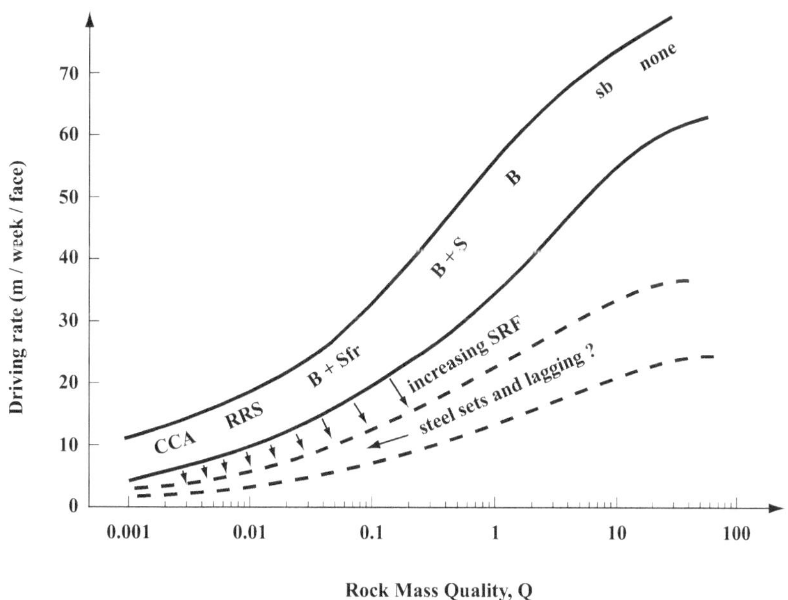

Figure 2.74 Approximate tunnel advance rates in meters per week per face for 70 to 90m^2 tunnels driven by NMT (modified from Grimstad, 1981). Curves for steel sets and lagging are tentative (Grimstad and Barton, 1993).

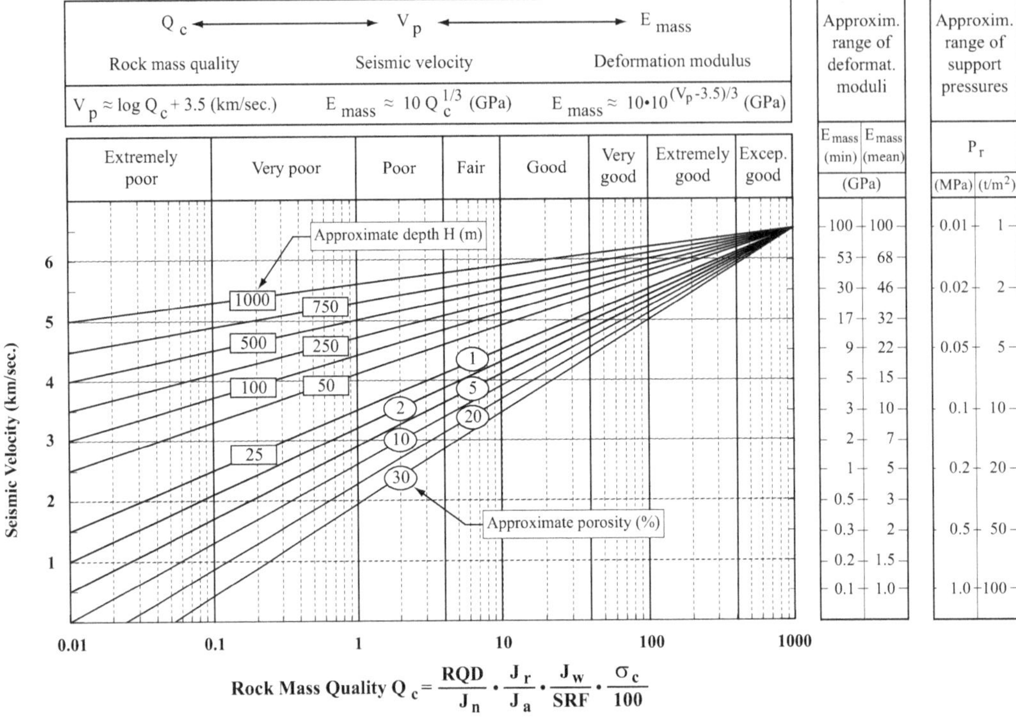

Figure 2.75 Rock mass quality, seismic velocity, and deformation modulus correlations for design (from Barton, 2002). (Reproduced by permission of Elsevier.)

As mentioned in the discussion of rating ratios, the ratio J_r/J_a can be used to represent the "overall frictional resistance," i.e. a combination of friction and cohesion. Barton (2002) refined this by defining a frictional component FC and a cohesive component CC as:

$$FC = \tan^{-1}\left(\frac{J_r}{J_a} J_w\right)$$
$$CC = \frac{RQD}{J_n} \frac{1}{SRF} \frac{\sigma_c}{100} \tag{2.45}$$

The inclusion of J_w roughly represents the softening effect of water or effective stresses. These two components are discussed here for reasons of completeness. It is not quite clear how these components should be used since Barton (2002) warns to directly associate them with the "c" and "ϕ" of continuum modeling. They may give an idea on the stress dependent and the stress independent components of shearing resistance. However, the latter is somewhat questionable given that SRF implicitly includes the stress level. If shearing resistance is derived from Q, it may be best to stay with the original "overall friction angle" = $\tan^{-1}(J_r/J_a)$.

In addition to the relations described, the Q value has also been correlated with the Lugeon test. The Lugeon value is an empirical expression of permeability that describes the water flow in liters/min from a 5 m packered-off section under 10 atmospheres of pressure. Barton (2001 and 2002) proposed the following relation:

$$L = \frac{1}{Q_c} \tag{2.46}$$

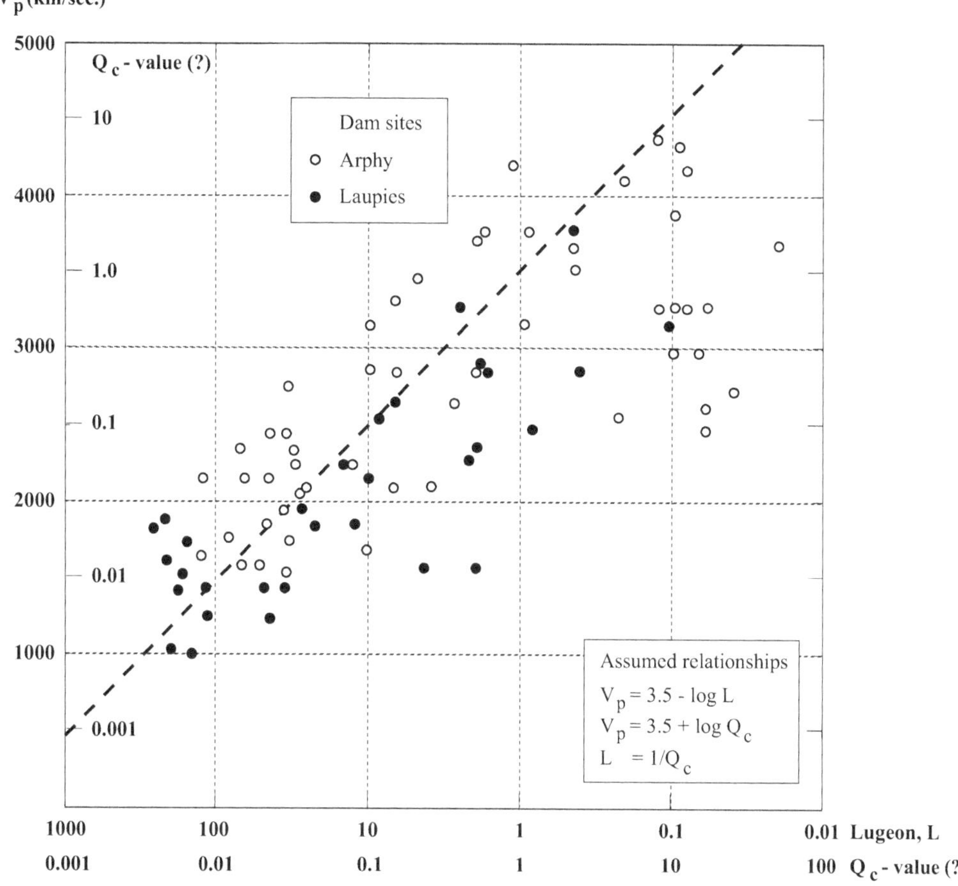

Figure 2.76 Dam site comparison of Lugeon values with P-wave velocities. Tentative Q_c – L correlation which will strictly correlate only to the nominal, near-Surface (25 m depth) sets of measurements. Low Lugeon values with low velocities may correspond to reduced injection pressures, therefore, reducing joint deformation effects (from Barton, 2002).

This relation was apparently developed from data from two French dam sites where originally the seismic relations were related to Lugeon values. Barton then used the V_p – Q_c relation to develop the Q_c – L relation (see Figure 2.76).

Finally, when using the Q-system in high stress environments, Peck (2000) suggested to use:

$$SRF = 34\left(\frac{\sigma_c}{\sigma_1}\right)^{-1.2} \tag{2.47}$$

Discussion

Similar to but possibly somewhat more than in other empirical rock mass classification systems, the Q-system is and has been subject to critique including both positive and negative comments. Probably the most thorough study is that by Palmström and Broch (2006) from which a number of points suggesting caution when using the Q-system are mentioned (paraphrased) here with additional comments (in cursive) by the authors of this book:

i. RQD is not an entirely reliable representation of fracturing. *While this is correct, and while the comments by Palmström and Broch regarding the effect of sampling line orientation on (Q) are also basically correct, these limitations apply to all fracture spacing measurements and not only to RQD.*
ii. J_n may not represent fracturing since it concentrates on joints sets.
iii. RQD/J_n can only, to a limited extent, represent block volume as shown by Palmström et al. (2002).
iv. The interblock shearing strength J_r/J_a usually works well but it might be difficult to determine how much shear movement has taken place since it affects this ratio.
v. J_w: This represents either inflow or water pressure. High values of either factor have detrimental effects on tunnel construction. However, it is not clear how high inflow affects tunnel support (except for installation problems).
vi. SRF. *Not surprisingly, this is the most controversial parameter.* As mentioned earlier, this parameter attempts to accommodate underlying factors which may (*or may not*) have similar effects on tunnel support but are physically quite different. In particular, categories (a) and (b) in Table 2.26f appear to represent the opposite end of rock conditions. Palmström and Broch (2006) mention a number of specific problems. The length (size) of the weakness zone is not included. It is questionable why in situ stress is only influenced by weakness zones but not the other parameters (RQD, J_n, J_r, J_a)? Palmström and Broch also question the limit where squeezing may occur; that is for H > 350 $Q^{1/3}$; see Equation (2.37) (e.g. since one needs first to know the Q and thus SRF before being able to decide if squeezing occurs or not, while this fact in turn needs to be known to determine SRF's. Nowhere is it said if this iterative process should actually be used.)

   While the comments above mainly relate to the underlying parameters, Palmström and Broch also make some general comments about the Q-system:
vii. Features not included:
   Orientation (*see earlier comments, including Barton's et al.'s comments).*
   Joint Size and Persistence *(these two characteristics are related to each other, they are difficult to measure and they are not included in other empirical classification systems either).*
   Joint Aperture *(one can argue that this is an effect included in* J_r, J_a, J_w).
   Rock Strength (as Palmström and Broch say, this is now considered through Q_c).
viii. General comments of caution and regarding applications. Figure 2.77 shows actual bolt spacing versus Q, which was originally published in Barton-Grimstad and to which Palmström and Broch added a line to project the support recommendations from the Q-support chart. *Quite clearly, there is substantial scatter. In the view of the authors of this book, one can argue that the support level (e.g. Figure 2.65) does not provide a single recommendation but a range thus reflecting the scatter of Figure 2.77.*
ix. Finally, Palmström and Broch argue that, practically speaking, the Q method "has its best application in jointed rock masses where instability is caused by rock falls. For most other types of ground behavior in tunnels, the Q system, like other empirical (classification) methods, has limitations." See Figure 2.78. The conclusion is based on the experience of the authors, as is said in Palmström and Stille (2007), where a table (reproduced here as Table 2.30) rates a variety of empirical methods with regard to different rock mass conditions.

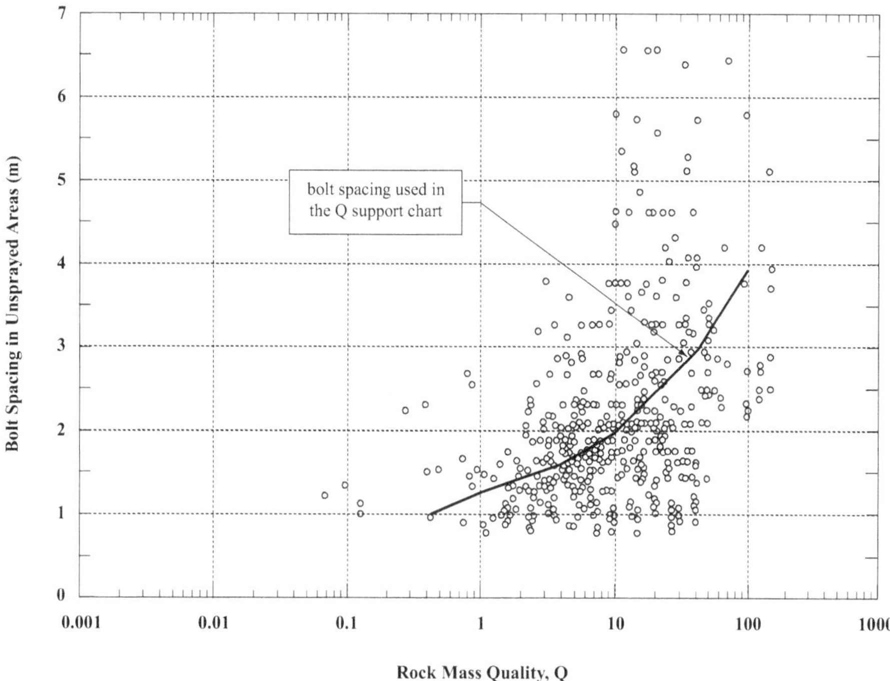

Figure 2.77 Bolt Spacing related to Q-value in unsprayed areas (modified from Grimstad and Barton, 1993). The line indicates the bolt spacing used in the Q-Support Chart (from Palmström and Broch, 2006). (Reproduced by permission of Elsevier.)

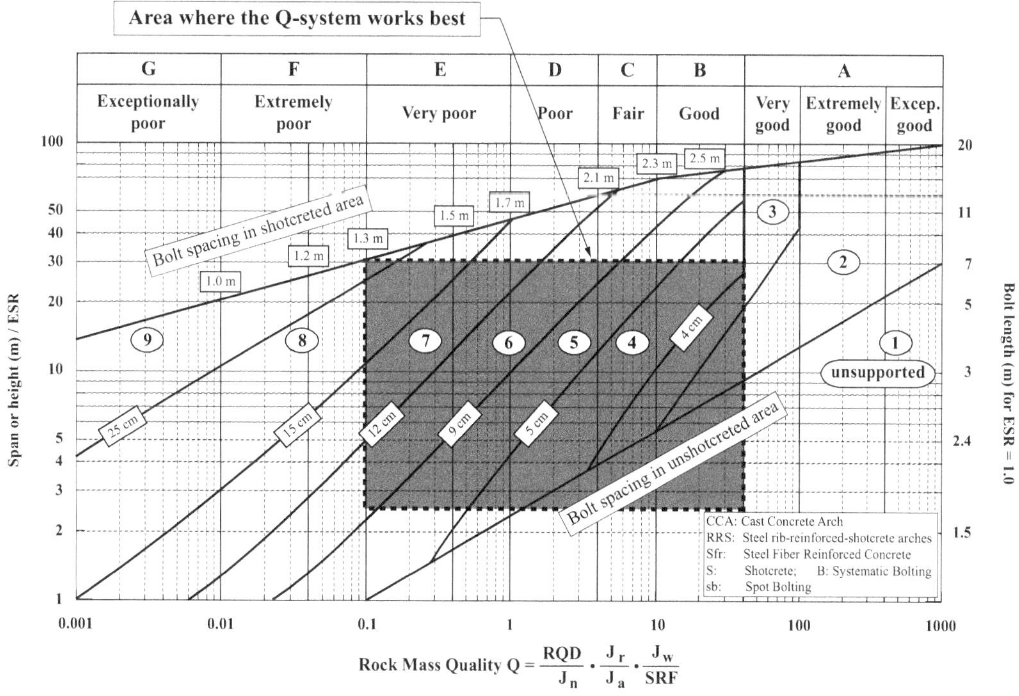

Figure 2.78 Limitations in the Q rock support diagram. Outside the unshaded area, supplementary methods/evaluations/calculations should be applied (from Palmström and Broch, 2006). (Reproduced by permission of Elsevier.)

Table 2.30 The fitness of various engineering design tools (from Palmstrom and Stille, 2007)

Ground behavior	Rock engineering and design tools								Triggering factor
	Classification systems								
	Judgment	Q	RMi support	NATM	Numerical modeling (for continuous ground)	Analytical calculations	Observational methods	Engineering judgment	
a Stable	2	2	1–2	1	1	2	1	1	Gravity driven
b Fall of block(s) or fragment(s)	1–2	1–2	1–2	1–2	2	2	2	1	
c Cave-in	3	2–3	2	3	3	2	3	2	
d Running ground	4	4	4	4	4	4	3	2	
e Buckling	4	3	3	3	2	2	2	2	Stress induced
f Rupturing from stresses	4	3	3	2	2	3	2	2	
g Slabbing, spalling	4	2	2	2–3	2	2	2	2	
h Rock burst	4	3–4	2	3	2	2	1–2	2	
i Plastic behavior (initial)	4	3–4	3	2–3	2	2	3	2	
j Squeezing ground	4	3	3	1–2	2	2	2	3	
k Ravelling from shaking or friability	4	4	4	3	4	4	2	2	Water influenced
l Swelling ground	4	3	3	3	3	3	2	2	
m Flowing ground	4	4	4	3–4	4	4	3	3	
n Water ingress	4	4	4	4	3	2	2	3	

Note: Fitness rating of the various tools: 1, suitable; 2, fair; 3, poor; 4, not applicable.

2.3.6.5.2 The Q_{TBM}

Starting with Barton (1999) and then in Barton (2000), the Q_{TBM} was proposed, based on "145 case records for over 1,000 km of tunnels." Q_{TBM} combines parameters in the original Q system (RQD, J_n, J_r, J_a J_w, SRF) with new parameters affecting the penetration rate of tunnel boring machines:

$$Q_{TBM} = \frac{RQD_o}{J_n}\frac{J_r}{J_a}\frac{J_w}{SRF}\frac{SIGMA}{F^{10}/20^9}\frac{20}{CLI}\frac{q}{20}\frac{\sigma_\theta}{5} \tag{2.48}$$

RQD_o is RQD interpreted in the tunnel direction, F is the average cutter load in tons force, SIGMA is the rock mass strength estimate (SIGMA = $5\gamma\ Q_{co}^{1/3}$), CLI is the Cutter Life Index as developed by NTH (e.g. Johannessen and Askilsrud, 1993; a few further comments will be made later), q is Quartz content in %, and σ_θ is the induced biaxial stress at the tunnel face.

Q_{TBM} can then be used to obtain the penetration rate PR:

$$PR = 5Q_{TBM}^{\frac{1}{5}} \tag{2.49}$$

As is well known, the advance rate (AR) is lower than the penetration rate (PR) reflecting effects such as cutter changes, other maintenance and repair, as well as other aspects of tunnel construction indirectly affecting the machine advance (muck removal problems, supply and installation of support, etc.). Barton relates AR and PR as shown in Figure 2.79. The lines in this figure and thus the AR/PR relation can be reflected in the simple equation:

$$AR = PR\,T^{m} \tag{2.50}$$

where T is the time in hours for completion of a particular length L of tunnel.

The exponent m has a negative value and expresses the decreasing utilization with time (lines 1, 2, 3 in Figure 2.79 have values of m = −0.17, −0.19, and −0.21, respectively). The value of m expresses the above-mentioned mix of influence in the expression shown below:

$$m = m_1\left(\frac{D}{5}\right)^{0.2}\left(\frac{20}{CLI}\right)^{0.15}\left(\frac{q}{20}\right)^{0.16}\left(\frac{n}{2}\right)^{0.05} \tag{2.51}$$

where D is the tunnel diameter, n the porosity, while q, CLI are as before. m_1 is never specifically defined in Barton (2000) but it appears to mainly reflect rock mass quality as shown in Figure 2.80 (note that this figure relates "m" and "m_1" to Q). Rather than using the relatively complex relation for m, one may simply use Table 2.31. Barton (2000) explains in detail that he obtained these factors essentially through a process of fine tuning (trial and error) to arrive at these relations. This explains for instance why two factors reflecting abrasion, namely, CLI and quartz content appear. The Q_{TBM}, PR, and AR relations are also shown in a graph (Figure 2.81).

Before critiquing the method, a comment needs to be made about CLI. This parameter has been developed by a group at NTNU (formerly called NTH) to predict TBM performance. Specifically, the index is (after Bruland, 1998):

$$CLI = 13.84\left(\frac{SJ}{AVS}\right)^{0.3847} \tag{2.52}$$

AVS is obtained from an abrasion test in which the steel (same type of steel as in the cutter) is pressed with 10 kg force against a rotating wheel on which the relevant crushed rock is

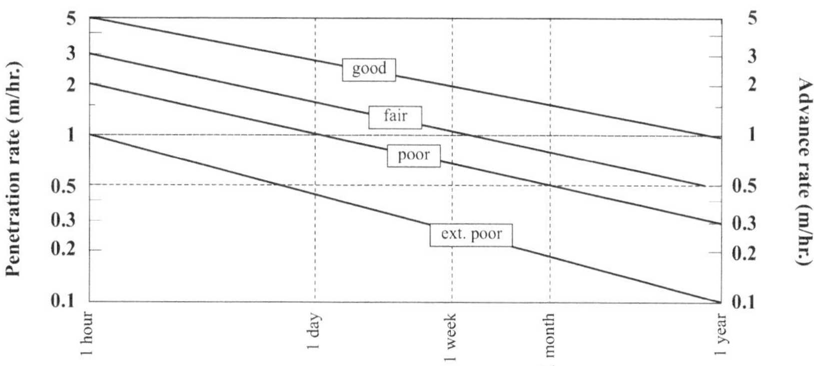

Figure 2.79 Declining average advance rate is seen as the unit of time (Day, Week, Month) and tunnel length increase, based on 145 TBM tunnels totaling >1000 km. (Barton, 2000).

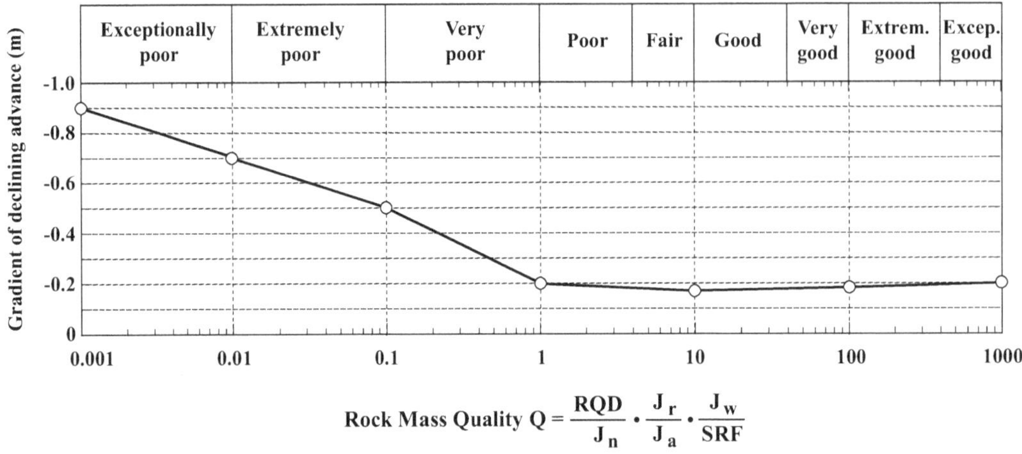

Figure 2.80 Preliminary estimate of declining advance rate gradient (-) m, as a function of Q-value (Barton, 2000).

Table 2.31 Advance rate, from Barton (1999)

(a) Deceleration gradient (-) m and its approximate relation to Q-value

Q	0.001	0.001	0.1	1	10	100	1000
m_1	~ −0.9	~ −0.7	~ −0.5	~ −0.22	~ −0.17	~ −0.19	~ −0.21

Unexpected events or expected bad ground. Many stability and support-related delays and gripper problems. Operator reduces PR This increases Q_{TBM}

Most variation of (−) m may be due to rock abrasiveness, i.e. cutter life index CLI, quartz content and porosity are important. PR depends on Q_{TBM}

Note: The subscript (1) is added to m for evaluation of Equation (2.51)

(b) Examples of declining advance rates for PR = 3 m/hr., m = (−) 0.2 Maximum hours (= real time) is assumed here

Period	PR	1 shift	1 day	1 week	1 month	3 month	1 year
Hours	1 hr.	10 hr.	24 hr.	168 hr.	720 hr.	2,160 hr.	8,760 hr.
U	1.00	0.63	0.53	0.36	0.27	0.22	0.16
AR (m/hr.)	3.0	1.9	1.6	1.1	0.8	0.6	0.5

(c) QTBM estimated from mean PR values, using Equation (2.49)

PR (m/hr.)	0.1	0.5	1.0	5	10
Q_{TBM}	$3.1\ 10^8$	10^5	3,125	1.0	0.03

placed. The steel loss in mg after 20 rotations is the AVS. SJ is obtained from Siever's miniature drill test. The miniature drill (8.5 mm diameter) is pressed against the rock with a 20 kg force and the SJ value is the drill depth after 200 rotations measured in 1/10 mm. Figure 2.82 from Bruland (1998) shows typical ranges of CLI values. Bruland states that 2000 samples were used to get the information shown in the figure.

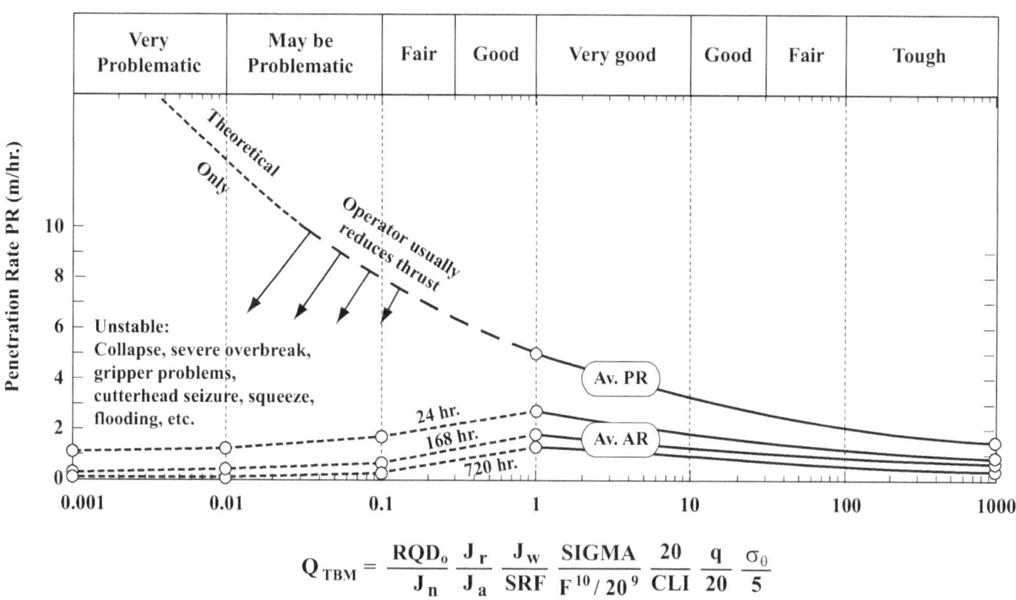

Figure 2.81 Suggested relation between PR, AR, and QTBM (Barton, 1999).

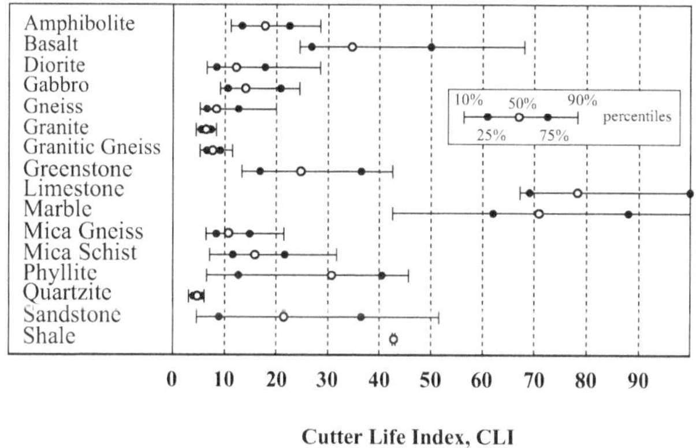

Figure 2.82 Typical CLI's for different rock types (Bruland, 1998).

Q_{TBM} has been strongly criticized by Blindheim (2005) to which Barton (2005) replied, and also criticized by Palmström and Broch (2006). One of the main points made by Palmström and Broch is reflected in Figure 2.83, from Sapigni et al. (2002), showing that AR plotted versus Q_{TBM} results in an extreme spread. Blindheim voices concerns regarding the ratios RQD_o/J_n and J_w/SRF, which are similar to concerns that have been mentioned earlier. Blindheim and Barton disagree regarding J_w, the former saying that water inflow generally reduces penetration rate while the latter states that at least for intermediate quantities, water may be helpful (cooling and positive effects of water pressure). Regarding the ratio, SIGMA/($F^{10}/20^9$), Sigma represents a double counting of Q and Blindheim questions if SIGMA can

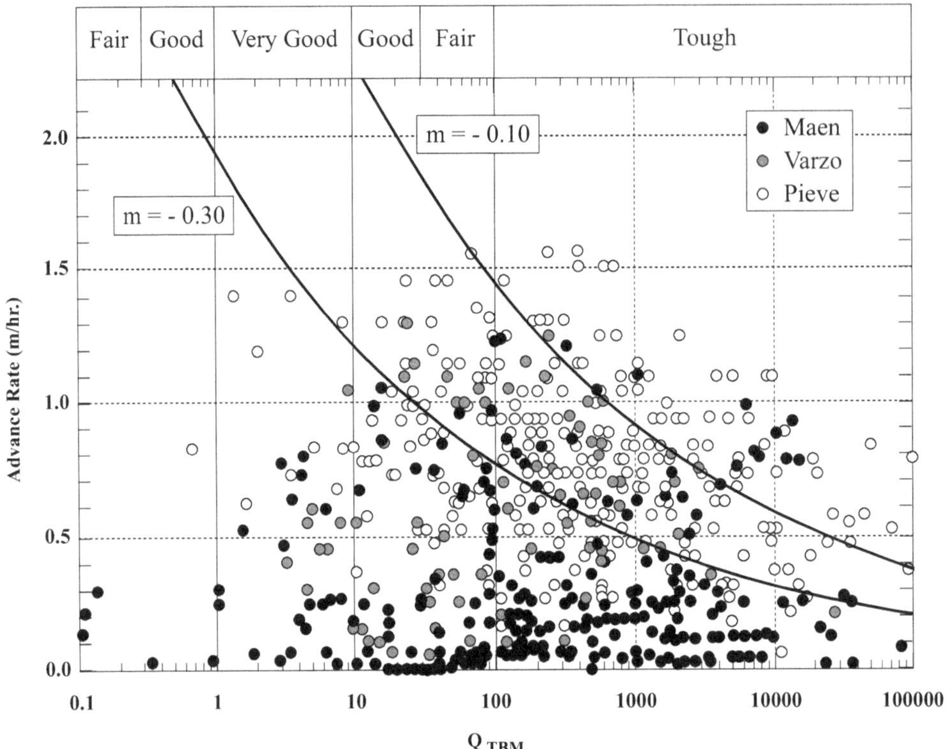

Figure 2.83 Advance rate versus QTBM. From Palmström and Broch (2006) (originally Sapigini et al., 2002).

represent anisotropy (as done by Barton). Also, the thrust is not the only machine parameter affecting PR (AR) but torque and machine type need to be considered. Barton justifies the use of SIGMA/($F^{10}/20^9$) by the fact that the NTH predictive method (Bruland, 1998) does not directly relate unconfined compression strength and cutter force, a statement that is only partially correct.

As already indicated above, both $\frac{20}{CLI}$ and $\frac{q}{20}$, in Equation (2.51) include some effects of abrasivity which, through its influence on cutter wear, will affect PR and through cutter changes will affect AR. Blindheim (2005) states that this is giving too much weight to abrasivity.

Finally, the factor in m or m_1 is questioned given that it is not clear how and why it relates to Q as applied in Figure 2.80, particularly given the fact that Barton (2000) implies that it is also affected by "level of site investigation, machine design, stability support during and tunnel management" (quote from Blindheim, 2005).

The authors of this book think that Q_{TBM} is an interesting example but that it may attempt to push a system intended for rock mass characterization in the context of support design too much in a direction for which it was not intended. This is particularly so since a number of methods for TBM penetration rate (advance rate) prediction and TBM wear exist, e.g. the NTH method (Bruland, 1998), the CSM method, etc. These methods consider the relevant factors such as compressive strength, jointing, abrasivity, more directly than Q_{TBM}. Particularly interesting is the fact that they use separate expressions for penetration rate and wear, i.e. they do the opposite of Q_{TBM} where "everything" is put into one expression.

2.4 EMPIRICAL METHODS – CONCLUSIONS

As discussed in Section 2.1, empirical methods can be classified (Figure 2.1) as using largely qualitative or quantitative descriptions of rock mass characteristics and then relating these characteristics, either directly or indirectly, to the design and possibly the construction characteristics of a tunnel. The indirect relations involve a transformation of rock mass characteristics either into rock loads, which are used in the structural analysis and design of tunnel supports, or into geotechnical parameters which can then be used to analyze and design the tunnel supports. In the direct relations, rock mass characteristics are directly related to support dimensions and other design and construction characteristics of a tunnel. The review in this chapter includes methods, which are not used much anymore but played an important role in the development of the methods that one uses today. Such methods of historical interest are the Terzaghi rock load method (qualitative – indirect) and the RSR-method (quantitative-direct). The reviewed methods which are most frequently used to date are:

1. Geomechanics Planning of Underground Construction Using Cyclic Advance or Continuous Advance; these are direct methods mostly used in conjunction with the application of the NATM.
2. The Hoek and Brown (GSI-based) Methods, a largely quantitative indirect method, producing geotechnical parameters.
3. RQD–rock support relation, a quantitative direct method which is used mostly in the United States.
4. The RMR (Geomechanics) and Q methods, which use largely quantitative descriptions of rock mass characteristics and relate them directly to tunnel support requirements (and some construction aspects).

When describing these methods in detail, attempts were made to point out where caution should be exercised when applying the particular method. These specific comments or "critiques" will not be repeated here but some general comments regarding the development and use of empirical methods will be made. This is best done by repeating the objectives that empirical methods should fulfill that were made in Section 2.1 and commenting on them (The objectives are printed in cursive type, the comments in normal type):

i. *Promote economical, yet safe, design.* Empirical methods by definition are based on preexisting cases, i.e. cases where the support materials and dimensions and geologic conditions have been observed but rarely where the detailed analysis underlying the design was available. By the fact that most of these cases have not failed, one can state that the designs are usually safe. On the other hand, since the degree of conservativism in the design is not known, the design may be overly conservative (not economical). It is important to note that this comment does not apply when geotechnical parameters are derived explicitly (Hoek and Brown) or implicitly (Geomechanics Planning) to design the tunnel. The problem is recognized in the other methods (RQD, RMR, Q), where extensive monitoring and updating to adapt to the particular ground conditions are recommended. The question arises if the users actually do so.
ii. *Generally applicable and robust* (insensitive to vagaries of use). Again, the issue of base cases enters. Given the derivation from base cases, the methods are strictly only applicable to geologic conditions, tunnel sizes, excavation methods and support types that were used to derive the methods. This is recognized to some extent when e.g. limiting size and stress ranges are mentioned but not all methods do so. One might say that the comments and suggestions on monitoring the performance and updating

the relations might compensate to some extent for the lack of stating limitations. In the opinion of the authors of this book, this is not sufficient – the application ranges should be explicitly stated! Regarding robustness, it appears that most methods are not very sensitive to being applied by different users. A number of cases, in which different users were involved, (e.g. Einstein, et al., 1983) attest to this. Nevertheless, it is fair to say that a larger number of parameters increase the possibility of differences.

iii. *Readily determinable parameters*. In principle, all the parameters are readily determinable if the rock mass if accessible. This means, however, that e.g. in cases of shielded TBM's, such information may be incomplete. Of greater concern is the use of the methods prior to construction, i.e. prior to exposing the rock mass. While intact rock properties and RQD are easily determined from borings, information on other rock mass properties may be more difficult to obtain. Much inference is then necessary. To be fair, one has to state that such limitations affect other methods (first principle based), similarly.

iv. *Subjectivity of the method*. The issue has been extensively discussed in the introductory Section 2.1 and one can repeat here that all empirical methods are essentially subjective be that in the assessment of parameters by the user, in the development of the underlying model, or both. As long as this is recognized, this is not at all a problem. As a matter of fact, it allows the user to bring his or her experience to bear when determining parameters. While subjectivity in parameter determination is easily recognized, this is not necessarily so regarding the subjectivity in the model development. Here, one relies on the subjective assessment of the model developer. This is, as just stated, acceptable, as long as it is recognized. Methods that require the user to formulate the relations such as the "Geomechanics Planning" force the user to explicitly formulate his/her subjective "model" rather than rely on that of the developer.

v. *Correctness of the method*. Again, and as stated in Section 2.1, this applies both to empirical and "first principle" approaches. In empirical methods, this is again related to the base cases and the often limited knowledge about the original design and analysis. Given that most cases involve tunnels that did not fail, one might say that something must be correct. This is, however, not entirely satisfactory. Actually, a failed case where one can backcalculate to some extent might be more informative. Continuous monitoring and associated backcalculating and case-specific updating may alleviate the lack of knowledge about the basic correctness.

In summary, the five requirements for satisfactory empirical methods are to a reasonable extent but not fully satisfied. The most important limitations are:

1. The fact that one does not know to what extent the resulting tunnel designs are overdesigned or just at the limit.
2. The range of applicability is strictly only correct for the range covered by the base cases.
3. Some methods evolved (changed) and continue to do so, reflecting the incorporation of additional information! (This has been mentioned in the specific reviews in Section 2.3, wherever applicable.)

While it may be difficult to provide more information to reduce limitation No. 1, it should be possible to do a better job regarding limitation No. 2. For limitation No. 3, this requires users to state very clearly what version they use and to be cognizant of the relevant literature.

Many of the developers of the empirical methods are aware of these and other limitations. They, therefore, recommend to observe and update the design (use of the observational method) and use several approaches in parallel, be they empirical or others.

Chapter 3

Direct and indirect methods for soil tunnels

With effects on the surface

3.1 INTRODUCTION

This chapter presents a number of empirical (direct method) and semi-empirical (i.e. mixture of analytical and empirical methods; indirect method) methods for tunnels excavated in soil. The distinction between tunnels in soil versus tunnels in rock is an old one and is based on the behavior of the ground that the tunnelers observed as it was excavated. One of the unintended consequences of such distinction is a bimodal view of geological materials where they are classified as soil, i.e. soft, and rock, i.e. hard. Such division has percolated the geotechnical/geomechanical communities, sometimes with adverse consequences. Of course, nature makes no such distinction, and geomaterials encountered in tunneling include soils, rocks, as well as a full gamut of transitional materials. Further, materials that could be initially classified as rock, i.e. when intact, can evolve with time and turn into what could be considered a soil, e.g. through processes of mechanical or chemical alteration and weathering.

Nevertheless, we have decided to preserve the distinction between tunnels in rock (Chapter 2) and tunnels in soil (this chapter), for two reasons: the first one is historical, with all its advantages and disadvantages, and the other for clarity of presentation, as different problems are found when dealing with the two extreme materials, with the natural result of development of different approaches in each case. Since, as mentioned, nature has the tendency of not following our conceptual models, the engineer may find that there is no method that applies to the case at hand. Thus any decision must be based on fundamental knowledge of geomechanics and sound experience, including that of others. Compilation of such experience, from a historical point of view, organized in such a way that provides some meaningful information, is one of the goals of Chapters 2 and 3.

The chapter has four more sections. Section 3.2 provides a summary of the empirical methods that have been used to estimate the loads on the support of tunnels excavated in soil. It is included for historical reasons only since they are no longer used in practice. Section 3.3 includes a short description of shield-based methods, which are the methods generally used for the construction of tunnels in soil. Section 3.4 contains a number of methods to estimate the deformations induced by tunneling in soil materials. Finally, Section 3.5 offers a number of recommendations regarding the use of the methods discussed in the Chapter.

3.2 METHODS TO ESTIMATE SUPPORT LOAD

Initial classifications of tunnel support were based simply on the capability of the tunneler to excavate the ground. Thus, tunnels were determined to be located in good, bad, or very bad ground. Good ground was found when the tunnel could be excavated without forepoling. If forepoling was needed, then the tunnel was in bad ground. If breasting was

DOI: 10.1201/9781003328940-3

necessary, then the tunnel was in very bad ground (Proctor and White, 1977). This classification was perhaps sufficient while timber was the support of choice. Once steel was introduced in tunneling, it was clear that a classification had to be more refined. Such classification, known as "Tunnelman's Ground Classification," is included in Table 3.1, after Terzaghi (1950).

Table 3.1 Tunnelman's ground classification. From Sinha (1989)

No.	*Classification*	*Tunnel working conditions*	*Representative soil types*
1	Hard	Tunnel heading may be advanced without roof support.	Very hard calcareous clay; cemented sand and gravel.
2	Firm	Tunnel heading may be advanced without roof support, and the permanent support can be constructed before the ground will start to move.	Loess above the water table; various calcareous clays with low plasticity such as the marls of South Carolina.
3	Slow Raveling	Chunks or flakes of material begin to drop out of the roof or the sides sometime after the ground has been exposed.	Fast raveling occurs in residual soils or in sand with clay binder below the water table. Above the water table the same soils may be slow raveling or even firm.
4	Fast Raveling	In fast raveling, the process starts within a few minutes; otherwise, it is referred to as slow raveling.	
5	Squeezing	Ground slowly advances into tunnel without fracturing and without perceptible increase of water content in ground surrounding the tunnel. (May not be noticed in tunnel but cause surface subsidence.)	Soft or medium-soft clay.
6	Swelling	Like squeezing ground, moves slowly into tunnel, but the movement is associated with a very considerable volume increase in the ground surrounding the tunnel.	Heavily precompressed clays with a plasticity index in excess of about 30; sedimentary formations containing layers of anhydrite.
7	Cohesive Running	The removal of the lateral support on any surface rising at an angle of more than 34° to the horizontal is followed by a "run," whereby the material flows like granulated sugar until the slope angle becomes equal to about 34°. If the "run" is preceded by a brief period of raveling, the ground is called cohesive running.	Cohesive running occurs in clean, fine, moist sand.
8	Running		Running occurs in clean, coarse, or medium sand above the water table.
9	Very Soft Squeezing	Ground advances rapidly into the tunnel as a plastic flow.	Clay and silts with high plasticity index.
10	Flowing	Flowing ground moves like a viscous liquid. It can invade the tunnel not only through the roof and the sides but also through the bottom. If the flow is not stopped, it continues until the tunnel is completely filled.	Any ground below the water table that has an effective grain size in excess of about 0.005 mm.
11	Bouldery	Problems incurred in advancing shield or in forepoling; blasting or hand-mining ahead of machine possibly necessary.	Boulder glacial till; rip-rap fill; some landslide deposits; some residual soils. The matrix between boulders may be gravel, silt, clay, or combinations thereof.

Tunnelman's classification is largely based on the observed behavior of soils, as they are excavated. It is divided into eleven classes (Proctor and White, 1977):

Class 1 and 2. Hard and Firm Ground: Tunnels can be constructed in firm ground without inducing noticeable ground deformations. The roof section of the tunnel can be left unsupported for several days without ground movement. These include sand and gravel soils with clay binder and fine-grained soils such as stiff, intact clay, as well as cemented soils.

Class 3 and 4. Raveling Ground: The term "raveling" is used to describe a soil that gradually breaks down into chunks or flakes. With time, more and more fragments of ground drop from the unsupported portions of the excavation and the cross section of the tunnel increases in size. Typical raveling soils include fine moist sand, sand and gravel mixtures with silt or clay as binder, and stiff fissured clays. The difference between "slow" and "fast" raveling is related to the time it takes for the phenomenon to take place. The difference was initially made based on the time it would take to place the steel ribs and lagging, as the support of the tunnel. If the stand-up time of the soil was about six hours, the time needed to place the steel ribs at a spacing of about 5 ft, the soil was classified as slow raveling; otherwise, as fast raveling.

Class 5 and 9. Squeezing and Very Soft Squeezing Ground: These are conditions associated with soft and medium clay, for squeezing ground, and with clays and silts with high plasticity, for very soft squeezing ground. In these soils, all the unsupported surfaces advance toward the excavation, slowly (squeezing) or rapidly (very soft squeezing), and continuously. An important characteristic of these soils is that they deform at constant moisture content and there is no change of volume during deformation.

Class 6. Swelling Ground: The behavior is very similar to that of squeezing ground, except that the movements are associated with an increase of volume. The volume change is either because there is an increase in water content, e.g. stiff clays that expand upon exposure, or due to chemical reactions, or because of combinations of water content changes and chemical reactions (see Section 6.5 for an in-depth discussion of tunnels in swelling ground).

Class 7 and 8. Running Ground: The term "running" describes the behavior of the soil as soon as it is left unsupported, and it is characteristic of clean granular soils above the water table. This can also be defined as raveling ground with zero stand-up time (conversely raveling ground is a running ground with a binder that provides stand-up time). Table 3.1 provides an indication that the soil stops movement once it reaches an angle of 34°, which is a typical angle of repose of granular soils. Very fine, moist sand above the water table has the properties of a cohesive running soil. If it is located below the water table, it behaves as flowing ground.

Class 10. Flowing Ground: Flowing Ground behaves as a thick liquid and invades the tunnel from all directions. If the flow is not stopped, it continues until the tunnel is completely filled. Soils in this category include granular soils and silts below the groundwater table. Even though the same soils, e.g. granular and silty soils, fall into "running" and "flowing," the behavior is completely different. Running ground stops once a cone of about 34° is formed, while flowing ground invades the tunnel as a fluid. The effects of running ground are local to the excavation, i.e. they only affect few feet around the tunnel, while the consequences of flowing ground can affect hundreds of feet beyond the tunnel. Finally, the bearing capacity of running ground may be high enough to support the roof load through steel sets and their (adequately sized) footings, while flowing ground has no bearing capacity.

Class 11: Bouldery Ground: As the name indicates, this class corresponds to a soil that is a mixture of particles with different sizes and includes boulders. The soil may present difficulties for excavation and in particular for forepoling if boulders are encountered.

Table 3.2 provides a relation between the Unified Soil Classification System and the Tunnelman's classification. Note that there is no one-to-one relation, as the same soil may correspond to more than one Tunnelman's class depending on consistency, heterogeneity and groundwater table position within the soil mass.

In soils, the support must be provided to the surrounding ground or the tunnel will collapse. In some soils, the support must be placed immediately, while in others the soil has a "stand-up time" that allows time to place the support. Past practice has been to erect an initial support to give immediate support to the soil and a final, or permanent, liner at a later stage. With TBM (Tunnel Boring Machine) construction, there is often only one liner, which is placed under the tail of the machine (further details about shield tunneling are given in Section 3.3). For initial support, three types have been used: steel ribs and lagging; liner plates (prefabricated steel plates) and shotcrete. For final lining: mass concrete, to provide long-term support and water-tightness, if needed; and segmental lining, made of brick, cast iron, steel segments, or precast concrete segments (Sinha, 1989). Modern support methods are discussed in Chapter 7.

Table 3.2 Correlation between the unified soil classification system and Tunnelman's classification (Sinha, 1989)

Tunnelman	*Tunnelman's ground classification*										
Standard soil classification	*Hard*	*Firm*	*Slow raveling*	*Fast raveling*	*Squeezing*	*Swelling*	*Cohesive running*	*Running*	*Very soft squeezing*	*Flowing*	*Bouldery*
Gravel and coarser											
Sand							A				
Silt		B							F		
Clay	C	D			E	D			F		
Gravel, with clay binder	C	D									
Gravel, with silt binder											
Sand, with clay binder	C	D									
Sand, with silt binder											
Cemented sand and gravel											
Highly organic soils											

A – moist, above water table
B – loess
C – stiff to very hard
D – stiff to hard
E – soft to medium
F – very soft

Notes: (1) The typical soil names refer to the dominant soil type with regard to their behavior in a tunnel; (2) the shaded areas indicate the soil types that usually cause the ground conditions described by the Tunnelman's terms.

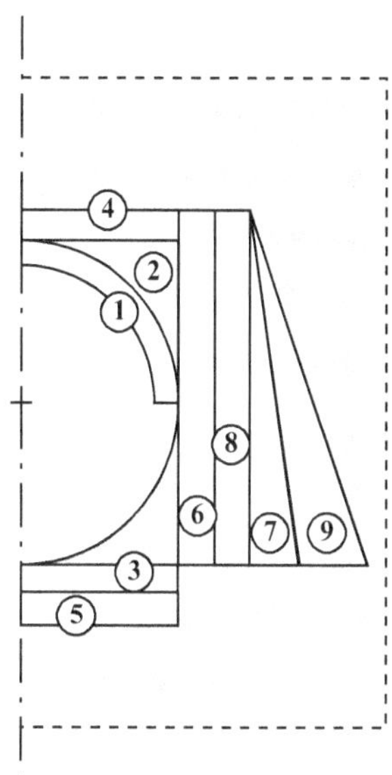

(1) Weight of the upper half of the tunnel

(2) Weight of the soil within the area marked 2

(3) Uniform upward force balancing 1 and 2

(4) Weight of the loading above the top of the tunnel

(5) Uniform upward reaction balancing 4

(6) Horizontal pressure due to water above the top of the tunnel

(7) Horizontal water pressure from top to bottom of the tunnel

(8) Horizontal pressure from soil above the top of the tunnel equal to the product of the weight of the soil (buoyant unit weight if sumerged) above the top of the tunnel and the factor K

(9) Horizontal pressure from soil between the top and the bottom of the tunnel. At any point, the pressure is the product of the weight of the soil between that point and the top of the tunnel and the factor K. Soil weighed as in 8.

Figure 3.1 Tunnel loading by Hewett and Johammesson (1922). Adapted from Sinha (1989).

Perhaps one of the earliest methods for the design of support in tunnels in soils is that of Hewett and Johammesson (1922), which is illustrated in Figure 3.1, after Sinha (1989). It is included here because it is one of the earliest attempts at identifying the origins of the loads on the support. The method consists of distinguishing potential sources of loading around the tunnel, as denoted in the figure, including both vertical and horizontal loads, and making a distinction between the loads coming from the soil and coming from the water. For the horizontal load, the factor K is taken between 0.5 and 0.6. The liner is represented as an elastic ring that must have the capacity to withstand all the loads.

In reality, the soil above the tunnel is only partially supported due to arching. Section 1.5 describes the concept of arching and the associated theory developed by Terzaghi (1943). Figure 2.17 illustrates the arching concept and Figure 2.18 the loads associated with it (see also Section 2.3.2.3 for details). As it has been discussed for rock, the magnitude of the vertical and horizontal loads is expressed in terms of the height H_p, of a mass of rock/soil above the tunnel.

Table 3.3 lists the loads that the support must withstand based on the type of soil and its Tunnelman's class. The table provides recommended values for H_p, H_p minimum ($H_{p,\,min}$) or maximum ($H_{p,\,max}$) for the design of the initial support. It is assumed that the soil, in all cases, is above the water table. If below the water table, air pressure may be used to prevent water from flowing into the tunnel. The loads for raveling and running ground are based on arching theory (Terzaghi, 1943) and are associated with a minimum displacement that the roof must have so full arching is developed. If these deformations are not produced, the loads will be higher. Thus, from the table, the pressures on the support, for raveling or running ground, are:

$$\text{Vertical: } P_v = \gamma H_p$$
$$\text{Horizontal: } P_h = 0.3\gamma(0.5H_t + H_p)$$

with γ being the unit weight of the soil, H_p the height of soil above the tunnel to be supported, and B and H_t the width and height of the tunnel (see Figures 2.17 and 2.18). Note that the

Table 3.3 Load on tunnel support (from Proctor and White, 1977)

(a) Raveling or Running Ground	
Dense Sand	$H_{p\ min} = 0.27\ (B + H_t)$ for yield of 0.01 $(B + H_t)$
	$H_{p\ max} = 0.60\ (B + H_t)$ for yield of 0.15 $(B + H_t)$
Loose Sand	$H_{p\ min} = 0.47\ (B + H_t)$ for yield of 0.02 $(B + H_t)$
	$H_{p\ max} = 0.60\ (B + H_t)$ for yield of 0.15 $(B + H_t)$
(b) Flowing Ground	
	$H_p = H$ or $2\ (B + H_t)$ whichever is smaller
(c) Squeezing Ground	
	$H_p = H - 2\ S_u\ H\ /[\gamma\ (B + 2H_t)]$
(d) Swelling Ground	
Intact	H_p = very small
Fissured	H_p = raveling ground

Rock load H_p in feet of rock on roof of support in tunnel with width B (ft) and height H_t (ft), at a depth H (ft) below the surface (see Figure 2.18). For circular tunnels, $H_t = 0$ and B = diameter of the tunnel.

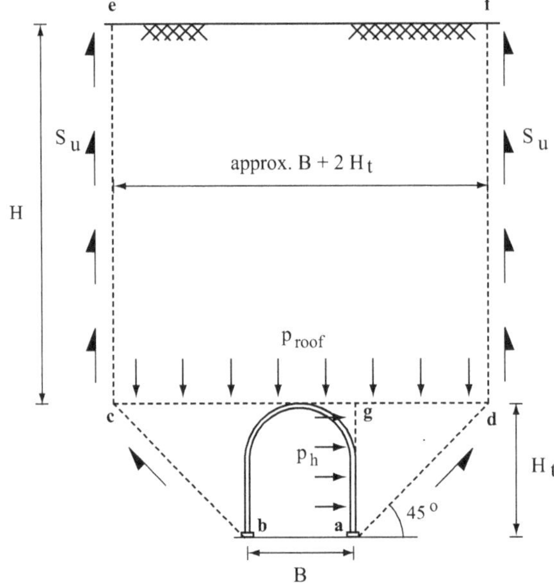

Figure 3.2 Support load in squeezing ground.

vertical pressure is that of a column of soil of height H_p with unit weight γ, and the horizontal pressure comes from limit equilibrium of the side wedge, e.g. wedge "bfd" in Figure 2.18, for typical values of the internal friction angle of sand. For squeezing ground, the values are obtained from limit equilibrium analysis (lower bound theorem) and for the failure mechanism depicted in Figure 3.2. Because the loads are obtained for initial support, i.e. short term, the undrained shear strength of the soil, S_u, is used. For the roof, equilibrium of the block "cdfe" results in the following pressure, $p_{roof} = \gamma H - 2\ S_u\ H/(B + 2\ H_t)$. For the lateral pressure, equilibrium of the wedge "adg" results in $p_h = p_{roof} + 0.5\gamma H_t - 2\ S_u$. Note that because of the assumption of short-term analysis, the slip lines from the bottom of the tunnel are at

45° with the horizontal. The table also lists the loads for swelling ground, but they are only included for completeness since they do not correspond well with reality (see Section 6.5).

As mentioned earlier, in Section 3.1, the empirical approaches discussed are no longer used. Instead, the support loads are estimated using analytical methods (see Chapter 4), or numerical methods (see Chapter 5), largely the Finite Element Method. A proper model not only requires the actual geometry of the tunnel, actual behavior of ground and support, but very importantly construction operations. In other words, the numerical model must capture the stress path that the soil and support undergo during construction and operation. Thus, knowledge of construction methods in soft soils is needed, which is the focus of Section 3.3. While experience shows that estimates of loads on support and ground are generally acceptable with current practice, deformations, particularly ground deformations, are not. For this reason, significant effort has been devoted to the development of different methods to estimate ground deformations, and in particular the settlements that may occur on the surface, above the tunnel, during construction. These are reviewed in Section 3.4.

3.3 CONSTRUCTION METHODS – SHIELD TUNNELING

Soil tunnels can be constructed by a variety of methods, such as shield tunneling with associated mechanical excavation and face support, pipe jacking, forepoling, and soil freezing with a number of different excavation and support strategies. Given the prevalence of shield-associated construction methods, this chapter will concentrate on these when discussing tunneling related deformations in Section 3.4. The following brief description of these methods is simply intended to provide the necessary background. Detailed descriptions of shield technology and related methods can be found in e.g. Maidl et al. (1996) and Zhang and Kieffer (2020).

A shield has the following components: the cutting edge, the trunk, and the tail (see Figure 3.3). The construction of the tunnel consists of repeated cycles that include the following steps: (1) excavation of a length of ground ahead of the face, typically one support ring long (Figure 3.3a); the excavation is done full face or partial face; (2) movement forward of the shield using hydraulic jacks and employing existing lining as reaction (Figure 3.3b); (3) retraction of the jacks and erection of a new ring of support under the shield (Figure 3.3c). The cycle is repeated. The major advantage of excavation with a shield is that the shield contains the ground, at its perimeter, until the initial or the final support is placed. In shield tunneling, the initial support may often be the final support. The shield thus has to withstand all the loads from the surrounding soil, construction loads, and prevent the flow of groundwater.

Shields can be classified based on the face support that they provide (for an in-depth description of the different types of shields, usage and history, see Maidl et al., 1996). The following types of shields can be distinguished (Figure 3.4):

Natural Face Support: The shield does not provide support to the ground at the face of the excavation. Thus, the face is stable because the ground is competent enough to provide sufficient stand-up time or the ground reaches a stable slope inside the tunnel (Figure 3.4a). Clearly, this type of support can only be used in tunnels with small diameter, or with intermediate shelves to contain the soil at the face for larger diameters, and mostly for soils above the groundwater table. Historically, this is the type of shield originally used to excavate in soft ground, and was first proposed by Brunel in 1806 and then used to excavate a tunnel under the Thames River in London (1825–1843).

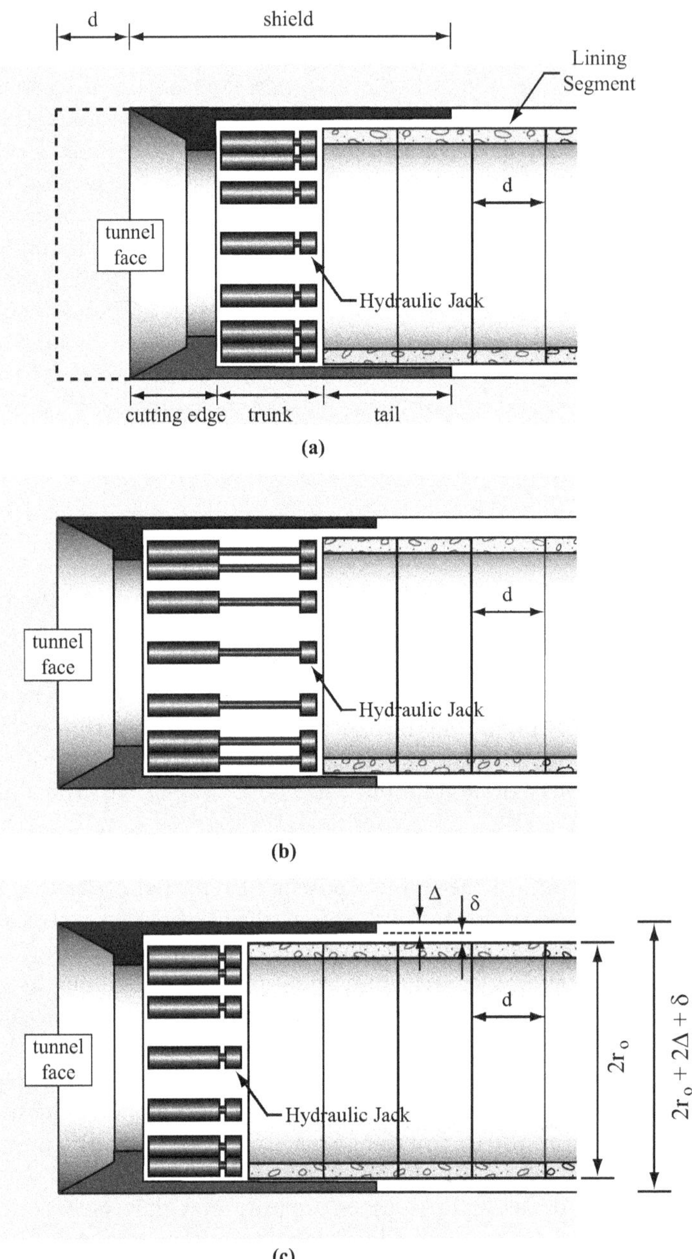

Figure 3.3 Shield tunneling concept. (a) Start position. (b) Excavation, Jacks extended. (c) New liner segment, Jacks retracted. (adapted from Tunnelling Technology, 1976).

Mechanical Face Support: The simplest method is to provide some face support with moveable face plates. The tunnel is excavated from top to bottom, and at the same time support is provided at the portion of the face just excavated. This is often done manually and is time consuming, with advance rates very small. An alternative is to provide mechanical support through a rotating wheel equipped with cutters that simultaneously supports and excavates the ground at the face (Figure 3.4b). As with the natural

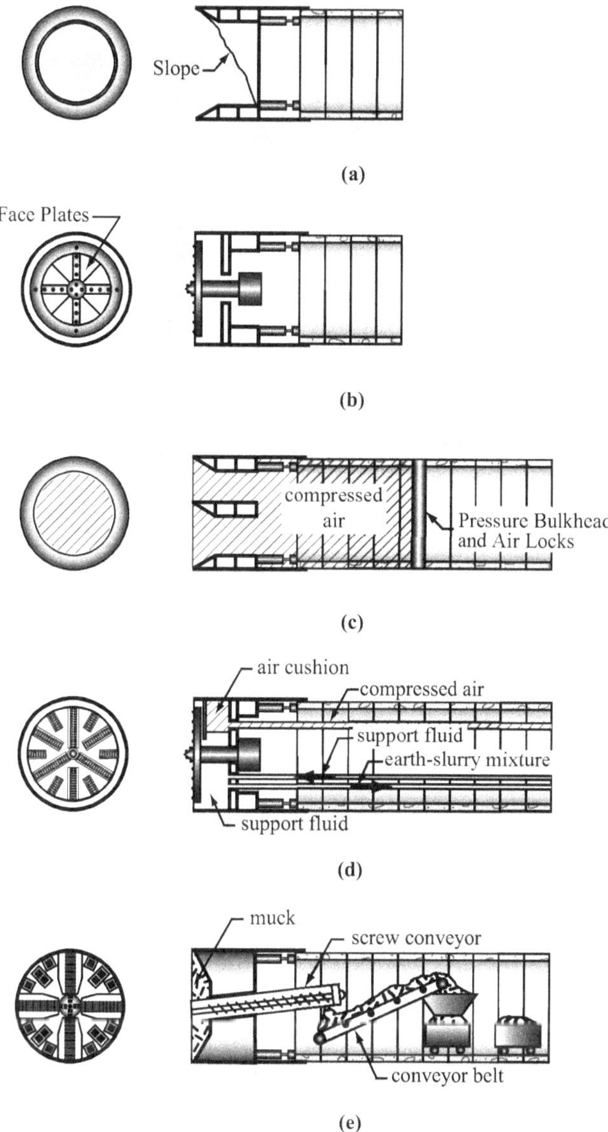

Figure 3.4 Types of face support. (a) Natural support. (b) Mechanical support. (c) Compressed air support. (d) Slurry support. (e) Earth pressure balance support. (after Maidl et al., 1996).

support, the method provides stability for the soil only (not for the water) and thus, in cohesionless soils, can be used for tunnels above the water table.

Compressed Air Support: This support is used to support the ground at the face and, most importantly, to prevent groundwater from entering the tunnel (Figure 3.4c). The principle is based on the application of an air pressure that is higher than the water pressure; thus, the magnitude of the air pressure applied is typically the hydrostatic pressure at the bottom of the tunnel. Since the water pressure (assuming hydrostatic conditions) increases linearly with depth, the result is an air pressure at the crown higher than the water pressure, which carries the danger of leakage or blowouts if the air pressure inside the tunnel is not carefully controlled. To be able to apply and maintain air pressure close to the face of the tunnel, a chamber needs to be created and sealed close to the face of the tunnel. This requires the installation of a pressure bulkhead integrated into the shield that can be moved as the shield advances. The bulkhead needs to have airlocks to allow the removal of the soil from the face and the supply of materials. An independent lock is needed for personnel to move in and out of the compressed air chamber, which needs to be equipped for decompression. Needless to say, and for safety and health reasons, exposure of personnel to air pressure is generally regulated, with limits on the maximum pressure, the time working under pressure and the decompression procedure. The combination of a shield and compressed air revolutionized underwater tunnel construction at the beginning of the 20th century, with earliest reports of using such combination for the construction of the London Underground. The design of compressed air support in tunnels is discussed in e.g. Jenny (1983). Although this method is not used as much as it was in the past, it is often relied upon to access the face of tunnels constructed by the slurry method or the earth pressure balance method.

Slurry Support: The face of the tunnel is supported by pressurized slurry. The shields are known as "bentonite," "suspension," or most commonly known as "slurry" shields. The method was patented by Haag in 1896, and the first slurry shield with a cutting wheel and hydraulic mucking was used in Japan in 1967. With the slurry shield, a fluid, typically bentonite, sometimes with polymers used for conditioning the slurry (or only polymers), is pumped to the face to provide a pressure equal to the total pressure existing at the face, earth and water pressure. Face stability is accomplished through the interaction of the bentonite (and/or polymers) suspension with the ground. The suspension, under pressure, penetrates the ground and within a few seconds creates a thin impermeable film (filter cake) that provides the face support. The fluid also serves as the transport medium of the materials removed from the face (Figure 3.4d). By controlling the addition or removal of the fluid at the face, the support pressure is adjusted. Slurry shields are most commonly used in sandy or silty soils. There are two main reasons: one is the requirement to create the filter cake, and the other economic, as the type of ground must allow the separation of the bentonite from the soil for reuse, typically in a separation plant on the surface.

Earth Pressure Balance (EPB) Support: The technique was first used in Japan in 1963. The fundamental idea behind the method is that the material excavated by the cutting wheel provides support to the face (Figure 3.4e). As the material at the face is excavated, it can (but does not have to) be modified by the addition of water, bentonite or clay suspensions and chemical additives. Removal of the material is typically done by a screw conveyor. The pressure at the face is controlled by balancing the rate of advance of the shield and removal of the material. Modification of the excavated material at the face is often necessary to obtain a consistency that provides adequate support pressure at the face and allows the soil to be carried by the screw conveyor. "Earth pressure balance" shields are best suited in clayey ground or in soils with low permeability. Other types of soils, such as sandy or gravely soils, can be excavated with the EPB shields with the addition of conditioners that decrease permeability and provide adequate consistency (the latter is important to prevent ground loss through the screw conveyor).

The need for face support can be related to the face stability number, N. It is defined, after Broms and Bennermark (1967), and given also in Equation (4.93):

$$N = \frac{\sigma_v - \sigma_i}{S_u} \tag{3.1}$$

where σ_v is the total vertical stress at the tunnel axis depth, σ_i is the internal support pressure (e.g. compressed air or face support), and S_u is the undrained shear strength of the soil. For stability numbers N between 4 and 6, a shield is usually required, if large settlements are acceptable. For N near 6, the displacements tend to increase and face support is often needed, especially when surface settlements need to be limited. For N values larger than 6 the face might be unstable and full-face support is usually required. This may be accomplished mechanically with EPB, slurry or compressed air. For more moderate N values, there are a number of quite sophisticated methods to determine earth or slurry pressure, e.g. Clough and Schmidt (1981), Lee et al. (1992), Anagnostou and Kovari (1994, 1996a), Vermeer et al. (2002), Mollon et al. (2010).

3.4 SOIL DEFORMATIONS

3.4.1 Introduction

There are three fundamental requirements for a satisfactory tunnel (Peck, 1969b): The first one is that the tunnel should be able to be built; the second one is that construction of the tunnel should not excessively damage below- or above-ground structures adjacent to the tunnel; and the third requirement is that the tunnel should withstand all the influences to which it may be subjected during its lifetime.

Terzaghi's (1936, 1943) recommendations on support pressures (see Sections 3.2 and 2.3.2) were based on the assumption that the ground around the tunnel deforms and takes load. Thus, ground deformations may be desirable consequences of tunneling since they are part of the supporting mechanism. Because the strength of the soil surrounding the tunnel is mobilized with its deformations, the soil contributes to support itself and reduce the structural demands on the liner (see discussion on ground structure interaction in Chapter 1). In other words, the shear stresses in the ground may increase and thus its deformations. Movements induced by tunneling operations may be significant and may affect a considerable volume of soil around the tunnel and may even reach the ground surface. Thus, preexisting structures such as buildings, pipes, and other tunnels may be affected by the construction of an adjacent tunnel and may experience undesirable deformations. It is not realistic to expect that tunnel excavation can be done with zero deformations, as it is not realistic for any other geostructure. What the design has to accomplish is to keep the extent and magnitude of the deformations within a tolerable range to prevent, or at least limit, potential damage to the surrounding built environment. This cannot be accomplished without reliable predictions of the deformations associated with the construction procedure, type of soil, groundwater conditions, geometry, and depth of the tunnel. When surface settlements are not accurately predicted, substantial damage can be caused to adjacent structures (Cording et al., 1978; Attewell et al., 1986).

Evaluation of the potential damage is a two-phase process. In the first phase the ground deformations induced by tunneling operations are predicted (Section 3.4.2). The second phase involves an estimate of the structure deformations and induced damage given the ground deformations (Section 3.4.3). Limited work has been done on estimating the induced structure deformations and damage, and typically involves consideration of soil–structure interaction (Selby, 1999), which requires the combined expertise of geotechnical and structural engineers. The reader is referred to the excellent work done by Cording and co-workers (Cording and Hansmire, 1975; Cording et al., 1978; Boscardin and Cording, 1989; Son and Cording, 2007) and others (Burland and Wroth, 1974; Attewell, 1977; Loganathan et al., 2000; Klar et al., 2005; Vorster et al., 2005; ITA/AITES Report 2006, 2007).

One can relate all this to what has been said about ground deformations around and ahead of a tunnel (be that in soil or rock) in Chapter 1. When the ground is significantly deformable and/or close to the surface, this will lead to settlements on the surface as schematically shown in Figure 3.5 (see also Figures 1.5, 1.7, and 1.9). Similar to what was discussed in Chapter 1, the deformation takes place around the tunnel in all directions and, consequently, the settlement troughs on the surface also extend beyond the footprint of the tunnel.

Close to the excavation, the soil displacements are large and toward the opening, with the largest deformations occurring above the tunnel. The soil deformations decrease as the distance from the tunnel increases. If the tunnel is shallow or the deformations in the soil are very large, the soil displacements may reach the surface, where they are observed as a settlement trough. The settlements are the largest above the center of the tunnel and decrease

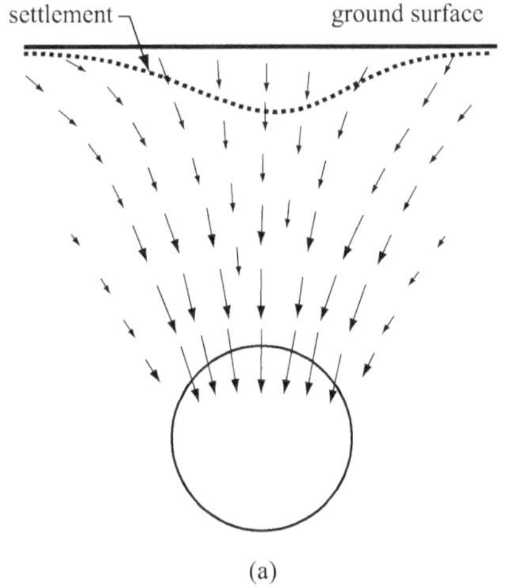

(a)

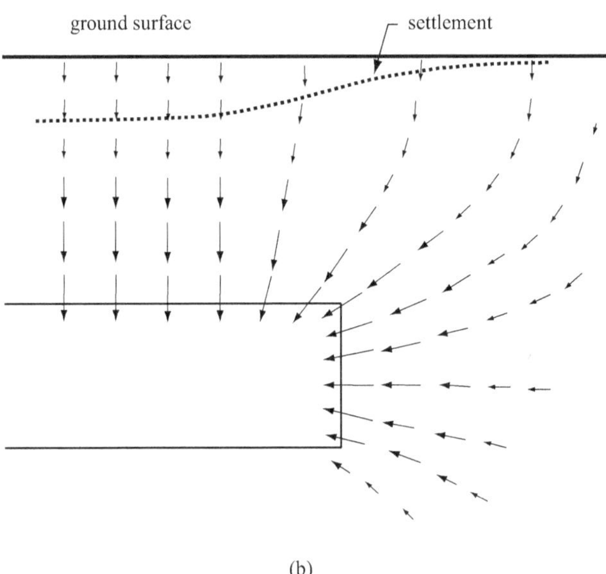

(b)

Figure 3.5 Conceptual model of ground movements toward tunnel. (a) Vertical cross section. (b) Vertical longitudinal section.

as the distance from the excavation increases. A similar phenomenon can be observed along a vertical section along the tunnel axis, as depicted in Figure 3.5b (see also Figure 1.5d). The excavation not only changes the stresses in the soil around the tunnel (Figure 3.5a), but also ahead of the opening, and so a point in the soil undergoes deformations well before the tunnel reaches its position (see also Figures 1.5 and 1.7). As the distance from the tunnel increases, the soil displacements decrease. Behind the tunnel face, and as the distance from the face increases, the soil movements are mostly radial and similar to those shown in Figure 3.5a (see also Figure 1.5d). If soil deformations reach the surface, they produce settlements which typically have the shape shown in the figure, with small magnitudes ahead of the face, increasing toward the face until they become constant at some distance behind the face.

This is shown three-dimensionally and with detailed longitudinal and transverse cross sections in Figure 3.6. The settlement trough, in horizontally bedded ground, is symmetric with respect to the tunnel axis. Figure 3.6a shows, as already mentioned, that ground deformations start at some distance ahead (along the direction of excavation) of the tunnel face and increase as points on the ground surface approach the position of the face of the tunnel (see also Figures 3.5b and 1.5d); simultaneously, the settlement trough widens. At some distance behind the tunnel face, the magnitude of the settlements and width of the settlement trough stabilize. Figure 3.6b is a plot of the settlements on a cross section far behind the face. The largest settlements occur at a point of the surface above the tunnel centerline. As the distance from the tunnel increases, the settlements decrease. Figure 3.6c shows the settlements on a vertical section along the tunnel axis.

The deformations toward the tunnel can be characterized or classified as movement at the face and at the perimeter (Figure 3.7). Very often these movements toward the tunnel are called ground loss, which is the difference between the actual and theoretical volume being excavated. One can use the schematic in Figure 3.7a to visualize a number of possibilities of movements (including no movement) and counteractions at the face and perimeter;

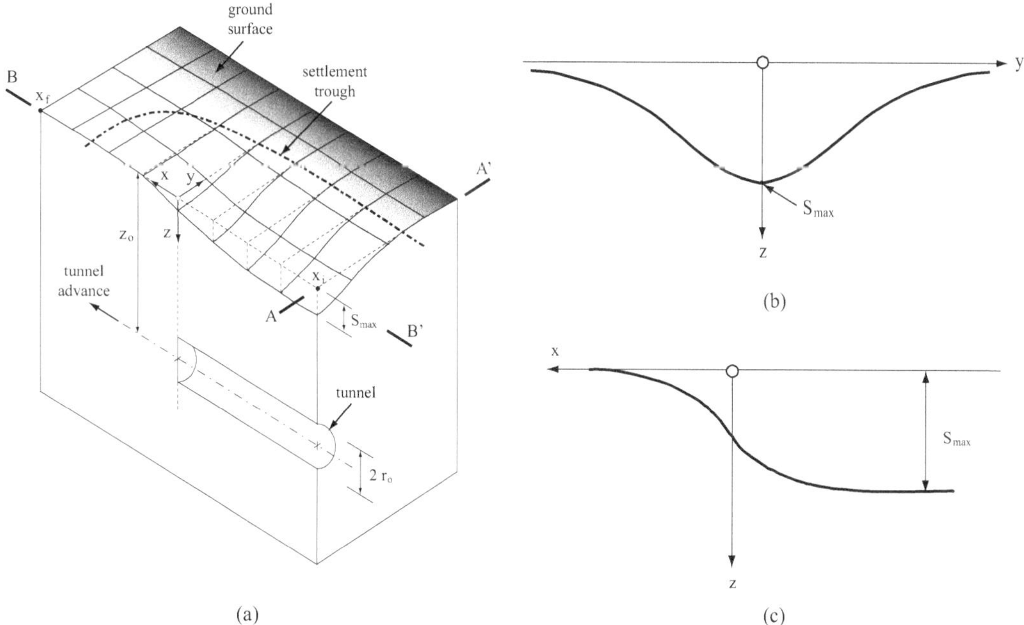

Figure 3.6 Settlement trough on ground surface due to tunnel excavation. (a) 3 D settlement trough. After Attewell (1977). (b) Cross-section A-A′. (c) Longitudinal section B-B′.

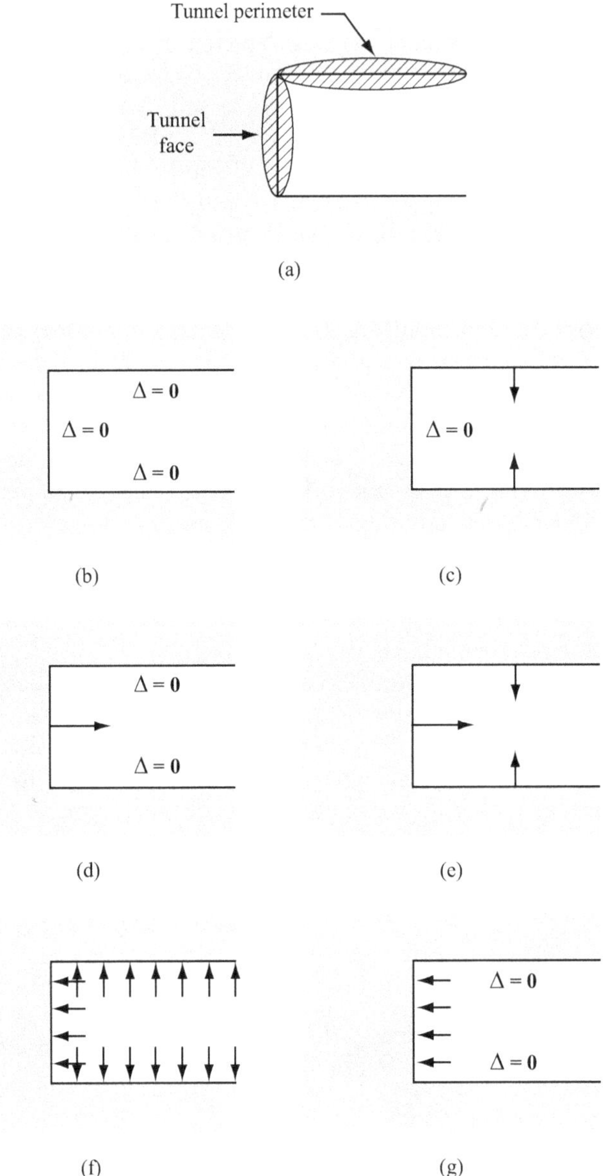

Figure 3.7 Different constraints at face and perimeter of the tunnel. (a) Tunnel face and perimeter. (b) Fixed face and fixed perimeter. (c) Fixed face and free perimeter. (d) Free face and fixed perimeter. (e) Free face and free perimeter. (f) Pressure at face and perimeter. (g) Pressure at face and fixed perimeter.

Figures 3.7b–e show various combinations of no movement and free movement, while Figures 3.7f–g show combinations of counterpressures and free movements. Clearly these schematic cases are simplifications but they can be related to real tunnel excavations:

Undoubtedly, other cases are possible.

The total ground loss can be decomposed into two: face loss and radial loss, also called face take and radial take. Face loss is caused by the deformations that the ground experiences ahead of the face of the tunnel, as the stress field inside the ground changes with tunnel

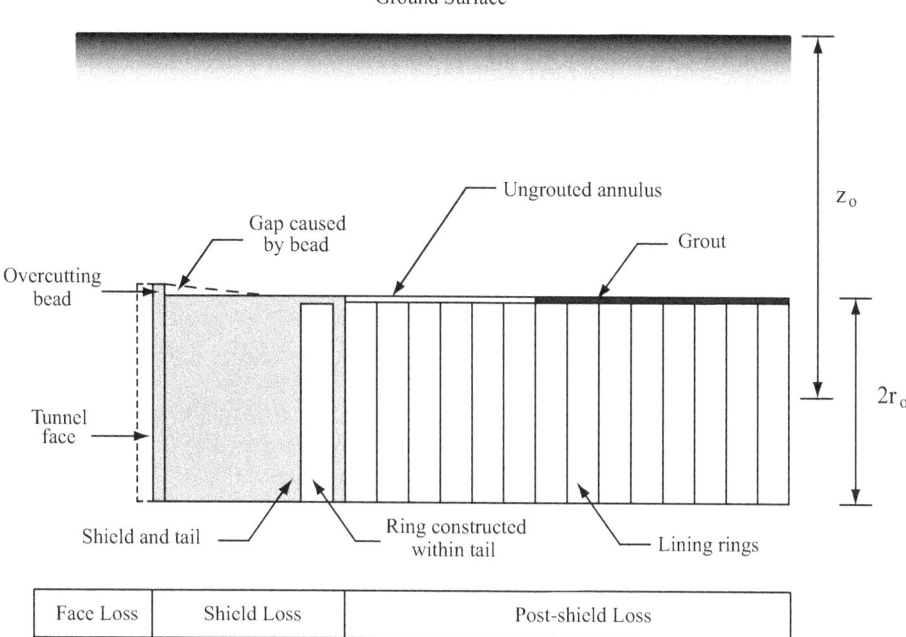

Figure 3.8 Sources of ground loss.

excavation (Figure 3.5b). These deformations have a strong axial component toward the face of the tunnel. Radial losses occur behind the face and, as the name indicates, have mostly a radial component (Figure 3.5a). Radial losses are further decomposed into shield loss and post-shield loss (Figure 3.8). Shield loss occurs because the shield, at least in its upper section, may not be in direct contact with the ground and this for several reasons: overcutting bead at front end of shield, shield weight causes downward driving, intentional upward pitch leaving a space behind the shield. Post-shield ground losses are caused by the delay and deformability of pea-gravel placement and grouting (for more details on these causes see e.g. Suwansawat and Einstein, 2006), and related to what was said in Chapter 1, the stress reduction at the opening perimeter leading to deformation and ovaling. Although the deformations related to stress changes are not ground losses in the strict sense, they are often included in them. Additional post-shield ground losses may be induced by consolidation, creep or other deformation mechanisms that may occur in the ground over time. In the methods discussed below, ground loss as a whole or in form of its just mentioned components is related to deformations. However, when using analytical and numerical methods all these factors are considered separately.

3.4.2 Estimate of ground deformations

Related to what has just been said, available methods to estimate ground deformations induced by tunnel excavation can be classified in a broad sense into direct and indirect methods. In direct methods the ground deformations, e.g. surface deformations, are estimated directly based on the mechanisms discussed. In indirect methods, the ground loss that occurs at the tunnel is computed first, and the result is then used to estimate deformations around the tunnel and/or surface settlements.

Techniques to estimate settlements induced by a single tunnel are presented first in Section 3.4.2.1 and by twin tunnels in Section 3.4.2.2. All methods assume shield excavation with open face, with no face support or air pressure.

3.4.2.1 Direct methods

This section includes those methods that provide an estimate of ground deformations, which are all developed from empirical observations of performance of existing tunnels. The methods should be used as reference only as their results are exclusively applicable within the range of cases compiled, and extrapolation to other tunnels, different ground conditions, or construction methods is questionable, at best. The methods are useful since they provide helpful guidelines, references, and an order of magnitude for the expected deformations. They also contain valuable experience accumulated over the years.

Peck (1969b) suggested that the vertical settlements at the surface along a cross section perpendicular to a single tunnel (Figure 3.6b) could be represented by the error function or normal probability (Gaussian) curve. Although there is no theoretical basis for such a function, experience shows that in many tunnels actual settlements follow the Gaussian distribution quite well, in particular in tunnels excavated in clay (Peck, 1969b; Attewell, 1977; Attewell and Farmer, 1974; Schmidt, 1974; Palmer and Belshaw, 1980; Mair et al., 1981; Clough and Schmidt, 1981; O'Reilly and New, 1982; Glossop and O'Reilly, 1982; Attewell et al., 1986; Mair et al., 1993; Fang et al., 1994; Klar et al., 2005; Clayton et al., 2006; Suwansawat and Einstein, 2007). Further comments about the applicability or non-applicability of the Gaussian curve will be made in conjunction with the discussion of Equation (3.8).

The equation for the settlement trough is (Peck, 1969b):

$$S(y) = S_{max}\, e^{\left(-\frac{y^2}{2i^2}\right)} \tag{3.2}$$

where S (y) is the vertical settlement along a section perpendicular to the tunnel axis (see Figure 3.9); S_{max} is the maximum settlement, which occurs above the tunnel centerline; y is the distance from the tunnel axis; see also Figures 3.6a and 3.6b; and i is the inflexion of the error curve, or settlement trough, and coincides with the standard deviation of the normal distribution function. Some geometric properties of the curve are: (1) the settlement is 0.61 S_{max} at the point of inflexion; (2) the point of maximum hogging curvature occurs at a distance $\pm\sqrt{3}\,i$, where the settlement is 0.22 S_{max} and the radius of curvature is $\rho = 2.24\, i^2/S_{max}$; and (3) the point of maximum sagging curvature is at the centerline, with $\rho = i^2/S_{max}$.

The settlement volume on the surface, V_s, is obtained by integration of Equation (3.2) and is given by:

$$V_s = \sqrt{2\pi}\, i S_{max} \approx 2.5\, i S_{max} \tag{3.3}$$

V_s is related to the ground loss and often expressed as a percentage of the cross-section area of the excavation per unit length of tunnel.

Horizontal deformations, U_y, can be estimated from Equation (3.4) with the assumption that the total ground displacements are radial toward the tunnel (O'Reilly and New, 1982).

$$U_y = S(y)\frac{y}{z_o} \tag{3.4}$$

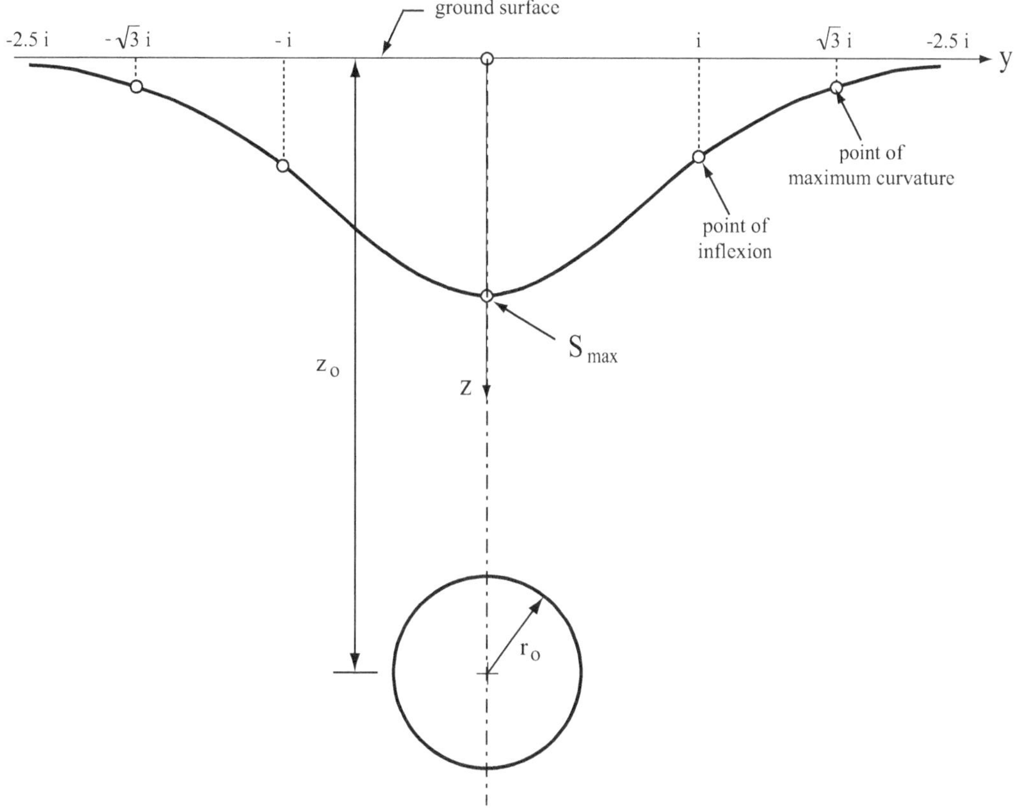

Figure 3.9 Settlement trough above the tunnel. Gaussian distribution.

Even though the Gaussian curve theoretically expands to infinity, in practice settlements are negligible beyond a distance of 2.5 i to 3 i from the centerline (e.g. New and O'Reilly, 1991; Mair et al., 1993). In Peck's method, the settlements are fully determined when the two parameters, S_{max} and i, are known. The parameters are obtained empirically. Peck (1969b) provided recommendations to estimate the width of the settlement trough based on observations in a number of tunnels. Figure 3.10 contains the original recommendations, where the normalized parameter i/r_0 is given as a function of the type of soil excavated and of the normalized tunnel depth z_0/r_0, where r_0 is the radius of the tunnel and z_0 is the depth of the tunnel axis below the ground surface (see Figures 3.8 and 3.9).

The relation between the inflexion parameter and tunnel depth has been updated over the years as more information has become available. Schmidt (1974) and Clough and Schmidt (1981), after expanding the tunnel database used by Peck (1969b), proposed the following relationship:

$$\frac{i}{r_0} = \left(\frac{z_0}{2\,r_0} \right)^{0.8} \tag{3.5}$$

The correlation proposed by O'Reilly and New (1982) has perhaps received the widest acceptance (e.g. Mair et al., 1993; Van der Berg et al., 2003; Clayton et al., 2006; Suwansawat and Einstein, 2007). It is:

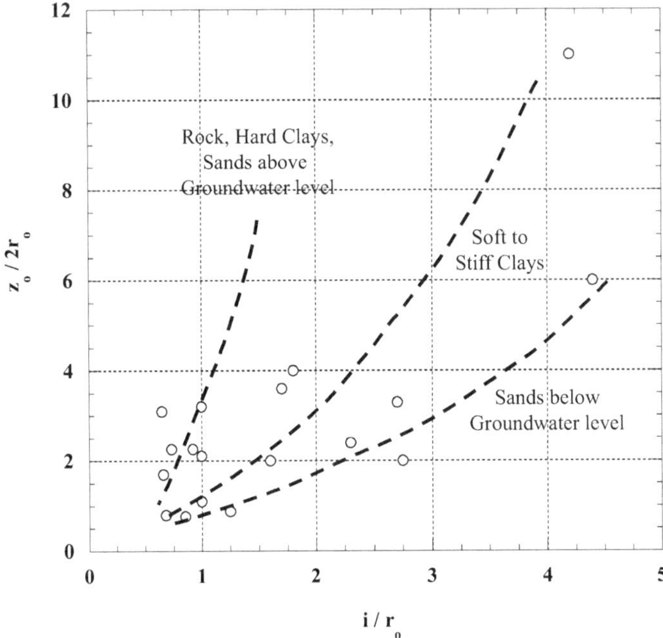

Figure 3.10 Relation between width and settlement trough. Data from Peck (1969b).

$$i = K z_o \tag{3.6}$$

where K is a constant that depends on ground conditions (see Figure 3.11). It is interesting to note that the width parameter i in the equation does not depend on the size of the tunnel, and only on the type of ground, given by the parameter K, and on tunnel depth. O'Reilly and New (1982) suggested K = 0.4 for stiff clays and 0.7 for soft, silty clays, with an average value of 0.5 for cohesive soils. The values are similar to those suggested by Mair and Taylor (1997), with K = 0.5 ± 0.1. For granular soils above the water table, O'Reilly and New (1982) suggested values ranging from 0.2 to 0.3, with 0.25 as the average. This is somewhat consistent with recommendations from Mair and Taylor (1997), with K = 0.35 ± 0.1, who reported no differences between tunnels above and below the water table, but found larger scatter for granular soils than for clayey soils. The recommendations were obtained from data on tunnels within the following limits: (a) tunnels with cover of at least one diameter; (2) tunnels with 0.5 to 2.5 m radius; (3) maximum depth of tunnel axis below ground surface of 10 m for granular soils and 30 m for cohesive soils.

Equation (3.6), which is applicable to the settlement trough on the surface, has been modified by Mair et al. (1993) for tunnels in clays to estimate settlements at depth. See Figure 3.12. The proposed equation takes the form:

$$\begin{aligned} i &= K(z_o - z) \\ K &= \frac{0.175 z_o + 0.325(z_o - z)}{z_o - z} \end{aligned} \tag{3.7}$$

where z is the depth below the surface where settlements are computed (see Figure 3.9). This equation assumes a Gaussian distribution of settlements and an increase of K with depth, which produces a comparatively wider settlement distribution. Mair et al. (1993) suggested

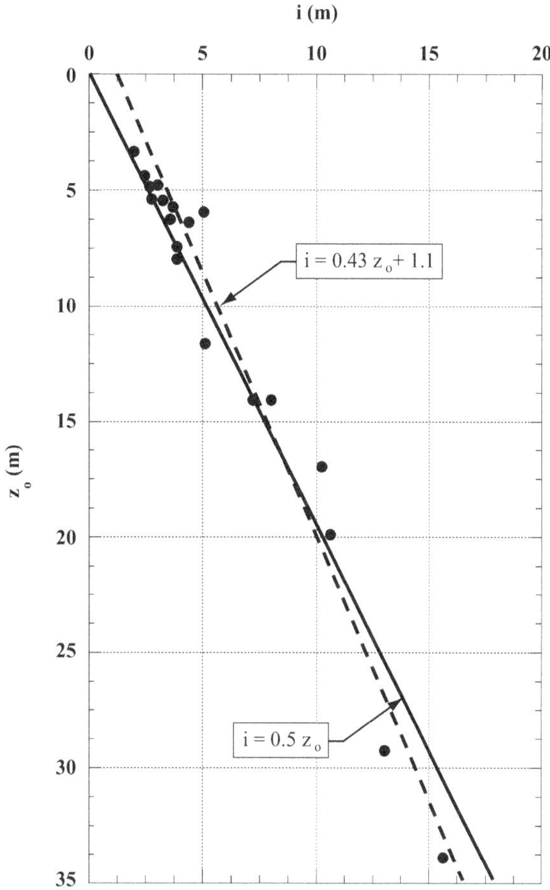

Figure 3.11 Relation between width of settlement trough and depth for tunnels in clay. From O'Reilly and New (1982).

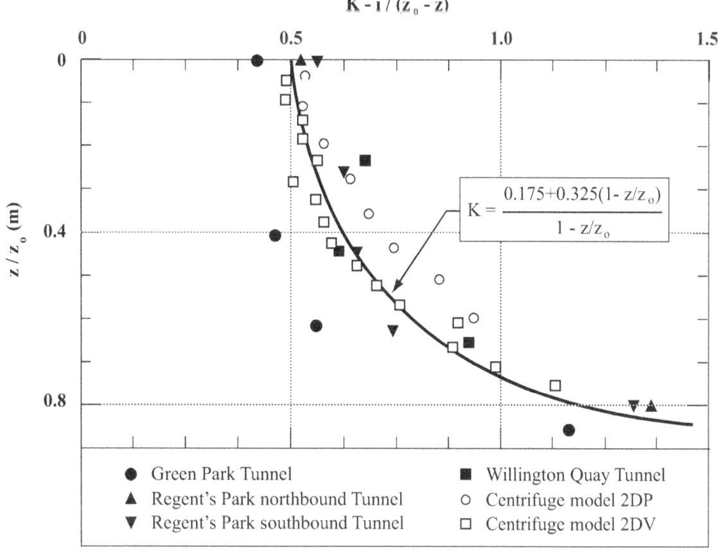

Figure 3.12 Variation of K with depth for tunnels in clays. From Mair et al. (1993).

that the equation should not be used at points closer than one diameter to the crown of the tunnel.

The second parameter needed in Equation (3.2), the maximum settlement, S_{max}, can be estimated from experience in similar tunnels. O'Reilly and New (1982) suggested the following values for the volume of the surface settlement trough:

For tunnels excavated in cohesive soils:

1. for stiff fissured clays, with or without shield, $V_s = 0.5–3\%$, with typical values between 1 and 2% (as mentioned earlier, the volume of the settlement trough is often expressed as a percentage of the cross-section area of the excavation per unit length of tunnel)
2. for glacial deposits, with shield in free air, $V_s = 2–2.5\%$; in compressed air, $V_s = 1–1.25\%$
3. for recent silty clay deposits, with shield in free air with the tunnel at failure or near-failure conditions, $V_s = 30–45\%$, and with shield in compressed air, $V_s = 5–20\%$.

Ground deformations in soft clay deposits strongly depend on tunnel stability, especially on face stability (Casarin and Mair, 1981; Clough and Schmidt, 1981; Kimura and Mair, 1981; Mair et al., 1981; Mitchell, 1983; Attewell et al., 1986; Macklin, 1999). A number of correlations have been proposed to relate the face stability of the tunnel directly to ground loss. This is done through the stability number (N) or through the Load Factor (LF). See Equation (3.1) for the definition of N. The Load Factor (LF) is defined as the ratio of the stability number over the stability number at collapse, i.e. $LF = N/N_{collapse}$. The following is a list of some of the relations proposed:

i. $V_s = m\,e^{(N-1)}$ for $N > 1$ and $V_s = m\,N$ for $N < 1$ (Clough and Schmidt, 1981), where $m = 0.002–0.006$. The correlation applies to clays with $E_u/S_u = 500–1500$ where E_u is the undrained Young's modulus.
ii. $V_s = \frac{S_u}{E_u} e^{N/2}$ (Mitchell, 1983), with $E_u/S_u = 200–700$. For poor workmanship, this volume loss should be increased by a factor of 3.
iii. $V_s = 1.33\,N - 1.4$ (Attewell et al., 1986), for $1.5 < N < 4$.
iv. $V_s = 0.23\,e^{4.4\,LF}$ (Macklin, 1999) for $LF \geq 0.2$ and for $z_o/r_o > 3$.

For tunnels excavated in granular soils:

Above the groundwater table, Attewell (1981) and Attewell et al. (1986) suggested a range of $V_s = 2–5\%$, and below the groundwater table, $V_s = 2–10\%$. For small diameter tunnels in man-made fills, Attewell (1981) recommended $V_s = 6\%$ for ancient fills, and 16% for fills less than 50 years old.

An expansion of the Gaussian distribution for settlements in a cross section to a three-dimensional distribution of the surface settlement trough has been proposed by Attewell et al. (1986). The distribution is given by the following function:

$$
\begin{aligned}
S(y) &= S_{max}\, e^{\left(-\frac{y^2}{2i^2}\right)} \left[G\left(\frac{x - x_i}{i}\right) - \left[G\left(\frac{x - x_f}{i}\right)\right]\right] \\
G(\eta) &= \frac{1}{\sqrt{2\pi}} \int_{-\infty}^{\infty} e^{\left(-\frac{\eta^2}{2}\right)} d\eta
\end{aligned}
\tag{3.8}
$$

where the variables and parameters S (y), S_{max} and i are the same as those used in Equation (3.2). The coordinate system is shown in Figure 3.6a; x_i is the initial or tunnel starting point and x_f

is the position of the face or final tunnel position. For y = 0 Equation (3.8) gives the surface settlement of a point above the tunnel axis, and it depends on the function G(η), which is a cumulative probability function. In Equation (3.8), the coordinate $x - x_f = 0$ and y = 0 corresponds to a point on the surface directly above the tunnel face; the value of the function is G (0) = 0.5. A point with coordinate $x - x_i = \infty$ is located very far from the tunnel face and $G(\infty) = 1$.

Equation (3.8) predicts a smooth trough on the surface along the tunnel axis, similar to that shown in Figure 3.6c. The settlements of a point on the surface vertically above the tunnel face are estimated as 50% of the total settlements, S_{max}. This should be compared with the value of 30% predicted in unsupported tunnels in elastic ground with axisymmetric loading (Panet and Guenot, 1982; Panet, 1995; Corbetta et al., 1991; Nam and Bobet, 2007) and with the value of 24–26% in elasto-plastic ground also with axisymmetric loading and unsupported tunnel (Lee et al., 1992). Experience shows however that at a point on the surface above the tunnel face the settlements may range between 10 and 50% of the total settlements (Attewell and Farmer, 1974; Schmidt, 1974; Van der Berg et al., 2003). Table 3.4 shows the estimates of the total settlement at the face of the shield and at the passage of the tail of the shield, as a function of the type of ground (Craig and Muir Wood, 1978). In addition, while the proposed curve predicts a smooth increase of settlements with the passage of the tunnel face, in reality the settlements increase very fast as the face moves below the point on the surface, with deformations of 65–70% of the total. Hence the proposed settlement trough predicted by Equation (3.8) should be merely viewed as a qualitative estimate of the settlements induced by the tunnel construction.

For clays, fine-grained materials and soils with low permeability, the tunnel excavation is sufficiently fast compared with the rate of dissipation of excess pore pressures induced in the soil such that the excavation process occurs in the soil under "undrained" conditions. In other words, due to the fast tunnel advance, there is no time for the water to flow out of the pores of the soil and so the volumetric strain in the soil is zero. Thus, the volume loss at the tunnel in low-permeability soils is equal the volume of the settlement trough at the surface, $V_s = V_f$. For granular soils this may not be the case. Loose granular soils tend to densify with changes in stress and thus the surface settlement volume may be larger than the volume loss at the tunnel. Cording and Hansmire (1975) observed that in tunnels excavated in granular soils there is a volume of soil at the springline with significant compression and a large volume of soil above the crown with large dilation. This phenomenon was also observed in the laboratory in model tests (Atkinson et al., 1977). In addition, fine-grained soils may experience long-term, time-dependent deformations (e.g. consolidation, as the excess pore pressures produced during excavation dissipate), while in granular soils the deformations occur mostly during construction. Hence, the Gaussian distribution seems to be appropriate for settlement data for most tunnels in normally consolidated clays, but it may not be satisfactory in overconsolidated clays, granular soils, and cases with large volumes of yielded ground

Table 3.4 Development of settlement profile (from Craig and Muir Wood, 1978)

	Percentage of total settlement completed	
Type of ground	*At face of shield (%)*	*At passage of tail of shield (%)*
Sand above water table	30–50	60–80
Stiff clays	30–60	50–75
Sand below the water table	0–25	50–75
Silts and soft clays	0–25	30–50[1]

[1] In case of tunnels under compressed air, these percentages are of the initial settlement

(New and O'Reilly, 1991; Loganathan and Poulos, 1998; Celestino et al., 2000; Vorster et al., 2005). Celestino et al. (2000) proposed the following three-parameter function:

$$S(y) = \frac{S_{max}}{1 + \left(\frac{|y|}{a}\right)^b}$$
$$\frac{a}{2r_o} = 0.39\frac{z}{2r_o} + 0.31 \text{ for stiff clay} \tag{3.9}$$
$$\frac{a}{2r_o} = 0.46\frac{z}{2r_o} + 0.37 \quad \text{for porous clay}$$

where S (y), S_{max}, y and z are defined as in Equation (3.2); r_o is the tunnel radius; and a and b are case-dependent parameters. Celestino et al. (2000) provide guidance for parameter a, for stiff and porous clays, as shown in the last two equations in (3.9). For parameter b, they suggest values in the range 2–3 for porous clays and 2–2.8 for stiff clays; however, the parameter b should be always larger than one. The volume of the settlement trough is:

$$V_s = \frac{2\pi a}{b}\frac{S_{max}}{\sin(\pi/b)} \tag{3.10}$$

3.4.2.2 *Indirect methods*

As mentioned, indirect methods estimate ground deformations from the ground loss. They rely on empirical estimates of the ground loss and use analytical and numerical methods to estimate ground deformations based on the deformations inside the tunnel.

Attewell (1977) provided an estimate of the distribution of total volume loss induced by tunneling operations (see also Attewell et al., 1986). In the method, the sources of volume loss are classified as: face loss, V_f; shield loss, V_b; and post-shield loss, V_p. See Section 3.4.1 and Figure 3.8. Their magnitudes are computed based on geometric considerations and as a percentage of the total theoretical volume excavated. Table 3.5 lists the contributions, from each source, to the total ground loss. The total volume loss is then $V_t = V_f + V_b + V_p$. The method is useful to estimate the sources of ground loss during construction. As the tunnel is excavated, the parameters needed to compute the different components of the ground loss can be measured and the relative importance of each of the sources can be estimated. With this information, appropriate decisions can be made to reduce, if necessary, those sources that contribute to the largest deformations.

The ground loss at the tunnel level and during construction (i.e. short term) has been obtained analytically by Lo et al. (1984) and Lee et al. (1992). It is applicable to cohesive saturated soils in undrained conditions. The concept of the "gap" parameter, g, is introduced (see also Section 4.2.1.2). The magnitude of the parameter g is the difference between the actual and the theoretical diameter of the tunnel excavated. Thus, the gap parameter is related to the ground loss as:

$$V_t = 100\frac{\pi(r_o + 0.5g)^2 - \pi r_o^2}{\pi r_o^2} \approx 100\frac{4gr_o + g^2}{4r_o^2} \tag{3.11}$$

The total gap is:

$$g = g_p + u_{3D} + \omega \tag{3.12}$$

Table 3.5 Total volume loss induced by shield operations (after Attewell, 1977; Attewell et al., 1986), $V_t = V_f + V_b + V_p$

Source	*Ground loss (%)*	*Remarks*
Face loss (V_f)	$V_f = 100k_1 \frac{m'}{l'}$ or $V_f = 200k_1 \frac{m}{r_o}$ (Cording et al., 1976)	m′ is the average rate at which the material moves unrestrained into the excavation; l′ is the average rate of tunnel advance; m is the average amount of soil movement into the tunnel; r_o the radius of the tunnel; and k_1 is a factor between 0 and 1, and is a measure of the doming deformation of the soil at the face; $k_1 \simeq 0.5$;
Shield loss ($V_b = V_{bb} + V_{bp} + V_{by} + V_{bc}$)	Overcutting due to the shield bead, V_{bb}; $V_{bb} = 200 l_s \frac{k_2}{r_o} \frac{m'}{l'}$	l_s is the length of the shield; l′ is the average rate of shield advance; and k_2 is a factor between 0 and 1 and represents the percentage of the bead cut closed by soil intrusion. The equation should be divided by a factor of 2 if the bead extends only 180° around the shield.
	Pitching of the shield, $V_{bp} = 50 \frac{p}{r_o}$ (Cording et al., 1976)	p is the difference in elevation between the front and rear of the shield minus the design grade of the tunnel.
	Yawing, V_{by} $V_{by} \approx 0$	These movements have been generally found to be much smaller than those produced by shield plowing (Hansmire and Cording, 1972), and are generally neglected. However poor workmanship in steering the shield could produce significant ground losses.
	Curve negotiation, V_{bc} $V_{bc} = 50 \frac{L^2}{r_o (r_c + r_o)}$ (Schmidt, 1974)	L is the length of the shield; and r_c is the radius of the curve.
Post shield loss (Vp)	$V_p = 200 \frac{l_u}{r_o} \frac{m'}{l'}$,	l_u is the unsupported ungrouted length behind the tail of the shield.

In Equation (3.12), g_p represents the physical gap between the crown of the excavated surface and the crown of the tunnel lining, u_{3D} is related to the ground loss that occurs at the face of the tunnel, and ω is the workmanship parameter and is a function of the steering of the shield and construction quality (note that while g_p and u_{3D}, in the method, are obtained analytically or numerically, ω is largely estimated empirically). The gap parameter can be considered as the maximum settlement at the tunnel crown, and it can be measured in the field.

The physical gap g_p is the geometric clearance between the outer skin of the shield and the lining, and may be obtained by the addition of the thickness of the shield tail, Δ, and the clearance needed for erection of the lining, δ. See Figure 3.13a. The figure shows the gap at the crown. This is because the weight of the shield and lining makes them rest on the invert of the tunnel where the gap is zero. Thus,

$$g_p = 2\Delta + \delta \tag{3.13}$$

The ground loss due to displacements of the soil ahead of the tunnel face, u_{3D} (Figure 3.13b), is obtained from three-dimensional analysis, for an elastic-perfectly plastic soil (Lee et al., 1992). The Tresca failure criterion was adopted in the analyses to simulate undrained conditions. The results from the numerical simulations were obtained for Young's modulus of the soil E_u/S_u = 200-800, which was either constant or increasing

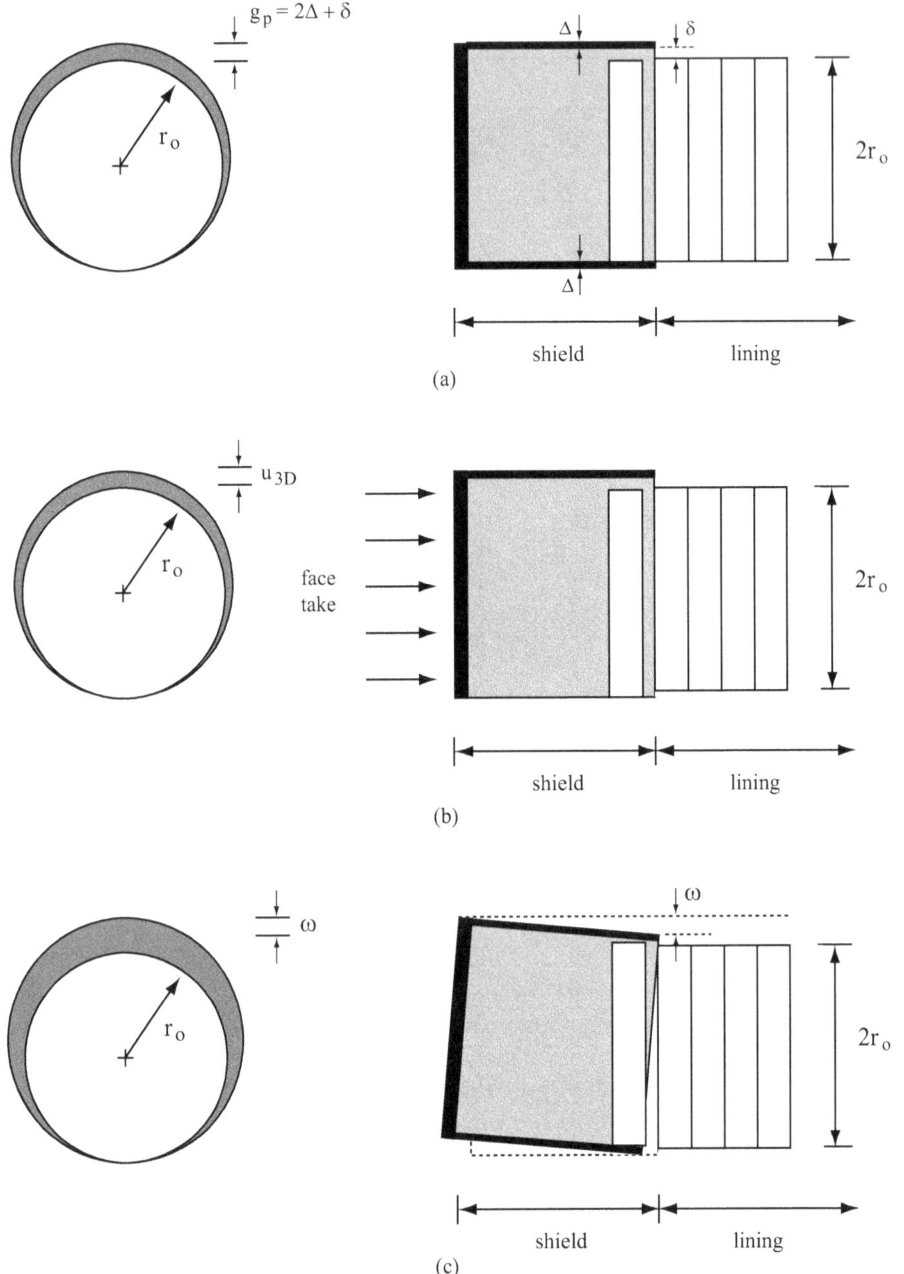

Figure 3.13 Sources for the gap parameter. (a) Physical gap g_p. (b) Face take u_{3D}. (c) Workmanship. (after Lee et al., 1992).

linearly with depth. The tunnel depth ranged from $z_o/r_o = 3$ to 9. The values of u_{3D} are given by:

$$u_{3D} = \frac{1}{2}\frac{r_o \sigma_{hr}}{E_u}\Omega \tag{3.14}$$

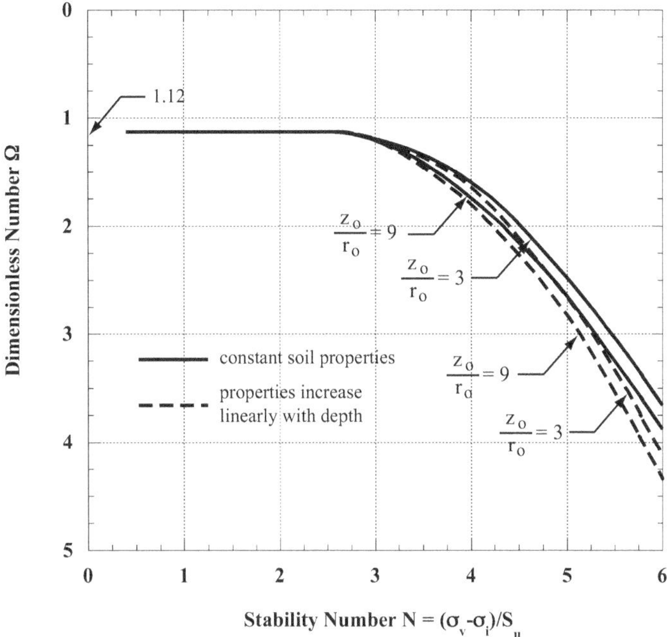

Figure 3.14 Dimensionless number Ω (after Lee et al., 1992).

In Equation (3.14), σ_{hr} is the total stress removed at the tunnel face, $\sigma_{hr} = K_o\,\sigma'_v + u_o - \sigma_i$, where K_o is the coefficient of earth pressure at rest, σ'_v is the effective vertical stress and u_o is the pore pressure, all computed at the tunnel axis and before tunnel excavation; σ_i is the support pressure at the face (compressed air pressure, pressure of the TBM at the face, etc.). Also in the equation, E_u is the undrained Young's modulus of the soil in extension. The factor Ω is given in Figure 3.14, which plots values for $K_o = 1$ and for soil properties that are either constant or increasing linearly with depth. The factor strongly depends on the stability number N and, to a much lesser extent, on tunnel depth. For stability numbers N less than about 2.5 the values of Ω are constant and equal to Ω = 1.12, which indicates that the soil ahead of the tunnel face remains elastic. Results from the numerical simulations (Lee et al., 1992) show that Ω is quite insensitive to K_o (results not shown in Figure 3.14), at least within the range considered $0.6 \le K_o \le 1.4$. For shields that provide support to the face such as slurry or earth pressure balance shields, face deformations, i.e. the value of u_{3D}, will decrease and may become zero if the pressure of the soil at the face is not released during construction.

Recall that the ground losses over the shield are due to radial movements of the soil caused by over-excavation, overcutting by the bead, and pitching and yawing of the shield. Over-excavation may be produced by intermittent alignment and steering problems that allow the soil to move radially and fill the void created. The radial displacements at the face due to over-excavation of the tunnel cross section are (Lee et al., 1992):

$$\omega = \frac{1}{3} r_o \left[1 - \sqrt{\frac{1}{1 + \frac{2(1+\nu_u)S_u}{E_u}\left(e^{\frac{N-1}{2}}\right)^2}} \right] \quad \text{if } \omega \le 0.6\,g_p\text{; otherwise:} \tag{3.15}$$

$$\omega = 0.6\,g_p$$

where ν_u is the undrained Poisson's ratio (i.e. $\nu_u \approx 0.5$). The first equation represents the radial deformations at the face if there is no support. The radial deformations along a longitudinal section of an unsupported tunnel, as illustrated in Figure 3.6c, start in the soil ahead of the face and increase toward the tunnel. At the tunnel face they have a magnitude of 50–60% of the maximum (Lee et al., 1992). The maximum radial deformations are limited by the physical gap, since for larger deformations the soil would be supported by the lining. Hence the value ω should be limited to a fraction of the physical gap, in this case a value of 0.6 is assumed.

Additional gap magnitudes may come from the shield bead, if there is one, and yawing and pitching (see also estimates in Table 3.5 from geometry considerations). If the shield has an overcutting bead, the additional gap is given by:

$$\omega = nt \tag{3.16}$$

where n = 0 if there is no bead, n = 1 when the bead is on the upper 180° of the shield and n = 2 for the entire 360°; t is the bead thickness. Yawing movements are in general reported to be small and are directly related to workmanship, and so they are very difficult to estimate a priori. The gap due to pitching is caused by the excess grade of the shield with respect to the theoretical grade (Figure 3.13c). It is given by:

$$\omega = pL \tag{3.17}$$

where p is the excess pitch and L is the length of the shield.

The ground losses after erection of the lining (post-shield losses) are due to the physical gap because the soil squeezes into the void left behind the shield as the shield moves forward. The magnitude of the ground loss may be reduced by backfilling the void with grout or pea gravel or by using an expanded lining. The loads on the lining will typically induce inward deformations at the crown and outward deformations at the springline (i.e. ovaling of the liner). For most cases, the magnitude of the ground loss associated with the lining ovalization is small compared with the other sources and may be neglected. An additional source for ground loss, not included in the derivation, is long-term deformations of the soil. If the volume of soil disturbed during construction is small and the tunnel does not act as a drain, the long-term deformations are typically much smaller than the short term. If disturbance of the soil is significant (e.g. large soil heave due to excessive pressure at the face) or if the tunnel acts as a drain (e.g. permeable or semi-permeable tunnels below the water table), long-term settlements can be substantial as well as the loads induced in the liner (Attewell, 1977; Finno and Clough, 1985).

In summary the total gap, given by Equation (3.12), is computed as the sum of: the physical gap, obtained from Equation (3.13); face loss, from Equation (3.14); and workmanship, obtained by the sum of Equations (3.15), (3.16), and (3.17).

Two-dimensional models have been used with the gap parameter to predict soil deformations and surface settlements (Rowe and Kack, 1983; Lo et al., 1984; Rowe, 1986; Rowe and Lee, 1992). In essence a two-dimensional simulation is employed to represent a three dimensional phenomenon by lumping face effects and face operations into the gap parameter. The idea is to use the gap computed in Equation (3.12) as a void distributed between the liner and the soil (see for example Figure 3.13), which represents the deformations induced by the construction procedure, shield geometry, quality of construction, etc. The soil can freely deform until the radial distance between the excavation and the liner is filled, in which case further soil displacements are prevented by the liner, which in turn takes load (see Sections 4.2.1.2 and 4.3.5). Elasto-plastic models for the soil are typically used, and should be preferred to elastic analysis. First because in soft soils there might be a substantial plastic volume of ground around the tunnel, and second because elastic models tend to underpredict the magnitude of the settlements and overpredict the width of the settlement trough (Rowe and Kack, 1983; Chou and Bobet, 2002). A substantial number of comparisons between two-dimensional models using the gap parameter and measurements from tunnel construction show that the method produces reasonable predictions (Rowe and Lee, 1992). Figure 3.15a shows a comparison between the method and measurements performed at Green Park Underground, a tunnel driven in heavily overconsolidated London clay. The tunnel was excavated at a depth of 29.3 m and had a radius of 2.07 m. The gap parameter estimated for the tunnel was 24–27 mm, which was used for the simulations. The soil Young's modulus adopted was E_u = 53 MPa. Figure 3.15a plots observed and predicted vertical settlements inside the soil above the tunnel centerline. The predictions are reasonable.

As an alternative to the method discussed, a relation between the crown settlement and the surface settlement can be employed. The gap parameter can be considered as the vertical displacement above the crown, and so the maximum surface settlement S_{max} can be estimated using the empirical correlation suggested by Ng (1991):

$$\begin{aligned} S_{max} &= \frac{1}{3}g && \text{for soft to firm clays} \\ S_{max} &= \left(0.95 - 0.22\frac{z_o}{2Nr_o}\right)g && \text{for stiff to hard clays} \end{aligned} \tag{3.18}$$

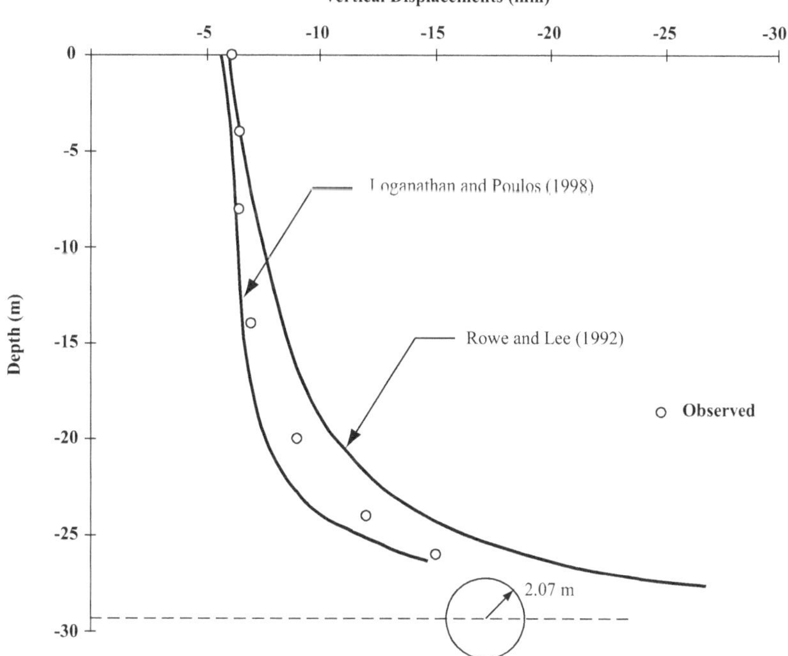

Figure 3.15a Green park underground tunnel. Subsurface settlements above centerline. Comparison between observations and Rowe and Lee (1992) and Loganathan and Poulos (1998) methods.

The concept of the gap parameter was used by Loganathan and Poulos (1998) to propose a close form solution to predict the ground displacements around a tunnel. The analytical solution derived by Verruijt and Booker (1996) was modified to capture better the magnitude and width of the settlement trough observed in tunnels. Verruijt and Booker (1996) obtained an analytical solution for the stress and strain fields of a tunnel excavated in an elastic medium subjected to uniform radial displacement and ovalization of the tunnel perimeter. The solution was an expansion of the formulation previously derived by Sagaseta (1987). The new approach provides vertical and horizontal displacements at a cross section well behind the tail of the shield, and is applicable, as the Lee et al. (1992) gap parameter, to undrained (i.e. short-term) deformations in tunnels in clay. The solution is given by the following equations:

$$U_z = r_o^2\left\{-\frac{z-z_o}{y^2+(z-z_o)^2}+(3-4\nu)\frac{z+z_o}{y^2+(z+z_o)^2}-\frac{2z[y^2-(z+z_o)^2]}{[y^2+(z+z_o)^2]^2}\right\}$$
$$\frac{4r_o g+g^2}{4r_o^2}e^{-\left(\frac{1.38y^2}{(z_o+r_o)^2}+\frac{0.69z^2}{z_o^2}\right)}$$
$$U_y = -r_o^2 y\left\{-\frac{1}{y^2+(z-z_o)^2}+\frac{3-4\nu}{y^2+(z+z_o)^2}-\frac{4z(z+z_o)}{[y^2+(z+z_o)^2]^2}\right\}$$
$$\frac{4r_o g+g^2}{4r_o^2}e^{-\left(\frac{1.38y^2}{(z_o+r_o)^2}+\frac{0.69z^2}{z_o^2}\right)} \qquad (3.19)$$

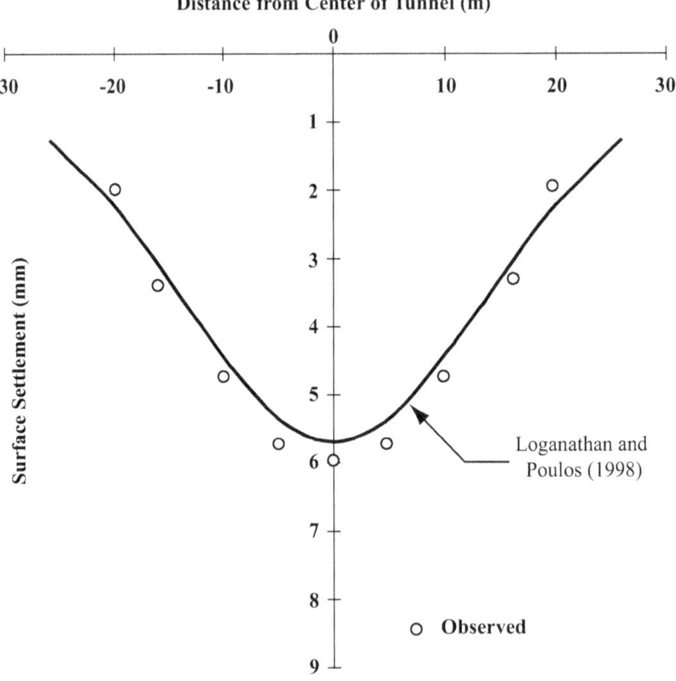

Figure 3.15b Green park underground tunnel. Surface settlements. Comparison between observations and Loganathan and Poulos (1998) method.

U_z are the vertical displacements and U_y are the horizontal displacements at coordinates y, z, which are defined in Figure 3.6; g is the gap parameter, which is obtained after Lee et al. (1992); e.g. Equation (3.12). Figure 3.15a shows a comparison between predictions made with the model and observations at Green Park Underground. The predictions are reasonable and of the same order as those obtained by Rowe and Lee (1992). Figure 3.15b shows observations and predictions of surface settlements for the same tunnel. Again, the differences are reasonable and within the level of accuracy expected for this type of prediction. Additional comparisons presented by Loganathan and Poulos (1998) and Loganathan et al. (2000) provide confidence in the method.

Pinto and Whittle (2014) extended the analytical solutions of Pinto (1999) and Verruijt and Booker (1996) to incorporate soil plasticity for the computation of ground deformations due to tunneling. As with Verruijt and Booker (1996), the sources of deformations considered are: u_ε, tunnel convergence; and u_δ, tunnel ovalization (Figure 3.16a). For a plastic, dilating soil with Poisson's ratio $\nu = 0.5$, the deformations are:

Due to Convergence:

$$\begin{aligned}
\frac{U_y}{u_\varepsilon r_o^{2\alpha-1}} &= \left\{\frac{y}{[y^2+(z-2z_o)^2]^\alpha} + \frac{y}{[y^2+z^2]^\alpha} - \frac{4(z-z_o)yz}{[y^2+z^2]^{\alpha+1}}\right\} \\
\frac{U_z}{u_\varepsilon r_o^{2\alpha-1}} &= -\left\{\frac{3z}{[y^2+z^2]^\alpha} + \frac{z-2z_o}{[y^2+(z-2z_o)^2]^\alpha} - \frac{4zy^2-2z_o[y^2-z^2]}{[y^2+(z-z_o)^2]^{\alpha+1}}\right\}
\end{aligned} \tag{3.20}$$

Due to Ovalization:

$$\begin{aligned}
\frac{U_y}{u_\delta r_o^{2\alpha-1}} &= y\left\{\begin{array}{l}
\dfrac{[y^2+(z-2z_o)^2]^2+[y^2-3(z-2z_o)^2][y^2+(z-2z_o)^2-r_o^2]}{[y^2+(z-2z_o)^2]^{\alpha+2}} \\
-\dfrac{[y^2+z^2]^2+[y^2-3z^2][y^2+z^2-r_o^2]}{[y^2+z^2]^{\alpha+2}} + 4\dfrac{y^2+(z-z_o)^2-z_o^2}{[y^2+z^2]^{\alpha+1}} \\
-8(z-z_o)\dfrac{(z-z_o)[y^2+(z-z_o)^2]-2z_o[z_o^2-y^2]-3(z-z_o)z_o^2}{[y^2+z^2]^{\alpha+2}}
\end{array}\right\} \\
\frac{U_z}{u_\delta r_o^{2\alpha-1}} &= \left\{\begin{array}{l}
-(z-2z_o)\dfrac{[y^2+(z-2z_o)^2]^2-[3y^2-(z-2z_o)^2][y^2+(z-2z_o)^2-r_o^2]}{[y^2+(z-2z_o)^2]^{\alpha+2}} \\
+z\dfrac{[y^2+z^2]^2-[3y^2-z^2][y^2+z^2-r_o^2]}{[y^2+z^2]^{\alpha+2}} - 4\dfrac{y^2(z+z_o)+(z-z_o)z^2}{[y^2+z^2]^{\alpha+1}} \\
-8z\dfrac{z_o(z-z_o)z^2+y^2[y^2+(z-z_o)^2-z_o(z-2z_o)]}{[y^2+z^2]^{\alpha+2}}
\end{array}\right\}
\end{aligned} \tag{3.21}$$

where x, y, z are the Cartesian coordinates, r_o is the radius of the tunnel, z_o is the depth of the tunnel (see Figure 3.6), and $\alpha = 1/(1\text{-}\sin\psi)$, with ψ, the dilation angle of the soil. Pinto et al. (2014) suggest calibrating the parameters in Equations (3.20) and (3.21) with three field measurements, as shown in Figure 3.16b, or alternatively by minimizing the least-squares differences between field measurements and predictions from the analytical solution. Estimates from the equations have been compared with field measurements from a number of tunnels (Pinto et al., 2014), and seem to provide good approximations of the ground deformations. Figure 3.16c shows the predicted and measured vertical deformations around the Second Heinenoord Tunnel, excavated with a slurry shield (Pinto et al., 2014).

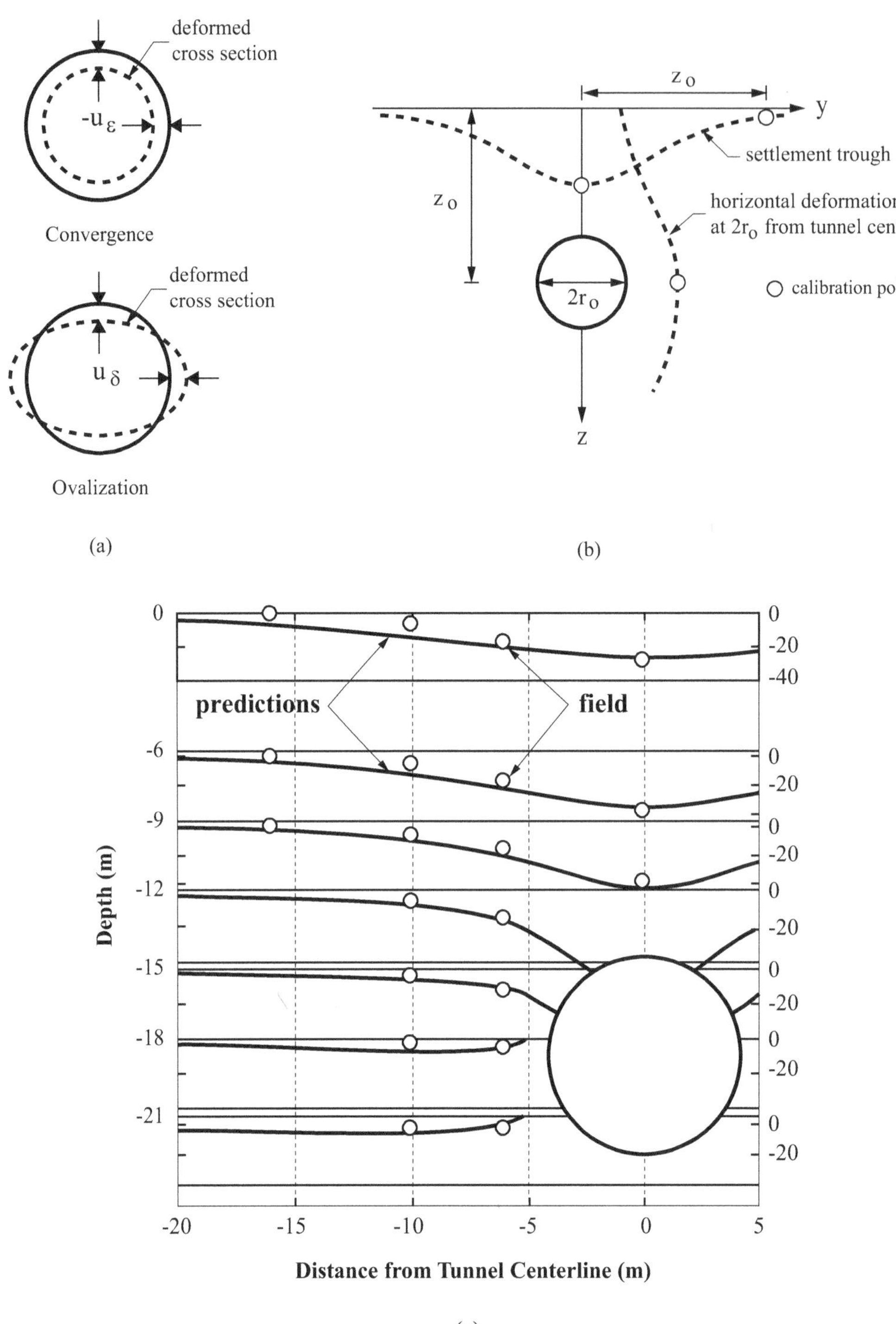

Figure 3.16 Calibration of analytical solution by Pinto and Whittle (2014) and comparison of predictions and measurements for the Heinenoord Tunnel. (a) Sources of tunnel deformation. (b) Model calibration. (c) Second Heinenoord Tunnel. Vertical displacements (mm). (after Pinto et al., 2014)

3.4.2.3 Twin tunnels

Figure 3.17 shows the surface settlement troughs observed during tunnel construction. The settlement trough depicted in Figure 3.17a is typical of single tunnels or of the first of the twin tunnels excavated. Construction of the second tunnel produces additional settlements as its excavation interacts with the first tunnel and induces distortions to the lining or additional deformations in the volume of soil affected by the first excavation. The resulting settlement trough is thus generally larger that the trough induced by any of the tunnels if excavated individually. The shape of the settlement trough can be symmetric and centered at the midpoint between the two tunnels or shifted toward one of the tunnels; see Figure 3.17b. Peck (1969b) suggested that if the tunnels are sufficiently distanced from each other, then the resulting settlement trough could be obtained by the superposition of the troughs created by the two tunnels if considered as single tunnels. The resulting curve can be also approximated with the Gaussian curve given by Equation (3.2). Peck (1969b) recommended, for the settlement trough width i, to use an "equivalent" radius, $r'_0 = r_0 + 1/2\,d$ where d is the distance between the tunnel centers. For settlement troughs shifted with the center (Figure 3.17b), New and Bowers (1993) suggested:

$$S(y) = S_{max} e^{\left[-\frac{(y-a)^2}{2i^2}\right]} \tag{3.22}$$

where a is a parameter that gives the shift with the center.

Suwansawat and Einstein (2007) investigated the settlement troughs produced during the construction of the Bangkok MRTA project. The project consisted of the construction of 20 km of twin tunnels and was divided into the north bound (NB) and south bound (SB) sections. The tunnels were excavated in stiff Bangkok clay about 15–25 m below the ground surface and had an outer diameter of 6.3 m and an inner diameter of 5.7 m. The tunnels were built below the groundwater table with Earth Pressure Balance (EPB) shields. It was observed that if the operation of the EPB shields was the same in both tunnels (i.e. same face pressure, grouting pressure, etc.) then the final settlement was symmetric. In contrast, when the face pressure, grout pressure, etc. were different, an asymmetric trough was produced (Figure 3.17c). An interesting observation was that the additional settlements induced by the passage of the second tunnel were smaller than the settlements created by the first tunnel and that they also conformed to a Gaussian distribution. The "additional" settlements were obtained by the difference between the final settlements measured after the passage of the second shield and those measured after passage of the first shield. Based on these observations, recommendations were made to obtain the settlement trough of twin tunnels for any type of settlement trough; i.e. symmetric, shifted symmetric or asymmetric (Figure 3.17). The procedure is illustrated in Figure 3.18 and is as follows:

1. Measure the maximum settlement at the centerline of the first tunnel and describe the settlement trough using a Gaussian distribution. The settlement trough width i can be obtained using the correlations proposed by O'Reilly and New (1982) given by Equation (3.6).
2. Measure the final settlement above the centerline of the second tunnel. This is point B in Figure 3.18.
3. Calculate the additional settlement produced by the passage of the second shield at the center of the second tunnel. This is point C in Figure 3.18, and is obtained by the

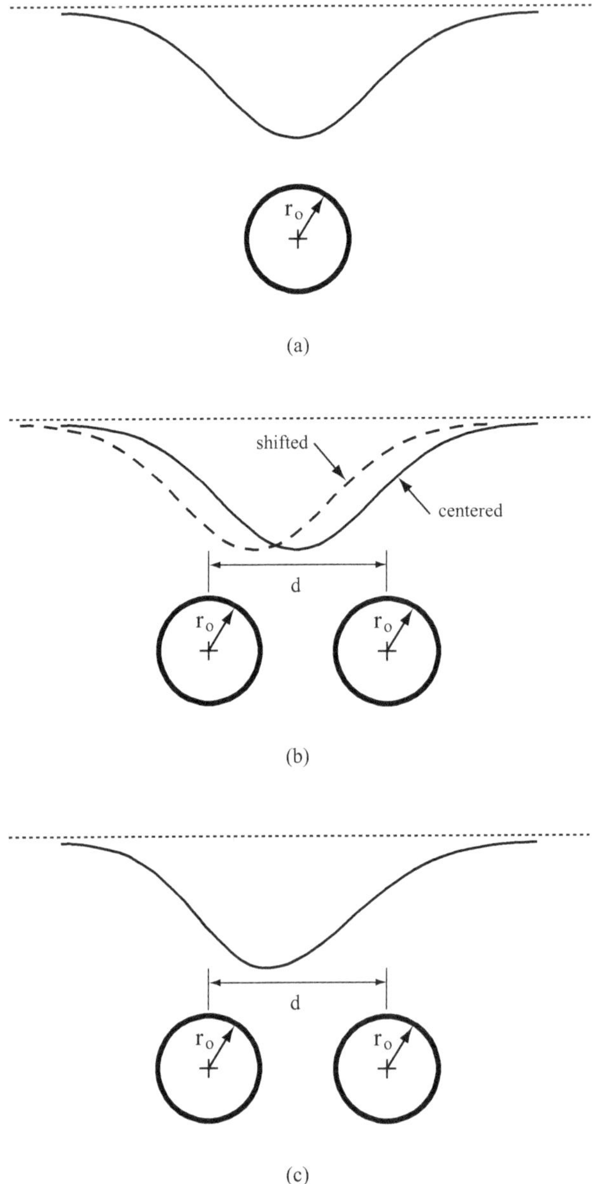

Figure 3.17 Types of settlement trough. (a) Single tunnel (b) Twin tunnel Symmetric settlement trough (c) Twin tunnel. Asymmetric settlement trough. (after Suwansawat and Einstein, 2007).

difference between the settlement from the first shield (point A) and the final settlement (point B).

4. Construct the Gaussian curve for the additional settlement. This curve has as maximum settlement the value obtained in step (3); that is, point C. The trough width i is also calculated with Equation (3.6).
5. The final settlement trough is obtained by superposition of the Gaussian curves for the first shield and for the additional settlements.

The method proposed by Suwansawat and Einstein (2007) is an empirical method and requires prior knowledge of the surface settlements above the centers of the tunnels. The authors

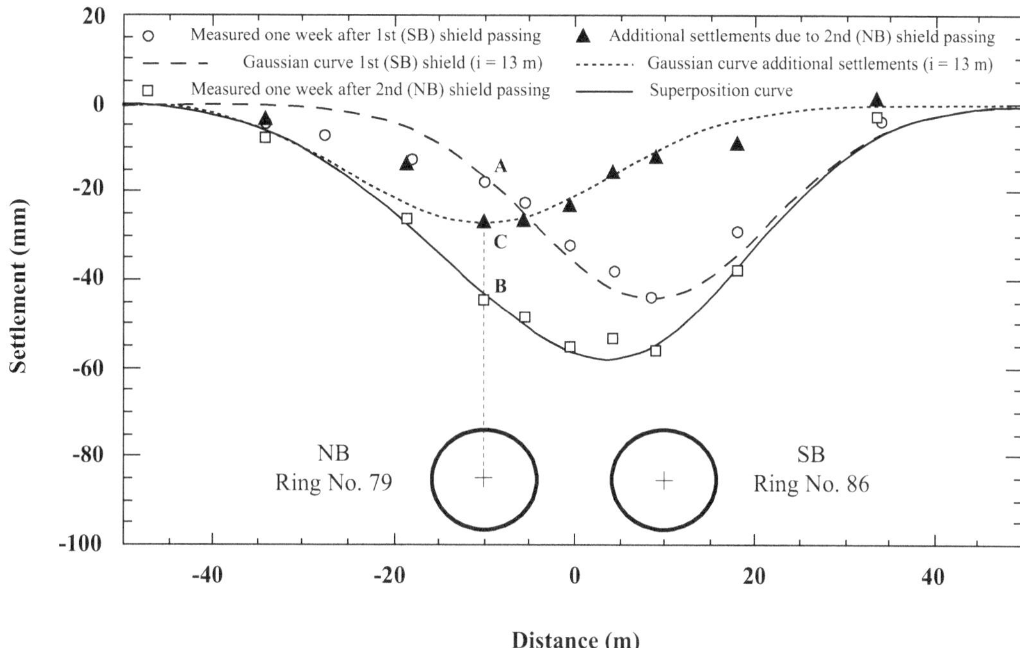

Figure 3.18 Surface settlements on 26-AR-001, Bangkok tunnel. Comparison between Gaussian curves and superposition curves (after Suwansawat and Einstein, 2007).

suggest that the method proposed can be used for twin tunnels located horizontally next to each other or stacked one on top of each other on a vertical plane.

3.4.3 Building damage due to tunneling

There are essentially two methods to estimate the damage to adjacent structures due to tunneling: the free-field method and the soil–structure interaction method. The free-field method is based on the notion that the structure will follow the deformations imposed to the ground due to the tunnel construction. This approach is acceptable for brick bearing wall and small frame structures (Burland and Wroth, 1974; Boscardin and Cording, 1989) since their stiffness is small enough such that no constraints are induced to the ground by the structure as it deforms. If the structure is stiff enough such that it limits the deformation of the ground, or not all ground deformations may be transmitted to the structure, then soil–structure interaction applies. Soil–structure interaction typically requires the use of numerical methods that can model the ground, the structure, and the construction process. These are discussed in Chapter 5, while this section is concerned with the free-field method.

Early studies on allowable settlements on structures include those of Peck et al. (1956), Skempton and MacDonald (1956) and Polshin and Tokar (1957). Skempton and MacDonald analyzed 98 buildings, 40 of them with damage. They introduced the notion of angular distortion, β (defined as the ratio between the differential settlement between two points and their distance), as an indicator of damage to buildings. They concluded that for values of β below 1/300, there was no cracking in walls and partitions, while for values larger than 1/150 there was structural damage. Subsequently Bjerrum (1963) expanded these recommendations for a range of different types of damage, as shown in Table 3.6. While the relation between damage and angular distortion seemed to work well in some instances, it was not always the case (Burland and Wroth, 1974).

Table 3.6 Damage criteria (Bjerrum, 1963)

Angular distortion, β	*Description of damage*
1/750	Limit where difficulties with machinery sensitive to settlements are to be feared.
1/600	Limit of danger for frames with diagonals.
1/500	Safe limit for buildings where cracking is not permissible.
1/300	Limit where first cracking in panel walls is to be expected.
	Limit where difficulties with overhead cranes are to be expected.
1/250	Limit where tilting of high, rigid buildings might become visible.
1/150	Considerable cracking in panel walls and brick walls, safe limit for flexible brick walls with H/L < ¼; limit where structural damage of general buildings is to be feared.

Burland and Wroth (1974) made the first attempt to relate damage with building geometry (given by the ratio of length of the building, L, to building height, H; i.e. L/H), critical tensile strain, ε_c (the limiting or critical tensile strain associated with the onset of cracking), building stiffness and building deformations. The first two concepts, e.g. L/H and ε_c, were borrowed from Polshin and Tokar (1957). Burland and Wroth applied the concept of critical strain to an elastic beam of length L, height H and unit thickness, which was taken, as a first approximation, as a representation of a building (see Figure 3.19). The beam undergoes two modes of deformation, both associated with a deflection Δ (Figure 3.19b): bending (Figure 3.19c) and shearing (Figure 3.19d). The equations that relate maximum bending strain, $\varepsilon_{b\,max}$, and maximum diagonal strain (or principal strain), $\varepsilon_{d\,max}$, as a function of beam deflection are:

When the neutral axis is in the middle of the beam:

$$\frac{\Delta}{L} = \left(0.167\frac{L}{H} + 0.65\frac{H}{L}\right)\varepsilon_{b\,max} \tag{3.23a}$$

$$\frac{\Delta}{L} = \left(0.25\frac{L^2}{H^2} + 1\right)\varepsilon_{d\,max} \tag{3.23b}$$

When the neutral axis is located at the lower extreme fiber of the beam

$$\frac{\Delta}{L} = \left(0.083\frac{L}{H} + 1.3\frac{H}{L}\right)\varepsilon_{b\,max} \tag{3.24a}$$

$$\frac{\Delta}{L} = \left(0.064\frac{L^2}{H^2} + 1\right)\varepsilon_{d\,max} \tag{3.24b}$$

The equations are not very sensitive to the type of loading, e.g. concentrated, distributed, etc. All relations are obtained for an isotropic beam with Poisson's ratio $\nu = 0.3$ (i.e. for a ratio E/G = 2.6). Note that Equation (3.24a) applies only to hogging since for sagging the tensile strain at the bottom of the beam would be zero. The maximum distortion of an isotropic elastic beam deforming in shear as well as in bending, with the neutral axis at the middle, is:

$$\frac{\Delta}{L} = \frac{1}{3}\beta\left[\frac{1 + 3.9(H/L)^2}{1 + 2.6(H/L)^2}\right] \tag{3.25}$$

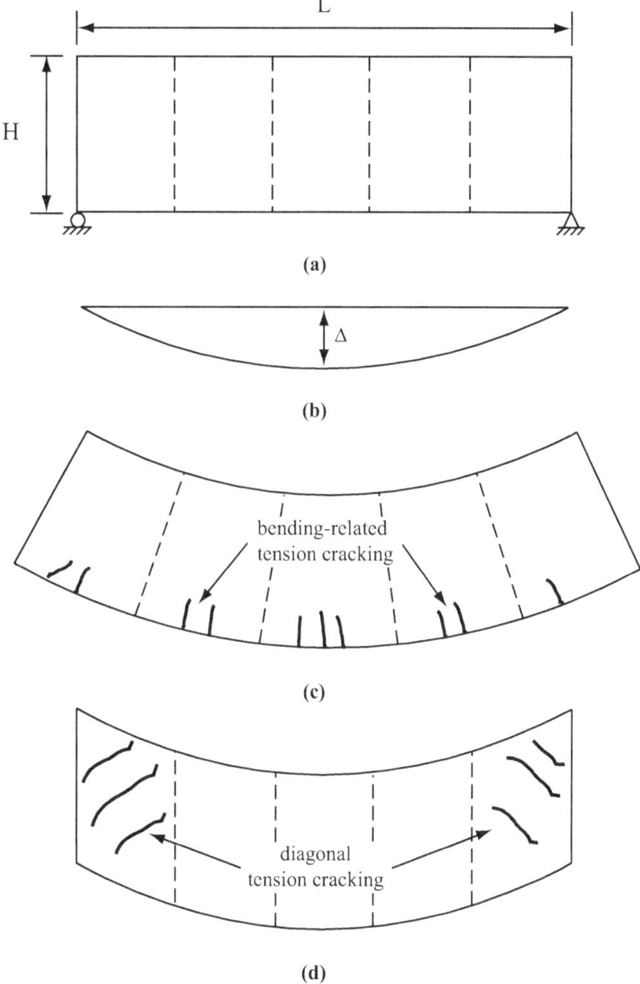

Figure 3.19 Deep beam model. (a) Simply supported beam. (b) Deflected shape. (c) Bending mode. (d) Shearing mode (after Burland and Wroth, 1974).

Figure 3.20 provides a comparison between the predictions for the onset of cracking obtained from the simple beam analysis and those observed from buildings (Burland and Wroth, 1974). The limiting curves in the figure were obtained from the beam equations and for a critical strain $\varepsilon_c = 0.075\%$ and $\beta = 1/500$. The following observations were made:

1. Determination of relative building deflection requires knowledge of: critical tensile strains of the building materials and finishes, length to height ratio of the structure, L/H, the approximate ratio between the equivalent longitudinal stiffness and the shear stiffness (E/G), and the degree of tensile restraint of elements in the building.
2. For structures with relatively low shear stiffness or a significant degree of tensile restraint such as frame buildings or reinforced bearing walls, the onset of cracking is due to diagonal tensile strain.
3. For structures with negligible tensile strength such as brick and masonry buildings, the onset of cracking is due to bending, particularly when the length to height ratio of the building is larger than two.

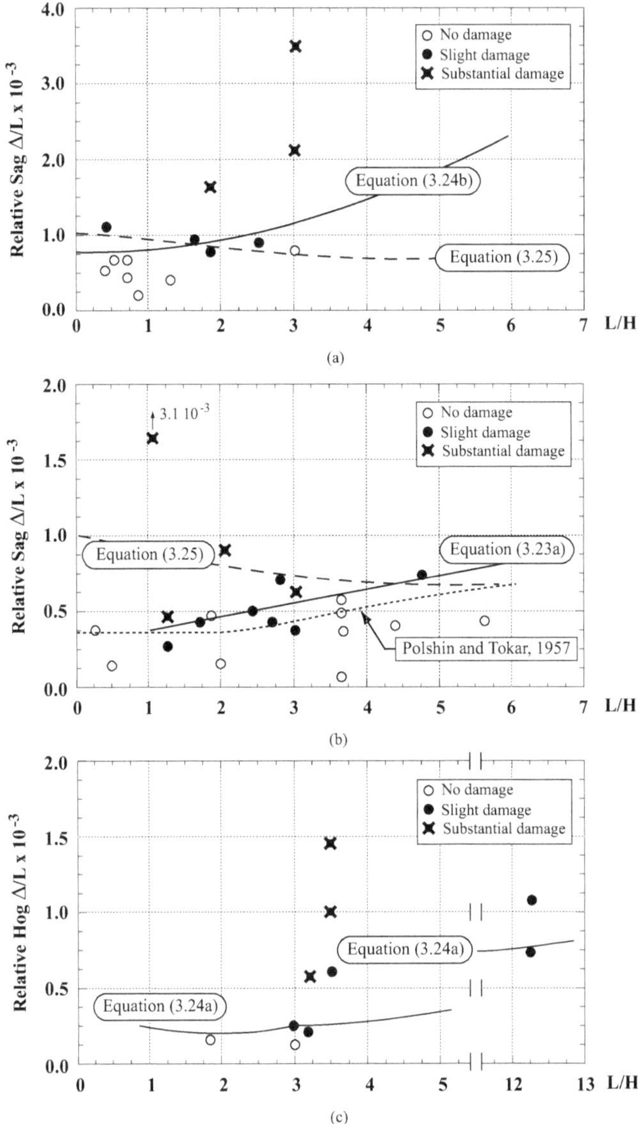

Figure 3.20 Relation between Δ/L and L/H for buildings showing various degrees of damage. ε_c = 0.075%, β = 1/500. (a) Frame buildings; (b) Load bearing walls; (c) Hogging of load bearing walls. From Burland and Wroth (1974).

4. Brick and masonry buildings subjected to hogging, cracking due to bending is likely to occur at very low deflection.
5. Figure 3.20 can be used as a very first estimate, even though the authors strongly suggest that additional data should be collected.

Boscardin and Cording (1989) expanded the concepts put forward by Burland and Wroth (1974). The beam undergoes two modes of deformation, both associated with a deflection Δ (Figure 3.19b): bending (Figure 3.19c) and shearing (Figure 3.19d). They related the angular distortion of the beam, subjected to a central point load and neutral axis at the lower edge of the beam, with the maximum deflection and horizontal strain. They obtained the following equations:

$$\beta = \frac{3\Delta}{L}\left[\frac{1+4\frac{E}{G}\left(\frac{H}{L}\right)^2}{1+6\frac{E}{G}\left(\frac{H}{L}\right)^2}\right]$$
$$\frac{\Delta}{L} = \left[0.064\left(\frac{L}{H}\right)^2+1\right]\left[\frac{\varepsilon_c - \varepsilon_h \cos^2\theta_{max}}{\sin 2\theta_{max}}\right] \tag{3.26}$$

where β is the angular distortion; Δ, H and L are as defined before (Figure 3.19a), E and G are the Young's modulus and shear modulus of the beam, ε_c is the critical strain associated with the onset of visible damage, ε_h is the horizontal strain and, when there is diagonal tension with horizontal extension, θ_{max} is the angle, measured from the horizontal, of maximum diagonal tensile strain.

Boscardin and Cording (1989) related building damage to angular distortion and horizontal strain. Figure 3.21a illustrates the determination of the angular distortion of a structure placed on the settlement trough induced by a tunnel. The horizontal strain, ε_h, is the

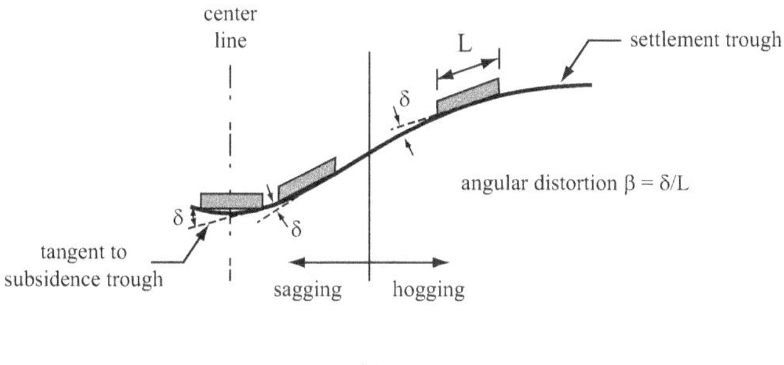

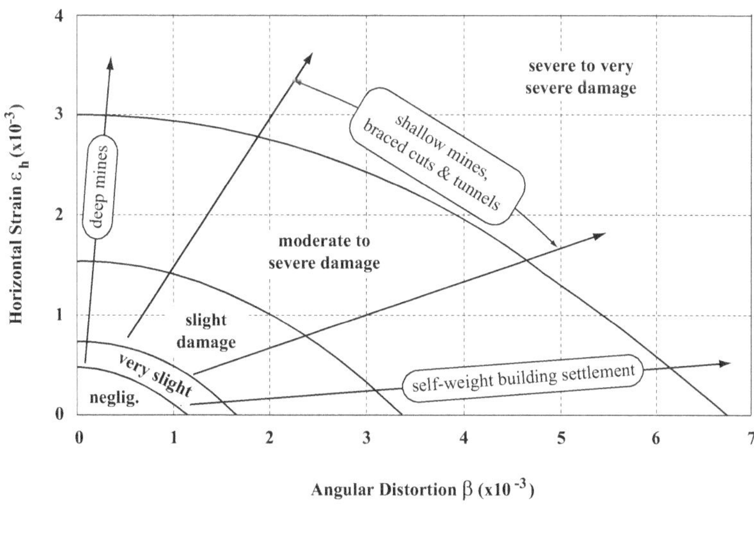

Figure 3.21 Relation between damage, angular distortion and horizontal extension strain for bearing wall and small frame structures. (a) Angular distortion within a settlement trough. (b) Damage in bearing wall and small frame structures. (adapted from Boscardin and Cording, 1989).

horizontal strain due to the relative horizontal movement of two reference points of the structure, and can be calculated either from the results of numerical methods when used to estimate surface deformations (Chapter 5), or from indirect methods, e.g. Equation (3.19), or from direct, empirical methods, e.g. Equation (3.4).

Figure 3.21b is a plot of Equation (3.26) with L/H = 1 and the neutral axis at one edge (Boscardin and Cording do not provide the values of E/G, but it seems that E/G = 2.6, the value for an isotropic beam, was used based on information from later publications, e.g. Son and Cording, 2005). The severity of damage is associated in the figure with the magnitude of the horizontal strain, ε_h, and the angular distortion, β. The damage is classified into five categories: negligible, very slight, slight, moderate to severe, and severe to very severe. The boundaries of the "very slight" damage zone are obtained with critical strains, ε_{crit}, 0.0005 and 0.00075, the magnitudes at which cracking is first noticeable (consistent with recommendations from Polshin and Tokar, 1957, and Burland and Wroth, 1974). The upper bound for "slight damage" is for a critical strain of 0.0015, which corresponds to an angular distortion of 1/300 and zero horizontal strain (as recommended by Skempton and MacDonald, 1956, and Bjerrum, 1963, for first cracking in panel walls and load-bearing walls settling under their own weight). The upper bound for the "moderate to severe damage" is obtained with the assumption of ε_{crit} = 0.0030, or for an angular distortion of 1/150 and zero horizontal strain (angular distortion for severe cracking in structures settling under their own weight, after Skempton and MacDonald, 1956, and Bjerrum, 1963). The radial lines in the figure indicate the regions where data from different excavations were obtained. A description of the type of damage, as observed in the field, is provided in Table 3.7 (from Boscardin and Cording, 1989, with modifications of the original descriptions by Burland et al., 1977). Later, Cording et al. (2001) generalized the concept of lateral strain (Figure 3.22b) and modified

Table 3.7 Classification of visible damage in buildings (from Boscardin and Cording, 1989; modified after Burland et al., 1977)

Class of damage	*Description of damage*[1]	*Approximate width*[2] *of cracks (mm)*
Negligible	Hairline cracks.	< 0.1
Very Slight	Fine cracks easily treated during normal redecoration. Perhaps isolated slight fracture in building. Cracks in exterior brickwork visible upon close inspection	< 1
Slight	Cracks easily filled. Re-decoration probably required. Several slight fractures inside building. Exterior cracks visible, some repointing may be required for weathertightness. Doors and windows may stick slightly	< 5
Moderate	Cracks may require cutting out and patching. Recurrent cracks can be masked by suitable linings. Tuck-pointing and possibly replacement of a small amount of exterior brickwork may be required. Doors and windows sticking. Utility service may be interrupted. Weathertightness often impaired.	5–15 or several cracks > 3 mm
Severe	Extensive repair involving removal and replacement of sections of walls, especially over doors and windows required. Windows and door frames distorted, floor slopes noticeable. Walls lean or bulge noticeably, some loss of bearing in beams. Utility service disrupted.	15–25, also depends on the number of cracks
Very Severe	Major repair required involving partial or complete re-construction. Beams lose bearing, walls lean badly and require shoring. Windows broken by distortion. Danger of instability.	Usually > 25, depends on the number of cracks

1 Location of damage in the building or structure must be considered when classifying the degree of damage.
2 Crack width is only one aspect of damage and should not be used alone as a direct measure of it.

somewhat the boundaries that describe the intensity of damage. The latest version of the damage criterion is given in Figure 3.22. The figure may be used to estimate the damage induced by tunneling in bearing-wall and small frame structures (Cording et al., 2001; Son and Cording, 2005; Cording et al., 2010). If a structure has enough capacity to withstand the ground strains, i.e. the deformations of the structure do not conform to the free-field ground deformations, usage of Figure 3.22b will result in a conservative estimate of the damage, and so numerical methods may be needed (Son and Cording, 2005, 2007, 2011). Figure 3.23 provides a comparison between observed settlements at the Mansion House in London (a 200 years old fragile building at the time of construction) and the "free-field" deformations predicted due to the excavation of a 3.05 m diameter tunnel at 15 m depth in London clay (Frischmann et al., 1994). The case study illustrates the potential benefits of the building stiffness, if they are properly considered.

Burland (1995) prepared an interactive diagram to determine the severity of the damage, based on the deep beam model (Burland and Wroth, 1974), analogous to the chart proposed by Boscardin and Cording (1989); that is, for L/H = 1 and E/G = 2.6. This is included in Figure 3.24. Damage is divided into five different categories that depend on (see Table 3.8 which provides also an equivalency with the degree of damage defined in Table 3.7) the

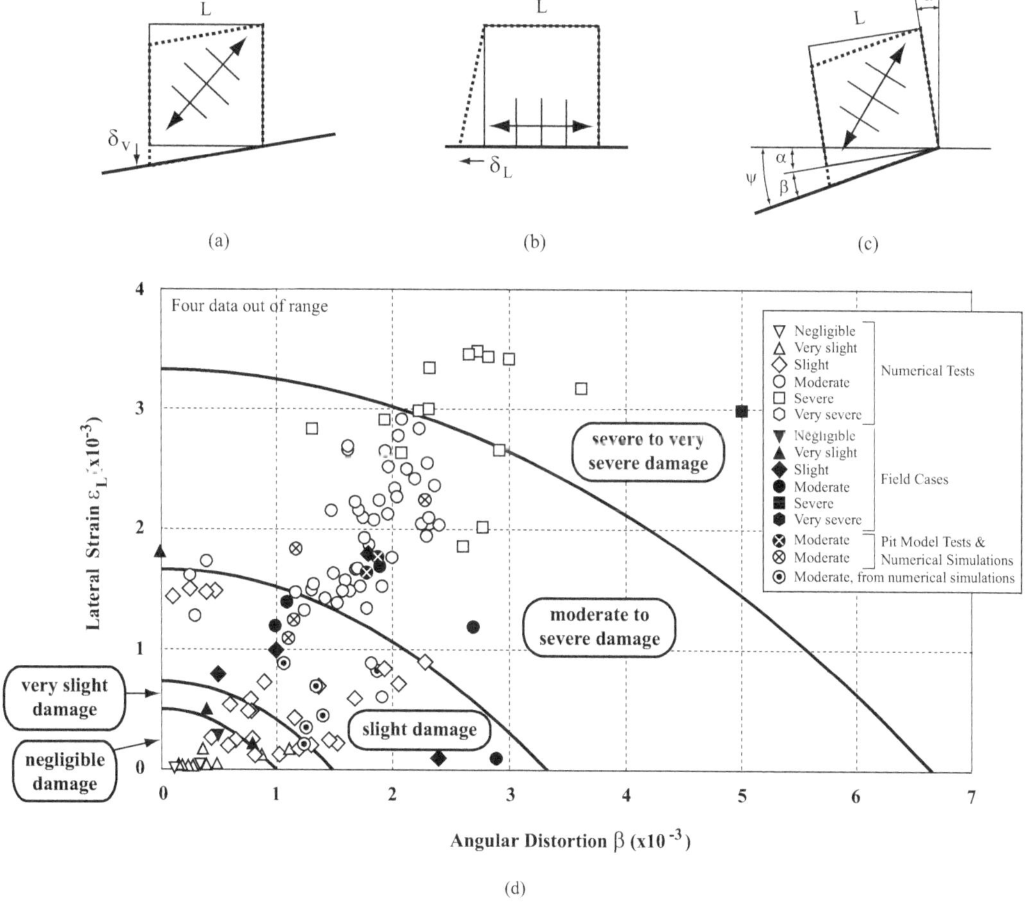

Figure 3.22 Relation between damage, angular distortion and horizontal extension strain (a) Angular distortion, $\beta = \delta_V/L$; (b) Lateral strain, $\varepsilon_L = \delta_L/L$; (c) Tilt, β = Slope (ψ) – Tilt (α); (d) building damage. (adapted from Son and Cording, 2005).

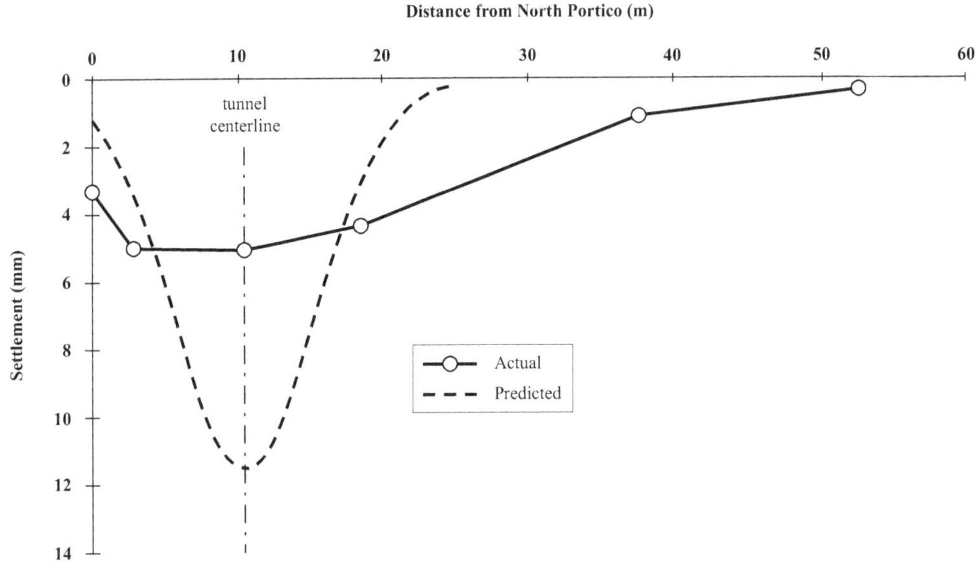

Figure 3.23 Influence on building stiffness on settlement profile. Mansion house in London (after Frischmann et al., 1994).

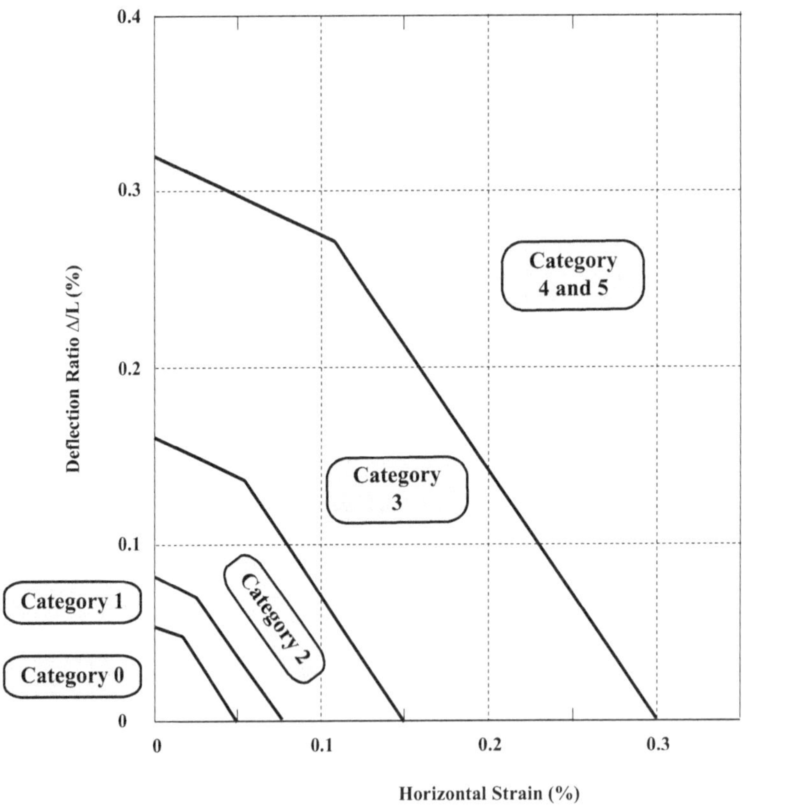

Figure 3.24 Damage category with deflection ratio and horizontal tensile strain for hogging deformations (L/H = 1) (from Burland, 1995).

Table 3.8 Relation between category of damage and limiting tensile strain, ε_{lim}[1] (from Burland, 1995)

Category of damage	*Normal degree of severity*	*Limiting tensile strain, ε_{lim}, (%)*[1]
0	Negligible	0 – 0.05
1	Very Slight	0.05 – 0.75
2	Slight	0.075 – 0.15
3	Moderate	0.15 – 0.3
4 to 5	Severe to Very Severe	> 0.3

[1] While e_{lim} is found in Burland (1995), Burland and Wroth (1974) use the concepts of limiting, e_{lim}, and critical tensile strain, e_c, interchangeably

limiting tensile strain, ε_{lim} (while ε_{lim} is found in Burland, 1995, Burland and Wroth, 1974, use the concepts of limiting, ε_{lim}, and critical tensile strain, ε_c, as defined before, interchangeably). The horizontal strain in the figure is the maximum of the tensile strains induced in the beam by flexion, shear, and/or horizontal deformations of the ground.

An alternative and yet, to some extent, complementary method to determine damage to buildings due to ground deformations is that of the maximum crack width (Boone, 1996; and Boone et al., 1999). The method is based on the deep elastic beam approximation for buildings (Burland and Wroth, 1974 and Boscardin and Cording, 1989) and utilizes the geometric changes in the building induced by ground deformations to compute the crack width due to the total tensile strain, ε_t (computed as the sum of the contributions from bending, shear and elongation due to ground movements) and the maximum principal tensile strain, ε_p. The strains are compared with the critical strain for the onset of cracking of the material, ε_c (see Table 3.9). If the strains imposed by the ground deformations are larger than the critical strains, the crack width is obtained by multiplying the corresponding strain with the associated length. Detailed equations to obtain the strains and crack widths associated with open

Table 3.9 Summary of critical strain, ε_c, at which cracking first occurs (from Boone, 1996)

Test conditions	*Mode of deformation*	*Critical strain ε_c (%)*
Brick Buildings with L/H>3	Tensile from flexure	0.05
Full scale frames with brick infill	Diagonal-tensile	0.081 – 0.137
	Shear approximation	0.16 – 0.27
Hollow tile and clinker block, brickwork	Shear distortions	0.22 – 0.33
	Diagonal-tensile	0.11 – 0.16
Full scale brick walls with supporting concrete beams, 1.2 < L/H < 3.0	Tensile from flexure	0.038 – 0.06
Concrete beams supporting brick walls	Tensile from flexure	0.035
Fiberboard or polywood on wood frame	Shear strain	0.6 – 1.66
Gypsum/fiberboard/plaster on wood frame	Shear strain	0.37 – 0.7
Structural clay tiles with cement-line	Shear strain	0.1
Clay brick with cement-lime mortar	Shear strain	0.1 – 0.2
Cement-line mortared concrete blocks	Shear strain	0.1
Core samples of brick and mortar	Tension	0.001 – 0.01[1]
Full-scale brick walls in field tests	Tension	0.02 – 0.03
Reevaluation of full-scale wall panel tests	Principal tensile	0.02 – 0.03

[1] Lower values represent failure along poorly mixed mortar joints

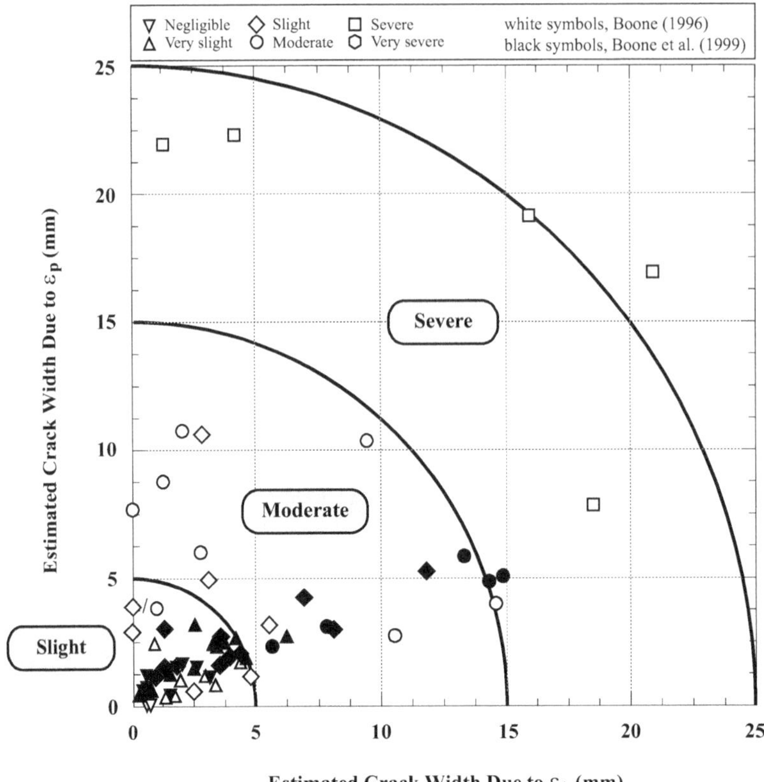

Figure 3.25 Damage estimation using the Boone (1996) method. From Boone et al. (1999).

excavations are given by Boone et al. (1999). Finally, the severity of the damage is estimated from Figure 3.25. There are four levels of damage: slight (crack width less than 5 mm), moderate (between 5 and 15 mm), severe (between 15 and 25 mm), and very severe (crack width larger than 25 mm).

Due to the effort required for accurate estimates of potential damage to buildings due to tunnel construction, and in particular in urban areas where the number of structures potentially affected may be very large, Burland (1995) suggests a three-stage method to assess the risk of damage:

Preliminary Assessment: It consists of drawing along the tunnel path contours of expected settlements and, according to recommendations given by Rankin (1988), eliminate those where the maximum settlement is less than 10 mm and the maximum rotation or slope (defined as the change in gradient of a line connecting two points in the building) is less than 1/500.

Second Stage Assessment: Further analysis is conducted on those buildings that did not pass the first stage. The building is represented as a deep beam and the potential damage is estimated based on the simple formulations obtained from elasticity, e.g. Equations (3.23) to (3.25) and chart in Figure 3.24 (an alternative to Burland's Figure 3.24, could be using Figures 3.22 or 3.25; note from the authors). Burland (1995) mentions that this is a conservative estimate because the building is assumed to conform to the ground deformations, i.e. it does not introduce constraints to the free-field ground displacements.

Detailed Evaluation: This is done to those buildings that, after the second stage, fall under Moderate (category 3 in Burland, 1995) or greater damage. The evaluation consists of a refinement of the second stage and includes: (1) the sequence and method of tunnel excavation; (2) structural details and structural continuity, i.e. consideration of steel or concrete frames; (3) building foundation, e.g. continuous foundation, piles; (4) orientation of the building; (5) soil–structure interaction; and (6) previous building movements, as those may affect the tolerance of the structure to sustain additional deformations.

3.5 CONCLUSIONS AND DISCUSSION

This chapter concentrates on tunneling in soil and the related direct and indirect methods (hence the title of the chapter). As has become clear, the design of tunnel supports in soil, be that for initial or final support, should be done with the analytical and numerical methods discussed in Chapters 4 and 5. What is presented in this chapter (Section 3.2) is meant to provide a brief historical perspective and, most importantly, to describe typical behavior of soils in tunnels and relate this to standard soil classification.

The main purpose of this chapter is to discuss methods that can be used to determine soil deformation with particular emphasis on the effect of soil deformation on the surface. To understand the methods described in this chapter but also the analytical and numerical procedures described later, basic ground behavior is presented first. In an expansion of what has been discussed regarding ground support interaction in Chapter 1, it is shown that ground deformation is caused by movement of the soil toward the face and toward the tunnel perimeter, the latter mostly in the section close to the face. The direct and indirect methods relate this soil movement into the tunnel, often called ground (volume) loss, to the deformation (settlement) at the surface. The difference between the two approaches is that direct methods provide recommendations based only on field observations, while the indirect methods are based on a combination of analytical and/or numerical methods and empirical approximations. The best known of these methods is the one originally developed by Peck and then further expanded by a number of British authors that involves relating tunnel geometry and soil types to sizing the settlement trough in form of the Gauss distribution. These approaches are entirely acceptable for a quick initial estimation of ground deformation during construction.

Field data, albeit still scarce, appear to support the extension of the methods to modern construction techniques that include EPB and slurry shields (Fang et al., 1994; Cooper et al., 2002; Van der Berg et al., 2003; Clayton et al., 2006; Suwansawat and Einstein, 2007). Surface settlements appear to correlate well with shield operation (Kasper and Meschke, 2006; Suwansawat and Einstein, 2007). Many case histories show that an appropriate face pressure can effectively limit ground losses.

More advanced methods attempt to relate the different components of ground movement into the tunnel such as movements of the face, and particularly, different movements around shield perimeter together with a limited set of soil parameters to deformation at the surface. Although the relations are quite detailed, one has to be aware of the fact that they also should only be applied for initial estimates during construction.

Clearly, one of the main purposes of determining ground deformations is to relate them to potential damage to structures through differential settlements and distortion. For this the approach by Burland and co-workers and Cording and co-authors is described in some detail but the reader is also made aware of other approaches.

If settlements cannot be reduced within the allowable limits, then preventive or remedial techniques may be employed. Such techniques are classified as follows (ITA/AITES 2006, 2007):

1. Improvement of the overall project conditions. This involves changing the project such that settlements are sufficiently small. This may be accomplished by increasing the tunnel depth cover, by placing the tunnel in a soil layer with good mechanical properties, reducing the tunnel cross section (i.e. twin tunnels instead of one single larger tunnel), and by removing curves on the tunnel alignment.
2. Improvement of ground conditions. These include compaction grouting, jet grouting, ground freezing, drainage (e.g. lowering the groundwater table), and compensation (consolidation) grouting (see also Cording et al., 1989).
3. Structural improvement of buildings. Underpinning of buildings affected by the tunnel construction and reinforcement of the structures are possible options.
4. Improvement while tunneling. This includes operations and construction procedures that can be applied to reduce ground deformations. They include: (a) staged excavation, where a reduction of the cross section usually results is small deformations, but at a cost of smaller advance rates; (b) face support, e.g. face bolting, shotcrete; (c) crown support to diminish settlements caused by inappropriate support installation; (d) pre-support of the ground ahead of the excavation with forepoling, umbrella vaults, pre-vaults, etc.; (e) underpinning the upper liner when using the bench and heading excavation method; and (f) closure of the lining with an invert arch as soon as possible.
5. Improvement of shield tunneling. This consists of methods to reduce ground deformations due to face or radial losses associated with shield operation. They consist of: (a) reduction of decompression ahead of the face by appropriate application of face pressure; (b) reduction of overcutting and shield length; and (c) backfill the void at the tail of the shield quickly and efficiently.

In short, the methods for estimating soil deformation in this chapter are described to provide designers with approaches to make initial estimates of potential soil tunnel construction effects at the surface. In most cases, it will then be necessary to do more detailed analytical and/or numerical work for more refined predictions. The initial estimates will, however, indicate where such work is necessary. Finally, it has to be said that support design and long-term deformation prediction has to be done with the analytical and numerical methods discussed in the following two chapters.

Chapter 4

Analytical methods

4.1 INTRODUCTION

The support of a tunnel must sustain the load induced by the ground as it is excavated and any other loads that may arise during the life of the tunnel. Chapter 1 provides the framework to understand, at least qualitatively, the interplay that exists between excavation and support behavior. Such interplay has been addressed in Section 1.2 utilizing the simple case of a deep tunnel with axial symmetry. While useful, the conclusions are limited since stress conditions at the depth of the tunnel are seldom hydrostatic. The effects of different far-field stresses, i.e. the coefficient of earth pressure at rest $K_o \neq 1$, may be significant and thus need to be considered for proper understanding of the interaction between ground, excavation and support. Such effects are illustrated in the following two cases, which are analogous to those discussed in Chapter 1 (Figures 1.5, 1.7 and 1.9), except that the far-field stresses are $\sigma_{vertical}$ = 1 MPa, $\sigma_{horizontal}$ = 0.5 MPa, and σ_{axial} = 0.5 MPa; that is, K_o = 0.5 in both axial and horizontal directions. The tunnel excavation (Figure 1.4) is the same, as well as the tunnel geometry, r_o = 2m, properties of the elastic medium, E = 500MPa, ν = 0.25, as those used in Chapter 1 (again, the values are not expected to be representative of any ground conditions; in particular, the ground stiffness is low to better illustrate ground response). Note that in this example the tunnel axis is parallel to the far-field principal stress σ_{axial}, but this may not be always the case. The Finite Element Method (FEM) code ABAQUS is used for the analysis. In the model, the ground is first subjected to the far-field stresses in such a way that the displacements obtained due to the geostatic stresses are zero. This is done to obtain displacements associated with tunnel excavation and not due to the initial stresses. In other words, the displacements computed are those that an observer would measure after the opening is built; the displacements associated with the initial stresses would have occurred before the construction of the tunnel. This is an important concept that will be further discussed in Section 4.2.1. In the FE analysis, the tunnel excavation is modeled by deactivating the appropriate elements. The boundaries are placed far enough from the tunnel such that their presence has no influence on the tunnel response. The results from the analysis are shown in Figure 4.1 and illustrate the effects of the excavation on the ground surrounding the tunnel. Figure 4.1a is a contour plot of the axial stresses, 4.1b of the vertical stresses, 4.1c of the horizontal stresses, and 4.1d of the displacements. Stress contours are plotted on vertical and horizontal planes through the tunnel axis and on a cross section perpendicular to the tunnel axis. The cross section is located far from the face of the tunnel such that there is no influence of the face and so it corresponds to plane strain conditions. As with the axisymmetric case (Figure 1.5), the presence of the tunnel has little influence on the axial stresses except immediately in front of the face. The excavation of the tunnel produces a net unloading, from zero axial stresses at the face to far-field stresses at a distance of about 1.5 to 2 r_o inside the ground, analogous to what is shown in Figure 1.5. Similarly, there is a redistribution of stresses around the tunnel.

DOI: 10.1201/9781003328940-4

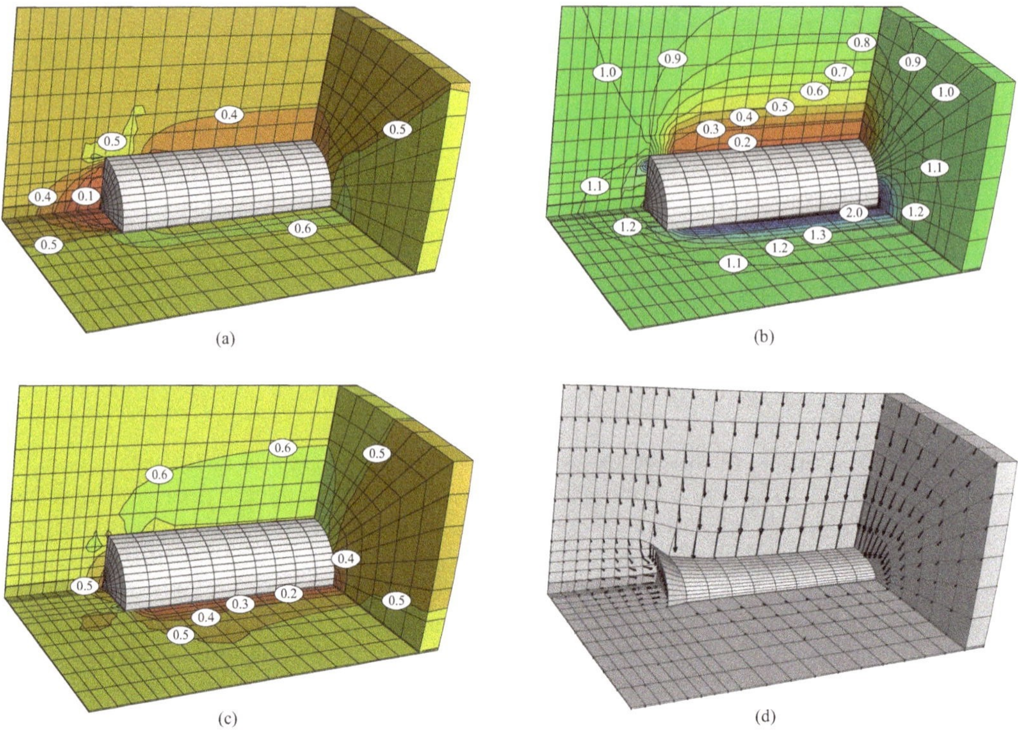

Figure 4.1 Three-dimensional view of stresses and displacements for an unsupported tunnel with r_o = 2 m, σ_v = 1.0 MPa, K_o = 0.5. (a) σ_{axial}. (b) $\sigma_{vertical}$. (c) $\sigma_{horizontal}$. (d) Dsiplacements.

There is an increase in vertical stresses ahead of the face over a length slightly larger than one radius, while the horizontal stresses are not much affected. Conditions corresponding to plane strain are also obtained at a distance of approximately 4 r_0 behind the face and apply to the stresses plotted on the cross sections. What is different, and shows the effects of K_0, is the distribution of stresses around the tunnel. For the case with K_o = 0.5, the excavation has a much larger effect on the vertical and horizontal stresses than for K_o = 1, with larger changes of stresses close to the tunnel. The tunnel deformations are also quite different. While close to the face, the ground displacements have a strong axial component and behind the tunnel, far from the face, mostly radial, similar to the axisymmetric case of Figure 1.5d, close to the tunnel, the differences are significant. The tunnel in Figure 1.5d moves inward with a uniform displacement at any given section; that is, due to the axial symmetry, the deformed shape of any cross section is circular. The deformed tunnel in Figure 4.1d adopts an oval shape with displacements toward the opening larger at the crown than at the springline (displacement direction and relative magnitude depends on the stress field). Figure 4.2 is a plot of the vertical settlements along the crown of the tunnel. The settlements are normalized by the factor E/σ_v, and the axial distance from the face by the tunnel radius, r_0, which makes the results comparable to those in Figure 1.7. As with the case with K_0 = 1, ahead of the face of the tunnel, beyond about 4 radii from the face, the settlements are small; the deformations increase significantly very close to the face and reach a maximum magnitude at about four to five radii behind the face, where plane strain conditions apply. Figure 4.2 clearly shows that the displacements at the crown, and thus the deformed shape of the tunnel, strongly depend on the value of K_0, with magnitudes much larger for K_o = 0.5 than for K_o = 1.

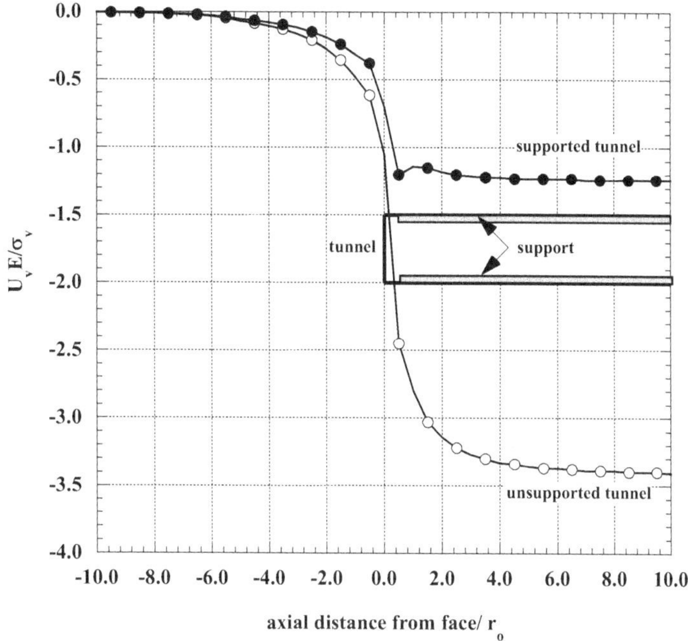

Figure 4.2 Vertical displacements at the crown of supported and unsupported tunnels, r_0 = 2 m, K_0 = 0.5.

Analogous effects are found in a tunnel where a liner is installed to provide support. The construction sequence is idealized in Figure 1.8. As with the unsupported tunnel case, the stresses induced in the ground are analyzed using the FEM code ABAQUS, with the condition that both the ground and the liner are elastic. The same tunnel dimensions, ground properties, and far-field stresses are used. For the liner, the following properties are employed: Young's modulus, E = 24000 MPa; Poisson's ratio, ν = 0.25, and liner thickness, t = 0.2 m. An unsupported length d = 1 m is assumed. Figure 4.3 shows the results. Figure 4.3a is a contour plot of the axial stresses, Figure 4.3b of the vertical stresses, Figure 4.3c of the horizontal stresses, and Figure 4.3d of the displacements. As with the axisymmetric case, Figure 1.9, the lined tunnel induces a significant reduction of axial stresses ahead of the tunnel face with magnitude similar to that observed for the unlined tunnel (Figure 4.1). The vertical stress contours show loading at the springline and unloading at the crown, as with the unlined tunnel; this is due to the redistribution of stresses around the tunnel, which is induced by the same mechanisms as those described for the unsupported tunnel (Figures 4.1b and c; see also Figure 1.6). The magnitudes of the stresses are much smaller with the lined tunnel than with the unlined tunnel, as with the axisymmetric case. The stresses in the liner however depend on the value of K_0. For K_0 = 1, the tangential stresses in the liner at any given section are constant (i.e. the liner is subjected only to compression in the tangential direction), as they should. For K_0 = 0.5, the tangential stresses in the liner are not constant along a section; in other words, the liner is subjected to both compression and bending, which results in larger tangential stresses for K_0 = 0.5 than for K_0 = 1. This observation is consistent with the ground and liner deformations. For the axisymmetric case, the liner has a uniform inward deformation at any cross section, while for the case analyzed here, the liner has an oval shape, with the major axis horizontal. The most important difference between the lined and unlined tunnels is that in the unlined tunnel the deformations at the crown and springline are both toward the excavation while in the lined tunnel the crown deformations are inward while

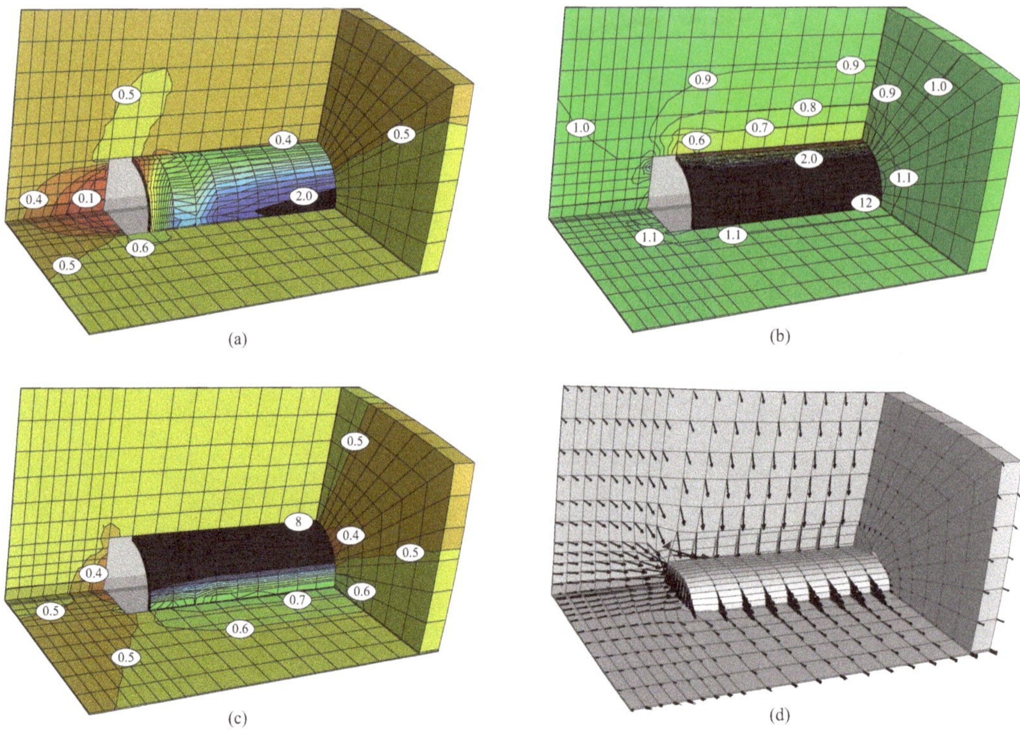

Figure 4.3 Three-dimensional view of stresses and displacements for a supported tunnel with r_o = 2 m, σ_v = 1.0 MPa, K_o = 0.5. (a) σ_{axial}. (b) $\sigma_{vertical}$. (c) $\sigma_{horizontal}$. (d) Dsiplacements.

the springline displacements are toward the ground. This is due to the stiffness of the liner, which in this case is relatively flexible but not very compressible, and so the settlements at the crown require an outward displacement at the springline. Such behavior impacts the horizontal stresses of the ground. Figure 4.3c shows that at the springline the horizontal stresses increase with respect to the far field. This is quite different than what happens with the unsupported tunnel where there is unloading (Figure 4.1c), and also for the axisymmetric lined tunnel (Figure 1.9b) that shows radial (i.e. vertical or horizontal) unloading. This is a favorable result for stability because, with a liner, the ground confinement at the crown and at the springline is larger than for the unsupported tunnel. A comparison of displacements for the supported and unsupported tunnels is provided in Figure 4.2. Similar to the unsupported tunnel, the displacements at distances larger than about two diameters ahead of the tunnel face are negligible. As the tunnel approaches, the displacements increase, and are significant immediately after passage of the tunnel face. At about two to three diameters behind the face, the displacements do not increase anymore and plane strain conditions are reached. This is analogous to what was found in Figure 1.7 for the K_o = 1 case. The most substantial effect of the liner in both cases (Figures 1.7 and 4.2) is the significant decrease of displacements of the ground. However, the deformations plotted in Figure 4.2 are larger than those in Figure 1.7.

Hence, the installation of a liner reduces the change of stresses in the ground compared with the unsupported tunnel, and so the ground deformations are also smaller. This occurs at the expense of an increase of load to the liner, which then needs to be designed to withstand the stresses transferred from the ground. The load that is transferred from the ground to the support depends on many factors, including the geometry and depth of the tunnel, initial stresses, properties of the ground and support and construction method. Perhaps there is no other problem in geotechnical engineering/geomechanics where ground, support and construction factors are so inter-related. This makes the problem very complex with many of the variables not fully known before construction. It is perhaps for this reason that there exists a large number of empirical and numerical methods that are available to estimate the support loads and the ground response. Numerical methods (Chapter 5) have the potential to account for some of the complexity of the problem, provided that sufficient information is available. Empirical methods (Chapters 2 and 3) bypass this requirement by relating directly the support requirements with easily determined ground properties or qualitative geologic descriptions (Einstein and Schwartz, 1979). Both approaches have applications to which they are best suited, but there exists the need for solutions that can be used to interpret data quantitatively and quickly within a rational analytical framework. Analytical solutions must satisfy three criteria (Einstein and Schwartz, 1979): (1) they must be simple to use; (2) they must be able to incorporate the most significant variables; and (3) they must correctly model the loading conditions and the ground–structure interaction around the tunnel.

Analytical methods are presented in Section 4.2, where some of the key concepts introduced in Chapter 1, and revisited using Figures 4.1 to 4.3, are quantitatively analyzed. While the discussion in this chapter will necessarily focus on two-dimensional considerations, due to the difficulties in finding close-form solutions for full three-dimensional problems, Section 4.3 includes a discussion on how to incorporate face effects, e.g. three-dimensional effects, into the two-dimensional assumption. Finally, Section 4.4 summarizes the most important points and provides recommendations for the use of the analytical solutions presented in the Chapter.

4.2 ANALYTICAL METHODS

The analytical methods can be broadly divided into two: continuum and discontinuum methods. The first method assumes that the ground is continuous, and so the stress and displacement fields in the medium are also continuous. Soils and heavily weathered rocks or heavily jointed rocks would fall into this category. The continuum methods can be further divided into two: (1) methods based on the assumption of elasticity; and (2) methods that consider plastic deformations of the ground medium. For the discontinuum method, it is assumed that instabilities are generated by the potential movement of blocks or wedges that can detach from the perimeter of the excavation. Moderately jointed rock in a low to medium stress field would be an example. The following sections provide an overview of solutions available for each of the three methods: elastic and plastic methods for the continuous medium and the wedge method for the discontinuous medium.

4.2.1 Elastic methods

Analytical solutions are available in only very few cases, where the following assumptions apply: (1) plane strain conditions in a direction perpendicular to the cross section of the tunnel; (2) the ground is either dry or fully saturated, homogeneous, and isotropic (limited solutions have been found for orthotropic and transversely anisotropic ground); (3) deformations of ground and liner remain within their respective elastic regimes. While limited in their applicability due to some of the assumptions made in their derivation, closed-form solutions provide the means for a quick quantitative estimate of the interaction between ground, excavation and support, and provide a simple framework where such interaction can be evaluated and presented in a rational manner. It has to be noted that such solutions are useful for a first-order estimate and may not replace more detailed analyses, usually requiring numerical methods. In addition, analytical solutions include all the fundamental variables and thus they are useful to identify those factors that are the most critical, and to verify numerical methods.

The solution of any elasticity problem must satisfy the equilibrium equations, the strain compatibility equations, the boundary conditions and, of course, Hooke's law. The following is a fairly general derivation of the governing equations in elasticity as applied to underground cavities, which includes saturated and dry media, short- and long-term conditions.

In polar coordinates the equilibrium equations are (Figure 4.4, where compression is positive):

$$\begin{aligned} &\frac{\partial \sigma_r}{\partial r} + \frac{1}{r}\frac{\partial \tau}{\partial \theta} + \frac{\sigma_r - \sigma_\theta}{r} = F_r \\ &\frac{1}{r}\frac{\partial \sigma_\theta}{\partial \theta} + \frac{\partial \tau}{\partial r} + 2\frac{\tau}{r} = F_\theta \end{aligned} \tag{4.1}$$

which are written, as they should, in total stresses. σ_r, σ_θ, and τ are the radial, tangential and shear stresses in polar coordinates, respectively; r and θ are the polar coordinates with origin at the center of the tunnel. F_r and F_θ are the body forces, which, in polar coordinates are:

$$\begin{aligned} F_r &= -\gamma \sin\theta \\ F_\theta &= -\gamma \sin\theta \end{aligned} \tag{4.2}$$

where γ is the total unit weight of the ground. Assuming that the principle of effective stresses applies:

$$\sigma_{ii} = \sigma'_{ii} + u \qquad \text{for } i = 1,2 \tag{4.3}$$

where u is the pore pressures. In effective stresses, the equilibrium equations are:

$$\begin{aligned} &\frac{\partial \sigma'_r}{\partial r} + \frac{1}{r}\frac{\partial \tau}{\partial \theta} + \frac{\sigma_r - \sigma_\theta}{r} + \frac{\partial u}{\partial r} = -\gamma \sin\theta \\ &\frac{1}{r}\frac{\partial \sigma'_\theta}{\partial \theta} + \frac{\partial \tau}{\partial r} + 2\frac{\tau}{r} + \frac{1}{r}\frac{\partial u}{\partial \theta} = -\gamma \cos\theta \end{aligned} \tag{4.4}$$

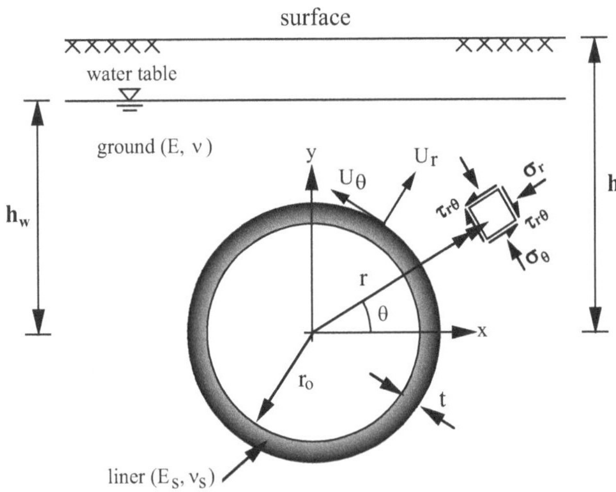

Figure 4.4 Tunnel.

The compatibility equation is, in plane strain:

$$\left(\frac{\partial^2}{\partial r^2}+\frac{1}{r}\frac{\partial}{\partial r}+\frac{1}{r^2}\frac{\partial^2}{\partial \theta^2}\right)\left[(\sigma'_r+\sigma'_\theta)+\frac{1}{1-\nu}u\right]=\frac{1}{1-\nu}\left(\frac{\partial F_r}{\partial_r}+\frac{1}{r}\frac{\partial F_\theta}{\partial_\theta}+\frac{F_r}{r}\right) \tag{4.5}$$

where ν is the Poisson's ratio. The right term of (4.5), given (4.2), is zero. The equation takes the final form:

$$\nabla^2(\nabla^2\phi)=\frac{1-2\nu}{1-\nu}\nabla^2 u \tag{4.6}$$

where ∇^2 is the Laplacian operator and ϕ the Airy stress function. Stresses are related to the Airy stress function through the following equations, which depend on the coordinate system, in this case polar and Cartesian (e.g. Timoshenko and Goodier, 1970):

In polar coordinates:

$$\begin{aligned}\sigma_r&=\sigma'_r+u=\frac{1}{r}\frac{\partial\phi}{\partial r}+\frac{1}{r^2}\frac{\partial^2\phi}{\partial\theta^2}-\gamma r\sin^3\theta\\ \sigma_\theta&=\sigma'_\theta+u=\frac{\partial^2\phi}{\partial r^2}-\gamma r\sin\theta\cos^2\theta\\ \tau&=-\frac{\partial}{\partial r}\left(\frac{1}{r}\frac{\partial\phi}{\partial\theta}\right)-\gamma r\cos\theta\sin^2\theta\end{aligned} \tag{4.7a}$$

In Cartesian coordinates:

$$\begin{aligned}\sigma_x&=\sigma'_x+u=\frac{\partial^2\phi}{\partial y^2}\\ \sigma_y&=\sigma'_y+u=\frac{\partial^2\phi}{\partial x^2}-\gamma y\\ \tau&=-\frac{\partial^2\phi}{\partial x\partial y}\end{aligned} \tag{4.7b}$$

Pore pressures are governed by the following field equation:

$$\frac{K}{\gamma_w}\left(\frac{\partial^2}{\partial r^2}+\frac{1}{r}\frac{\partial}{\partial r}+\frac{1}{r^2}\frac{\partial^2}{\partial\theta^2}\right)u=-\frac{\partial\varepsilon_{vol}}{\partial t} \tag{4.8}$$

where K is the coefficient of permeability, γ_w is the unit weight of water, and ε_{vol} is the volumetric strain. Equation (4.8) can be written, for plane strain, as:

$$\frac{EK}{(1-\nu)(1-2\nu)\gamma_w}\left(\frac{\partial^2}{\partial r^2}+\frac{1}{r}\frac{\partial}{\partial r}+\frac{1}{r^2}\frac{\partial^2}{\partial\theta^2}\right)u=-\frac{\partial(\sigma'_r+\sigma'_\theta)}{\partial t} \tag{4.9}$$

In general, mechanical-flow problems are coupled so Equations (4.6) and (4.9) must be solved simultaneously. There are few cases however where the equations can be uncoupled.

The first case is for dry ground, where pore pressures are zero. For this problem Equation (4.8) has the trivial solution u = 0 and Equation (4.6) recovers the classical form (Timoshenko and Goodier, 1970):

$$\nabla^2(\nabla^2 \phi) = 0 \tag{4.10}$$

The second case corresponds to problems with steady-state conditions or for long-term analysis. In those cases, $\frac{\partial}{\partial t} = 0$, and thus $\nabla^2 u = 0$. The analysis can be de-coupled if the flow problem is solved first. Once the pore pressures are known, Equation (4.10) is solved with appropriate boundary conditions. Note that because $\nabla^2 u = 0$, Equation (4.6) changes into (4.10). The effective stresses are obtained from (4.7) from the Airy function and the pore pressures. This is the approach used to obtain the solution to the problem of a pressure tunnel in Section 6.3.2.

The third case is for short-term analysis (i.e. undrained analysis). In this case there is no change of volume in the ground, thus $\varepsilon_{vol} = 0$. For plane strain conditions,

$$\varepsilon_{vol} = \frac{(1-2\nu)(1+\nu)}{E}(\sigma'_r + \sigma'_\theta) = 0 \tag{4.11}$$

which requires:

$$(\sigma'_r + \sigma'_\theta) = 0 \tag{4.12}$$

Combining Equation (4.12) with (4.9) results in $\nabla^2 u = 0$, and so again Equation (4.6) transforms into (4.10). Once the Airy stress function is known, stresses can be computed from (4.7). Strains, in plane strain, can be obtained from the following equations:

In polar coordinates:

$$\begin{aligned}
\varepsilon_r &= \frac{1+\nu}{E}[\sigma'_r - \nu(\sigma'_r + \sigma'_\theta)] \\
\varepsilon_\theta &= \frac{1+\nu}{E}[\sigma'_\theta - \nu(\sigma'_r + \sigma'_\theta)] \\
\gamma_{r\theta} &= \frac{\tau}{G}
\end{aligned} \tag{4.13a}$$

In Cartesian coordinates:

$$\begin{aligned}
\varepsilon_x &= \frac{1+\nu}{E}[\sigma'_x - \nu(\sigma'_x + \sigma'_y)] \\
\varepsilon_y &= \frac{1+\nu}{E}[\sigma'_y - \nu(\sigma'_x + \sigma'_y)] \\
\gamma_{ry} &= \frac{\tau}{G}
\end{aligned} \tag{4.13b}$$

Note that the strains are expressed in terms of effective stresses. Strains and displacements are related by:

$$
\begin{aligned}
\varepsilon_r &= \frac{\partial U_r}{\partial r} \\
\varepsilon_\theta &= \frac{1}{r}U_r + \frac{1}{r}\frac{\partial U_\theta}{\partial \theta} \\
\gamma_{r\theta} &= \frac{1}{r}\frac{\partial U_r}{\partial \theta} + \frac{\partial U_\theta}{\partial r} - \frac{1}{r}U_\theta
\end{aligned}
\tag{4.14a}
$$

in polar coordinates. In Cartesian coordinates,

$$
\begin{aligned}
\varepsilon_x &= \frac{\partial U_x}{\partial x} \\
\varepsilon_y &= \frac{\partial U_y}{\partial y} \\
\gamma_{xy} &= \frac{\partial U_x}{\partial y} + \frac{\partial U_y}{\partial x}
\end{aligned}
\tag{4.14b}
$$

Displacements are computed by integration of (4.14). Note that with the sign convention used of compression positive, compressive strains are positive and so integration of (4.14) gives displacements positive in the negative direction of the coordinate system. What is done in the following is change the sign of the displacements that are obtained after integration of (4.14) to have displacements positive in the positive axes directions of the coordinate system. Note that this gives displacements that are not consistent with stresses. This has been done to provide results with stresses positive in compression and displacements positive along the positive direction of the coordinate system.

The three cases: dry ground, steady-state and short-term conditions are very relevant for geotechnical engineering/geomechanics. It is important to note that in all three cases the problem resides in solving the same equation: Equation (4.10). This equation has received significant attention in the past resulting in different methods for its solution (Mindlin, 1940, 1948; Muskhelishvili, 1954; Sokolnikoff, 1956; Savin, 1961; Timoshenko and Goodier, 1970; Soutas-Little, 1973). Two are the methods that have been used the most: the complex variable method and the Airy stress function method (relative stiffness method). The methods are briefly introduced in the following sections.

4.2.1.1 Complex variable method

Complex variable theory and conformal mapping techniques have been extensively used for the solution of problems of openings in infinite elastic media (Mindlin, 1940, 1948; Muskhelishvili, 1954; Sokolnikoff, 1956) where water is not present; in other words when pore pressures are zero anywhere in the medium. The fundamental theories of complex variable and conformal mapping have been extensively described by Muskhelishvili (1954), and later on by Savin (1961) and Timoshenko and Goodier (1970). A number of researchers have successfully implemented this technique into different disciplines of engineering: (Theocaris and Petrou, 1989; Theocaris, 1991; Motok, 1997; Gerçek, 1991, 1997; Exadaktylos and Stavropoulou, 2002; Exadaktylos et al., 2003; Huo et al., 2006; Li and Wang, 2008). For circular tunnels the method has been very effectively used by Verruijt and coworkers (Verruijt,

1997, 1998; Strack and Verruijt, 2002) and Bobet and coworkers (Bobet, 2011, 2016a, b; Vitali et al., 2018b, 2019, 2020).

According to complex variable theory, any biharmonic function, for example the Airy stress function ϕ in Equation (4.10), can be expressed as:

$$\phi = \mathrm{Re}[\bar{z}\varphi(z) + \chi(z)] \tag{4.15}$$

where z is a complex variable (z = x + iy; with $i = \sqrt{-1}$) and $\bar{z}$ is the complex conjugate of $z(\bar{z} = x - iy); \varphi(z)$ and $\psi(z) = \chi'(z) = \dfrac{d\chi(z)}{dz}$ are two analytic complex functions, also known as "complex potential functions". The stress components, for example in a Cartesian coordinate system (Figure 4.4), can be expressed in terms of the complex potentials as:

$$\begin{aligned} \sigma_x + \sigma_y &= 2[\varphi'(z) + \overline{\varphi'(z)}] \\ \sigma_y - \sigma_x + 2i\tau &= 2[\bar{z}\varphi''(z) + \psi'(z)] \end{aligned} \tag{4.16}$$

The displacements in plane strain are:

$$2G(U_x + iU_y) = (3 - 4v)\varphi(z) - z\overline{\varphi'(z)} - \overline{\psi(z)} \tag{4.17}$$

where U_x and U_y are horizontal and vertical displacements, respectively; G is the shear modulus of the medium, $G = E/[2(1 + \nu)]$.

Complex variable theory is used to determine stresses and deformations of the ground, while structural theory is used to find stresses and deformations of the tunnel support, if any. Compatibility of normal and shear stresses and displacements is invoked at the interface between the ground and the structure to solve the problem of a supported tunnel. For an unsupported tunnel, the solution needs to satisfy the boundary conditions imposed at the perimeter of the opening, which typically are of the form of imposed displacements or of imposed tractions; for example zero radial and shear tractions at the perimeter of an unsupported opening.

The solution requires finding the two complex potentials $\varphi(z)$ and $\psi(z)$, which, given (4.16) and (4.17), satisfy the boundary conditions. It is convenient however to apply a coordinate transformation to map the opening, which can be of any shape, into a unit circle. Conformal mapping allows for the transformation (Churchill, 1960; Savin, 1961). The variable z maps into ζ through a single-valued analytic function, $z = \omega(\zeta)$, such that a point in the z-plane corresponds to a unique point in the ζ-plane and vice-versa. A number of functions have been used in the literature that have different expressions depending on the shape of the opening. The following are typical examples:

$$z = \omega(\zeta) = \frac{a_o}{\zeta} + a_1\zeta + a_2\zeta^2 + \ldots. \tag{4.18a}$$

$$z = \omega(\zeta) = -i a_o \frac{1+\zeta}{1-\zeta} \tag{4.18b}$$

where a_o, a_1, a_2, etc. are complex constants that depend on the shape of the opening. Equation (4.18a) has been used to map openings in an infinite medium with elliptical, rectangular, and with other conventional cross sections such as "D" shape, double arch, arched roof and parabolic floor (Savin, 1961; Exadaktylos and Stavropoulou, 2002; Gerçek, 1991, 1997; Huo et al., 2006). Equation (4.18b) has been used to map a circular shallow tunnel in a semi-infinite medium (Verruijt, 1997, 1998; Strack and Verruijt, 2002). Figure 4.5 illustrates the transformation of a rectangular opening in the z-plane into a unit circle in the ζ-plane. The values of the constants a_0, a_1, a_2, etc. in (4.18a) depend on the dimensions of the rectangle given by a and b. The rectangle ABCD in the z-plane, shown in Figure 4.5, is transformed into a unit circle A'B'C'D' in the ζ-plane. Point A' in the unit circle corresponds to point A in the rectangle, and so on. Note that ABCD is in a clockwise direction, whereas A'B'C'D' is counter-clockwise. This is so because the conformal mapping function transforms infinity in the z-plane into the origin in the ζ-plane, and vice versa. In other words, the area of interest (the ground) in the z-plane is outside the rectangle, which corresponds to the area enclosed by the unit circle in the ζ-plane.

With the transformation, the complex functions take the form:

$$\begin{aligned} \varphi(z) &= \varphi(\omega(\zeta)) = \varphi(\zeta) \\ \psi(z) &= \psi(\omega(\zeta)) = \psi(\zeta) \end{aligned} \tag{4.19}$$

The boundary conditions for the stresses are imposed in terms of the resultant force at the boundaries rather than through Equation (4.16). The resulting equation is (Muskhelishvili, 1954):

$$i(F_x + iF_y) = i\int_s (t_x + it_y)\,ds = \varphi(z) + z\overline{\varphi'(z)} + \overline{\psi(z)} + C \tag{4.20}$$

where t_x and t_y are the surface tractions acting at the boundary (e.g. along the perimeter of the rectangular opening in Figure 4.5) and C is an integration constant. With the conformal mapping transformation, the boundary conditions for stresses given by Equation (4.20) take the form:

$$\varphi(\sigma) + \frac{\omega(\sigma)}{\overline{\omega'(\sigma)}}\overline{\varphi'(\sigma)} + \overline{\psi(\sigma)} = i(F_x + iF_y) \tag{4.21}$$

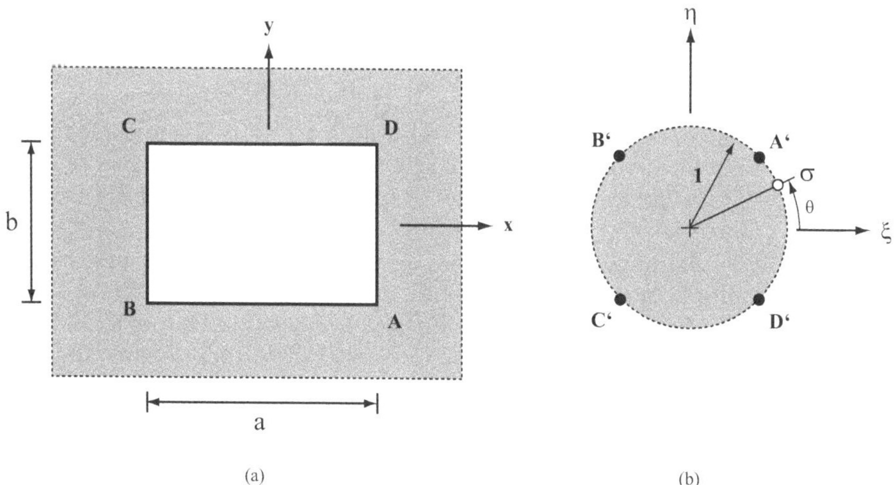

Figure 4.5 Conformal mapping transformation of a rectangular opening in the z-plane into a unit circle in the ζ-plane. (a) z plane. (b) ζ plane.

where $\sigma = e^{i\theta}$, which is a point along the surface of the unit circle (Figure 4.5) and $\varphi(\sigma)$, $\varphi'(\sigma)$, $\psi(\sigma)$ are the boundary values of these functions for $\zeta \to \sigma$ inside the circle. Muskhelishvili (1954) has shown that from (4.21) the following two integral equations are obtained, from which the complex potentials $\varphi(\zeta)$ and $\psi(\zeta)$ can be found:

$$\begin{aligned} 2\pi i\varphi(\zeta) + \int_s \frac{\omega(\sigma)}{\overline{\omega'(\sigma)}} \overline{\varphi'(\sigma)} \frac{d\sigma}{\sigma - \zeta} &= i\int_s \frac{F_x + iF_y}{\sigma - \zeta} d\sigma \\ 2\pi i\psi(\zeta) + \int_s \frac{\overline{\omega(\sigma)}}{\omega'(\sigma)} \varphi'(\sigma) \frac{d\sigma}{\sigma - \zeta} &= -i\int_s \frac{F_x - iF_y}{\sigma - \zeta} d\sigma \end{aligned} \tag{4.22}$$

The equation for displacements, Equation (4.17), after conformal mapping is:

$$(3 - 4\nu)\varphi(\sigma) - \frac{\omega(\sigma)}{\overline{\omega'(\sigma)}} \overline{\varphi'(\sigma)} - \overline{\psi(\sigma)} = 2G(U_x + iU_y) \tag{4.23}$$

Similar to what is done for the stress boundary conditions, Equation (4.23) can be expressed in integral form (Muskhelishvili, 1954) as:

$$\begin{aligned} 2\pi i\,(3 - 4\nu)\varphi(\zeta) - \int_s \frac{\omega(\sigma)}{\overline{\omega'(\sigma)}} \overline{\varphi'(\sigma)} \frac{d\sigma}{\sigma - \zeta} &= 2G\int_s \frac{U_x + iU_y}{\sigma - \zeta} d\sigma \\ 2\pi i\psi(\zeta) + \int_s \frac{\overline{\omega(\sigma)}}{\omega'(\sigma)} \varphi'(\sigma) \frac{d\sigma}{\sigma - \zeta} &= -2G\int_s \frac{U_x - iU_y}{\sigma - \zeta} d\sigma \end{aligned} \tag{4.24}$$

The complex potentials $\varphi(z)$ and $\psi(z)$ are obtained from (4.21) or (4.22) for stress-based boundary conditions or from (4.23) or (4.24) for displacement-based boundary conditions. As an alternative, the complex potentials functions may be expressed as a Laurent series as (Muskhelishvili, 1954):

$$\begin{aligned} \varphi(z) &= \sum_{k=1}^{k=\infty} A_k \log(z - z_k) + a_o + \sum_{k=1}^{k=\infty} a_k z^k + \sum_{k=1}^{k=\infty} b_k z^{-k} \\ \psi(z) &= \sum_{k=1}^{k=\infty} B_k \log(z - z_k) + c_o + \sum_{k=1}^{k=\infty} c_k z^k + \sum_{k=1}^{k=\infty} d_k z^{-k} \end{aligned} \tag{4.25}$$

where A, B, a_k, and b_k are complex constants and z_k is an arbitrary point inside the opening. The constants are obtained from the boundary conditions, i.e. from Equations (4.21) or (4.22) if stresses are imposed to the boundaries or from (4.23) or (4.24) if displacements are prescribed.

A number of problems have been solved using the complex variable method and they include single or multiple openings in an infinite medium with or without support (Savin, 1961; Exadaktylos and Stavropoulou, 2002; Gerçek, 1991, 1997; Huo et al., 2006; Li and Wang, 2008), deep circular tunnels in transversely-anisotropic rock (Hefny and Lo, 1999; Bobet, 2011, 2016a, b; Vitali et al., 2020) and in a semi-infinite medium with and without gravity forces (Verruijt, 1997, 1998; Strack and Verruijt, 2002; Vitali et al., 2018b, 2019).

The equations required for the boundary conditions often adopt an analytically intractable form due to the conformal mapping transformation and may require the use of a computer to reach a solution. Because of this, the results are often provided in the form of charts. For some simple problems, a close-form solution can be obtained. The best known of those is that of a deep unsupported circular opening in an infinite medium. The solution is given by the following equations (the solution can also be obtained using the method described in the next section):

$$\sigma_r = \frac{1}{2}\sigma_v(1+k)\left(1-\frac{r_o^2}{r^2}\right) - \frac{1}{2}\sigma_v(1-k)\left(1-4\frac{r_o^2}{r^2}+3\frac{r_o^4}{r^4}\right)\cos 2\theta$$

$$\sigma_\theta = \frac{1}{2}\sigma_v(1+k)\left(1+\frac{r_o^2}{r^2}\right) + \frac{1}{2}\sigma_v(1-k)\left(1+3\frac{r_o^4}{r^4}\right)\cos 2\theta$$

$$\tau = \frac{1}{2}\sigma_v(1-k)\left(1+2\frac{r_o^2}{r^2}-3\frac{r_o^4}{r^4}\right)\sin 2\theta \tag{4.26}$$

$$U_r = -\frac{1+\nu}{E}\left\{\frac{1}{2}\sigma_v(1+k)\left(1-2\nu+\frac{r_o^2}{r^2}\right) - \frac{1}{2}\sigma_v(1-k)\left(1+4(1-\nu)\frac{r_o^2}{r^2}-\frac{r_o^4}{r^4}\right)\cos 2\theta\right\} r$$

$$U_\theta = -\frac{1+\nu}{E}\frac{1}{2}\sigma_v(1-k)\left[1+2(1-2\nu)\frac{r_o^2}{r^2}+\frac{r_o^4}{r^4}\right] r\sin 2\theta$$

where σ_v is the far-field vertical stress acting at the depth of the center of the opening ($\sigma_v = \gamma h$, where γ is the unit weight of the ground and h is the depth of the center of the opening measured from the ground surface; see Figure 4.4), and k is the ratio of horizontal to vertical stresses at the tunnel center ($k = K_o = \sigma_h/\sigma_v$, where σ_h is the far-field horizontal stress at the depth of the tunnel center); r and θ are the polar coordinates (Figure 4.4) and E and ν are the elastic properties of the ground; r_o is the radius of the opening. Stresses and displacements are defined as shown in Figure 4.4. The displacements in (4.26) include the deformations due to the initial far-field stresses σ_v and σ_h. These displacements would have occurred well before the tunnel excavation. The displacements that an observer would see after the opening is completed are (these displacements are referred as "net" displacements):

$$U_r^{net} = -\frac{1+\nu}{E}\left\{\frac{1}{2}\sigma_v(1+k)\frac{r_o^2}{r^2} - \frac{1}{2}\sigma_v(1-k)\left[4(1-\nu)\frac{r_o^2}{r^2}-\frac{r_o^4}{r^4}\right]\cos 2\theta\right\} r$$

$$U_\theta^{net} = -\frac{1+\nu}{E}\frac{1}{2}\sigma_v(1-k)\left[2(1-2\nu)\frac{r_o^2}{r^2}+\frac{r_o^4}{r^4}\right] r\sin 2\theta \tag{4.27}$$

Note that the displacements in (4.26) increase with radial distance. This is due to the far-field stresses which extend to infinity. The net displacements in Equation (4.27) vanish at infinity as the disturbance in the ground produced by the tunnel decreases with distance.

Figure 4.6 is a contour plot of the stresses induced by the opening in the surrounding medium. The geometry and elastic properties used are the same as those employed in the calculations for Figure 4.1 and are: $r_o = 2$ m, E = 500 MPa, $\nu = 0.25$, $\sigma_v = 1$ MPa, k = 0.5. Figure 4.6a shows radial and tangential stresses. Note that there are two axes of symmetry in the problem: a vertical axis through the center of the tunnel and a horizontal axis also through the center, and so the results are symmetric with respect to the two axes. Figure 4.6a plots, on the half right, the radial stresses σ_r and on the half left the tangential stresses σ_θ. The radial stresses at the perimeter of the tunnel are zero since there is no support; they increase as the radial distance from the tunnel increases toward the far-field vertical stress, on the vertical axis, and toward the far-field horizontal stress on the horizontal axis. The tangential stresses are the largest at the springline and they decrease toward the far-field stresses with radial distance. At the perimeter of the excavation, i.e. $r = r_o$, Equation (4.26) gives:

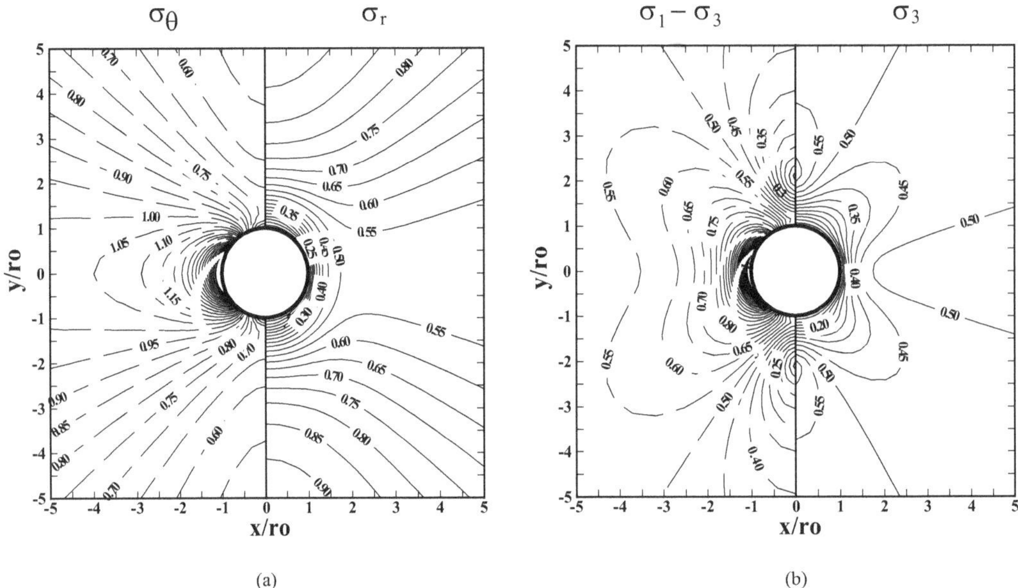

Figure 4.6 Stresses around unsupported deep circular opening. (a) Radial (right) and tangential (left) stresses. (b) Minor principal (right) and deviatoric (left) stresses.

$$\sigma_\theta \,|_{r=r_o} = \sigma_v(1+k) + 2\sigma_v(1-k)\cos 2\theta \tag{4.28}$$

With the input data used to plot Figure 4.6, Equation (4.28) gives values for σ_θ equal to 0.5 MPa at the crown and 2.5 MPa at the springline. See Figure 4.6a.

Figure 4.6b plots the minor principal stress σ_3 and the deviatoric stress $\sigma_1 - \sigma_3$. The minor principal stress at the perimeter of the opening is zero and coincides with the radial stress. With increasing radial distance, σ_3 increases quickly. The deviatoric stress is the largest close to the opening and decreases very fast as the distance from the tunnel increases. At the perimeter of the excavation the deviatoric stress is equal to the tangential stress since the minor principal stress is zero. This is an important observation that indicates that the ground around the excavation is subjected to a state of uniaxial compression. As the radial distance increases, confinement, σ_3, rapidly increases. Hence it is expected that the ground may yield first close to the excavation. Since confinement increases rapidly with radial distance, the plastic zone, if any, would be limited to portions of the ground close to the tunnel. This issue is further discussed in Section 4.2.2.

The tangential stresses at the perimeter of a closed curve that can be expressed in parametric form as (Figure 4.7),

$$\begin{aligned} x &= a\cos\beta + c\cos 3\beta \\ y &= b\sin\beta - c\sin 3\beta \end{aligned} \tag{4.29}$$

are given by (Obert and Duvall, 1967):

$$\begin{aligned} &[(a^2+6bc)\sin^2\beta + (b^2+6ac)\cos^2\beta - 6c(a+b)\cos^2 2\beta + 9c^2]\sigma_\theta \\ &= (\sigma_h+\sigma_v)(a^2\sin^2\beta + b^2\cos^2\beta - 9c^2) - \tau_{hv}(a+b)^2\frac{a+b+6c}{a+b+2c}\sin 2\beta \\ &-\frac{(a^2-b^2)(\sigma_h+\sigma_v)-(a+b)^2(\sigma_h-\sigma_v)}{a+b-2c}[(a-3c)\sin^2\beta - (b-3c)\cos^2\beta] \end{aligned} \tag{4.30}$$

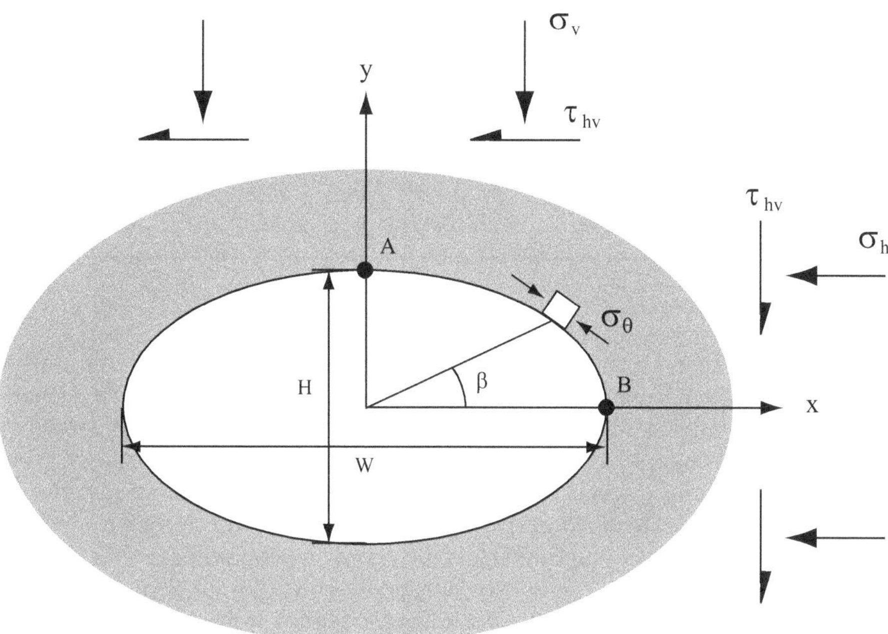

Figure 4.7 Curved opening in infinite medium.

In Equations (4.29) and (4.30) a, b, and c are parameters and β is an angle. The condition for the curve is that it must be simple; i.e. it must not cross itself. The width and height of the curve axes are W = 2(a + c) and H = 2(b + c), respectively. σ_h, σ_v, and τ_{hv} are the applied far-field stresses in the Cartesian coordinate system used to describe the curve. Examples of curves include circles, ellipses, and approximate ovaloids and rectangles with rounded corners. An ellipse is obtained taking c = 0 in (4.29) and (4.30). The major and minor axis of the ellipse are given by 2a and 2b. For a biaxial stress field that is normal and parallel to the ellipse axes (i.e. $\tau_{hv} = 0$), the tangential stress is:

$$\sigma_\theta = (\sigma_h + \sigma_v) + 2(\sigma_h - \lambda\sigma_v)\frac{(\lambda+1)\sin^2\beta - 1}{(\lambda^2-1)\sin^2\beta + 1} \tag{4.31}$$

where λ = a/b = W/H is the width to height ratio. What is interesting is that the tangential stresses do not depend on the absolute values of the axes dimensions, but on the aspect ratio of the ellipse given by λ = a/b.

Table 4.1 shows values of the tangential stresses σ_θ at points A and B (Figure 4.7) along the perimeter of the ellipse with different height to width ratios and different far-field stresses (note that results for λ = a/b = 1 are those of a circular opening).

Equation (4.31) shows that when $\sigma_h/\sigma_v = \lambda$, the tangential stresses are constant around the opening and equal to $\sigma_\theta = (\lambda + 1)\,\sigma_v$. When λ = 1 the ellipse transforms into a circle. Note that at the springline, point B, the stresses increase substantially as the ellipse becomes slender and the main axis is oriented perpendicular to the largest far-field compression stress σ_v. The contrary happens at the crown, point A, where the tangential stresses decrease and become more negative also as the larger axis of the ellipse increases in size. The problem of elliptical openings has also been solved using methods other than the complex variable method (e.g. using curvilinear coordinates; see Timoshenko and Goodier, 1970).

Table 4.1 Values of σ_θ/σ_v for elliptical holes with different aspect ratios λ = a/b

	a/b = 1/4		*a/b* = 1/3		*a/b* = ½		*a/b* = 1		*a/b* = 2		*a/b* = 4	
σ_h/σ_v	*A*	*B*	*A*	*B*	*A*	*B*	*A*	*B*	*A*	*B*	*A*	*B*
0	−1	1.5	−1	1.66	−1	2	−1	3	−1	5	−1	9
1/3	2	1.17	1.33	1.33	0.67	1.67	0	2.67	−0.33	4.67	−0.5	8.67
1	8	0.5	6	0.67	4	1	2	2	1	4	0.5	8

Complex variable theory has also been used to obtain stresses in the medium when multiple circular openings are placed. Of particular interest is the problem of two identical circular holes in an infinite medium (a different approach for the solution of the problem of the twin tunnels has been suggested by Kooi and Verruijt, 2001, through an iterative procedure). Figure 4.8 shows the two circular openings, with identical radius r_o and distance between centers d. The tunnels are assumed deep and subjected to far-field stresses σ_h and σ_v, which are perpendicular and parallel, respectively, to the plane that contains the axes of the tunnels. Table 4.2 contains values of the normalized tangential stress σ_θ/σ_v or σ_θ/σ_h for three different loading cases. In the first case, the twin openings are loaded in biaxial compression with a uniform far-field stress $\sigma_h = \sigma_v$; in the second case, a uniaxial compression is applied with the direction of loading perpendicular to the plane containing the axes of the tunnels, $\sigma_h = 0$ and $\sigma_v \neq 0$; and in the last case they are also loaded in uniaxial compression but in a direction parallel to the plane of symmetry, $\sigma_h \neq 0$ and $\sigma_v = 0$. The stresses at the perimeter of the openings are maximum at the springline, $\theta = 0$ or $\theta = \pi$, in the first two cases, and at the crown and invert for the last case, $\theta = \pm\pi/2$ (Obert and Duvall, 1967).

Table 4.2 shows that the stresses at the perimeter of the openings increase as the width of the pillar between the tunnels decreases. The presence of the second tunnel has negligible influence when the distance between the tunnels is between four to six radii, d/r_o > 4–6, depending on the loading conditions, or when the thickness of the pillar is between one to two diameters.

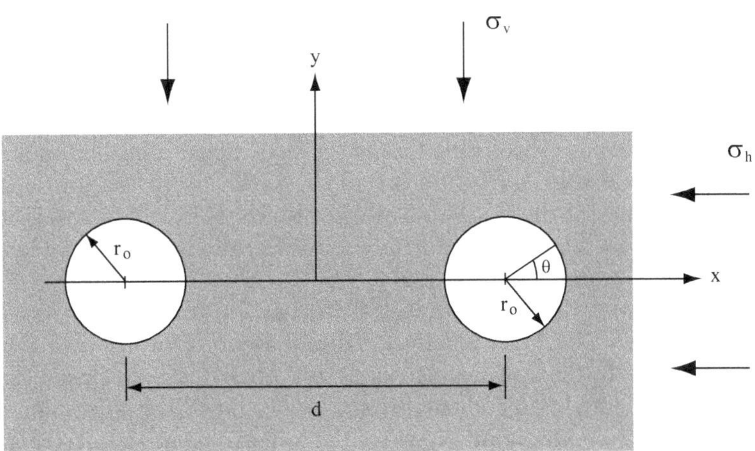

Figure 4.8 Twin circular tunnels in infinite medium.

Table 4.2 Values of σ_θ/σ_v for twin tunnels

	$\sigma_h = \sigma_v \neq 0$		$\sigma_h = 0$ and $\sigma_v \neq 0$		$\sigma_h \neq 0\ \sigma_v = 0$
d/r_o	$\theta = 0$	$\theta = \pi$	$\theta = 0$	$\theta = \pi$	$\theta = \pm\, \pi/2$
2	2.894	0.000	3.869	0.000	2.569
3	2.255	2.887	3.151	3.264	2.623
4	2.158	2.411	3.066	3.020	2.703
6	2.080	2.155	3.020	2.992	2.825
10	2.033	2.049	3.004	2.997	2.927
16	2.014	2.018	3.001	2.999	2.970
22	2.000	2.000	3.000	3.000	3.000

Source: Data from Obert and Duvall (1967)

4.2.1.2 Relative stiffness method

The relative stiffness method can be traced as far back as 1969 when Peck first introduced the concept of flexible and rigid liners (Peck, 1969b). According to the concept, a flexible liner has a pressure distribution and a deflected shape such that the bending moments at all points in the liner are negligible. A rigid liner deflects insignificantly under the loads imposed by the soil. Figure 4.9 illustrates the concept; see also discussion in Section 1.3. Before the tunnel is excavated, the initial stresses in the ground are σ_v and $\sigma_h = K_o\, \sigma_v$; K_o is the coefficient of earth pressure at rest. Upon excavation, the flexible liner deforms and since it is not able to

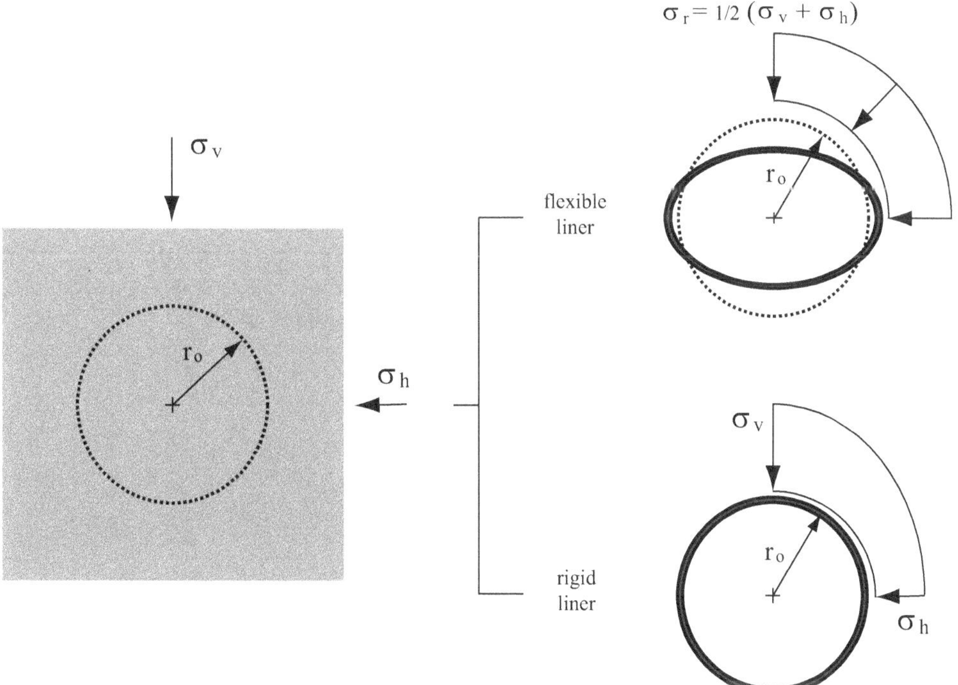

Figure 4.9 Flexible versus rigid liner.

withstand any bending moments, the distribution of radial stresses, σ_r, and axial thrust, T, of the liner must be constant and equal to (Peck et al., 1972):

$$\begin{aligned}\sigma_r &= \frac{1}{2}(1+K_o)\sigma_v \\ T &= \frac{1}{2}(1+K_o)\sigma_v r_o\end{aligned} \tag{4.32}$$

The rigid liner does not deform. Without a change of displacements in the ground, the stresses do not change. So the liner must support the initial stresses of the ground (Figure 4.9). No displacements of the liner occur. The maximum moment M and axial force T are (Peck et al., 1972):

$$\begin{aligned}M &= \pm\frac{1}{4}\,(1-K_o)\,\sigma_v\,r_o^2 \\ T_{springline} &= \sigma_v\,r_o \\ T_{crown} &= K_o\,\sigma_v\,r_o\end{aligned} \tag{4.33}$$

In both cases, it is assumed that full slip conditions exist between the liner and the ground (i.e. no shear stresses exist at the contact). Equation (4.33) results in a load demand on rigid liners that is too high. Liners have a finite stiffness and deflect with loading, and so in reality liners will fall between the two extreme cases. Peck et al. (1972), based on the work by Burns and Richard (1964), provided the thrust, moment and radial displacements at the crown and springline of a deep tunnel with a support with finite stiffness. Plane strain conditions were assumed at a cross section perpendicular to the tunnel axis; also the ground and the liner were elastic, homogeneous and isotropic. The liner was treated as a thin shell. The results were given as a function of the compressibility and flexibility of the liner relative to the surrounding ground (thus the name relative stiffness method). The expressions of the relative compressibility and flexibility were obtained assuming that the ground inside the tunnel contributed to the deformations. An alternative definition was later provided by Einstein and Schwartz (1979).

The problem analyzed by Einstein and Schwartz (1979) is that shown in Figure 4.4 except that the tunnel is placed very deep (i.e. $h \gg r_o$) in a weightless ground (i.e. $\gamma = 0$) and there are no pore pressures in the medium (i.e. $u = 0$). The solution of the ground, that is the solution of Equation (4.10), resides in finding an appropriate Airy stress function ϕ. A particular solution of Equation (4.10) was given by Timoshenko and Goodier (1970) as:

$$\begin{aligned}\phi = {} & a_o \ln r + b_o r^2 + c_o r^2 \ln r + d_o r^2 \theta + a'_o \theta \\ & + \frac{1}{2} a_1 r \theta \sin\theta + (b_1 r^3 + a'_1 r^{-1} + b'_1 r \ln r)\cos\theta \\ & - \frac{1}{2} c_1 r \theta \cos\theta + (d_1 r^3 + c'_1 r^{-1} + d'_1 r \ln r)\sin\theta \\ & + \sum_{n=2}^{\infty} (a_n r^n + b_n r^{n+2} + a'_n r^{-n} + b'_n r^{-n+2})\cos n\theta \\ & + \sum_{n=2}^{\infty} (c_n r^n + d_n r^{n+2} + c'_n r^{-n} + d'_n r^{-n+2})\sin n\theta\end{aligned} \tag{4.34}$$

where the parameters a_o, b_o, c_o, etc. are determined from the boundary conditions.

Stresses are obtained from Equation (4.7) where the pore pressures and unit weight are zero; i.e. u = 0, γ = 0. Strains are obtained from stresses and the displacements from integration of the strains.

The liner is assumed as a closed thin shell. The stress-displacement relations are (Flügge, 1966):

$$
\begin{aligned}
&\frac{d^2U_\theta^s}{d\theta^2}+\frac{dU_r^s}{d\theta}=-\frac{C(1-\nu^2)}{E}r_o\tau^s\\
&\frac{dU_\theta^s}{d\theta}+U_r^s+\frac{C}{F}\left(\frac{d^4U_r^s}{d\theta^4}+2\frac{d^2U_r^s}{d\theta^2}+U_r^s\right)=\frac{C(1-\nu^2)}{E}r_o\sigma_r^s
\end{aligned}
\tag{4.35}
$$

with the assumption that $t << r_o$ (the liner thickness is much smaller than the tunnel radius and so $r_o + t \approx r_o$); σ^s_r and τ^s are the radial and tangential stresses, respectively, applied to the liner and located at the liner-ground interface; U^s_r and U^s_θ are the radial and tangential displacements of the liner. C and F are, respectively, the compressibility and flexibility ratios defined as:

$$
\begin{aligned}
C&=\frac{Er_o(1-\nu_s^2)}{E_sA_s(1-\nu^2)}\\
F&=\frac{Er_o^3(1-\nu_s^2)}{E_sI_s(1-\nu^2)}
\end{aligned}
\tag{4.36}
$$

where E_s and ν_s are the elastic constants of the liner, and A_s and I_s are the area and moment of inertia of the cross section of the liner per unit length of tunnel (e.g. $A_s = t$ and $I_s = 1/12\ t^3$ for a liner of constant thickness t).

The solution is reached by imposing the boundary conditions at infinity and at the ground–liner interface. At infinity,

$$
\begin{aligned}
\sigma_r\,|_{r\to\infty}&=\frac{1}{2}(\sigma_v+\sigma_h)-\frac{1}{2}(\sigma_v-\sigma_h)\cos 2\theta\\
\sigma_\theta\,|_{r\to\infty}&=\frac{1}{2}(\sigma_v+\sigma_h)+\frac{1}{2}(\sigma_v-\sigma_h)\cos 2\theta\\
\tau\,|_{r\to\infty}&=\frac{1}{2}(\sigma_v-\sigma_h)\sin 2\theta
\end{aligned}
\tag{4.37}
$$

Equation (4.37) establishes that far from the tunnel the stresses coincide with the far-field stresses σ_v and σ_h, which are expressed, as shown in (4.37), in polar coordinates. At the ground–liner interface the boundary conditions are, for no-slip conditions between the liner and the ground:

$$
\begin{aligned}
\sigma_r\,|_{r=r_o}&=\sigma_r^s\\
\tau\,|_{r=r_o}&=\tau^s\\
U_r^{net}\,|_{r=r_o}&=U_r^s\\
U_\theta^{net}\,|_{r=r_o}&=U_\theta^s
\end{aligned}
\tag{4.38}
$$

Full slip between support and ground requires:

$$\begin{aligned} \sigma_r \,|_{r=r_0} &= \sigma_r^s \\ \tau \,|_{r=r_0} &= \tau^s = 0 \\ U_r^{net} \,|_{r=r_0} &= U_r^s \end{aligned} \tag{4.39}$$

No slip conditions means that a point of the ground in contact with the support moves exactly the same as the corresponding point of the support. Full slip implies that the shear stress at the ground–support contact is zero, and so the tangential displacements of the ground may not be the same as those of the support. The above expressions (4.38) and (4.39) apply at the interface between support and ground. However, because of the assumption $t << r_0$, $(r_0 + t) \approx r_0$, the radius of the tunnel can be taken as the radius of the excavation and the radius of the support. In typical tunnels, the thickness of the liner is of the order of $t \approx 0.1\ r_0$, and so the assumption is reasonable. For thicker liners, the errors may not be negligible and they increase with the thickness of the liner.

Note that net displacements are used in (4.38) and (4.39) since these are the displacements that the support needs to withstand. This is because the displacements produced by the far-field stresses occurred before excavation and support installation, and so they have no effect on the ground–support interaction. Only the change of displacements due to excavation, support, etc. should be considered (Einstein and Schwartz, 1979). This issue has been also discussed in detail by Pender (1980). Net displacements are the total displacements minus the displacements produced by the far-field stresses given in (4.37), which are:

$$\begin{aligned} U_r &= -\frac{1+\nu}{E}\left[\frac{1}{2}(\sigma_v + \sigma_h)(1-2\nu)r - \frac{1}{2}(\sigma_v - \sigma_h)r\cos 2\theta\right] \\ U_\theta &= -\frac{1+\nu}{E}\frac{1}{2}(\sigma_v - \sigma_h)r\sin 2\theta \end{aligned} \tag{4.40}$$

Thus the displacements from Equation (4.40) should be subtracted from those obtained using the Airy stress function in (4.34).

The final solution is given by the following expressions for stresses and displacements:

$$\begin{aligned} \sigma_r &= \frac{1}{2}\sigma_v(1+k) + a_o r^{-2} - \left[\frac{1}{2}\sigma_v(1-k) + 6a_2' r^{-4} + 4b_2 r^{-2}\right]\cos 2\theta \\ \sigma_\theta &= \frac{1}{2}\sigma_v(1+k) - a_o r^{-2} + \left[\frac{1}{2}\sigma_v(1-k) + 6a_2' r^{-4}\right]\cos 2\theta \\ \tau &= \left[\frac{1}{2}\sigma_v(1-k) - 6a_2' r^{-4} - 2b_2' r^{-2}\right]\sin 2\theta \\ U_r^{net} &= -\frac{1+\nu}{E}\{-a_o r^{-1} + [2a_2' r^{-3} + 4(1-\nu)b_2' r^{-1}]\cos 2\theta\} \\ U_\theta^{net} &= -\frac{1+\nu}{E}[2a_2' r^{-3} - 2(1-2\nu)b_2' r^{-1}]\sin 2\theta \end{aligned} \tag{4.41}$$

The parameters are, for the no-slip condition:

$$
\begin{aligned}
a_o &= -\frac{(1-\nu)CF}{C+F+(1-\nu)CF}\frac{1}{2}\sigma_v(1+k)r_o^2 \\
a_2' &= -\beta b_2' r_o^2 \\
b_2' &= \frac{(1-\nu)C}{4\nu-6\beta+(1-\nu)(1-3\beta)C}\frac{1}{4}\sigma_v(1-k)r_o^2 \\
\beta &= \frac{2\nu F+(1-\nu)CF+6(1-\nu)C}{3(C+F)+2(1-\nu)CF}
\end{aligned}
\tag{4.42}
$$

For the full slip condition:

$$
\begin{aligned}
a_o &= -\frac{(1-\nu)CF}{C+F+(1-\nu)CF}\frac{1}{2}\sigma_v(1+k)r_o^2 \\
a_2' &= \frac{(1-\nu)(6+F)}{3(5-6\nu)+(1-\nu)F}\frac{1}{4}\sigma_v(1-k)r_o^4 \\
b_2' &= -\frac{3+2(1-\nu)F}{3(5-6\nu)+(1-\nu)F}\frac{1}{4}\sigma_v(1-k)r_o^2
\end{aligned}
\tag{4.43}
$$

The loads acting on the liner are related to the interface stresses through the following equations (e.g. Flügge, 1966):

$$
\begin{aligned}
r_o\frac{dT}{d\theta}-\frac{dM}{d\theta} &= -r_o^2\tau^s \\
r_oT+\frac{d^2M}{d\theta^2} &= r_o^2\sigma_r^s
\end{aligned}
\tag{4.44}
$$

where T and M are the axial force and moment on the cross section of the liner. Note that because the geomechanics sign convention is used, the axial force is positive in compression and the moment is positive clockwise. Their values are given by:

For no slip:

$$
\begin{aligned}
T &= \frac{1}{2}\sigma_v(1+k)r_o + a_o r_o^{-1} + \left[\frac{1}{2}\sigma_v(1-k)r_o - 2a_2' r_o^{-3}\right]\cos 2\theta \\
M &= \left[\frac{1}{4}\sigma_v(1-k)r_o^2 + a_2' r_o^{-2} + b_2'\right]\cos 2\theta
\end{aligned}
\tag{4.45}
$$

For full slip:

$$
\begin{aligned}
T &= \frac{1}{2}\sigma_v(1+k)r_o + a_o r_o^{-1} + \frac{1}{3}\left[\frac{1}{2}\sigma_v(1-k)r_o + 6a_2' r_o^{-3} + 4b_2' r_o^{-1}\right]\cos 2\theta \\
M &= \frac{1}{3}\left[\frac{1}{2}\sigma_v(1-k)r_o^2 + 6a_2' r_o^{-2} + 4b_2'\right]\cos 2\theta
\end{aligned}
\tag{4.46}
$$

For $k = 1$, the far-field stresses are $\sigma_v = \sigma_h$ and the problem is axisymmetric. The solution is independent of the angular coordinate θ and thus the displacements of the ground and liner are only radial and there are no moments in the liner. For values of k ≠ 1 the deformed cross section acquires an elliptical shape with the major and minor axes coinciding with the vertical and horizontal axes (Figure 4.4) because these are the axes of symmetry. Whether the springline or the crown move inward or outward depends on the values of C and F, i.e. on the ratio E/E_s or on the relative flexibility of the liner, and on the value of k. Note that taking $C = F = \infty$ in Equations (4.42) and (4.43), the solution for an unsupported tunnel is obtained. This solution coincides, as it should, with Equations (4.26) for stresses and (4.27) for net displacements.

Figures 4.10 and 4.11 are contour plots of the stresses around a tunnel with support for the no-slip and for the full-slip conditions, respectively. The results are obtained for the same input parameters as those used for Figure 4.6. That is, $r_o = 2$ m, $\sigma_v = 1$ MPa, $\sigma_h = 0.5$ MPa (i.e. k = 0.5), ground properties: E = 500 MPa, $\nu = 0.25$, and liner properties: t = 0.2 m, $E_s = 24000$ MPa, $\nu_s = 0.25$. This makes the results comparable with those of Figure 4.6, obtained for the same tunnel but without support. A comparison of Figure 4.10 with 4.6 indicates that the radial stresses around the tunnel are larger for the tunnel with support than for the unsupported tunnel, while the opposite is true for the tangential stresses: they are higher in the unsupported tunnel than in the supported tunnel. This is caused by the liner taking loads from the ground. For the unsupported tunnel, the radial stresses at the perimeter of the opening are zero because there is no support. For the tunnel with the liner, the radial stresses are compressive because there is a load transfer from the ground to the support. The fact that there is no support in the tunnel of Figure 4.6 means that the ground must support itself, and so the tangential stresses are much larger than those of the supported tunnel. These differences can also be observed by comparing the plots of the minor principal stress and the deviatoric stress. In the unsupported tunnel the confinement is smaller and the deviatoric stress is larger than in the tunnel with support. The liner provides confinement to the ground, as the radial stresses at the ground–liner contact are not zero. The confinement results then in smaller load, i.e. smaller deviatoric stress, in the ground. Although not shown in the figures, the displacements induced in the ground for the case of the tunnel with support are smaller

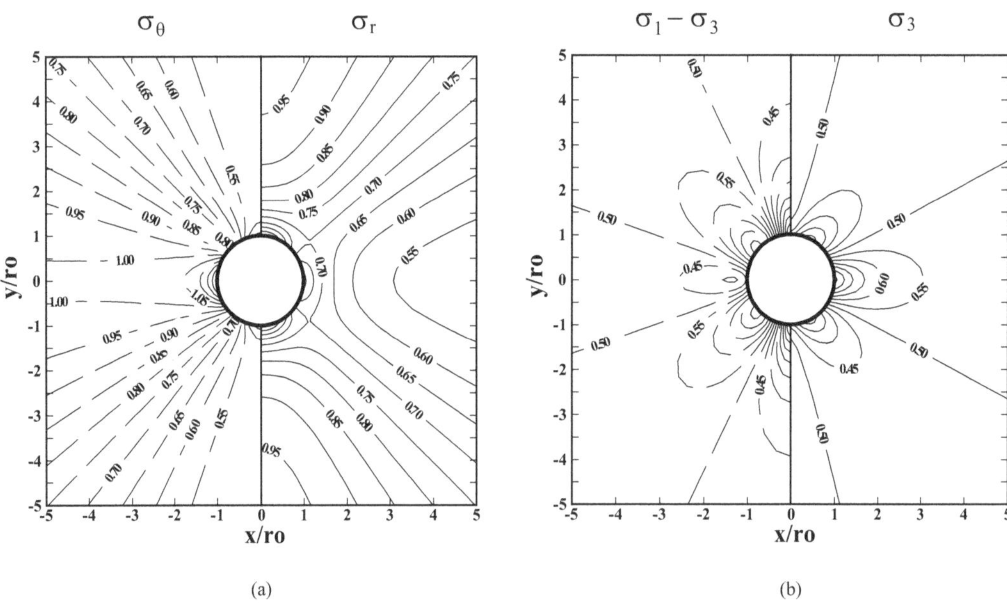

Figure 4.10 Stresses around a supported deep circular opening. No slip. (a) Radial (right) and tangential (left) stresses. (b) Minor principal (right) and deviatoric (left) stresses.

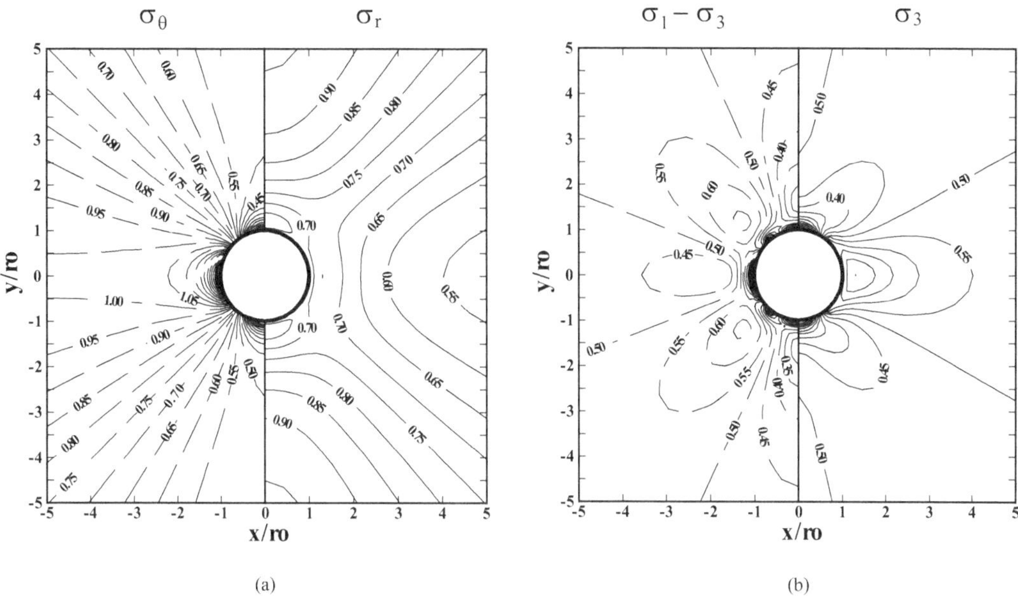

Figure 4.11 Stresses around a supported deep circular opening. Full slip. (a) Radial (right) and tangential (left) stresses. (b) Minor principal (right) and deviatoric (left) stresses.

than those of the unsupported tunnel. This is the result of the smaller loads taken by the ground when there is support.

The effects of friction at the interface can be observed by comparing Figures 4.10 and 4.11. The differences are not as significant as those between the supported and unsupported tunnels, but the tendency is for higher stresses in the ground for the full slip case than for the no slip. The exception is for the tangential stresses at the crown, which appear to be smaller for the case with full slip. This observation can again be explained with soil–liner interaction. The absence of shear stresses at the contact between the ground and the liner results in a decrease of ground support. There are two mechanisms of load transfer from the ground to the liner, in terms of stresses: through radial stresses and through shear stresses. The absence of shear stresses results in an increase of load transfer through radial stresses, and so the radial stresses in the ground are larger for the full slip case. When there is full slip, the ground at the interface is less constrained than when there is no slip and so it can deform more. With larger deformations the ground takes more load, and so the tangential stresses (also the deviatoric stresses) increase.

The response of the liner was explored by Einstein and Schwartz (1979), who provided a detailed discussion of the effects of liner stiffness, coefficient of earth pressure at rest, Poisson's ratio, etc. In essence, Equations (4.41), (4.45), and (4.46) indicate that the response of the ground and liner depend on the relative stiffness between the medium and the support, which is given by the parameters C, which is a measure of the relative compressibility, and F, which is a measure of the relative flexibility. The stiffer the liner is, the smaller the stresses in the ground and the larger the loads of the liner. This is an intuitive conclusion because stiffer supports contribute more to resist the ground. The relative stiffness is defined to a large extent by the geometry of the liner and by the ratio E/E_s, the ratio between the stiffness of the ground and the liner. The effect of the Poisson's ratio on the results is modest. Figures 4.12, 4.13 and 4.14 are plots of the normalized thrust, $T/(\sigma_v r_o)$, moment, $M/(\sigma_v r_o^2)$, and displacements, $U_s\ E/[(1 + \nu)\sigma_v r_o]$, of the liner, respectively, as a function of the relative stiffness between the ground and the liner, E/E_s. The results are obtained at the springline

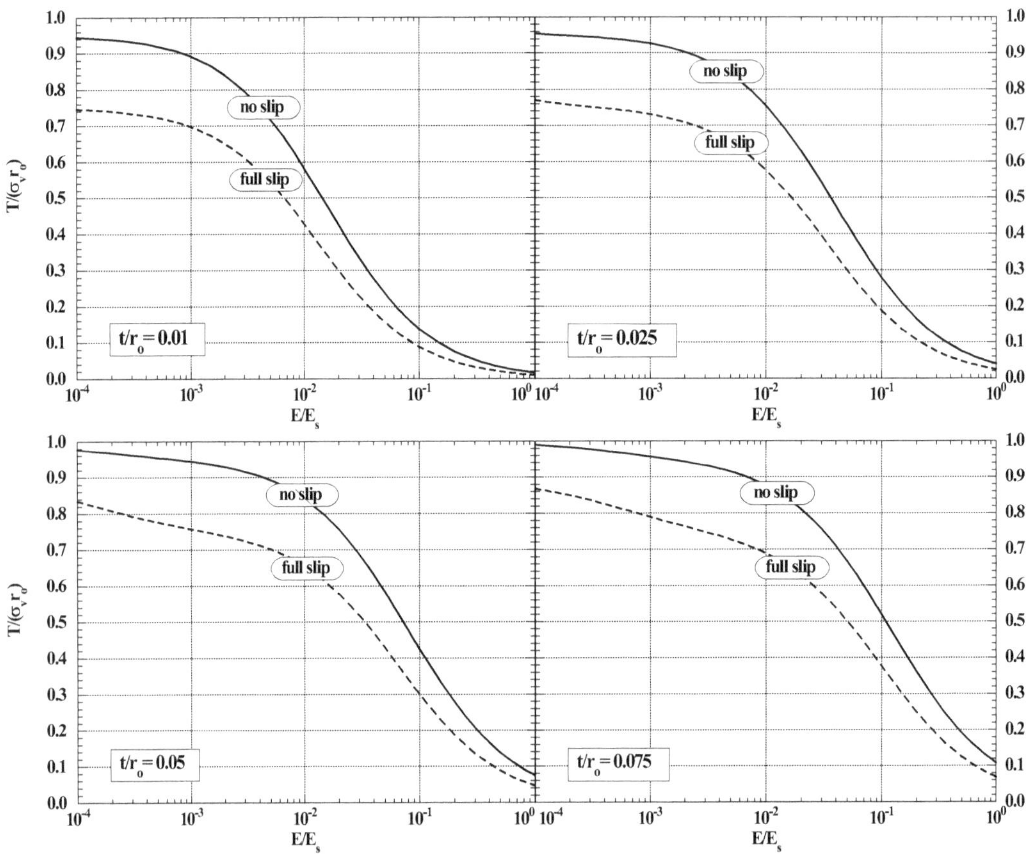

Figure 4.12 Normalized thrust at springline. $K_o = 0.5, \nu = \nu_s = 0.25$.

of the tunnel, i.e. at $\theta = 0°$, and for $k = K_o = 0.5$, $\nu = \nu_s = 0.25$, and for no slip and full slip conditions at the ground–liner interface. As one can see, the relative stiffness ratio, E/E_s, has a major effect on the response of the liner, which exhibits a large reduction in normalized thrust and bending moments as the stiffness ratio increases, i.e. as the ground becomes stiffer (or the liner becomes softer). The thickness of the liner also has an effect on the response of the support, albeit smaller than the relative stiffness. As expected, the thicker the liner, the larger the load on the liner and the smaller the radial displacements. The figures also can be used to ascertain the effect of the ground–liner interface. No slip conditions result in larger thrust, but smaller moments and radial displacements than full slip conditions.

A useful exercise that provides further insight into the ground–liner interaction phenomenon is to compute Equations (4.41) and (4.45) and (4.46) for a perfectly flexible and incompressible liner (Peck's liner; see Figure 4.9). This is often the case for actual tunnels, in particular for tunnels in soft to medium stiff soils and in soft rock. Mathematically, this is accomplished by setting $C \rightarrow 0$ and $F \rightarrow \infty$ in the equations. The result is:

For no slip:

$$a_o = 0$$

$$a_2' = \frac{\nu}{3-2\nu}\frac{1}{2}\sigma_v(1-k)r_o^4 \tag{4.47a}$$

$$b_2' = -\frac{3}{3-2\nu}\frac{1}{4}\sigma_v(1-k)r_o^2$$

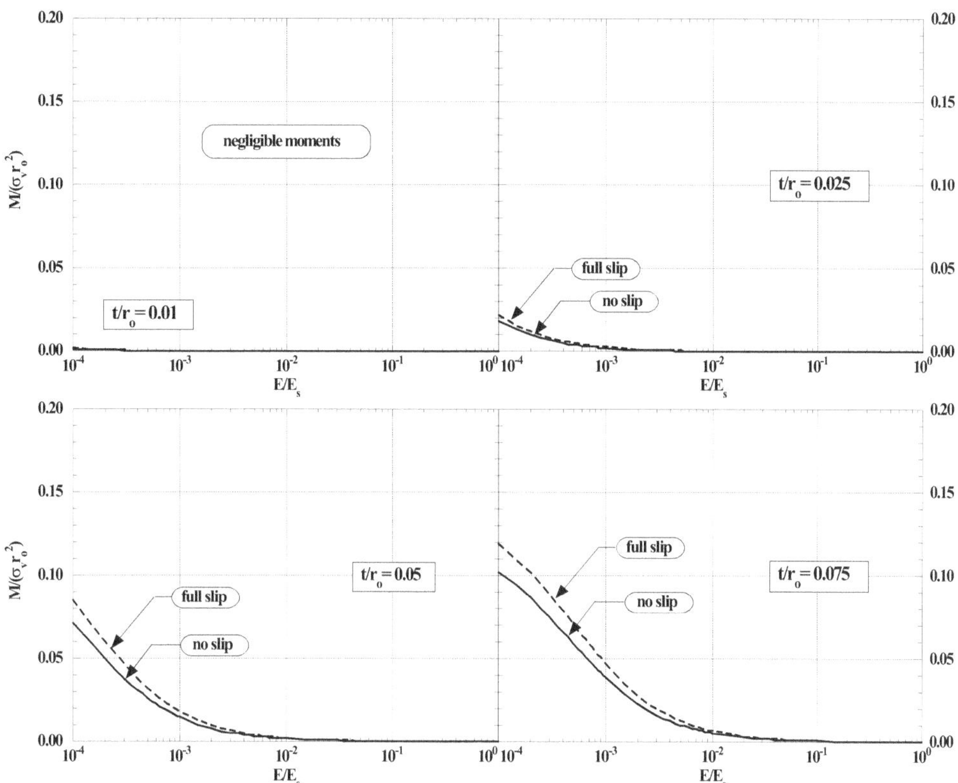

Figure 4.13 Normalized moment at springline. $K_o = 0.5$, $\nu = \nu_s = 0.25$.

and for full slip:

$$\begin{aligned} a_o &= 0 \\ a_2' &= \frac{1}{4}\sigma_v(1-k)r_o^4 \\ b_2' &= -\frac{1}{2}\sigma_v(1-k)r_o^2 \end{aligned} \tag{4.47b}$$

Stresses and displacements can be obtained from (4.41) and (4.47). At the tunnel perimeter, for the no slip case:

$$\begin{aligned} \sigma_r|_{r=r_o} &= \frac{1}{2}\sigma_v(1+k) + \frac{1}{2}\frac{3-4\nu}{3-2\nu}\sigma_v(1-k)\cos 2\theta \\ \sigma_\theta|_{r=r_o} &= \frac{1}{2}\sigma_v(1+k) + \frac{1}{2}\frac{3+4\nu}{3-2\nu}\sigma_v(1-k)\cos 2\theta \\ \tau|_{r=r_o} &= \frac{3-4\nu}{3-2\nu}\sigma_v(1-k)\sin 2\theta \\ U_r^{net}|_{r=r_o} &= \frac{1+\nu}{E}\frac{3-4\nu}{3-2\nu}\sigma_v(1-k)r_o\cos 2\theta \\ U_\theta^{net}|_{r=r_o} &= -\frac{1+\nu}{E}\frac{1}{2}\frac{3-4\nu}{3-2\nu}\sigma_v(1-k)r_o\sin 2\theta \end{aligned} \tag{4.48a}$$

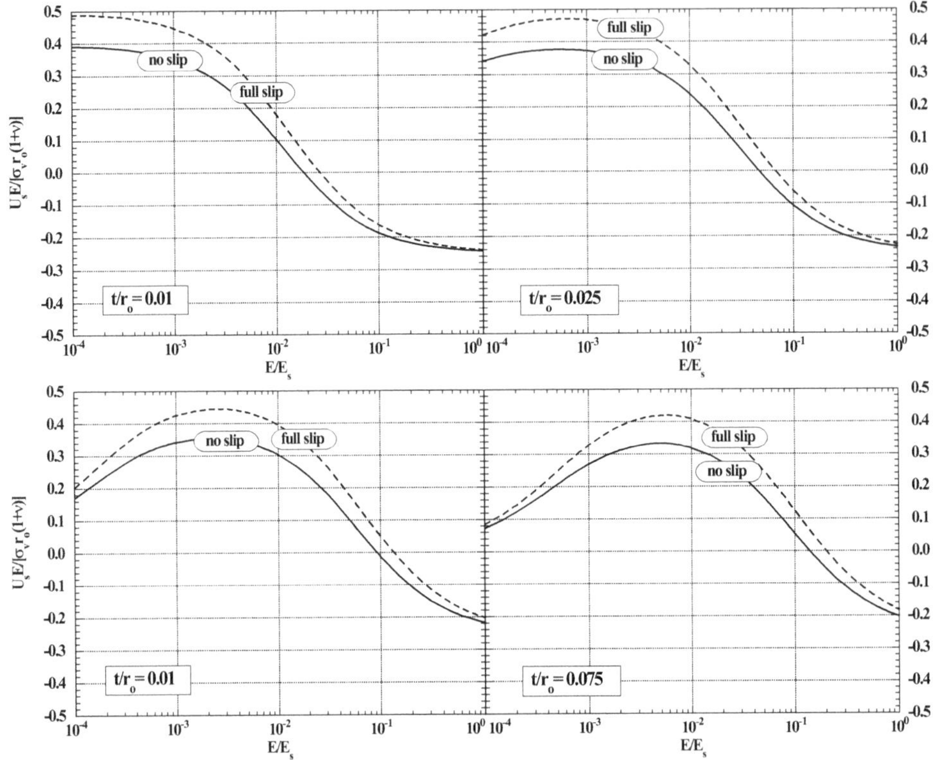

Figure 4.14 Normalized liner radial displacements at springline. $K_o = 0.5, \nu = \nu_s = 0.25$.

For the full slip case:

$$\begin{aligned}
\sigma_r \mid_{r=r_o} &= \frac{1}{2}\sigma_v(1+k) \\
\sigma_\theta \mid_{r=r_o} &= \frac{1}{2}\sigma_v(1+k) + 2\sigma_v(1-k)\cos 2\theta \\
\tau \mid_{r=r_o} &= 0 \\
U_r^{net} \mid_{r=r_o} &= \frac{1+\nu}{E}\frac{1}{2}(3-4\nu)\sigma_v(1-k)r_o\cos 2\theta \\
U_\theta^{net} \mid_{r=r_o} &= -\frac{1+\nu}{E}\frac{1}{2}(3-4\nu)\sigma_v(1-k)r_o\sin 2\theta
\end{aligned} \tag{4.48b}$$

Note that the radial stress for the full slip case is the same as that proposed by Peck et al. (1972) and given in Equation (4.32). Inspection of Equations (4.48a) and (4.48b) indicate that the tangential stresses are the largest for the full slip case, provided that k<1. The radial stresses are largest at the springline for the no slip case and are largest at the crown for the full slip case, again for k<1. The radial displacements for the full slip case are always larger than for the no slip case. These findings are consistent with the soil–structure interaction discussion regarding the plots in Figures 4.10 to 4.14. The axial load and moment in the liner are:

$$\begin{aligned}
T^{noslip} &= \frac{1}{2}\sigma_v(1+k)r_o + \frac{1}{2}\frac{3-4\nu}{3-2\nu}\sigma_v(1-k)r_o\cos 2\theta \\
T^{fullslip} &= \frac{1}{2}\sigma_v(1+k)r_o \\
M^{noslip} &= M^{fullslip} = 0
\end{aligned} \tag{4.49}$$

The axial force for the full slip case coincides with that proposed by Peck et al. (1972), given in Equation (4.32). The moments in both cases are zero because of the infinitely large flexibility of the liner. Note also that the axial load for the no slip case is larger than for the full slip at the springline and smaller at the crown, provided that k<1.

The relative stiffness method (i.e. the Airy stress function method) can be used to approximate stresses and displacements of a shallow tunnel in a heavy ground where stresses increase linearly with depth due to the ground's unit weight. Bobet (2001) found a solution for this problem with the additional assumption that the soil around the tunnel has deformed by a magnitude "w" before the liner is installed. This assumption tries to address the ground loss that occurs during shield excavation in soft ground. Section 3.4 contains a discussion about the ground movements induced by shield excavation as well as a quantification of their sources and magnitudes. The parameter "w" can be taken as an "equivalent ground loss" which is related to deformations due to any physical gap between the ground and the liner, due to the unsupported tunnel face, or due to the quality of workmanship. All these deformations constitute the "ground loss" (see Section 3.4). Hence,

$$w = \frac{\text{ground loss}(\%)}{100}\frac{\pi r_o^2}{2\pi r_o} = \frac{\text{ground loss}(\%)}{200} r_o \tag{4.50}$$

Because of the movements of the ground prior to liner installation, the additional assumption of full slip between the ground and the liner is made.

Equations (4.34) to (4.36) still apply, but now with the boundary conditions:

$$\begin{aligned}
\tau \,|_{\theta=\pm\frac{\pi}{2}} &= 0 \\
U_\theta \,|_{\theta=\pm\frac{\pi}{2}} &= 0 \\
U_r^{net} \,|_{r=r_o} &= U_r^s \,|_{r=r_o} - w \\
\sigma_r \,|_{r=r_o} &= \sigma_r^s \,|_{r=r_o} \\
\tau \,|_{r=r_o} &= \tau^s \,|_{r=r_o} = 0 \\
\sigma_y \,|_{r\to\infty} &= \gamma y = \gamma(h - r\sin\theta) \\
\sigma_y \,|_{r\to\infty} &= k\sigma_y
\end{aligned} \tag{4.51}$$

The solution is given by:

$$\begin{aligned}
\sigma_r &= \frac{a_o}{r^2} + \frac{1}{2}\gamma h(1+k) + \left[-\frac{1}{4}\gamma r(k+3) - 2c_1' r^{-3} + c_1 r^{-1} + d_1' r^{-1}\right]\sin\theta \\
&\quad - \left[\frac{1}{2}\gamma h(1-k) + 6a_2' r^{-4} + 4b_2' r^{-2}\right]\cos 2\theta + \left[\frac{1}{4}\gamma r(1-k) - 12c_3' r^{-5} - 10d_3' r^{-3}\right]\sin 3\theta \\
\sigma_\theta &= -\frac{a_o}{r^2} + \frac{1}{2}\gamma h(1+k) + \left[-\frac{1}{4}\gamma r(3k+1) + 2c_1' r^{-3} + d_1' r^{-1}\right]\sin\theta \\
&\quad + \left[\frac{1}{2}\gamma h(1-k) + 6a_2' r^{-4}\right]\cos 2\theta + \left[-\frac{1}{4}\gamma r(1-k) + 12c_3' r^{-5} + 2d_3' r^{-3}\right]\sin 3\theta \\
\tau &= \left[-\frac{1}{4}\gamma r(1-k) + 2c_1' r^{-3} - d_1' r^{-1}\right]\cos\theta \\
&\quad - \left[-\frac{1}{2}\gamma h(1-k) + 6a_2' r^{-4} + 2b_2' r^{-2}\right]\sin 2\theta + \left[\frac{1}{4}\gamma r(1-k) + 12c_3' r^{-5} + 6d_3' r^{-3}\right]\cos 3\theta
\end{aligned} \tag{4.52}$$

$$U_r^{net} = -\frac{1+\nu}{E}\left\{-\frac{a_o}{r} + [c_1' r^{-2} + c_1(1-\nu)\ln r + d_1'(1-2\nu)\ln r]\sin\theta \right.$$
$$\left. + [2a_2' r^{-3} + 4(1-\nu)b_2' r^{-1}]\cos 2\theta + [3c_3' r^{-4} + (5-4\nu)d_3' r^{-2}]\sin 3\theta\right\}$$
$$U_\theta^{net} = -\frac{1+\nu}{E}\left\{-[c_1' r^{-2} - c_1((1-\nu)\ln r + \nu) - d_1'(1-2\nu)(\ln r - 1)]\cos\theta \right.$$
$$\left. -\frac{1}{2}[-4a_2' r^{-3} + 4(1-2\nu)b_2' r^{-1}]\sin 2\theta + [-3c_3' r^{-4} + (1-4\nu)d_3' r^{-2}]\cos 3\theta\right\}$$

with

$$\begin{aligned}
a_o &= -\frac{1}{2}\frac{\gamma h(1+k)(1-\nu^2)CF + 2E(C+F)\frac{w}{r_o}}{(C+F)(1+\nu)+(1-\nu^2)CF} r_o^2 \\
c_1 &= \gamma r_o^2 \\
c_1' &= -\frac{1}{8}\left(k - \frac{\nu}{1-\nu}\right)\gamma r_o^4 \\
d_1' &= -\frac{1}{4}\frac{1-2\nu}{1-\nu}\gamma r_o^2 \\
a_2' &= \frac{1}{4}\frac{(F+6)(1-\nu)}{(1-\nu)F+3(5-6\nu)}\gamma h(1-k)r_o^4 \\
b_2' &= -\frac{1}{4}\frac{2(1-\nu)F+3}{(1-\nu)F+3(5-6\nu)}\gamma h(1-k)r_o^2 \\
c_3' &= -\frac{1}{12}\frac{(1-\nu)F+4(5-4\nu)}{(1-\nu)F+8(7-8\nu)}\gamma(1-k)r_o^6 \\
d_3' &= \frac{1}{8}\frac{(1-\nu)F+8}{(1-\nu)F+8(7-8\nu)}\gamma(1-k)r_o^4
\end{aligned} \tag{4.53}$$

The axial force and moment acting on a cross section of the liner, T and M, are as follows:

$$\begin{aligned}
T = &-\frac{1}{2}\frac{\left[2E\frac{w}{r_o} - \gamma h(1+k)(1+\nu)\right](C+F)}{(C+F)(1+\nu)+(1-\nu^2)CF} r_o \\
&+ \frac{3}{2}\frac{3-4\nu}{(1-\nu)F+3(5-6\nu)}\gamma h(1-k)r_o\cos 2\theta \\
&- \frac{3-4\nu}{(1-\nu)F+8(7-8\nu)}\gamma(1-k)r_o^2\sin 3\theta \\
M = &\frac{3}{2}\frac{3-4\nu}{(1-\nu)F+3(5-6\nu)}\gamma h(1-k)r_o^2\cos 2\theta \\
&- \frac{3-4\nu}{(1-\nu)F+8(7-8\nu)}\gamma(1-k)r_o^3\sin 3\theta
\end{aligned} \tag{4.54}$$

Figure 4.15 shows contour plots of a shallow tunnel with the following characteristics: $r_o = 2$ m, depth $h = 20$ m, ground unit weight $\gamma = 20$ kN/m³, $k = 0.5$, $E = 50$ MPa, $\nu = 0.25$,

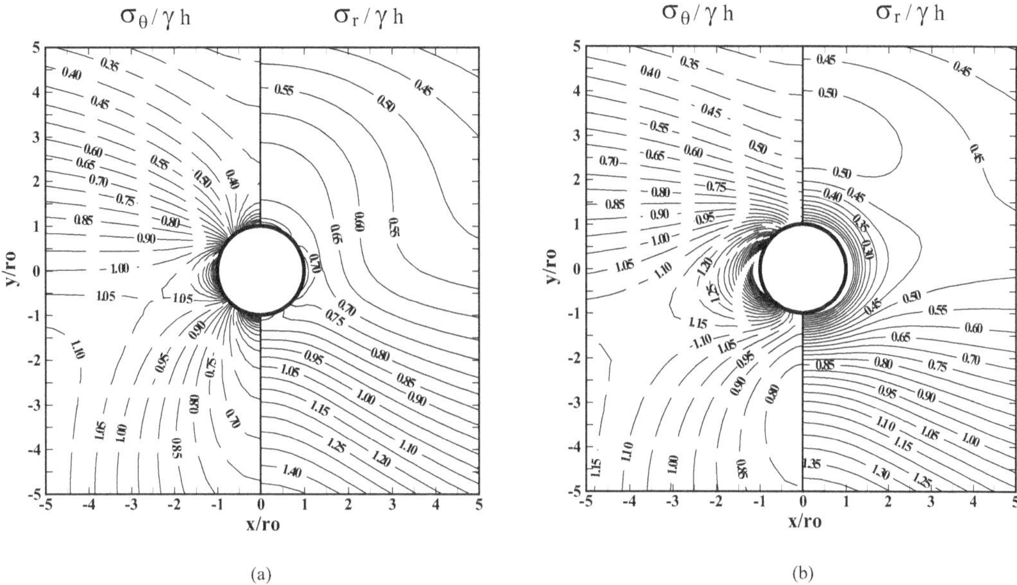

Figure 4.15 Stresses around a supported shallow circular opening. (a) Ground loss w = 0% normalized radial (right) and tangential (left) stresses. (b) Ground loss w = 2% normalized pore pressures (right) and tangential (left) stresses.

and for the liner, t = 0.2 m, E_s = 24000 MPa, ν_s = 0.25. The major difference between this example and the previous ones is that the Young's modulus of the ground has been reduced to 50 MPa (to enhance results and thus help with the discussion). The stresses in Figure 4.15 are normalized by the vertical stress at the center of the unexcavated tunnel, which should make the results somewhat comparable to the previous figures. Figure 4.15a plots contours of radial and tangential stresses for a tunnel where there is no ground loss, i.e. w = 0, while Figure 4.15b plots the stresses when w = 2%. The most important difference between Figures 4.11a, which corresponds to a deep tunnel with full slip, and Figure 4.15a, apart from the magnitude of the stresses, which depends on the relative stiffness of the ground and liner, is that the stresses around the shallow tunnel have lost the radial symmetry that had for the deep tunnel. This is the result of the gravity field that makes stresses increase with depth. The effects of ground loss are quite dramatic. A relatively small displacement of the ground previous to liner installation carries a significant reduction of radial stresses and a large increase of tangential stresses. This is an expected result because as the ground deforms more, it has to carry a larger load. A direct consequence of this is larger displacements of the ground and smaller loads on the support (not shown in the figure).

It is interesting to note that in Equation (4.52) displacements are proportional to the logarithm of the radial coordinate. This is due to the gravity field and the assumption of plane strain. In this case, the excavation of the tunnel, as seen by a point far from the tunnel (invoking the Saint Venant's principle), results in a point load applied at the center of the tunnel and distributed uniformly in the tunnel's axial direction, with magnitude equal to the unit weight of the excavated cross section times the unit weight of the ground. This is the well-known problem of a line load in an infinite medium, which has a similar increase of displacements with radial distance from the point of application of the load. Hence, Equation (4.52) should be used to obtain relative displacements between two points or absolute displacements by making the displacements at a given point zero. The location of zero displacements may be located on the surface far enough from the tunnel (Strack and Verruijt, 2002) or at a stiff layer (Chou and Bobet, 2003). The depth of the stiff layer is not significant for the evaluation of the stresses in the ground or in the liner, as long as the layer is deeper than about five times the tunnel radius, but it is critical for a proper evaluation of absolute ground displacements. It is also important to mention that the solution for a shallow tunnel presented is not exact in the sense that it does not fulfill the requirements of zero normal and shear stresses on the surface. The exact solution can be obtained using complex variable theory (Strack and Verruijt, 2002). The solution included here, however, provides acceptable results, with errors smaller than about 10%, for depth to radius ratios larger than 1.5; $h/r_o > 1.5$.

Equations (4.52) to (4.54), together with the results plotted in Figure 4.15, indicate that the magnitude of the gap between the shield and the ground "w" is one of the most important sources for ground stresses and deformations. They also show that deformations increase with the tunnel radius and decrease with the depth of the tunnel.

For incompressible and perfectly flexible liners ($C = 0$, $F = \infty$), which is representative of a large number of tunnels, particularly segmental tunnels in soft ground, the stresses and displacements of the ground at the tunnel perimeter, and the axial force and moment of the liner are:

$$
\begin{aligned}
&\sigma_r \left.\right|_{r=r_o} = \frac{1}{2}\gamma h(1+k) - \frac{E}{1+\nu}\frac{w}{r_o} \\
&\sigma_\theta \left.\right|_{r=r_o} = \frac{1}{2}\gamma h(1+k) + \frac{E}{1+\nu}\frac{w}{r_o} - \frac{1}{2}\gamma r_o\left[2k + \frac{1-2\nu}{1-\nu}\right]\sin\theta \\
&\qquad + 2\gamma h(1-k)\cos 2\theta - \gamma r_o(1-k)\sin 3\theta \\
&\tau \left.\right|_{r=r_o} = 0 \\
&U_r^{net} \left.\right|_{r=r_o} = -\frac{1+\nu}{E}\left\{\frac{E}{1+\nu}w - \frac{1}{8}\frac{1}{1-\nu}\gamma r_o^2[k-(1+k)\nu - 2(3-4\nu)\ln r_o]\sin\theta\right. \\
&\qquad \left. -\frac{1}{2}(3-4\nu)\gamma h r_o(1-k)\cos 2\theta + \frac{1}{8}(3-4\nu)\gamma r_o^2(1-k)\sin 3\theta\right\} \\
&U_\theta^{net} \left.\right|_{r=r_o} = -\frac{1+\nu}{E}\left\{\frac{1}{8}\frac{1}{1-\nu}\gamma r_o^2[(1-\nu)k + 2 - \nu + 2(3-4\nu)\ln r_o]\cos\theta\right. \\
&\qquad \left. +\frac{1}{2}(3-4\nu)\gamma h r_o(1-k)\sin 2\theta + \frac{1}{8}(3-4\nu)\gamma r_o^2(1-k)\cos 3\theta\right\} \\
&T = \frac{1}{2}\gamma h(1+k)r_o - \frac{E}{1+\nu}w \\
&M = 0
\end{aligned}
\tag{4.55}
$$

Note that for a deep tunnel, $h>>r_o$, Equation (4.55) coincides with (4.48b), except for the term in "w," which arises due to the ground loss. This can be ascertained by making in (4.55) $\sigma_v = \gamma h$ and $\gamma r_o = 0$ since $r_o << h$. A tunnel can be considered deep when the depth of the tunnel below the surface, if no excess pore pressures exist or there is no drainage toward the tunnel, is larger than about ten times the tunnel radius (Bobet, 2003).

As the ground loss increases, the stresses in the ground increase very fast, as shown in Figure 4.15. In soft ground, it is expected that significant yielding may occur with even small deformations. If this is the case, Equations (4.51) to (4.55) should be used with caution or only qualitatively to estimate the contribution of each parameter to the tunnel response. Section 3.4 includes other methods that are better suited to estimate ground deformations and stresses when yielding of the ground occurs.

4.2.1.3 Tunnels below the water table

It is quite common to have tunnels excavated below the water table (Figure 4.4). The solution then has to account for the response of the saturated soil to the changes of stresses imposed by the tunnel construction. Such response will depend on the permeability of the soil, in addition to other engineering properties and geometry parameters. If the permeability of the ground is high and the construction rate (i.e. the rate at which the change of stresses is imposed to the ground) is low, then no noticeable excess pore pressures will develop and so a long-term analysis is appropriate. If however the permeability of the ground is low and the construction rate is high, then excess pore pressures will occur during construction, which will dissipate long after the tunnel excavation has been completed. In this case both short-term and long-term calculations are necessary. For example, Anagnostou and Kovari (1996a) suggested that drained conditions would prevail in those tunnels excavated in materials with a permeability higher than 10^{-6} to 10^{-7} m/sec, with advance rates of 0.1 to 1 m/hr. or less.

Figures 4.16 to 4.19 show the results of a three-dimensional analysis with ABAQUS of an unsupported deep tunnel below the water table. The tunnel geometry, ground properties, and loading are similar to those of Figure 4.1. The tunnel radius is $r_0 = 2$m, and the ground properties: E = 500 MPa, $\nu = 0.25$. The far-field loading applied is $\sigma'_{vertical} = 1$ MPa, $K_0 = 0.5$, and u = 0.5 MPa. The pore pressures around the perimeter of the excavation are imposed as atmospheric, i.e. u = 0. Thus the only difference between the results presented in Figures 4.1 and 4.16 to 4.19 is the addition of the pore pressures. Note, however, that with pore pressures the total vertical stress at the depth of the tunnel axis is 50% larger and the total horizontal stresses 100% larger. Figure 4.16 is a plot of contours of effective axial, vertical, and horizontal stresses and the pore pressures generated for short-term conditions. Figure 4.17 is a plot of the contours for long-term conditions. Similar to what happens in the tunnel with no water pressure (Figure 4.1) the tunnel does not change much the axial stresses, except ahead of the face where there is substantial unloading. The case of the tunnel with water pressure (Figure 4.16a) shows much larger unloading than the tunnel with no water pressure. The vertical stresses in the two cases do not differ much. Both show unloading at the crown and loading at the springline. There is a bit more loading ahead of the tunnel in Figure 4.16b. The differences are much larger for the horizontal stresses. The tunnel below the water table has much more unloading at the springline and larger loading at the crown. Given that the results for the vertical stress are similar, this means that the shear stresses in the ground for the tunnel with pore pressures are larger than in the tunnel with no water. Figure 4.16d shows the pore pressures around the tunnel. Since the far-field pore pressures in the example are 0.5 MPa, the figure indicates that the excess pore pressures around the tunnel are negative (note that the pore pressures around the opening are smaller than the free-field pore pressures). This is an expected result because the excavation of the opening in essence induces unloading in the ground. Since the ground cannot change volume, the mean effective stresses are maintained. Thus the tendency of the ground to expand because of the unloading is counterbalanced by negative excess pore pressures. This is an important observation because the negative excess pore pressures help the ground support itself. This is so because the negative pore pressures maintain confinement and thus the strength of the soil is higher (this is discussed further in Section 4.2.2). Also because smaller displacements in the ground are induced (this is shown later in Figure 4.19) the shear stresses are also reduced. A comparison between Figures 4.16 and 4.17 provides an indication of the changes in behavior due to the dissipation of the excess pore pressures and the change of pore water flow that is induced, as the tunnel is in essence a drain. The pore pressures shown in Figure 4.17d

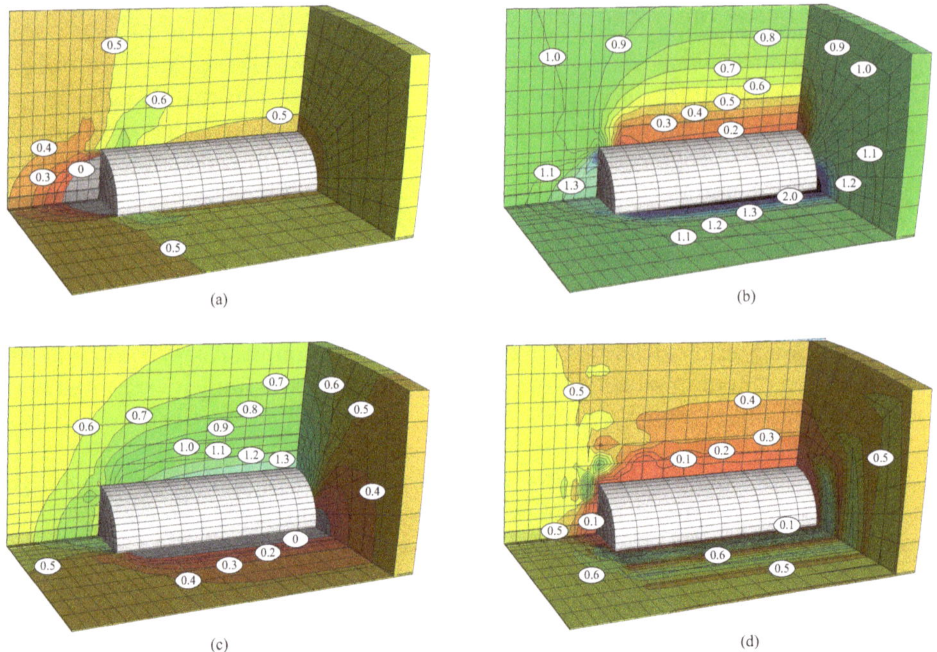

Figure 4.16 Three-dimensional view of stresses and pore pressures for an unsupported deep tunnel with $r_o = 2$ m, $\sigma'_v = 1.0$ MPa, $K_o = 0.5$, u = 0.5 MPa. Short-term analysis. (a) σ'_{axial}. (b) $\sigma'_{vertical}$. (c) $\sigma'_{horizontal}$. (d) Pore pressures.

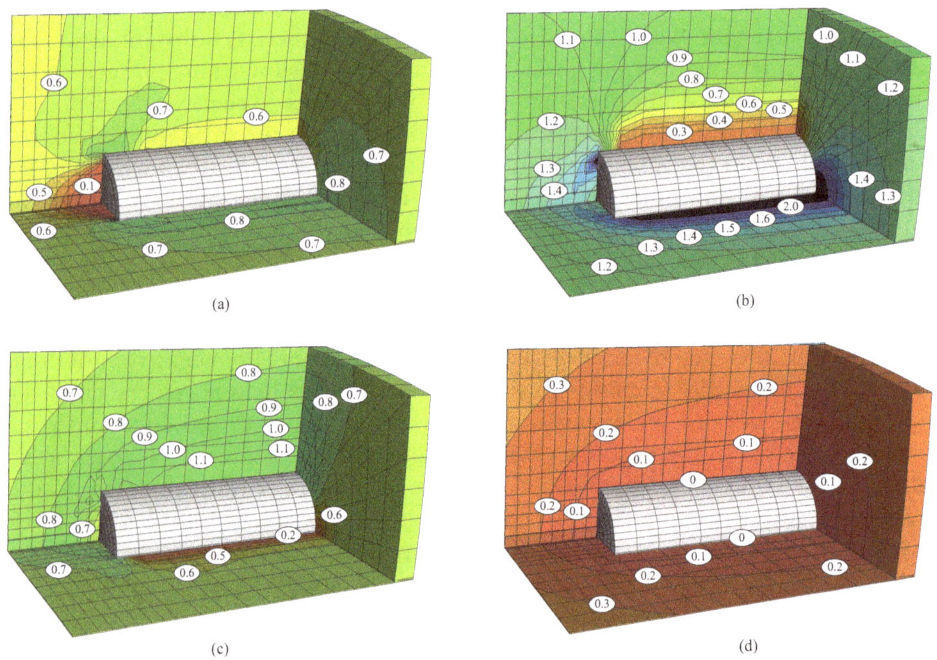

Figure 4.17 Three-dimensional view of stresses and pore pressures for an unsupported deep tunnel with $r_o = 2$ m, $\sigma'_v = 1.0$ MPa, $K_o = 0.5$, u = 0.5 MPa. Long-term analysis. (a) σ'_{axial}. (b) $\sigma'_{vertical}$. (c) $\sigma'_{horizontal}$. (d) Pore pressures.

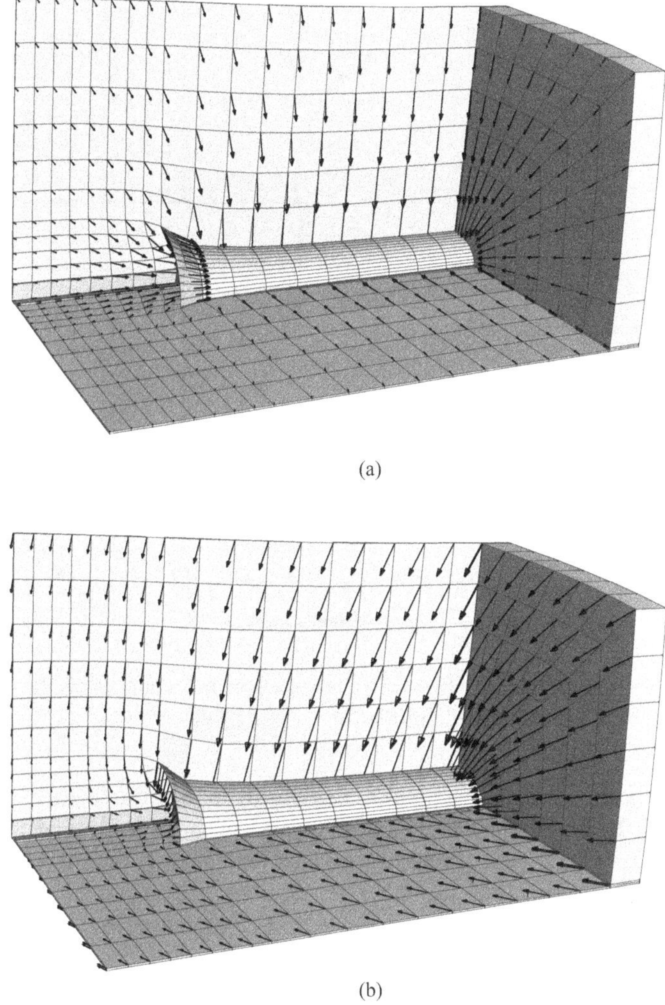

(a)

(b)

Figure 4.18 Three-dimensional view of displacements for an unsupported deep tunnel with r_o = 2 m, σ'_v = 1.0 MPa, K_o = 0.5, u = 0.5 MPa. (a) Short term. (b) Long term.

correspond to steady-state conditions. The tunnel indeed acts as a drain, with the pore pressures slowly increasing with increasing distance from the tunnel toward the far field. It is important to note the large area of influence of the tunnel. The figure shows results within a distance of about four radii from the center of the tunnel. While this distance was sufficient in the previous cases to show the volume of influence of the tunnel, in this case it is not. Inspection of Figures 4.16 and 4.17 shows that the long-term analysis is associated with an increase of all stresses: axial, vertical, and horizontal, both in magnitude and in extent beyond the tunnel.

Conceptually, the response of a tunnel below the water table can be decomposed into the response that results from the ground alone and the response due to the water. This decomposition is strictly true with the assumption of elasticity. The effects of the ground are due to its buoyant unit weight (in the case analyzed due to the far-field effective stresses). Since the applied stresses are the same for the cases of Figures 4.1 and 4.17, the component of the ground alone is exactly that shown in Figure 4.1 (note that the results in Figure 4.16

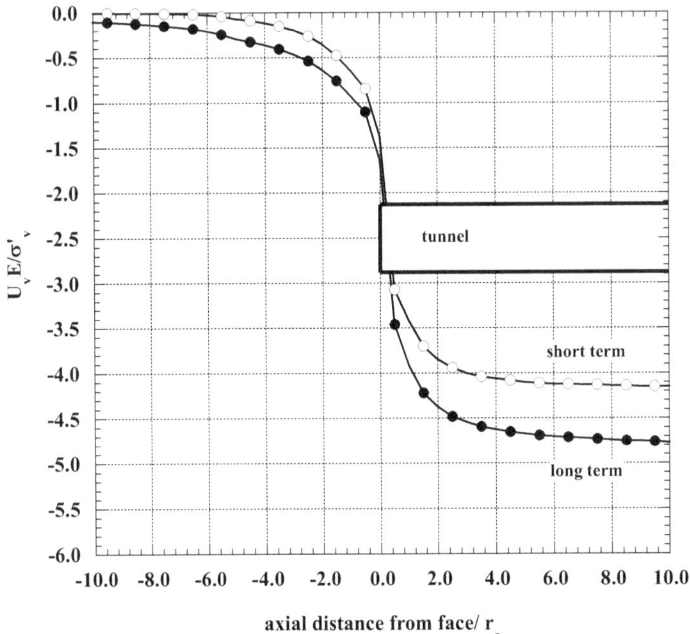

Figure 4.19 Vertical displacements at the crown of a deep unsupported tunnel below the water table, r_0 = 2 m, K_0 = 0.5 for short- and long-term analysis.

are not those of Figure 4.1 because the assumption of no volume change is not satisfied in Figure 4.1). The differences between Figures 4.1 and 4.17 are only due to the water seepage toward the excavation. The increase of effective stresses in Figure 4.17 compared with Figure 4.1 is caused by the water flow. In the literature, the concept is also referred to as "seepage forces" (e.g. Lambe and Whitman, 1969). The increase in stresses is associated with an increase of displacements.

Displacements are plotted in Figure 4.18 for short term and for long term. The deformed shapes of the opening are similar to that of the tunnel with no water. There is an ovalization of the tunnel with larger movements at the crown than at the springline. For the short term, the displacements are larger near the tunnel and quickly decrease away from the tunnel. The deformations ahead of the tunnel are toward the face and then slowly rotate behind the face until at a section far behind the tunnel face where the displacements are mostly radial. The pattern is quite different for long term. The displacements have a significant component toward the face of the tunnel. Also, the displacements do not decrease much with distance from the excavation. This is a somewhat counterintuitive result, which can be explained using the decomposition of behavior between ground and water. The displacements shown in Figure 4.1 are those of the ground component. Hence the water component produces the counterintuitive movements shown in Figure 4.18. In Section 4.2.1.3.2, an expression for the displacements induced around a circular opening with drainage is presented. The result is that the pore pressures change with radial distance following a logarithmic function. Mathematically, this results in a monotonic increase of displacements from the point of drainage. The outcome is the counterintuitive result of larger displacements with increasing radial distance. Hence, the displacements shown along the vertical plane are the sum of displacements of decreasing magnitude with radial distance for the ground component and of increasing magnitude with distance for the water component. The displacements on the horizontal plane, with the exception of those close to the tunnel, where the radial distance is relatively small, are in essence due to the water component. The water flow around the tunnel can be visualized as perpendicular to the pore pressure contours in Figure 4.17d, which for a deep tunnel are the equipotentials. The flow induces seepage forces, which in turn produce displacements. Hence it would be expected to have horizontal displacements with a component toward the right in Figure 4.18b, following the direction of seepage, and increasing with horizontal distance due to the logarithmic function. Figure 4.18b, however, shows the horizontal displacements in the opposite direction. This is just the result of the boundary conditions used for the simulation where zero horizontal displacements are imposed at a cross section placed far in front of the face (i.e. on the left hand side in the figure). Since this is a fixed location, it requires that the rest of the model moves toward the left. In summary, the counterintuitive displacements for the long-term case are the result of displacements increasing with radial distance and the boundary conditions used in the simulation. This is not an issue when displacements decrease with radial distance because the boundary conditions imposed far from the tunnel have no relevance in the solution. The boundary conditions in terms of displacements are, however, important when displacements increase with distance and care should be taken interpreting the results. It must be realized that in the long-term case, absolute movements have no physical meaning and one should use relative movements between two points.

The magnitude of the vertical displacements at the crown is shown in Figure 4.19. The displacements follow a sigmoidal curve similar to that observed for the tunnel with no water table (Figure 4.2). What is interesting is the difference between short- and long-term displacements. Because the negative excess pore pressures help support the ground, the displacements are smaller for the short term than for the long term. Also, long-term displacements occur at distances from the face larger than for the short term. This is a direct consequence of the large volume of ground affected by the pore water pressure depression around the tunnel for long-term conditions. The increase of displacements from those in Figure 4.2 to the long-term displacements in Figure 4.19 is, as with the stresses, due to the seepage forces that exist in the ground as a result of the water flow toward the tunnel. The increase may be significant, in this particular case about 40%.

An analytical solution can be obtained for the problem of a tunnel below the water table using superposition (Bobet, 2001) and considering two stages in the construction process: (1) immediately after construction; (2) long after construction. In the first stage, it is assumed that construction operations are completed very rapidly so no excess pore pressures dissipate; that is, a short-term analysis is performed. In the second stage, it is assumed that a long time has passed after construction so all excess pore pressures produced have dissipated; hence, a long-term analysis is performed. In addition, different drainage may exist at the interface between the liner and the ground. The following two extreme cases are considered: no drainage or full drainage. In the no-drainage case the long-term pore pressures at the ground–liner interface are those of the free field. In the full drainage case, the pore pressures are zero.

4.2.1.3.1 Short-term analysis

The Airy function in (4.34) still applies. Total stresses are found from (4.7), and effective stresses and pore pressures from (4.7) and (4.12). The boundary conditions are very similar to (4.51), with the fundamental difference being that they should be written in terms of effective stresses. They are as follows:

$$
\begin{aligned}
\tau\,|_{\theta=\pm\frac{\pi}{2}} &= 0\\
U_\theta\,|_{\theta=\pm\frac{\pi}{2}} &= 0\\
U_r^{net}\,|_{r=r_o} &= U_r^s\,|_{r=r_o} - w\\
\sigma_r' + u\,|_{r=r_o} &= \sigma_r^s\,|_{r=r_o}\\
\tau\,|_{r=r_o} &= \tau^s\,|_{r=r_o} = 0\\
\sigma_y'\,|_{r\to\infty} &= \gamma_b y = \gamma_b(h - r\sin\theta)\\
\sigma_x'\,|_{r\to\infty} &= k\sigma_y'\\
u\,|_{r\to\infty} &= \gamma_w(h_w - r\sin\theta)
\end{aligned}
\tag{4.56}
$$

The solution is given by the following expressions:

$$
\begin{aligned}
\sigma_r' =\ & \frac{a_o}{r^2} + \frac{1}{2}\gamma_b h(1+k) - \frac{(1-v)CF}{C+F+(1-v)CF}\gamma_w h_w\frac{r_o^2}{r^2}\\
& -\left[\frac{1}{4}\gamma_b r(k+3) - \frac{1}{2}\gamma_w\frac{r_o^2}{r} + 2c_1' r^{-3} - \frac{1}{2}c_1 r^{-1}\right]\sin\theta\\
& -\left[\frac{1}{2}\gamma_b h(1-k)6a_2' r^{-4} + 2b_2' r^{-2}\right]\cos 2\theta + \left[\frac{1}{4}\gamma_b r(1-k) - 12c_3' r^{-5} - 6d_3' r^{-3}\right]\sin 3\theta\\
\sigma_\theta' =\ & -\frac{a_o}{r^2} + \frac{1}{2}\gamma_b h(1+k) + \frac{(1-v)CF}{C+F+(1-v)CF}\gamma_w h_w\frac{r_o^2}{r^2}\\
& -\left[\frac{1}{4}\gamma_b r(3k+1) + \frac{1}{2}\gamma_w\frac{r_o^2}{r} - 2c_1' r^{-3} + \frac{1}{2}c_1 r^{-1}\right]\sin\theta\\
& +\left[\frac{1}{2}\gamma_b h(1-k) + 6a_2' r^{-4} + 2b_2' r^{-2}\right]\cos 2\theta + \left[-\frac{1}{4}\gamma_b r(1-k) + 12c_3' r^{-5} + 6d_3' r^{-3}\right]\sin 3\theta\\
\tau =\ & \left[-\frac{1}{4}\gamma_b r(1-k) + 2c_1' r^{-3}\right]\cos\theta - \left[-\frac{1}{2}\gamma_b h(1-k) + 6a_2' r^{-4} + 2b_2' r^{-2}\right]\sin 2\theta\\
& +\left[\frac{1}{4}\gamma_b r(1-k) + 12c_3' r^{-5} + 6d_3' r^{-3}\right]\cos 3\theta\\
u =\ & \gamma_w(h_w - r\sin\theta) + \frac{1}{2}\left(\gamma_w\frac{r_o^2}{r} + c_1 r^{-1}\right)\sin\theta - 2b_2' r^{-2}\cos 2\theta - 4d_3' r^{-3}\sin 3\theta\\
U_r^{net} =\ & -\frac{1+v}{E}\left\{-\frac{a_o}{r} + \frac{(1-v)CF}{C+F+(1-v)CF}\gamma_w h_w\frac{r_o^2}{r} + \left[c_1' r^{-2} + \frac{1}{2}\left(c_1 + \gamma_w r_o^2\right)\ln r\right]\sin\theta\right.\\
& \left. + \left[2a_2' r^{-3} + 2b_2' r^{-1}\right]\cos 2\theta + \left[3c_3' r^{-4} + 3d_3' r^{-2}\right]\sin 3\theta\right\}\\
U_\theta^{net} =\ & -\frac{1+v}{E}\left\{-\left[c_1' r^{-2} - \frac{1}{2}\left(c_1 + \gamma_w r_o^2\right)(1+\ln r)\right]\cos\theta\right.\\
& \left. + 2a_2' r^{-3}\sin 2\theta - \left[3c_3' r^{-4} + d_3' r^{-2}\right]\cos 3\theta\right\}
\end{aligned}
\tag{4.57}
$$

where

$$
\begin{aligned}
a_o &= -\frac{1}{2}\frac{\gamma_b h(1+k)(1-\nu^2)CF + 2E(C+F)\dfrac{w}{r_o}}{(C+F)(1+\nu)+(1-\nu^2)CF} r_o^2 \\
c_1 &= \gamma_b r_o^2 \\
c_1' &= \frac{1}{8}\gamma_b(1-k)\, r_o^4 \\
a_2' &= \frac{1}{4}\frac{(1-\nu)F+3}{(1-\nu)F+6}\gamma_b h(1-k) r_o^4 \\
b_2' &= -\frac{1}{4}\frac{2(1-\nu)F+3}{(1-\nu)F+6}\gamma_b h(1-k) r_o^2 \\
c_3' &= -\frac{1}{12}\frac{(1-\nu)F+12}{(1-\nu)F+24}\gamma_b(1-k) r_o^6 \\
d_3' &= \frac{1}{8}\frac{(1-\nu)F+8}{(1-\nu)F+24}\gamma_b(1-k) r_o^4
\end{aligned}
\tag{4.58}
$$

The axial force and moment in the liner are:

$$
\begin{aligned}
T = &-\frac{1}{2}\frac{\left[2E\dfrac{w}{r_o} - \gamma_b h(1+k)(1+\nu) - 2\gamma_w h_w(1+\nu)\right](C+F)}{(C+F)(1+\nu)+(1-\nu^2)CF} r_o \\
&+\frac{3}{2}\frac{1}{(1-\nu)F+6}\gamma_b h(1-k) r_o \cos 2\theta \\
&-\frac{1}{(1-\nu)F+24}\gamma_b(1-k) r_o^2 \sin 3\theta \\
M = &\ \frac{3}{2}\frac{1}{(1-\nu)F+6}\gamma_b h(1-k) r_o^2 \cos 2\theta \\
&-\frac{1}{(1-\nu)F+24}\gamma_b(1-k) r_o^3 \sin 3\theta
\end{aligned}
\tag{4.59}
$$

For short-term conditions, whether there is full drainage at the ground–liner interface or the interface is impermeable, the results are not affected.

Equations (4.58) and (4.59) depend on the Poisson's ratio of the ground. This is in contrast to solutions found in the literature where expressions for short-term analysis are taken from long-term analysis with the Poisson's ratio equal to 0.5. This is done because the volumetric strain in the soil is zero if $\nu = 0.5$; see Equation (4.11). This approach is not correct. First, because the Poisson's ratio, in elasticity, is a material property and cannot change with the loading conditions; and second because this assumption violates the constitutive behavior of the material since deformations are a function of effective stresses. Because of these two reasons, the practice of taking $\nu = 0.5$ should be abandoned.

For incompressible and perfectly flexible liners (C = 0, F = ∞), the stresses and displacements of the ground at the tunnel perimeter and the axial force and moment of the liner are:

$$
\begin{aligned}
\sigma'_r\,|_{r=r_o} &= \frac{1}{2}\gamma_b h(1+k) - \frac{E}{1+\nu}\frac{w}{r_o} - \frac{1}{2}(\gamma_b-\gamma_w)r_o\sin\theta \\
&\quad -\gamma_b h(1-k)\cos 2\theta + \frac{1}{2}\gamma_b r_o(1-k)\sin 3\theta \\
\sigma'_\theta\,|_{r=r_o} &= \frac{1}{2}\gamma_b h(1+k) + \frac{E}{1+\nu}\frac{w}{r_o} + \left[\frac{1}{2}(\gamma_b-\gamma_w)r_o - \gamma_b r_o(1+k)\right]\sin\theta \\
&\quad +\gamma_b h(1-k)\cos 2\theta - \frac{1}{2}\gamma_b r_o(1-k)\sin 3\theta \\
\tau\,|_{r=r_o} &= 0 \\
u\,|_{r=ro} &= \gamma_w(h_w - r_o\sin\theta) + \frac{1}{2}(\gamma_b+\gamma_w)r_o\sin\theta + \gamma_b h(1-k)\cos 2\theta - \frac{1}{2}\gamma_b r_o(1-k)\sin 3\theta \\
U_r^{net}\,|_{r=r_o} &= -\frac{1+\nu}{E}\left\{\frac{E}{1+\nu}w + \frac{1}{8}r_o^2[\gamma_b(1-k)+4(\gamma_b+\gamma_w)\ln r_o]\sin\theta\right. \\
&\quad \left. -\frac{1}{2}\gamma_b h\, r_o(1-k)\cos 2\theta + \frac{1}{8}\gamma_b r_o^2(1-k)\sin 3\theta\right\} \\
U_\theta^{net}\,|_{r=r_o} &= -\frac{1+\nu}{E}\left\{-\frac{1}{8}r_o^2[\gamma_b(1-k)-4(\gamma_b+\gamma_w)(1+\ln r_o)]\cos\theta\right. \\
&\quad \left. +\frac{1}{2}\gamma_b h\, r_o(1-k)\sin 2\theta + \frac{1}{8}\gamma_b r_o^2(1-k)\cos 3\theta\right\} \\
T &= \frac{1}{2}[\gamma_b h(1+k) + 2\gamma_w h_w]r_o - \frac{E}{1+\nu}w \\
M &= 0
\end{aligned}
\tag{4.60}
$$

If the tunnel is deep, stresses and displacements of the ground can be obtained by making in Equations (4.57) and (4.58) $h >> r_o$ and $\sigma'_v = \gamma_b h$. Taking $w = 0$, the expressions are:

$$
\begin{aligned}
\sigma'_r &= \frac{1}{2}\sigma'_v(1+k) - \frac{(1-\nu)CF}{C+F+(1-\nu)CF}\left[\frac{1}{2}\sigma'_v(1+k)+\gamma_w h_w\right]\frac{r_o^2}{r^2} \\
&\quad -\frac{1}{2}\sigma'_v(1-k)\left[1+3\frac{(1-\nu)F+3}{(1-\nu)F+6}\frac{r_o^4}{r^4} - \frac{2(1-\nu)F+3}{(1-\nu)F+6}\frac{r_o^2}{r^2}\right]\cos 2\theta \\
\sigma'_\theta &= \frac{1}{2}\sigma'_v(1+k) + \frac{(1-\nu)CF}{C+F+(1-\nu)CF}\left[\frac{1}{2}\sigma'_v(1+k)+\gamma_w h_w\right]\frac{r_o^2}{r^2} \\
&\quad +\frac{1}{2}\sigma'_v(1-k)\left[1+3\frac{(1-\nu)F+3}{(1-\nu)F+6}\frac{r_o^4}{r^4} - \frac{2(1-\nu)F+3}{(1-\nu)F+6}\frac{r_o^2}{r^2}\right]\cos 2\theta \\
\tau &= \frac{1}{2}\sigma'_v(1-k)\left[1-3\frac{(1-\nu)F+3}{(1-\nu)F+6}\frac{r_o^4}{r^4} + \frac{2(1-\nu)F+3}{(1-\nu)F+6}\frac{r_o^2}{r^2}\right]\sin 2\theta \\
u &= \gamma_w h_w + \frac{1}{2}\frac{2(1-\nu)F+3}{(1-\nu)F+6}\sigma'_v(1-k)\cos 2\theta \\
U_r^{net} &= -\frac{1+\nu}{E}\left\{\frac{(1-\nu)CF}{C+F+(1-\nu)CF}\left[\frac{1}{2}\sigma'_v(1+k)+\gamma_w h_w\right]\frac{r_o^2}{r}\right. \\
&\quad \left. +\frac{1}{2}\sigma'_v(1-k)\left[\frac{(1-\nu)F+3}{(1-\nu)F+6}\frac{r_o^4}{r^3} - \frac{2(1-\nu)F+3}{(1-\nu)F+6}\frac{r_o^2}{r}\right]\cos 2\theta\right\} \\
U_\theta^{net} &= -\frac{1+\nu}{E}\frac{(1-\nu)F+3}{(1-\nu)F+6}\frac{1}{2}\sigma'_v(1-k)\frac{r_o^4}{r^3}\sin 2\theta
\end{aligned}
\tag{4.61}
$$

The axial force and moment in the liner are given by, from (4.59):

$$T = \frac{1}{2}\frac{[\sigma'_v(1+k)+2\gamma_w h_w](C+F)}{C+F+(1-\nu)CF} r_o + \frac{3}{2}\frac{1}{(1-\nu)F+6}\sigma'_v(1-k)r_o\cos 2\theta$$
$$M = \frac{3}{2}\frac{1}{(1-\nu)F+6}\sigma'_v(1-k)r_o^2\cos 2\theta \tag{4.62}$$

Equations (4.61) and (4.62) can be used to obtain expressions for a deep tunnel with an incompressible and perfectly flexible liner. The expressions can also be obtained from (4.60) taking the terms in $\gamma\ r_o$ and $\gamma_w\ r_o$ equal to zero since $r_o << h$ and $r_o << h_w$.

For a deep tunnel with no slip, also with w = 0, the stresses and displacements for the ground are:

$$\sigma'_r = \frac{a_o}{r^2} + \frac{1}{2}\sigma'_v(1+k) - \frac{(1-\nu)CF}{C+F+(1-\nu)CF(1-r_o^2/R^2)}u_w\frac{r_o^2}{r^2}$$
$$-\left[\frac{1}{2}\sigma'_v(1-k)+6a'_2r^{-4}+2b'_2r^{-2}\right]\cos 2\theta$$

$$\sigma'_\theta = -\frac{a_o}{r^2} + \frac{1}{2}\sigma'_v(1+k) + \frac{(1-\nu)CF}{C+F+(1-\nu)CF(1-r_o^2/R^2)}u_w\frac{r_o^2}{r^2}$$
$$+\left[\frac{1}{2}\sigma'_v(1-k)+6a'_2r^{-4}+2b'_2r^{-2}\right]\cos 2\theta$$

$$\tau = \left[\frac{1}{2}\sigma'_v(1-k)-6a'_2r^{-4}-2b'_2r^{-2}\right]\sin 2\theta \tag{4.63}$$

$$u = u_w + \frac{(1-\nu)CF}{C+F+(1-\nu)CF(1-r_o^2/R^2)}u_w\frac{r_o^2}{R^2} - 2b'_2r^{-2}\cos 2\theta$$

$$U_r^{net} = -\frac{1+\nu}{E}\left\{-\frac{a_o}{r} + \frac{(1-\nu)CF}{C+F+(1-\nu)CF(1-r_o^2/R^2)}u_w\frac{r_o^2}{r} + 2(2a'_2r^{-3}+2b'_2r^{-1})\cos 2\theta\right\}$$

$$U_\theta^{net} = -\frac{1+\nu}{E}2a'_2r^{-3}\sin 2\theta$$

where R is the radial distance from the center of the tunnel to where far-field pore pressures are recovered. In many cases $R >> r_o$, which can be implemented in (4.63) by taking the ratio $r_o/R = 0$. The parameters in (4.63) are given by:

$$a_o = -\frac{1}{2}\frac{(1-\nu)CF}{C+F+(1-\nu)CF}\sigma'_v(1+k)r_o^2$$

$$a'_2 = \frac{1}{4}\frac{[3C+F+(1-\nu)CF](1-\nu)}{[2+(1-\nu)C][6+(1-\nu)F]}\sigma'_v(1-k)r_o^4 \tag{4.64}$$

$$b'_2 = -\frac{1}{4}\frac{[3(C+F)+2(1-\nu)CF](1-\nu)}{[2+(1-\nu)C][6+(1-\nu)F]}\sigma'_v(1-k)r_o^2$$

The axial force and moment in the liner are:

$$T = \frac{a_o}{r_o} + \frac{1}{2}\sigma'_v(1+k)r_o - \frac{(1-\nu)CF}{C+F+(1-\nu)CF(1-r_o^2/R^2)}\left(1-\frac{r_o^2}{R^2}\right)u_w r_o + u_w r_o$$

$$+\left[\frac{1}{2}\sigma'_v(1-k)r_o - 6a'_2 r_o^{-3} + 2b'_2 r_o^{-1}\right]\cos 2\theta \qquad (4.65)$$

$$M = \frac{1}{4}\sigma'_v(1-k)r_o^2\cos 2\theta$$

For an incompressible and perfectly flexible liner, the results take the form:

$$\sigma'_r|_{r=r_o} = \frac{1}{2}\sigma'_v(1+k) - \frac{1}{2}\sigma'_v(1-k)\cos 2\theta$$

$$\sigma'_\theta|_{r=r_o} = \frac{1}{2}\sigma'_v(1+k) + \frac{1}{2}\sigma'_v(1-k)\cos 2\theta$$

$$\tau|_{r=r_o} = \frac{1}{2}\sigma'_v(1-k)\sin 2\theta$$

$$u|_{r=r_o} = u_w + \frac{3}{4}\sigma'_v(1-k)\cos 2\theta$$

$$U_r^{net}|_{r=r_o} = \frac{1+\nu}{E}\sigma'_v(1-k)r_o\cos 2\theta \qquad (4.66)$$

$$U_\theta^{net}|_{r=r_o} = -\frac{1+\nu}{E}\frac{1}{4}\sigma'_v(1-k)r_o\sin 2\theta$$

$$T = \frac{1}{2}\sigma'_v(1+k)r_o + u_w r_o - \sigma'_v(1-k)r_o\cos 2\theta$$

$$M = \frac{1}{4}\sigma'_v(1-k)r_o^2\cos 2\theta$$

Comparison of Equations (4.60), that correspond to full slip, when applied to a deep tunnel and $w = 0$, and Equations (4.66), for no slip and also for a deep tunnel, shows that at the ground–liner interface, and for $k<1$: (1) at the springline, the effective radial stresses for the full-slip case are smaller than for the no-slip case, the effective tangential stresses are larger, and the pore pressures are smaller; (2) the opposite is true at the crown; (3) the radial displacements for the full-slip case are always smaller than for the no-slip case, while the tangential displacements are always larger; and (4) the axial load for the full-slip case is larger than for the no-slip case at the springline and smaller at the crown.

4.2.1.3.2 Long-term analysis

As with the short-term analysis, the solution is found by decomposing the problem into the contributions of the ground and the water. For the ground, the solution is given by Equations (4.52), (4.53) and (4.54) where the buoyant unit weight should be used (i.e. replace in the equations γ by γ_b). For the contribution of the water, two extreme cases are possible: there is full drainage at the liner-ground interface, and so the pore pressures at this location are zero, or the interface is impermeable and so the pore pressures at the contact are equal to the far field. The rest of the boundary conditions are the same as those in (4.56), taking $\sigma'_x = \sigma'_y = 0$ when only the water is considered. The following provides a close-form solution for a shallow tunnel with ground loss w.

If the interface is impermeable, the solution (water contribution only) is given by:

$$\begin{aligned}
\sigma'_r &= \frac{a_o}{r^2} + (-2c'_1 r^{-3} + c_1 r^{-1} + d'_1 r^{-1})\sin\theta \\
\sigma'_\theta &= -\frac{a_o}{r^2} + (2c'_1 r^{-3} + d'_1 r^{-1})\sin\theta \\
\tau &= (2c'_1 r^{-3} - d'_1 r^{-1})\cos\theta \\
u &= \gamma_w (h_w - r\sin\theta) \\
U_r^{net} &= -\frac{1+\nu}{E}\left\{-\frac{a_o}{r} + [c'_1 r^{-2} + c_1(1-\nu)\ln r + d'_1(1-2\nu)\ln r]\sin\theta\right\} \\
U_\theta^{net} &= \frac{1+\nu}{E}\{c'_1 r^{-2} - c_1[(1-\nu)\ln r + \nu] - d'_1(1-2\nu)(\ln r - 1)\cos\theta\}
\end{aligned} \tag{4.67}$$

with

$$\begin{aligned}
a_o &= -\frac{(1-\nu)CF}{C+F+(1-\nu)CF}\gamma_w h_w r_o^2 \\
c_1 &= \gamma_w r_o^2 \\
c'_1 &= -\frac{1}{8}\frac{1-2\nu}{1-\nu}\gamma_w r_o^4 \\
d'_1 &= -\frac{1}{4}\frac{1-2\nu}{1-\nu}\gamma_w r_o^2
\end{aligned} \tag{4.68}$$

For the liner:

$$\begin{aligned}
T &= \frac{C+F}{C+F+(1-\nu)CF}\gamma_w h_w r_o \\
M &= 0
\end{aligned} \tag{4.69}$$

For an incompressible, perfectly flexible liner ($C = 0$, $F = \infty$) Equations (4.67) and (4.69) take the following form:

$$\begin{aligned}
\sigma'_r|_{r=r_o} &= \gamma_w r_o \sin\theta \\
\sigma'_\theta|_{r=r_o} &= -\frac{1}{2}\frac{1-2\nu}{1-\nu}\gamma_w r_o \sin\theta \\
\tau|_{r=r_o} &= 0 \\
u|_{r=r_o} &= \gamma_w (h_w - r_o \sin\theta) \\
U_r^{net}|_{r=r_o} &= \frac{1+\nu}{E}\frac{1}{8}\frac{1}{1-\nu}\gamma_w r_o^2[1-2\nu-2(3-4\nu)\ln r_o]\sin\theta \\
U_\theta^{net}|_{r=r_o} &= -\frac{1+\nu}{E}\frac{1}{8}\frac{1}{1-\nu}\gamma_w r_o^2[3-2\nu+2(3-4\nu)\ln r_o]\cos\theta \\
T &= \gamma_w h_w r_o \\
M &= 0
\end{aligned} \tag{4.70}$$

In the case of a deep tunnel, stresses and displacements of the ground are given by (Bobet, 2001):

$$
\begin{aligned}
\sigma'_r &= \frac{(1-\nu)CF}{C+F+(1-\nu)CF}\frac{r_o^2}{R^2-r_o^2}\left(1-\frac{R^2}{r^2}\right)u_w \\
\sigma'_\theta &= \frac{(1-\nu)CF}{C+F+(1-\nu)CF}\frac{r_o^2}{R^2-r_o^2}\left(1+\frac{R^2}{r^2}\right)u_w \\
\tau &= 0 \\
u &= u_w \\
U_r^{net} &= -\frac{1+\nu}{E}\frac{(1-\nu)CF}{C+F+(1-\nu)CF}\frac{r_o^2}{R^2-r_o^2}\left(1-2\nu+\frac{R^2}{r^2}\right)r\,u_w \\
U_\theta^{net} &= 0 \\
T &= \frac{C+F}{C+F+(1-\nu)CF}u_w\,r_o \\
M &= 0
\end{aligned}
\tag{4.71}
$$

which apply to both full slip and no-slip conditions.

If the liner is incompressible and perfectly flexible,

$$
\begin{aligned}
\sigma'_r &= \sigma'_\theta = \tau = 0 \\
u &= u_w \\
U_r^{net} &= U_\theta^{net} = 0 \\
T &= u_w\,r_o \\
M &= 0
\end{aligned}
\tag{4.72}
$$

Equation (4.72) indicates that the liner must support the full water pressure, in addition to the loads due to the ground. Note that if the liner is not incompressible, the liner and the ground share the water pressure. This must be so because, if the liner is compressible, it will move inward with the water pressure. This is associated with analogous movements of the ground (unless the ground detaches from the liner), and so the ground takes some of the load from the water pressure. Hence the assumption generally made to design the liner for the full hydrostatic water pressure may be conservative.

If the interface is permeable, a seepage flow is established in the ground with the water flowing toward the tunnel. The flow of water induces seepage forces in the ground, which in turn induce effective stresses. The solution (water component only) is given by:

$$
\begin{aligned}
\sigma'_r &= \frac{a_o}{r^2}+2b_o+c_o+2c_o\ln r-\gamma_w h_w+[c_1r^{-1}-2c'_1r^{-3}+d'_1r^{-1}]\sin\theta \\
&\quad +(-6a'_2r^{-4}-4b'_2r^{-2})\cos 2\theta+\gamma_w(h_w-r_o\sin\theta)\frac{\ln\left[1+\frac{4h_w}{r}\left(\frac{h_w}{r}-\sin\theta\right)\right]}{\ln\left[1+\frac{4h_w}{r_o}\left(\frac{h_w}{r_o}-\sin\theta\right)\right]}
\end{aligned}
$$

$$
\sigma'_\theta = -\frac{a_o}{r^2}+2b_o+3c_o+2c_o\ln r-\gamma_w h_w+[2c'_1r^{-3}+d'_1r^{-1}]\sin\theta
$$

$$+6a_2' r^{-4}\cos 2\theta + \gamma_w (h_w - r_o \sin\theta)\frac{\ln\left[1+\frac{4h_w}{r}\left(\frac{h_w}{r}-\sin\theta\right)\right]}{\ln\left[1+\frac{4h_w}{r_o}\left(\frac{h_w}{r_o}-\sin\theta\right)\right]}$$

$$\tau = (2c_1' r^{-3} + d_1' r^{-1})\cos\theta - (6a_2' r^{-4} + 2b_2' r^{-2})\sin 2\theta$$

$$u = \gamma_w (h_w - r\sin\theta) - \gamma_w (h_w - r_o \sin\theta)\frac{\ln\left[1+\frac{4h_w}{r}\left(\frac{h_w}{r}-\sin\theta\right)\right]}{\ln\left[1+\frac{4h_w}{r_o}\left(\frac{h_w}{r_o}-\sin\theta\right)\right]}$$

$$\begin{aligned}
U_r = -\frac{1+\nu}{E}\Bigg\{ & -\frac{a_o}{r} + 2(1-2\nu)b_o r + c_o(1-4\nu)r + 2(1-2\nu)c_o r(\ln r - 1) - (1-2\nu)\gamma_w h_w r \\
& +[(1-\nu)c_1 \ln r + c_1' r^{-2} + (1-2\nu)d_1' \ln r]\sin\theta + (2a_2' r^{-3} + 4(1-\nu)b_2' r^{-1})\cos 2\theta \\
& + \frac{(1-2\nu)\gamma_w (h_w - r_o\sin\theta)}{\ln\left[1+\frac{4h_w}{r_o}\left(\frac{h_w}{r_o}-\sin\theta\right)\right]}\Bigg\{ r\ln\left[1+\frac{4h_w}{r_o}\left(\frac{h_w}{r_o}-\sin\theta\right)\right] \\
& -2h_w \sin\theta \ln(r^2 + 4h_w^2 - 4h_w r\sin\theta) + 4h_w \cos\theta \,\mathrm{atan}\left(\frac{r\cos\theta}{2h_w - r\sin\theta}\right)
\end{aligned} \qquad (4.73)$$

$$\begin{aligned}
U_\theta = -\frac{1+\nu}{E}\Bigg\{ & 4c_o(1-\nu)r\theta - [-(1-\nu)c_1 \ln r - \nu c_1 + c_1' r^{-2} + (1-2\nu)d_1'(1-\ln r)]\cos\theta \\
& +[2a_2' r^{-3} - 2(1-2\nu)b_2' r^{-1}]\sin 2\theta \\
& - \frac{(1-2\nu)\gamma_w (h_w - r_o\sin\theta)}{\ln\left[1+\frac{4h_w}{r_o}\left(\frac{h_w}{r_o}-\sin\theta\right)\right]}\Bigg\{ 2h_w \cos\theta \ln(r^2 + 4h_w^2 - 4h_w r\sin\theta) \\
& +4h_w \sin\theta \;\mathrm{atan}\left(\frac{r\cos\theta}{2h_w - r\sin\theta}\right) + 2r0 - r\;\mathrm{atan}\left(\frac{4h_w r\cos\theta}{4h_w^2 - r^2}\right) \\
& - r\;\mathrm{atan}\left(\frac{r^2\sin 2\theta}{4h_w^2 + r^2\cos 2\theta}\right)\Bigg\}
\end{aligned}$$

with,

$$\begin{aligned}
a_o = & \frac{h_w^2 r_o^2}{[C+F+(1-\nu)CF]h_w^2 + [(C+F)(1-2\nu) - (1-\nu)CF]r_o^2} \\
& \Bigg\{[(C+F)(1-2\nu) - (1-\nu)CF]\left[\gamma_w h_w - \frac{c_1}{h_w} + 2\frac{c_1'}{h_w^3} - \frac{d_1'}{h_w}\right] \\
& +c_o\left[2(1-\nu)(C+F) - \frac{3}{2}(1-\nu)\frac{r_o^2}{h_w^2}C + 2(1-2\nu)\ln\frac{r_o}{h_w}(C+F) - 2(1-\nu)\ln\frac{r_o}{h_w}CF\right]\Bigg\}
\end{aligned}$$

$$
\begin{aligned}
b_o &= -\frac{1}{2}\left[\frac{a_o}{h_w^2} + c_o + 2c_o \ln h_w + \left(-\gamma_w h_w + \frac{c_1}{h_w} - 2\frac{c_1'}{h_w^3} + \frac{d_1'}{h_w}\right)\right] \\
c_o &= \frac{(1-2\nu)}{2(1-\nu)} \frac{\gamma_w h_w}{\ln\left[1 + \frac{4h_w^2}{r_o^2}\right]} \\
c_o &= \gamma_w r_o^2 \\
c_1' &= -\frac{1-2\nu}{8(1-\nu)} \gamma_w r_o^4 \\
d_1' &= -\frac{1-2\nu}{4(1-\nu)} \gamma_w r_o^2 \\
a_2' &= \frac{1}{6} \frac{1-\nu}{(1-\nu)F + 3(5-6\nu)} \frac{c_o r_o^6}{h_w^2} \\
b_2' &= -\frac{1}{2} \frac{1-\nu}{(1-\nu)F + 3(5-6\nu)} \frac{c_o r_o^4}{h_w^2}
\end{aligned}
\tag{4.74}
$$

The thrust and moment in the liner are:

$$
\begin{aligned}
T &= \left(\frac{a_o}{r_o^2} + 2b_o + c_o + 2c_o \ln r_o\right) r_o + \left(2\frac{a_2'}{r_o^3} + \frac{4}{3}\frac{b_2'}{r_o}\right)\cos 2\theta \\
M &= \left(2\frac{a_2'}{r_o^2} + \frac{4}{3} b_2'\right)\cos 2\theta
\end{aligned}
\tag{4.75}
$$

Equations (4.73) and (4.74) do not fully satisfy the condition of zero vertical and shear stresses at the surface. Also, in the derivations of displacements, it has been assumed that $\partial(h_w - r_o \sin\theta)/\partial\theta = 0$, which is not entirely correct. Thus the results should be taken only as approximate. The above derivation gives satisfactory results (errors smaller than 10–15%) for openings with a depth to radius ratio larger than 1.5 ($h/r_o > 1.5$) and in the vicinity of the tunnel, within 2–5 radii from the center of the tunnel. The accuracy of the solution increases with depth of the tunnel.

The water inflow, Q, into the excavation per unit length of tunnel is given by the following expression (El Tani, 2003):

$$
\begin{aligned}
Q &= 2\pi K \frac{\lambda^2 - 1}{\lambda^2 + 1} \frac{h_w}{\ln\lambda} \\
\lambda &= \frac{h_w}{r_o} - \sqrt{\frac{h_w^2}{r_o^2} - 1}
\end{aligned}
\tag{4.76}
$$

where K is the permeability of the ground medium.

For a very deep tunnel, a solution has been obtained with the condition that $R \gg r_o$ by Bobet (2001), and is given by the following expressions:

$$
\sigma_r' = \frac{r_o^2}{R^2 - r_o^2}\left[\frac{(1-\nu)CF}{C + F + (1-\nu)CF} - \frac{1-2\nu}{2(1-\nu)}\right]\left(1 - \frac{R^2}{r^2}\right) u_w + \frac{u_w}{2(1-\nu)} \frac{\ln\frac{R}{r}}{\ln\frac{R}{r_o}}
$$

$$\sigma_\theta' = \frac{r_o^2}{R^2 - r_o^2}\left[\frac{(1-\nu)CF}{C+F+(1-\nu)CF} - \frac{1-2\nu}{2(1-\nu)}\right]\left(1+\frac{R^2}{r^2}\right)u_w + \frac{u_w}{2(1-\nu)}\frac{\ln\frac{R}{r}+1-2\nu}{\ln\frac{R}{r_o}}$$

$$\tau = 0$$

$$u = u_w \frac{\ln\frac{r}{r_o}}{\ln\frac{R}{r_o}}$$

$$U_r^{net} = -\frac{1+\nu}{E}\left\{\frac{r_o^2}{R^2-r_o^2}\left[\frac{(1-\nu)CF}{C+F+(1-\nu)CF} - \frac{1-2\nu}{2(1-\nu)}\right]\left(1-2\nu+\frac{R^2}{r^2}\right) + \frac{1-2\nu}{2(1-\nu)}\frac{\ln\frac{R}{r}+1-\nu}{\ln\frac{R}{r_o}}\right\}u_w r \tag{4.77}$$

$$U_\theta^{net} = 0$$

$$T = \frac{C+F}{C+F+(1-\nu)CF}u_w r_o$$

$$M = 0$$

The water inflow into the tunnel is:

$$Q = 2\pi K \frac{u_w/\gamma_w}{\ln R/r_o} \tag{4.78}$$

By comparing the expressions in (4.73) and (4.77), it can be shown (Bobet, 2001) that the radius of influence of a deep tunnel, R, is equal to the depth of the tunnel below the water table; i.e. $R = h_w$. A similar conclusion was also reached by Schleiss (1986) and Fernández et al. (1994). Because the water flow toward the tunnel changes the stresses around the tunnel over a large volume (see Figure 4.17d), a tunnel can only be considered deep when the ratio of the depth of the tunnel to the radius of the tunnel is larger than 20; i.e. $h/r_o > 20$. This represents an increase by a factor of two compared with the case of a tunnel without pore pressures.

The expressions for the axial load and moment in (4.77) should be compared with (4.69), which provides the liner loads when the ground–liner interface is impermeable. Since both expressions coincide, the interesting result of identical liner loads in a deep tunnel for a perfectly permeable and impermeable liner is obtained. With the assumption of elasticity this observation can be taken further to conclude that the liner loads of a deep tunnel will not change if the drainage conditions at the ground–liner contact change from impermeable to permeable and vice versa. Note that this is not correct for shallow tunnels. The expressions for the stresses and displacements of the ground (except for displacements at the tunnel perimeter) are completely different (Bobet, 2003; Nam and Bobet, 2006). The stresses and the displacements of the ground are much larger, as it will be shown later, when discussing Figures 4.20 and 4.21.

Figures 4.20 to 4.22 show results of a tunnel with $r_o = 2$ m, $h = h_w = 20$m, $\gamma = 20$kN/m^3, E = 50 MPa, $\nu = 0.25$, $E_s = 24000$ MPa, $\nu_s = 0.25$, t = 0.2, and no gap between the ground and the liner (w = 0). The results are comparable to those of Figure 4.15a, which are obtained for

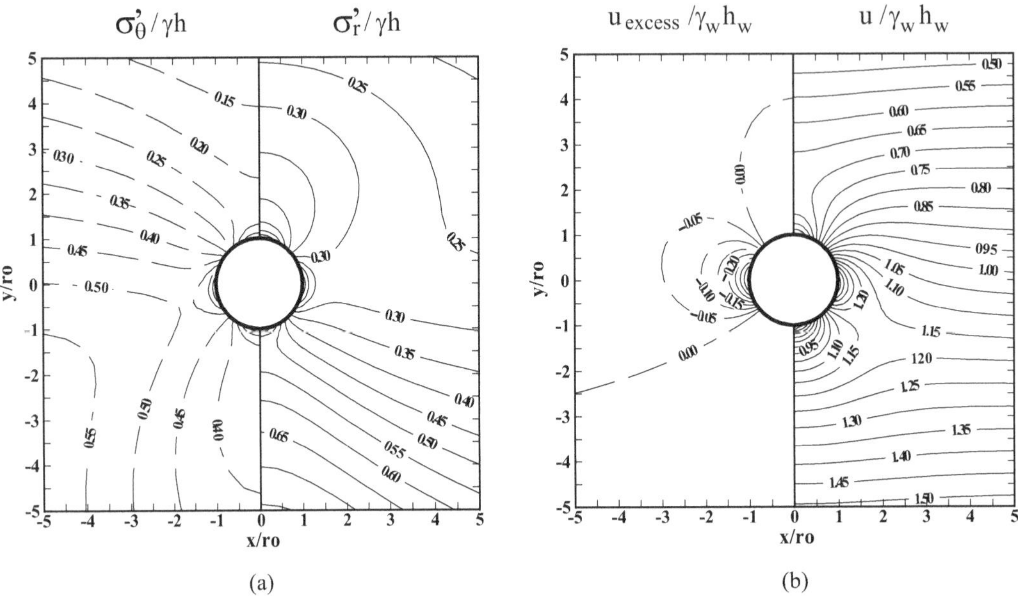

Figure 4.20 Stresses and pore pressures around a supported shallow circular opening. Short-term analysis. (a) Normalized effective radial (right) and tangential (left) stresses. (b) Normalized pore pressures (right) and excess pore pressures (left).

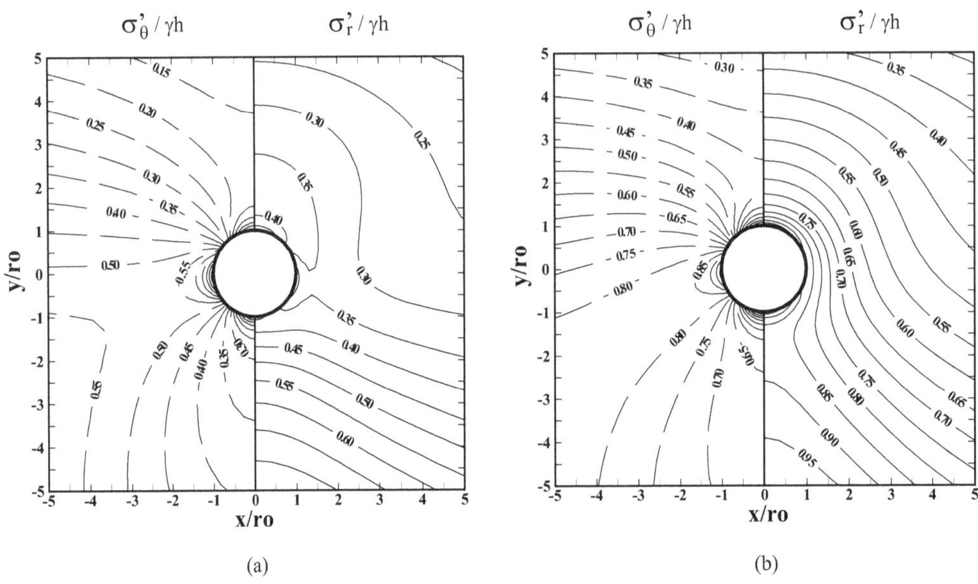

Figure 4.21 Stresses around a supported shallow circular opening. Difference between drainage and no drainage at liner–ground contact. Long-term analysis. (a) Normalized radial (right) and tangential (left) stresses. Long term. no drainage. (b) Normalized radial (right) and tangential (left) stresses. Long term. Full drainage.

the same tunnel but without water. An assumption has been made that the total unit weight of the soil is the same in both cases. A comparison of effective radial and tangential stresses, plotted in Figure 4.20a for short-term analysis results, and in Figure 4.15a for a medium without water, reveals that the stresses from the short-term analysis are much smaller (note that in the two cases stresses are normalized by the total vertical stress at the center of the tunnel). This is due to two reasons: (1) the buoyancy effect since the tunnel is submerged below the water table; and (2) the negative excess pore pressures originated because of the unloading of the ground. The pore pressures are shown in Figure 4.20b that presents total pore pressures and excess pore pressures. In the example, the largest negative excess pore pressures occur at the springline of the tunnel.

Figure 4.21 shows the results for the long-term analysis. There are two sets of results: one for impermeable liner, Figure 4.21a, and the other for permeable liner, Figure 4.21b. The stresses in the ground for long-term conditions with the impermeable liner are slightly larger than those for short-term conditions but still substantially lower than for the tunnel in the medium without water, Figure 4.15a. The difference between the results shown in Figures 4.15a and 4.21a are due, to a large extent, to the buoyancy effects of the water that decrease the effective stresses in the ground. The stresses in the liner, however, are the same. In the case with no water, the liner withstands the full total weight of the ground while in the case with water the liner must support the reduced weight of the ground due to buoyancy, but it also has to support the water pressure. Full drainage at the ground–liner interface has a dramatic effect when compared with the case of impermeable liner. Both the tangential stresses and the radial stresses increase. This is the result of the development of seepage forces in the ground as the water flows toward the tunnel. The stresses slowly recover as the radial distance increases. This is due to the logarithm term in Equation (4.73) that determines the rate of change of stresses and pore pressures, particularly far from the tunnel. The area of influence is also very large, of the order of the depth of the tunnel below the water table. A similar logarithmic term is found also in the expression for displacements, which produces the counterintuitive result of increasing displacements as the radial distance from the tunnel increases.

The response of the ground in contact with the liner and of the liner itself can be found in Figure 4.22. The figure plots the effective tangential stresses in the ground at the tunnel perimeter, normalized by the total vertical stress at the depth of the tunnel. It also plots the tangential stresses in the liner in the fiber in contact with the ground, normalized by the vertical total stress. Three cases are plotted: medium with no water, medium saturated short-term, and medium saturated long-term analyses with full drainage. The tangential stresses in the ground obtained for the short term are small compared with the other two cases. As discussed before, this is due to the buoyancy effects of the water and the negative excess pore pressures. With time, the excess pore pressures dissipate and flow toward the tunnel is established. The result is larger effective tangential stresses in the ground and an increase of the liner loads. The plot for the tunnel with no water shows tangential stresses in the ground and liner stresses larger than the tunnel with drainage and long-term conditions. This is the consequence of the buoyancy of the ground.

4.2.2 Plastic methods

Figure 4.23 is a plot of the results obtained for a deep unsupported tunnel with radius $r_0 = 2$ m, subjected to the far-field stresses, $\sigma'_v = 1$ MPa, and $K_0 = 0.5$. The ground is assumed to follow the Coulomb failure, with the following properties: Young's modulus E = 500 MPa, Poisson's ratio $\nu = 0.25$, internal frictional angle $\phi = 30°$, cohesion, c = 0.15 MPa, and dilation angle $\psi = 20°$. The properties are chosen to make the results comparable to those of Figure 4.1 and to cause yielding of the ground. The comparison between the two figures indicates that the axial stresses are not much affected by plasticity. The horizontal stresses are somewhat larger at the springline in Figure 4.23. The most important difference is observed

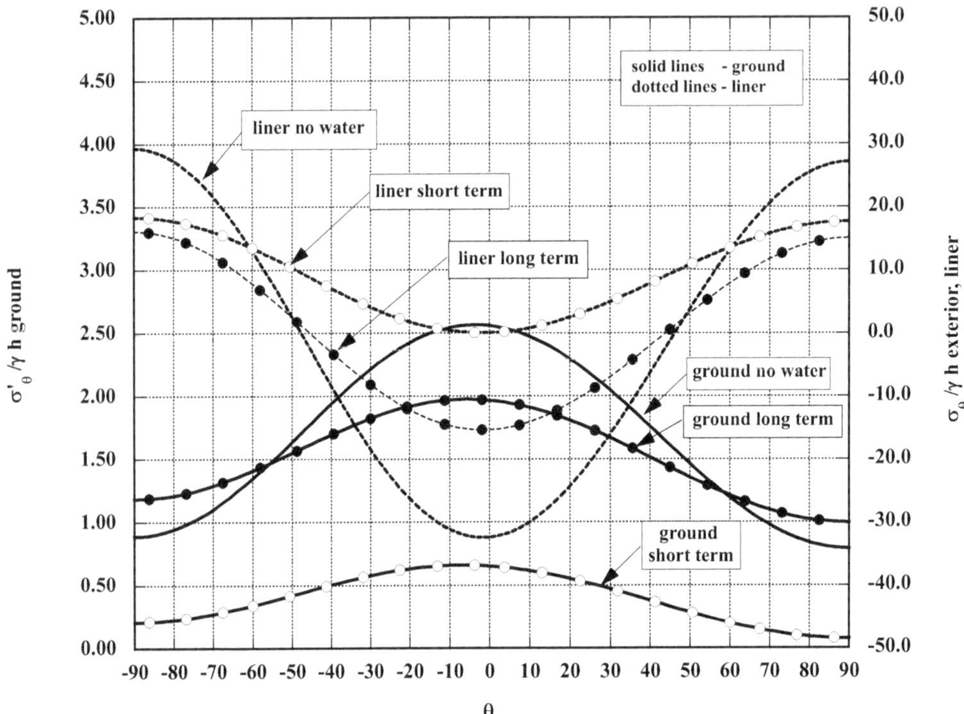

Figure 4.22 Normalized effective tangential stresses for the ground and exterior tangential stresses for the liner, for a shallow tunnel. r_0 = 2 m, h = h_w = 20 m, γ = 20 kN/m^3, E = 50 MPa, ν = 0.25, E_s = 24000 MPa, ν_s = 0.25, t = 0.2 m.

in the vertical stresses. Vertical stress concentrations in Figure 4.23 occur at some distance ahead of the face of the tunnel and also further into the medium at the springline. Due to redistribution of stresses, the ground ahead of the face and at the springline should have stresses larger than the free field, as one can see in Figure 4.1b. However at these locations the ground yields and so the stresses are transferred further into the ground. This is possible because as the distance from the opening increases, axially for the face and radially for the excavation, the confinement in the ground increases, and so the shear strength of the ground increases. With an elastic ground, the larger stresses occur at the perimeter of the opening, Figure 4.1c, while in the elasto-plastic ground, the largest stresses may be found at some distance from the excavation, Figure 4.23b. Similar observations have been made by Lee and Rowe (1990) and Eberhardt (2001). Figure 4.23c, which is a plot of the volume of ground that yields, confirms that yielding extends into the ground in front of the face and around the springline. There is no yielding at the crown. This is because, in this case, there is significant unloading in the vertical direction and moderate loading in the horizontal direction, which are not sufficient to induce yielding.

Yielding of the ground is associated, as one would expect, with larger displacements. The shape of the deformed opening (not shown in Figure 4.23) is similar to that of the tunnel in the elastic ground, except that the magnitude of the deformations is larger. Figure 4.24 shows the longitudinal distribution of the vertical displacements at the crown of the tunnel. In addition to the expected result of larger deformations, the figure also shows that the distance ahead of the tunnel, where ground deformations start to occur, is similar in both cases, while for the elasto-plastic ground, plane strain conditions appear to occur at slightly larger distances behind the face.

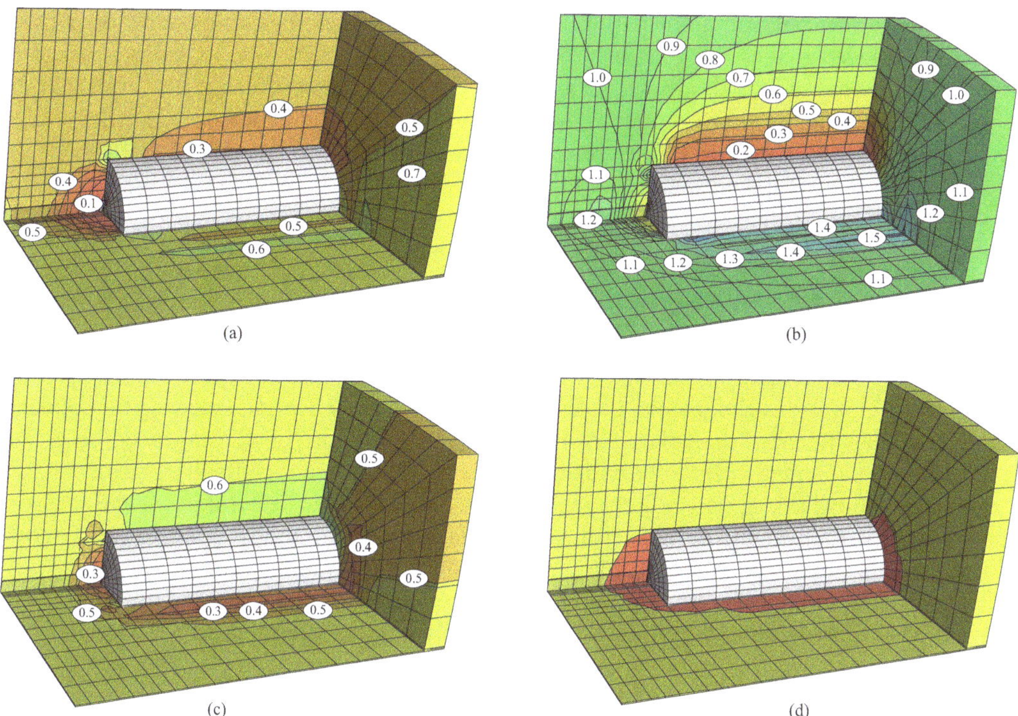

Figure 4.23 Three-dimensional view of stresses and plastic zone for an unsupported deep tunnel with r_o = 2 m, σ'_v = 1.0 MPa, K_o = 0.5, ϕ = 30°, ψ = 20°, c = 0.15 MPa. (a) σ'_{axial}. (b) $\sigma'_{vertical}$. (c) $\sigma'_{horizontal}$. (d) Plastic zone.

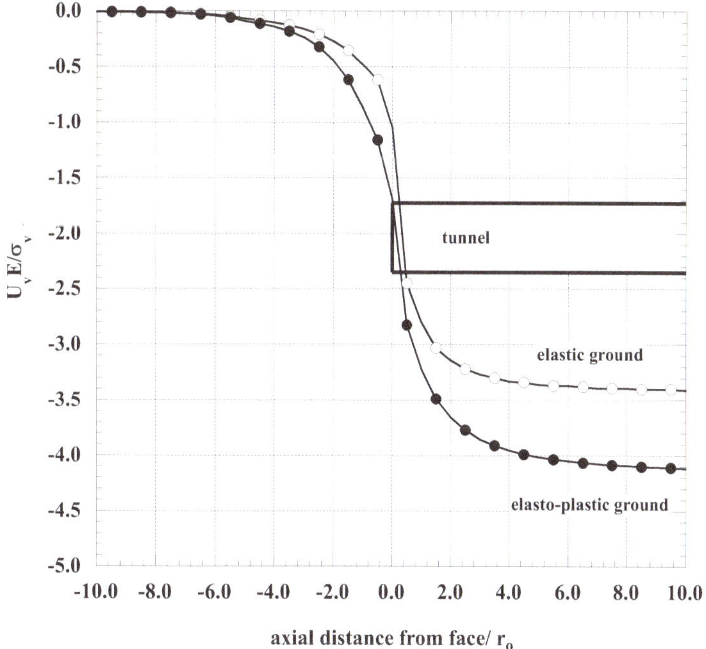

Figure 4.24 Vertical displacements at the crown of an unsupported tunnel, for elastic and elasto-plastic ground, r_o = 2 m, K_o = 0.5.

Consideration of plastic deformations of the medium where the tunnel, with or without support, is placed makes the general problem intractable analytically. Its solution must be found using numerical methods (see Chapter 5). There are two problems, however, that are of interest to tunneling where some progress has been made: the behavior of a cross section far behind the face, where plane strain conditions can be assumed, and the stability of the cross section of the tunnel and of the tunnel face. These two problems are discussed in the next two sections.

4.2.2.1 Plane strain analysis

The tunnel is assumed deep in a homogeneous isotropic medium, with circular cross section, and subjected to a hydrostatic far-field stress and to a constant internal stress. See Figure 4.25. Under these conditions the problem is axisymmetric. In other words, stresses and displacements are independent of the polar coordinate θ. The medium can be decomposed into a plastic and an elastic zone, with the boundary defined by the radial coordinate r_p (Figure 4.25). Close to the tunnel (i.e. $r_o \leq r \leq r_p$), the medium undergoes plastic deformations, and far from the opening ($r \geq r_p$) the medium is within its elastic regime. The solution has to satisfy equilibrium, compatibility, the material constitutive model, and boundary conditions.

The equation of equilibrium, Equation (4.1), takes a simple form because of the assumptions of deep tunnel and axial symmetry, which imply $F_r = F_\theta = 0$ and $\partial/\partial\theta = 0$. Hence,

$$\frac{\partial \sigma_r}{\partial r} + \frac{\sigma_r - \sigma_\theta}{r} = 0 \tag{4.79}$$

In the elastic zone, stresses can be computed from the Airy stress function in (4.34) and have the following expressions:

$$\begin{aligned} \sigma_r &= \sigma_o + \frac{C_1}{r^2} \\ \sigma_\theta &= \sigma_o - \frac{C_1}{r^2} \\ \tau &= 0 \end{aligned} \tag{4.80}$$

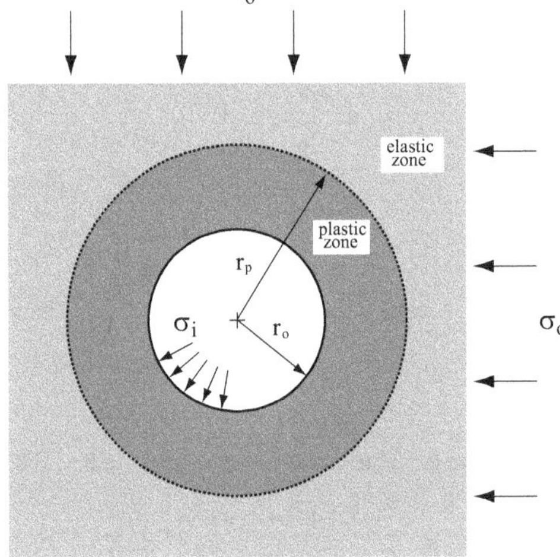

Figure 4.25 Circular deep tunnel with axisymmetric loading.

where the condition that $\sigma_r = \sigma_o$ for $r \to \infty$ has been imposed. C_1 is a constant that is obtained from continuity of radial stresses across the elastic–plastic interface. In other words the radial stresses in the elastic zone at $r = r_p$ must be the same as the radial stresses in the plastic zone also at $r = r_p$. Since r_p is unknown, a second equation is needed, which is obtained by imposing the condition that stresses obtained from (4.80) at $r = r_p$ must satisfy the yield condition.

In the plastic region, the material must obey the yield condition, which can be defined as:

$$\begin{aligned} \sigma_1 &= Y(\sigma_3) \quad \text{or} \\ \sigma_\theta &= Y(\sigma_r) \end{aligned} \tag{4.81}$$

where $Y(\sigma_3)$ is the yield function. Because the excavation produces unloading in the radial direction and loading in the tangential direction, the major principal stress is the tangential stress and the minor principal stress is the radial stress. Equation (4.81) and (4.79) produce a differential equation that can be integrated. The variable in this equation is typically σ_r. Integration of the differential equation gives an expression for the radial stress and one integration constant. The constant is obtained by imposing the boundary condition at the tunnel perimeter. That is, $\sigma_r|_{r=ro} = \sigma_i$.

Once the radial stresses in the plastic zone are known, the tangential stresses are obtained from (4.81).

Displacements are computed from the integration of strains. Note that due to the symmetry of the problem, the tangential displacements are always zero. Computation of radial displacements in the elastic zone is straightforward, and it is done by integration of the radial strains produced by the stresses in (4.80). In the plastic zone, for small deformation analysis, the strains are decomposed into elastic and plastic, as follows:

$$\begin{aligned} \varepsilon_r &= \varepsilon_r^e + \varepsilon_r^p \\ \varepsilon_\theta &= \varepsilon_\theta^e + \varepsilon_\theta^p \end{aligned} \tag{4.82}$$

The superscripts e and p denote, respectively, elastic and plastic strains. The plastic strains are obtained from a plastic potential $G(\sigma_r, \sigma_\theta) = 0$, from the flow rule:

$$\begin{aligned} d\varepsilon_r^p &= d\lambda \frac{\partial G(\sigma_r, \sigma_\theta)}{\partial \sigma_r} \\ d\varepsilon_\theta^p &= d\lambda \frac{\partial G(\sigma_r, \sigma_\theta)}{\partial \sigma_\theta} \end{aligned} \tag{4.83}$$

When the functions Y and G are the same, the flow rule is associated. Otherwise it is non-associated. The parameter $d\lambda$ is obtained as part of the calculations. Since the associated flow rule gives volumetric strains that are too large compared with actual observations of geomaterials, a non-associated flow rule is typically assumed. The one most used has the following form:

$$G(\sigma_r, \sigma_\theta) = \sigma_\theta - N_\psi \sigma_r = \sigma_\theta - \frac{1+\sin\psi}{1-\sin\psi}\sigma_r \tag{4.84}$$

where ψ is the dilation angle. Combining (4.83) and (4.84) a relation is found for the plastic strains:

$$\varepsilon_r^p + N_\psi \varepsilon_\theta^p = 0 \tag{4.85}$$

Displacements are finally obtained combining (4.85) with the relation between strains and displacements given by Equation (4.14a). Hence,

$$\frac{dU_r}{dr}+N_\psi\frac{U_r}{r}=\varepsilon_r^e+N_\psi\varepsilon_\theta^e=\frac{1+\nu}{E}\{\sigma_r+N_\psi\sigma_\theta-(1+N_\psi)[\nu(\sigma_r+\sigma_\theta)+(1-2\nu)\sigma_o]\} \tag{4.86}$$

In (4.86) the strains produced by the initial stress σ_o have been subtracted, so the displacements computed are the net displacements, only due to the internal pressure, σ_i. The procedure has been applied to obtain closed-form solutions for different material models. The results for two models are included in this section because of their interest to the geomechanics community: Coulomb, often used for soils, and Hoek-Brown, used for rocks.

The yield function for the Coulomb model in two dimensions is:

$$\sigma_\theta=\frac{1+\sin\phi}{1-\sin\phi}\sigma_r+\frac{2c\cos\phi}{1-\sin\phi}=N_\phi\sigma_r+N_c \tag{4.87}$$

Note that when $\psi=\phi$ the associated flow rule is recovered. A number of authors have worked on this problem, either analytically (Wang, 1996; Carranza-Torres and Fairhurst, 1997; Yu, 2000; Carranza-Torres, 2003; Park and Kim, 2006) or analytically and/or numerically (Anagnostou and Kovari, 1993; Yu, 2000). We follow here the approach suggested by Park and Kim (2006). The solution is:

$$\begin{aligned}
\sigma_r^e &= \sigma_o-\frac{(N_\phi-1)\sigma_o+N_c}{N_\phi+1}\frac{r_p^2}{r^2}\\
\sigma_\theta^e &= \sigma_o+\frac{(N_\phi-1)\sigma_o+N_c}{N_\phi+1}\frac{r_p^2}{r^2}\\
U_r^e &= -\frac{1+\nu}{E}\frac{(N_\phi-1)\sigma_o+N_c}{N_\phi+1}\frac{r_p^2}{r}\\
\sigma_r^p &= \left(\sigma_i+\frac{N_c}{N_\phi-1}\right)\left(\frac{r}{r_o}\right)^{N_\phi-1}-\frac{N_c}{N_\phi-1}\\
\sigma_\theta^p &= N_\phi\left(\sigma_i+\frac{N_c}{N_\phi-1}\right)\left(\frac{r}{r_o}\right)^{N_\phi-1}-\frac{N_c}{N_\phi-1}\\
U_r^p &= -\frac{1+\nu}{E}\left\{2(1-\nu)\frac{(N_\phi-1)\sigma_o+N_c}{N_\phi+N_\psi}\left(\frac{r_p}{r}\right)^{N_\psi+1}-(1-2\nu)\frac{(N_\phi-1)\sigma_o+N_c}{N_\phi-1}\right.\\
&\quad\left.+\left[(1-\nu)\frac{(1+N_\psi)(1+N_\phi)}{N_\phi+N_\psi}-1\right]\frac{(N_\phi-1)\sigma_i+N_c}{N_\phi-1}\left(\frac{r}{r_o}\right)^{N_\phi-1}\right\}r\\
\frac{r_p}{r_o} &= \left[\frac{2}{N_\phi+1}\frac{(N_\phi-1)\sigma_o+N_c}{(N_\phi-1)\sigma_i+N_c}\right]^{\frac{1}{N_\phi-1}}
\end{aligned} \tag{4.88}$$

The superscripts e and p correspond to the elastic and plastic zones, respectively. Additional closed-from solutions have been found for a number of interesting cases, which include: short-term or undrained excavation (Yu, 2000), undrained excavation with the ground following the Modified Cam Clay model with non-linear elasticity (Yu, 2000; Cao et al., 2002), opening in strain-softening rock (Brown et al., 1983; Carranza-Torres and Fairhurst, 1997), tunnel with seepage forces due to flow toward the excavation (Lee et al., 2006b).

The Hoek and Brown yield criterion (Hoek and Brown, 1980a, b) was derived empirically for rocks, and has the following expression (see Section 2.3.4.2 for further details):

$$\sigma_0 = \sigma_r + \sqrt{m\sigma_c\sigma_r + s\sigma_c^2} \tag{4.89}$$

where σ_c is the unconfined compression strength of the intact rock, and m and s are properties of the rock (s = 1 for intact rock and s<1 for fractured rock). A number of authors have investigated the problem of a deep circular opening excavated in a Hoek-Brown rock mass with axisymmetric loading (Wang, 1996; Carranza-Torres and Fairhurst, 1999; Sofianos and Nomikos, 2006; Park and Kim, 2006). The following is the complete solution for the circular opening with the Hoek-Brown failure criterion. The solution is taken from Park and Kim (2006) for consistency with the previous formulation with the Coulomb yield criterion:

$$\sigma_r^e = \sigma_o + \left(\sigma_i - \sigma_o + N_m \ln\frac{r_p}{r_o} + N_c \ln^2\frac{r_p}{r_o}\right)\frac{r_p^2}{r^2}$$

$$\sigma_\theta^e = \sigma_o - \left(\sigma_i - \sigma_o + N_m \ln\frac{r_p}{r_o} + N_c \ln^2\frac{r_p}{r_o}\right)\frac{r_p^2}{r^2}$$

$$U_r^e = -\frac{1+\nu}{E}(\sigma_o - \sigma_y)\frac{r_p^2}{r}$$

$$\sigma_r^p = \sigma_i + N_m \ln\frac{r}{r_o} + N_c \ln^2\frac{r}{r_o}$$

$$\sigma_\theta^p = \sigma_i + N_m + (N_m + 2N_c)\ln\frac{r}{r_o} + N_c \ln^2\frac{r}{r_o}$$

$$\begin{aligned} U_r^p = -\frac{1+\nu}{E}\Bigg\{ & \left[\frac{N_\psi - \nu N_\psi - \nu}{1+N_\psi} N_m + (1-2\nu)(\sigma_i - \sigma_o)\right]\left[1 - \left(\frac{r_p}{r}\right)^{N_\psi+1}\right] \\ & + \left[(1-2\nu)N_m - 2(1-\nu)\frac{1-N_\psi}{1+N_\psi} N_c\right]\left[\frac{1}{1+N_\psi}\left(\left(\frac{r_p}{r}\right)^{N_\psi+1} - 1\right)\right. \\ & \left. + \ln\frac{r}{r_o} - \left(\frac{r_p}{r}\right)^{N_\psi+1} \ln\frac{r_p}{r_o}\right] \\ & + (1-2\nu)N_c\left[\ln^2\frac{r}{r_o} - \left(\frac{r_p}{r}\right)^{N_\psi+1} \ln^2\frac{r_p}{r_o}\right] + (\sigma_o - \sigma_y)\left(\frac{r_p}{r}\right)^{N_\psi+1}\Bigg\} \end{aligned} \tag{4.90}$$

$$\frac{r_p}{r_o} = e^{\frac{-N_m + \sqrt{N_m^2 - 4N_c(\sigma_i - \sigma_y)}}{2N_c}}$$

$$\sigma_y = \sigma_o + \frac{1}{8}m\sigma_c - \frac{1}{2}\sigma_c\sqrt{\left(\frac{m}{4}\right)^2 + m\frac{\sigma_o}{\sigma_c} + s}$$

$$N_m = \sqrt{m\sigma_c\sigma_i + s\sigma_c^2}$$

$$N_c = \frac{1}{4}m\sigma_c$$

The solution has been expanded to include the case of softening of the rock mass after yielding (Sharan, 2003, 2005; Park and Kim, 2006). This is accomplished by using the following yield function for the rock after undergoing plastic deformations:

$$\sigma_\theta = \sigma_r + \sqrt{m_r\,\sigma_c\,\sigma_r + s_r\,\sigma_c^2} \tag{4.91}$$

where the parameters m_r and s_r are for the yielded rock. A common assumption is made that the unconfined compression strength of the intact rock remains unchanged.

A solution has been proposed for the generalized Hoek-Brown criterion. It is not a closed-form solution since it requires numerical integration (Carranza-Torres, 2004). The criterion is expressed as (Hoek et al., 2002):

$$\sigma_\theta = \sigma_r + \sigma_{ci}\left(m_b\,\frac{\sigma_r}{\sigma_{ci}} + s\right)^a \tag{4.92}$$

In Equation (4.92) σ_{ci} is the unconfined compression strength of the intact rock, and m_b, s and a are semi-empirical parameters, which can be obtained through correlations with the Geological Strength Index (GSI); see Carranza-Torres and Fairhurst (2000); Hoek et al. (2002) and Carranza-Torres (2004). See also Section 2.3.4.2 for details.

Figure 4.26 shows the results of Equation (4.90) applied to a deep tunnel with radius, r_o = 2 m, subjected to far-field stresses σ_o = 1 MPa, and interior stresses σ_i = 0. The Hoek-Brown failure criterion is assumed for the ground. Two cases are considered: (1) "hard" rock with m_i = 0.5, s = 0, σ_c = 50 MPa; and (2) "soft" rock with m_i = 0.2, s = 0, σ_c = 25 MPa. In both cases ν = 0.25. The terms "hard" and "soft" are used here in a relative sense since in both cases the material could be considered as a soft rock. The values have been chosen mostly to highlight differences in the results. Radial displacements for the hard rock are computed assuming no dilation, ψ = 0. For the soft rock, two dilations have been chosen: ψ = 0 and ψ = 20°. Again the values are not intended to be representative of a particular material but to illustrate their influence on the results. Figure 4.26a plots the radial and tangential stresses around the tunnel. Note that the radial coordinate is normalized by the tunnel radius. The tunnel in the hard rock has higher radial stresses than in the soft rock, and it also has higher tangential stresses close to the tunnel. The tangential stresses show a discontinuity at a distance of about 1.15 r_o. This corresponds to the extent of the plastic zone around the excavation. For the softer material, yielding occurs within a radial distance of about 1.7 r_o. Since within the yield zone the material cannot take large stresses, a load transfer occurs to the rock placed at larger radial distances from the tunnel. This explains the smaller tangential stresses near the tunnel and the larger tangential stresses in the elastic zone, compared to the case of hard rock. In both cases, as the distance from the tunnel increases, the radial and tangential stresses tend to converge to the value of the far-field stresses. The displacements plotted in Figure 4.26b show the expected trend of larger values for the soft rock than for the hard rock. Dilation, given by the dilation angle ψ, has important effects. As the dilation angle increases, the displacements near the tunnel (within the plastic zone, where strains are a function of ψ) increase significantly. It is important to note that for the axisymmetric problem stresses are independent of dilation. Hence, as dilation increases, the volumetric strain of the material increases, and so the radial displacements increase. This result has also been noticed by Wang (1996) and Park and Kim (2006).

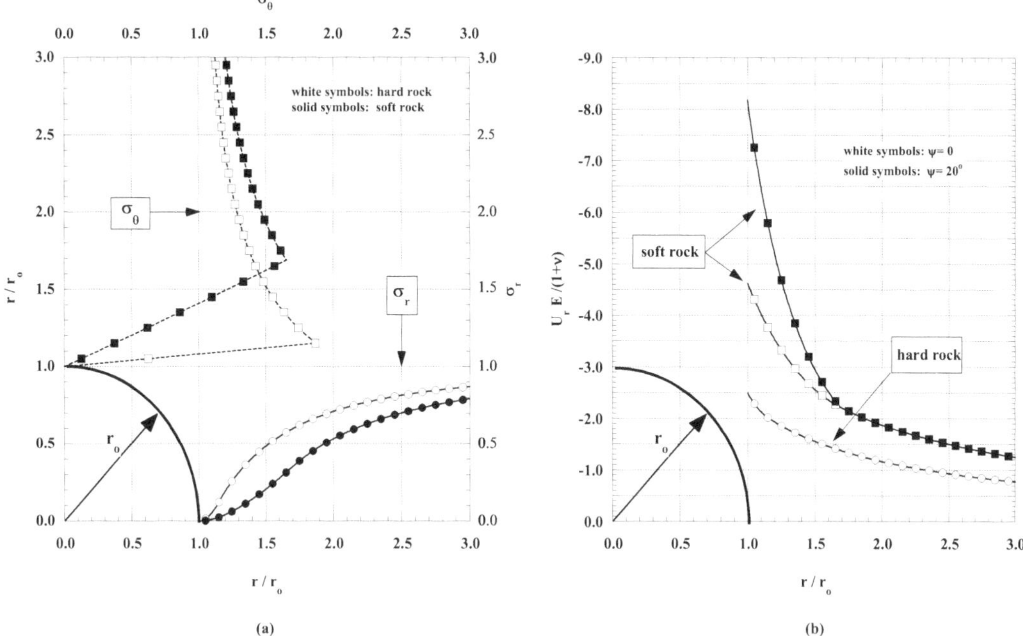

Figure 4.26 Stresses and displacements of a deep tunnel with axisymmetric loading. Hoek-Brown failure criterion. r_0 = 2 m, σ_0 = 1 MP, σ_i = 0, ν = 0.25. (a) Radial and tangential stresses. (b) Radial Displacements.

The analytical solutions presented in this section have the serious limitations that are only applicable to plane strain conditions, to axisymmetric loading and circular cross sections. This renders the solutions with very little practical applications, even though some effort has been made to approximate face effects and obtain the ground reaction curve for the characteristic curves method (Carranza-Torres and Fairhurst, 2000). As can be seen in Figure 4.23 a general, non-uniform far-field stress has a significant influence on the stresses and displacements of the ground around the tunnel. There can be significant differences at the springline and the crown, depending on the coefficient of earth pressure at rest K_0. Nevertheless, these, as all analytical solutions, provide insight into the problem, provide valuable qualitative information, and can be used to interpret field observations and to calibrate and verify numerical methods.

A limited number of comparisons have been made using the Coulomb failure criterion and the Hoek-Brown failure criterion for the same material, using equivalent strength properties to define cohesion and internal friction angle for the Coulomb failure (Wang, 1996 and Park and Kim, 2006). The equivalent properties are obtained by linearization of the Hoek-Brown envelope within the stress range of interest. The comparisons show that the results can be quite different, and the magnitude of the differences is a function of the applied stresses σ_0 and σ_i. Sofianos and Nomikos (2006) provide recommendations to adjust the equivalent properties for the axisymmetric problem.

4.2.2.2 Tunnel stability

The previous section has described the elasto-plastic response of the ground surrounding the tunnel as the internal pressure decreases. With further reduction of the internal support, the size of the plastic zone around the opening increases, and so the displacements of the ground increase. There will be a critical support pressure where the rate of increase of displacements is very large compared to the rate of decrease of the internal pressure, at which point the tunnel collapses.

Collapse is a stability phenomenon that can happen at the perimeter of the tunnel and also at the face of the tunnel. Laboratory experiments have been conducted to observe this phenomenon (Casarin and Mair, 1981; Chambon and Corté, 1994; Kamata and Mashimo, 2003; Sterpi and Cividini, 2004). The collapse of a circular tunnel in granular media is illustrated in Figure 4.27a. The tests were conducted in the laboratory under plane strain conditions of a tunnel section with circular shape located far from the face (Sterpi and Cividini, 2004). The granular media surrounding the opening was simulated with aluminum rods. The collapse was induced by decreasing the internal support of the opening. The thick dashed

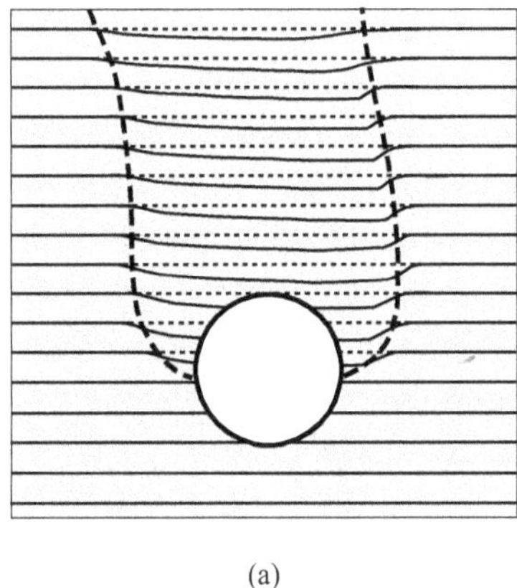

(a)

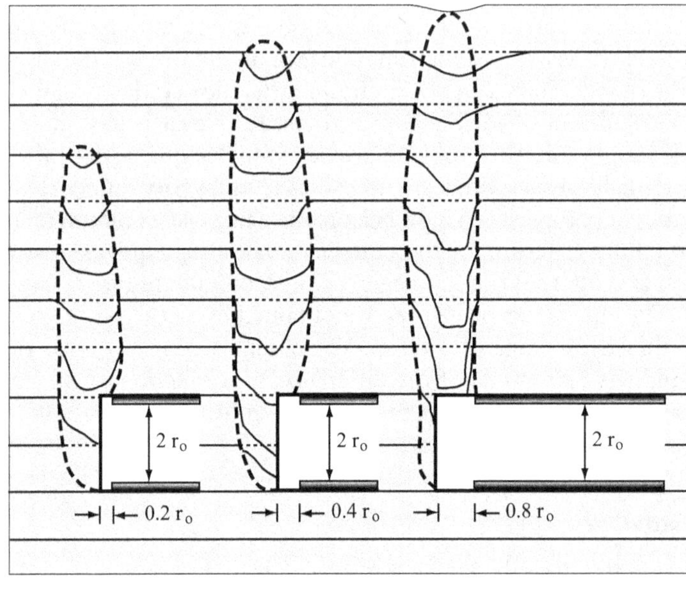

(b)

Figure 4.27 Tunnel collapse. (a) Collapse of the cross section. Adopted from Sterpi and Cividini (2004). (b) Face collapse. Adopted from Chambon and Corté (1994).

lines in the figure denote the limit of the failure, which consisted of two narrow bands of shear localization. Localization started at the tunnel springline and propagated toward the surface. The material inside the bands moved downward almost as a rigid body since no appreciable deformation was noticed. It was reported that the shape of the bands depended on the tunnel depth and on the rod density. The non-symmetrical collapse shape reported in Figure 4.27a was due to initial non-homogeneities in the rod distribution, which increased as the collapse progressed.

Figure 4.27b describes the collapse induced at the face of a circular opening excavated in dry sand. The tests were conducted in a centrifuge to simulate different tunnel depths. A pressure latex membrane placed inside a box of Fontenebleau sand (Chambon and Corté, 1994) kept the shape of the tunnel. Failure of the face was induced by decreasing the pressure inside the membrane. A number of tests were performed with different lengths between the support and the face, as shown in Figure 4.27b. Failure occurred in three stages. In the first stage the pressure at the face was reduced but no significant movements were detected (elastic stage). Below a certain critical pressure, further decrease of support involved plastic deformations at the face (plastic stage). In this stage, holding or increasing the internal pressure stabilized the face. Failure (failure stage) was sudden and took the form of free flow into the tunnel. The failure envelope had the shape of a cylindrical bulb, as shown in Figure 4.27b. The thick dashed lines in the figure denote the limits of the failure, which as in Figure 4.27a, correspond to a narrow zone of strain localization. The failure started at or near the invert of the tunnel, extended about a half diameter in front of the face and propagated upward almost vertically. The other limit of the failure started at the end of the support and propagated vertically. Collapse occurred when the two shear bands met each other isolating the failure bulb. For deep tunnels, the failure bulb did not reach the surface. For shallow tunnels, failure could reach the surface and induce significant subsidence. As with the plane strain experiments, the deformations of the soil inside the shear bands were reported as very small, with the soil moving as a rigid block. Experiments with different unsupported lengths indicated that for unlined lengths smaller than 0.2 r_0 the collapse pressures were similar to those of a tunnel with support up to the face, and the failure patterns were also similar. For longer unlined tunnels (unsupported length larger than 0.8 r_0), the collapse affected the unlined part first and the failure bulb was much larger, with a failure pattern similar to that reported for plane strain, Figure 4.27a; collapse occurred also earlier than for the lined tunnel. The observations made by Chambon and Corté (1994) have been reported by others (e.g. Casarin and Mair, 1981; Kamata and Mashimo, 2003; Sterpi and Cividini, 2004).

As the experiments indicate, the collapse of the tunnel is a progressive three-dimensional mechanism, which can only be approximated numerically with suitable material models (Sterpi and Cividini, 2004). There are, however, two cases where a solution has been attempted: (1) collapse of the plane strain cross section; and (2) collapse of the face. See Figure 4.28.

Stability solutions are found using the limit equilibrium theorems of plasticity, where the soil is idealized as a homogeneous, isotropic, elastic, and perfectly plastic material. The collapse mechanism is unique, from plasticity theory, and it can be approximated using the lower and upper bound solutions. The lower bound theorem states that if a stress field is found that satisfies equilibrium and does not violate the yield condition, then the solution is a lower bound of the exact solution. The upper bound theorem states that the loads obtained from calculations of the work done by a collapse mechanism that is kinematically admissible are larger than the exact loads. The approach then consists of finding the largest lower bound and the lowest upper bound solutions to approximate the exact solution, since it has to be found between the two bounds. Clearly an upper bound solution provides an unsafe estimate and a lower bound a safe estimate.

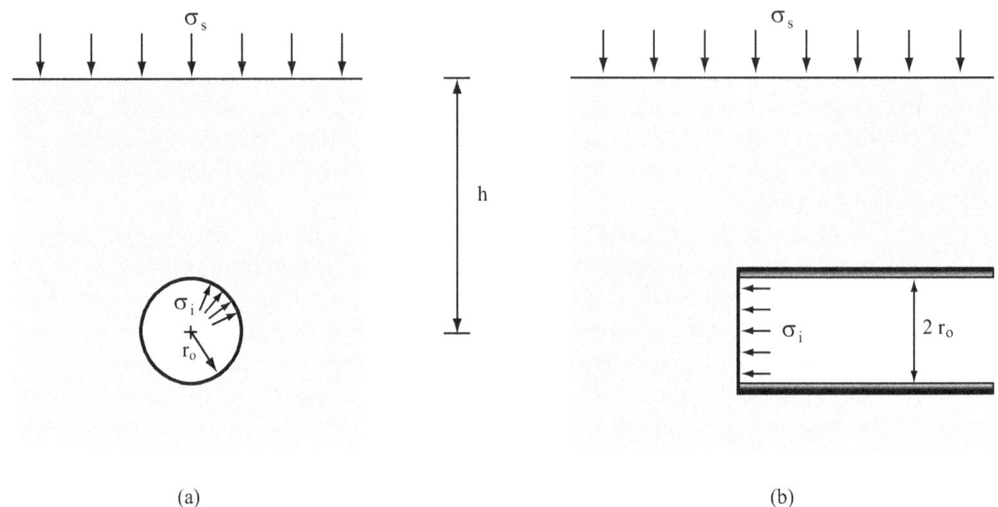

Figure 4.28 Tunnel stability. (a) Plane strain cross section. (b) Tunnel face.

Two analyses are possible: short term and long term. For the short term, it is assumed that the stress change rate induced in the ground due to tunnel operations is very fast and so there is no time for excess pore pressure dissipation. The strength of the material is given by its undrained shear strength. For the long-term analysis, the assumption made is that all excess pore pressures have dissipated and steady-state conditions have been reached. The strength of the material is given by its cohesion and internal friction angle. In the following, it is assumed that the soil follows the Coulomb failure envelope.

Lower and upper bound solutions for short-term analysis have been proposed by Davis et al. (1980) for a shallow tunnel with circular cross section.

The stability of the plane strain cross section, Figure 4.28a, is given in Figure 4.29. The figure is a plot of the stability number N with depth of the tunnel, Figure 4.4, and the ratio $\gamma\ r_0/S_u$ where γ is the total unit weight of the soil, r_0 is the radius of the opening, and S_u is the undrained shear strength of the soil, which is assumed constant with depth. The stability number N was first defined by Broms and Bennermark (1967), after extruding clay under pressure through circular openings, as:

$$N = \frac{\sigma_v - \sigma_i}{S_u} \tag{4.93}$$

In (4.93) σ_v is the total vertical stress at the tunnel axis, as if the tunnel does not exist (i.e. $\sigma_v = \gamma\ h + \sigma_s$); σ_i is the support pressure at the perimeter of the tunnel (e.g. air pressure); and S_u is the shear strength of the soil at the tunnel axis. See Figure 4.28a. The stability number is also used to determine surface settlements and ground deformations due to tunnel excavation; see Section 3.4. Broms and Bennermark (1967) found from their experiments that for stability ratios less than 6 the opening was stable, irrespective of the depth of the tunnel. Note that the value of the stability ratio N in Figure 4.29 increases with the depth of the tunnel, and only for very deep tunnels, a value of 6 or larger is found.

The results presented in Figure 4.29 were obtained by Davis et al. (1980) from numerical analyses. Similar results for the upper bound solution were also found by Klar et al. (2007) using the elastic displacement equations proposed by Verruijt and Booker (1996).

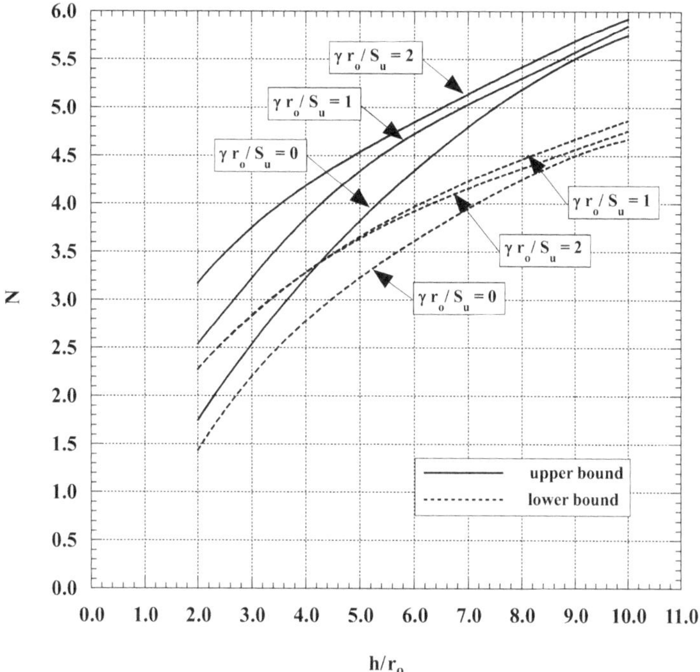

Figure 4.29 Plane strain stability. Short-term analysis. Upper and lower bound for circular tunnels. Adopted from Davis et al. (1980).

An analytical solution exists for the lower bound for a weightless ground ($\gamma\ r_o/S_u = 0$), and it is given by:

$$N = 2\ln\left(\frac{h}{r_o}\right) \tag{4.94}$$

The stability of the face is investigated assuming that the perimeter of the tunnel is fixed and so only the face can have displacements. The face is supported by an internal pressure σ_i as shown in Figure 4.28b. The short-term stability number N for the face is given in Figure 4.30 (Davis et al., 1980). The upper bound was found numerically. The lower bound is given by the following equations:

$$N = 2 + 2\ln\left(\frac{h}{r_o}\right), \text{ for } \frac{h}{r_o} < 2.72 \tag{4.95a}$$

$$N = 4\ln\left(\frac{h}{r_o}\right), \quad \text{for } \frac{h}{r_o} > 2.72 \tag{4.95b}$$

A comparison of Figures 4.29 and 4.30 shows that the stability ratio for the plane strain cross section is much smaller than for the face stability. Given the definition of the stability ratio, this represents a much larger support pressure for the plane strain case. This is not a surprising result due to the three-dimensional effects at the face that help with the support. The validity of the short-term stability equations has been checked by comparing its predictions with experimental results (Davis et al., 1980).

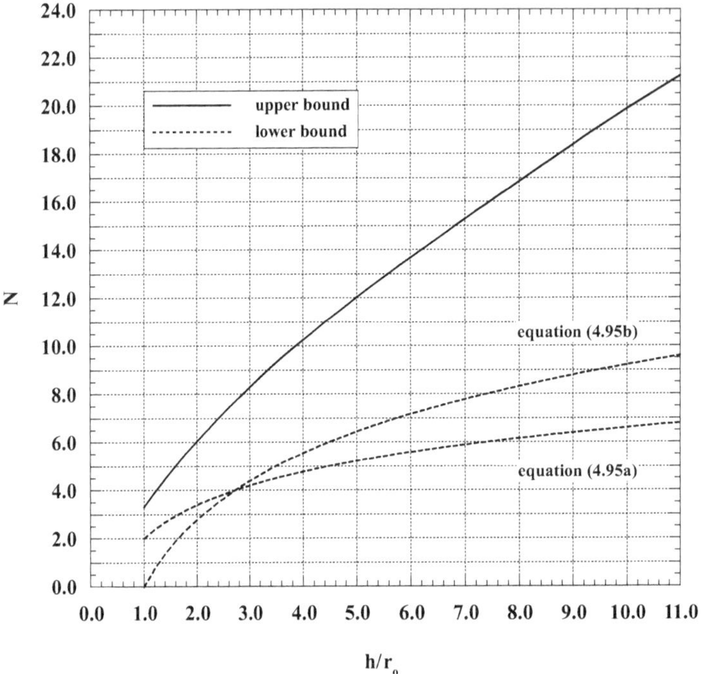

Figure 4.30 Face stability. Short-term analysis. Upper and lower bound for circular tunnels. Adopted from Davis et al. (1980).

For long-term analysis, a lower bound estimate for the support pressure of the plane strain cross section was provided by Terzaghi (1943), who developed the rock load theory based on the silo formulation (Janssen, 1895). Section 1.5 provides a detailed discussion of arching theory and on Terzaghi's findings. The following gives a summary of the main concepts, and, while it duplicates some of the information in Section 1.5, it is included here for completeness. Terzaghi's derivation is based on equilibrium of a thin horizontal slice of the ground above the tunnel of width 2B (Figure 4.31; see also Figure 1.36) subjected to its own weight and to fully mobilized friction along its vertical sides. An assumption is made that the horizontal stress on the vertical sides is proportional to the vertical stress, $\sigma_h = K\,\sigma_v$. Hence the shear stress at the side of the slice is $\tau = c + K\,\sigma_v \tan\phi$, where c and ϕ are the cohesion and internal friction angle of the ground. The width 2B is estimated assuming that the limit of the failure starts at the invert of the tunnel at an angle of $45 + \phi/2$ with the horizontal until it reaches the crown of the tunnel, where the failure is delimited by a vertical shear band. Thus, as indicated in the figure, $2B = b + 2H/\tan(45 + \phi/2)$, where b and H are the actual width and height of the tunnel. Equilibrium of the slice produces a differential equation in terms of σ_v, which can be integrated. The result is:

$$\sigma_v = \frac{\gamma B - c}{K\tan\phi}\left(1 - e^{-K\tan\phi\frac{h_2}{B}}\right) + \gamma h_1 e^{-K\tan\phi\frac{h_2}{B}} \tag{4.96}$$

Note that Equation (4.96) is the same as Equation (1.3) taking in (1.3) $z = h_2$. Terzaghi found in his experiments that the value of K increased gradually from 1 to 1.5 over a height of about B. He recommended, for practical purposes, to use a value of K = 1. In Equation

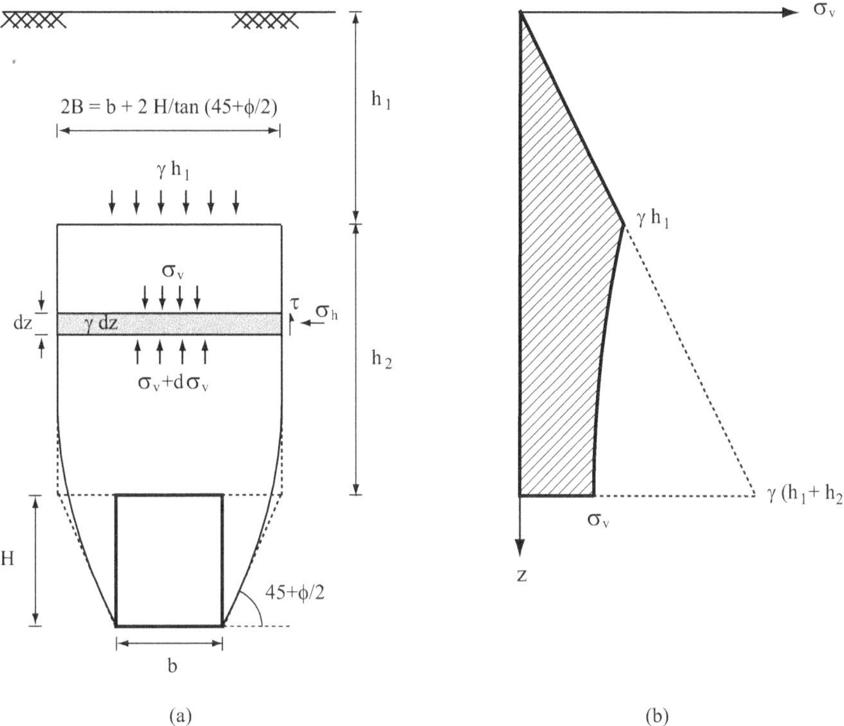

Figure 4.31 Terzaghi's rock pressure theory. (a) Model. (b) Vertical stress with depth.

(4.96) h_2 is the height above the tunnel where arching occurs. As h_2 increases, the magnitude of the vertical stress σ_v at the tunnel required to provide support decreases. For heights larger than about 5 to 6 B full arching has developed and σ_v does not decrease much further (Figure 4.31b). Equation (4.96) indicates that for a deep tunnel ($h_2 \to \infty$, or for practical purposes $h_2 > 5$ to 6 B) the vertical stress reaches a minimum value equal to $\sigma_v \to (\gamma B\text{-}c)/(K \tan \phi)$. This value is negative (theoretically no support pressure is needed) when $\gamma B < c$; however tensile failure is still possible at the crown.

Trapdoor experiments appear to validate the predictions obtained from Equation (4.96), as observed by Iglesia et al. (1999) and Adachi et al. (2003), and as discussed in Section 1.5.

The long-term upper- and lower-bound stability for the tunnel face in a cohesive and frictional media was analyzed by Leca and Dormieux (1990) with the assumption, similar to the undrained analysis, that the liner reaches the face of the tunnel. The upper bound solution is given by the following equation:

$$(c + \sigma_s \tan\phi)N_s + 2\gamma r_o \tan\phi N_\gamma = (c + \sigma_i \tan\phi) \tag{4.97}$$

In (4.97) c and ϕ are the cohesion and internal friction angle of the soil, γ is the unit weight of the soil, σ_s is the surcharge on the ground surface, if any, and σ_i is the support pressure applied at the face of the tunnel (Figure 4.28b). The parameters N_s and N_γ are given in Figure 4.32. The values have been computed from the solution provided by the authors.

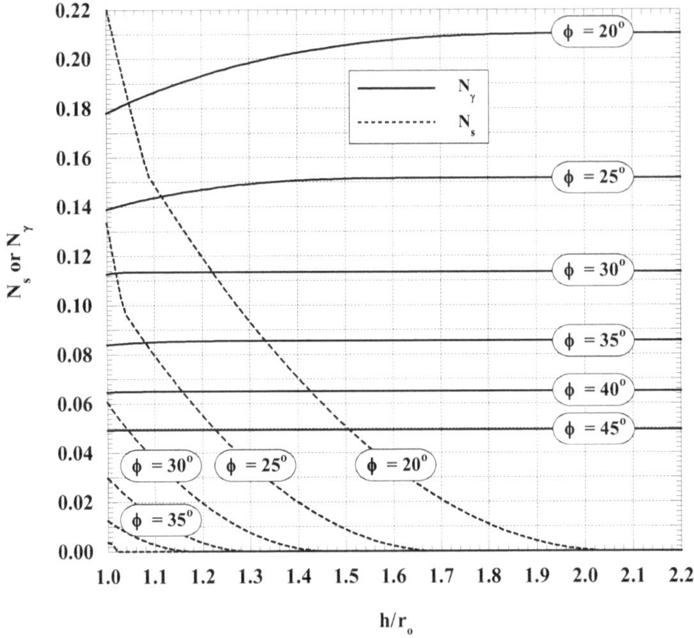

Figure 4.32 Face stability. Long-term analysis. Parameters N_s and N_γ for the upper bound solution. Adopted from Leca and Dormieux (1990).

For the lower bound, Equation (4.97) still applies, but with the following parameters:

$$N_s = \frac{1-\sin\phi}{1+\sin\phi}\left(\frac{h}{r_o}\right)^{-\frac{2\sin\phi}{1-\sin\phi}} \quad \text{or}$$

$$N_s = \left(\frac{h}{r_o}\right)^{-\frac{4\sin\phi}{1-\sin\phi}} \tag{4.98}$$

$$N_\gamma = \frac{1}{2}\frac{1-\sin\phi}{1+\sin\phi}\left(\frac{h}{r_o}+1\right)$$

The equation for N_s is chosen such that it gives the largest result.

The solutions provided for long-term analysis have been satisfactorily compared with experimental results (Leca and Dormieux, 1990; Chambon and Corté, 1994).

The stability equations presented have been expanded by a number of authors to more complex scenarios. The effects of seepage forces on the stability of the tunnel face have been investigated by Lee and Nam (2001), (2004), and Lee et al. (2004a). The seepage forces may increase significantly the force required to support the face of the tunnel (i.e. the stability of the tunnel is reduced), which is a function of the head difference between the tunnel and the free field. In addition, the seepage forces may increase as the rate of advance of the tunnel increases, reducing the stability of the face. A numerical parametric study conducted by Lee and Nam (2004) suggested that such reduction should be evaluated for tunnels excavated in ground with permeability lower than 10^{-6} m/sec. Anagnostou and Kovari (1994) provided recommendations for the face stability of tunnels excavated with slurry shield as a function of the depth of penetration of the slurry into the ground ahead of the face. The same authors (Anagnostou and Kovari, 1996a, b) provided an upper bound solution for the face stability of tunnels excavated below the water table using an earth-pressure balance shield.

The number of closed-form solutions is understandably limited due to the complexity of the problem. Small-scale laboratory experiments (e.g. Casarin and Mair, 1981; Chambon and Corté, 1994; Iglesia et al., 1999; Yoo and Shin, 2003; Kamata and Mashimo, 2003; Sterpi and Cividini, 2004; Lee et al., 2006a) and numerical modeling (Eisenstein and Ezzeldine, 1994; Eisenstein and Rossler, 1995; Kamata and Mashimo, 2003; Sterpi and Cividini, 2004; Kolymbas, 2005; Klar et al., 2007) have been successfully used to estimate the stability of tunnels for more complex problems.

4.2.3 Wedge methods

The elastic and elasto-plastic methods reviewed in Sections 4.2.1 and 4.2.2 are applicable, within the limitations of the assumptions made, to a continuous medium. Most soils and heavily weathered rocks may fall within this category, as well as heavily fractured rock. However for blocky rock, the failure mechanisms are dominated by rigid movements of blocks or wedges delimited by discontinuities. Failure of the blocks can be produced by falling, sliding along one face or sliding along two faces, or a combination of sliding and rotation. Failure is generally sudden and it only takes a very small displacement to mobilize the peak friction angle along the sliding plane(s). It is thus advisable to provide support to potentially unstable blocks as soon as possible. Failure of even relatively small blocks may remove support from other blocks, thus contributing to additional unraveling; the process will continue until the rock supports itself by arching.

The geometry of the blocks is controlled by the structure of the rock mass, which is defined by the type, orientation and number of discontinuities, their spacing and size or intensity. Intersection of discontinuities such as bedding, joints, and faults determines the shape of the blocks, while the size and spacing of the discontinuities define the volume of the block. At least three discontinuities plus the perimeter of the tunnel are needed to form a tetrahedral block (Figure 4.33). It is thus an obvious requirement for any stability analysis to have access to detailed geological characterization of the rock mass.

Stability analysis of block failures is done in two steps using limit equilibrium analysis. In the first step the blocks that are kinematically possible are identified. In "Block Theory" (Goodman and Shi, 1985; Goodman, 1989), this consists in dividing the blocks into "removable" and "non-removable." Removable blocks are those that potentially can slide toward the excavation. Such identification is done using stereographic projection and Shi's theorem (Goodman and Shi, 1985). In the second step, a static equilibrium analysis of those wedges that are kinematically possible (or removable) is conducted to either obtain the safety factor of the block or to determine the support load needed for an adequate factor of safety. The stereographic projection method is also used for the stability computations. In the stability calculations, all forces acting on the block need to be considered. These include the weight of the block, the normal and shear forces on each sliding plane, water pressures, external support, seismic load, and any other loads that may arise during construction and operation. Note that the stresses in the rock induced by the excavation should be incorporated through the resultant normal and shear loads on each discontinuity (Mauldon et al., 1997; Sofianos et al., 1999; Nomikos et al., 2006). It is customary to assume that friction along the sliding planes is governed by the Coulomb friction law and the joints have no cohesion. It is also assumed that the discontinuities are fully persistent (i.e. they are very large compared to the size of the opening) and thus stability calculations are performed on the largest block that can be formed given the dimensions of the tunnel. In addition, the largest block can appear

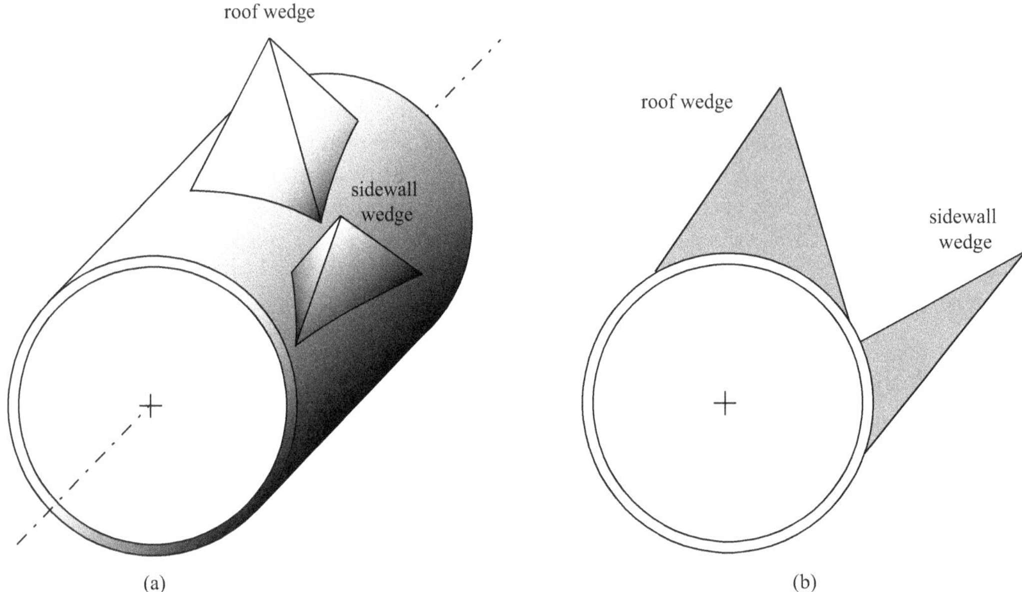

Figure 4.33 Wedge failure. (a) 3D view. (b) Cross section.

anywhere along the tunnel. These assumptions provide a conservative estimate of the safety of the excavation. A factor of safety of 1.3 is recommended for temporary excavations and 1.5 for permanent excavations.

With the introduction of affordable and fast computers and the wide availability of technical software, the stability analysis method using the stereographic projection method has become obsolete. In practice, software programs for block analysis are routinely used. The graphical method and block theory are, however, still valuable tools for education and visualization purposes. Review of the method is beyond the scope of this book, and the interested reader is referred to the publications of Goodman and co-workers (Goodman and Shi, 1985; Goodman, 1989).

Support to potentially unstable blocks should be provided by a stiff system to limit deformations as the transition in the joints from peak to residual strength requires very small sliding. Rockbolts are typically used, either grouted, or mechanically anchored. The latter are tensioned to provide a stiff response (see Section 7.4 for an in-depth discussion of rockbolts). The bolts should be long enough such that they are anchored behind the potential sliding plane of the block; a minimum distance of one meter is recommended between the block limit and the anchor of the bolt. Shotcrete also provides support because of its shear strength. However, because blocks need to be supported quickly and the shotcrete requires time to harden, it is not a solution often used. Besides, support is provided by the shotcrete only at the intersections of the block faces and the tunnel and so not all the shotcrete surface is effective. For tunnels in rock masses with closely jointed rocks, the use of shotcrete may be more appropriate as it helps prevent the initial failure of small wedges, which untreated, could induce a progressive failure. See Section 7.5 for more information on shotcrete.

The wedge or block method should be used in cases where isolated wedges are formed, and so the stability of the opening is controlled by individual blocks. For a jointed rock mass where the response of the tunnel is determined by the relative movements of interconnected blocks (see Section 5.1), the assumption of a single wedge cannot be made, and so the method should not be employed. Instead the use of discrete element methods is appropriate (see Sections 5.5 to 5.7). The response of the tunnel is then a function not only of the engineering properties of the intact rock and discontinuities, but also of the orientation, number, spacing of the discontinuities, and block connectivity (Shen and Barton, 1997).

4.3 THREE-DIMENSIONAL CONSIDERATIONS

The concept of the characteristic curves method (discussed in Section 1.3) is useful in providing some insight into the ground–liner response obtained using the two-dimensional analytical methods discussed. In particular, it is instructive to observe ground–structure interaction, as shown in Figure 4.34 (see also e.g. Figures 1.18 and 1.19). Both the relative stiffness method and the plasticity models, with the assumption of plane strain, are based on the premise that excavation and liner installation occur at the same time. This would correspond to the curves marked as 1 and 2 in Figure 4.34. However, as shown in Figures 4.2, 4.19, and 4.24, which plot ground displacements along a vertical section along the tunnel axis, ground displacements start to occur well ahead of the face of the tunnel. At the face, the ground displacements may be between 30 and 50% of the maximum displacements that take place at a cross section far behind the face (where plane strain conditions occur and thus the two-dimensional assumption applies). The curve in Figure 4.34 that would represent better this behavior is that shown as 3 (support delay). Inspection of the figure indicates that the two-dimensional assumption of simultaneous excavation and support is conservative for the support since the stresses computed are too high. The displacements are too low, and the tangential stresses in the ground are also too low. As a reference, and considering the unsupported tunnel case with elastic ground, the errors would be of the order of 30–50%, which are too high for any practical purposes.

There are a number of methods that have been proposed to estimate the actual, three-dimensional, liner loads based on two-dimensional results (Kielbassa and Duddeck, 1991; Möller, 2006; Karakus, 2007). The following five methods are perhaps the most discussed in the literature: (1) the Simplified Analysis Method; (2) the Characteristic Curves Method; (3) the β method; (4) the α method; and (5) the gap method.

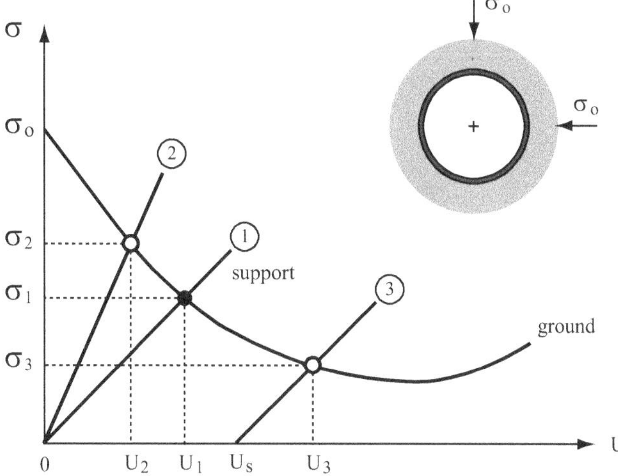

Figure 4.34 Characteristic curves method.

4.3.1 The simplified analysis method

The Simplified Analysis Method (SAM) was proposed by Schwartz and Einstein (1979) and (1980). It is based on the relative stiffness method. The support loads are computed with the following equation:

$$\begin{aligned} T &= \lambda_y \lambda_d T_{rel\,stiff} \\ M &= \lambda_y \lambda_d M_{rel\,stiff} \end{aligned} \tag{4.99}$$

where $T_{rel\ stiff}$ and $M_{rel\ stiff}$ are the axial load and moment in the liner obtained from the relative stiffness method, i.e. Equations (4.45) or (4.46); λ_y is the yield factor and is a coefficient that accounts for yielding of the ground; and λ_d is the support delay factor and accounts for the reduction of support load due to the support delay. The coefficients were obtained after a parametric study using a Coulomb failure criterion and different unsupported lengths (dimension d in Figure 1.8). In general, the moment should have different scale factors than the thrust, but the authors of the method stated that this approach provides a conservative estimate.

The parameter λ_d is given by:

$$\lambda_d = 0.98 - 0.57 \frac{d}{r_o} \tag{4.100}$$

The parameter depends only on the ratio of the unsupported length to the radius of the tunnel. Interestingly no other properties of the ground, liner or loading (e.g. K_o) are significant.

The parameter λ_y is obtained by solving the following equation twice:

$$F_g(\sigma) - F_s(\sigma) - F_d(\lambda_d) = 0 \tag{4.101}$$

F_g (σ) is the ground characteristic curve, F_s (σ) is the support characteristic curve, and F_d (λ_d) represents the support delay. The ground characteristic curves in the SAM method are obtained from a plane strain analysis of a cross section of the tunnel assuming axisymmetric loading. In other words, the ground characteristic curves are the radial displacements in equation (4.88), with the assumption of a Coulomb failure, or in (4.90) with the assumption of a Hoek-Brown failure. The support characteristic curve is also obtained from a plane strain analysis with axisymmetric loading. For example, for a linear elastic support, the characteristic curve is:

$$F_s(\sigma) = -\frac{(1 - \nu_s^2)}{E_s A_s} \sigma r_o^2 \tag{4.102}$$

where E_s, ν_s, and A_s are the Young's modulus, Poisson's ratio, and cross-section area, respectively, of the support. All other variables are as defined previously. The support delay function is:

$$F_d(\lambda_d) = -\frac{(1 + \nu)}{E}(1 - \lambda_d)\sigma r_o \tag{4.103}$$

where λ_d is the support delay factor from (4.100).

The coefficient λ_y is obtained by solving the non-linear Equation (4.101) twice in terms of σ. The first time the equation is solved with the characteristic function of the ground that accounts for yielding (e.g. from the radial displacements in (4.88) or (4.90)). Let's assume the solution is σ_1. The second time, the equation is solved assuming that the ground does not yield. For plane strain and axisymmetric loading, the ground characteristic curve is:

$$F_s(\sigma) = -\frac{(1+\nu)}{E}(\sigma_o - \sigma)r_o \tag{4.104}$$

Let's assume that the solution is σ_2. The yield factor is:

$$\lambda_y = \frac{\sigma_1}{\sigma_2} \geq 1 \tag{4.105}$$

If the yield factor is equal to one, no yielding occurs. The authors recommend using the method for $0.5 \leq K_o \leq 1.5$ and for values of $\lambda_y \leq 2$.

4.3.2 The characteristic curves method

The method is discussed in Section 1.3. As mentioned in the Section, the method is based on obtaining the radial deformations (convergence) of the ground and the support when subjected to a uniform radial pressure; see Figures 1.12 and 4.34. The two curves are obtained independently. Analytical or numerical solutions have been often employed considering a wide range of elasto-plastic behavior for the ground and support (e.g. Daemen, 1975, 1977; Schwartz and Einstein, 1979; Hoek and Brown, 1980a; Brown et al., 1983, provide a good review of the material behavior assumed until about 1980; see also Eisenstein and Branco, 1991; El-Nahhas et al., 1992; Nguyen-Minh and Guo, 1993, 1996; Bernaud and Rousset, 1996; Carranza-Torres and Fairhurst, 2000; Oreste, 2003b; Bobet, 2010a, c). The curves can also be obtained from field observations (Verman et al., 1995). The two most important limitations of the method are that calculations are performed with the assumption of plane strain and axial symmetry with respect to the tunnel axis. The first assumption, that of plane strain, can be assumed in certain cases (e.g. unsupported tunnel, pressure tunnel). Within certain limitations, the results can be modified to incorporate three-dimensional effects due to the presence of the tunnel face. The second assumption, that of axial symmetry, is quite limiting and is only applicable to circular tunnels with their axis parallel to one of the far-field principal stresses, $K_o = 1$, and homogeneous and isotropic ground. Due to this limitation, the method is mostly used qualitatively, as a tool to conceptually analyze soil–structure interaction.

A proposal using the Characteristic Curves Method (CCM) has been made by Carranza-Torres and Fairhurst (1999) and (2000) to account for three-dimensional, e.g. face, effects. The procedure is similar to the Simplified Analysis Method (SAM). The most important difference is how the support delay is computed. In the CCM, the ground reaction curve, as with the SAM, is obtained from equations (4.88) or (4.90). The proposal was originally developed for rocks following the Hoek-Brown failure criterion, but there is no conceptual limitation to extend it to other materials. The curve for the liner is obtained, also as with the SAM, from the stress-radial displacement of the plane strain cross section of the liner. The support delay, however, is computed differently. In the SAM, it is given by Equation (4.100). In the CCM, the support delay is obtained from a longitudinal plot of the radial displacements of a point of the ground at the perimeter of the excavation with the assumption of no support; e.g. the displacement curve in Figure 4.2. The displacements are obtained from measurements made in the field, from elasto-plastic numerical models or from the following empirical relation (based on convergence measurements made at the Mingtam Power Cavern project):

$$\frac{U_r}{U_{r,max}} = \left[1 + e^{-\frac{x/r_o}{1.10}}\right]^{-1.7} \tag{4.106}$$

where U_r is the radial displacement at a distance x behind the face, and $U_{r,max}$ is the displacement that occurs far from the face (i.e. where plane strain conditions apply).

Figure 4.35 illustrates the process. For example, if the support is placed at a distance $2r_o$ behind the face, the ground would have deformed already U_s. The magnitude U_s is the delay used in Figure 4.34 to translate the support reaction curve from the origin.

If the ground can be assumed elastic, and for axisymmetric loading, the longitudinal radial deformation curves can be approximated as:

For ground without water, Panet (1995) suggested the following equation:

$$\frac{U_r}{U_{r,max}} = 0.25 + 0.75\left[1 - \left(\frac{0.75}{0.75 + x/r_o}\right)^2\right] \tag{4.107}$$

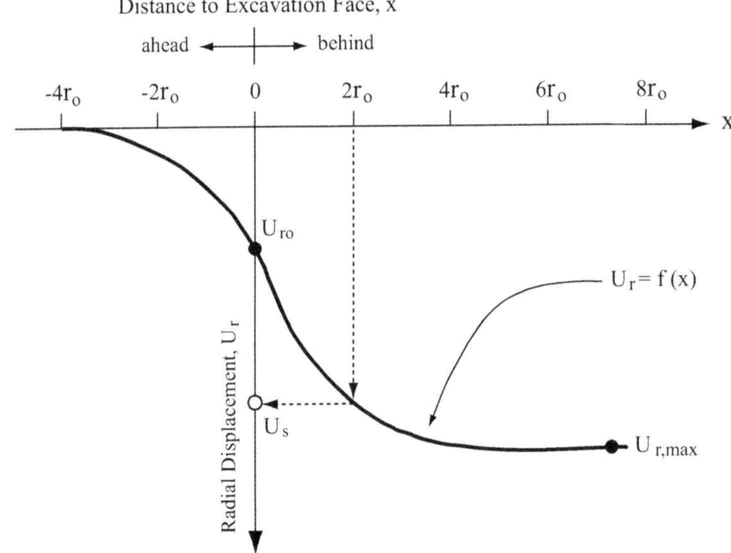

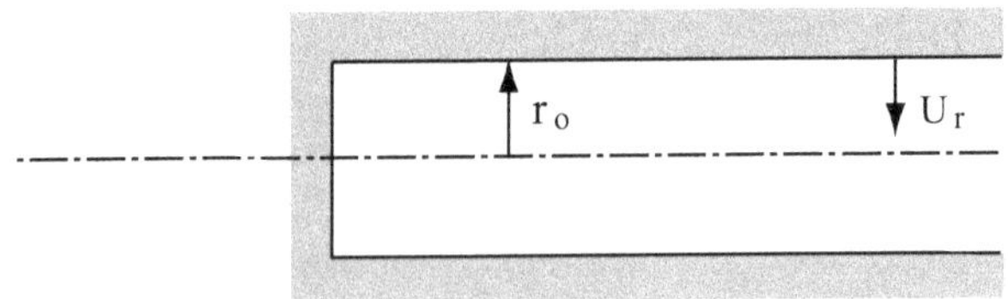

Figure 4.35 Longitudinal radial displacements.

where $x \geq 0$ is the distance behind the tunnel face, and r_o is the tunnel radius (see Figure 4.35). A somewhat similar equation was proposed by Corbetta et al. (1991), also for $x \geq 0$:

$$\frac{U_r}{U_{r,max}} = 0.29 + 0.71[1 - e^{-1.5(x/r_o)^{0.7}}] \tag{4.108}$$

For $x \leq 0$ Nam and Bobet (2007) suggested the following relation:

$$\frac{U_r}{U_{r,max}} = 0.28\, e^{1.05\, x/r_o} \tag{4.109}$$

In (4.107) to (4.109), $U_{r,max}$ is given by:

$$U_r = -\frac{1+\nu}{E}\sigma_o r_o \tag{4.110}$$

Expressions (4.107) to (4.109), although different, provide similar results. At the tunnel face ($x = 0$), the displacements predicted range between 25 and 29% of the maximum displacements. The differences are reasonably acceptable and well within the precision that could be expected from field measurements. The maximum radial displacement ($U_{r,max}$) occurs at approximately 5–6 tunnel radii behind the tunnel face, and the displacements are negligible at approximately 4 tunnel radii ahead of the face. The rate of displacement change is very large close to the tunnel face.

For a tunnel excavated below the water table with drainage toward the opening, Nam and Bobet (2007) proposed:

Ahead of the face, i.e. for $x > 0$,

$$\frac{U_r}{U_{r,max}} = \frac{U_{ro}}{U_{r,max}} + \left(1 - \frac{U_{ro}}{U_{r,max}}\right)\left[1 - \left(\frac{0.75}{0.75 + x/r_o}\right)^2\right] \tag{4.111}$$

or

$$\frac{U_r}{U_{r,max}} = \frac{U_{ro}}{U_{r,max}} + \left(1 - \frac{U_{ro}}{U_{r,max}}\right)[1 - e^{-1.5(x/r_o)^{0.7}}] \tag{4.112}$$

where U_{ro} is the radial displacement at the tunnel face and $U_{r,max}$, as before, the maximum, or plane strain, radial displacement. They are:

$$\frac{U_{ro}}{U_{r,max}} = 0.28 + 0.19(1 - e^{-u_o/\sigma'_{v0}}) \tag{4.113}$$

where σ'_{vo} and u_o are the far-field effective vertical stress and pore pressure; i.e. $\sigma'_{vo} = \sigma_o - u_o$. $U_{r,max}$ is given by (4.110) where σ_o is the total far-field stress.

For the displacements ahead of the tunnel face, $x < 0$:

$$\frac{U_r}{U_{r,max}} = 0.28\, e^{1.05\, x/r_o} + 0.2\, e^{0.1 x/r_o}(1 - e^{-u_o/\sigma'_{v0}}) \tag{4.114}$$

For a tunnel below the water table, maximum radial displacements are obtained at a distance of 5–6 radii behind the tunnel face; ahead of the tunnel face the influence of the excavation is negligible at a distance of 15 to 30 tunnel radii. With water seepage, the normalized ground deformations are much larger ahead of the tunnel face while they are similar to those of the dry ground behind the tunnel face. These observations compare well with the numerical results shown in Figures 4.17 and 4.19.

Equations (4.106) to (4.114) indicate that the radial displacement ratios ahead and behind the face of the tunnel depend only on the distance along the tunnel axis measured from the face of the tunnel. Interestingly, they are independent on the Young's modulus of the ground, the Poisson's ratio, and the size of the tunnel. An investigation on the effects of the Poisson's ratio by Unlu and Gercek (2003) suggests that the radial displacements at the face have a stronger dependency on ν than provided by previous formulations. They suggested the following relation:

$$\frac{U_{r0}}{U_{r,max}} = 0.22\nu + 0.19 \tag{4.115}$$

For an elastic response of the ground and liner, the radial stresses and displacements of the liner, $\sigma_{r,\,delay}$ and $U_{r,delay}$, considering a support delay U_s, are given by:

$$\begin{aligned} \sigma_{r,delay} &= \frac{U_{r,max} - U_s}{U_{r,max}} \sigma_{r,plane\ strain} \\ U_{r,delay} &= \frac{U_{r,max} - U_s}{U_{r,max}} \sigma_{r,plane\ strain} \end{aligned} \tag{4.116}$$

where $\sigma_{r,\,plane\ strain}$ and $U_{r,plane\ strain}$, are obtained from the relative stiffness solution where a plane strain, 2D, model is considered. The final radial displacements of the ground are those of the liner plus the support delay U_s.

The previous expressions have been obtained for cases with elastic ground and axial symmetry. Tonon and Amadei (2002) showed that the equations are also valid when the rock mass is transversely anisotropic with the plane of transverse anisotropy parallel to the tunnel axis and/or the far-field stress is non-uniform. This means that any point along the perimeter of the tunnel displaces by the same fraction of its maximum plane strain displacement, as the tunnel advances. This fraction is only a function of the distance to the tunnel face. Note, however, that with an anisotropic rock or with a non-isotropic far-field stress, the ground–liner interaction curves cannot be obtained with radial displacements only if the ground–support interface is not frictionless. This is because the tangential component of the ground displacements at the tunnel perimeter induces shear stresses at the interface, which in turn affect the response of the liner. In this case three-dimensional, numerical models need to be used.

4.3.3 The β method

This is the method perhaps most used in practice. In the β method, or stress reduction method, the presence of the face is approximated by a stress applied to the perimeter of the tunnel. The method consists of two steps (see Figure 4.36a). In the first step, the tunnel is excavated and a pressure equal to $\beta\,\sigma_o$ is applied to the perimeter of the opening; σ_o is the far-field stress and $0<\beta<1$ is the stress reduction factor. In the second step, the liner is placed and the stress $\beta\,\sigma_o$ is shared by the ground and the liner. The magnitude of β is related to the "delay" in placing the support. Thus a large unsupported length of the tunnel is associated with a small β; and

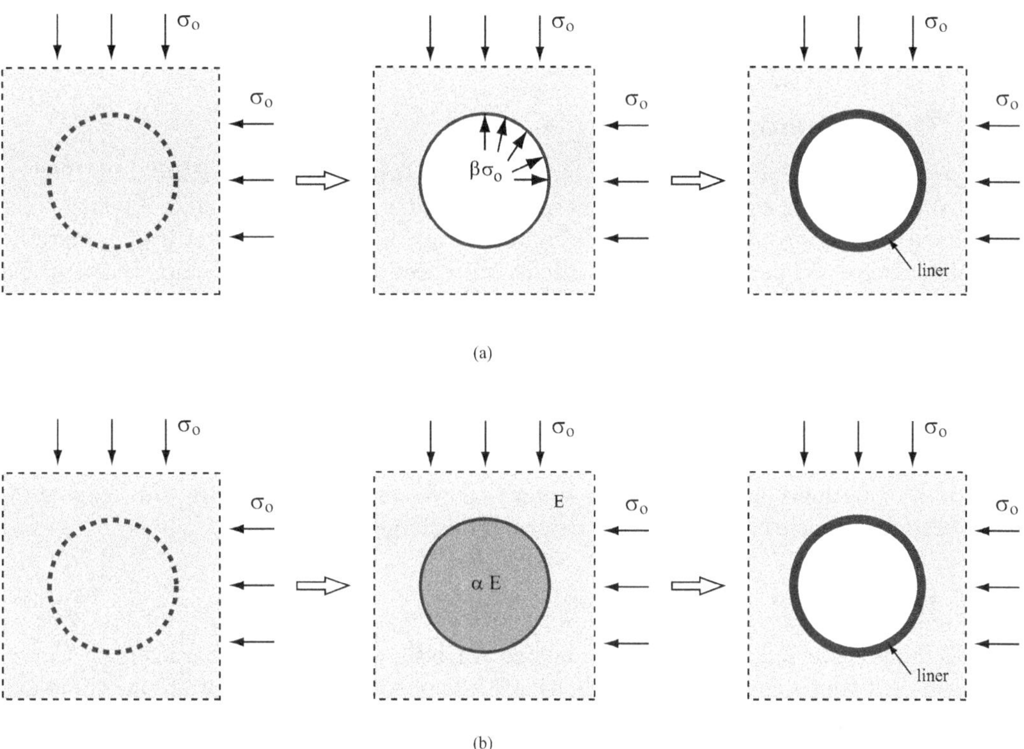

Figure 4.36 2D approximation to 3D analysis: β and α methods. (a) β Method: Stress reduction method. (b) α Method: Stiffness reduction method.

vice versa, a short unsupported length would have a large β. A large factor β results in small displacements of the ground, small radial displacements at the perimeter of the excavation and large loads of the support. Contrary, a small factor β results in large ground displacements and small support loads. In the method, the magnitude of β must be known a priori. A typical value of β is 0.5, but it can range from 0.2 to 0.8. It has to be noted that approximations to different response may require different stress reduction factors; i.e. the values of β for the same problem are different if one wants to match axial forces in the liner, moments, or ground settlements. In addition, the values of β strongly depend on the construction method, ground stiffness, plasticity model, K_0, unsupported tunnel length, etc. (Möller, 2006).

4.3.4 The α method

The α method or stiffness reduction method, Figure 4.36b, is applied in two steps. In the first step the stiffness of the ground that is excavated is reduced to a value α E, where $0 < \alpha < 1$ is the stiffness reduction factor and E is the Young's modulus of the ground (Swoboda, 1990). After equilibrium is reached, the ground inside the opening is removed and the liner is placed. As with the β method, large values of α are consistent with small unsupported length (or small support delay) and are associated with small ground displacements and large support loads. Small values of α apply to large unsupported lengths (large support delay) and result in large ground displacements and small liner loads. Typical values range between 0.2 and 0.5. The magnitude of α must be known to perform the calculations. As with the β method, an estimate of the magnitude of the parameter requires prior knowledge of the deformations induced in the ground during excavation, which can only be estimated, without complex three-dimensional numerical modeling, by experience.

4.3.5 The gap method

The gap method (Lo et al., 1984; Lee et al., 1992; Bobet, 2001; Chou and Bobet, 2002) is applicable to tunnels excavated in cohesive saturated soils to determine stresses and displacements of the ground and liner during construction, assuming undrained conditions. Computation of the gap parameter is discussed in Section 3.4.2.2. The gap represents the radial distance that the ground has to displace before contact with the liner; see for example Figure 3.13. The gap is a measure of all the ground displacements, due to all sources, that occur before liner installation. The magnitude of the gap is a function of the ground properties, loading, and construction method. The analysis is done with the assumption of plane strain, either numerically or analytically. Close-form solutions for stresses and displacements using the gap method are given in Equations (4.57) to (4.59). The liner does not carry any load until the ground deformations are larger than the gap. In other words, the loads of the liner are only a function of the ground deformations larger than the gap.

4.3.6 Discussion

The methods discussed, namely the Simplified Analysis Method, the Characteristic Curves Method, the β, the α, and the gap Methods, are all approximate. To illustrate the complexities that arise from a 3D analysis, a tunnel deep inside a soft rock, a shale, is analyzed using the FEM ABAQUS. The tunnel has a circular cross section with inside radius $r_o = 2$ m. The boundaries of the discretization are placed at 20 times the tunnel radius from the axis of the tunnel or the face of the excavation, which is a distance large enough such that the boundaries do not influence the results (Bobet, 2003). Due to the symmetry of the problem, only one quarter of the discretization is used. The discretization is done with 8-node brick-type elements with pore pressures at the nodes. The center of the tunnel is placed at a depth of 700 m, with a saturated unit weight of the shale $\gamma = 25$ kN/m^3. The water table is assumed at the surface. The far-field effective stresses and pore pressures are applied at the boundaries, with a magnitude equal to the stresses and pressures computed at the axis of the tunnel. In other words the effects of rock weight within the modeled region are neglected. This is an acceptable assumption for tunnels deeper than ten times the tunnel radius (Bobet, 2003).

The construction process corresponds to a full face tunnel excavation without face support. This process is simulated in ABAQUS by a series of steps, where in each step the rock elements that are excavated are deactivated and the liner elements that are placed are activated. This sequential process is needed because the material model used for the rock is elasto-plastic, and thus the solution is stress-path dependent. The material model used in the simulations was developed by Bellwald (1991) and Aristorenas (1992) and it is based on drained and undrained laboratory tests on Opalinus clay shale. The following is a summary description of the model. Further details can be found in Aristorenas (1992).

The incremental stress-strain equation can be written in matrix form as:

$$\{d\sigma'\} = [C^{ep}]\{d\varepsilon\} \tag{4.117}$$

where $\{d\sigma'\}$ and $\{d\varepsilon\}$ are the incremental effective stress and strain tensors, respectively, and $[C^{ep}]$ is the elasto-plastic stiffness matrix, $[C^{ep}] = [C^e] - [C^p]$; $[C^e]$ is the elastic stiffness matrix and $[C^p]$ is the plastic stiffness matrix.

The elastic properties are given by the shear modulus, G, the bulk modulus, K, and the coupling modulus, H, which are defined as follows:

$$G = \frac{\sigma'_{oct}}{a}; \quad K = \frac{2.3026\,\sigma'_{oct}}{C_{B\varepsilon}}; \quad H = \frac{\sigma'_{oct}}{m_1} \tag{4.118}$$

In equation (4.118), σ'_{oct} is the effective octahedral stress, and a, $C_{B\varepsilon}$, and m_1 are material elastic properties. The coupling modulus couples the elastic deviatoric and volumetric behavior, and is a measure of the material's anisotropy. The plastic stiffness matrix is obtained from:

$$\begin{aligned} [C^p] &= \frac{[C^e]\{a\}\{a\}^T[C^e]}{A + \{a\}^T[C^e]\{a\}} \\ \{a\}^T &= \left[\frac{\partial F}{\partial \sigma'_{11}}\ \frac{\partial F}{\partial \sigma'_{22}}\ \frac{\partial F}{\partial \sigma'_{33}}\ \frac{\partial F}{\partial \sigma_{12}}\ \frac{\partial F}{\partial \sigma_{13}}\ \frac{\partial F}{\partial \sigma_{23}}\right] \end{aligned} \tag{4.119}$$

F is the yield function and A is a hardening term, given by:

$$\begin{aligned} F &= \frac{2}{\sqrt{3}} b\sqrt{J_2} + m_2 \frac{I_1}{3} - c^* \\ A &= \frac{4}{3}\frac{b(\sigma'_{oct})^2}{\sqrt{3J_2}} \end{aligned} \tag{4.120}$$

I_1 is the first invariant of the effective stress tensor and J_2 is the second invariant of the deviatoric stress tensor; c^* is a state variable that depends on the current stress state; b and m_2 are material plastic properties. Failure is defined by the Drucker-Prager failure criterion.

The parameters used for the model are based on conventional triaxial compression and extension tests from Bellwald (1991) and are as follows: total unit weight, $\gamma = 25$ kN/m^3; internal friction angle, $\phi = 30°$; cohesion, c = 0.8 MPa; parameters a = 0.002, b = 0.002, $C_{B\varepsilon}$ = 3 MPa, $m_1 = 0.004$, and $m_2 = -0.005$; and permeability, $K = 5 \times 10^{-10}$ m/s.

The liner is assumed elastic, with properties: Young's modulus, E = 24,000 MPa and Poisson's ratio, $\nu = 0.25$. The thickness of the liner is 0.2 m, which corresponds to a radius of excavation of 2.2 m. Three advance rates are considered 1 m/day, 4 m/day, and 16 m/day. The influence of the lateral coefficient of earth pressure at rest, K_o, is also investigated. Three values are explored, K_o = 0.5, 1.0, and 1.5; it is assumed that the two principal horizontal stresses are equal. In addition, the effects of the unsupported tunnel length, d, are studied using d = 0 (no unsupported length), d = 1 m (0.5 r_o), and d = 2 m (one tunnel radius).

Figure 4.37 shows contour plots of effective stresses and pore pressures for K_o = 1, advance rate of 4 m/day, and for an unsupported length d = 1 m. This is taken as the base case. The far-field total stresses are $\sigma_{vertical} = \sigma_{horizontal} = \sigma_{axial}$ = 17.5 MPa, and pore pressures u = 6.87 MPa, and so the far-field effective stresses are 10.63 MPa. The contours are computed after an excavation of 20 r_o and are representative of the stresses induced in the rock and support after each excavation and support cycle. The figure shows that the excavation induces a significant unloading of the axial stresses inside the rock ahead of the tunnel excavation, which extends to a distance of about five tunnel radii from the tunnel face. Behind the face, the axial stress is not much affected. The tunnel excavation has a large effect on the horizontal and vertical effective stresses, as shown in the figure. There is significant unloading at the springline for the horizontal stress and at the crown for the vertical stress. This is expected since in

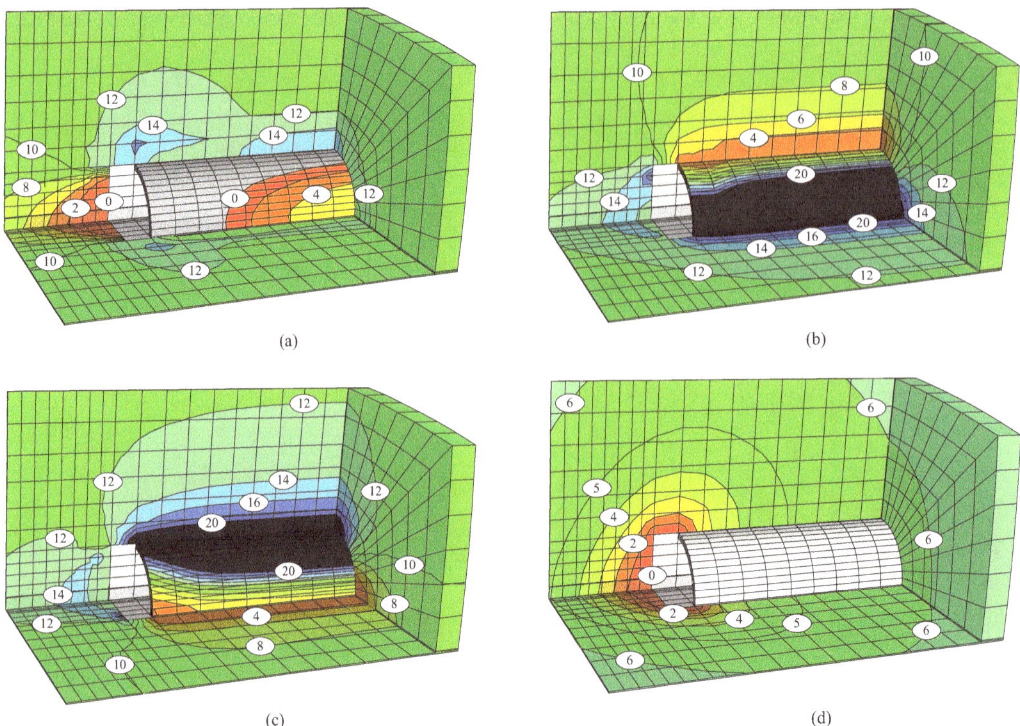

Figure 4.37 Three-dimensional view of stresses and pore pressures of a supported tunnel excavated in shale. Tunnel at 700 m depth, with water table at the surface. r_o = 2 m, isotropic far-field stresses. (a) σ_{axial}. (b) $\sigma_{vertical}$. (c) $\sigma_{horizontal}$. (d) Pore pressures.

the direction of a particular stress, say the vertical stress, there is a transfer of stresses from the springline, which unloads, toward the crown, which is loaded. Stress redistribution also occurs, for similar reasons, along the axial direction. The figure shows an increase of both the vertical and horizontal stresses, compared to the far-field values, in the rock ahead of the tunnel face. The extent of the rock affected is about five tunnel radii, similar to the extent observed for axial stresses. Thus the rock at the face of the tunnel is subjected to a state of stress similar to the unsupported rock at the perimeter of the tunnel. This state of stress has no load in one direction and may have a significant load, several times the in situ stress, in the other two directions. While not evident in Figure 4.37 (it is shown better in Figure 4.38), the stress distribution around the tunnel is not axisymmetric, as one would expect with K_o = 1. This is a consequence of the material model, in which it is assumed that bedding of the rock is horizontal and thus the material is more deformable in the vertical direction than in the horizontal direction. Figure 4.37 also shows the pore pressures generated. The pore pressures at the face and in the unsupported rock at the tunnel perimeter are zero, as they should since the rock is exposed to atmospheric pressure. The pore pressures quickly increase inside the rock mass, which is related to the stress gradient induced by the excavation, and affect a volume of rock within a distance of five to six tunnel radii from the face. The pore pressures equilibrate with time to the ambient pore pressures (6.87 MPa for this case), as will become clear in the following discussion. The drainage toward the tunnel, given the material properties and tunnel size, has little effect on the magnitude of the pore pressures; while the tunnel opens a drainage path toward the opening, the excavation rate is much faster than the time it would take for drainage to be significant. Instead, the unloading induced by the excavation

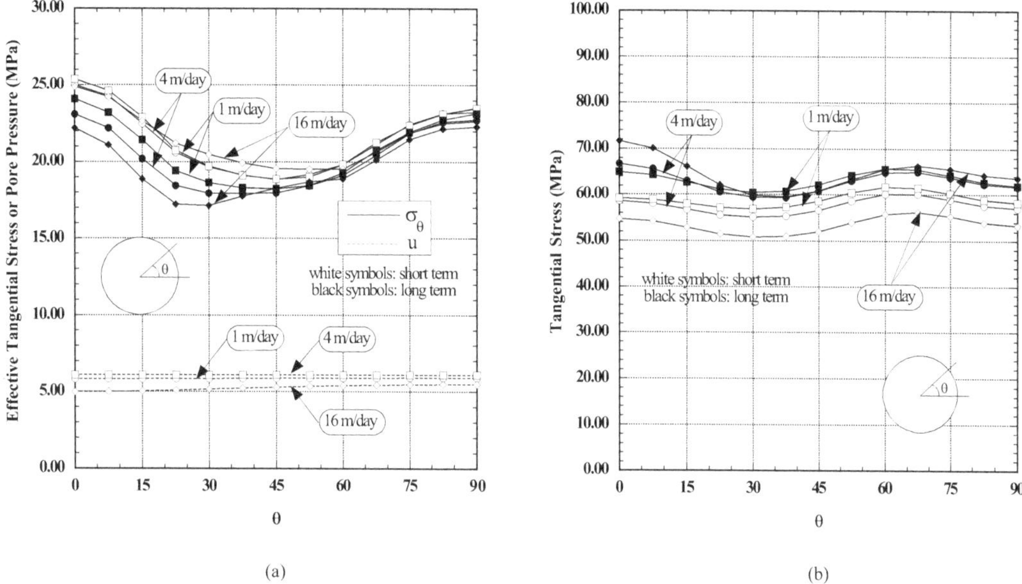

Figure 4.38 Tangential stresses of the liner and effective tangential stresses and pore pressures of the ground at the liner–ground interface. Deep tunnel in shale. Influence of excavation rate (from Bobet and Einstein, 2008). (a) Ground. (b) Liner. (Republished with permission of American Society of Civil Engineers.)

generates negative excess pore pressures, which reduce the pore pressures in the vicinity of the tunnel. This is an interesting phenomenon because during short-term (i.e. during excavation) negative excess pore pressures help maintain the mean effective stresses of the ground, thus helping the ground support itself. The result is an increased stability of the excavation and reduced loads of the support. With time, the negative excess pore pressures dissipate, the mean effective stresses in the ground increase, and the loads on the support increase. See complementary discussion in Section 4.2.1.3.

The interplay between excavation, pore pressures, and stresses in the ground and support is shown in Figure 4.38a. The figure plots effective tangential stresses and pore pressures for the rock in contact with the liner (the cylindrical coordinates used in Figure 4.38 have, as cylinder axis, the axis of the tunnel, and the origin of the angular coordinates coincides with the horizontal axis; thus in the figure $\theta = 0°$ corresponds to the springline and $\theta = 90°$ to the crown). The results apply to a cross section far from the face and are shown for different excavation rates: 1 m/day, 4 m/day, and 16 m/day. During excavation (short term) the difference in stresses is small, less than 5%, although the excavation rates differ by more than one order of magnitude. This is due to the induced change in excess pore pressures, which is the largest with the faster excavation rate. With time, the negative excess pore pressures dissipate and the effective stresses in the ground decrease. The largest change corresponds to the 16 m/day excavation rate case, which is the one that generates the largest excess pore pressures. The response of the liner is quite interesting. Figure 4.38b is a plot of the tangential stresses of the liner computed at the contact with the ground. The largest excavation rate induces the smallest loads during excavation (short term). With time, negative excess pore pressures dissipate, the effective stresses in the ground decrease and there is a load transfer from the ground to the liner, and so the stresses in the liner increase. The increase is the largest with the fastest excavation rate. However, the long-term stresses in the liner are not significantly affected by the excavation rate.

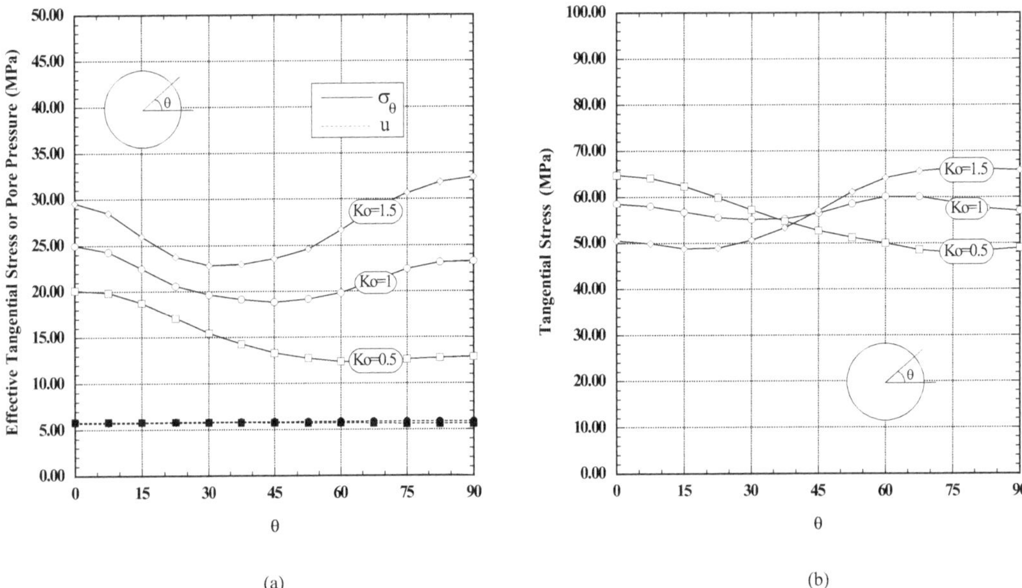

Figure 4.39 Tangential stresses of the liner and effective tangential stresses and pore pressures of the ground at the liner–ground interface. Deep tunnel in shale. Influence of K_0 (from Bobet and Einstein, 2008). (a) Ground. (b) Liner. (Republished with permission of American Society of Civil Engineers.)

Figure 4.39 can be used to evaluate the effects of K_0 on the tunnel response. Figure 4.39a plots the effective tangential stresses and pore pressures of the rock in contact with the liner for different values of the coefficient of earth pressure at rest, for an advance rate of 4 m/s, and for an unsupported length d = 1 m. The results are computed at the end of an excavation round (i.e. short term). One can see that as K_0 increases the (short term) effective tangential stresses increase. It is also apparent from the figure that $K_0 = 1$ does not produce uniform results. This is a consequence of the material model anisotropy, which makes the rock more deformable in the vertical than in the horizontal direction. Thus even for $K_0 = 1$ there is more deformation (unloading) at the crown than at the springline. The pore pressures, however, are not much affected by K_0. Figure 4.39b shows the tangential stresses at the liner fiber in contact with the ground. As observed with the ground, a far-field isotropic loading ($K_0 = 1$) does not induce uniform stresses in the liner. As expected, the stresses at the springline are the largest with $K_0 = 0.5$ and are the smallest with $K_0 = 1.5$. The influence of K_0 on the liner stresses is moderate, with differences at most of 30% between the stresses at the springline and at the crown.

Figure 4.40 shows plots similar to those of Figures 4.38 and 4.39, where the influence of the unsupported length of the tunnel is explored. For the ground, the short-term effective tangential stresses slightly change with increasing unsupported length. This is associated with an increase of excess pore pressures (decrease of total pore pressures) with increasing unsupported length. This is expected because there is a larger unloading of the ground as d increases. The increase of the unsupported length has a very large influence on the liner stresses. The smallest loads are registered for the largest unsupported length. This is understandable because, as the unsupported length increases, the ground deforms more, the pore pressure decrease, and so the loads on the liner decrease.

Figures 4.41 and 4.42 show a comparison between the results obtained from the three-dimensional model and results from a two-dimensional plane strain analysis obtained using the β method and assuming simultaneous excavation and installation of the liner (i.e. the relative stiffness method). The base case is used for the comparison; i.e. $K_0 = 1$, d = 1m,

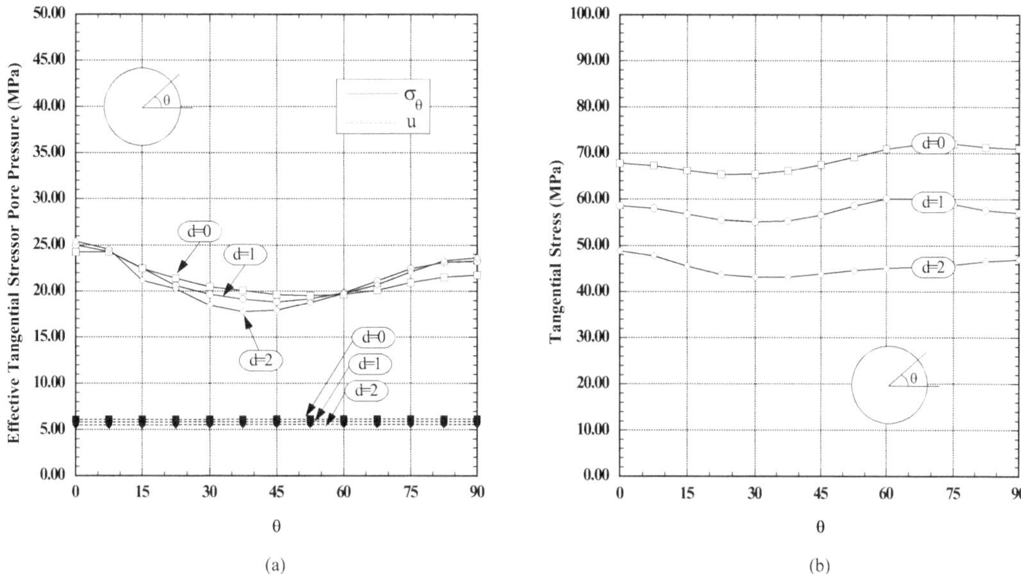

Figure 4.40 Tangential stresses of the liner and effective tangential stresses and pore pressures of the ground at the liner–ground interface. Deep tunnel in shale. Influence of unsupported length. (a) Ground. (b) Liner.

advance rate of 4 m/day. For the β method, the short term has been divided into two stages of equal duration, each half of the duration used in the three-dimensional computation. In the first stage the elements corresponding to the excavation are removed and a total stress equal to the stress reduction factor times the total far-field stress is applied to the perimeter of the excavation. In the second stage the liner elements are activated and the applied stress is removed. Figure 4.41 shows the results for short term. Figure 4.41a plots the effective tangential and pore pressures of the ground in contact with the liner, and Figure 4.41b the tangential stresses of the fiber of the liner in contact with the ground. Analogous plots for the long term are included in Figure 4.42. Three values of the stress reduction factor β are used: β = 0.3, 0.5, and 0.6. The objective is not to calibrate the method and thus find the optimum value of β, but to explore how the solution changes with different magnitudes of the load reduction factor. For short term, Figure 4.41a, the effective tangential stresses of the ground are somewhat insensitive to β. The assumption of simultaneous excavation and liner installation also gives comparable results. Note, however, that the differences between the two-dimensional and the three-dimensional results are significant. The differences are more important for the pore pressures. A large magnitude of the stress reduction factor induces smaller excess pore pressures. This is expected since with larger β the ground deformations are smaller. Interestingly, the assumption of simultaneous excavation and liner installation gives the best results. Similar to what happens for the ground, as the stress reduction factor decreases, the stresses in the liner decrease. This is also expected because the ground deforms more. The results obtained with β = 0.6, as average, are not far from the three-dimensional results, but the shape of the stress distribution in the liner is not captured. The relative stiffness assumption (simultaneous excavation and liner placement), as expected, over-predicts the liner stresses. For the long-term analysis, the trends are similar, except that the differences between the 2D and the 3D results are much larger.

The comparisons presented in Figures 4.41 and 4.42 illustrate the difficulties in obtaining accurate results with a two-dimensional analysis. As discussed by Möller (2006), there is no

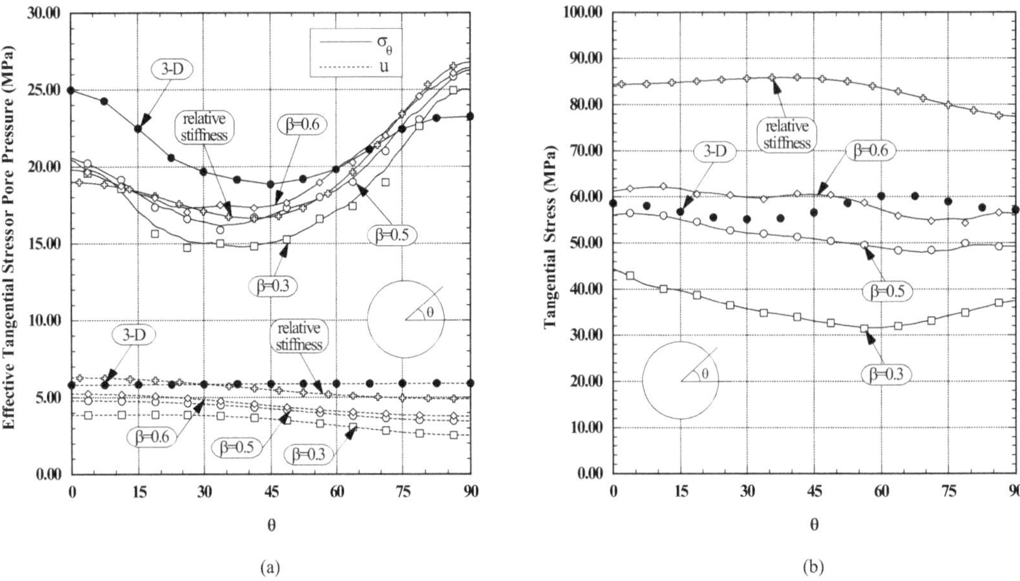

Figure 4.41 Tangential stresses of the liner and effective tangential stresses and pore pressures of the ground at the liner–ground interface. Deep tunnel in shale. 2D vs 3D analysis. β Method. Short-term analysis. (a) Ground. (b) Liner.

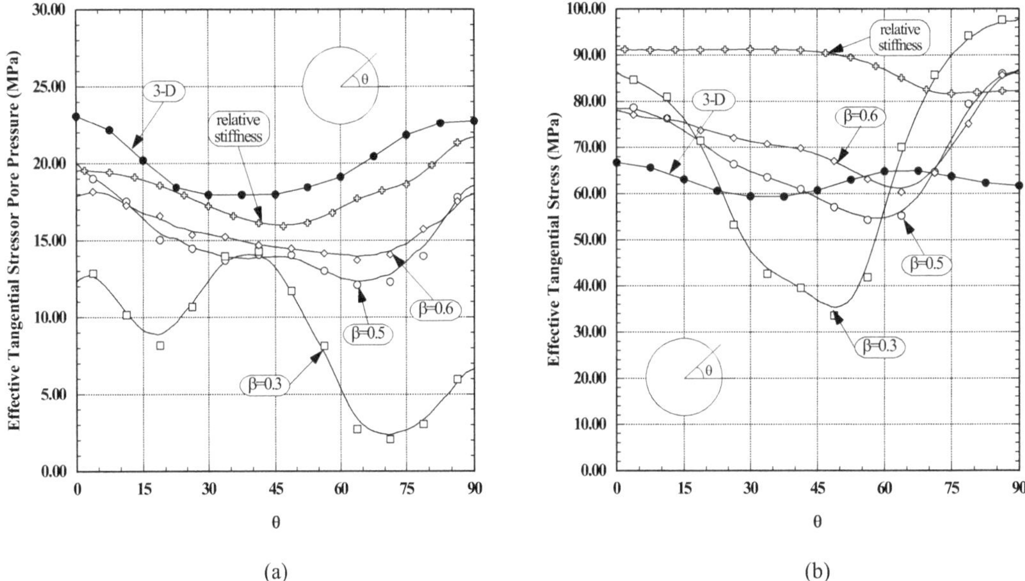

Figure 4.42 Tangential stresses of the liner and effective tangential stresses and pore pressures of the ground at the liner–ground interface. Deep tunnel in shale. 2D vs 3D analysis. β Method. Long-term analysis. (a) Ground. (b) Liner.

single value of β that can be used to match three-dimensional analysis. The stress reduction factor can be calibrated to approximate either the stresses of the ground, or the liner, either for short term or long term, etc. In addition, the values of β depend on the construction process, loading, material model and properties, etc. For example, Figure 4.40 shows that the results are quite different depending on the unsupported length, and so β will change with d. Since the magnitude of the stress reduction factor must be estimated, and unless a three-dimensional analysis is conducted, which would negate the purpose of performing the simpler two-dimensional analysis, the quality of the results from 2D analyses cannot be ascertained or measured. The same discussion applies to all the other methods. Where the two-dimensional approximations are potentially useful is to monitor the tunnel response at a site where the ground conditions and the construction method are uniform. The method (i.e. β, α method, etc.) can be calibrated once from field measurements and used afterward to identify deviations from expected conditions and for monitoring.

4.4 DISCUSSION AND RECOMMENDATIONS

The analysis presented in Figures 4.37–4.42 may be considered as of moderate complexity since it is a three-dimensional model with an elasto-plastic ground and includes both short-term and long-term conditions. However, the model does not make an attempt to analyze the excavation and the liner installation processes. Also a simple elastic model is assumed for the liner. The example, however, illustrates well the differences between two- and three-dimensional simulations.

It is clear than without a "correction" applied to the outcome from two-dimensional, either plane strain or axisymmetric models, the results can fall very far from those obtained from three-dimensional simulations. The drawback is that the magnitude of the correction to be applied to the predictions obtained from 2D models is not known beforehand and strongly depends on many factors, including construction process, size of the tunnel, loading, material models used, presence of water, short term or long term, to name a few. Even when the magnitude of the correction can be identified for a particular result, e.g. liner stresses, the same factor may not be appropriate for other results such as stresses of the ground and displacements.

It also has to be considered the different modes of loading of the ground along the excavation of the tunnel. The stress and displacement patterns obtained from a three-dimensional analysis around a tunnel face are very different from those obtained from a plane strain assumption (Lee and Rowe, 1990). The stress path at the crown and invert of a tunnel far from the face, where plane strain conditions would apply, follow a triaxial extension, but ahead of the face it is closer to triaxial compression. In addition, Cantieni and Anagnostou (2009) show that a point in the ground unloads as the tunnel approaches and, after the liner/support is installed, it reloads. These different loading paths, which are followed by a 3D model, cannot be obtained with a 2D model.

It is apparent that, given the availability and good quality of numerical codes (see Chapter 5 for a description of the numerical methods most used for tunneling), it is advisable to conduct three-dimensional simulations for design. As with all simulations and discretizations, engineering judgment is needed to include those assumptions and simplifications that make the model workable and yet provide a sufficient approximation to reality. This does not imply that the analytical, two-dimensional methods included in this chapter have no use or no practical application. Analytical solutions can be an effective tool for practitioners. Even though closed-form solutions may not be able to cover all possible scenarios and they incorporate assumptions that may be too restrictive for some problems (elastic or elastic-perfectly plastic behavior, homogeneity, isotropy, plane strain, axial symmetry, etc.), they are extremely useful to identify those parameters that have the largest influence on the solution and to provide the means to quickly obtain estimates of stresses and deformations which can be used for preliminary design or as a first step for more elaborate numerical models. The insight gained from the analysis can be used to optimize potential numerical remodeling efforts by identifying pertinent parameters for sensitivity analysis. Estimates obtained from analytical relationships can also be used in pre-feasibility and feasibility studies to obtain early and relatively accurate assessments of excavation and support requirements. Finally, analytical relationships can also be used to check the validity of the results obtained from numerical simulations. Most importantly, analytical solutions provide a rational framework to understand and quantify the observed response of ground and support during and after construction, which in tunnels it is intimately coupled with the construction process.

Chapter 5

Numerical methods

5.1 INTRODUCTION

Analytical methods, as described in Chapter 4, are very useful because they provide results with limited effort and highlight the most important variables that determine the solution of a problem. Analytical solutions have often a limited application since they must be used within the range of assumptions made for their development. Such assumptions often include elastic behavior, homogeneous, isotropic material, time-independent behavior, quasi-static loading, etc. Soils and rock masses display non-linear behavior, either because it is inherent to the material or because it has been externally induced (e.g. past stress history). Rocks and soils may not be isotropic or homogeneous, and the loading may not be static, or the geometry of the problem may be complex. In these cases, solutions can only be obtained numerically.

Numerical methods give only approximations to the correct or exact mathematical solution. This is so because some simplifications are made to solve the system of differential equations either inside the continuum or at the boundaries of the discretization. It has to be mentioned also that the problem that is solved is a conceptualization (simplified model) of the actual physical problem. The conceptualization applies to the geometry of the problem, the loading process or history, and the response of the geomaterials to loading. The better the approximation to the actual problem, through this conceptualization process, the more accurate the solution will be with respect to the response observed in the field or in the laboratory.

Numerical methods have been extensively used in the past several decades due to advances in computing power. In a broad sense, numerical methods can be classified into continuum and discontinuum methods (Jing and Hudson, 2002; Jing, 2003). Continuum methods may incorporate the discontinuities in the medium, if present, explicitly or implicitly, while in discontinuum methods discontinuities are incorporated explicitly. The need to use, for a particular problem, continuum or discontinuum methods depends on the size, or scale, of the discontinuities with respect to the size, or scale, of the problem that needs to be solved. There are no quantitative guidelines to determine when one method should be used instead of the other. Figure 5.1 (following Brady, 1987) provides some qualitative guidance. For example Figure 5.1a illustrates an opening in a medium without discontinuities; in this case the displacement field is continuous and thus continuum numerical methods are appropriate. Figure 5.1b shows a tunnel excavated in a medium with a small number of discontinuities, which divide the medium into a small number of continuous regions. The displacement field will be continuous inside each region but may be discontinuous across the discontinuities. If a continuum model is used, the model should be able to consider the specific discontinuities. The medium depicted in Figure 5.1c is determined by the number of discontinuities with spacing and continuity such that the blocks defined are within the scale of the opening. In this case, displacements may be determined by the slip and opening or closing of the

DOI: 10.1201/9781003328940-5

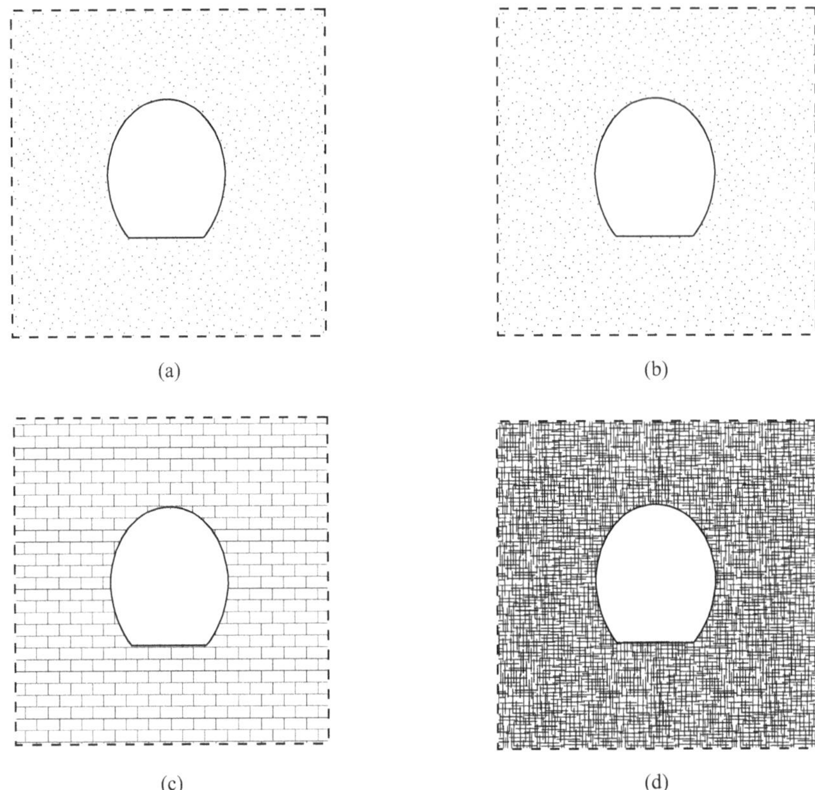

Figure 5.1 Continuum and discrete models (Bobet et al., 2009). (a) Continuum. (b) Discrete-continuum. (c) Discrete. (d) Pseudo-continuum. (Republished with permission of American Society of Civil Engineers.)

discontinuities and rotation of the blocks. Thus a discontinuum numerical method seems appropriate. If the medium is heavily jointed such that the blocks defined by the discontinuities have a size much smaller than the opening, Figure 5.1d, a pseudo-continuous displacement field is produced and the use of a continuum model seems reasonable.

There is quite a large number of numerical methods that have been used in the literature to estimate the behavior of underground structures. The most important, or at least the most used methods, are for: Continua, the Finite Difference Method, Finite Element Method and Boundary Element Method; and Discontinua, the Distinct Element Method, Discontinuous Deformation Analysis, and Bonded Particle Model. Two other methods are included which do not follow this classification: Meshless Methods and Artificial Neural Networks (even though one could argue that Artificial Neural Networks is not a numerical method in the sense used in this chapter, it is included here for completeness). The following sections provide the fundamental assumptions and the mathematical framework for each method and an overview of the range of problems where each method has been successfully used.

5.2 CONTINUUM: THE FINITE DIFFERENCE METHOD

The Finite Difference Method (FDM) is based on the premise that governing differential equations can be adequately represented by finite differences. The method is the oldest among the numerical methods presented here and was used even before the arrival of computers.

Timoshenko and Goodier (1970) attribute the first application of the method to Runge, who in 1908 used it for the solution of torsion problems. With the FDM, the set of differential equations is reduced to a system of linear equations, which can be solved by any of the classical methods. Southwell (1946) developed the relaxation method, which provides a fast solution of the system of equations; this promoted a much wider use of the FDM. The method really took off with the advent of computers.

The Finite Difference Method is discussed in this book first for historical reasons, second because it is still widely applied in elastostatics (e.g. computer code FLAC), and third because this is the method most used to solve dynamic problems.

The following presents a concise derivation of the system of equations that results from the approximation of the governing differential equations for 2D, plane strain, by finite differences. This is intended to provide a fundamental understanding of the method, which will be useful to assess its advantages, disadvantages, and limitations.

With the method, a grid is superimposed to the domain, as shown in Figure 5.2. For clarity it is assumed that grid points in the horizontal direction are spaced by a constant distance h and in the vertical direction by k. The derivatives of a continuous function f at a point with coordinates x_i, x_j in the domain are approximated as:

$$
\begin{aligned}
\left.\frac{\partial f}{\partial x}\right|_{x_i,x_j} &= \frac{1}{2h}[-f(x_i-h,x_j)+f(x_i+h,x_j)]+O(h^2)\\
\left.\frac{\partial f}{\partial y}\right|_{x_i,x_j} &= \frac{1}{2k}[-f(x_i,x_j-k)+f(x_i,x_j+k)]+O(k^2)\\
\left.\frac{\partial^2 f}{\partial x^2}\right|_{x_i,x_j} &= \frac{1}{h^2}[f(x_i-h,x_j)-2f(x_i,x_j)+f(x_i+h,x_j)]+O(h^2)\\
\left.\frac{\partial^2 f}{\partial y^2}\right|_{x_i,x_j} &= \frac{1}{k^2}[f(x_i,x_j-k)-2f(x_i,x_j)+f(x_i,x_j+k)]+O(k^2)\\
\left.\frac{\partial^2 f}{\partial x\partial y}\right|_{x_i,x_j} &= \frac{1}{4hk}[f(x_i-h,x_j-k)-f(x_i-h,x_j+k)\\
&\quad -f(x_i+h,x_j-k)+f(x_i+h,x_j+k)]+O(h^2,k^2)
\end{aligned}
\tag{5.1}
$$

The error of the derivatives is of the order of h^2 or k^2, and it is represented by the function O() at the end of each equation. Equation (5.1) gives the central finite difference formulas, which should be preferred because they produce errors equal or smaller than the forward or backward finite difference formulas (Boresi et al., 2003).

There are two approaches that can be followed: the stress approach, and the displacement approach. In the stress approach, the governing equations are written in terms of stresses (or in terms of the Airy stress function in elasticity). For a 2D, homogeneous, isotropic, and elastic body, the governing equation is given by (Timoshenko and Goodier, 1970):

$$
\left(\frac{\partial^2}{\partial x^2}+\frac{\partial^2}{\partial y^2}\right)\left(\frac{\partial^2\phi}{\partial x^2}+\frac{\partial^2\phi}{\partial y^2}\right)=0 \tag{5.2}
$$

In (5.2) ϕ is the Airy stress function. For interior points, the equation can be approximated with central finite difference formulas as (Wang, 1953; Timoshenko and Goodier, 1970; Fenner, 1986; Xu, 1992; Boresi et al., 2003):

$$
\begin{aligned}
&20\phi_{i,j}-8(\phi_{i+1,j}+\phi_{i-1,j}+\phi_{i,j-1}+\phi_{i,j+1})+2(\phi_{i+1,j+1}+\phi_{i+1,j-1}+\phi_{i-1,j+1}+\phi_{i-1,j-1})\\
&+(\phi_{i+2,j}+\phi_{i-2,j}+\phi_{i,j-2}+\phi_{i,j+2})=0
\end{aligned}
\tag{5.3}
$$

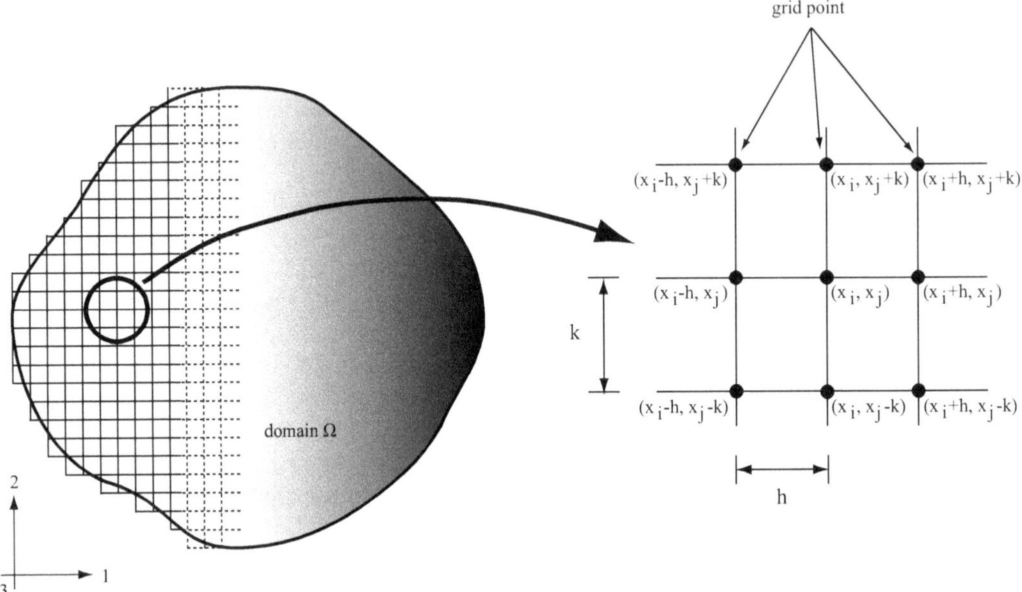

Figure 5.2 Finite difference grid in 2D (Bobet, 2010c). (Republished with permission of Springer nature VB.)

In the equation, the sub-indices represent the position of the point in the grid; for example i, j represents a point with coordinates (x_i, x_j), i + 1, j the point $(x_i + h, x_j)$, i, j + 1 the point $(x_i, x_j + k)$ and so on. Prescribed displacements or stresses at the boundary can be written in terms of derivatives of the Airy stress function. If there is a total of N grid points, the end result is a system of N equations, one per point given by (5.3) for interior points and one per boundary condition at the boundary, and N unknowns, which are the values of the Airy stress function at each point. Stresses and displacements at the grid points can be expressed in terms of the derivatives of the Airy stress function (e.g. Timoshenko and Goodier, 1970), which in turn can be approximated with finite differences following Equation (5.1).

The displacement approach provides more versatility than the stress approach, and it should in general be preferred. For non-homogeneous bodies, for problems involving displacement boundary conditions, or when body forces are not constant, the displacement approach is more efficient. In fact there are problems where the stress approach may not be applicable such as those with non-constant body forces.

For the 2D plane strain problem in a homogeneous isotropic medium, the governing equation is (Timoshenko and Goodier, 1970):

$$\begin{aligned} 2(1-\nu)\frac{\partial^2 u}{\partial x^2} + (1-2\nu)\frac{\partial^2 u}{\partial y^2} + \frac{\partial^2 v}{\partial x \partial y} = 0 \\ 2(1-\nu)\frac{\partial^2 v}{\partial y^2} + (1-2\nu)\frac{\partial^2 v}{\partial x^2} + \frac{\partial^2 u}{\partial x \partial y} = 0 \end{aligned} \quad (5.4)$$

where u and v are the displacements in the x and y directions, respectively, and ν is the Poisson's ratio. For clarity no body forces are included in (5.4). Using the central finite difference approximation, the equations are written as:

$$
\begin{aligned}
&2(1-\nu)\frac{1}{h^2}[u_{i-1,j}-2u_{i,j}+u_{i+1,j}]+(1-2\nu)\frac{1}{k^2}[u_{i,j-1}-2u_{i,j}+u_{i,j+1}]\\
&+\frac{1}{4hk}[v_{i-1,j-1}-v_{i-1,j+1}-v_{i+1,j-1}+v_{i+1,j+1}]=0\\
&2(1-\nu)\frac{1}{k^2}[v_{i,j-1}-2v_{i,j}+v_{i,j+1}]+(1-2\nu)\frac{1}{h^2}[v_{i-1,j}-2v_{i,j}+v_{i+1,j}]\\
&+\frac{1}{4hk}[u_{i-1,j-1}-u_{i-1,j+1}-u_{i+1,j-1}+u_{i+1,j+1}]=0
\end{aligned}
\tag{5.5}
$$

Stresses are related to the derivatives of the displacements, which in turn can be approximated by finite differences. If stresses are prescribed at the boundaries, they can be obtained using forward or backward finite difference approximations of the displacement derivatives. As with the stress approach, a linear system of equations is obtained; in this case 2N equations, since there are two equations per grid point. The solution of the system results in the displacements at the grid points.

Even though the process has been presented for a uniform grid, it can be easily applied to grids with different spacing, and to curved boundaries. As one would expect the results improve as the grid distance decreases.

Discontinuities can be incorporated in the model by using grid points on each side of the discontinuity. The relative displacement between corresponding grid points determines the slip along the discontinuity, and frictional laws (e.g. Coulomb) can be enforced by adding new equations to the system of equations that relate shear stress with normal stress. Normal and shear displacements can also be related to the shear and normal stiffness of the discontinuity.

The method can also be readily used to solve dynamic problems, where displacements are a function of position and time. Explicit time integration techniques are often used to provide solutions using small time increment steps. Dynamic problems require a maximum time step Δt to ensure stability of the solution, which is given by:

$$
\begin{aligned}
\Delta t &= \min\left(\frac{\Delta x}{C_p}\right)\\
C_p &= \sqrt{\frac{K+4/3G}{\rho}}
\end{aligned}
\tag{5.6}
$$

C_p is the compressional or P-wave velocity in the medium, Δx is the grid spacing; K is the bulk modulus, G the shear modulus, and ρ the density of the medium. Equation (5.6) indicates that the maximum time step is controlled by the stiffer material in the medium (i.e. the larger the K or G of the material, the larger the C_p, thus the smaller Δt). It is not unusual to run tens of thousands of steps to complete a numerical analysis. While the number of steps is quite large, the time required to conclude each step and the memory required to store the solution is small, and thus complex dynamic problems can be analyzed in a reasonable period of time.

The finite difference approach is very well suited to incorporate non-linear behavior. The solution is then obtained on a stepwise process involving sufficiently small loading increments until the desired final state is reached. At the end of each loading step, displacements at the grid points are obtained; stresses are then updated based on the non-linear behavior of the material, and another small loading increment is added. The new increment starts with the updated stress field from the previous increment. This is a forward scheme that does not require iteration, unlike other techniques such as Finite Element Methods that use implicit solution methods.

Figure 5.3 shows an example of a finite difference grid for a shallow tunnel subjected to earthquake loading. A similar grid was used by Kirzhner and Rosenhouse (2000) for the analysis of the dynamic response of tunnels. Since the interest from this type of simulations is typically stresses and displacements around the tunnel, the grid spacing is smaller in this area. A support can be modeled by adding grid points at the perimeter of the tunnel with a beam-type connection between the nodes. The seismic motions are input at the bottom of the discretization usually in the form of acceleration time history. A key feature of discretizations for dynamic analyses, also common to other numerical methods, is that the artificial boundaries (right and left in Figure 5.3) should not affect the results. It is necessary then that the Finite Difference code used has features that make the boundaries non-reflective.

The Finite Difference Method has found multiple applications in tunneling, from support and ground loading (Oreste and Peila, 1997; Alejano et al., 1999; Xie et al., 1999; Mohan et al., 2001; Boidy et al., 2002; Jayanthu et al., 2004; Graziani et al., 2005; Karakus and Fowell, 2005; Li et al., 2005; Yasitli and Unver, 2005), to interaction between tunnels and other structures (Kitiyodom et al., 2005). It is, however, in dynamic analysis where the FD method has found its widest use (Kirzhner and Rosenhouse, 2000; Sandoval and Bobet, 2020a, b).

The FDM is still the focus of intensive research, where significant contributions are being made to improve the flexibility of the method by developing schemes capable of discretizing the medium with unstructured grids of arbitrary shape (e.g. Voronoi grids, Du et al., 2003).

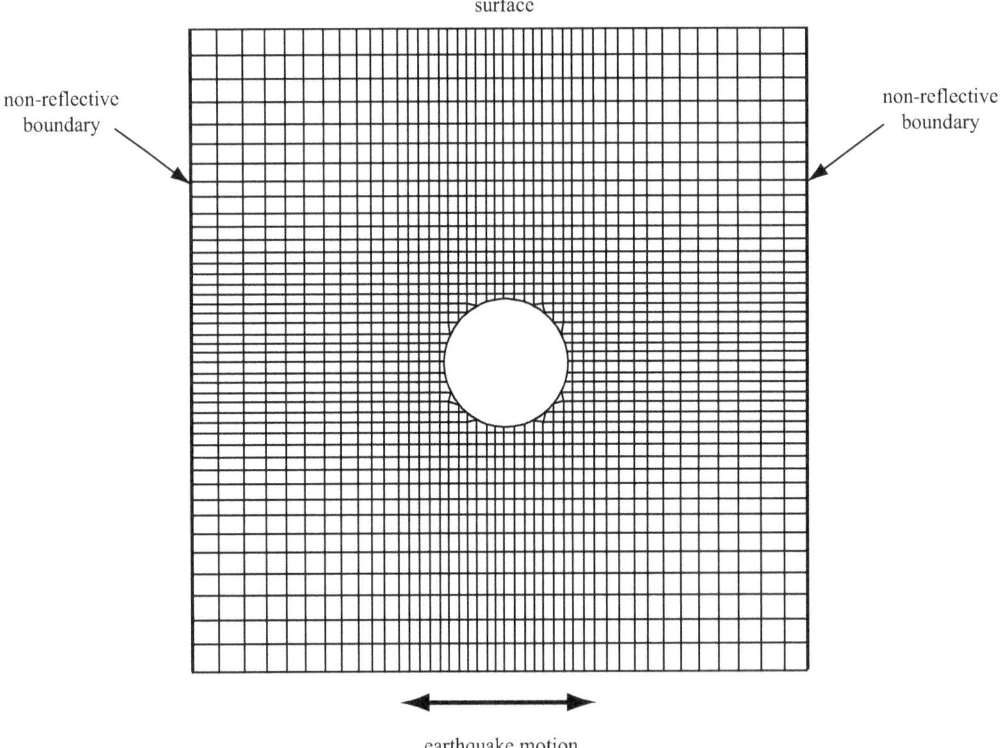

Figure 5.3 Example of finite difference grid.

Two promising methods are the result of this effort: The Finite Volume Method (FVM) and the Generalized Finite Difference Method (GFDM). The FVM is based on the solution of the integral form of the field partial differential equations over control volumes of polyhedral shape that are used to divide the continuum (Lahrmann, 1992; Oñate et al., 1994; Jasak and Weller, 2000). The GFDM is capable of solving any second-order partial differential equation over an irregular arrangement of nodes with any type of boundary condition (Cocchi, 2000; Benito et al., 2001). These developments are the precursors of the Meshless Methods (MM), which will be introduced with some detail in Section 5.8

5.3 CONTINUUM: THE FINITE ELEMENT METHOD

The Finite Element Method (FEM) is by far the method most used for the analysis of tunnels in continuous or quasi-continuous media. The term "Finite Element," according to Bathe (1982), was first introduced by Clough (1960), after seminal contributions by Argyris (1955) and Turner et al. (1956). The method consists of the discretization of the continuum into small elements that intersect at their nodes (Figure 5.4). The method relies on the assumption that, through appropriately chosen interpolation functions, displacements at any point within the element can be accurately obtained from the displacements at the nodes.

The following is intended to provide a summary of the fundamentals upon which the FEM numerical algorithms originate. The formulation is presented for the displacement-based FEM, which is the most commonly used method.

The method is based on the principle of virtual displacements, which states that, for a body in equilibrium, any compatible (i.e. satisfies boundary conditions) small virtual displacements applied to the body, the total internal work associated with the virtual displacement field must be equal to the total virtual external work. In mathematical form,

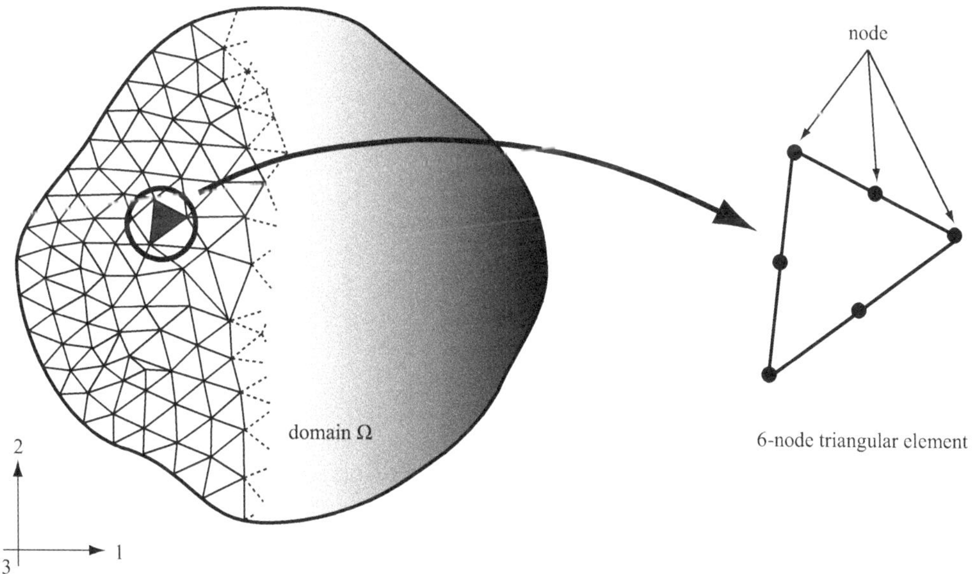

Figure 5.4 Finite element discretization in 2D (Bobet et al., 2009). Republished with permission of American Society of Civil Engineers.

$$\int_V \bar{\varepsilon}_{ij}\sigma_{ij}\,dV = \int_V \bar{u}_i f_i^B\,dV + \int_S \bar{u}_i^S f_i^S\,dS + \bar{u}_i^C f_i^C \tag{5.7}$$

In (5.7), the indices i, j = 1,2,3 are the directions of the three coordinate system; $\bar{\varepsilon}_{ij}$ and $\bar{u}_i$ denote the virtual strains and virtual displacements, respectively; σ_{ij} are the (unknown) stresses in the body, and f^B, f^S, and f^C are body forces, surface forces, and concentrated forces applied to the body. The stresses are related to the strains through the material model

$$\sigma_{ij} = C_{ijkl}\,\varepsilon_{kl} + \sigma_{ij}^o \tag{5.8}$$

C_{ijkl} is the compliance matrix and $\sigma_{ij}{}^o$ are the initial stresses.

The FEM is based on the assumption that the displacements of any point inside an element, u_i, can be approximated through interpolation (or shape) functions, h_k, conveniently selected, and the displacements of the nodes of the element, $\hat{u}_k$. The interpolation functions will be discussed later. Displacements and strains at any point inside an element can then be expressed as:

$$\begin{aligned} u_i &= h_k\,\hat{u}_{ik} \\ \varepsilon_{ij} &= \frac{1}{2}\left(\frac{\partial u_i}{\partial x_j} + \frac{\partial u_j}{\partial x_i}\right) = \frac{1}{2}\left(\frac{\partial h_k}{\partial x_j}\hat{u}_{ik} + \frac{\partial h_k}{\partial x_i}\hat{u}_{jk}\right) \\ \varepsilon &= B\hat{U} \end{aligned} \tag{5.9}$$

where $\hat{u}_{ik}$ is the displacement along direction i of node k. The derivatives can be lumped into a matrix B which factors the nodal displacements $\hat{U}$, appropriately arranged.

Hence, stresses are written, from (5.8) and (5.9), in matrix form:

$$\sigma = CB\hat{U} + \sigma^o \tag{5.10}$$

Substitution of (5.9) and (5.10) into (5.7) gives:

$$\begin{aligned} \hat{\bar{U}}^T\left[\left(\sum_m \int_{V^m} B^T C B\,dV^m\right)\hat{U} + \sum_m \int_{V^m} B^T\sigma^o dV^m\right] &= \hat{\bar{U}}^T \sum_m \int_{V^m} H^m f^{mB} dV^m \\ &\quad + \hat{\bar{U}}^T \sum_m \int_{S^m} H^m f^{mS} dS^m + \hat{\bar{U}}^T f^C \end{aligned} \tag{5.11}$$

where matrix notation is used instead of the index notation. External forces f^C are applied at the nodes of the elements. Volume and surface integration is accomplished by summation of the contributions of all elements. Note that nodal displacements can be taken out of the integrals, and $\bar{U}$ and $\hat{U}$ are matrices that contain all the virtual and actual nodal displacements of the body, ordered from node 1 to node N, where N is the total number of nodes. In (5.11) the index m represents the element number, from m = 1 to M, where M is the total number of elements.

Note that all the terms in (5.11) are multiplied by the virtual displacements $\hat{\bar{U}}^T$, which can be eliminated from the equation. Re-arranging,

$$K\hat{U} = R \tag{5.12}$$

Equation (5.12) provides a system of equations that can be solved in terms of the nodal displacements. K is the stiffness matrix and is given by:

$$K = \sum_m \int_{V^m} B^T C B dV^m \tag{5.13}$$

R is the load vector, which from (5.11) is:

$$R = \sum_m \int_{V^m} H^m f^{mB} dV^m + \sum_m \int_{S^m} H^m f^{mS} dS^m - \sum_m \int_{V^m} B^T \sigma^o dV^m + f^C \tag{5.14}$$

For a dynamic analysis, Equation (5.12) is written as:

$$M\ddot{\hat{U}} + D\dot{\hat{U}} + K\hat{U} = R \tag{5.15}$$

where $\dot{\hat{U}}$ and $\ddot{\hat{U}}$ are the first- and second-time derivatives of the nodal displacements; M is the mass matrix and D is the damping matrix, which are:

$$\begin{aligned} M &= \sum_m \int_{V^m} \rho H^T H dV^m \\ D &= \sum_m \int_{V^m} \kappa H^T H dV^m \end{aligned} \tag{5.16}$$

where ρ is density and κ is damping.

The flexibility of the method in (5.11) is evident. Material properties and material behavior can change from element to element, and the input loading, both internal and external, can vary through the domain. While such extreme variation in properties and loading across the body and boundaries is not practical, it is not uncommon to have several materials and complex loading in the simulations. It can also be noticed in (5.14) that stepwise loading can be easily implemented by making σ^o, the stresses at the beginning of the current loading increment, equal to the stresses at the end of the previous loading increment.

The accuracy of the final solution depends on the number of nodes (or number of elements) used for the simulations. Figure 5.5 shows 2D elements commonly used in tunneling. The 1D elements in Figures 5.5a and 5.5d can be used for rockbolts and anchors, and the 2D elements for the ground and support (the support, in 2D is often modeled with beam elements and in 3D with shell elements, when the thickness of the support is small compared with the size of the tunnel; otherwise, solid elements should be used). Instead of using elements of any

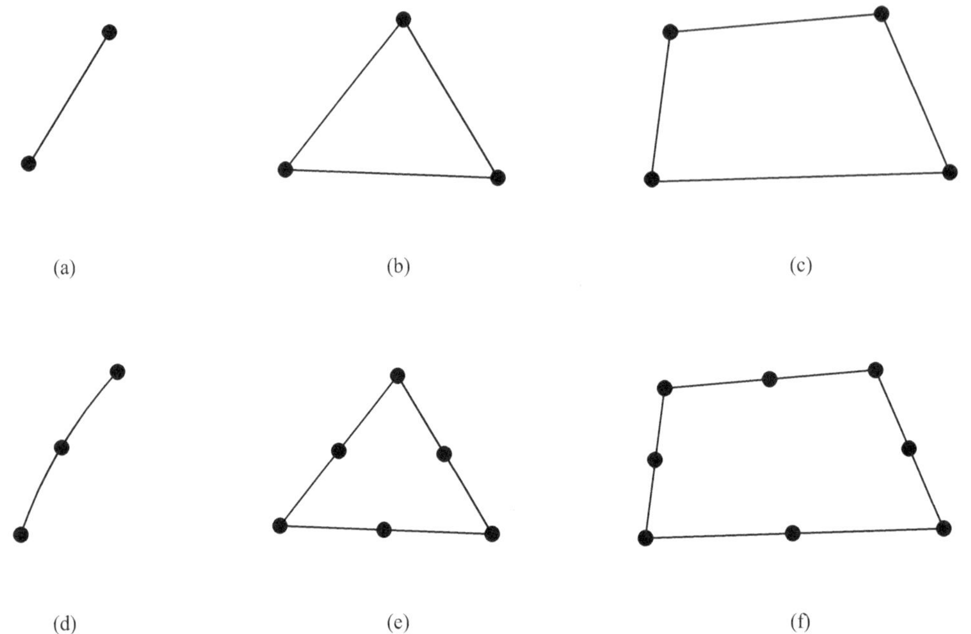

Figure 5.5 Examples of two-dimensional isoparametric elements. (a) 2-node line element. (b) 3-node triangular element. (c) 4-node cuadrilateral element. (d) 3-node line element. (e) 6-node triangular element. (f) 8-node cuadrilateral element.

arbitrary shape, the numerical integration of Equations (5.13), (5.14), and (5.16) is better accomplished by mapping the actual elements into new elements with simpler shapes and with their own coordinate system. Figure 5.6 shows a 4-node and an 8-node isoparametric element mapped into 2 × 2 unit square elements.

Coordinates and displacements of any point inside the actual element are expressed in terms of the coordinates and displacements of the nodes through the interpolation or shape functions h_k.

$$\begin{aligned} x_i &= h_k\, \hat{x}_{ik} \\ u_i &= h_k\, \hat{u}_{ik} \end{aligned} \tag{5.17}$$

The shape functions depend on the type of element, i.e. linear, triangular, quadrilateral, etc., and on the number of nodes of the element. For example the shape functions for a 4-node element are:

$$\begin{aligned} h_1 &= \frac{1}{4}(1+\xi_1)(1+\xi_2) \\ h_2 &= \frac{1}{4}(1-\xi_1)(1+\xi_2) \\ h_3 &= \frac{1}{4}(1-\xi_1)(1-\xi_2) \\ h_4 &= \frac{1}{4}(1+\xi_1)(1-\xi_2) \end{aligned} \tag{5.18}$$

where ξ_1 and ξ_2 are the coordinates expressed in a system local to the element, as shown in Figure 5.6.

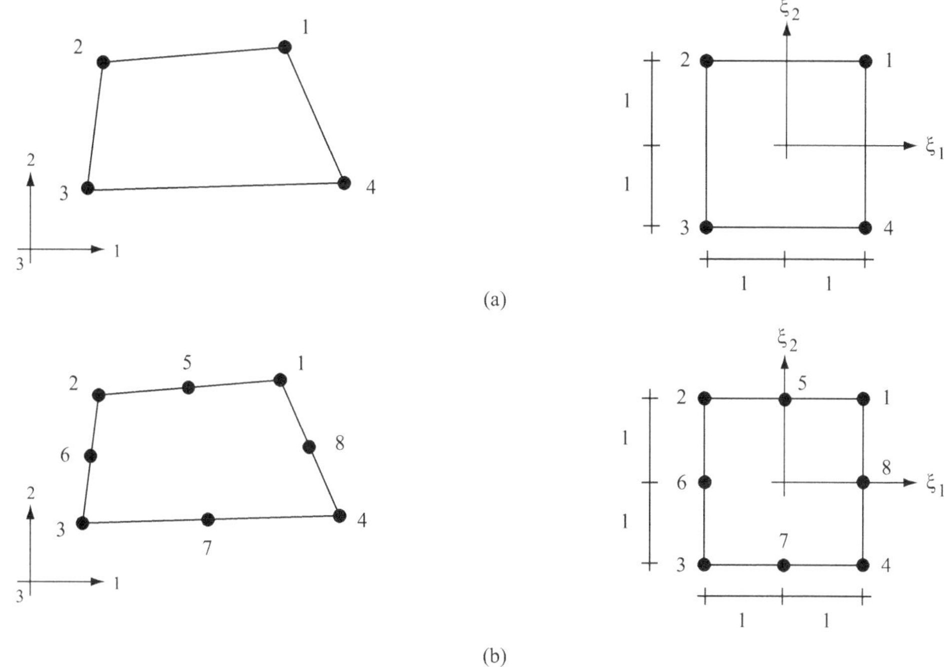

Figure 5.6 Mapping of elements from global to local coordinate system. (a) 4-node cuadrilateral element. (b) 8-node cuadrilateral element.

And for an 8-node element,

$$\begin{aligned}
h_1 &= -\frac{1}{4}(1+\xi_1)(1+\xi_2)(1-\xi_1-\xi_2) \\
h_2 &= -\frac{1}{4}(1-\xi_1)(1+\xi_2)(1+\xi_1-\xi_2) \\
h_3 &= -\frac{1}{4}(1-\xi_1)(1-\xi_2)(1+\xi_1+\xi_2) \\
h_4 &= -\frac{1}{4}(1+\xi_1)(1-\xi_2)(1-\xi_1+\xi_2) \\
h_5 &= \frac{1}{2}(1-\xi_1{}^2)(1+\xi_2) \\
h_6 &= \frac{1}{2}(1-\xi_1)(1-\xi_2^2) \\
h_7 &= \frac{1}{2}(1-\xi_1^2)(1-\xi_2) \\
h_8 &= \frac{1}{2}(1+\xi_1)(1-\xi_2^2)
\end{aligned} \tag{5.19}$$

The interpolation functions h_k return a value of one when applied at the element node k, and zero at all other nodes. This is required, since from (5.17) the coordinates and displacements of the node should be recovered. Low-order elements (e.g. 3-node elements) have linear interpolation functions. In other words, the displacements vary linearly across the element. As a consequence, from Equations (5.9) and (5.10) strains and stresses are constant throughout the element. In contrast, the 8-node element has a parabolic interpolation, which results in a linear variation of strains and stresses across the element. Even though elements with parabolic interpolations (bottom elements in Figure 5.5) are computationally more expensive than elements with linear interpolations (top elements in Figure 5.5), they should be preferred because of the increased level of accuracy, in particular in areas where stress gradients are expected to be high; e.g. around the opening of the tunnel. Elements with linear interpolations can of course be used, but they would require a much denser mesh.

The integration in (5.13), (5.14), and (5.16), because of the coordinate change that maps elements into simple geometries, needs to be expressed in terms of the coordinate system local to the element. Hence,

$$
\begin{aligned}
K &= \sum_m \int_{V^m} B^T C B \det J \, dV^m \\
R &= \sum_m \int_{V^m} H^m f^{mB} \det J \, dV^m + \sum_m \int_{S^m} H^m f^{mS} \det J^S \, dS^m - \sum_m \int_{V^m} B^T \sigma^o \det J \, dV^m + f^C \\
M &= \sum_m \int_{V^m} \rho H^T H \det J \, dV^m \\
C &= \sum_m \int_{V^m} \kappa H^T H \det J \, dV^m
\end{aligned}
\tag{5.20}
$$

The terms in the integrals are now expressed in terms of the coordinate system local to the element (Figure 5.6), and J is the Jacobian operator for the coordinate transformation.

$$
j_{ij} = \frac{\partial x_j}{\partial \xi_i} \tag{5.21}
$$

where ξ_i represents the coordinate system local to the element (Figure 5.6).

There are many types of elements that have been developed, similar to the two discussed here. Relevant to tunneling are contact or joint elements, and infinite elements. Contact elements are intended to model frictional discontinuities. However, the tendency seems to be to use "contact surfaces" rather than contact elements. In either case, the element or contact surface can open under tension and close under compression. In compression, slip is produced when the shear stress reaches the shear strength, which is defined by a suitable behavior model, e.g. Coulomb. Also normal and shear displacements can be related to normal and shear stresses through the corresponding normal and shear stiffness properties. Infinite elements impose the condition of zero displacements at infinity (Kumar, 2000 provides an application of infinite elements for the numerical analysis of underground excavations). These contacts, elements, and many others are typically found in the element library of most commercially available codes.

Figure 5.7 shows an example of a 3D mesh for a circular unsupported tunnel (see also Figures 1.5, 1.9, 4.1, 4.3 and 4.37). In the discretization the element density increases toward the excavation since this is where the stress gradients are expected to be the largest. The boundaries are placed far from the opening such that their presence should not affect the results close to the tunnel (a number of specific details are given later in Section 5.9).

The Finite Element Method has been and is currently heavily used for geotechnical engineering analysis in general and for tunneling in particular. The range of cases where the method is used spans the entire gamut of possibilities, from elasticity to plasticity, from static

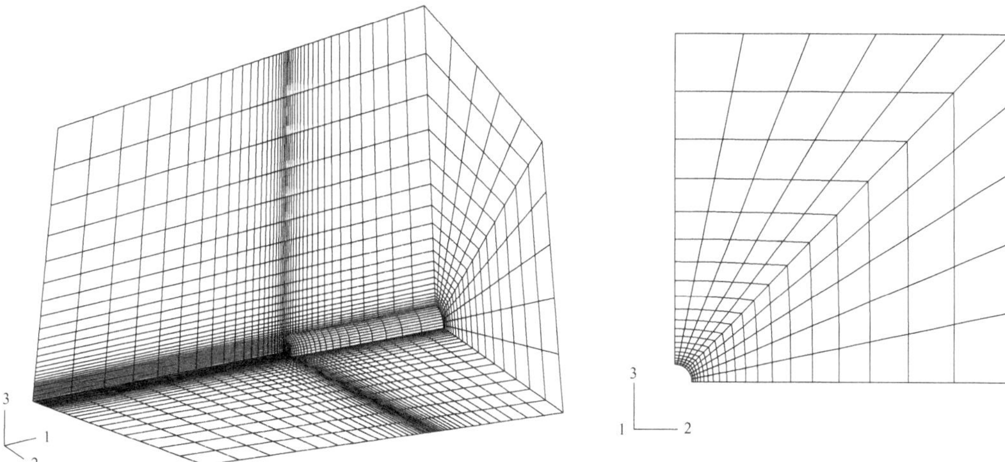

Figure 5.7 3D FEM discretization of unsupported tunnel.

to dynamic analyses, from ground behavior to soil–structure interaction. Many researchers and practitioners routinely use the method, as one can see after a cursory review of the technical literature. In tunneling, the FEM is used to ascertain stresses in the support and the ground due to tunnel excavation (Swoboda and Abu-Krisha, 1999; Pelizza et al., 2000; Eberhardt, 2001; Dhawan et al., 2002; Öttl et al., 2003; Zhu et al., 2003; Javadi and Snee, 2002; Lackner et al., 2002; Sterpi and Cividini, 2004; Boldini et al., 2005), to examine the effects of groundwater seepage on ground and liner stresses and deformations (Yoo, 2005; Nam and Bobet, 2006, 2007), to predict ground deformations due to tunneling (Addenbrooke and Potts, 2001; Rodriguez-Roa, 2002; Karakus and Fowell, 2003; De Farias et al., 2004; Franzius and Potts, 2005; Karakus and Fowell, 2005; Ieronymaki et al., 2017; Vitali et al., 2021), to investigate the effects of tunneling operations on the response of preexisting structures such as piles, buildings, and tunnels (Addenbrooke and Potts, 2001; Lee and Ng, 2003) or the response of existing tunnels due to surface construction (Abdel-Meguid et al., 2002), and to estimate the response of underground structures to dynamic loading due to blasting or seismic events (Wang, 1993; Law and Lam, 2003; Dhawan et al., 2004; Huo et al., 2005, 2006a, b; Sandoval and Bobet, 2020b; Savigamin and Bobet, 2021). The list is by no means intended to be exhaustive, but to provide references to illustrate the wide range of application of the method.

5.4 CONTINUUM: THE BOUNDARY ELEMENT METHOD

With the Boundary Element Method (BEM), only the boundaries of the continuum need to be discretized. This is in contrast to the other two continuum methods, the Finite Difference and the Finite Element methods, where the entire medium has to be discretized. Also, if the medium extends to infinity, which is common in problems involving underground excavations, no artificial boundaries such as those needed in FDM and FEM are required. The BEM automatically satisfies far-field conditions. In the BEM, the solution is approximated at the boundaries while equilibrium and compatibility are exactly satisfied in the interior of the medium. In FDM and FEM the approximations are made inside the medium. The advantage of limiting the discretization to the boundaries is that the problem is reduced by one order: from 3D to a 2D surface problem at the boundary, and from 2D to a line problem. Thus the method is very attractive for those problems where the volume to boundary surface ratio is large.

The technique used in the BEM consists in essence of transforming the governing differential equations, which apply to the entire medium, to integral equations, which only consider boundary values (Venturini, 1983; Brebbia et al., 1984; Crouch and Starfield, 1983). In a boundary value problem, some parameters such as stresses and displacements are known while others are not, which then are part of the solution. There are two approaches to solve for the unknown parameters. In the first approach (Direct BEM) the unknowns are solved directly, and once they are obtained, stresses and displacements at any point in the continuum can be obtained directly from the solution. In the second approach (Indirect BEM), the solution is in terms of some "fictitious" quantities, typically stresses or displacements. The fictitious quantities are obtained first and the actual stresses, and displacements at any point in the medium are expressed in terms of these fictitious quantities.

5.4.1 The direct Boundary Element Method

The method started with the pioneer work by Rizzo (1967) and later on by Cruse (1969) who developed the fundamental equations for the 2D and 3D elastostatic problem, respectively. Since then, the method has expanded significantly and can be used to solve problems involving plasticity, dynamic loading, groundwater flow, soil–structure interaction, soil-structure-fluid interaction, to name a few.

The formulation of the method is based on Betti's reciprocity theorem (Betti, 1872), which states that for a linear-elastic solid subjected to two quasi-static loading states I and II, the work done by the forces of case I applied to the displacements of state II is equal to the work done by the forces of case II applied to the displacements of state I.

Figure 5.8 shows the conceptual stages. Figure 5.8a depicts the tractions and displacements at the boundaries of state I. State II is obtained by the solution of Kelvin's problem, a concentrated unit load in an infinite medium. In equation form, Betti's theorem can be expressed as:

$$u_i^I(\xi) = \int_S t_j^I(x)u_{ij}^{II}(\xi, x)dS(x) - \int_S t_{ij}^{II}(\xi, x)u_j^I(x)dS(x) \tag{5.22}$$

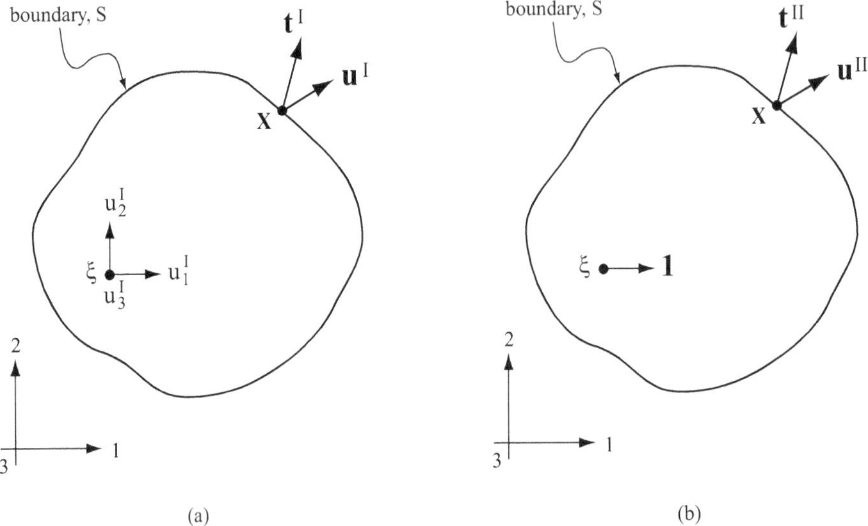

Figure 5.8 Loading states. (a) State I Loading. (b) State II Loading.

In Equation (5.22), and for clarity, body forces and any initial state of stress have been neglected. The left-hand side of the equation is the work done by the unit force in state II, applied along the direction i at point ξ, with the displacements of the same point in state I. On the right-hand side of the equation, $t_j^I(x)$ are the tractions at the boundary at point X with coordinates x due to state I along the direction j; $u_{ij}^{II}(\xi, x)$ is the displacement component j at point X due to the unit concentrated load applied at point ξ along direction i; $t_{ij}^{II}(\xi, x)$ is the state II traction at the boundary at point X, along direction j due to a unit load applied at point ξ along the i direction; $u_j^I(x)$ is the state I displacement along direction j of point X; and dS is the elemental boundary area for 3D problems or elemental boundary length for 2D problems. The notation used follows that from Venturini (1983) and Bittnar and Sejnoha (1996).

The solution of Kelvin's problem (state II loading) is given by (see e.g. Timoshenko and Goodier, 1970; Soutas-Little, 1973):

$$
\begin{aligned}
u_{ij}^{II} &= \frac{1}{16\pi(1-\nu)Gr}\left[(3-4\nu)\delta_{ij} + \frac{x_i}{r}\frac{x_j}{r}\right] \\
\sigma_{ijk}^{II} &= -\frac{1}{8\pi(1-\nu)r^2}\left[(1-2\nu)\left(-\frac{x_i}{r}\delta_{jk} + \frac{x_j}{r}\delta_{ik}\frac{x_k}{r}\delta_{ij}\right) + 3\frac{x_i}{r}\frac{x_j}{r}\frac{x_k}{r}\right]
\end{aligned}
\tag{5.23}
$$

In the equation, u_{ij}^{II} is the displacement in the j direction (Figure 5.9) at point X due to a unit load in direction i. The point X has coordinates x_1, x_2, and x_3 relative to the point of application of the load and r is the distance between point X and the point of application of the load; see Figure 5.9 that depicts a unit load applied along the direction i = 3. σ_{ijk}^{II} is the stress σ_{jk} at point X due to a unit load in direction i. δ_{ij} is the Kronecker delta ($\delta_{ij} = 1$ when i = j, $\delta_{ij} = 0$ when i ≠j). For 2D, plane strain, Equations (5.23) are:

$$
\begin{aligned}
u_{ij}^{II} &= \frac{1}{8\pi(1-\nu)G}\left[-(3-4\nu)\ln r\,\delta_{ij} + \frac{x_i}{r}\frac{x_j}{r}\right] \\
\sigma_{ijk}^{II} &= -\frac{1}{4\pi(1-\nu)r}\left[(1-2\nu)\left(-\frac{x_i}{r}\delta_{jk} + \frac{x_j}{r}\delta_{ik}\frac{x_k}{r}\delta_{ij}\right) + 2\frac{x_i}{r}\frac{x_j}{r}\frac{x_k}{r}\right]
\end{aligned}
\tag{5.24}
$$

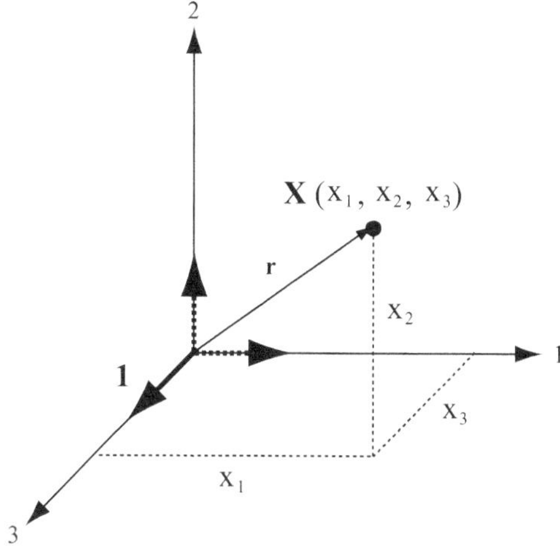

Figure 5.9 Coordinate system used for Kelvin's problem.

The traction components on a surface that has an outward normal n_k are given by:

$$t_{ij} = \sigma_{ijk} n_k \tag{5.25}$$

Thus all terms with superscript II in Equation (5.22) are known. The unknowns correspond to terms with superscript I. Note that Equation (5.22) gives three equations (in 3D), one for each of the components i = 1,2,3 (two equations in 2D). If the tractions and displacements at the boundary due to loading stage I are known, then displacements at any point ξ inside the medium can be obtained using Equation (5.22). Stresses can be computed from displacements, given the fundamental relation between stresses, strains, and displacements. For example in elasticity (e.g. Timoshenko and Goodier, 1970):

$$\sigma_{ij}^{I} = \frac{2\nu G}{1-2\nu}\delta_{ij}\frac{\partial u_k^I}{\partial x_k} + G\left(\frac{\partial u_i^I}{\partial x_j} + \frac{\partial u_j^I}{\partial x_i}\right) \tag{5.26}$$

It is at the boundaries where tractions or displacements are imposed. Thus, it is convenient to apply Equation (5.22) to points ξ and X at the boundary. As the point of application of the load ξ approaches point X a singularity occurs in the integrals. The integrals can be evaluated by dividing the boundary into two domains: the domain close to the load ε, and the rest of the boundary S-ε, as depicted in Figure 5.10a. Taking the limit $\varepsilon \to 0$ in the integrals results in the following expression of Equation (5.22):

$$c_{ij}(\xi)u_j^I(\xi) = \int_S t_j^I(x)u_{ij}^{II}(\xi,x)dS(x) - \int_S t_{ij}^{II}(\xi,x)u_j^I(x)dS(x) \tag{5.27}$$

In (5.27) the coefficient c_{ij} depends on the shape of the boundary around point ξ. In plane strain the coefficient is (Venturini, 1983; Bittnar and Sejnoha, 1996):

$$c_{ij}(\xi) = \frac{\psi(\xi)}{2\pi}\delta_{ij} \tag{5.28}$$

$\psi(\xi)$ is the angle, in radians, shown in Figure 5.10b, which corresponds to the interior angle between the tangents to the boundary in the vicinity of point ξ. For a smooth boundary, $\psi = \pi$ and the coefficient is equal to 1/2. For an interior point, $\psi = 2\pi$, the coefficient is one and Equation (5.22) is recovered. For an exterior point, c = 0.

In the BEM, the boundary of the continuum is divided into elements. For a 3D continuum the boundary elements are surface, 2D elements and for a 2D continuum the elements at the boundary are line, 1D elements. Thus the BEM method reduces by one the dimension of the problem. Figure 5.11 illustrates a 2D continuum with line elements.

The integrals in Equation (5.27) are then evaluated for each element, and their contribution over the boundary S is obtained as a discrete sum of element integrals. Equation (5.29) is the result of such discretization.

$$c_{ij}(\xi)u_j^I(\xi) = \sum_{m=1}^{M}\int_{S^m} t_j^I(x)u_{ij}^{II}(\xi,x)dS^m(x) - \sum_{m=1}^{M}\int_{S^m} t_{ij}^{II}(\xi,x)u_j^I(x)dS^m(x) \tag{5.29}$$

In the equation, the integrals over the entire boundary are obtained as the summation of M integrals over each element. S^m is the area (2D) or length (1D) of element m.

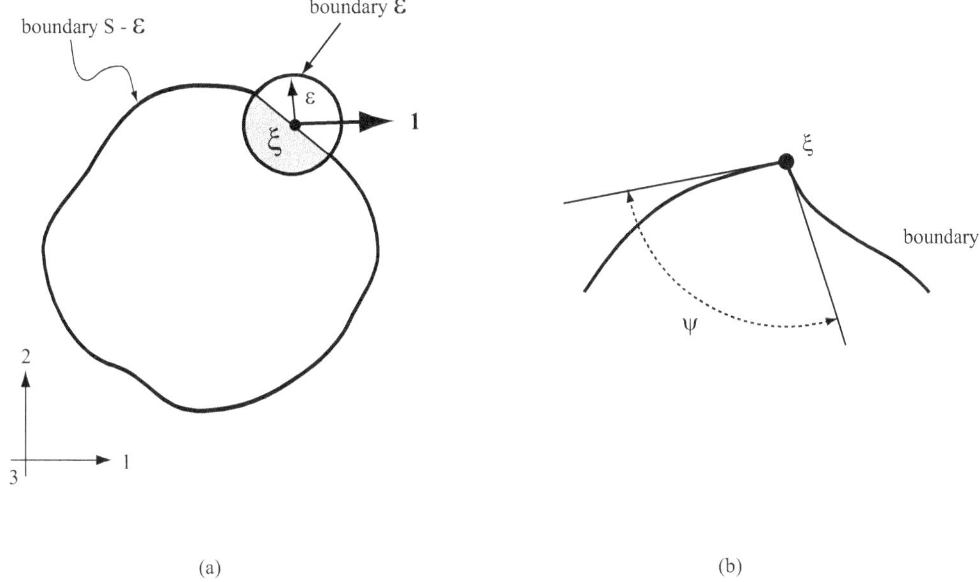

Figure 5.10 Determination of coefficients C_{ij}. (a) Integration around singularity point. (b) Angle ψ at a boundary point.

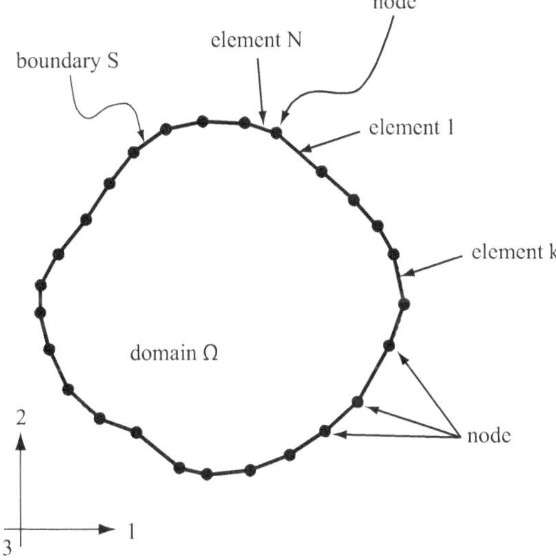

Figure 5.11 Example of discretization with boundary elements in 2D (Bobet, 2010c). (Republished with permission of Springer nature VB.)

Equation (5.29) is applied to a characteristic point in each element (also called collocation point); e.g. the point at the center of the element. For a 2D problem, there are two equations that can be generated at each point; one equation per direction of application of the load, i = 1,2 (three equations in 3D). Since there are M elements, the total number of equations generated is 2M (3M in 3D). At the boundaries, a normal and a shear traction (a normal and two shear tractions in 3D) or a displacement normal or parallel to the plane of the element (one normal and two parallel displacements in 3D) can be applied. For a well-posed problem, tractions and displacements cannot be imposed at the same point. Let's assume, for clarity of the argument, that tractions are imposed and displacements are unknown and both are constant over the length of the element (or the length of the element is small enough such that the parameters can be considered constant). In this case the stresses t_j^I are known while the displacements u_j^I are unknown. Since there are M elements and one collocation point per element, the total number of unknowns is 2M (3M in 3D). The problem can be solved since there are the same number of equations as unknowns.

The final form of the Boundary Element Method is based on the idea of interpolation functions, which is borrowed from the Finite Element Method formulation. Coordinates, displacements, and tractions at any point inside an element are expressed, through the interpolation functions, in terms of the corresponding values at the nodes of the elements, as it is done in (5.17).

The interpolation functions are generally the same as those used for the Finite Element Analysis (Section 5.3), and are expressed in a coordinate system local to the element (Figure 5.6). In terms of nodal parameters and element local coordinate system, Equation (5.29) is written, in matrix form, as:

$$C(x)\hat{U}^I(x) = \sum_{m=1}^{M}\int_{S^m} H^m(x)\hat{T}^I(x)U^{II}(\xi,x)\det J\, dS^m - \sum_{m=1}^{M}\int_{S^m} T^{II}(\xi,x)H^m(x)\hat{U}^I(x)\det J\, dS^m \tag{5.30}$$

where now points X and ξ at the boundary are expressed in terms of the local coordinates of the element. J is the Jacobian of the coordinate transformation from the global system to the system local to the element, i.e. Equation (5.21). In matrix form, Equation (5.30) is written as:

$$A\hat{U} = B\hat{T} \tag{5.31}$$

The matrix A groups the left-hand side of Equation (5.30) and the second integral on the right-hand side. The matrix B is the first integral on the right-hand side. Note that since u and t are nodal values, they are constant over the element and can be moved outside the integral. The integrals are singular and are computed numerically (Davies, 1993). The result is a system of linear equations, which in general is non-symmetric. The system is of order 2M for 2D or 3M for 3D problems, which can be used to determine displacements at the boundary given the tractions, or in more general terms, to obtain unknown displacements and tractions at the boundary given the known displacements and tractions imposed at the boundary.

Displacements at any point inside the medium can be obtained from Equation (5.30) with the point ξ being the point of interest. In other words, the functions u_{ij}^{II} and t_{ij}^{II} are evaluated at point ξ. Since this is an interior point, $c_{ij} = 1$.

Stresses are obtained from Equation (5.26), which after some algebraic manipulation takes the form:

$$\sigma(\xi) = \sum_{m=1}^{M}\int_{S^m} D^{II}(\xi,x)H^m(x)\hat{T}^I(x)\det J dS^m - \sum_{m=1}^{M}\int_{S^m} S^{II}(\xi,x)H^m(x)\hat{U}^I(x)\det J dS^m \tag{5.32}$$

where the components of the matrices $D^{II}(\xi,x)$ and $S^{II}(\xi,x)$ are, respectively:

$$d^{II}_{ijk}(\xi,x) = \frac{2\nu}{1-2\nu}\frac{\partial u^{II}_{kl}(\xi,x)}{\partial x_l}G\delta_{ij} + G\left(\frac{\partial u^{II}_{ki}(\xi,x)}{\partial x_j} + \frac{\partial u^{II}_{kj}(\xi,x)}{\partial x_i}\right)$$

$$s^{II}_{ijk}(\xi,x) = \frac{G}{2\pi\alpha(1-\nu)r^{\beta}}\left\{\beta\left[(1-2\nu)\delta_{ij}\frac{x_k}{r} + \nu\left(\delta_{ik}\frac{x_j}{r} + \delta_{jk}\frac{x_i}{r}\right) - \gamma\frac{x_i}{r}\frac{x_j}{r}\frac{x_k}{r}\right]\frac{x_i}{r}n_i\right.$$
$$\left. + \beta\nu\left(n_i\frac{x_j}{r}\frac{x_k}{r} + n_j\frac{x_i}{r}\frac{x_k}{r}\right) + (1-2\nu)\left(\beta n_k\frac{x_i}{r}\frac{x_j}{r} + n_j\delta_{ik} + n_i\delta_{kj}\right) - (1-4\nu)n_k\delta_{ij}\right\} \quad (5.33)$$

In Equation (5.33) $\alpha = 2$, $\beta = 3$ and $\gamma = 5$ in 3D, and $\alpha = 1$, $\beta = 2$ and $\gamma = 4$ in 2D; x_i is the relative coordinate between points x and ξ; and n_i are the components of the normal to the plane where stresses are computed.

Figure 5.12 shows an example of a mesh that could be used to address 3D problems regarding a tunnel opening without support. In this particular case, the boundary conditions consist of the far-field stresses and zero normal and shear stresses at the tunnel perimeter and face. Figure 5.12 illustrates a common technique of increasing the density of elements where the solution is of interest (e.g. near the face in this case) or where the stress gradient is the largest. As the distance from the face of the tunnel increases, the size of the elements may be increased as the precision required decreases. The discretization should reach far enough from the tunnel face such that it does not influence the results. Special elements such as infinite elements can also be used for this purpose.

The approach formulated for homogeneous materials can be applied to problems involving regions with different elastic properties, or to problems with frictional discontinuities. In these cases, the domain is divided into as many sub-domains as homogeneous materials exist, or in sub-domains bound by discontinuities. Figure 5.13a shows a tunnel in a homogeneous elastic medium that is crossed by a major frictional discontinuity. The continuum is sub-divided in two regions, regions 1 and 2 at each side of the discontinuity (Figure 5.13b). Equations (5.30) or (5.31) in matrix form can be directly applied to each of the regions using the discontinuity as part of their boundary. Hence Equation (5.31), for each region, is:

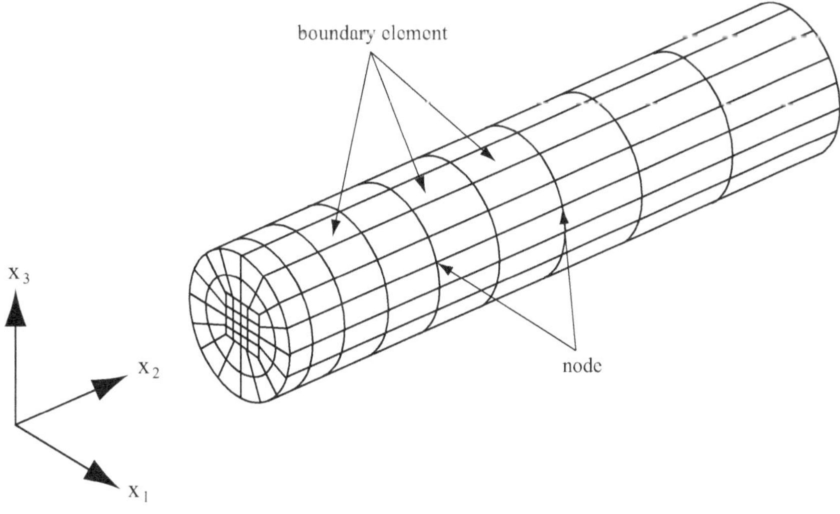

Figure 5.12 Example of discretization with boundary element methods.

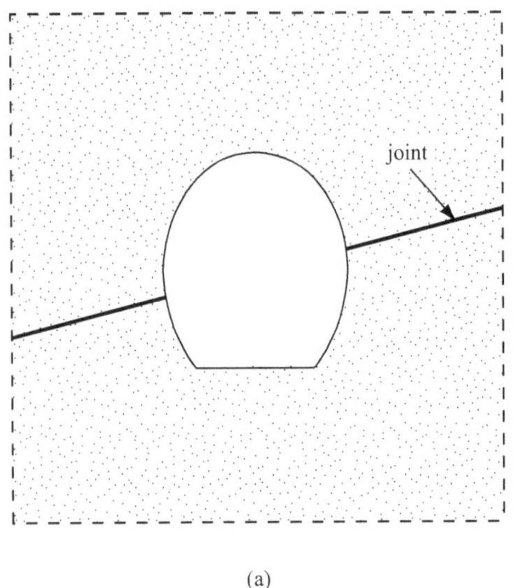

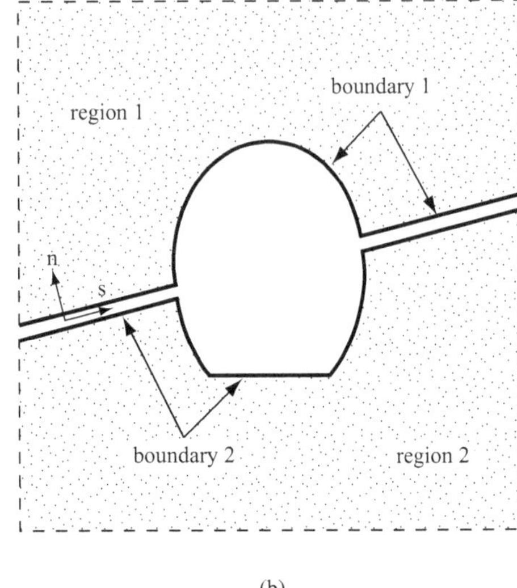

Figure 5.13 BEM applied to a tunnel in continuum medium with joints. (a) Continuum with a Joint. (b) Discretization.

$$A^1\hat{U}^1 = B^1\hat{T}^1$$
$$A^2\hat{U}^2 = B^2\hat{T}^2 \quad (5.34)$$

The superscripts 1 and 2 correspond to each region. Equilibrium of forces at the interface requires:

$$t_n^1 = t_n^2$$
$$t_s^1 = t_s^2 \quad (5.35)$$

where t_n and t_s are the tractions acting at the boundary of each region in the direction normal and parallel to the boundary, respectively. Note that the positive n and s directions may change from boundary to boundary, in which case the appropriate sign should be used in (5.35). Normal and shear tractions can be related to displacements through the normal and shear stiffness of the discontinuity as follows:

$$t_n = K_n(u_n^1 - u_n^2) + t_n^0$$
$$t_s = K_s(u_s^1 - u_s^2) + t_s^0 \quad (5.36)$$

K_n and K_s are the normal and shear stiffness, respectively, of the joint; u_n and u_s are the displacements normal and parallel to the discontinuity, and t_n^0 and t_s^0 are the initial normal and shear tractions. If there is no slip along the discontinuity, Equations (5.34), (5.35), and (5.36) provide the complete solution of the problem. If there is slip, the maximum shear traction along the discontinuity is determined by the friction law chosen. For a Coulomb-type friction,

$$|t_s| \le c + t_n \tan\phi \quad (5.37)$$

c is the cohesion and ϕ is the friction angle; in the equation, t_n is compression.

Boundary Element Methods are particularly well suited to address static continuum problems with small boundary to volume ratio, with elastic behavior, and with stresses or displacements applied to the boundaries. Actual problems may not always conform to these limitations. For example, rocks and soil deposits may undergo significant yielding under moderate stresses, gravity forces may be significant for shallow tunnels, and inertia may play an important role with dynamic loading (e.g. blasting, earthquake).

Significant work has been done to include within the BEM body forces (e.g. Venturini, 1983; Bittnar and Sejnoha, 1996) plastic behavior (e.g. Banerjee et al., 1979, 1980; Telles and Brebbia, 1981a, b; Venturini, 1983), and inertia forces (e.g. Mansur and Brebbia, 1983, 1985); a comprehensive historical review of development and implementation of the numerical techniques is provided by Manolis et al. (1993).

Dynamic and body forces require integration over the entire volume domain which leads to the need for discretization of the entire continuum. The plasticity algorithms require integration at least over the volume of the material that undergoes yielding and convergence of the solution, as with FEM, is attained through iteration. With plastic deformations and with cases where integration needs to be extended over part or the entire volume, the advantage that the BEM offers regarding limited discretization of the continuum may be lost. Efficient hybrid BEM-FEM solutions are possible, where a FEM discretization is used for those parts of the continuum where plastic deformations occur (i.e. in tunnels, near the opening), while Boundary Elements are used in elastic regions (i.e. far from the tunnel). The advantage of the coupled (hybrid) FEM-BEM is reduced discretization and automatic satisfaction of boundary conditions at infinity. The challenge of the hybrid approach is the generation of nodal forces and displacements from the BEM that are consistent with those of FEM, and also that the resulting stiffness matrix is non-symmetric (in contrast with FEM where the stiffness matrix is symmetric). Figure 5.14 shows an example of a hybrid discretization, where the tunnel liner and a volume of the ground next to the tunnel, where plastic deformations occur, are discretized with Finite Elements. Far from the tunnel, and where the deformations are elastic, Boundary Elements are used. As mentioned, boundary elements automatically satisfy far-field stresses and thus a hybrid method that combines both FEM and BEM may have the advantage of a simplified mesh and more economic computation time and smaller storage. Similar benefits, however, can be obtained by using FEM and infinite elements at the boundaries rather than with the Boundary Elements.

The BEM has not been used as extensively as other methods such as the FEM for the analysis of tunnels and other underground excavations. Perhaps one of the first instances where the method was used was by Gioda (1984), who computed stresses around a deep unlined tunnel in an infinite elastic medium. Since then the method has been used for tunnels in isotropic elastic rock masses with a limited number of discontinuities (Brady, 1987; Pande et al., 1990; Beer, 1993), in elasto-plastic isotropic media (Carrer and Telles, 1993), in anisotropic media (Pan et al., 1998; Tonon and Amadei, 2002; Chuhan et al., 2004), and in tunnels subjected to dynamic loading in an elasto-plastic medium (Carrer and Telles, 1993; Beskos, 1993). Where the BEM appears to find application is in conjunction with the FEM through a hybridized combination of the two methods. Hybrid BEM-FEM has been utilized to address tunnel problems in elastic and elasto-plastic ground medium for static (Gioda and Carini, 1985; Ushijima and Einstein, 1985; Beer, 1993; Eberhardsteiner et al., 1993; Yuan et al., 2000) and dynamic loading (Esmaeili et al., 2006).

5.4.2 The indirect Boundary Element Method

The method received considerable attention after the work of Crouch and Starfield (1983) and others (e.g. Wiles and Curran, 1982; Chan et al., 1990), even though earlier applications exist, mostly in mining (Berry and Sales, 1962). The work by Crouch and Starfield and others

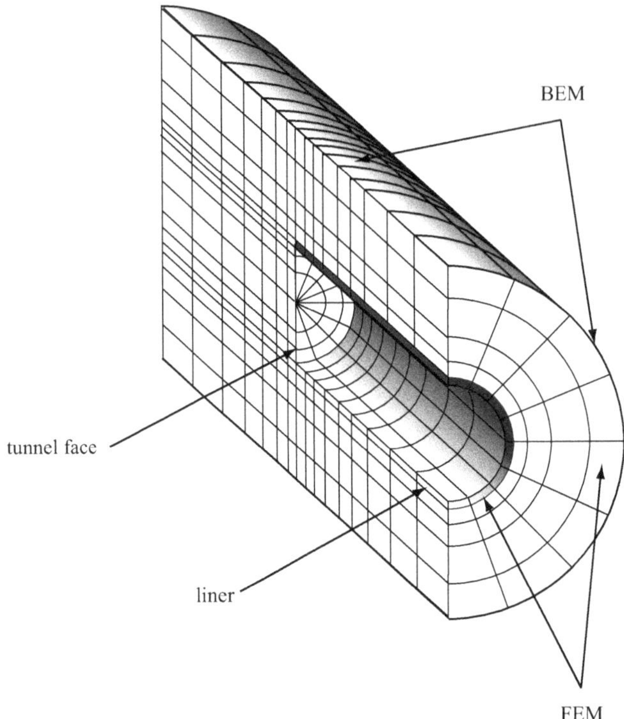

Figure 5.14 Hybrid BEM – FEM mesh (Bobet, 2010c). Republished with permission of Springer nature VB.

provided a full mathematical formulation and a framework to implement the methodology into usable code. Since then, the method has found a niche in the excavation of tabular bodies and in problems involving joints and cracks, even though some examples can be found in the literature involving tunneling (Crouch and Starfield, 1983; Shou, 2000).

The method is based on a solution in terms of fictitious parameters which then can be used to obtain stress and displacement fields anywhere in the continuum. The method is particularly suited for problems with features that have one dimension much smaller than the other two, such as discontinuities. It is exclusively used in linear elastic and homogeneous media. In these problems, the direct Boundary Element Method may result in ill-conditioned sets of equations as the two sides of the discontinuity are very close to each other.

There are two types of Indirect Boundary Element Methods: the Displacement Discontinuity Method (DDM) and the Stress Discontinuity Method (SDM). Both methods are based on the premise that the boundaries of the continuum can be discretized into small segments or elements, and there exists an analytical formulation that provides stresses and displacements anywhere in the continuum due to the tractions or displacements imposed at the boundaries. Figure 5.15 shows the discretization with discontinuity elements of a 2D rectangular body with a fracture.

Figure 5.16 shows a Displacement Discontinuity Element (DDE) and a Stress Discontinuity Element (SDE). The DDE can have relative displacements between its upper and lower surfaces. In other words, there is a displacement jump (discontinuity) across the element. Figure 5.16a shows a constant displacement discontinuity element. The displacement discontinuity has magnitude D_1 in the direction parallel to the element (shear direction), and of magnitude D_2 perpendicular to the element (normal direction). In SDEs, a constant line load with components P_1 and P_2 (shear and normal) is applied along the element, which results in a traction jump between the upper and lower surface equal to the magnitude of the applied

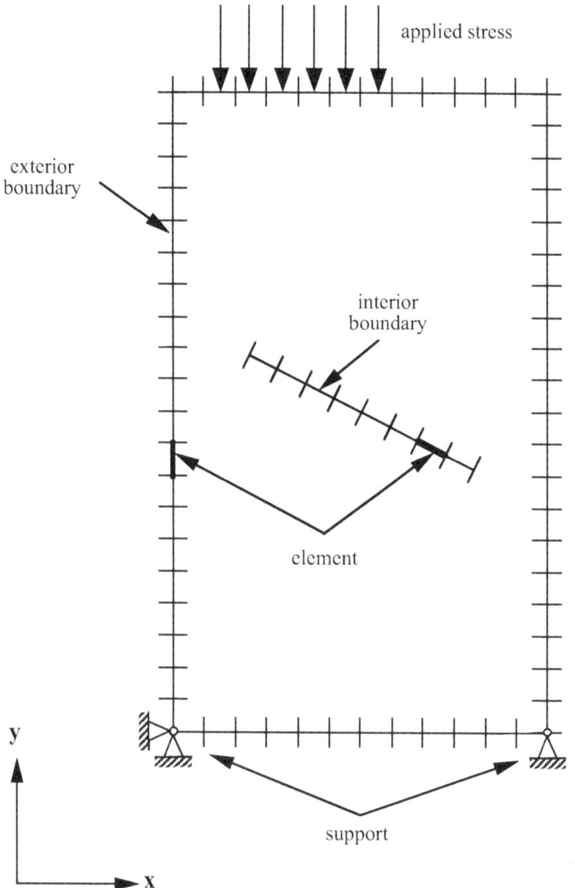

Figure 5.15 Example of 2D discretization with discontinuity elements.

force (Figure 5.16b). The magnitudes of the displacement discontinuities D_1 and D_2 and of the force components P_1 and P_2 are the fictitious parameters used for the solution. Note that the parameters have physical meaning, as they represent the closure and shear displacements of the element and the tractions applied to the element.

The stresses and displacements produced by a constant displacement discontinuity element (CDDE) are obtained from the solution of a semi-infinite dislocation (Figure 5.17a). For the semi-infinite dislocation, stresses and displacements are expressed in terms of the fictitious parameters as (only non-zero values are reported):

$$
\begin{aligned}
\sigma_{ss} &= CD_1 n \frac{3s^2+n^2}{(s^2+n^2)^2} + CD_2 s \frac{n^2-s^2}{(s^2+n^2)^2} \\
\sigma_{sn} &= -CD_1 s \frac{s^2-n^2}{(s^2+n^2)^2} + CD_2 n \frac{n^2-s^2}{(s^2+n^2)^2} \\
\sigma_{nn} &= -CD_1 n \frac{s^2-n^2}{(s^2+n^2)^2} - CD_2 s \frac{3n^2+s^2}{(s^2+n^2)^2} \\
u_s &= \frac{D_1}{2\pi}\left[\operatorname{atan}(n,s) - \pi + \frac{sn}{2(1-\nu)(s^2+n^2)}\right] + \frac{D_2}{2\pi}\left[\frac{1-2\nu}{4(1-\nu)}\ln(s^2+n^2) - \frac{s^2-n^2}{4(1-\nu)(s^2+n^2)}\right] \\
u_n &= -\frac{D_1}{2\pi}\left[\frac{1-2\nu}{4(1-\nu)}\ln(s^2+n^2) + \frac{s^2-n^2}{4(1-\nu)(s^2+n^2)}\right] + \frac{D_2}{2\pi}\left[\operatorname{atan}(n,s) - \pi - \frac{sn}{2(1-\nu)(s^2+n^2)}\right]
\end{aligned}
\tag{5.38}
$$

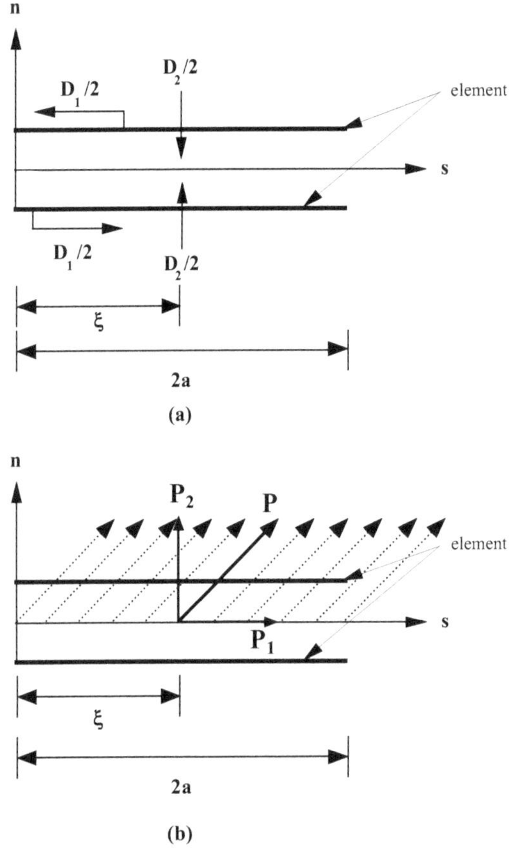

Figure 5.16 Constant discontinuity elements. (a) DDE. (b) SDE.

where $C = \frac{E}{4\pi(1-\nu^2)}$, E is the Young's modulus, and ν the Poisson's ratio; s and n are the coordinates of the point of interest with respect to the coordinate system attached to the dislocation (Figure 5.17a). The function "atan (n,s)" is the inverse tangent of the ratio of "n" over "s"; the angle is measured counterclockwise with origin at the "s" axis and takes values between "0" and 2π. Equation (5.38) applies to plane strain conditions, which is the problem of interest since it is assumed that one of the dimensions of the crack is much larger than the other two.

Figure 5.17b shows how to obtain the stress and displacement influence functions for a DDE. A finite dislocation of length "2a" (the total length of a crack is usually denoted as "2a") is obtained from two semi-infinite dislocations: one with discontinuities D_1 and D_2 and the other one with the same magnitudes but with opposite sign. The two semi-infinite dislocations are placed along the "s" axis and offset by a distance "2a" from each other. For $s < 2a$, the displacement magnitudes D_1 and D_2 are recovered; for $s > 2a$, the values of D_1 and D_2 are zero since the two semi-infinite dislocations cancel out.

Following the same reasoning, stresses and displacements can be obtained for elements with distributions of displacement discontinuities other than constant. For example the discontinuity distribution for an element with displacement discontinuities increasing linearly (LDDE) from its origin (D_{i0} at $s = 0$; $i = 1,2$) to its end (D_{if} at $s = 2a$; $i = 1,2$) can be expressed as:

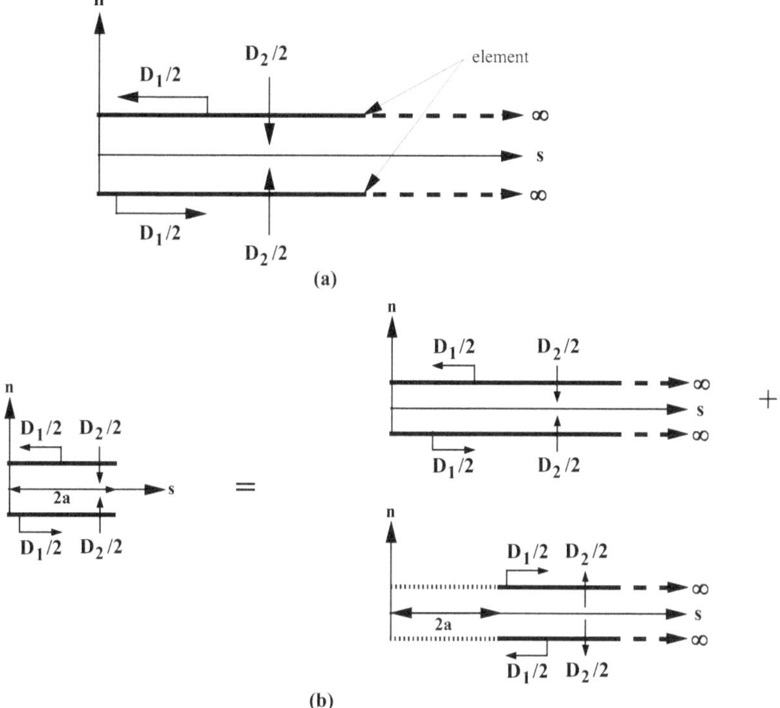

Figure 5.17 Dislocations. (a) Semi-infinite dislocation. (b) Finite dislocation from two semi-infinite dislocations.

$$D_i = D_{i0} + \frac{D_{if} - D_{i0}}{2a}\xi \text{ for } 0 < \xi < 2a \tag{5.39}$$

where ξ is the distance from the origin to the point of interest along the discontinuity (Figure 5.16). Stresses and displacements induced by a LDDE can be found by the superposition of three problems: (1) a semi-infinite dislocation with discontinuity D_{i0} with origin at the origin of coordinates; (2) an infinite number of semi-infinite dislocations with discontinuity $dD_i = \frac{D_{if} - D_{i0}}{2a} d\xi$ with origin at a distance $s = \xi$ from the origin of coordinates; and (3) a semi-infinite dislocation with discontinuity $-D_{if}$ with origin at $s = 2a$.

Since the DDM is often used for problems involving cracks and crack propagation, it is useful to have elements that can reproduce the singularity created at the tips. This can be done with an element (RRE) with a square root distribution of displacement discontinuity of the form:

$$D_i = D_{i0}\sqrt{\frac{\xi}{a}} \quad \text{for } 0 < \xi < 2a \tag{5.40}$$

This element reproduces the $\frac{1}{\sqrt{r}}$ stress singularity at the tip of a crack (r is the radial distance from the tip). Similar to the LDDE, the solution is found by superposition of two problems: (1) An infinite number of semi-infinite dislocations with discontinuity $dD_i = \frac{D_{i0}}{2\sqrt{a}}\frac{d\xi}{\sqrt{\xi}}$

at a distance s = ξ; and (2) a semi-infinite dislocation with discontinuity $-\sqrt{2}D_{i0}$ with origin at s = 2a.

Stresses and displacements induced by a constant SDE can be found by integration of the solution of a line load distributed over the length of the element (Figure 5.16b). Stresses and displacements due to a line load are given in (5.24). As with DDEs, stresses and displacements for elements with load distributions other than constant can be obtained. For example the fictitious parameter for elements with a linear distribution of load (LSDE) is:

$$P_i = P_{i0} + \frac{P_{if} - P_{i0}}{2a}\xi \quad \text{for} 0 < \xi < 2a \tag{5.41}$$

where P_{i0} is the force acting at the origin of the element along the direction i, P_{if} is the force along direction i at the end of the element, and ξ is the distance from the origin of the element. The solution is found by direct integration of (5.24) over the length of the element with $dP_i = \left[P_{i0} + \frac{P_{if} - P_{i0}}{2a}\xi\right]d\xi$.

If the boundary (both exterior and interior) is divided into elements, and each element has one or more collocation (reference) points ordered from 1 to M (e.g. CDDE, RRE and SDE have one collocation point because the stress or displacement discontinuity distribution along the element is defined with only one parameter, while LDDE and LSDE have two collocation points since the distribution is defined with two parameters: the value of the stress or displacement at the origin and at the end of the element), then the stresses and displacements at collocation point "m" in the coordinate system local to the element (see Figure 5.16) are given by:

$$\sigma_{m,k} = \{\sigma_{m,s}, \sigma_{m,n}, u_{m,s}, u_{m,n}\} \quad \text{for} m = 1, M \quad \text{and} k = 1, 4 \tag{5.42}$$

For multiple elements, the stresses or displacements at any given point are obtained by linear superposition of the contributions of all the elements. Thus,

$$\sigma_{m,k} = A_{m,k,j} V_j. \tag{5.43}$$

where V_j is the fictitious parameter "j," with j ranging from 1 to 2M (there are two fictitious parameters for each collocation point; here V_j denotes either the displacement discontinuity D_j or the stress discontinuity P_j). $A_{m,k,j}$ is the influence function of the fictitious parameter "j," of the stress "k" of the collocation point "m." The influence functions are obtained from integration of the fundamental solutions for an infinite dislocation (5.38) for DDEs or for a line load (5.24) with the assumptions appropriate to each element, e.g. constant displacement or stress distribution, linear, parabolic, square root.

Friction along discontinuities can be easily implemented by adding to the system of equations the following relations:

$$\begin{aligned} D_{m,n} &= 0 \\ |\sigma_{m,s}| &= \mu \sigma_{m,n} \end{aligned} \tag{5.44}$$

which indicate that the normal displacement discontinuity must be zero at collocation point m, since the discontinuity is closed, and that the shear stress is equal to the coefficient of friction, μ, times the normal stress. Further, the magnitude of D_n can be monitored to determine whether the discontinuity opens or closes with loading.

In 2D, Equation (5.43) provides a linear system of 2M equations (3M in 3D) that apply to points at the boundary. At these points, since they are located at the boundary, tractions, displacements, or a suitable combination of the two must be known. This results in 2M unknowns (3M in 3D), and thus the system can be solved in terms of the fictitious parameters V_j. Once V_j are known, stresses and displacements at any point in the continuum can be obtained by superposition of the influence of each of the elements.

Codes have been written with SDEs or DDEs (e.g. Crouch and Starfield, 1983). Stress Discontinuity Elements are very well suited to discretize exterior boundaries where stresses are applied, while Displacement Discontinuity Elements are more efficient in modeling the body and the tips of cracks. Both SD and DD elements can be combined into a single, hybridized method that can take advantage of the particular characteristics of the two types of elements (Chan et al., 1990; Kuriyama et al., 1995; Shou, 2000; Bobet, 2001c; Bobet and Mutlu, 2005).

The method and associated algorithms have been expanded to three dimensions (Kuriyama et al., 1995; Cayol and Cornet, 1997; Shou et al., 1997; Vijayakumar et al., 2000), to inhomogeneous bodies and anisotropic media (Crouch and Starfield, 1983), and to dynamic analysis (Mack, 1993). The method has also been combined with other numerical procedures such as the Direct BEM (Cayol and Cornet, 1997).

Where the Indirect Boundary Element Method has been perhaps used with more success is in the investigation and modeling of fractured media (Shen and Stephansson, 1994; Shen et al., 1995; Bobet and Einstein, 1998; Vásárhelyi and Bobet, 2000; Bobet, 2000), mining (Bruneau et al., 2003), geological processes in discontinuous rock masses (Olson and Pollard, 1988, 1989, 1991; Pollard et al., 1990; Zeller and Pollard, 1992), and in slope stability (Scavia, 1995).

Figure 5.18 shows a comparison between the crack pattern obtained from experiments and the pattern simulated by a hybrid DD-SD method. The simulations have been done with the code FROCK (Chan et al., 1990; Bobet, 2000). In the example, two open fractures in a prismatic gypsum specimen are loaded in uniaxial compression (Bobet, 2001a). As a result, wing (tensile) and secondary (shear) cracks are produced at the tips of the flaws. With increasing loading, the cracks propagate and coalescence is produced by the linkage of the two internal secondary cracks with a tensile crack (Figure 5.18b). The flaws are modeled with 20 LDD elements with one RR element at each tip. New cracks are simulated by adding new DD elements along the path of propagation of the cracks.

The Indirect Boundary Element Method has definite advantages over other numerical methods such as the Finite Element Method (FEM) or the Finite Difference Method (FDM) in that only the boundary needs to be discretized and no remeshing is necessary as new cracks are formed during the analysis. However, the method can be used only on elastic materials where small-scale yielding at the tip of the cracks occurs. Similar to the Direct BEM, the solution is only obtained at the boundaries and additional computation is required to find stresses and displacements at other points.

5.5 DISCONTINUUM: THE DISTINCT ELEMENT METHOD

The Distinct Element Method (DEM) was introduced by Cundall (1971) as a model to simulate large movements in blocky rock masses, and then used for soils, which were modeled as discs (Cundall and Strack, 1979). Later on the method has been applied to spherical and polyhedral blocks (Cundall, 1987; Cundall, 1988; Hart et al., 1988; Pande et al., 1990; Cundall and Hart, 1992; Potyondy and Cundall, 2004) for both soils and rocks. The DEM belongs to the family of Discrete Element Methods, which Cundall and Hart (1992) define as

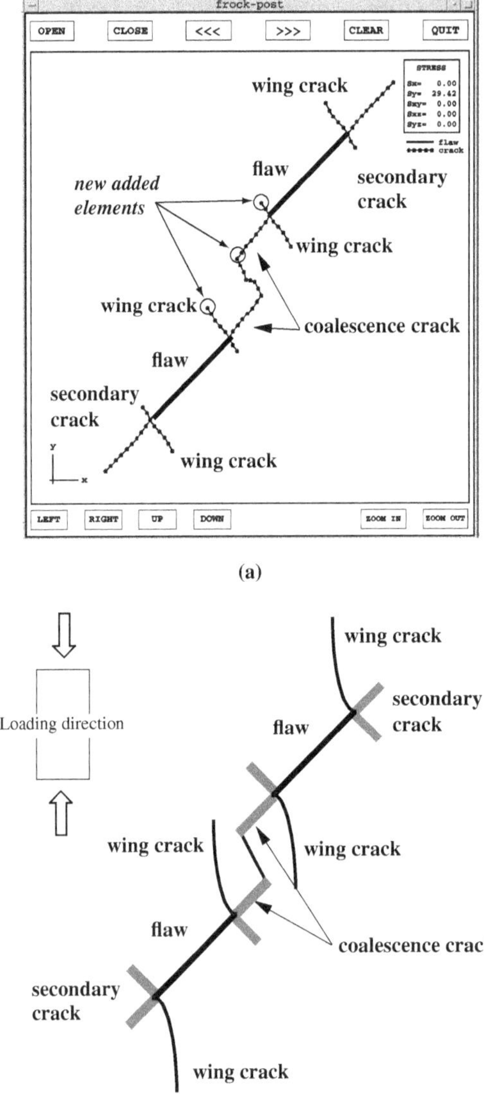

Figure 5.18 Model with DDM of coalescence between two open flaws in uniaxial compression. (a) Model. (b) Experiments.

those that: (1) allow finite displacements and rotations of discrete bodies, including detachment; and (2) automatically recognize new contacts between bodies during calculations. Discrete Element Methods need to address three key issues: (1) representation of contacts; (2) representation of solid material; and (3) detection and revision of contacts during execution. An in-depth discussion of these issues is provided by Cundall and Hart (1992).

In the DEM, it is assumed that the medium is divided by fully persistent discontinuities that delimit through their intersections a finite number of blocks, which in turn are interconnected through the discontinuities. Figure 5.19 provides an idealization of a discretization with the DEM of a medium with two sets of discontinuities. The following provides key concepts for the formulation of the DEM. For clarity, the discussion is restricted to two-dimensional discretizations with rigid bodies.

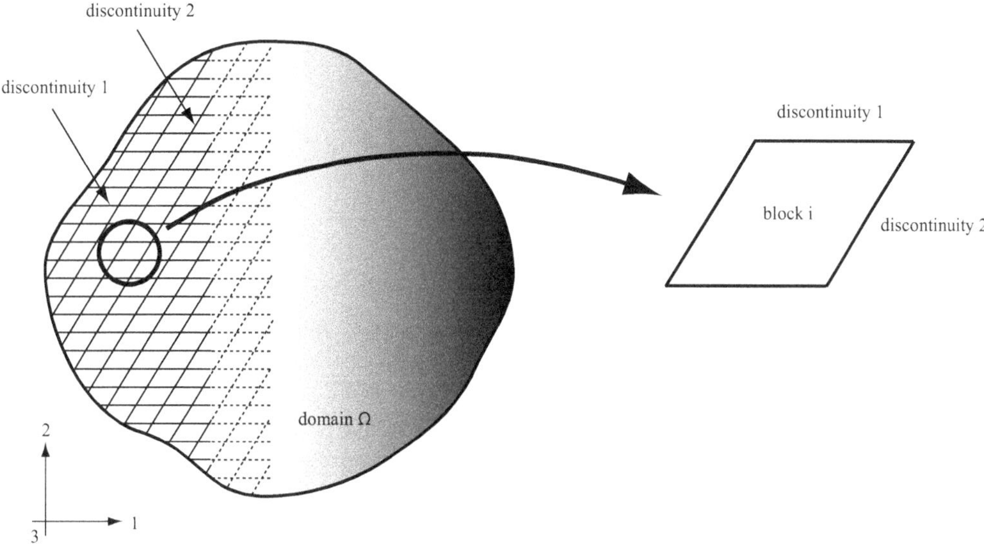

Figure 5.19 Distinct element method discretization (Bobet et al., 2009). (Republished with permission of American Society of Civil Engineers.)

A single block is subjected to forces arising from the contacts, if any, from the surrounding blocks and from internal forces (e.g. gravity). The displacement of the block is governed by Newton's second law of motion:

$$\begin{aligned} m\ddot{u}_i^t + c\dot{u}_i^t &= F_i^t \\ I\ddot{\theta}_i^t + c\dot{\theta}_i^t &= M_i^t \end{aligned} \tag{5.45}$$

where t is time, m is the mass, and I is the moment of inertia of the element, u_i is the displacement of the gravity center of the element in the direction i, $\ddot{u}_i$ and $\dot{u}_i$ are the acceleration and the velocity of the gravity center, $\ddot{\theta}$ and $\dot{\theta}$ are the angular acceleration and angular velocity of the element; c is the viscous damping, and F_i and M are the resultant force and moment applied at the center of gravity. In the DEM, Equation (5.45) is solved in the time domain using an explicit finite difference method. Using the central finite difference approximation, velocities and displacements are given by:

$$\begin{aligned} \dot{u}_i^{t+\Delta t/2} &= \left[D_1 \dot{u}_i^{t-\Delta t/2} + \frac{F_i^t}{m}\Delta t \right] D_2 \\ \dot{\theta}^{t+\Delta t/2} &= \left[D_1 \dot{\theta}^{t-\Delta t/2} + \frac{M^t}{I}\Delta t \right] D_2 \\ D_1 &= 1 - \frac{c}{m}\frac{\Delta t}{2} \\ D_2 &= \frac{1}{1 + \frac{c}{m}\frac{\Delta t}{2}} \\ u_i^{t+\Delta t} &= u_i^t + \dot{u}_i^{t+\Delta t/2}\Delta t \\ \theta^{t+\Delta t} &= \theta^t + \dot{\theta}^{t+\Delta t/2}\Delta t \end{aligned} \tag{5.46}$$

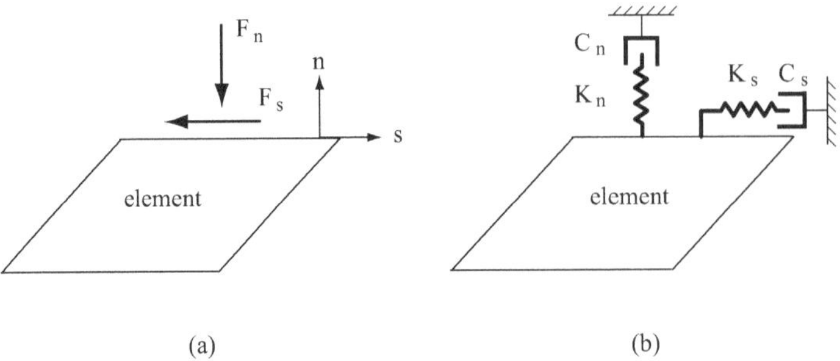

Figure 5.20 Forces at the boundary of DEM elements (Bobet, 2010c). (a) Contact forces. (b) Contact forces idealization. (Republished with permission of Springer nature VB.)

The forces acting at each element boundaries are originated by the interaction of the element with the surrounding elements. At each boundary, a normal and a shear force appear as the result of the relative movements between the two elements that share the discontinuity. The forces at the interface are obtained using a penalty method where the magnitude of the forces is related to the relative movements between the two elements and the stiffness of the discontinuity. Figure 5.20a shows the positive forces at the top of the element, and Figure 5.20b shows an idealization of the contact between blocks. The normal force is proportional to the relative movement of the two blocks across the contact and along the normal direction. The shear force is proportional to the relative movement along the direction of the contact. Expressions for the forces are:

$$\begin{aligned} F_n^{t+\Delta t} &= F_n^t - K_n \Delta u_n^{\Delta t} A_c - \beta K_n \Delta \dot{u}_n^{\Delta t} A_c \\ F_s^{t+\Delta t} &= F_s^t - K_s \Delta u_s^{\Delta t} A_c - \beta K_s \Delta \dot{u}_s^{\Delta t} A_c \end{aligned} \tag{5.47}$$

K_n and K_s are the normal and shear stiffness of the contact (subscripts n and s refer to the directions normal and parallel to the discontinuity, respectively); Δu_n and Δu_s are relative displacements between the two elements, and A_c is the contact area. A damping factor, the third term on the right-hand side of the equation, is normally included to attenuate or prevent "rattling" of the contact between blocks. Damping (C_n and C_s in Figure 5.20) is often expressed as proportional to the normal and shear stiffness (βK_n and βK_s in (5.47)), but other expressions for damping have been proposed (e.g. damping proportional to the rate of change of the kinetic energy of the element; Cundall, 1987).

The magnitude of the shear force is limited by the constitutive relation used for the contact surface. For a Coulomb-type friction law,

$$F_s^{t+\Delta t} \le c A_c + F_n^{t+\Delta t} \tan\phi \tag{5.48}$$

where c and ϕ are the cohesion and friction angle of the contact surface. If the shear force obtained from (5.47) is larger than that from (5.48), it is reduced to the limiting magnitude given by (5.48), where the value of cohesion may be reduced to zero if the bonds along the discontinuity are deemed broken.

The calculations are performed from one state, where the solution is fully known, to another state in small time increments. The procedure is as follows (Hart et al., 1988): the law of motion is applied through Equation (5.46) with current forces to update the position of each element. As a result, the relative displacements and velocities at the contacts between elements are obtained. From the relative displacements, contact forces are updated using Equation (5.47) and new resultant forces and moments at the center of gravity of each element are computed. The cycle is repeated with small increments until the final solution is obtained. In the formulation, time can represent actual time when performing a dynamic analysis, or a fictitious parameter to represent loading increment from one loading stage to the next.

As with the Finite Difference Method, numerical stability requires a time increment smaller than the critical time step, which is given by (Hart et al., 1988);

$$\Delta t_{crit} = \kappa\sqrt{\frac{m_{min}}{2K_{max}}} \tag{5.49}$$

where m_{min} is the smallest element mass, K_{max} is the largest normal or shear stiffness in the discretization, and κ is a factor that takes into account the fact that an element may be in contact with more than one element. Hart et al. (1988) suggest a value for κ equal to 0.1.

Typical runs are completed with thousands of cycles involving very small time increments. The solution of Equations (5.46) and (5.47) is a forward process, and thus the computer time required in each cycle is very small; also, the storage information needed for each element is small. So the process discussed so far does not require intensive computation power or large storage capabilities. Where such requirements become significant is for the algorithm to recognize and keep track of all the contacts between elements during execution. A very simple procedure would be to compare the position of each element with the rest of the elements at the end of each cycle. For a discretization with n elements, this would require of the order of n^2 operations in each cycle, which would make the entire method impractical.

Considerable effort has been devoted to develop efficient algorithms, which on the one hand need to accurately describe the interaction between elements and on the other hand are not computationally intensive. The problem is complex as the algorithms need to identify not only what elements are in contact but also the type of contact: corner to corner, corner to edge, or edge to edge, since the magnitude and direction of the contact forces depend on the type of contact. A number of approaches have been proposed to identify contacts, such as global searching algorithms, buffer zone definition, contact or field zone, binary tree structures, space decomposition, and alternating digital tree (Hocking, 1992; Dowding et al., 1983; Cundall, 1988; O'Connor and Dowding, 1992; Ghaboussi and Barbosa, 1990). A comprehensive review of these methods is provided by Mohammadi (2003).

The following description applies to a procedure for convex elements proposed by Cundall (1988), which is very efficient and has been applied to a number of computer codes (e.g. UDEC, Universal Distinct Element Code; Lemos et al., 1985; Cundall, 1987; and LDEC, Livermore Distinct Element Code; Morris et al., 2002a, 2002b and 2004). The first step consists of the identification of the neighbors of each element. The space is divided into rectangular cells (Figure 5.21a provides a conceptual two-dimensional representation), and each element is mapped into the cells that occupies. Neighboring blocks can then be identified easily by looking at the entries into the cells, which are the identifiers of the blocks occupying the cells. As an element moves through the simulation, it is re-mapped and tested for new contacts with the new neighbor elements. Contact between two elements is determined by finding first the "common plane" that bisects the space between the two elements, and then test each block separately for contact with the common plane. Point C in Figure 5.21b is on the common plane, and it is initially placed at the midpoint between the centroids of elements A and B. The common plane (Figure 5.21b) is found by an iteration process in which the gap (or overlap) between the common plane and the closest corner of each element is maximized. This is done by translations and rotations of the contact plane about point C.

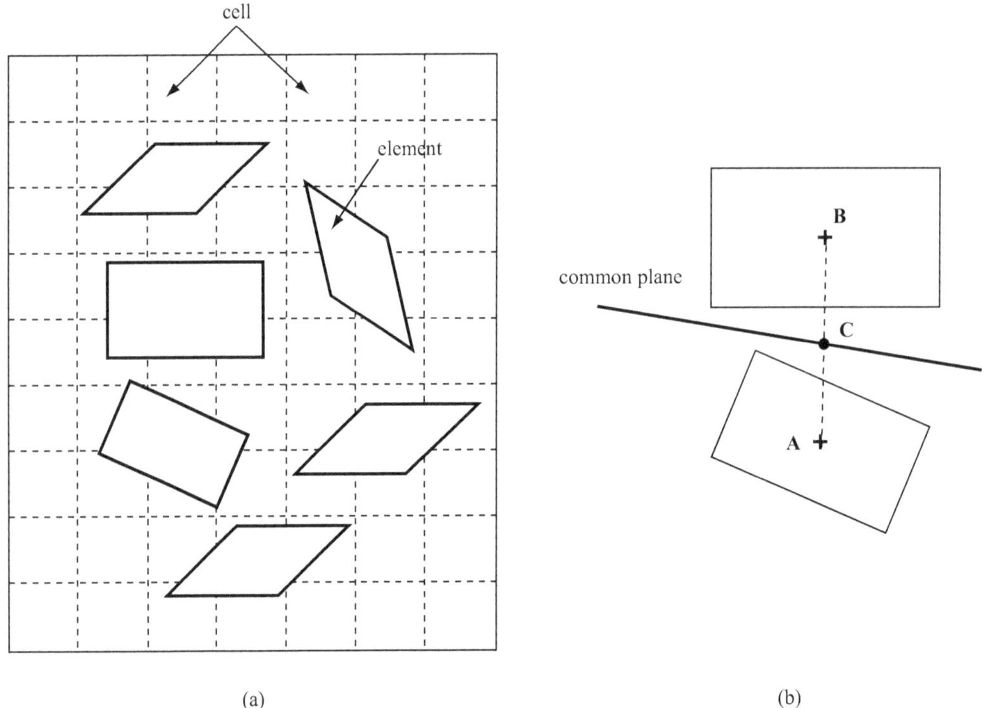

Figure 5.21 Contact detection between elements. (a) Element mapping. (b) Common plane.

Once the common plane is identified, simple tests between corners and the common plane can be done to check whether the coordinates of the corner are contained in the plane, cross the plane (overlap) or not (gap). An edge contact is produced when two corners lie on the plane, and a face contact when three corners are contained in the common plane. Point C is where the contact forces between the two elements (Equation (5.47)) are applied with the positive normal direction for element A, as the unit vector perpendicular to the common plane toward block B.

Contact identification and update is computationally very intensive and it is still an active topic of research. A method that is also computationally efficient involves the representation of elements as three-dimensional superquadratics (Williams and Pentland, 1992). The expressions for superquadratics are simple, and substitution of the coordinates of any point into the mathematical expressions provides a very simple means to identify the position of the point with respect to the element since a return value of zero means that the point is on the surface of the element, a negative value means that the point is inside and a positive value that the point is outside.

The Distinct Element Method is nowadays a very versatile and extensively validated procedure. It has been developed for full three-dimensional problems, and by discretizing the elements with Finite Difference or Finite Element meshes, can be applied to deformable bodies (Pande et al., 1990; Cundall and Hart, 1992; Eberhardt et al., 2004) and to fragmentation of discontinua (Hocking, 1992; Eberhardt et al., 2004). It can be used for static and for dynamic calculations (Taylor and Preece, 1992; Chen and Zhao, 1998).

Discrete Element Methods are extensively used in soil mechanics to model granular assemblies (e.g. Cook and Jensen, 2002; Konietzky, 2002; O'Sullivan et al., 2002, 2004, 2006) and in rock mechanics where the rock mass is modeled as an assembly of blocks connected through fully persistent joints (e.g. Cundall, 1987; Ishida et al., 1987; Pande et al., 1990) and to investigate stress changes due to the presence of discontinuities (Cundall, 1987; Su and Stephansson, 1999). The method has also found widespread application to tunnels excavated in discontinuous rock masses (O'Connor an Dowding, 1992; Bhasin et al., 1996; Shen and Barton, 1997; Bhasin and Hoeg, 1998; Konietzky, 2002; Dowding et al., 2000; MacLaughlin and Clapp, 2002; Morris et al., 2004; Gandhe et al., 2005; Hao and Azzam, 2005; Kemeny, 2005; MacLaughlin et al., 2006).

Figure 5.22a shows the discretization used to investigate the response of a tunnel in a discontinuous rock mass subjected to blast loading (Morris and Block, 2006; Bobet et al., 2009). Figures 5.22b and 5.22c show the response of the tunnel immediately after detonation and

Figure 5.22 DEM simulation of an underground structure subjected to dynamic loading (Morris and Block, 2006; Bobet et al., 2009). (a) Discretization of the rock mass. (b) Block displacement at t = 0 ms. (c) Block displacement at t = 30 ms. (Republished with permission of American Society of Civil Engineers.)

30 ms later. The simulations were run using parallel processing and the Livermore Distinct Element Code (LDEC), and consisted of 8 million blocks with approximately 100 million contacts.

5.6 DISCONTINUUM: DISCONTINUOUS DEFORMATION ANALYSIS

The Discontinuous Deformation Analysis (DDA) is a Discrete Element Method following the definition by Cundall and Hart (1992); see Section 5.5. The method started with the work of Shi and Goodman (1984), (1985), and since then it has received considerable attention by the geoengineering community.

The method is fully described in Shi (1992a) and (1993). In essence, the medium is discretized into elements or blocks that are in contact with each other only through the discontinuities. The discretization used in Figure 5.19 to illustrate the DEM could perfectly apply to the DDA. There are fundamental differences between the DEM and DDA. In the DEM, each block is treated separately while in the DDA the total potential energy of the system is minimized to find the solution. In the DEM, stresses and forces are unknowns while displacements are computed from stresses; in DDA the displacements are the unknowns. In the DEM, the contacts are resolved using a penalty method which results in the definition of the contact forces, while in the DDA interpenetration of blocks is prevented by adding springs to the contacts. The DEM uses an explicit procedure to solve the equilibrium equations, while the DDA is an implicit method. While the DDA is a fully discontinuous analysis, it resembles and follows the procedures developed for FEM.

The DDA, similar to the DEM, needs to address three key issues (Section 5.5): (1) representation of contacts; (2) representation of solid material; and (3) detection and revision of contacts during execution. The elements can be convex or non-convex, and their shapes are determined by the location of their contacts with the neighboring elements. Thus blocks are represented by polyhedra, with the contacts between blocks consisting of edge to face, edge to edge, or face to face.

It is assumed that any large displacements or deformations are the result of the accumulation of small displacements and deformations after a sufficiently large number of steps. Within each step the displacements of any block are small, and thus they can be given by a first-order approximation of the form:

$$\begin{aligned} u &= u_o + (x - x_o)a_1 + (y - y_o)a_2 \\ v &= v_o + (x - x_o)b_1 + (y - y_o)b_2 \end{aligned} \tag{5.50}$$

where u and v are the x- and y-axis displacements of a point with coordinates x and y; u_o and v_o are the rigid body motions at point x_o, y_o and a_i and b_i i = 1,2 are constants. Strains can be computed from (5.50). In turn displacements can be expressed as a function of strains as follows:

$$\begin{aligned} u &= u_o + (x - x_o)\varepsilon_{xx} + (y - y_o)\left(\frac{1}{2}\gamma_{xy} - r_o\right) \\ v &= v_o + (y - y_o)\varepsilon_{yy} + (x - x_o)\left(\frac{1}{2}\gamma_{xy} + r_o\right) \end{aligned} \tag{5.51}$$

where ε_{xx}, ε_{yy}, and γ_{xy} are the axial strains and the shear strains in the x and y axis, respectively, and r_o is the rigid block rotation, in radians, about point x_o, y_o. Equation (5.51) is expressed in matrix notation:

$$U = TD \tag{5.52}$$

where $U = (u, v)$, $D^T = (u_o, v_o, r_o, \varepsilon_{xx}, \varepsilon_{yy}\ \gamma_{xy})$ and T have the appropriate coefficients from (5.51). The matrix D represents the unknowns for each element; thus there are a total of 6 degrees of freedom or unknowns. Note that strains in each element are constant. For a system of N elements or blocks, the total number of unknowns is 6N. Minimization of the potential energy of the system of blocks, following FEM convention, is expressed as:

$$K_{ij}D_j = F_i \tag{5.53}$$

D_j is made of 6 × 1 sub-matrices that contain the 6 unknowns of each element j; K_{ij} is composed of 6 × 6 stiffness sub-matrices associated with the corresponding degrees of freedom of element j, and F_i is a set of 6 × 1 force sub-matrices of element i. The diagonal terms of the matrix K_{ij}, i.e. K_{ii}, depend on the material properties of element i (shown later) and the non-diagonal terms K_{ij} ($i \neq j$) on the contacts between elements. All the sub-matrices K_{ij} are obtained by minimizing the potential energy associated with strain energy, initial stresses, concentrated and distributed loads, body forces, inertia forces, viscosity, displacement constraints at the element contacts, etc. Shi (1993) provides full derivation of the equations. Derivation of the 2D stiffness sub-matrices associated with strain energy is included to illustrate the process. Strain energy Π for a two-dimensional single element is given by:

$$\Pi = \frac{1}{2}\int_S \varepsilon_{ij}\sigma_{ij}\,dxdy = \frac{1}{2}\iint D^T C D\,dxdy = \frac{1}{2}D^T C D S \tag{5.54}$$

where C is the 6 × 6 compliance matrix and S is the area of the element. The members of the stiffness matrix for element i are given by:

$$K_{ij} = \frac{\partial^2 \Pi}{\partial d_{ik}\,\partial d_{jk}} = \delta_{ij} S C \tag{5.55}$$

where d_{ik} is the degree of freedom k, k = 1,6, of element i. Note that (5.55) are the diagonal terms of (5.53). In a similar manner new terms originated by initial stresses, forces, etc. are added to the appropriate terms of the stiffness matrix.

In the DDA, no tension and no penetration between blocks are allowed. The kinematics of the block system are incorporated into the equations of equilibrium (5.53) by adding very stiff springs between appropriate elements to lock the movement in the corresponding direction. Tension between two elements can be modeled by applying a lock in the direction where tension is permitted; once the lock is removed (i.e. a critical tensile threshold is reached) the elements can separate. Hence by adding or removing locks along appropriate directions, movements between blocks can be prevented, thus preventing penetration. Within a certain loading step (load increment) an iteration process is applied where locks are added or removed as appropriate until all kinematic constrains (e.g. no penetration) are satisfied. Imposing the kinematics of the problem requires addressing two issues: (1) determine contacts between blocks; and (2) add to the global equilibrium equations the appropriate stiff springs.

The contact identification process starts after definition of the elements where some threshold distance ρ is established such that only elements within the threshold distance are checked for contact. As the simulation proceeds, potential contacts between elements are updated. If within a single step the relative displacement between two elements is smaller than their initial distance, no contact check is performed. The distance between blocks i and j is determined by:

$$\eta_{ij} = \min\left[\sqrt{(x_A - x_B)^2 + (y_A - y_B)^2}\right] \tag{5.56}$$

where A is a vertex of block i (or j) and B is on the boundary of block j (or i); see Figure 5.23a.

If interpenetration between two elements is detected, then stiff springs are placed between the two elements and the system is recalculated. Figure 5.23b shows an initial point P_1 from element i which has moved passed the edge P_2P_3 of element j to position P'_1. A stiff spring is added to the global equilibrium matrix to prevent such interpenetration. The potential energy associated with the spring is given by:

$$\prod_k = \frac{1}{2} p d^2 \tag{5.57}$$

where p is a very large positive number, large enough such that the displacements of the spring are three to four orders of magnitude smaller than the total displacement of the element; d is the interpenetration distance between elements i and j, as shown in Figure 5.23b. The distance d is given by:

$$d = \frac{1}{\sqrt{(x_2 - x_3)^2 + (y_2 - y_3)^2}} \begin{vmatrix} 1 & x_1 + u_1 & y_1 + v_1 \\ 1 & x_2 + u_2 & y_2 + v_2 \\ 1 & x_3 + u_3 & y_3 + v_3 \end{vmatrix} \tag{5.58}$$

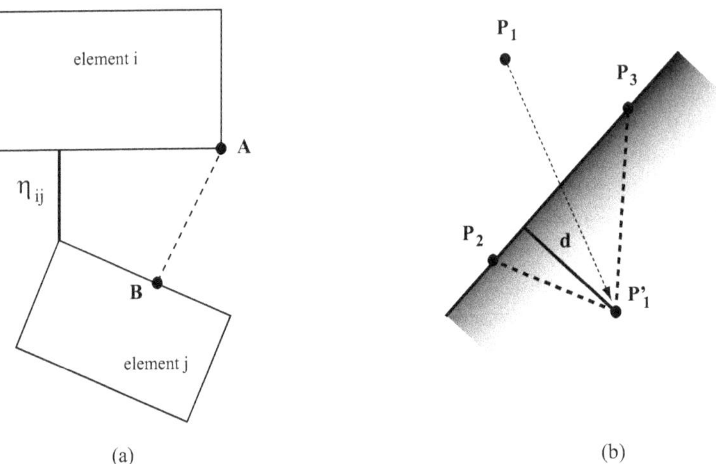

Figure 5.23 Contact between elements in DDA. (a) Distance between two elements. (b) Penetration distance.

The coordinates x_i, y_i of each point are obtained in terms of the six degrees of freedom of each element, after Equation (5.51). Minimization of the potential energy gives the terms to be added to the global stiffness matrix in (5.53). They are obtained by taking the derivatives of (5.57), as it is done in (5.55).

The procedure of solving the equilibrium equations, determining interpenetration, and adding stiff springs is repeated until no interpenetration occurs. At the end of each iteration, the spring force is calculated. If the component of the force normal to the contact is tensile, the normal spring is removed. If the component of the force parallel to the contact is larger than the maximum allowed by the constitutive model (e.g. $F_s > \mu F_n$, Coulomb), a normal spring and a pair of frictional forces are applied at the contact to allow for sliding and prevent penetration in the normal direction. If the component of theforce parallel to the contact is smaller than the maximum allowed, springs in both the normal and parallel directions are applied to prevent any relative movement at the contact.

The method, which was originally developed for 2D problems (Shi and Goodman, 1984, 1985; Shi, 1993), has been expanded to 3D (Shi, 2001; Jiang and Yeung, 2004). The method has been coupled with FE methods to address both continuum and discontinuum problems (Cheng and Zhang, 2002). Because the DDA method prevents interpenetration of elements, which is in contrast to what is done in other discontinuum methods such as the DEM and BPM (discussed later), it is well suited for coupled flow-mechanical problems. Progress has been made to develop models for flow through discontinuities, which depends on the connectivity of the discontinuities and on their behavior due to mechanical loading (Kim et al., 1999). The limitation of the original DDA that the blocks could not break has been overcome by new developments in modeling, where blocks are divided into sub-blocks when tensile or shear stresses reach the strength of the material; thus the DDA has been extended to fragmentation and fracture propagation problems (Lin et al., 1996; Koo and Chern, 1998).

The procedure described applies to "forward" analysis where, given the geometry and input loading, a solution in terms of displacements, strains, and/or stresses is reached. A similar procedure can be established for a "backward" analysis, where the material constants are obtained from deformation measurements (Shi, 1993).

Validation of the Displacement Discontinuity Analysis has been done extensively by comparing predictions from the method with analytical solutions, with other numerical methods, with laboratory and with field measurements (e.g. Hatzor and Feintuch, 2001; MacLaughlin et al., 2001; MacLaughlin and Berger, 2003; Yeung et al., 2003; Tsesarsky et al., 2005; Wu, 2007; MacLaughlin and Doolin (2006) provide an extensive review). So DDA can be viewed as a mature and reliable method. DDA has found a niche in rock mechanics to investigate slope stability (Hatzor et al., 2004; Sitar et al., 2005; Wu, 2007; Kveldsvik et al., 2009) and tunneling in discontinuous rock masses (Shi, 1993; Wu et al., 2004; Tsesarsky and Hatzor, 2006). Additional case studies are presented in Lu (2003).

Figure 5.24 illustrates an example application of the DDA method (Jing, 1998). In the figure, a shallow rectangular tunnel in a rock mass medium with two joint sets is subjected to a vertical load on the surface. The figure shows the different stages of the failure, from initial conditions, Figure 5.24a to final failure, Figure 5.24f.

5.7 DISCONTINUUM: BONDED PARTICLE METHOD

The Bonded Particle Method (Potyondy and Cundall, 2004) originates from the application of the DEM (Section 5.5) to a discontinuous medium modeled as discs in two dimensions or spheres in three dimensions. The key idea of the method is that the rock material can

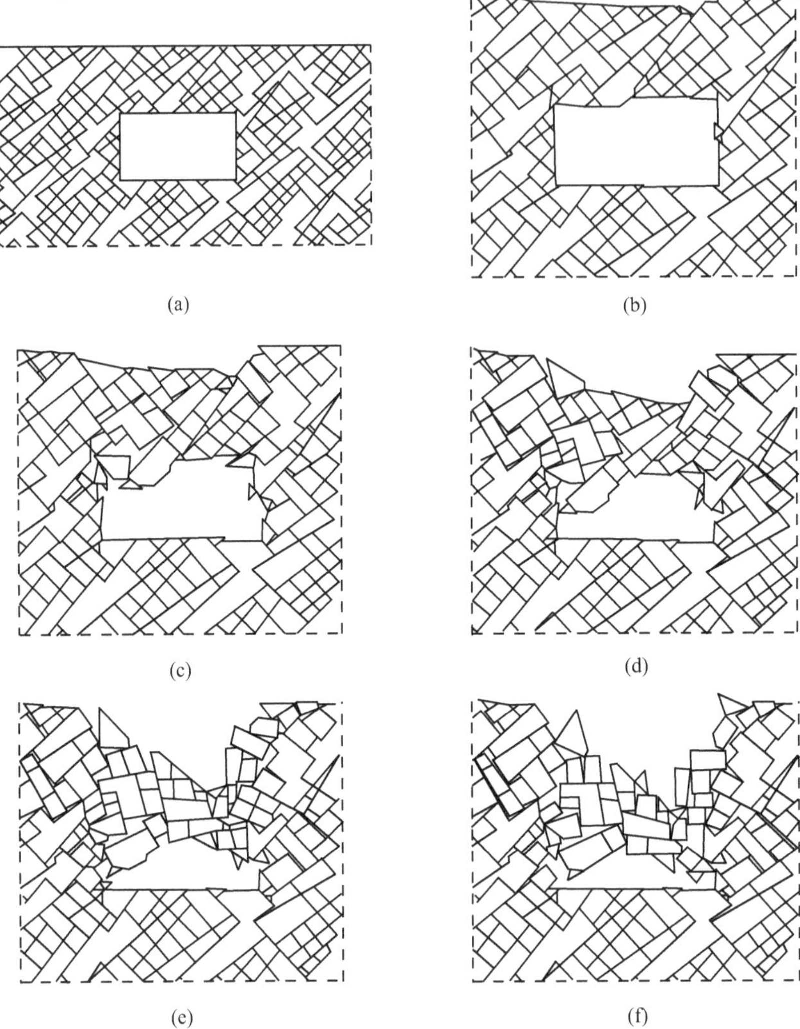

Figure 5.24 Example application of DDA method. 8 × 5 tunnel with a vertical load applied at the surface. (a) initial geometry; (b) at time, t = 0.001 s; (c) t = 0.002 s; (d) t = 0.003 s; (e) t = 0.004 s; and (f) t = 0.005 s. Adapted from Jing, (1998). (Reproduced by permission of Elsevier.)

be approximated by an agglomerate of cemented grains; see Figure 5.25a. The grains or particles are assumed rigid with circular or spherical shape. The particles interact with each other through their contacts such that deformation is produced at the particle contacts and by relative displacements between particles (Figure 5.25b). Tensile and shear cracks between particles occur when the tensile or shear strength of the contact is reached.

As with the DEM, Newton's second law of motion is solved through a central finite difference algorithm to determine the displacements and velocities of each particle due to the forces acting on the particle. The forces arise from the weight of the particle and from the contact forces between particles. Equations (5.45) and (5.46) are used to determine the motions of any particle. The solution of a problem with static or dynamic loading is done incrementally with very small time steps (for static loading, time is an auxiliary variable related to the load increment during each step). The procedure follows that of the Distinct Element Method (Section 5.5): Displacements and velocities of each grain are computed using Equation (5.46) with the magnitude of the forces equal to those at the end of the previous step. From absolute

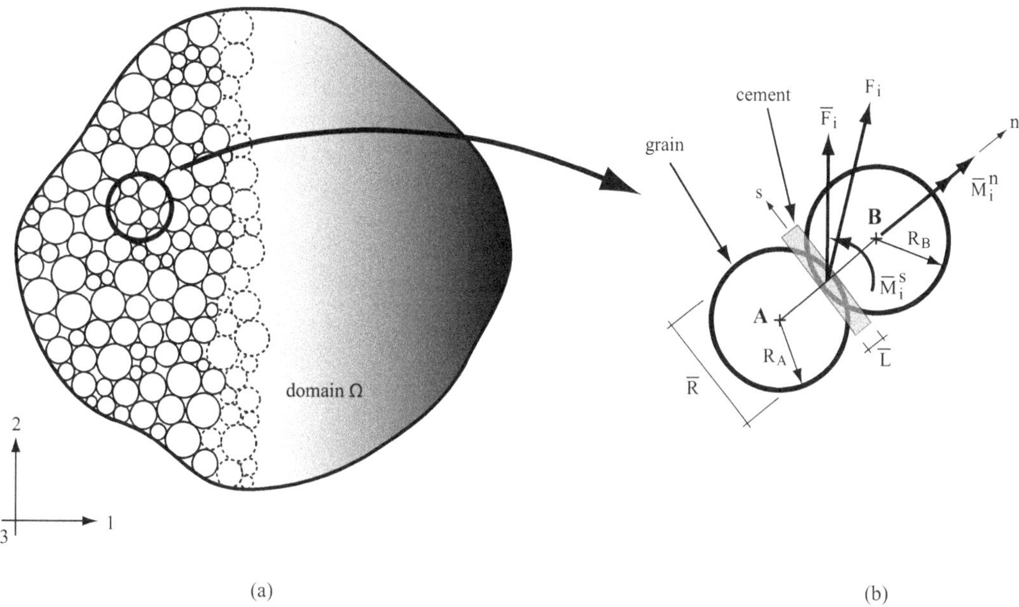

Figure 5.25 Bonded particle method discretization (Bobet, 2010c). (a) Discretization. (b) Cemented grains. (Republished with permission of Springer Nature VB.)

displacements, the relative motions between particles in contact is obtained, which in turn are used to determine the magnitude of the forces and moments acting between particles. The updated loads are then used to compute motions for the next time increment. The process is repeated until the complete solution of the problem is obtained. During the process, contact between particles is reviewed and updated as new contacts may be formed or old ones are destroyed, as bonds between particles break. Inter-particle forces and moments are obtained based on the relative motions between particles and on the properties of the particles and bond. The magnitude of the forces and moments (Figure 5.25b) is given by:

$$\begin{aligned} F_i &= F_i^n \vec{n} + F_i^s \vec{s} \\ \bar{F}_i &= \bar{F}_i^n \vec{n} + \bar{F}_i^s \vec{s} \\ \bar{M}_i &= \bar{M}_i^n \vec{n} + \bar{M}_i^s \vec{s} \end{aligned} \tag{5.59}$$

where F_i is the inter-particle force between particle A and particle B (Figure 5.25b), with components F_i^n and F_i^s in the directions normal and parallel, respectively, to the contact between the two particles; $\bar{F}_i$ and $\bar{M}_i$ are the force and moment carried by the bond between the two particles. The magnitude of the loads is given by (Potyondy and Cundall, 2004):

$$\begin{aligned} \Delta F_i^n &= \frac{k_n^A k_n^B}{k_n^A + k_n^B} \Delta U^n \\ \Delta F_i^s &= -\frac{k_s^A k_s^B}{k_s^A + k_s^B} \Delta U^s \\ \Delta \bar{F}_i^n &= \bar{k}_n A \Delta U^n \\ \Delta \bar{F}_i^s &= -\bar{k}_s A \Delta U^s \\ \Delta \bar{M}^n &= -\bar{k}_s J \Delta \theta^n \\ \Delta \bar{M}^s &= -\bar{k}_n I \Delta \theta^s \end{aligned} \tag{5.60}$$

k_n^A, k_s^A, k_n^B and k_s^B are the normal and shear stiffnesses of particles A and B, and $\bar{k}_n$ and $\bar{k}_s$ are the normal and shear stiffness of the bond between particles; ΔU^n and ΔU^s are the incremental normal and shear displacements between particles, and $\Delta\theta^n$ and $\Delta\theta^s$ are the incremental rotational angles also in the normal and shear directions (note that there is no rotation in the normal direction in two-dimensional problems); A, I, and J are the area, moment of inertia, and polar moment of inertia of the bond between the two particles, and are given by:

$$A = \begin{cases} 2\bar{R} & \text{in 2D} \\ \pi\bar{R}^2 & \text{in 3D} \end{cases}$$
$$I = \begin{cases} \frac{2}{3}\bar{R}^3 & \text{in 2D} \\ \frac{1}{4}\pi\bar{R}^4 & \text{in 3D} \end{cases} \tag{5.61}$$
$$J = \begin{cases} \text{n/a} & \text{in 2D} \\ \frac{1}{2}\pi\bar{R}^4 & \text{in 3D} \end{cases}$$

$\bar{R}$ is the bond radius, and R_A and R_B are the radii of particles A and B, respectively, as shown in Figure 5.25b. $\bar{R}$ is usually expressed as the product of a factor $\bar{\lambda}$ times the smallest of R_A and R_B. The maximum tensile and shear stresses acting on the bond are calculated as:

$$\bar{\sigma}_{max} = -\frac{\bar{F}_i^n}{A} + \frac{|\bar{M}_i^s|\bar{R}}{I}$$
$$\bar{\tau}_{max} = \frac{\bar{F}_i^s}{A} + \frac{|\bar{M}_i^n|\bar{R}}{J} \tag{5.62}$$

When the maximum tensile or shear stress reaches the tensile strength of the bond, $\bar{\sigma}_c$, or shear strength, $\bar{\tau}_c$, the bond breaks and it is removed from the model.

The shear force F_i^s in (5.60) is limited by the constitutive law used for inter-particle friction (e.g. Coulomb with $F_i^s \leq \mu F_i^n$; μ is the coefficient of friction between particles). If the relative displacement between two particles is negative, there is a gap between the two particles and the normal and shear forces are set to zero; if it is positive, the two particles overlap and thus there are normal and shear forces between the particles.

Thus the following microproperties are needed for the model: k_n, k_s, and μ which are associated with the grains, and $\bar{R}$ (or $\bar{\lambda}$), $\bar{k}_n$, $\bar{k}_s$, $\bar{\sigma}_c$ and $\bar{\tau}_c$, which depend on the bond.

Even though the Bonded Particle Model is relatively new, it has been already used for a wide range of applications within geotechnical engineering. The model has been applied to investigate the strength of soils and rock materials (Boutt and McPherson, 2002; Wanne, 2002; Potyondy and Cundall, 2004; Holt et al., 2005), slope stability (Wang et al., 2003a), damage to the rock mass during tunnel excavation and tunnel support (Fakhimi and Labuz, 2002; Tannant and Wang, 2002, 2004; Maynar and Rodríguez, 2003; Potyondy and Cundall, 2004), fracture mechanics (Konietzky et al., 2002; Potyondy and Cundall, 2004), blasting and dynamic analysis (Cundall et al., 2002; Olson et al., 2002; Hazzard and Young, 2004), behavior of granular materials and powders (Li and Holt, 2002; Kleier and Kleinschrodt, 2002; Bwalya and Moys, 2002)

The list of applications of the method is not exhaustive, and it is intended to provide a measure of the wide range of fields where the method is used. The method has been the focus

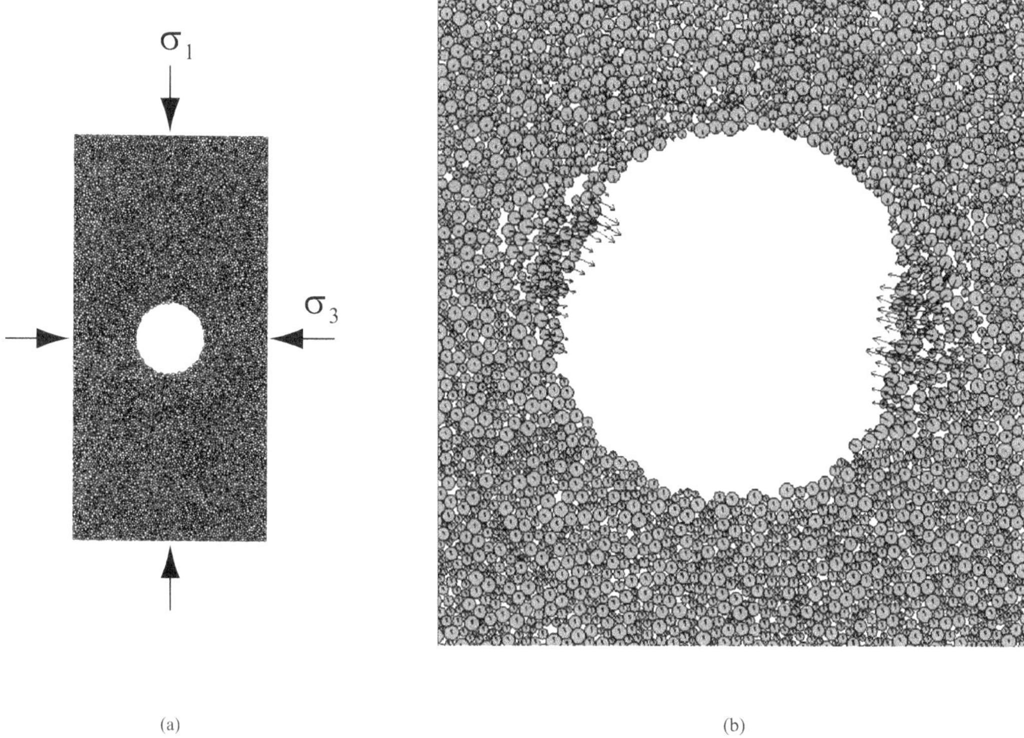

Figure 5.26 Example of BPM model. Circular tunnel subjected to biaxial compression with σ_3 = 7.5 MPa. From Fakhimi et al. (2002). Reproduced by permission of Elsevier. (a) Model. (b) Detail at failure. (Republished with permission of Elsevier.)

of recent conferences where a large number of cases and applications have been presented even in fields beyond civil engineering (e.g. Konietzky, 2002).

Figure 5.26 illustrates the use of the model (Fakhimi et al., 2002) to determine the damage zone around a circular opening, in the form of tensile and shear cracks. The model reproduces the results of experiments conducted on Berea sandstone where an opening of 14 mm diameter was placed into a prismatic block that was loaded in plane strain with 7.5 MPa confinement. A uniform particle size distribution was used to model the rock, with average particle size 0.2 mm, similar to the actual size of the Berea sandstone grains. Figure 5.26b shows the final stage of failure of the opening with significant cracks and notches between grains.

5.8 OTHER METHODS

There are other numerical methods, beyond those discussed in the previous sections, that are or have been used in geomechanics and particularly in tunneling (key block method, fracture network, fracture flow, hybrid methods, etc.). Inclusion of all the methods would not be practical, and it was decided that this chapter would focus on the methods arguably with clear application to and with long tradition in tunneling. Of course the outcome will change with time as new methods are developed and existing methods become obsolete or are superseded by new ones. The discussion, however, could not be complete without the addition of two other methods: meshless methods and Artificial Neural Network methods. These two methods cannot be classified into continuum or discontinuum due to their fundamental nature, as it will be explained in the next sections, and yet they need to be included due to their high interest in tunneling. Meshless methods, even though still under development, have the potential to revolutionize the field of numerical methods by removing the constraints that physical discretizations introduce. Artificial Neural Network methods use a completely different paradigm. They are based on experience from previous outcomes to predict expected results following an approach that does not have to consider the mechanics of the problem. The two methods are briefly introduced and discussed in the next two sections.

5.8.1 Meshless methods

Meshless methods' (MM) origins can be traced to the work by Lucy (1977), in an attempt to study astrophysical phenomena. It was not until the early 1990s when MM saw their most important development after the work of Nayroles et al. (1992) and Belytschko et al. (1994). The driving force behind these methods is the reduction of the dependency of numerical schemes on the physical discretization of the medium. The approximation is based entirely in terms of nodes. The idea is that the medium can be represented by a number of nodes (Figure 5.27a) placed without the need to follow any particular structure, and new nodes can be incorporated during computation. The key advantage of meshless methods is a significant reduction of time for mesh generation, which is today by far the most time-consuming task during modeling. Because the methods do not depend on an existing discretization, discontinuities can be incorporated easily, which makes the methods both continuum and discontinuum, following the classification presented in Section 5.1.

The first meshless method proposed was the smooth particle hydrodynamics method (SPH; Lucy, 1977), which was followed by the diffuse element method (DEM; Nayroles et al., 1992) and the element-free Galerkin method (EFG; Belytschko et al., 1994), and then reproducing kernel particle methods (RKPM; Liu et al., 1995), partition of unity finite element

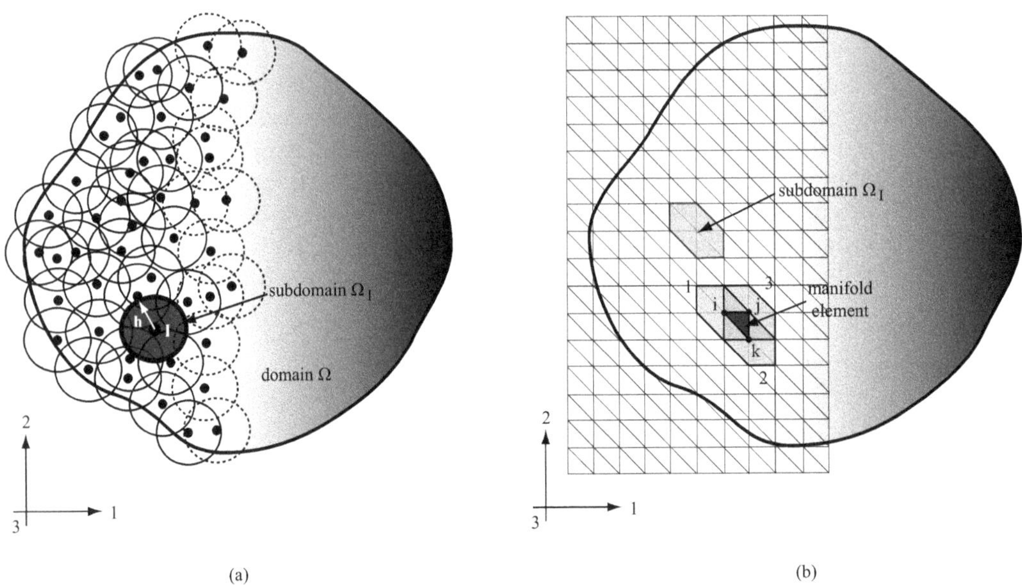

Figure 5.27 Meshless methods. (a) Linear square method. (b) Manifold method.

methods (PUFEM; Babuska and Melenk, 1996), h-p clouds (Duarte and Oden, 1996), the finite point method (Oñate et al., 1996a, 1996b), the generalized finite difference method (GFD; Perrone and Kao, 1975; Gavete et al., 2003), and others. Belytschko et al. (1996) and Chen et al. (2006) provide and excellent review of meshless methods. It has to be mentioned that this is a field of rapid growth and what it is included in this section may become obsolete in few years.

All meshless methods, although different, share fundamental assumptions such as the use of nodes for the solution of the problem and weight functions. The following is a brief summary of the element-free Galerkin method (EFG; Belytschko et al. 1994; Pan et al. 2005) in two dimensions, which on the one hand is used to illustrate the fundamentals of the approach, and on the other hand it is arguably the meshless method that has produced the most considerable impact and still receives significant attention from the research community.

The EFG method is based on a moving least-square procedure (MLS) and on weight functions. In the method, an arbitrary set of nodes x_I, I = 1 to N is placed in the domain Ω (Figure 5.27a). A weight function is defined in a subdomain Ω_I associated with node I. The weight function is nonzero within the subdomain and zero outside the subdomain. Each subdomain is associated with a node I and has the shape of discs or squares in two-dimensional problems with size h (a subdomain is highlighted in Figure 5.27a). The subdomains overlap and there are usually five to ten discs overlapping a particular node I (for clarity fewer discs are drawn in the figure).

The displacements at any point are given by:

$$u^h(x) = \sum_{i=1}^{m} p_i(x)a_i(x) = P^T(x)A(x) \tag{5.63}$$

where $u^h(x)$ is the approximation of the displacement function u(x), at spatial coordinate x, using subdomains of size h; $p_i(x)$ are monomial basis functions and $a_i(x)$ are their coefficients. The total number of terms in the basis is m. The summation in (5.63) is also expressed as the product of the matrix P, which is composed of the monomial basis $p_i(x)$, and A, which contains the coefficients $a_i(x)$. Examples of the basis are:

$$\begin{aligned} &\text{Linear:} \quad P^T = (1, x) \quad \text{in 1D and } P^T = (1, x, y) \quad \text{in 2D} \\ &\text{Quadratic:} \ P^T = (1, x, x^2) \text{in 1D and } P^T = (1, x, y, x^2, xy, y^2) \text{in 2D} \end{aligned} \tag{5.64}$$

Other forms can be used, e.g. singular functions to better approximate stresses at the tips of cracks.

The coefficients $a_i(x)$ are obtained by minimizing the difference between the approximated and actual displacements at nodes I. This gives the function:

$$\begin{aligned} J &= \sum_I w(x - x_I)(u^h(x_I) - u(x_I))^2 = \sum_I w(x - x_I)\left[\sum_I p_i(x_I)a_i(x) - u(x_I)\right]^2 \\ &= (PA - U)^T W(PA - U) \end{aligned} \tag{5.65}$$

where $w(x - x_I)$ is a weight function, W is the matrix of the weight functions, and U is the matrix of the displacements at nodes I. The weight functions suggested are exponential, cubic spline, quartic spline, and SPH spline (Belytschko et al., 1996). For example the cubic spline function is defined as:

$$w(s) = \begin{cases} \frac{2}{3} - 4s^2 + 4s^3 & \text{for } s \le \frac{1}{2} \\ \frac{4}{3} - 4s + 4s^2 - \frac{4}{3}s^3 & \text{for } \frac{1}{2} < s \le 1 \\ 0 & \text{for } s > 1 \end{cases} \tag{5.66}$$

where s = $(x\text{-}x_I)/h$ is the argument of the weight function. Minimizing the function J gives:

$$\frac{\partial J}{\partial a_i} = P^T WPA - P^T WU = 0 \tag{5.67}$$

The coefficients of the basis $a_i(x)$, contained in the matrix A, are obtained from (5.67). Substitution in (5.63) provides the final form of the approximation:

$$u^h(x) = \sum_{I=1}^{N} (P^T A^{-1} P^T W)U = \sum_{I=1}^{N} \Phi U \tag{5.68}$$

This expression is similar to the relation used in FEM (Section 5.3) to relate displacements inside the elements with nodal displacements which is done through interpolation functions h_k; i.e. Equation (5.17). A weak form of the variational principle (Galerkin method) can be used to satisfy the equations of equilibrium and boundary conditions; e.g. Equation (5.7). A procedure similar to that used for the FEM can be followed by expressing displacements as a function of nodal displacements, as in Equation (5.68), and stresses as a function of strains (Equation (5.8) given the constitutive model. The final result is a set of equations similar to (5.12) where a "stiffness" matrix should be inverted to obtain the displacements at the nodes x_I. Note, however, that in FEM the values of the interpolation functions at the nodes are equal to one, but this is not the case in (5.68). In other words, the functions Φ are not interpolation functions. For this reason, imposition of specified displacement conditions at the boundary is not as straightforward as in FEM. Different approaches have been proposed such as Lagrange multipliers, modified variational principle, penalty methods, and perturbed Lagrangian (Belytschko et al., 1996). This results in a non-symmetric stiffness matrix, which requires additional computation time, as compared with the FEM. Also computation of the integrals needed to build the stiffness matrix is more complex than FEM and requires also additional computer time. An attractive approach to reduce the difficulties associated with the displacement boundary conditions is to couple the meshless method with a FEM, where the interior of the domain is discretized with nodes following the procedure outlined for the MM, while the boundary region is discretized with finite elements (Belytschko et al., 1996).

While computational costs associated with meshless methods can be more than ten times higher than with FEM, it is expected that given the rapid evolution of the field, the method will attain a level of efficiency comparable to the FEM.

Meshless methods are beginning to be applied to discontinuous media, including jointed rock masses (Zhang et al., 2000). Particular mention needs to be given to a variation of MM applied to blocky systems: the Numerical Manifold Method (NMM). The NMM was first proposed by Shi (1992b), (1996), and (1997) and then further developed by others (Luo et al., 2001; Li et al., 2005; Zhang and Zhou, 2006). As in MM the domain is divided into subdomains, but with the NMM the subdomains are hexagons (Figure 5.27b). The intersection of three hexagons is a triangle, which is called a manifold element. The entire domain is composed of the common intersections of the subdomains.

Displacements within an element are approximated with the weighted displacements of the subdomains that form the element.

$$u(x) = \sum_{i=1}^{n} w_i(x)u_i(x) \tag{5.69}$$

where u(x) is the displacement of any point x inside the element, $w_i(x)$ are the weight functions of each of the subdomains from i = 1, n which intersection forms the element (there are typically three hexagonal subdomains per element), and u_i are the displacements in each subdomain. The weight functions are defined such as:

$$\begin{aligned} & w_i(x) \geq 0 && \text{in subdomain i} \\ & w_i(x) = 0 && \text{outside subdomain i} \\ & \sum_{i=1}^{n} w_i(x) = 1 && \text{inside the element} \end{aligned} \tag{5.70}$$

The simplest weight function is the linear function. Inside the element, it takes the form:

$$w_i(x) = a_i + b_i x + c_i y \tag{5.71}$$

The coefficients a_i, b_i, and c_i, i = 1, 2, 3 corresponding to each of the hexagons that form the triangular manifold element, are obtained by satisfying (5.70). Thus,

$$\begin{gathered} \begin{pmatrix} w_1(x) \\ w_2(x) \\ w_3(x) \end{pmatrix} = \frac{1}{\Delta} \begin{pmatrix} x_j y_k - x_k y_j & y_j - y_k & x_k - x_j \\ x_k y_i - x_i y_k & y_k - y_i & x_i - x_k \\ x_i y_j - x_j y_i & y_i - y_j & x_j - x_i \end{pmatrix} \begin{pmatrix} 1 \\ x \\ y \end{pmatrix} \\ \Delta - (x_j - x_i)(y_k - y_i) - (x_k - x_i)(y_j - y_i) \end{gathered} \tag{5.72}$$

where x_i and y_i are the coordinates of the vertices of the triangular element (Figure 5.27b).

The subdomain displacements $u_i(x)$ in Equation (5.69) are expressed in term of basis functions $p_i(x)$, similar to the meshless method.

$$u_i(x) = \sum_{i=1}^{m} p_i(x)a_i \tag{5.73}$$

where m is the number of basis functions and a_i are unknown constant coefficients. The basis functions are the same as those defined for the MM; i.e. Equation (5.64). Thus displacements take the final form:

$$\begin{gathered} u(x) = (T_1(x) \quad T_2(x) \quad T_3(x))(D_1 \quad D_2 \quad D_3)^T \\ T_i(x) = \begin{pmatrix} w_i(x)p_i(x) & 0 & 0 \\ 0 & w_i(x)p_i(x) & 0 \\ 0 & 0 & w_i(x)p_i(x) \end{pmatrix}, i = 1, 2, 3 \end{gathered} \tag{5.74}$$

The matrices D_i are the constant coefficients a_i of the basis functions. In the original form of the NMM (Shi, 1992), the displacements in (5.74) are used to compute the potential energy associated with the problem, which is then minimized following the same procedure as in the DDA (Section 5.6). Another approach has been proposed by Li et al. (2005) with a weak form of the variational principle (Galerkin method), as it is done in the meshless methods. Both procedures yield the same result, but the Galerkin method is more robust and more general.

The structure of the NMM consists of three main parts (Shi, 1997): the mathematical procedure described, block kinematics, and the simplex integration method. Block kinematics are borrowed from DDA and follow the same tracking and contact techniques described in Section 5.6. The simplex integration method is based on analytical solutions of the integrals over triangular elements, so the results are of high precision and the computation time is small.

As with the MM, the NMM is well suited to address both continuum and discontinuum problems. It has some of the deficiencies associated with MM in which the computation time is still larger than other methods such as FEM or DDA, and it is a method still under development. However, the method will receive steady improvements over the next few years, and it is expected that it will be brought to a level comparable to that of other well established methods. In fact the method has started being used for fracture problems, where cracks can propagate over the domain (e.g. Zhou et al., 2003; Zhang et al., 2003; Wang et al., 2003b; Kourepinis et al., 2003; and Su et al., 2003).

5.8.2 Artificial Neural Networks

Artificial Neural Networks (ANN) belong to the broader family of Deep Learning (DL) methods, which in turn are part of Machine Learning (ML) or Artificial Intelligence (AI) methods (ML and AI are often used interchangeably). ML and AI methods try to imitate how humans learn. This is a rather different approach that what has been discussed so far, where all methods included in the previous Sections reach a solution addressing the mechanics of the problem where equilibrium, constitutive model, strain compatibility, and boundary conditions are rigorously satisfied; what distinguishes one method from the other is how this is mathematically accomplished. DL methods, however, are based on a completely different paradigm, which includes statistics (learning) and predictive capabilities. Learning can be supervised or unsupervised (Shinde and Shah, 2018; Jong et al., 2021). In supervised learning, predictions from the model can be compared with data and the model can be adjusted to improve output, i.e. the method "learns," while in unsupervised learning, patterns are learned by the model, but predictions cannot be checked against measurements (Jong et al., 2021). In tunneling, supervised learning is commonly used.

The origin of Neural Networks can be traced back to McCulloch and Pitts (1943) who wrote the first mathematical model for a neuron, but it was not until the 1980s when the introduction of new architectures and learning processes made ANNs useful and practical tools. After that, steady advances in the field took place until 2006, where the breakthrough work by Hinton et al. (2006) revolutionized the subject and the term "Deep Learning" was established. DL is a neural network with a large number of layers and parameters (see Shinde and Shah, 2018 and Emmert-Streib et al., 2020 for a time-history of the origin and evolution of DL). As with the human brain, ANNs are composed of a number of interconnected units called neurons. Each neuron receives information, processes the information, and sends the results to other neurons (see Figure 5.28a). The characteristics of ANNs are that the information is stored over the entire network, are massively parallel processing systems, are fault-tolerant, can reach a solution with ill-defined or imprecise information, and can learn and adapt. The disadvantages are that ANN systems operate as "black boxes" in that there is no

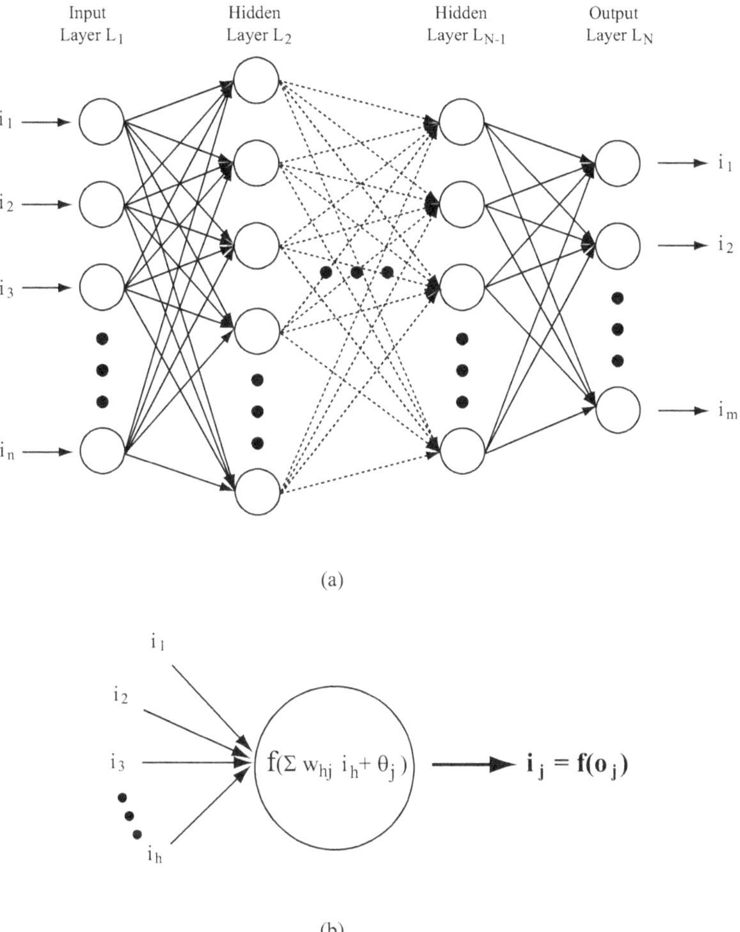

Figure 5.28 Artificial neural network (Bobet, 2010c). (a) Feedforward Neural Network. (b) Operation in Neuron j. (Republished with permission of Springer Nature VB.)

possibility of assessing how they work internally, their design guidelines and operation are somewhat arbitrary, training may be difficult or impossible, and their performance may not be easily predicted (Schalkoff, 1997). On the one hand, the input parameters my reflect the underlying mechanical relations and represent the key variables controlling tunnel behavior. On the other hand, ANNs may be perceived as highly sophisticated curve fitting techniques. However, they have proven to provide reasonable solutions to imprecisely formulated problems or to phenomena only described through observations (Jing and Hudson, 2002).

There are several types of Artificial Neural Networks depending on the characteristics of each neuron, the learning or training scheme, network architecture, and network function (e.g. De Wilde, 1996; Schalkoff, 1997; Shinde and Shah, 2018; Emmert-Streib et al., 2020): Feedforward Neural Networks (FFNN); Recurrent Neural Networks (RNN); Deep Belief Networks (DBN); Autoencoder; and Long-Term Memory Networks (LSTM). FFNNs, also known as Multilayer Perceptron (MLP), have a number of neuron layers, with all neurons in a layer connected with all neurons in the following layer, but the information is only passed forward (Figure 5.28a). A special FFNN is the Convolutional Neural Network (CNN), where a neuron in a given layer may not be connected to all the neurons in the next layer, thus reducing the number of parameters and computation time. In RNNs, the information may propagate through a layer of neurons more than once. DBNs combine RNN and FFNN, typically using RNNs for input and FFNN for output. Autoencoder networks are similar to DBNs and are unsupervised neural network models. They have the same size of the input and output layers, and they have a symmetric structure. LSTM networks are a type of RNNs, but with a memory cell included in the network architecture to store information over extended times (as opposed to RNNs where the information is stored for a short time).

In tunneling, most of the work, so far, has been done using Feedforward Neural Networks, even though an increasing number of publications seem to indicate that other DL techniques may be better suited to predict tunneling operations and tunnel performance. See e.g. Jong et al. (2021) for a review of AI techniques used in underground construction and Zhang et al. (2020) for shield-tunneling. As mentioned, DL or ML is a field that is rapidly evolving and it is expected that significant advancements will occur in the next few years that may render the information provided here outdated. It is beyond the scope of the book to explore in detail all algorithms proposed. What is done is provide in the following a summary review of the FFNN method; first, because of its simplicity, to better illustrate key concepts and second, because, as mentioned, there is a significant body of work done with the method.

The Feedforward Neural Network (FFNN) has two or more layers of neurons (Figure 5.28a). The first layer receives the input applied to the network and the last layer contains the output. As mentioned, the units or neurons in each layer are forward connected only to the units or neurons in the next layer. There is no connection between neurons in the same layer. Thus FFNNs are connected to the exterior by the input and output layers only. The layers between the input and output are called hidden layers. As shown in Figure 5.28a, the input consists of n units, each corresponding to an input parameter, and m output units, each corresponding to a requested result parameter. There can be any number of hidden layers, and each layer can have any number of units.

The information stored in each neuron, often called the state of the neuron, is passed forward to its connected neuron in the next layer and modified by a connection weight and a bias or threshold value. The resulting value is further modified in the receiving neuron by a function called the activation or transformation function (Figure 5.28b). For example, neuron j in layer L_k receives input from the neurons in layer L_{k-1}. If the state of neuron j is denoted by i_j

$$i_j = f\left(\sum_{h\in L_{k-1}} (W_{hj} i_h + \theta_j)\right) = f(o_j), \quad j \in L_k \tag{5.75}$$

where f is the activation function; w_{hj} is the weight associated with the connection between neuron h in layer L_{k-1} and neuron j in layer L_k (note that w_{jh} does not exist since there is no connection back from neuron j to neuron h); θ_j is the bias or threshold value associated with neuron j; and o_j is the argument of the function.

The process in the network works as follows: an array of input values is defined as the state of the neurons in the input layer. These values are transmitted to the second, hidden layer, following the protocol defined in Equation (5.75); the state of the neurons in the second layer is transmitted to the third layer where new calculations are performed to obtain the state of the neurons in this layer. The process is repeated until the output layer is reached. The state of the neurons in the output layer constitutes the output of the system. The weights and biases are not known, which requires training of the FFNN; the activation function, however, is defined within the code. Several functions are possible (linear, multiplicative, rectifier, softmax, etc.). The sigmoidal function is an example of an activation function, and is expressed as:

$$f(o_j) = \frac{1}{1+e^{-oj}} \tag{5.76}$$

The sigmoidal function is part of the family of squashing functions, which constrain the output to values in the range 0 to 1. It is a continuous function and its first derivative exists, which is necessary for the training of the FFNN.

Training of the network (i.e. to obtain the values of weights and biases) is done by comparing the output provided by the FFNN with actual results, t_m, associated with a given input. The strategy normally used is to minimize the difference between actual and predicted results using the error norm:

$$E = \sum_{m \in L_N} [t_m - f(o_m)]^2 \tag{5.77}$$

There are different strategies to minimize the error E in (5.77) by changing the values of the weights and the biases. The most common strategy is the backpropagation algorithm or delta rule (De Wilde, 1996; Schalkoff, 1997), where the derivatives of the error function E with respect to the weights or biases are set to zero; i.e. the error norm in (5.77) is minimized. Other methods are available; e.g. the Levenberg-Marquardt approximation (e.g. Singh and Singh, 2005) or the Stochastic Gradient Descent, which is often preferred because it results in faster learning (Bottou, 2010).

The delta rule results in the following equations:

$$\begin{aligned}
w_{pq}(t+1) &= w_{pq}(t) + \Delta w_{pq} + \alpha[w_{pq}(t) - w_{pq}(t-1)] \\
\theta_p(t+1) &= \theta_p(t) + \Delta\theta_p + [\theta_p(t) - \theta_p(t-1)] \\
\Delta w_{pq} &= 2\eta[t_p - f(o_p)]f'(o_p)i_q, && p \in L_N, q \in L_{N-1} \\
\Delta\theta_p &= 2\eta[t_p - f(o_p)]f'(o_p), && p \in L_N, q \in L_{N-1} \\
\Delta w_{pq} &= 2\eta \sum_{m \in L_N} [t_m - f(o_m)]f'(o_m)w_{mp}f'(o_p)i_q, && p \in L_{N-1}, q \in L_{N-2} \\
\Delta\theta_p &= 2\eta \sum_{m \in L_N} [t_m - f(o_m)]f'(o_m)w_{mp}, && p \in L_{N-1}, q \in L_{N-2}
\end{aligned} \tag{5.78}$$

The parameter t corresponds to the number of passes or epoch of the training data through the network; η is the learning rate coefficient, typically between 0 and 1; and α is the momentum coefficient, also between 0 and 1. The magnitude of the learning rate determines how fast convergence is expected, and it should depend on the shape of the error function. A larger value can be used with a smooth error function, but a smaller value should be used with rapidly changing error surfaces. The parameter α is introduced to help with convergence and to minimize oscillations in the solution. Both parameters, η and α, are pre-defined within the code.

There are no rules to design FFNNs. The input and output neurons, in terms of numbers and characteristics, are defined by the user. Thus, the user needs to decide what are the variables that may affect the results and what are the results needed. The number of hidden layers and the number of neurons per layer are problem-dependent. Increasing the number of neurons and/or hidden layers does not necessarily result in better predictions. In fact overfitting the FFNN is a real danger, which may induce erroneous results. The strategy often followed consists of dividing the available data into two sets: one for training and the other for validation. A number of strategies can be tested with different number of hidden layers, different number of neurons per hidden layer, and number of epochs for training. Each trained FFNN is then tested against the validation data, selecting the FFNN with the smallest differences (Suwansawat and Einstein, 2006). There is no guarantee, however, that the process described will result in at least one of the FFNNs providing satisfactory results. Once the FFNN is trained and selected it can be used for predictive purposes. It is very important to realize that the FFNN should not be used to make predictions outside the range of cases within which it has been trained.

Despite the shortcomings of the FFNNs, they have been successful in giving accurate predictions to problems that cannot be solved following a mechanistic approach because some of the inputs or conditions needed are not well defined or the input data may be not completely reliable. FFNNs are being used in many fields of geoengineering. For example FFNNs have been applied to obtain soil and rock properties (Meulenkamp and Alvarez Grima, 1999; Singh et al., 2001; Yang and Rosenbaum, 2002; Kahraman et al., 2006; Singh et al., 2006; Sonmez et al., 2006) including soil liquefaction (Goh, 1994; Young-Su and Byung-Tak, 2006), slope stability (Deng and Lee, 2001; Sakellariou and Ferentinou, 2005), deep excavation deformations (Chua and Goh, 2005), mining and tunneling support (Feng et al., 1996; Yang and Zhang, 1997; Deng et al., 2003; Feng and An, 2004; Deb et al., 2006). Within tunneling, FFNNs have found a particular niche in estimating ground deformations due to tunneling (Shi et al., 1998; Lueke and Ariaratnam, 2006; Neaupane and Adhikari, 2006; Suwansawat and Einstein, 2006) and predicting tunneling operation performance, where FFNNs were used by Gajewski and Jonak (2006) and Jonak and Gajewski (2006) to estimate cutting tools status during excavation, by Javadi (2006) to predict air loss during excavation, and by Singh and Singh (2005) and Kabiesz (2006) to estimate ground vibrations during mining operations. FFNNs have also been coupled with FEM, where the Finite Element Method is used to solve the mechanics of the problem or to produce the data for training the FFNN, or the FFNN is used to obtain input parameters for the FEM from back-calculation or to make predictions based on input data from the FEM (Pichler et al., 2003).

As mentioned, other DL methods are increasingly being used in tunneling, and it is expected that they will become dominant in the near future. This is shown in Jong et al. (2021), who compiled the number of citations on AI applications to underground structures, for the period 2011–2021. While the largest number of citations went to ANNs (25% of the total), the numbers steadily decreased over the time period investigated.

5.9 DISCUSSION AND RECOMMENDATIONS

Numerical methods are tools that the engineer has to evaluate qualitatively and quantitatively the effects of geology on the design and the consequences of the design on geology. The methods can be used both in a forward analysis where, given geometry and properties, results are obtained (e.g. stresses, displacements), or on a backward analysis where, given results or measurements, ground properties or ground behavior are approximated.

In any analysis, the following needs to be determined: geometry, geology in terms of layers, depth, size, etc.; appropriate boundary conditions; material properties and behavior such as elastic, plastic, visco-elastic, etc.; and construction process. Without exception, all the details and complexities of the problem cannot be introduced into the numerical model. This is so because in many cases the geology and material behavior are not fully known, the actual construction process cannot be predicted, or the numerical model is necessarily applied to a limited volume of the entire domain. In any case, assumptions and decisions need to be made. The goal is to create a model that is simple enough such that it can be implemented and interpreted within a reasonable amount of time, and yet it is accurate enough such that the results sufficiently approximate the performance of the design.

All numerical models visited in this chapter are capable of providing reasonable results when sound engineering judgment is employed with their use. A word of caution needs to be added for Artificial Neural Networks since their use should be confined within the range of the database employed for their training.

The largest portion of time spent in modeling is during pre-processing or discretization and post-processing or results analysis. It is perhaps for this reason that the most used numerical methods in practice are those that include user-friendly pre- and post-processing capabilities. These are almost exclusively commercial codes. The following is a list of the codes most referenced in the literature: Finite Difference Method: FLAC and FLAC^{3D} (ITASCA Consulting Group, Inc.); Finite Element Method: ABAQUS (Hibbit, Karlson and Sorensen, Inc.), ADINA (ADINA R & D Inc.), ANSYS (ANSYS, Inc.), MIDAS (MIDAS IT) PENTAGON-2D and -3D (Emerald Soft), PHASE^2 (Rockscience), PLAXIS (Plaxis BV); Boundary Element Method: BEFE (coupled BEM-FEM, Computer Software and Services (CSS)), EXAMINE ^{2D} and EXAMINE^{3D} (Rockscience); Distinct Element Method: UDEC, 3DEC (ITASCA Consulting Group, Inc.), EDEM (DEM Solutions) and LDEC (Licensed by Lawrence Livermore National Laboratory); and Bonded Particle Method: PFC^{2D} and PFC^{3D} (ITASCA Consulting Group, Inc.). All codes are based on the principles of mechanics and they rigorously solve (in the context of numerical solutions) equilibrium equations, boundary conditions, strain compatibility, and the constitutive material model (note that EXAMINE can only consider elastic analysis). The choice between one code and another, within the realm of continuum or discontinuum, is often based on personal or company preferences. All codes have a very steep learning curve and it may take significant time and effort for a company/institution to train engineers in any one particular software. Thus there is a tendency to keep the expertise within a very reduced number of numerical codes.

The codes listed can be divided into Continuum (FLAC, ABAQUS, ADINA, ANSYS, MIDAS, PENTAGON, PHASE, PLAXIS, EXAMINE, BEFE) and Discontinuum (UDEC, 3DEC, EDEM, LDEC, PFC). For soils, it is often assumed that a continuum approach is appropriate. For rocks however, there are no guidelines to decide when a continuum or a discontinuum model should be used. If very few discontinuities are present in the medium, a continuum model can still be efficient; with a large number of discontinuities (e.g. the size of the blocks determined by the discontinuities is much smaller than the size of the opening) a pseudo-continuum model can still be applied. When the size of the blocks is of the same order as the opening, a discontinuum model seems more reasonable. This issue is still under debate; on the one hand there is a large experience-based on continuum models successfully used in rock masses, but on the other hand there is mounting evidence that in discontinuous media the stress field obtained with a continuous model does not compare well with the stress jumps across discontinuities predicted by discontinuous models (Cundall, 1987; Barton, 2006).

With all the codes, an idealization and a discretization of the medium is required. While each problem is different, and thus it requires unique decisions, there are some common issues that can be analyzed. The following provides a discussion and some recommendations regarding 2D versus 3D model and discretization.

5.9.1 2D versus 3D

The problems to address can be divided into construction and operation. During construction, a tunnel is a 3D structure and thus requires a 3D model. During operation the tunnel has its axial dimension much larger than the other two dimensions, and thus it may be modeled as 2D plane strain, with the plane strain direction along the axis of the tunnel; examples include internal fluid pressure in pressure tunnels, seismic design, and secondary support. Note that this does not include caverns, which are 3D structures during construction and operation; it also neglects stress history, which may or may not be significant.

Three-dimensional models are still time consuming to prepare and to interpret. The tendency until recent years has been to rely on computations performed with 2D models and allow for some approximations for the model and/or interpretation of the results. Today very efficient pre- and post-processing capabilities available with commercial codes make it possible to attempt 3D modeling. Indeed, an increasing number of publications with 3D models have been observed in the technical literature in recent years.

In 2D models, an assumption often made is to place the liner at the same time as the ground is excavated (Figure 5.29a; see discussion in Section 4.3). The result is that all the deformations of the ground are transferred to the liner; this produces stresses in the tunnel support that tend to be too large and deformations in the ground that tend to be too small. This is because, as the tunnel is excavated, the soil ahead of the tunnel deforms and so only a fraction of the ground deformations are taken by the liner. Two methods have been used to account for three-dimensional effects in two-dimensional models: the load reduction method and the stiffness reduction method (e.g. Swoboda, 1990; Golser and Schubert, 2003). Further comments are provided in Section 4.3. With the load reduction method (Figure 5.29b), a fictional internal stress support of magnitude $\beta\ \sigma_0$ (σ_0 is the far-field stress in Figure 5.29a) is applied to the perimeter of the tunnel to simulate the longitudinal arching provided by the unexcavated portion of the tunnel. With the stiffness reduction method (Figure 5.29c) the excavated portion of the ground retains a stiffness with magnitude α E, where E is the stiffness of the ground. The methods are equivalent to assuming a delayed support installation, where the liner is placed after some displacements of the ground have occurred. The parameters α and β are chosen such that the magnitudes of the final stresses in the liner and ground displacements correspond to those obtained in the field. In problems without symmetry, complex excavation and support process, or with plastic behavior, an estimate of the magnitude of the parameters requires prior knowledge of the deformations induced in the ground during excavation; a certain degree of conservatism and sound engineering judgment should be exercised when estimating the parameters α or β (see discussion and criticism of the methods in Section 4.3.6).

A note of caution is needed when using a 2D (typically plane strain) approach because of the inability of the two-dimensional model to follow the stress history that occurs in the ground ahead and behind the face of the tunnel (see Section 1.2). Cantieni and Anagnostou (2009) show that a point on the perimeter of the excavation undergoes a full cycle of stress reversal, which cannot be captured by a 2D model where the radial stress follows a monotonic unloading path. In general, the errors induced by the plane strain assumption increase with the stiffness of the liner and with the length of the unsupported tunnel because of the larger stress reversal associated with these situations (Cantieni and Anagnostou, 2009).

In three-dimensional analyses, the tunnel is placed in a ground volume delimited by the boundaries of the discretization (see next section for a discussion about placement of the boundaries), as shown in Figure 5.30a. Modeling should be conducted following the actual construction process. This is necessary when an elasto-plastic ground model is introduced in the simulation as the results are stress path-dependent; i.e. the results depend on the stress history, which is defined by the construction process. This is not the case for elastic analyses where the solution is independent of the construction process and depends only on the initial and final states. Tunnel construction in the 3D model may be simulated by a successive removal/deactivation of elements at the face of the tunnel to simulate excavation and addition/activation of elements behind the face to simulate liner installation. Figure 5.30b

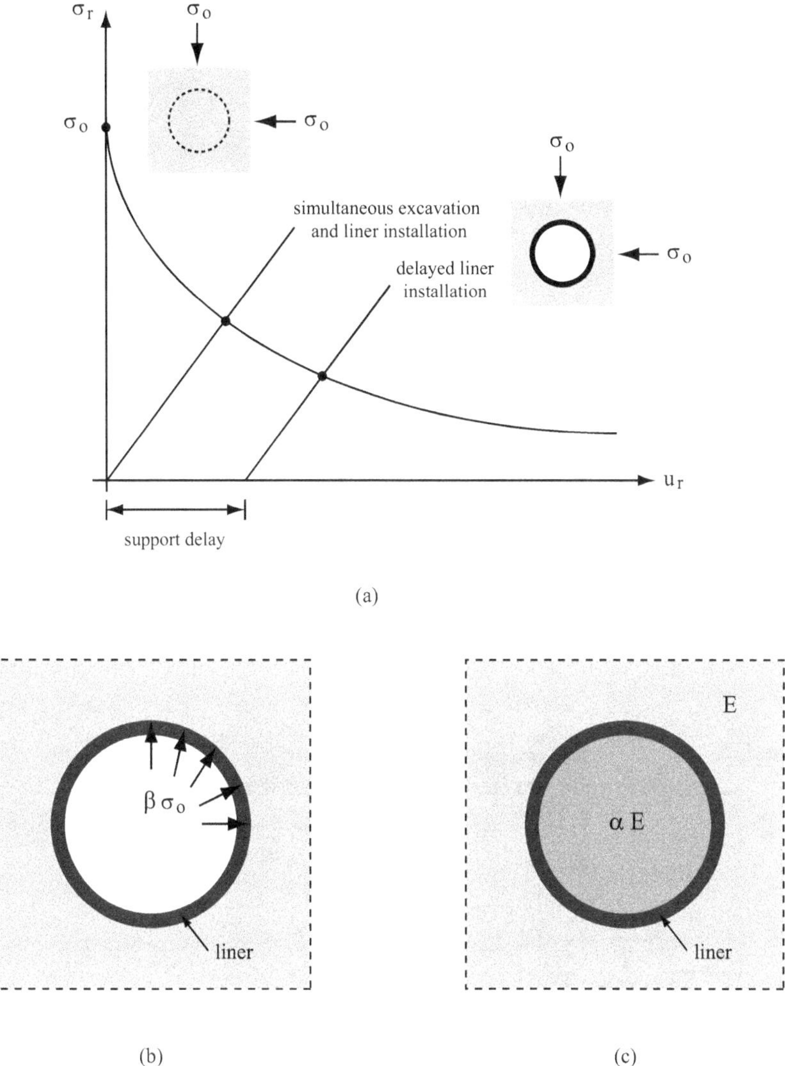

Figure 5.29 2D approximation to 3D analysis. (a) Characteristic curves. (b) Load reduction method. (c) Stiffness reduction method.

illustrates the construction sequence described. The process of excavation and support installation is followed from start of construction to the end. For steady-state conditions, the start of the simulation can be at an arbitrary location and the process of construction is simulated until steady-state conditions are found; in other words, until the results obtained at the end of an excavation-support cycle do not differ from the results obtained at the end of the previous cycle (Franzius and Potts, 2005). Figure 5.31 is an example of the results obtained following the process described. The figure plots the radial and face displacements of the ground around the tunnel once steady-state conditions are reached. The solution corresponds to a tunnel with a circular cross section of 8 m diameter, a liner 0.4 m thick, unsupported length of 2 m, and with an advancement rate of 5 m per day. The tunnel is assumed deep with in situ stresses of 20 MPa and is modeled assuming axisymmetric conditions. The Cam-clay model is used for the ground, a normally consolidated clay, with the following properties:

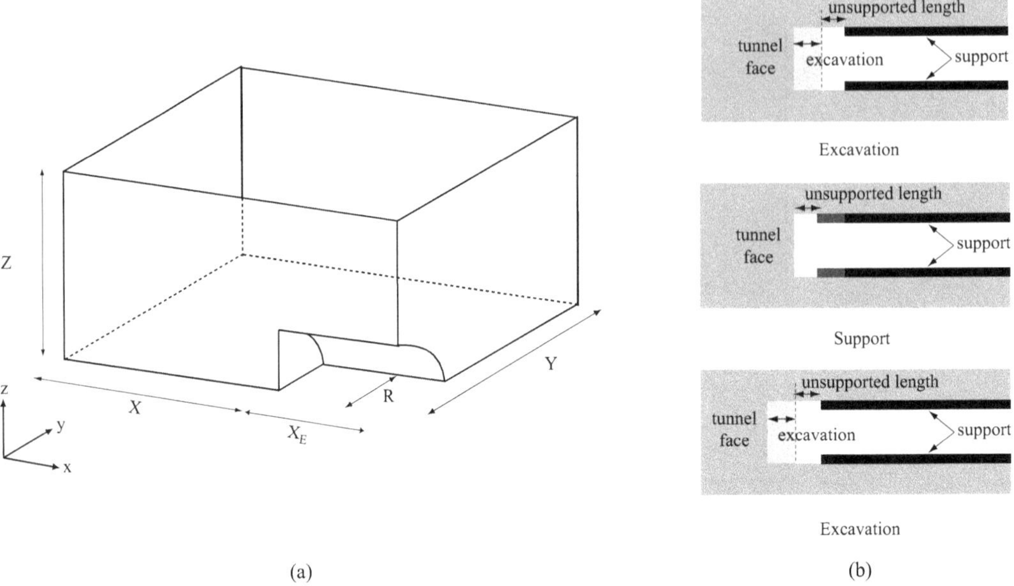

Figure 5.30 3D modeling with continuum models. (a) Discretization dimensions. (b) Excavation and support cycle.

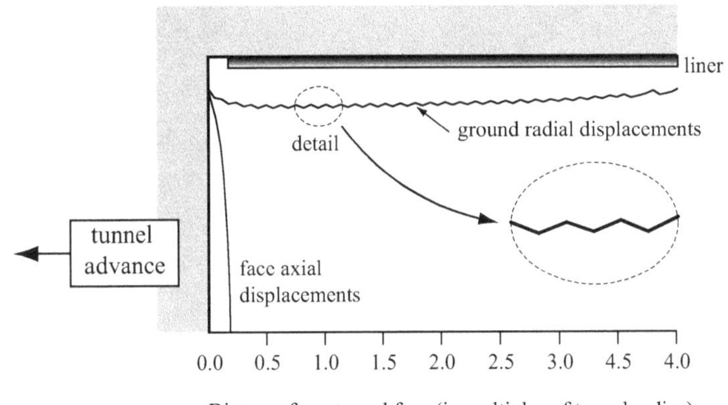

Figure 5.31 Ground displacements at tunnel face and tunnel perimeter from FEM analysis (displacements increase by a factor of 10).

compression index $C_c = 0.22$; swell index $C_s = 0.044$, internal friction angle $\phi = 23°$, permeability $K = 10^{-8}$ cm/sec, initial void ratio $e_o = 0.44$, coefficient of earth pressure at rest $K_o = 0.5$, and for the concrete liner, Young's modulus E = 21.7 GPa and Poisson's ratio $\nu = 0.25$ (Einstein and Bobet, 1997). What is interesting to see in the figure is the saw-tooth shape of the radial displacements (see detail). This is observed even when the step increments used in the finite element model are relatively small, 0.5 m per step. Note that the distance from peak to peak corresponds to the step used in the simulation, i.e. 0.5 m. This is caused by the "hard contact" between the liner and the soil at the end of the liner closer to the face (left end of the liner in the figure). The saw-tooth shape is also observed for the distribution of radial and tangential stresses in the liner since they are related, as they should, to displacements.

The amplitude and wave length of the saw-tooth shape decrease as the length of the step used in the simulations decreases. An alternative method for homogeneous problems and steady-state simulations has been developed by Anagnostou (Anagnostou, 2007a,b, 2008) using a Lagrangian coordinate system attached to the face of the tunnel. The advantage of the approach is that a solution for problems with steady-state conditions can be reached in one step; also, the solution provides smooth results, i.e. the saw-tooth shape of the results due to the discrete step process disappears. The disadvantage is that it is quite difficult to implement the method in commercial codes because the Lagrangian coordinate requires modification of Equation (5.12), which results in a non-symmetric system of equations.

The material models found in the literature are many and quite different, which is expected given the different materials encountered in tunneling. The most common material models used are Cam-Clay, Drucker-Prager, Mohr-Coulomb for soils, and Mohr-Coulomb and Hoek-Brown for rocks, with associated and non-associated flow rules. For discontinuities, the model most used is Mohr-Coulomb with constant or stress-dependent shear and normal stiffness. These models are the most popular, in part because they are available in the standard material libraries of commercial numerical codes. It has to be mentioned that important recent advances have been made on the development of coupled thermo-hydro-mechanical models for geomaterials (THM models) for nuclear repositories (e.g. Tsang et al., 2005). It is expected that such models will eventually become routinely used in tunneling. The support is often assumed elastic, with properties either constant or changing with time to simulate hardening of concrete or shotcrete materials (Oreste and Peila, 1997). Recently, material models for shotcrete based on chemo-plasticity have been proposed and successfully used that incorporate time-dependent behavior and early-age shrinkage and creep (Hellmich et al., 1999, 2000; Boldini et al., 2005).

5.9.2 Discretization

With any numerical method, only a portion of the entire domain is discretized for analysis. The exception may be Boundary Element Methods where only the boundaries are discretized. Since realistically with BEM only elastic materials can be considered, this method is often coupled with FEM (e.g. BEFE code), and thus a discretization is still required.

The question arises as to where the boundaries should be placed and how to discretize the volume of ground inside the boundaries. Boundaries are artificial limits imposed by the model to the entire domain, and thus the overriding requirement is that the presence of the boundaries should not change the results.

Considerable experience exists with continuum models, but there are no guidelines that can be applied to all problems. For static analysis and for problems that do not involve pore pressures or groundwater flow (i.e. dry ground conditions), the boundaries perpendicular to the tunnel axis (see Figure 5.30a) should be placed at a distance of about eight to ten tunnel radii, $Y = Z = 8\text{-}10R$, and the boundaries along the tunnel axis at six to eight times the tunnel radius behind the face ($X_E = 6\text{-}8R$) and five to six times the radius ahead of the face ($X = 5\text{-}6R$). For problems with coupled pore pressures (e.g. water flow toward the tunnel, short-term analysis) the boundaries should be placed much farther. It has been recommended $X = Y = 20R$ and $X_E = 6\text{-}8R$ and $X = 15\text{-}30R$ (Bobet, 2003; Nam and Bobet, 2006) for elastic analysis and for elasto-plastic, $X_E > 20R$ and $X = 26R$ (Franzius and Potts, 2005). Numerical results of tunnels below the water table with drainage toward the tunnel indicate that the region investigated should be as large as the actual region to obtain realistic results for the consolidation time, i.e. when steady-state conditions are reached (Einstein et al., 1995a). Again, the dimensions and recommendations provided should be taken as indicative and may depend on the dimensions of the tunnel, depth, loading, and material model used. For an interesting discussion on boundaries and element types and sizes, applied to tunneling, see Vitali et al. (2018a).

Dynamic analyses often involve imposing the earthquake motions at the base of the discretization. The lateral boundaries should not reflect the waves impinging on them. There are different approaches that can be taken to minimize the effects of the lateral boundaries such as free boundary far from the tunnel, dashpots, special algorithms for "quiet" boundaries, which are usually code-dependent, or infinite elements. Depending on the type of boundary chosen, and the size of the tunnel, the distance at which the boundary needs to be placed will change.

It is recommended that, in both static and dynamic analyses, different runs be performed with the boundaries at increasing distances from the tunnel. The boundaries can then be placed at a distance beyond which the solution does not change.

There is not much experience yet with discontinuum models. As a rule of thumb one may expect that the boundaries should be placed at a distance larger than that recommended for continuum models and that includes a sufficient number of blocks.

The far-field stresses, if imposed at the boundaries, have in general a triangular (if the discretization reaches the surface) or a trapezoidal shape. For tunnels with an overburden equal or larger than ten times the tunnel radius ($Z \geq 10R$, with Z measured from the surface) the results can be sufficiently approximated with constant stress distributions such that the vertical stress computed at the center of the tunnel is applied to the top boundary and the horizontal stress, also computed at the center of the tunnel, is applied to the lateral boundaries (Einstein and Schwartz, 1979; Bobet, 2003).

For discontinuum models, the discretization is determined by the existing discontinuities or by the size of the ground grains. For continuum models, the discretization is somewhat arbitrary. A larger density of elements or nodes should be placed in regions where the solution changes rapidly (i.e. large stress gradients) or where results are needed (e.g. liner and near the liner). In continuum models the solution tends to improve as the number of elements increases (i.e. the size of the elements decrease). There are, however, no guidelines as for the optimum number or size of elements, so similar to what is recommended for the location of the boundaries, several trials should be run with increasing number of elements. The discretization with the smallest number of elements that produces the same solution as discretizations with increasing number of elements can be chosen for the analyses. As for the type of elements in FEM, there are reports of volumetric locking with low-order elements in elasto-plastic analysis (e.g. constant strain, three-node triangular elements), so it is advisable to use, if possible, higher-order isoparametric elements (BTS and ICE, 2004).

Chapter 6

Special cases

6.1 INTRODUCTION

The preceding chapters provide a general framework, either based on empiricism or on first principles through analytical or numerical formulations. They are general enough such that their range of applicability is broad. There are, however, a number of particular applications of tunneling where the information provided, albeit still applicable, may not be sufficient given the specialized nature of the tunnel and ground conditions. The objective of this chapter is to cover a few of those special cases. Clearly, it is not possible in a single chapter, or even a single volume, to cover all possibilities, and so it was decided to include these special cases that are either common in practice, in the authors' opinion, or scenarios where the authors have a particular interest.

It was decided to include the following special cases in this chapter: Seismic Design of Tunnels, in Section 6.2; Pressure Tunnels, in Section 6.3; Tunnels in Rockburst areas, in Section 6.4; and Tunnels in Swelling Ground in Section 6.45. At the end of each section, a short summary and discussion/recommendations from the authors are provided.

6.2 SEISMIC DESIGN

6.2.1 Introduction

Underground structures are considered safer than aboveground structures. This idea is supported by overwhelming data, which show significant differences of damage produced during recent earthquakes on structures placed on the ground surface and structures placed underground (Ichikawa et al., 1982; Abdel Salam and Abdel Fattah, 1994; Law and Lam, 1999; Gazetas et al., 2005). However, during earthquakes in Japan in 1995, in Turkey and Taiwan in 1999, severe damage and even collapse occurred in a number of underground structures. The Hyogoken-Nambu earthquake of January 17, 1995, in Japan, caused the failure of the Daikai station, a cut and cover structure in the subway system in Kobe, and produced severe damage to the Kosaku Nagata and Sannomiya stations on the same line and to an underground shopping mall near the Shinkaichi station (Iida et al., 1996; Shawky and Maekawa, 1996; Huo et al., 2005), as well as significant damage to the Shinkansen (the Japanese high-speed train) tunnels (Asakura and Sato, 1996). The November 1999 Turkey earthquake contributed to the failure of the Bolu tunnel that was under construction (Dalgiç, 2002), the Chi-chi earthquake in Taiwan in September 1999 caused damage to a large number of mountain tunnels (Wang et al., 2001), as well as to underground gas and water pipelines (Chen et al., 2002), while the 2008 Wenchuan earthquake was responsible for the collapse of the Longxi tunnel (Yu et al., 2016). Widespread lifeline damage (oil, gas, water, electricity)

DOI: 10.1201/9781003328940-6

was also experienced during the San Fernando earthquake in 1971. Also, during the 1994 Northridge earthquake, extensive lateral crushing occurred in the corrugated metal pipe used for the outlet of the reconstructed San Fernando Dam (Bardet and Davis, 1995), which was caused by large pore pressures in the soil around the pipe.

Dowding and Rozen (1978) studied the response of 71 tunnels to earthquake motions. They recorded different levels of damage ranging from cracking to closure in 42 of the observations (59%). The number of cases was later increased by Owen and Scholl (1981) with 56 additional tunnels. The database was further expanded by Sharma and Judd (1991) who compiled information on 192 tunnels from 85 earthquakes throughout the world; in 94 cases (49%) the tunnels suffered from small to heavy damage. Wang et al. (2001) compiled the damage observed in mountain tunnels after the 1999 Chi-chi earthquake. Of a total of 57 tunnels investigated, 49 (86%) suffered some damage. The data show that shallow tunnels are at greater risk during an earthquake than deeper tunnels; roughly 60% of the total cases with overburden depths less than 50 m suffered some damage. The type of ground is also a factor; 79% of the openings in soils were reported to have suffered some damage while excavations in rock were much less likely to suffer damage. The data also show that openings in low-strength soils are at greater risk than in high strength soils and rocks. Earthquake parameters, such as magnitude and epicentral distance are also important. More than half the damage reports were for events that exceeded magnitude 7 in the Richter scale. Additionally, nearly 75% of the cases where damage was reported were within 50 km of the earthquake epicenter. The data analyzed was conclusive in that there was no damage in tunnels with Peak horizontal Ground Accelerations (PGA) up to 0.19 g (accelerations were measured on the ground surface, above the tunnel). In most of the cases where damage was reported, the peak ground accelerations were larger than 0.4 g. Total collapse of the structure was more likely at PGA larger than 0.5–0.55g. While the databases are useful to identify critical parameters and provide a qualitative assessment of tunnel response during earthquakes, they do not contain sufficient information to quantitatively determine the relative importance of each parameter or to predict damage or lack of damage (Owen and Scholl, 1985).

The field evidence compiled shows that tunnels are vulnerable to seismic damage. As discussed, roughly 50–70% of the inspected tunnels have suffered moderate to heavy damage during a strong earthquake. Clearly, this level of damage is unacceptable and shows a systematic, widespread problem with the seismic design of tunnels. The apparent discrepancy between the general perception of safety and the observed damage has been rooted in the poor understanding of the load transfer mechanisms between the ground and the structure and the effects of seismic-induced distortions on the structure's behavior, which in the past has led engineers to ignore earthquake loadings for tunnel design in seismic areas.

6.2.2 Dynamic loading and structure response

The difference between above-ground and below-ground structures is that structures placed on the surface and subjected to ground motions have a response determined by inertia forces and their displacements can be significantly different from those imposed by the ground. Design methods for aboveground structures typically involve the application of pseudo static forces, which approximate dynamic-induced forces. Underground structures, however, are constrained by the surrounding ground, and thus it is unlikely that their displacements differ significantly from those of the ground. Hence, their analysis should be based on the displacements imposed by the ground rather than on inertia loading.

Dynamic-induced deformations can be imposed on an underground structure by three different mechanisms: (1) ground failure, (2) direct shearing displacements of active faults intersecting the structure, and (3) ground shaking.

Ground failure includes rock or soil sliding at the tunnel portals or liquefaction of the ground surrounding the structure. Liquefaction results in a large reduction of the shear strength of the soil and extensive ground deformations. It is not realistic to design underground structures to withstand or accommodate such deformations, and it is generally advisable to avoid soils that can liquefy or take measures during construction to prevent liquefaction such as soil densification, soil treatment or by providing isolation mechanisms to the structure to prevent flotation (Hashash et al., 2001).

It is not always possible to avoid crossing active faults. The solution for fault crossing of large openings consists of designing the tunnels such that the deformations imposed by the fault can be accommodated without much distress of the structure, rather than trying to provide support and prevent deformations. For relatively small steel pipelines (1 to 2 m diameter) an increase of thickness of the pipe has been sufficient to resist imposed displacements without structural damage. For larger structures, there are two methods that have been successfully used: (1) placement of a frangible material between the tunnel support and the surrounding ground (backpacking); and (2) construction of a "seismic chamber" that is large enough to absorb the total accumulated displacements of the fault. A frangible material has a very low stiffness compared to the tunnel support or has stress-strain response with a long-yield plateau. Examples include low-strength concrete and granular materials (Desai et al., 1989). The fault deformations are absorbed by the frangible material without transferring much of the stresses to the tunnel support. The width and length of the seismic chamber are a function of the amount of displacement expected from the fault and the maximum curvature allowed to road, track, or pipe contained in the chamber (Hashash et al., 2001). Each method requires an over-excavation at the fault crossing and steel reinforcement of the ground support to resist bending. In addition, systematic maintenance of the section crossing the fault as well as periodic realignment of the inner tunnel is necessary. Ideally, a tunnel should cross a fault perpendicularly. If this is not possible, compression and extension may be induced in the liner, which in water-bearing areas may cause unacceptable water inflow if enough ductility has not been provided to the tunnel support; e.g. by a system of closely spaced joints in the liner. The two methods described have been successfully used. For example seismic chambers proposed by the U.S. Department of Transportation to cross a number of active faults for the California High-Speed Train Program are expected to accommodate several meters of fault displacement. A somewhat different scenario is the Claremont tunnel, a water conveyance tunnel that crosses the Hayward Fault in Oakland. The fault creeps and causes damage. The tunnel was built in 1928, and the concrete liner was damaged and eventually repaired in 1966. At that time, a large shaft was placed at the crossing location with the concept to more easily lower construction equipment for repairs wherever they are necessary. However, in 1994, it was decided to provide additional safety also to sudden fault displacement through a bypass tunnel. This bypass tunnel is larger (5.2 m diameter) compared to the original tunnel (2.7 m diameter) and, most importantly, includes a 30-m-long "steel carrier pipe" of 2.6 m diameter where the fault is crossed. This pipe is placed on movable bearings such that it can be shifted during a major fault movement. Bypass construction started in 2004 and was completed in 2007. Further details can be obtained in e.g. Caulfield et al. (2005).

Ground shaking refers to oscillating displacements produced by propagating waves through the ground. Damage to the tunnel may include cracking, spalling, and failure of the liner. The response of an underground structure to ground shaking depends on (St John and Zahrah, 1987): the shape, dimensions, and depth of the excavation, the properties of the ground where the structure is located, the properties of the support system, and the severity of the ground shaking, which is determined by the magnitude of the earthquake, the intensity of the earthquake, the frequency content, and the duration of the event (St John and Zahrah, 1985).

The characteristics of the ground shaking depend on the source of the dynamic loading: fault rupture/earthquake or explosion, and location of the tunnel with respect to the source: near- or far-field effects. Near-field motions are short in duration compared with far-field motions and involve high frequencies, of the order of tens of cycles per second and relatively few significant pulses (Pratt et al., 1979; Dowding, 1985). As a result, the particle velocities are low and the accelerations are high (Dowding, 1985). Far-field motions are characterized by longer durations and lower frequency content, between 0.1 and 10 Hz, which result in lower accelerations than near-field and higher particle velocities. Explosions close to the tunnel may have frequency content similar to that of near-field earthquake motions, but a smaller number of cycles (Dowding, 1985). The response of the tunnel changes dramatically with the frequency content of the dynamic excitation. High frequency compressive waves, which have a small wavelength (wavelength $\lambda = V/f$, where V is the wave velocity and f the frequency), can be reflected from the surface of the opening and cause slabbing when the tensile strength of the liner or of the unsupported tunnel surface is reached. Stress waves reflection from the cavity and entrapment and circulation of the waves around the tunnel may occur when the wavelengths are smaller than the tunnel radius. This is possible when a tunnel is located near the explosion or seismic source, but it may not occur in tunnels located in the far-field due to the longer wavelengths associated with the ground motion (Dowding, 1985; St John and Zahrah, 1987).

Large diameter underground structures can be classified as: (1) bored or mined tunnels; (2) cut and cover tunnels; and (3) immersed tube tunnels. The dynamic loads can be grouped, based on their source, as: (1) seismic; and (2) explosive. The following discussion applies to bored or mined and cut and cover tunnels subjected to body waves induced by earthquakes; that is, the effects of surface waves, e.g. Rayleigh and Love waves are not discussed. While there is no question that the effects of surface waves on shallow structures should also be investigated, current knowledge is very limited and much more work needs to be done. Explosions, on the one hand, pipes or other utilities and immersed tunnels, on the other hand, are beyond the scope of this work.

Underground structures, mined or cut and cover, in a seismic area may be dimensioned following a two-level design event criterion, as proposed by Wang (1993):

1. The Operating Design Earthquake (ODE). This is the earthquake event that can reasonably be expected to occur at least once during the design life of the facility. In other words, the earthquake has a probability of occurrence of 40–50%. The design goal is that the overall system shall continue to operate during and after an ODE with little to no damage, with the underground structure response remaining within its elastic regime.
2. The Maximum Design Earthquake (MDE). This is the earthquake event that has a small probability of exceedance of 3–5% during the facility life. The MDE design goal is that public safety shall be maintained during and after the event. For some facilities such as lifelines and critical transportations systems, the public safety requirement may dictate that the structure remains operational after a MDE event.

Wang (1993) proposed the following loading criteria, based on load factor design.

For ODE:

For bored or mined tunnels: $U = 1.05D + 1.3L + \beta(EX + H) + 1.3EQ$

For cut and cover structures: $U = 1.05D + 1.3L + \beta(E1 + E2) + 1.3EQ$

For MDE:

For bored or mined tunnels: $U = D + L + EX + H + EQ$

For cut and cover structures: $U = D + L + E1 + E2 + EQ$

Where U is the required structural strength capacity, D is the dead load of the structural components, L is the live load, EX is the load produced by excavation of the tunnel, H is the hydrostatic water pressure, E1 is the vertical static load due to earth and water, E2 is the horizontal static load due to earth and water, EQ is the load due to earthquake motion, β is 1.05 if extreme loads are assumed with little uncertainty, otherwise β is 1.3. Note that in the above equations, the condition L = 0 (no live load) should also be considered.

FHWA (2009) prescribes, for road tunnels, different load combinations (see Section 7.2.3). It also considers a two-level design: first level, where the tunnel survives an earthquake with a return period of 2,500 years (2% probability of exceedance in 50 years, or 3% probability of exceedance in 75 years); and second level, where the tunnel suffers minimal damage, i.e. it can be put in service immediately after the event, for an earthquake with a 50% probability of exceedance in 75 years, which corresponds to a return period of 108 years. For the second level, return periods of 500 years are also used.

There are two basic approaches in present seismic design. One approach is to carry out dynamic, non-linear soil–structure interaction analysis using finite element or finite difference methods, where inertia forces are included. The input motions in these analyses are time histories emulating design response spectra. Input motions are applied to the boundaries of a soil island to represent propagating motion waves. In the second approach, the pseudo-static approach, inertia forces are neglected. The earthquake loading is simulated as a static far-field stress, strain or displacement applied to the ground in which the structure is embedded.

Mow and Pao (1971) investigated the stress concentrations produced by impinging compressional and shear waves propagating normal to the tunnel axis. They found that the peak dynamic stress concentrations produced by compressional waves, with a wavelength 25 times larger than the size of the cavity, were 10–15% higher than stresses resulting from static analysis with far-field applied stresses equal to the peak free-field. Similar results were found for shear waves. These conclusions were supported by Hendron and Fernández (1983), Merritt et al. (1985), and Monsees and Merritt (1988) who showed that the dynamic amplification of stress waves impinging on a tunnel is negligible if the rise time of the pulse is larger than about two times the transit time of the pulse across the opening; in other words, when the wave length (λ) of peak velocities is at least eight times larger than the width (B) of the opening (Figure 6.1). The finding is in general agreement with the observations by Sandoval and Bobet (2020a) who found the response of deep circular tunnels insensitive to frequency content for λ/B ratios larger than 10 when the ground has a non-linear response (as opposed to the λ/B ratio of 8 when the response is elastic) and for drained loading. For undrained loading with accumulation of excess pore pressures in the ground and also for a non-linear behavior of the ground, the λ/B ratio should be larger than 8 to 9. In those cases, the seismic load can be considered as a pseudo static load. This is an important conclusion, which has been used to derive simple analytical formulations for the seismic design of underground structures. It is important to note that a pseudo-static analysis may be used for tunnels placed in the far field, where frequencies of the ground motion are within the 0.1 to 10 Hz range. For example, following Hendron's and Fernández's (1983) criterion, the size of a tunnel in a relatively soft soil, with a shear wave velocity of 200 m/s, with a 5 Hz earthquake frequency should be at most 5 m. For stiffer soils and rocks, the size of the opening could be much larger. For

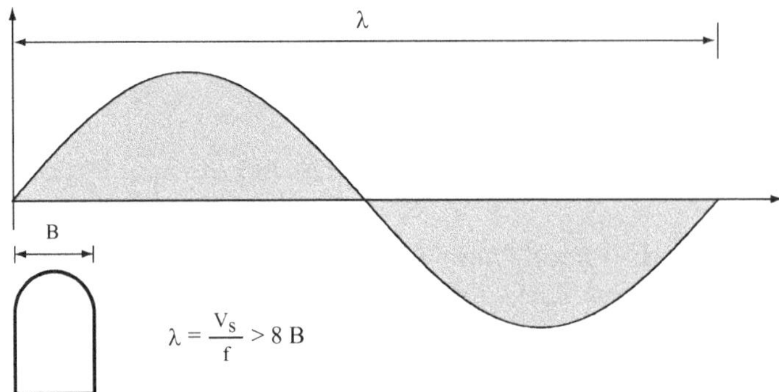

Figure 6.1 Comparison of wave length of peak velocities and size of opening. Adapted from Hendron and Fernández (1983).

the near-field, higher frequencies require a very small tunnel size (for the previous case, one to two orders of magnitude smaller) and the pseudo static approximation may not be accurate. In this case, numerical analyses should be conducted (e.g. Hwang and Lysmer, 1981; Labreche, 1983; Kirzhner and Rosenhouse, 2000; Esmaeili et al., 2006; Sandoval and Bobet, 2020b; Tsinidis et al., 2020). This issue is particularly important for tunnels in a discontinuous medium, where the stability of the blocks surrounding the excavation strongly depends on the input frequency and earthquake duration, in addition to the motion amplitude, and stability is particularly sensitive to the orientation of the discontinuities (Dowding and Belytschko, 1983; Dowding, 1985). Chapter 5 provides information regarding the numerical methods used for dynamic analysis of underground structures.

In the pseudo-static approach, soil–structure interaction may or (erroneously) may not be considered. When the interaction is included in the analysis, analytical solutions, finite element or finite difference methods are typically used and provide the magnitude of seismic-induced displacements or strains in the tunnel. If soil–structure interaction is neglected, it is assumed that the structure follows the deformations of the ground. This is known as the free-field approach, where relationships are obtained based on the premise that the structure must accommodate the free-field deformations without loss of its integrity. This approach is presented here because it is still used, given its simplicity but, as will be shown, should be either abandoned or, at most, applied with extreme caution, and only as a first step before using the more complex, and realistic, methods discussed in the following sections. The free-field method could be useful as "back of the envelope" calculation to have a sense of the "order of magnitude" of the tunnel deformations induced by the earthquake.

6.2.3 Free-field deformations

There are two types of seismic waves in an infinite medium: compressional or P-waves and shear or S-waves. Compressional waves produce motions parallel to the direction of propagation while shear waves produce motions perpendicular to the direction of propagation. See Figure 6.2. Near the surface, two other waves are observed: Rayleigh and Love waves.

Figure 6.3 shows an S-wave in an infinite medium propagating at an angle ϕ with the axis of the tunnel and inducing motions along a plane that makes an angle β with the tunnel axis. The displacement field is given by (here we follow the work by Kouretzis et al., 2006; see also St John and Zahrah, 1987):

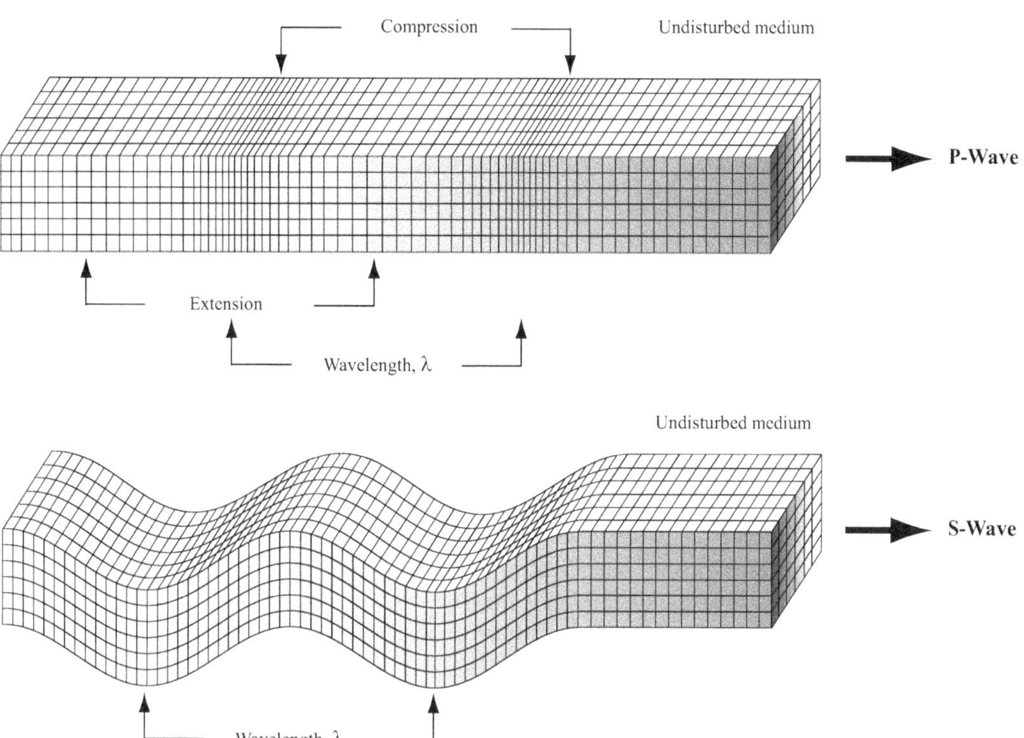

Figure 6.2 Body waves.

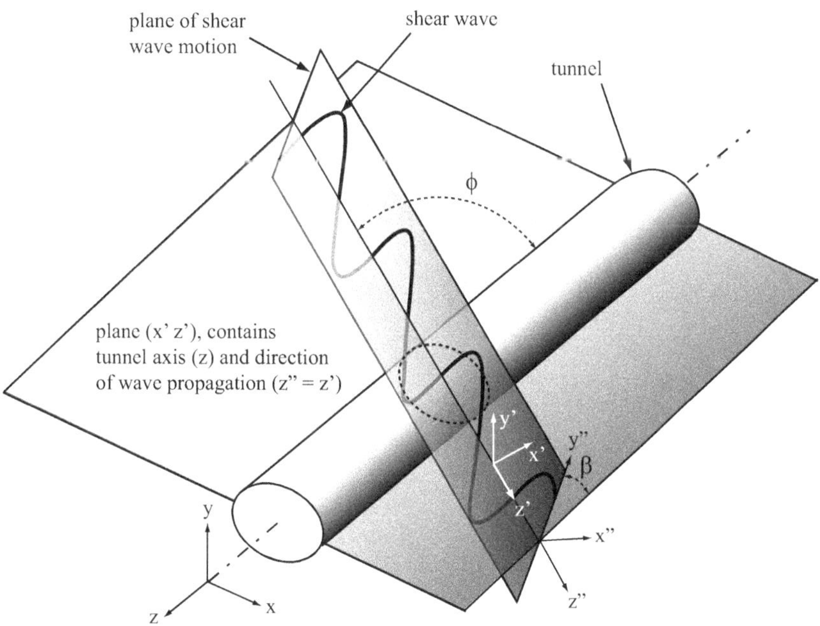

Figure 6.3 Shear wave impinging on a tunnel. Adapted from Kouretzis et al. (2006).

$$U_{x''} = 0$$
$$U_{y''} = U^{s}_{max} \sin\left[\frac{2\pi}{\lambda}(z' - V_s t)\right] \quad (6.1)$$
$$U_{z''} = 0$$

where U is the displacement at a point with coordinates x″, y″, z″ = z′, U^{s}_{max} is the amplitude of the displacement motion, λ is the wavelength, V_s is the velocity of the shear wave, and t is time. Note that Equation (6.1) satisfies the equation of S-wave propagation in an infinite medium, given by (e.g. Kramer, 1996):

$$\frac{\partial^2}{\partial t^2}\left(\frac{\partial U_{z''}}{\partial y''} - \frac{\partial U_{y''}}{\partial z''}\right) = V_s^2\left(\frac{\partial^2}{\partial x''^2} + \frac{\partial^2}{\partial y''^2} + \frac{\partial^2}{\partial z''^2}\right)\left(\frac{\partial U_{z''}}{\partial y''} - \frac{\partial U_{y''}}{\partial z''}\right)$$
$$V_s = \sqrt{\frac{G}{\rho}} \quad (6.2)$$

where G is the shear modulus of the medium and ρ its density. The following discussion assumes that the presence of the tunnel does not affect the ground motions and thus strains induced in the tunnel can be computed directly from the free-field motions. Strains in a cylindrical coordinate system are given by (Figure 6.4):

$$\text{Axial strain:} \quad \varepsilon_{zz} = \frac{\partial U_z}{\partial z}$$
$$\text{Tangential strain:} \quad \varepsilon_{\theta\theta} = \frac{1}{r}\frac{\partial U_\theta}{\partial \theta} + \frac{U_r}{r} \quad (6.3)$$
$$\text{Shear strain:} \quad \gamma_{\theta z} = \frac{1}{r}\frac{\partial U_z}{\partial \theta} + \frac{\partial U_\theta}{\partial z}$$

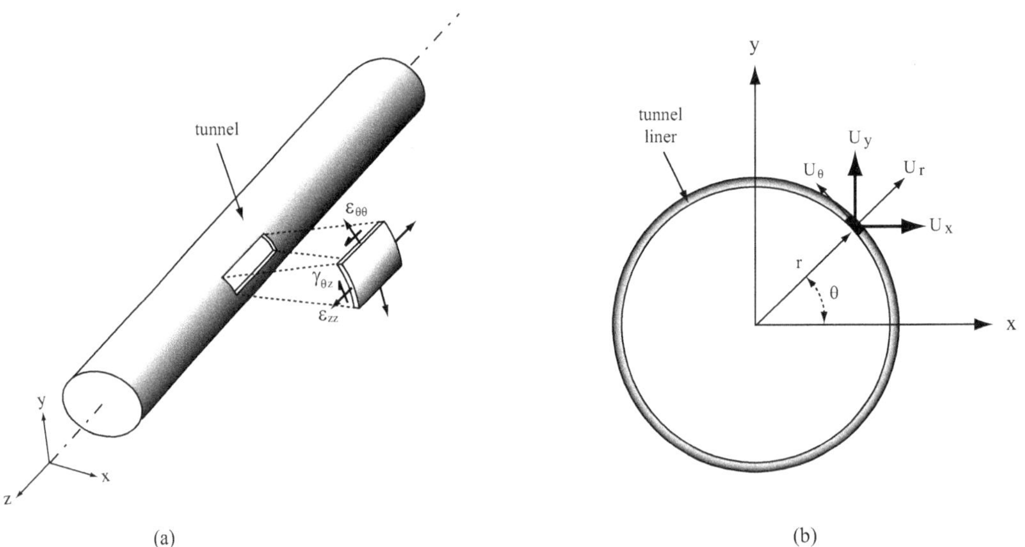

Figure 6.4 Strains and displacements at the tunnel liner. (a) Strains of the tunnel liner. (b) Displacements in cartesian and polar coordinate system.

The radial strains and the other two shear strains are not computed; they would not be significant because the radial and shear stresses at the perimeter of the opening are zero.

The shear wave given by (6.1) is decomposed into a SH wave with motions in the x'z' plane and propagating in the z' direction, and into a SV wave with motions in the y'z' plane propagating also in the z' direction.

The SH wave is defined as (Figure 6.3):

$$
\begin{aligned}
U_{x'} &= U^{s}_{max} \cos\beta \sin\left[\frac{2\pi}{\lambda}(z' - V_z t)\right] \\
U_{y'} &= 0 \\
U_{z'} &= 0
\end{aligned}
\tag{6.4}
$$

The SH wave is decomposed along the z and x axes (Figure 6.5) into (Kouretzis et al., 2006): (1) an apparent SH wave propagating along the z-axis; (2) an apparent P wave propagating along the z-axis; (3) an apparent P wave propagating along the x-axis; and (4) an apparent SH wave propagating along the x-axis. The equations of motion of each wave are:

$$
\begin{aligned}
&1.\ U_x = U^{s}_{max} \cos\beta \cos\phi \sin\left[\frac{2\pi}{\lambda/\cos\phi}\left(z - \frac{V_s}{\cos\phi} t\right)\right] \\
&2.\ U_z = -U^{s}_{max} \cos\beta \sin\phi \sin\left[\frac{2\pi}{\lambda/\cos\phi}\left(z - \frac{V_s}{\cos\phi} t\right)\right] \\
&3.\ U_x = U^{s}_{max} \cos\beta \cos\phi \sin\left[\frac{2\pi}{\lambda/\sin\phi}\left(x - \frac{V_s}{\sin\phi} t\right)\right] \\
&4.\ U_z = -U^{s}_{max} \cos\beta \sin\phi \sin\left[\frac{2\pi}{\lambda/\sin\phi}\left(x - \frac{V_s}{\sin\phi} t\right)\right]
\end{aligned}
\tag{6.5}
$$

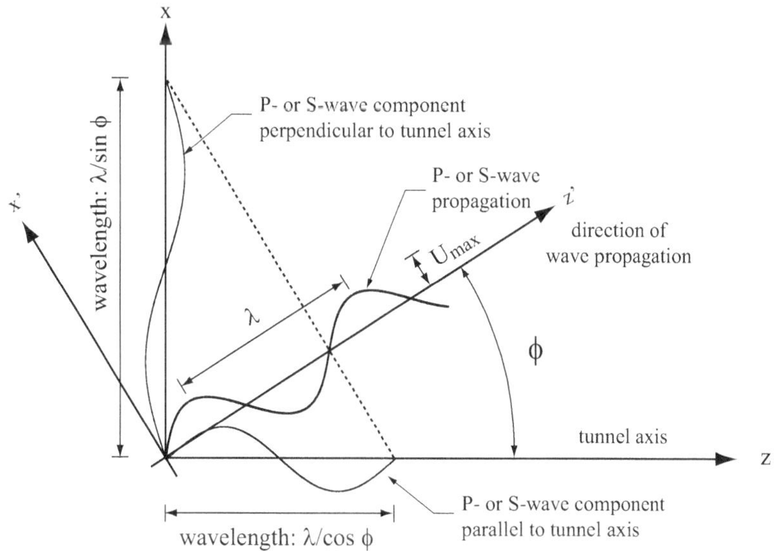

Figure 6.5 Wave decomposition into components parallel and perpendicular to tunnel axis.

The SV wave is (Figure 6.3):

$$
\begin{aligned}
&U_{x'} = 0 \\
&U_{y'} = U^{s}_{max} \sin\beta \sin\left[\frac{2\pi}{\lambda}(z' - V_s t)\right] \\
&U_{z'} = 0
\end{aligned}
\tag{6.6}
$$

The SV wave is decomposed along the z and x axes (Figure 6.5) into: (1) an apparent SV wave propagating along the z-axis; and (2) an apparent SV wave propagating along the x-axis. The equations of motion are:

$$
\begin{aligned}
&1.\ U_y = U^{s}_{max} \sin\beta \sin\left[\frac{2\pi}{\lambda/\cos\phi}\left(z - \frac{V_s}{\cos\phi} t\right)\right] \\
&2.\ U_y = U^{s}_{max} \sin\beta \sin\left[\frac{2\pi}{\lambda/\sin\phi}\left(x - \frac{V_s}{\sin\phi} t\right)\right]
\end{aligned}
\tag{6.7}
$$

The strains are computed from (6.3), (6.5), and (6.7), noting that (Figure 6.4b):

$$
\begin{aligned}
U_r &= U_x \cos\theta + U_y \sin\theta \\
U_\theta &= -U_x \sin\theta + U_y \cos\theta \\
x &= r\cos\theta
\end{aligned}
\tag{6.8}
$$

The maximum strains are (Kouretzis et al., 2006):

$$
\begin{aligned}
\varepsilon_{zz} &= \pm \frac{V^{s}_{max}}{2V_s} \cos\beta \sin 2\phi \\
\varepsilon_{\theta\theta} &= \pm \frac{V^{s}_{max}}{2V_s} (\cos\beta \sin 2\phi \sin^2\theta + \sin\beta \sin\phi \sin 2\theta) \\
\gamma_{\theta z} &= \pm \frac{V^{s}_{max}}{V_s} (\cos\beta \cos 2\phi \sin\theta + \sin\beta \cos\phi \cos\theta)
\end{aligned}
\tag{6.9}
$$

where V^{s}_{max} is the maximum particle velocity and is obtained taking the time derivative in (6.1); $V^{s}_{max} = \frac{2\pi}{\lambda} V_s U^{s}_{max}$.

A P-wave propagating at an angle ϕ with the axis of the tunnel induces motions along the direction of propagation:

$$
\begin{aligned}
&U_{x'} = 0 \\
&U_{y'} = 0 \\
&U_{z'} = U^{P}_{max} \sin\left[\frac{2\pi}{\lambda}(z' - V_p t)\right]
\end{aligned}
\tag{6.10}
$$

where the plane z′z contains the axis of the tunnel and the direction of propagation of the P-wave (Figure 6.5); U^{p}_{max} is the maximum displacement and V_p is the velocity of the compressional wave. Equation (6.10) satisfies the equation of P-wave propagation in an infinite medium, given by (Kramer, 1996):

$$\frac{\partial^2}{\partial t^2}\left(\frac{\partial U_{x'}}{\partial x'} - \frac{\partial U_{y'}}{\partial y'} + \frac{\partial U_{z'}}{\partial z'}\right) = V_p^2\left(\frac{\partial^2}{\partial x'^2} + \frac{\partial^2}{\partial y'^2} + \frac{\partial^2}{\partial z'^2}\right)\left(\frac{\partial U_{x'}}{\partial x'} + \frac{\partial U_{y'}}{\partial y'} + \frac{\partial U_{z'}}{\partial z'}\right)$$
$$V_p = \sqrt{\frac{2G(1-v)}{\rho(1-2v)}} \tag{6.11}$$

The P wave is decomposed along the z and x axes (Figure 6.5) into: (1) an apparent P wave propagating along the z axis; (2) an apparent S wave propagating along the z axis; (3) an apparent P wave propagating along the x axis; and (4) an apparent S wave propagating along the x axis. The equations of motion of each wave are:

$$\begin{aligned}
&1.\ U_z = U^P_{max}\cos\phi\sin\left[\frac{2\pi}{\lambda/\cos\phi}\left(z - \frac{V_p}{\cos\phi}t\right)\right]\\
&2.\ U_x = U^P_{max}\sin\phi\sin\left[\frac{2\pi}{\lambda/\cos\phi}\left(z - \frac{V_p}{\cos\phi}t\right)\right]\\
&3.\ U_x = U^P_{max}\sin\phi\sin\left[\frac{2\pi}{\lambda/\sin\phi}\left(x - \frac{V_p}{\sin\phi}t\right)\right]\\
&4.\ U_z = U^P_{max}\cos\phi\sin\left[\frac{2\pi}{\lambda/\sin\phi}\left(x - \frac{V_p}{\sin\phi}t\right)\right]
\end{aligned} \tag{6.12}$$

Following the same procedure as that used for S waves, the maximum strains are:

$$\begin{aligned}
\varepsilon_{zz} &= \pm\frac{V^P_{max}}{V_p}\cos^2\phi\\
\varepsilon_{\theta\theta} &= \pm\frac{V^P_{max}}{V_p}\sin^2\phi\cos^2\theta\\
\gamma_{\theta z} &= \pm\frac{V^P_{max}}{V_p}\sin 2\phi\cos\theta
\end{aligned} \tag{6.13}$$

where $V^P{}_{max}$ is the maximum particle velocity and is obtained taking the time derivative in (6.10); $V^P_{max} = \frac{2\pi}{\lambda}V_p U^P_{max}$.

The far-field displacements induce the following deformations to a tunnel (Figure 6.6): (1) axial (longitudinal) compression and extension (Figure 6.6a); (2) transverse compression and extension (Figure 6.6b); (3) axial (longitudinal) bending or snaking (Figure 6.6c); (4) axial (longitudinal) shear (Figure 6.6d); and (5) ovaling or racking (Figures 6.6e and f). Axial compression and extension is produced by components of seismic waves that produce motions parallel to the tunnel axis. Following the decomposition of the randomly oriented shear and compression waves discussed in Equations (6.1) to (6.13), axial compression and extension is caused by wave 2 of the shear wave SH decomposition given in Equation (6.5) and by wave 1 of the P-wave decomposition of Equation (6.12). Transverse compression is produced by waves with components that produce motion perpendicular to the tunnel cross

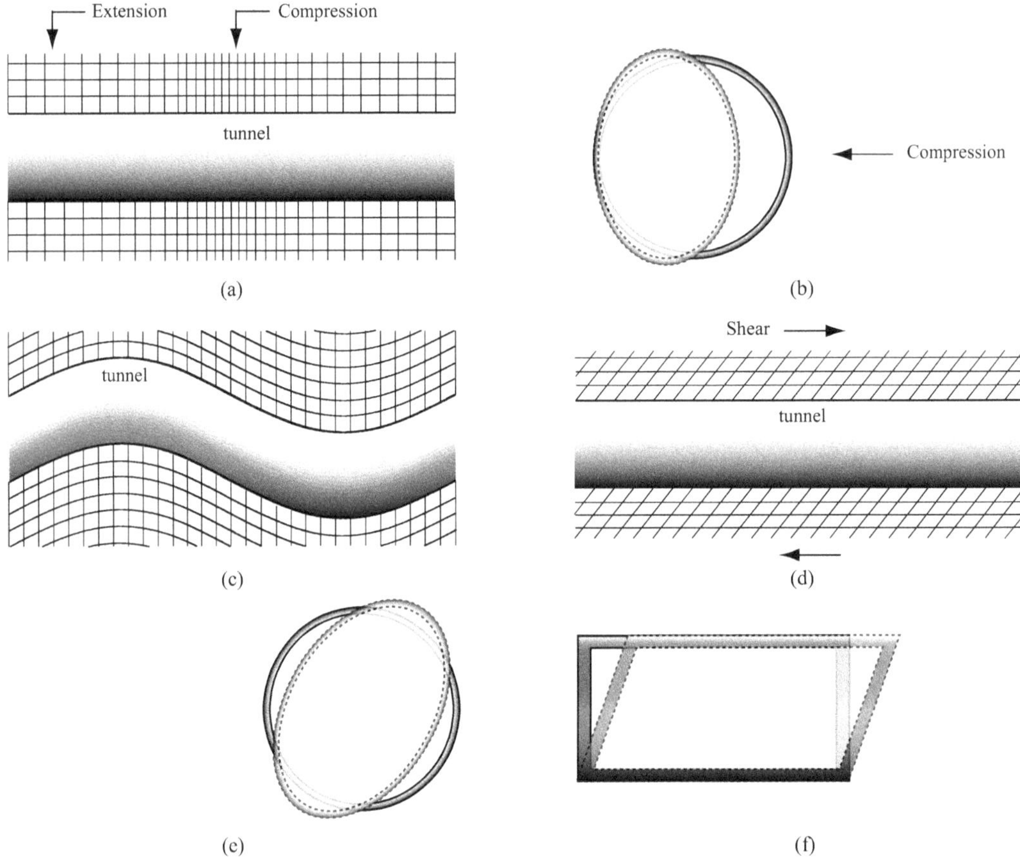

Figure 6.6 Deformation modes of tunnels due to seismic waves. (a) Axial compression-extension. (b) Transverse compression-extension. (c) Axial bending - Snaking. (d) Axial shear. (e) Ovaling. (f) Racking. Adapted from Owen and Scholl (1981).

section. This is produced by waves 3 both in Equations (6.5) and (6.12). Axial bending or snaking is caused by components of the seismic waves that produce motion perpendicular to the tunnel axis. It is caused by waves 1 in (6.5) and (6.7), and by wave 2 in (6.12). Axial shear is the result of shear waves traveling perpendicular to the tunnel axis with motions also parallel to the tunnel axis. These motions are produced by waves 4 in both (6.5) and (6.12). Finally ovaling and racking are caused by shear waves traveling perpendicular to the tunnel axis. They are the result of wave 2 in (6.7). If soil–structure interaction is considered, different modes of deformation can be produced by the same wave; e.g. a shear wave traveling along the axis of the tunnel will produce both snaking and transverse compression. In general, ovaling and racking deformations are the most critical (Wang, 1993).

Until very recently, seismic design of tunnels, when included, was done considering that the tunnel would move following the free-field deformations of the ground (Kuesel, 1969; Owen et al., 1979; Owen and Scholl, 1981; Hendron and Fernández, 1983; Merritt et al., 1985; St John and Zahrah, 1985, 1987; Monsees and Merritt, 1988; Hashash et al., 2001; Anderson et al., 2008). With this assumption, the strains of the liner are given by Equations (6.9) and (6.13). The rationale behind this was based on the idea that the approach was conservative if the stiffness of the tunnel was larger than the surrounding ground. In this case the tunnel would deform less, as the tunnel would resist the deformations of the ground, and thus the assumption of free-field displacements would overestimate the actual tunnel displacements. For flexible tunnels, in particular for tunnels with a flexibility ratio, $F = \frac{(1-v_s^2)E\,r_o^3}{6(1+v)E_s\,I_s}$ (Hendron and Fernández, 1983), larger than 20, i.e. F > 20, the liner is considered perfectly flexible, and it was believed that the tunnel would just conform to the displacements imposed by the surrounding ground. In the definition of the flexibility ratio, E and ν are the Young's modulus and Poisson's ratio of the ground, E_s and I_s are the Young's modulus and moment of inertia of the support, and r_o is the radius of the tunnel. The accuracy of this approach has been apparently supported by a number of successful comparisons between predictions of structure distortions obtained from free-field displacements and results from Finite Element dynamic analyses, where a deep structure was subjected to seismic motions applied in the far field (Merritt et al., 1985; St. John and Zahrah, 1985; Monsees and Merritt, 1988; Ichikawa et al., 1982; Pakbaz and Yareevand, 2005). It will be shown that this approach need to be used with extreme caution as it may lead to unsafe design.

The seismic design of tunnels is usually done independently for longitudinal bending (snaking) and ovaling or racking, using in each case the direction of the seismic wave (see Figure 6.3) that produces the largest strains in the liner (limited research seems to indicate that such assumption, i.e. two-dimensional analyses, may underestimate the response and that full three-dimensional analyses are needed; e.g. Yu et al., 2016, 2017. Further research is needed to reach a more definitive conclusion). For snaking, this occurs with a shear wave (shear waves produce larger strains than compressional waves) with motions in the plane defined by the axis of the tunnel and the direction of propagation (i.e. $\beta = 0°$ in Figure 6.3) and at an angle of 45° with the axis of the tunnel ($\phi = 45°$); this maximizes the axial strain ε_{zz} in (6.9). Ovaling of the tunnel is the largest for a shear wave traveling perpendicular to the tunnel axis ($\beta = 90°$ and $\phi = 90°$ in Figure 6.3), which maximizes the tangential strain $\varepsilon_{\theta\theta}$ in (6.9), while the maximum axial strains occur when $\beta = 0°$ and $\phi = 45°$. Thus the maximum strains are:

$$\begin{aligned} \varepsilon_{zz,max} &= \pm\frac{V_{max}^s}{2V_s} \\ \varepsilon_{\theta\theta,max} &= \pm\frac{V_{max}^s}{2\,V_s} \quad \text{or} \quad \gamma_{max} = \pm\frac{V_{max}^s}{V_s} \end{aligned} \tag{6.14}$$

Strains due to the curvature induced at the location of the tunnel by the traveling waves are added to those in (6.9) and (6.13). The additional strains are computed as $\varepsilon_{zz} = r_o\rho_{curv}$, or as the product of the tunnel radius, r_o, and the curvature ρ_{curv}. The curvature is given by $1/\rho_{curv} = \partial^2 U / \partial x^2$, or as the second derivative of displacements, U, with respect to the direction of motion of the seismic wave (St John and Zahrah, 1985; Wang, 1993; Hashash et al., 2001). The maximum strains recommended are:

$$\varepsilon_{zz,max} = \pm\frac{V_{max}^s}{V_s}\sin\phi\cos\phi \pm r_o\frac{a_s}{V_s^2}\cos^3\phi \tag{6.15}$$

where a_s is the peak particle acceleration and ϕ is the angle that maximizes the axial strains. The second term in the equation is the added term from curvature. Kouretzis et al. (2006) have shown that this approach is conservative.

Most soils exhibit a pronounced nonlinear behavior under significant shear loading with their shear stiffness decreasing rapidly with increasing shear strain. With load reversal, the soil experiences hysteretic behavior with the percentage of the material damping (i.e. the size of the stress-strain loop) being largely independent of frequency (e.g. Kramer, 1996). Figure 6.7 shows experimental data on soils from Seed et al. (1986) and Vucetic and Dobry (1991). The degradation of shear modulus with strain and the hysteretic behavior and damping are critical characteristics of the behavior of soils during cyclic loading, which influences the soil–structure interaction. This was early recognized by Merritt et al. (1985) and Monsees and Merritt (1988) who recommended using in the preceding calculations the shear wave velocity for the in situ conditions and for the level of strain expected. These authors even provided recommendations for seismic-induced shear strains in different soils for different earthquake magnitudes. While shear stiffness degradation with strain may not be significant for hard rocks, it is a very important issue for soils. Shear wave velocities are typically measured in the field with geophysical methods, which induce very small strains in the ground, and thus the soil's shear modulus is close to the maximum value G_{max} (Figure 6.7). Earthquakes may induce large strains in the soil, which will produce a significant reduction of the soil's shear stiffness (Figure 6.7), reducing the shear wave velocity V_s. Thus the far-field shear strains may be much larger than those obtained from the preceding equations using shear wave velocities measured at small strains.

Seismic design of underground structures, as it is already done for above ground structures, requires site-specific ground response analysis for the ODE and MDE. The computer programs SHAKE (Schnabel et al., 1972) or DEEPSOIL (Hashash et al., 2010) may be employed to estimate the free-field soil deformations for various types of ground and for any given

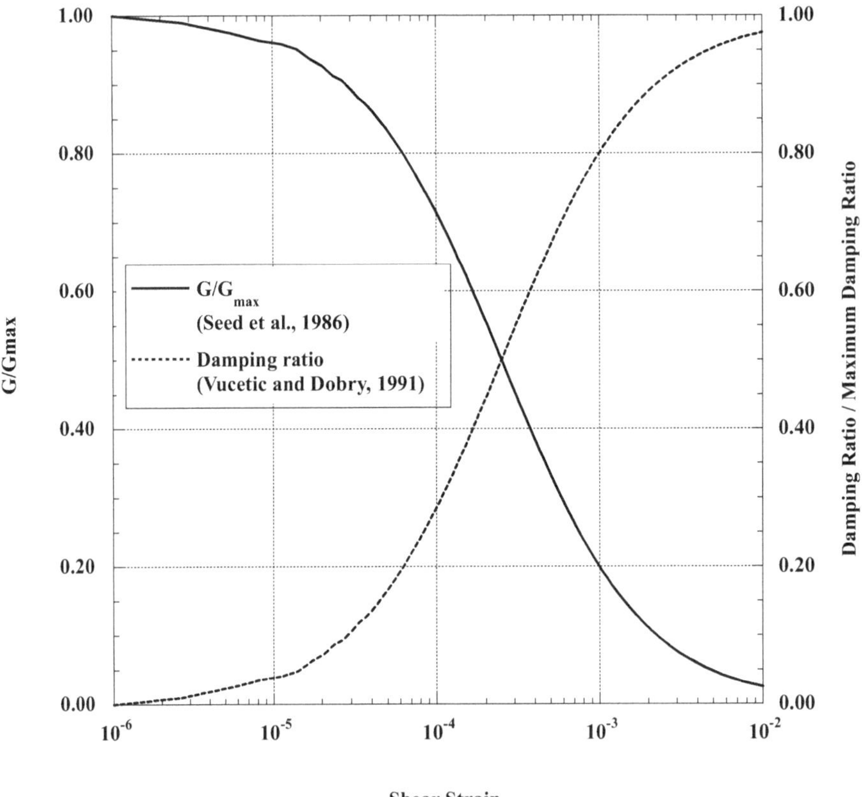

Figure 6.7 Shear modulus and damping ratio with shear strain.

ground motion. Following the rationale behind the approach of free-field ground deformations, the first step would be to obtain site-specific ground deformations using either ODE or MDE (Figure 6.8a). The structure distortion would then be the difference between the ground displacement at the top and bottom of the structure (Figure 6.8b). For circular tunnels, $\Delta r_o/r_o = \frac{1}{2}\gamma_{max}$, where r_o is the radius of the tunnel and γ_{max} is the maximum shear strain of the ground at the location of the tunnel also obtained from site-specific ground response analysis.

While the free-field method approach is very attractive because it is simple, it is an approximate method and it should be used, at best, as a first step. Caution should be used to account for the degradation of the liner and surrounding soil stiffness as cyclic ground displacements take place. The reason for the shortcomings of the free-field method is that it does not consider soil–structure interaction. If the structure is much stiffer than the surrounding ground, the design may be too conservative. As a side note and yet a very important one, it has not been uncommon for cut and cover structures to use, for their seismic design, the free-field method where the free-field shear strain is obtained using the second equation in (6.14) with the small-strain shear wave velocity. The results may seem reasonable because of two errors that somehow compensate each other. First, by using small-strain shear wave velocities, the resulting far-field shear strain is unrealistically small. Second, because cut and cover structures are much stiffer than the surrounding ground, they deform less than the free field, so the actual free-field deformations would be too large for the structure. The method, if considered, should be used with extreme caution because the factor of safety is not known from the calculations, and it is possible that the approach results in an unsafe design. If the structure is softer than the surrounding ground, the assumption that it deforms following the ground is not correct and errs on the unsafe side. Figure 6.9 shows three extreme cases corresponding to a circular deep tunnel subjected to seismic-induced deformations in the form of a quasi-static shear strain imposed in the far field. Figure 6.9a illustrates the deformation of a tunnel that has the same stiffness as the surrounding ground (i.e. the stiffness of the tunnel is

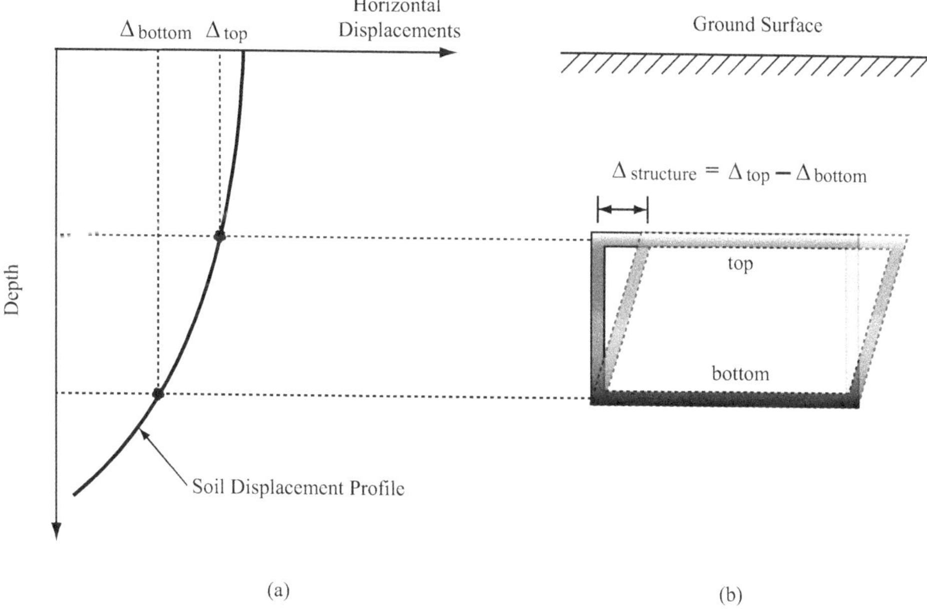

Figure 6.8 Structure distortion from free-field soil displacements. (a) Seismic-induced soil displacements. (b) Racking of a rectangular underground structure. Adapted from Wang (1993).

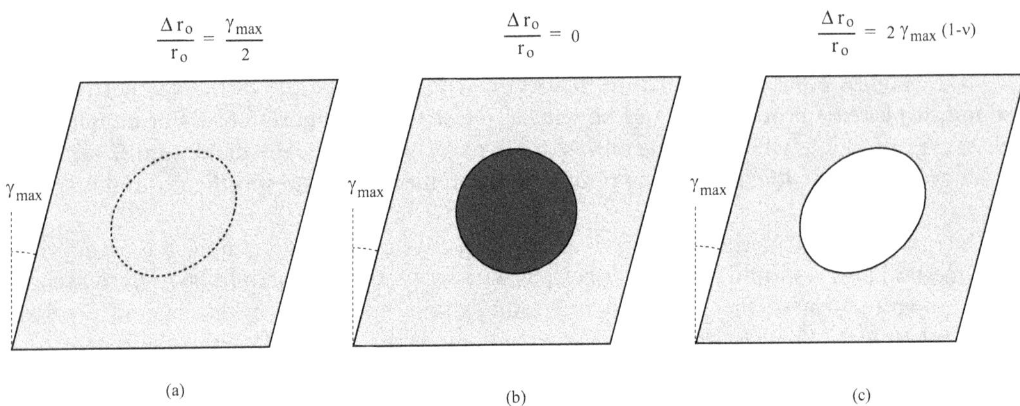

Figure 6.9 Deep circular tunnel subjected to far-field shear strain with different stiffness. (a) Tunnel with same stiffness as ground. (b) Stiff tunnel. (c) Tunnel opening.

that of the soil it replaces). Figure 6.9b plots the response of a tunnel which is infinitely stiff, and Figure 6.9c of a circular opening (i.e. there is no support). In the first case (Figure 6.9a), the tunnel does indeed follow exactly the deformations of the free field. In the second case (Figure 6.9b) the tunnel deformations are zero; this is at the expense of large stresses on the tunnel. In the third case, and assuming that the ground remains elastic (no degradation of the soil stiffness), the tunnel deformations are given by $\Delta r_o/r_o = 2\ \gamma_{max}\ (1 - \nu)$, where ν is the Poisson's ratio of the ground. The deformations can be three to four times those computed using the free-field approach, i.e. Figure 6.9a with $\Delta r_o/r_o = 1/2\ \gamma_{max}$. Further, because the strains around the opening are also larger, the shear stiffness degradation of the soil around the opening would not be considered inducing ever larger deformations!

In conclusion, the free-field method is not recommended. It has significant limitations and should only be used, at best, to provide a first estimate, i.e. the order of magnitude, of seismic distortions. Instead, the method introduced in the next section, which considers soil–structure interaction, should be employed.

6.2.4 Soil–structure interaction

The presence of the structure, because it has a different stiffness than the surrounding ground, changes the deformations of the ground. An accurate determination of the response of a structure placed below ground requires assessment of soil–structure interaction. The design is done by subjecting the structure to two independent modes of loading: one in the axial direction that maximizes the axial strains and one in the transverse direction (recall the comment in the preceding section regarding the notion that 2D analyses may be unconservative).

6.2.4.1 Axial design

There are two approaches that have been proposed for axial design: one is based on elastic-beam analysis, where the liner is assumed as an elastic beam in an elastic infinite space (St John and Zahrah, 1987), and the second one following the principles of the relative stiffness method (Savigamin and Bobet, 2021).

For the first method, based on elastic-beam analysis, in the axial direction, strains are computed using two shear waves: one with $\beta = 0°$ and at $\phi = 45°$ (Figure 6.3), which is the one that produces the largest axial strains (first equation in (6.14) or first term in (6.15)), and

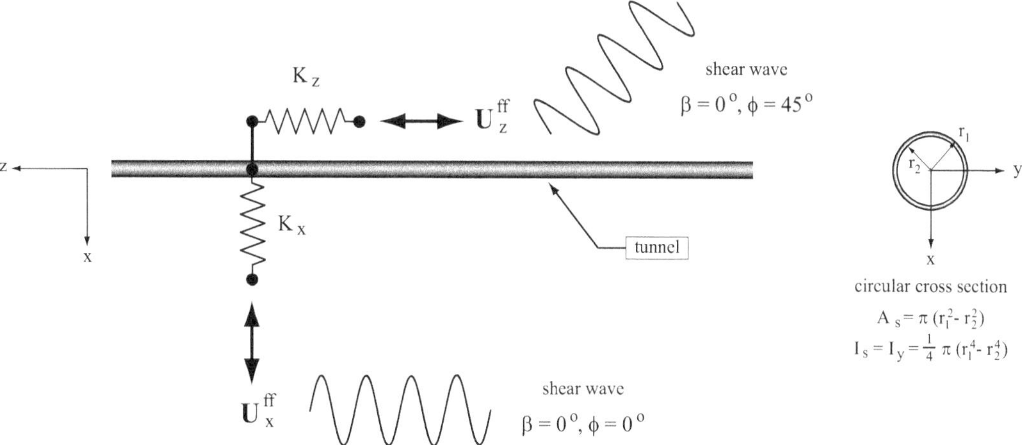

Figure 6.10 Soil–structure interaction for axial design (see Figure 6.3 for coordinates and definitions).

another one with β = 0° and at φ = 0°, which maximizes the bending strains (second term in (6.15)). See Figure 6.10. Axial strains or loads are evaluated by modeling the tunnel as an elastic beam supported on an elastic foundation (St John and Zahrah, 1987). The equations for the tunnel are:

$$
\begin{aligned}
E_s I_s \frac{d^4 U_x}{dz^4} &= K_x(U_x^{ff} - U_x) \\
E_s A_s \frac{d^4 U_z}{dz^4} &= K_z(U_z - U_z^{ff})
\end{aligned}
\tag{6.16}
$$

The first equation corresponds to motions induced by the shear waves perpendicular to the tunnel axis and the second equation to motions parallel to the tunnel axis. See Figure 6.10. In the equations, E_s, I_s, and A_s are the Young's modulus, moment of inertia of the liner along the axis perpendicular to the plane where bending occurs (plane x-z in (6.16) given the coordinates used), and the cross-section area of the liner (see Figure 6.4, where I_s and A_s are given for a circular cross section); U_x are the displacements normal to the axis of the tunnel (see Figures 6.3 and 6.4); U_x^{ff} are the free-field ground displacements due to the earthquake at the location of the tunnel, and are computed using site-specific ground response analysis (e.g. programs SHAKE or DEEPSOIL or similar); K_x is the transverse spring coefficient or foundation modulus in the transverse direction and K_z is the axial spring coefficient or axial foundation modulus (see Figure 6.10); both are given in terms of force per unit displacement per unit length of tunnel. Their magnitudes are (St John and Zahrah, 1987):

$$
K_x = K_z = \frac{32\pi G(1-\nu)}{(3-4\nu)} \frac{r_o}{\lambda}
\tag{6.17}
$$

where G is the shear modulus of the ground, ν the ground's Poisson's ratio, r_o the radius of the tunnel or half the width of the opening, and λ is the wavelength of the seismic-induced displacements.

Figure 6.11 shows the axial force, Q, the shear force V, the moment M, and the force between the ground and the structure, P, induced by the ground motions. The maximum moment M_{max} and associated shear force V_{max} and axial force Q_{max} are produced by the shear

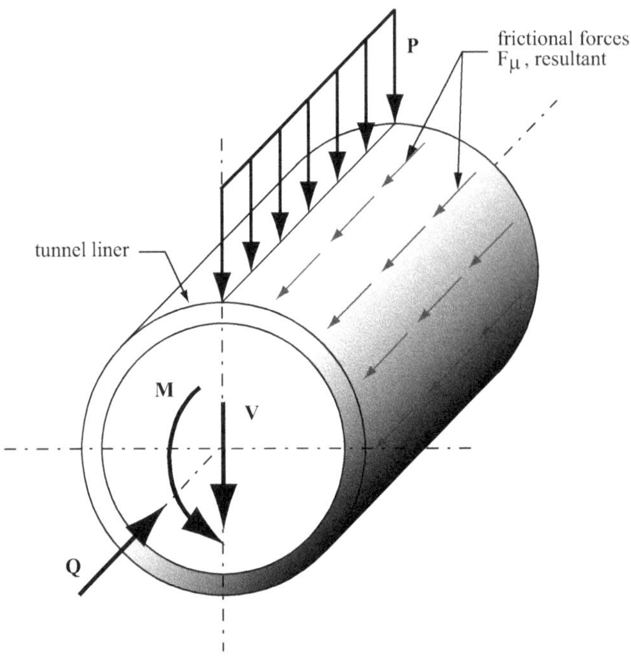

Figure 6.11 Forces and moment due to axial loading.

wave with $\beta = 90°$ and at $\phi = 0°$, and are given in (6.18). See Figure 6.11. The equations also provide the maximum pressure between the structure and the surrounding ground P_{max}.

$$
\begin{aligned}
M_{max} &= \pm \frac{\left(\frac{2\pi}{\lambda}\right)^2}{1+\frac{E_s I_s}{K_x}\left(\frac{2\pi}{\lambda}\right)^4} E_s I_s U_x \\
V_{max} &= \pm \frac{\left(\frac{2\pi}{\lambda}\right)^3}{1+\frac{E_s I_s}{K_x}\left(\frac{2\pi}{\lambda}\right)^4} E_s I_s U_x \\
Q_{max} &= \pm \frac{\frac{2\pi}{\lambda}}{2+\frac{E_s A_s}{K_z}\left(\frac{2\pi}{\lambda}\right)^2} E_s A_s U_x \\
P_{max} &= \pm \frac{\left(\frac{2\pi}{\lambda}\right)^4}{1+\frac{E_s I_s}{K_x}\left(\frac{2\pi}{\lambda}\right)^4} E_s I_s U_x
\end{aligned}
\qquad (6.18)
$$

The cross section of the tunnel is designed to support the combined forces from (6.18). Note that the moment and forces can be applied in any direction, since the shear wave can also impinge the tunnel in any direction. The equations depend on the wavelength of the shear waves λ. St John and Zahrah (1987) provided the following expressions for the forces and moment when their magnitude is maximized with respect to λ.

$$\begin{aligned}
M_{max} &= \pm\frac{1}{3}\left(4E_s I_s B^2\right)^{\frac{1}{3}} U_x \quad \text{when}\,\lambda = 2\pi\left(\frac{E_s I_s}{2B}\right)^{\frac{1}{3}} \\
V_{max} &= \pm B U_x \quad \text{when}\,\lambda = 0 \\
P_{max} &= \pm\frac{4}{5}\left(\frac{4B^4}{E_s I_s}\right)^{\frac{1}{3}} U_x \quad \text{when}\,\lambda = 2\pi\left(\frac{E_s I_s}{4B}\right)^{\frac{1}{3}} \\
Q_{max} &= \pm B U_x \quad \text{when}\,\lambda = 0 \\
B &= \frac{16(1-\upsilon)}{3-4\upsilon} G r_o
\end{aligned} \tag{6.19}$$

For a compressional wave, the maximum axial force Q_{max} acting on the cross section of the tunnel is for a wave parallel to the tunnel, i.e. $\phi = 0°$:

$$Q_{max} = \pm\frac{\left(\frac{2\pi}{\lambda}\right)}{2 + \frac{E_s A_s}{K_z}\left(\frac{2\pi}{\lambda}\right)^2} E_s A_s U_z \leq \frac{1}{4} F_\mu \lambda \tag{6.20}$$

As before, the axial force, if maximized with respect to the wavelength, is $Q_{max} = B\,U_z$, with the value B of given in (6.19) and for a wavelength, $\lambda = 0$. The axial force is limited by the magnitude of the shear force that can be developed at the contact between the liner and the ground, per unit length, and is given by F_μ.

In the second method, based on the relative stiffness method (Savigamin and Bobet, 2021), axial shear is produced by a shear wave that propagates perpendicular to the tunnel axis and has motion parallel to the tunnel axis (Figure 6.6d). Given that the tunnel is very long in its axis direction, it can be assumed that $\partial/\partial z = 0$, where the z-axis is parallel to the tunnel axis. Equilibrium in cylindrical coordinates can be simplified to:

$$\begin{aligned}
&\frac{\partial\tau_{rz}}{\partial r} + \frac{1}{r}\frac{\partial\tau_{\theta z}}{\partial\theta} + \frac{\tau_{rz}}{r} = 0 \\
&\sigma_{rr} = \sigma_{\theta\theta} = \tau_{r\theta} = 0 \\
&u_r = u_\theta = 0
\end{aligned} \tag{6.21}$$

The first expression can be written, in terms of axial displacements, as:

$$\frac{\partial^2 u_z}{\partial r^2} + \frac{1}{r^2}\frac{\partial^2 u_z}{\partial\theta^2} + \frac{1}{r}\frac{\partial u_z}{\partial r} = 0 \tag{6.22}$$

A solution is found using separation of variables. The stresses in the liner are (Savigamin and Bobet, 2021):

$$\begin{aligned}
\sigma_{rz}^{s} &= \gamma_{ff}\left[1-\left(\frac{r_i}{r}\right)^2\right]A_1 G_s \sin\theta \\
\sigma_{\theta z}^{s} &= \gamma_{ff}\left[1+\left(\frac{r_i}{r}\right)^2\right]A_1 G_s \cos\theta \\
A_1 &= \frac{2}{1+\dfrac{G_s}{G}+\left(1-\dfrac{G_s}{G}\right)\left(\dfrac{r_i}{r_o}\right)^2}
\end{aligned} \tag{6.23}$$

where γ_{ff} is the far-field shear strain induced by the earthquake, r_i is the interior radius of the liner, r_o is the exterior radius of the liner, i.e. the liner thickness is r_o-r_i, G_s is the shear modulus of the liner material and G the shear modulus of the ground, and r, θ, and z are the cylindrical coordinates. The solution in Equation (6.23) has been verified numerically (Savigamin and Bobet, 2021).

Axial bending, or snaking (Figure 6.6c), is produced by a shear wave that propagates parallel to the tunnel axis and has motion perpendicular to the tunnel axis. In the free field, the equations of motion are (see Figure 6.4 for coordinate system):

$$\begin{aligned}
u_{y,ff} &= u_{ff}\cos\frac{2\pi}{\lambda}(z-V_s t) \\
u_{x,ff} &= u_{z,ff} = 0
\end{aligned} \tag{6.24}$$

where z is the coordinate along the tunnel axis, u_{ff} is the far-field wave amplitude, λ is the wavelength, V_s is the shear wave velocity, and t is the time. A solution is found assuming quasi-static analysis, a thin-walled shell for the liner, and that the wavelength of the sinusoidal motion λ is much larger than the tunnel radius ($\lambda >> r_o$). Savigamin and Bobet (2021) found that all the forces and moments acting on the liner, for wavelength values representative of earthquake epicenters far from the tunnel, are small, except for the shear forces N_{rz} and $N_{\theta z}$. Those are given by:

$$\begin{aligned}
N_{rz}^{s} &= -\frac{t\left(r_o^2+\dfrac{t^2}{12}\right)}{(1+k)\dfrac{G_s}{G}t+r_o}\frac{1}{r_o}\frac{4\pi}{\lambda}G_s u_{ff}\sin\frac{2\pi z}{\lambda}\cos\theta \\
N_{\theta z}^{s} &= -\frac{r_o t}{(1+k)\dfrac{G_s}{G}t+r_o}\frac{4\pi}{\lambda}G_s u_{ff}\sin\frac{2\pi z}{\lambda}\cos\theta
\end{aligned} \tag{6.25}$$

where $k=\dfrac{1}{12}\left(\dfrac{t}{r_o}\right)^2$, G_s and G are the shear modulus of the liner material and ground, respectively, t is the liner thickness and r_o the radius of the tunnel.

Note that the solutions in (6.23) and (6.25) indicate that the liner (and also the ground) are loaded in shear (note that shear stresses in the ground and shear forces in the liner are the only non-zero variables), which is quite different than what is given in (6.19), based on elastic-beam theory, where the liner is loaded in bending (the solution is in terms of axial, shear force and moment). Comparisons between predictions from the analytical solutions and from finite element methods (Savigamin and Bobet, 2021) show that Equations (6.23) and (6.25) are correct and better represent the deformations of ground and support for axial shear (Figure 6.6d) and axial bending (Figure 6.6c). Interestingly, because the ground in loaded in shear only, Equations (6.23) and (6.25) apply for drained and undrained loading conditions. This is so because pure shear, in elasticity, does not produce volumetric strains and so it does not induce excess pore pressures.

Note also that the results are based on the assumption that the ground remains elastic, which may not be so in cases where the tunnel is excavated in soft soils. For a preliminary design, the magnitude of the shear modulus G of the ground should be compatible with the level of strain induced in the ground around the structure by the earthquake. For detailed design, a numerical method should be used. It is, however, usually found that the transverse design (Section 6.2.4.2) is more critical than the axial design, and thus detailed calculations for axial design may not always be done.

6.2.4.2 *Transverse design*

For the transverse design, the tunnel is subjected to a shear wave propagating perpendicular to the axis of the tunnel (β = 90° and ϕ = 90°). This is equivalent to impose to the far-field shear stresses $\tau_{ff} = G\,\gamma_{ff}$ (Figure 6.12), with G equal to the shear modulus of the ground and γ_{ff} the free-field shear strains, which, as before, should be computed from site specific response analyses for the ODE and MDE. The shear wave-induced deformations cause racking (rectangular cross section) or ovaling (circular cross section) of the structure. These deformations always seem to be more critical than the axial deformations discussed in Section 6.2.4.1.

Wang (1993) ran dynamic parametric analyses on rectangular structures with different dimensions and stiffnesses, with the assumption that both the ground and the structure remain elastic. He found that the flexibility ratio of the structure, F^w, correlates well with the ratio between the structure's deformations and the free-field ground deformations. The flexibility ratio is expressed as:

$$F^w = \frac{G\,a}{S_1\,b} \tag{6.26}$$

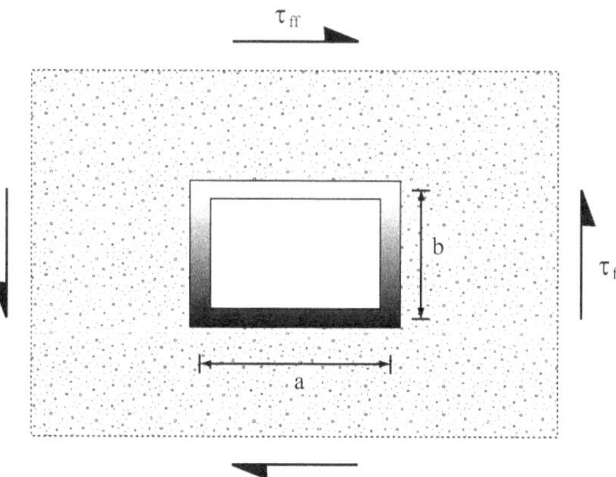

Figure 6.12 Rectangular structure in an infinite medium subjected to far-field shear stresses.

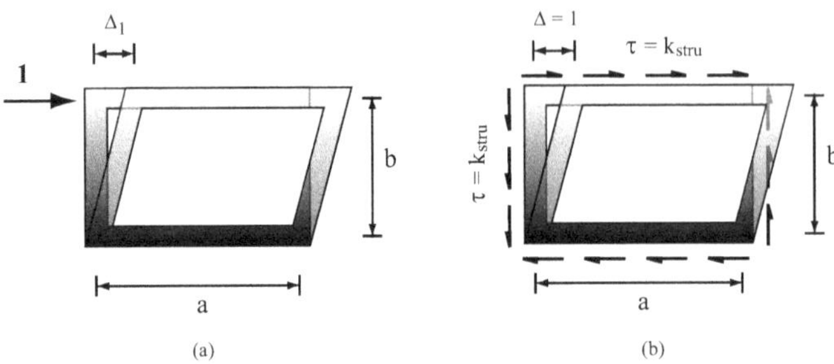

Figure 6.13 Structure stiffness according to Wang (1993) and Penzien (2000). (a) Structure stifness after Wang (1993). (b) Structure stifness after Penzien (2000).

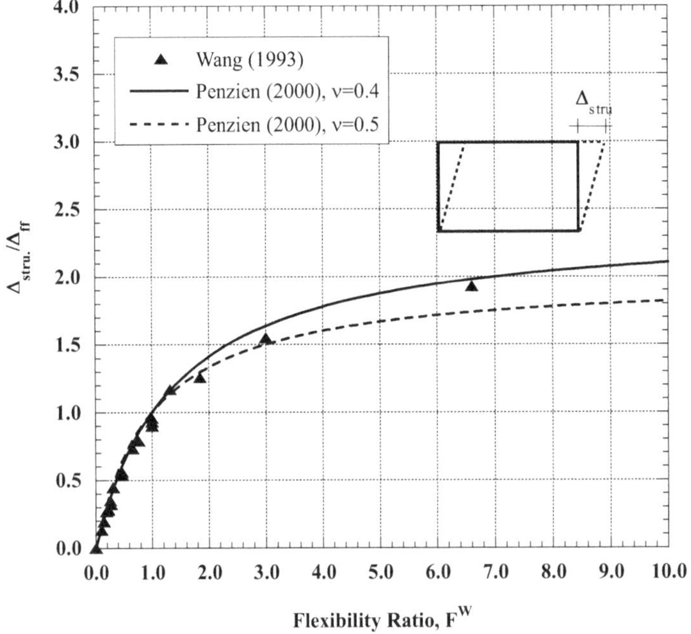

Figure 6.14 Wang (1993) and Penzien (2000) solutions.

where a and b are the dimensions of the structure (Figures 6.12, 6.13a), G is the shear modulus of the ground, and $S_1 = 1/\Delta_1$ with Δ_1 equal to the plane strain displacement produced on the structure by a unit horizontal concentrated force applied to the top of the structure (Figure 6.13b). Figure 6.14 shows the results from Wang (1993) in the form of the structure's racking ratio, defined as the ratio between the structure's Δ_{stru} and the free-field deformation Δ_{ff}.

If the structure is very rigid, $F^w = 0$, it will not deform regardless of the displacements imposed by the surrounding ground. If $F^w < 1$, the structure is stiffer than the ground and it will deform less than the free field (Figure 6.14); if $F^w = 1$ the structure deforms with the soil because it has the same stiffness; if $F^w > 1$ the structure is less stiff than the ground and it will deform more than the free field; in the limit, as $F^w \rightarrow \infty$, the structure deformations are those of the opening.

Penzien (2000) proposed an approximated method to evaluate the racking deformation of deep rectangular tunnels subjected to a far-field shear stress. It is an approximate solution because of the assumption that the load transfer between the ground and the structure takes place only through shear stresses at the interface, and the deformations of a rectangular opening are approximated by those of a circular opening. Penzien showed that the deformations of the structure depend on the relative stiffness, or the stiffness ratio, between the ground and the structure. The relative stiffness is defined with the parameter $\frac{k_{stru}}{k_{soil}}$, which is the ratio between k_{stru}, the stiffness of the structure and k_{soil}, the stiffness of the soil. k_{stru} is equal to the magnitude of a uniform shear stress applied to the perimeter of the structure that produces a unit displacement of the structure in plane strain (Figure 6.13b); and k_{soil} = G/b, where G is the shear modulus of the soil and b is the height of the structure. The relative stiffness is the inverse of Wang's flexibility ratio. The normalized deformation of the structure (racking ratio), or the ratio between the structure deformation (Δ_{stru}) and the free-field ground deformation (Δ_{ff}), $\Delta_{stru}/\Delta_{ff}$, is obtained from:

$$\frac{\Delta_{stru}}{\Delta_{ff}} = \frac{4(1-v)}{1+(3-4v)\dfrac{k_{stru}}{k_{soil}}} = \frac{4(1-v)}{1+(3-4v)\dfrac{1}{F^{w}}} \tag{6.27}$$

where ν is the Poisson's ratio of the ground. When the structure is much stiffer than the surrounding ground (i.e. $k_{stru} >> k_{soil}$), $\frac{k_{stru}}{k_{soil}}$ is very large and the deformation of the structure approaches zero. This corresponds to a very rigid structure embedded in a much softer material; the structure will keep its original shape no matter the deformation that the surrounding material undergoes. On the contrary, if the structure is much more flexible than the surrounding ground (i.e. $k_{stru} << k_{soil}$), $\frac{k_{stru}}{k_{soil}}$ is very small, and the deformation of the structure, $\Delta_{stru.}$, is $4(1 - \nu)\Delta_{ff}$. This is as if the structure does not exist and the solution corresponds to the deformation of the opening. Note that the value $4(1 - \nu)\Delta_{ff}$ is exact for circular openings subjected to a far-field constant shear stress, but it is an approximation for rectangular openings. Another important assumption made in developing Equation (6.27) is that ground displacements are imposed on the structure only through shear stresses at the interface, thus neglecting the contribution of any normal stresses. Despite those approximations, Penzien's approach provides results similar to those from Wang (1993), as one can see in Figure 6.14.

A closed-form solution for rectangular tunnels was developed by Huo et al. (2006), which addressed some of the shortcomings of previous solutions. The development of the solution is based on complex variable theory and conformal mapping techniques, and relies on the following assumptions: (1) a deep rectangular structure inside an infinite medium; (2) plane strain conditions in any transverse section perpendicular to the longitudinal axis of the structure; (3) homogeneous and isotropic ground; (4) elastic response of structure and surrounding ground; (5) pseudo-static analysis. The solution provides the normalized deformation of a rectangular structure (racking ratio) in an infinite elastic medium subjected to a far-field shear stress (Figure 6.12). Because of the symmetry of the problem, the stress distribution around the structure is assumed as that depicted in Figure 6.15a. The shear stress τ^i is assumed constant along the perimeter of the structure, and the normal stress has a linear distribution along the four sides of the rectangle with a maximum magnitude p_1^i along the side of length "a" and p_2^i along the height "b." Because of moment equilibrium, $p_1^i a^2 = p_2^i b^2$. A positive sign of the normal and shear stress indicates a distribution as shown in Figure 6.15a. Stresses of equal magnitude and opposite sign are applied to the ground at the interface between the structure and the ground. The closed-form solution is an approximation because: (1) the

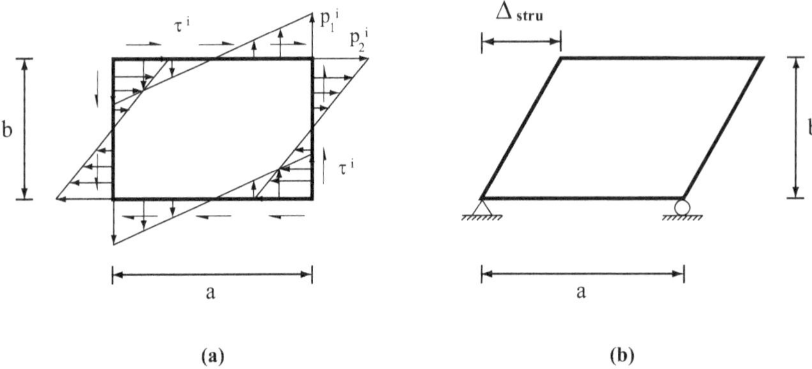

Figure 6.15 Structure deformations (Bobet et al., 2008). (a) loading. (b) deformations. Republished with permission of Canadian Science Publishing.

assumption made regarding the stress distribution at the structure–ground interface; and (2) the assumption that the thickness of the structure is small compared with the size of the opening.

The normalized structure deformations are given by:

$$\frac{\Delta_{stru.}}{\Delta_{ff}} = (1 - v_s^2)\{N\Delta_{p_2^i} + [M\Delta_{p_2^i} + \Delta_{\tau^i}]L\}\frac{G}{b}$$
$$\Delta_{stru.} = \frac{\Delta_{stru.}}{\Delta_{ff}}\gamma_{ff}\,b \tag{6.28}$$

where $\frac{\Delta_{stru.}}{\Delta_{ff}}$ is the ratio of the structure's racking deformation over the free-field ground deformation; ν_s is the Poisson's ratio of the structure; G is the shear modulus of the ground; γ_{ff} are the free-field shear strains computed at the center of the structure; and b is the height of the structure; M, N, and L are parameters that depend on the stiffness of the ground and the structure, the aspect ratio of the structure m = a/b, and the Poisson's ratio of the ground ν; $\Delta_\tau{}^i$ is the deformation of the structure due to a unit shear stress τ^i, and $\Delta_{p_2^i}$ is the deformation of the structure due to the linear normal stress distribution (Figure 6.15a) with $p_2^i = 1$ and $p_2^i = (b/a)^2$. Thus, the structure's deformation Δ_{stru}, which is the difference between the displacements of the top and bottom of the structure (see Figure 6.15b), is equal to:

$$\Delta_{stru} = (1 - v_s^2)(\tau^i\Delta_{\tau^i} + p_2^i\Delta_{p_2^i}) \tag{6.29}$$

The values of $\Delta_\tau{}^i$ and $\Delta_{p_2^i}$ can be obtained analytically or numerically from structural analysis. For the case of a rectangular structure with width "a" and height "b" with an interior central column, with members' stiffness: lateral walls, $(EI)_w$; top and bottom slabs, $(EI)_s$; central column, $(EI)_c$, the deformations are (Huo et al., 2006):

$$\Delta_{\tau^i} = \frac{1}{24}mb^4\,\frac{\frac{2}{(EI)_c}\left[\frac{m}{(EI)_s} + \frac{1}{(EI)_w}\right] + \frac{m}{2(EI)_s}\left[\frac{3m}{2(EI)_s} + \frac{2}{(EI)_w}\right]}{\frac{2}{(EI)_c} + \frac{1}{(EI)_w} + \frac{3m}{(EI)_s}}$$

$$\Delta_{p_2^i} = \frac{1}{4} b^4 \frac{\dfrac{1}{45(EI)_c}\left[\dfrac{m}{(EI)_s} + \dfrac{1}{(EI)_w}\right] + \dfrac{m}{480(EI)_s}\left[\dfrac{7m}{2(EI)_s} + \dfrac{5}{(EI)_w}\right]}{\dfrac{1}{3(EI)_c} + \dfrac{1}{6(EI)_w} + \dfrac{m}{2(EI)_s}} \tag{6.30}$$

where m = a/b is the aspect ratio of the structure (Figure 6.12). Note that $F^w = \frac{G}{b}\Delta_{\tau^i}$; in other words, Wang's flexibility ratio F^w can be obtained by multiplying the deformations of the structure due to a unit shear stress by the ratio of the ground's shear modulus to the height of the structure. If there is no central column, $(EI)_c = 0$, Equation (6.30) simplifies to:

$$\begin{aligned} \Delta_{\tau^i} &= \frac{1}{24} m b^4 \left[\frac{m}{(EI)_s} + \frac{1}{(EI)_w}\right] \\ \Delta_{p_2^i} &= \frac{1}{60} b^4 \left[\frac{m}{(EI)_s} + \frac{1}{(EI)_w}\right] \end{aligned} \tag{6.31}$$

The parameters M and N in Equation (6.28) depend on the Poisson's ratio of the ground ν and on the aspect ratio m of the structure (Huo et al., 2006), and are given in Figure 6.16a. The parameter L is defined as:

$$L = -\frac{(1 - v_s^2) N \Delta_{p_2^i} - \dfrac{a}{G} F_1}{(1 - v_s^2)(M \Delta_{p_2^i} + \Delta_{\tau^i}) - \dfrac{a}{G} F_2} \tag{6.32}$$

where F_1 and F_2 can be found in Figure 6.16b. F_1 and F_2 are also a function of ν and m.

Huo et al. (2006) provided extensive comparisons between predictions obtained with the analytical solution and results from numerical simulations using Finite Element Methods. Best

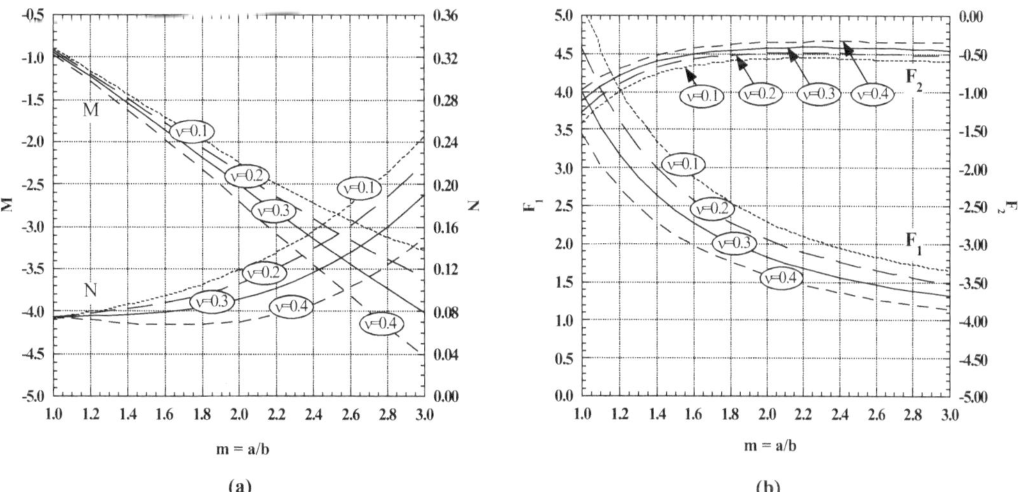

Figure 6.16 Parameters in Equations (6.28) and (6.32), (Bobet et al., 2008). (a) Parameters M and N. (b) Parameters F_1 and F_2. Republished with permission of Canadian Science Publishing.

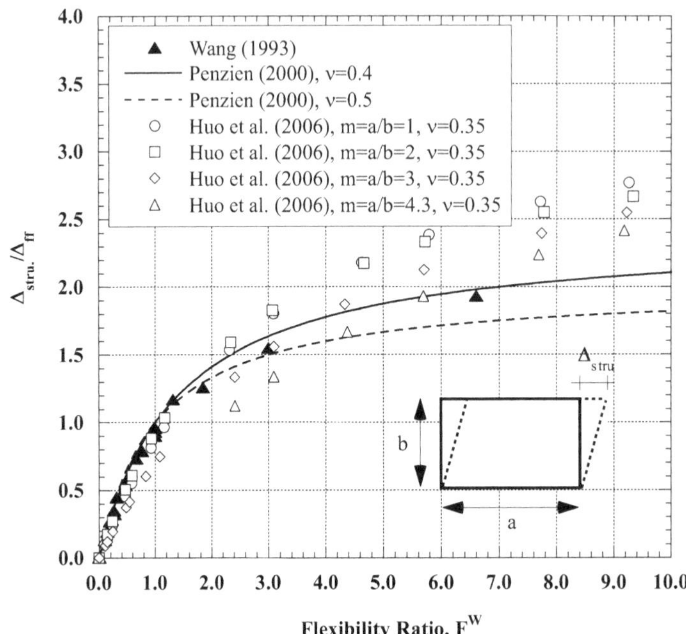

Figure 6.17 Comparison between Huo et al. (2006) Analytical solution and Wang (1993) and Penzien (2000) results.

results were found when in Equations (6.28) through (6.32) the height and width were taken as the interior dimensions of the structure; that is, the dimensions from wall to wall rather than the distance between axes of horizontal and vertical members. The comparisons showed that the analytical solution provided results with differences typically within a 2–8% range with respect to numerical methods. As an example, Figure 6.17 shows a comparison between Huo et al. solution and Wang and Penzien's solutions. Huo et al.'s results bracket Wang's and Penzien's results and compares well with those earlier results for flexibility ratios smaller than about 1.0 to 1.5, which correspond to structures stiffer than the surrounding ground or slightly softer. For softer structures, with flexibility ratios larger than 2, distortions predicted by Huo et al. increase much faster than those from Wang and Penzien. For very large flexibility ratios (i.e. no structure) Huo et al.'s analytical solution predicts normalized distortions equal to those of an unsupported opening (3.774, 3.469 and 3.844 for aspect ratios m = 1, 2, and 3, respectively, which agree with theoretical results; e.g. Savin, 1961), while Wang's and Penzien's are 2.4–2.8 and 2.6, respectively, which are too small (see Figure 6.17). In summary, for small flexibility ratios, smaller than 1.0 to 1.5, the three solutions give similar results. For larger flexibility rations, Huo et al.'s approach should be preferred.

Huo et al. (2006) showed that the normalized structure displacement is independent of the absolute size of the structure; in other words, the normalized displacement depends on m = a/b, the ratio of length to height of the structure, which is a measure of the shape of the structure. They also showed that the normalized displacement depends on the relative stiffness between the structure and the surrounding ground, and not on the absolute stiffness of the structure or the ground, which supports the findings from Wang (1993) and Penzien (2000).

The loads on the members of the structure can be estimated once the racking deformation is obtained. Wang (1993) recommended a pseudo-static structural analysis with the following two loading conditions: (1) a point load applied as shown in Figure 6.18a, with a magnitude such that it produces the computed racking deformation (Equations (6.28) or (6.29) or Figure 6.17); and (2) a linear load distribution (Figure 6.18b), also with a magnitude that

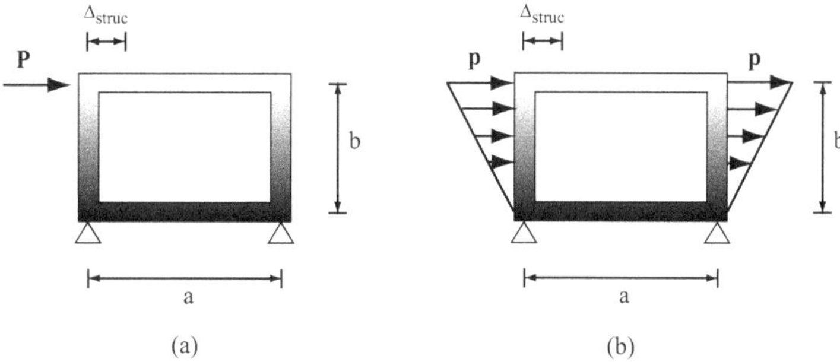

Figure 6.18 Structural analysis (adopted from Wang, 1993). (a) Point load. (b) Triangular load.

induces in the structure the computed racking deformation. Wang recommended using for design the worst loading condition.

The closed-form solution proposed by Huo et al. (2006) provides the stresses along the perimeter of the structure.

$$\begin{aligned} p_1^i &= \frac{p_2^i}{m^2} \\ p_2^i &= M\tau^i + N\frac{G}{b}\Delta_{ff} \\ \tau^i &= L\frac{G}{b}\Delta_{ff} \end{aligned} \tag{6.33}$$

Once the stresses are obtained, they can be used to compute the distribution of axial and shear forces, and bending moments along the members of the structure. Similar to what happens with displacements, comparisons between the analytical solution and numerical methods indicate that better results are obtained when using the interior dimensions (e.g. wall to wall dimensions) for the width and height of the structure in (6.33).

The two methods presented, Wang's (1993) and Huo et al. (2006), are approximations given the assumptions made. The finding that the two methods do not provide exactly the same results can be traced back to the different hypothesis and simplifications made for their derivation. Both are very useful for preliminary analysis and for pre-design of the structure. The two methods give a reasonable approximation of the structure's racking deformations, but they are not accurate for the magnitude of the structure members' loads. For the final design, a numerical analysis is recommended. This is particularly important to estimate the loads in the structure.

A closed-form analytical method has been developed for deep circular tunnels using the relative stiffness method (Einstein and Schwartz, 1979; Davis, 1999; Bobet, 2003). A far-field shear stress is equivalent to a compressive and a tensile stress rotated 45° with respect to the direction of the shear stress (Figure 6.19). From the relative stiffness method, and for the case of no slip between the liner and the ground, the thrust T, moment M, and radial displacements U_r of the liner are:

$$\begin{aligned} T &= \pm(1-C_2)\tau_{ff}\, r_o \cos 2\theta \\ M &= \pm\frac{1}{2}(1+C_1+C_2)\tau_{ff}\, r_o^2 \cos 2\theta \\ U_r &= \pm\frac{1}{2G}[1-2(1-\nu)C_1-C_2]\,\tau_{ff}\, \tau_o \cos 2\theta \end{aligned} \tag{6.34}$$

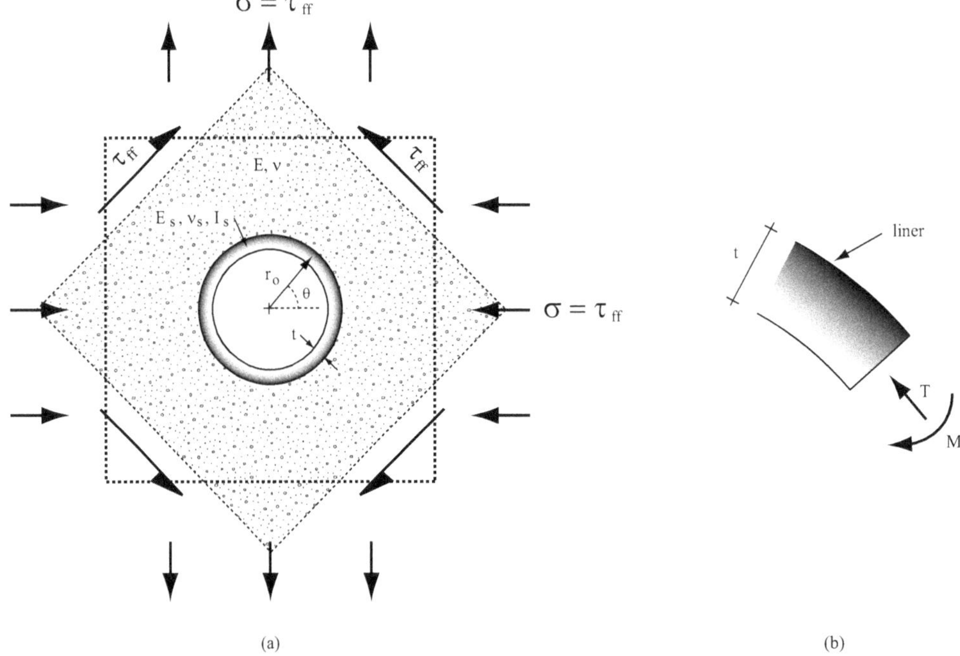

Figure 6.19 Circular tunnel subjected to far-field shear stresses. (a) Far-field seismic-induced shear stresses. (b) Thrust and moment in liner

where

$$C_1 = -2\frac{(1-\nu)^2C+(1-\nu)-[(1-\nu)C+4]3/F}{(1-\nu)^2C+(1-\nu)(3-2\nu)+[(1-\nu)(5-6\nu)C+4(3-4\nu)]3/F}$$
$$C_2 = \frac{1}{3}\frac{(1-\nu)C-2-C_1[(1-\nu)C+4\nu]}{(1-\nu)C+2} \quad (6.35)$$

For full slip between the ground and the tunnel,

$$T = \pm\frac{12(1-\nu)}{(1-\nu)F+3(5-6\nu)}\tau_{ff}\, r_o \cos 2\theta$$
$$M = \pm T\, r_o \quad (6.36)$$
$$U_r = \pm\frac{1}{G}\frac{2(1-\nu)^2F}{(1-\nu)F+3(5-6\nu)}\tau_{ff}\, r_o \cos 2\theta$$

In the above equations, τ_{ff} is the far-field shear stress, $\tau_{ff} = \gamma_{ff}$ G; G and ν are the shear modulus and Poisson's ratio of the ground, respectively; r_o is the tunnel radius; F and C are the flexibility and compressibility ratios, defined as (Einstein and Schwartz, 1979):

$$C = \frac{E\, r_o(1-\nu_s^2)}{E_s\, A_s(1-\nu^2)}$$
$$F = \frac{E\, r_o^3(1-\nu_s^2)}{E_s I_s(1-\nu^2)} \quad (6.37)$$

A_s is the cross-section area of the liner per unit length of tunnel (i.e. $A_s = t$), and I_s is the moment of inertia of the liner per unit length of tunnel (i.e. $I_s = 1/12\ t^3$; Figure 6.19). The tunnel distortion can be computed as:

$$\frac{\Delta r_o}{r_o} = \frac{1}{r_o} U_r\big|_{\theta=0^o}$$
$$\frac{\dfrac{\Delta r_o}{r_o}}{\left(\dfrac{\Delta r_o}{r_o}\right)_{freefield}} = \frac{\dfrac{1}{r_o} U_r\big|_{\theta=0^o}}{\dfrac{1}{2}\gamma_{ff}} = \frac{2G}{\tau_{ff}\, r_o} U_r\big|_{\theta=0^o} \qquad (6.38)$$

The second equation represents the tunnel distortion, i.e. $\Delta r_o/r_o$, with respect to that of the free field and provides a measure of the degree of constraint that the tunnel imposes to the surrounding ground.

If the ground is saturated, excess pore pressures may be produced during rapid loading (e.g. during a seismic event). If it is assumed that the duration of loading is very small relative to the permeability of the ground so no pore pressure dissipation occurs (i.e. undrained analysis), the ground does not change volume. Bobet (2003) expanded the relative stiffness method and found the solution for a deep circular tunnel in a saturated ground (see further discussion in Section 4.2.1.3), where the far-field stresses produce excess pore pressures. Following the principle of effective stresses, the solution for no slip between liner and ground is:

$$\begin{aligned}
T &= \pm(1-2C_4)\tau_{ff}\, r_o \cos 2\theta \\
M &= \pm\frac{1}{2}(1+2C_3+2C_4)\tau_{ff}\, r_o^2 \cos 2\theta \\
U_r &= \pm\frac{1}{2G}[1-2C_3-2C_4]\tau_{ff}\, r_o \cos 2\theta \\
C_3 &= -\frac{(1-\nu)^2C+(1-\nu)-[(1-\nu)C+4]3/F}{[2+(1-\nu)C][(1-\nu)+6/F]} \\
C_4 &= \frac{1}{2}\frac{(1-\nu)^2C-12/F}{[2+(1-\nu)C][(1-\nu)+6/F]}
\end{aligned} \qquad (6.39)$$

For the full slip case,

$$\begin{aligned}
T &= \pm\frac{6}{6+(1-\nu)F}\tau_{ff}\, r_o \cos 2\theta \\
M &= \pm T\, r_o \\
U_r &= \pm\frac{1}{G}\frac{(1-\nu)F}{6+(1-\nu)F}\tau_{ff}\, r_o \cos 2\theta
\end{aligned} \qquad (6.40)$$

Note that for cases where there is no slip between the liner and the ground, the tunnel distortions and the liner loading depend on both compressibility and flexibility ratios. If there is full slip between the ground and the liner, the results depend only on the flexibility ratio. Note also that due to the cyclic nature of the dynamic excitation, the sign of the thrust and moment can be either positive or negative.

Typical tunnels have a small compressibility ratio C compared with their flexibility ratio F. For the limit case, when the compressibility is negligible, i.e. $C = 0$, the following relations apply for the cases considered: no excess pore pressures (denoted as dry ground in the equations), undrained loading (saturated ground in the equations), and no slip or full slip between liner and ground:

$$\begin{aligned}
\frac{(U_r)_{dry,no\,slip}}{(U_r)_{dry,full\,slip}} &= \frac{(M)_{dry,no\,slip}}{(M)_{dry,full\,slip}} = \frac{2[(1-\nu)F+3(5-6\nu)]}{(1-\nu)(3-2\nu)F+12(3-4\nu)} < 1\\
\frac{(T)_{dry,no\,slip}}{(T)_{dry,full\,slip}} &= \frac{[(1-\nu)F+12][(1-\nu)F+3(5-6\nu)]}{3[(1-\nu)(3-2\nu)F+12(3-4\nu)]} > 1\\
\frac{(U_r)_{saturated,no\,slip}}{(U_r)_{saturated,full\,slip}} &= \frac{(M)_{saturated,no\,slip}}{(M)_{saturated,full\,slip}} = 1\\
\frac{(T)_{saturated,no\,slip}}{(T)_{saturated,full\,slip}} &= \frac{1}{6}[(1-\nu)F+12] > 1\\
\frac{(U_r)_{dry,no\,slip}}{(U_r)_{saturated,no\,slip}} &= \frac{(T)_{dry,no\,slip}}{(T)_{saturated,no\,slip}} = \frac{(M)_{dry,no\,slip}}{(M)_{saturated,no\,slip}} =\\
&\quad \frac{4(1-\nu)[(1-\nu)F+6]}{(1-\nu)(3-2\nu)F+12(3-4\nu)} > 1 \text{ when } F > \frac{12}{1-\nu}
\end{aligned} \tag{6.41}$$

The first two equations, related to dry ground, i.e. no excess pore pressures, show that radial displacements and moments are always the largest with full slip, while no-slip conditions carry the largest axial load. For saturated ground, where excess pore pressures develop, the third and fourth equations indicate that radial displacements and moments are independent of the interface conditions between the ground and the liner and that the axial load is the largest when there is no slip between the liner and the ground. The final expression in (6.41) can be used to ascertain the effects of undrained loading on the response of the liner. It shows that, when the flexibility ratio F is larger than about 12, dry loading conditions (no excess pore pressures, drained loading analysis) result in larger deformations and larger load demand to the liner than saturated conditions (excess pore pressures are generated during undrained loading). For stiffer tunnels (F less than about 12), the undrained loading condition may be more severe.

For rectangular structures, however, and for drained loading, full slip conditions always reduce the seismic demand on the structure compared with the no slip condition. Bobet et al. (2008) investigated the effect of friction between the tunnel support and the ground on the distortion of the tunnel. They run a series of simulations using a finite element program, which considered a rectangular structure in an infinite homogeneous, isotropic, and elastic medium, similar to that shown in Figure 6.12. They used interface elements between the structure and the surrounding ground to model slip. Interface elements allowed slip when the shear stress was larger than the shear strength, which was modeled following the Coulomb friction law. The interface elements also allowed the formation of gaps when the normal stress was tensile. Therefore, a small constant normal stress with magnitude 120 kPa was applied to the boundaries of the discretization to prevent the ground from detaching from the structure (i.e. to make the normal stresses at the interface compressive). The seismic-induced deformations were modeled by applying a constant shear stress at the far boundaries of the discretization with magnitude 50 kPa. Different interfaces were evaluated, with coefficients of friction μ ranging from 0 (full slip), to very large magnitudes (no slip, tied interface). Cohesion was always taken as zero. The properties of the structure were E_s = 24000 MPa,

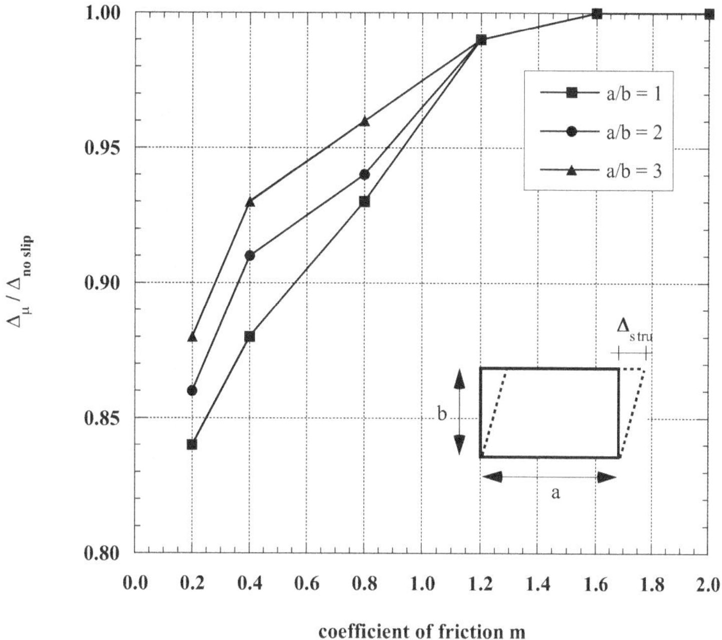

Figure 6.20 Effect of friction coefficient on normalized structure deformation (Bobet et al., 2008). Republished with permission of Canadian Science Publishing.

ν_s = 0.15, and of the ground, G = 80 MPa, ν = 0.35. The thickness of the slabs and walls of the structure was 0.5 m. Figure 6.20 shows the effects of the frictional interface. The structure deformations, Δ_μ, normalized with respect to the deformations obtained with a tied interface, $\Delta_{no\ slip}$ (no slip condition), are plotted as a function of the interface friction, and for three different structure aspect ratios, a/b = 1, square (4 × 4 m), 2 (4 × 8 m), and 3 (4 × 12 m). The structure deformations increase with the increase of the friction coefficient. The structure deformations with an interface coefficient of friction μ greater than 1.2 converge to the results of a tied interface. The structure deformations with tied interface were 12%, 9% and 7% larger than those with μ = 0.4 for a/b = 1, 2, and 3, respectively. The structure deformations were the smallest with μ = 0.2, which were about 84–88% of the case with a tied interface. Results with $0 < \mu < 0.2$ were not deemed realistic because openings between ground and structure appeared in the numerical model due to the small values of the friction coefficient; these results are shaded in the figure.

The depth of underground structures has a relatively small effect compared to other factors such as free-field ground movements, relative stiffness, and shape of the structure. Wang (1993) investigated the effects of depth on square structures under seismic loading. He conducted a number of finite element dynamic analyses with 4.5 m × 4.5 m square tunnels where the relative stiffness between the structure and the ground was kept constant. He concluded that the presence of the ground surface had a secondary effect on structure deformations when the ratio of overburden depth to structure height was larger than one. For overburden depths larger than two times the height of the structure, the normalized structure deformation was relatively independent of the burial depth and could be approximated by the deformation of a structure in an infinite medium. With no overburden, the normalized displacements were about 80% of those of a deep structure. For a typical cut-and-cover structure, the overburden to height ratio is between 0 and 1.0, which corresponds to structure deformations about 10–20% smaller than the deformations computed with the assumption of a deep structure. These conclusions were later supported by Huo (2005), who included in the analyses the shape of the structure, with values of length to height ratio, a/b = 1 (4 × 4 m), 2 (4 × 8 m), and 3 (4 × 12 m). He conducted static analysis using a Finite Element Method with the following properties: of the structure, E_s = 24000 MPa, ν_s = 0.15; of the ground G = 80 MPa, ν = 0.35; thickness of slabs and walls 0.5 m; overburden to height ratios, d/b, from 1.5 (deep structure) to 0 (no overburden). Figure 6.21 shows depth effects on structure deformations. The deformations are given by the ratio $\Delta_{shallow}/\Delta_{deep}$, where $\Delta_{shallow}$ is the deformation of the structure at the chosen depth and Δ_{deep} is the deformation of the same structure but placed very deep below the surface. The figure shows that depth effects depend on the shape of the structure. The figure also shows that shallow structures have smaller deformations than deeper structures, which is in agreement with the findings from Wang (1993). Very close to the surface, the reduction of deformations can be as much as 15–20%. This reduction decreases quickly with depth. For burial depths, given by the ratio d/b, greater than about 1.5, the structure deformations are those of a deep structure for all practical purposes. Huo (2005) also showed that these results are independent of the actual dimensions of the structure. In other words, two structures with different dimensions but with the same aspect ratio a/b have the same deformations, when all other variables are unchanged.

Both numerical and analytical results (e.g. Wang, 1993; Huo et al., 2006) show that for rectangular structures and for a given earthquake, the key factor that determines the response of the structure is the relative stiffness between the structure and the surrounding ground and the shape of the structure, given by the aspect ratio a/b. Interface friction and structure depth have smaller effects. For a structure embedded in stiff soil or rock, the change of ground stiffness with strain may not be significant, but relatively soft soils may have a very large stiffness reduction.

The presence of the underground structure modifies the deformations, and thus the stiffness reduction, of the surrounding soil. Away from the structure the ground moves freely

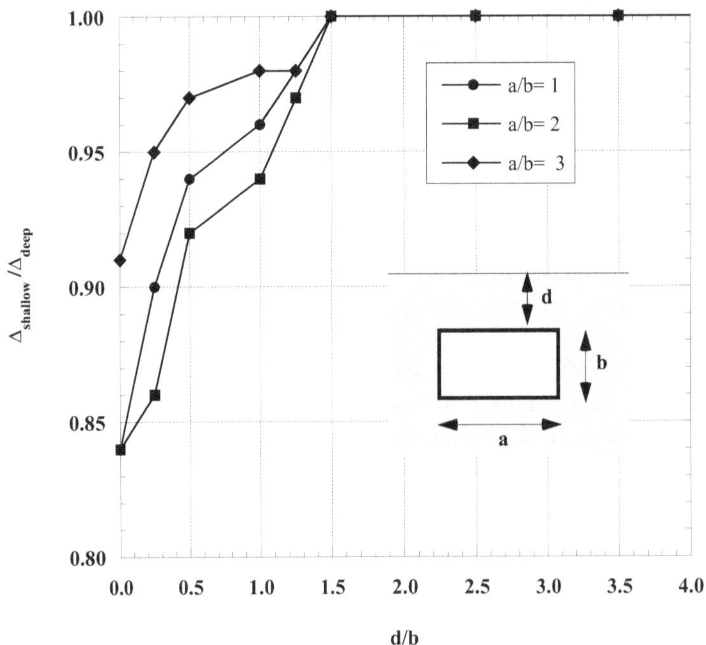

Figure 6.21 Effect of burial depth on normalized structure deformation.

under the input motions generated by the earthquake. Because of the considerable magnitude of the transient input, large soil strains with associated large degradation of the shear modulus develop within the ground materials resulting in significant soil deformations (and associated reduction of stiffness). Close to the structure, however, the soil movements are affected by the presence of the structure. Stiff structures reduce the deformations of the surrounding soil with respect to those of the free field (as if there was no structure) and, as a consequence, the degradation of the soil shear modulus is limited. On the other hand, in the vicinity of flexible structures, the surrounding ground can experience strains larger than the free field and the shear modulus of the soil will substantially degrade favoring larger deformations around the opening. Huo et al. (2005) investigated numerically this mechanism for their analysis of the failure of the Daikai station in Kobe, Japan, during the January 17, 1995, Kobe earthquake. Figure 6.22 is a contour plot of shear modulus degradation (G/G_{max}) in the ground surrounding the cross section of the Daikai station at peak structure distortion (at about four seconds after the start of the earthquake). Only the left half of the soil structure is shown because of symmetry. The shear modulus degradation of the soil shown in the figure is controlled by the presence of the structure. A small shear modulus degradation (effective modulus about 80–90% of the maximum shear modulus) develops in an area of soil near the structure within a distance of about the height of the opening. Beyond this area the shear modulus degradation increases rapidly until it reaches the far-field response (effective modulus about 30–40% of maximum shear modulus). This is an interesting observation because it indicates that the soil surrounding the structure displaces concurrently with the structure.

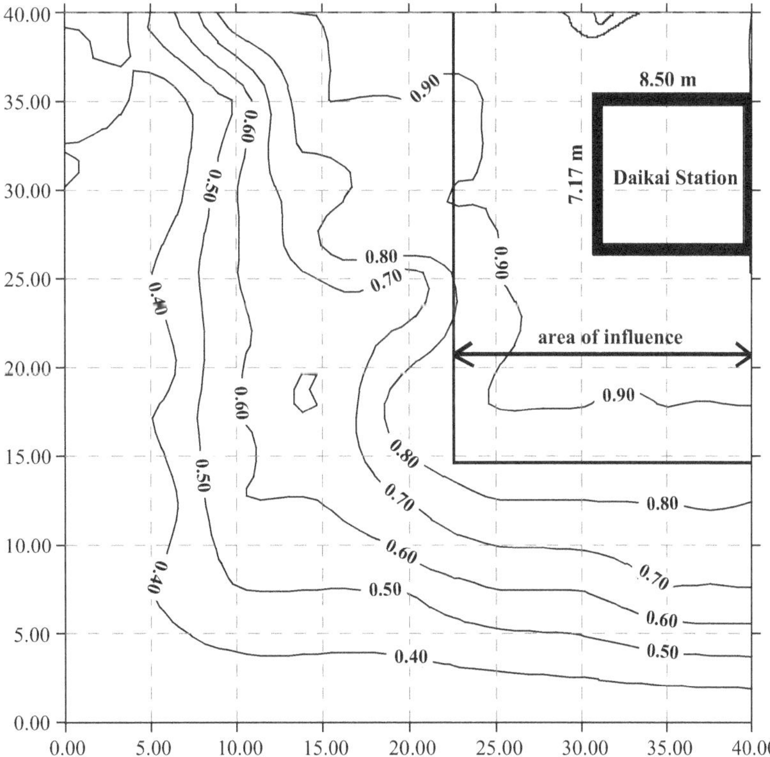

Figure 6.22 Contour plot of shear modulus degradation (G/Gmax) at maximum structure distortion. Labels in axes are distance in meters (Huo et al., 2005). Republished with permission of American Society of Civil Engineers.

Due to the significant effect of ground deformations on the response of rectangular structures embedded in soft soil, interface characteristics are also relevant. This is so because friction along the interface governs the nature of the loads and/or displacements transferred from the ground to the rectangular structure and thus affects the level of constraint imposed from the structure to the soil. During shaking, the structure deforms not only due to the normal stresses but also due to the shear stresses induced at the interface. For a given ground motion, the soil stiffness determines the magnitude of the free-field displacements that can potentially be imposed on the structure. The capacity of transferring such displacements to the rectangular structure is determined by the relative stiffness of the structure with respect to the surrounding soil and by the friction developed along the interface; in other words, the deformations of the soil around the structure may not be the same as those of the free field and are dependent on the relative stiffness between the structure and the soil and on the interface characteristics. These two factors are interrelated and have opposite effects. A strong interface can effectively transfer large shear stresses and thus could induce significant structural deformations. However, if the structure is stiff and the interface friction remains high the displacements of the surrounding soil are highly constrained, resulting in limited soil stiffness degradation and limited soil deformations around the rectangular structure. A weaker interface on the same structure will transmit smaller shear stresses to the structure, which would tend to limit its deformation. However, the reduced constraint on the attached soil mass will result in larger seismically induced deformations or perhaps in slip along the interface. As a consequence, the degradation of the soil shear modulus will increase generating in turn larger deformations in the surrounding soil and high normal stresses at the interface. Hence the potential to develop considerable soil displacements is increased although the capability to transmit them in shear to the rectangular structure is small.

Dynamic elasto-plastic numerical analyses, which included soil stiffness degradation with strain, conducted for the Daikai station and for the LA Civic Center station, both with a rectangular cross section (Bobet et al., 2008), showed that structure deformations were slightly larger with a frictional interface than with a tied interface. A tied interface decreased structure deformations by about 5–10% with respect to a frictional interface when the non-linear soil behavior model was used. This is in contrast to what happens with an elastic soil model where the structure deformations increase 10–15% with a tied interface (Figure 6.20). The reason for the difference is the interplay that exists between the constraint that the structure imposes on the soil and the soil's stiffness degradation with strain. With an elastic model, the soil's stiffness does not change. As shear stresses from the soil can be better transferred to the structure with a tied interface, the structure deformations are larger than with a frictional interface. However, with a strain-softening soil model the stiffness of the soil is reduced with soil deformations. A frictional interface has two opposing effects: on the one hand, it limits the magnitude of the shear stress that can be transmitted to the structure, and on the other hand, it allows larger deformations of the soil surrounding the structure since slip can occur at the interface. As the soil deforms more, the stiffness of the soil is reduced, which in turn induces further deformations in the soil with the imposed earthquake motions. This is consistent with numerical predictions that indicate a reduction of soil shear modulus of $G/G_{max} \sim 78\%$ around the Daikai structure with a frictional interface compared with a reduction of 85% with a tied interface (Huo et al., 2005). Thus the softer soil will deform more, but the soil's capacity to transmit deformations to the structure in shear is reduced. The final outcome from the inter-related behavior of soil and the cut and cover rectangular Daikai station is larger structure deformations with a frictional interface.

The same concepts regarding support stiffness and interface characteristics should apply to tunnels with circular cross sections. However, to the author's knowledge, no quantitative analysis has been done on this issue. As with rectangular structures, the relative stiffness between the tunnel and the surrounding ground determines largely the deformations imposed on the structure (Equations (6.34) and (6.36)). Contrary to what is observed for rectangular structures, full slip conditions may induce larger deformations in the ground and structure, as indicated in Equations (6.41). If so, the shear modulus degradation of the soil surrounding the tunnel will be much larger for full slip than for no slip conditions, which will further induce larger displacements in the soil and in the structure. Thus interface conditions for underground structures with circular cross section may have a much larger influence on soil and structure behavior than for structures with a rectangular cross section.

6.2.5 Recommendations

There is enough field evidence to conclude that underground structures are vulnerable to seismic deformations, and thus they must be designed to accommodate the ground motions imposed by the earthquake. If the structure is far from the epicenter of the earthquake, a static analysis may be sufficient to approximate the dynamic response of the structure. Any seismic analysis must include ground–structure interaction. Otherwise, the results may not be accurate. Thus, methods such as the free-field method should be abandoned or, at most, used as a very first approximation, as they may be unsafe.

The closed-form solutions presented are developed with the assumption that the stiffness of the ground and liner remain constant. This may be the case for stiff soils and rocks when the seismic-induced strains are small, and for liners where the damage produced by the earthquake loading is negligible. For soft soils this is not the case, and a significant reduction of the soil's stiffness occurs. A practical method is proposed for preliminary design in which Huo's et al. (2006) analytical solution and/or Wang's (1993) method are combined with an iterative scheme to account for the ground shear modulus degradation with cyclic loading (i.e. equivalent linear method). The method was first recommended by Bobet et al. (2008) who verified it by comparing its predictions with full dynamic numerical methods where the ground's stiffness depended on the level of strain. For cut and cover rectangular structures the approach consists of the following steps (Figure 6.23):

1. Obtain the free-field ground deformation Δ_{ff} at the center of the structure. The free-field ground deformation can be obtained from ground response analyses; e.g. SHAKE or DEEPSOIL or similar using as input horizontal accelerations induced by the design earthquake. The free-field ground deformations are: $\Delta_{ff} = \gamma_{ff}\, b = \tau_{ff}\, b/G$, where G is the ground shear modulus, b is the height of the structure, and τ_{ff} and γ_{ff} are the far-field ground shear stress and shear strain, respectively. The shear modulus, shear stress, and strain are those computed at the center of the structure.
2. Compute the structure compliances Δ_τ^i and $\Delta_{p_2^i}$ and the aspect ratio m = a/b (a is the width of the structure). This can be done with closed-form solutions such as those of Equations (6.30) or (6.31) or from structural analysis.
3. Compute parameters M, N, and L, from Figure 6.16 and Equation (6.32). As an alternative, compute Wang's flexibility ratio F^w from (6.26). Use the value of the ground's shear modulus obtained from step 6 in the previous iteration. An initial guess is needed to start the iteration process.
4. Find the structure distortion $\Delta_{stru.}$ from Equation (6.28) or from Figures 6.14 or 6.17, given Δ_{ff} from step 1.
5. Determine the shear strain, γ, of the ground close to the structure. The soil deformations around the structure are constrained by the presence of the structure. Use $\gamma = \Delta_{stru.}/b$.

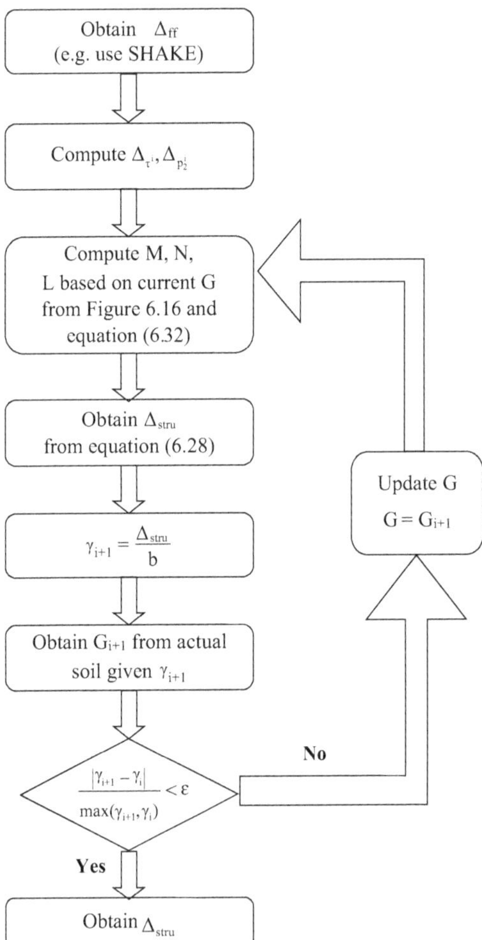

Figure 6.23 Iterative procedure for computing racking of rectangular underground structures (Bobet et al., 2008). Republished with permission of Canadian Science Publishing.

6. Update the ground shear modulus. The shear modulus is obtained from the G-γ curve of the soil (e.g. Figure 6.7 or from resonant column tests, Seed et al., 1986), given the magnitude of the shear strain found in step 5.
7. Evaluate the difference between strains computed in the current and previous iterations. If the differences are small (e.g. error, $\varepsilon < 1\%$), convergence has been reached and the iteration process is completed. The structure deformation is that obtained in step 4. If the differences are larger than the maximum error allowed, a new iteration is attempted with the updated shear modulus of the soil. The iteration should start at step 3, unless the structure stiffness is also updated, in which case the new iteration should start at step 2. Usually convergence is found after a reasonably small number of iterations.
8. Compute normal and shear stresses along the perimeter of the structure using Equation (6.33) with the final, updated, shear modulus of the ground.
9. Use results from step 8 to obtain normal, shear loads, and moments for the elements of the structure using structural analysis. As an alternative, use the larger results from the two loading conditions shown in Figure 6.18, where the magnitude of the concentrated force or the distributed load is such that they induce in the structure the computed racking deformations obtained at the end of step 7.

The method can be expanded to include changes of stiffness of the structure if it moves into its non-linear regime. The iterative process remains the same, with the structure's compliance computed in step 2 updated given the structure's deformations obtained at the end of step 4 in the previous iteration (while this naturally follows from the idea of iteration for the soil, it has not been verified yet).

The procedure has been validated by comparing displacement predictions with those using full dynamic elasto-plastic soil models. Huo (2005) and Bobet et al. (2008) computed the response of the Daikai station subjected to the 1995 Kobe earthquake and the response of the Los Angeles Civic Center Station during the 1994 Northridge earthquake. Both stations have a rectangular cross section. These authors used a full dynamic Finite Element method with a hysteretic soil model with the actual accelerations induced by the earthquake in the ground as input loading. They compared the structure distortions obtained with the numerical method and with the iterative procedure. The two methods gave comparable results, with differences of the order of 10–15%.

For tunnels with a circular cross section, a similar approach is proposed (Figure 6.24).

An alternative, or as complementary, to the analytical approach is a static numerical approach using one of the continuum methods discussed in Chapter 5 (typically Finite Difference of Finite Element Method). In this approach, both the ground and the structure are explicitly modeled, together with a credible interface. The earthquake response of the underground structure is approximated by solving the static problem numerically where at the boundaries of the discretization the free-field ground deformations induced by the earthquake are imposed. These free-field deformations are obtained from a 1D analysis (e.g.

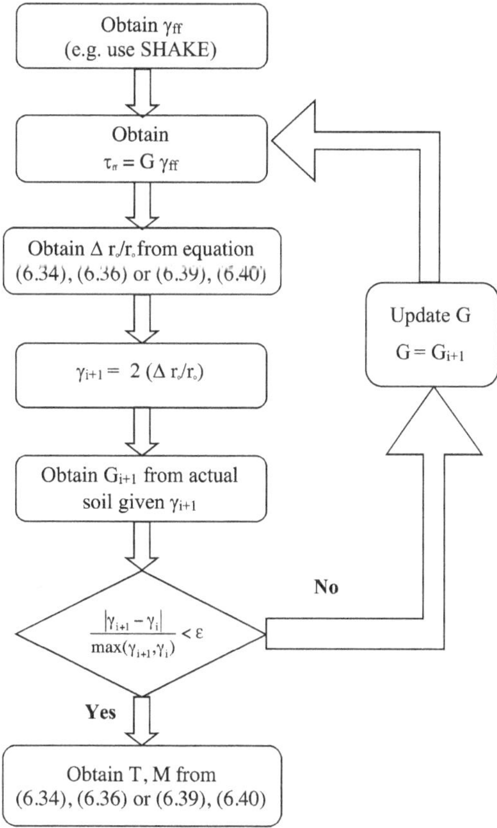

Figure 6.24 Iterative procedure for computing ovaling of circular underground structures.

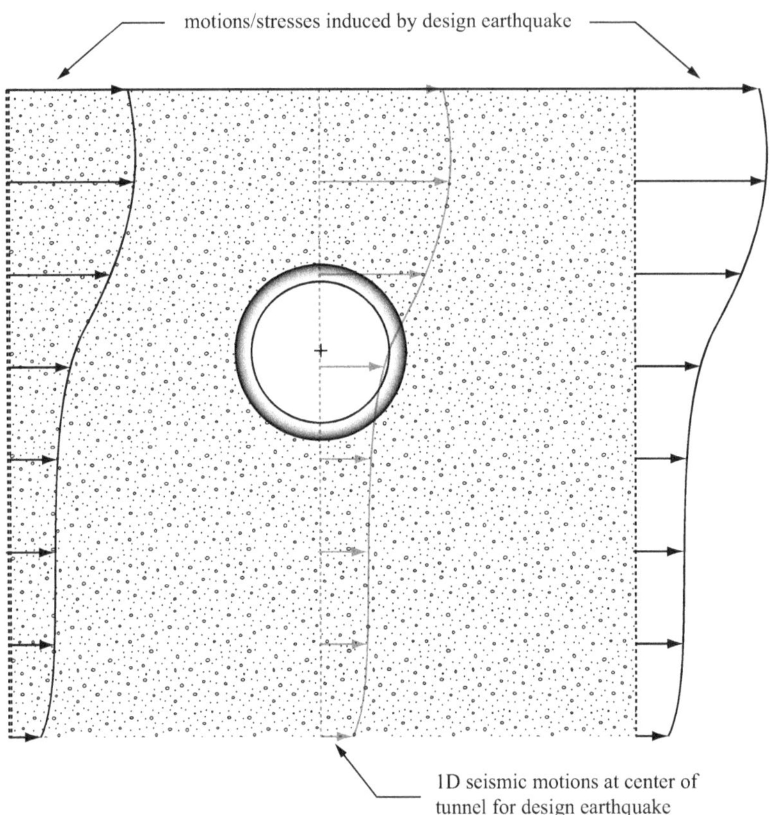

Figure 6.25 Input seismic demand for numerical quasi-static analysis.

SHAKE, DEEPSOIL or similar) of the site response, where a column of the soil, with the actual stratigraphy and ground properties are used with the desired earthquake loading at the base of the ground column (Figure 6.25). The free-field ground distortions obtained are those imposed at the boundaries of the numerical model. It is highly recommended to use an appropriate soil model that takes into account the soil stiffness degradation with strain; alternatively, an iterative approach similar to that used in Figures 6.23 and 6.24 can be attempted.

The two procedures, analytical or static numerical, give good approximations for a preliminary seismic design of underground structures in a continuous medium, especially where deformations need to be estimated. The solutions, taking into consideration the level of uncertainty that exists for predicting the response of ground and structure during cycles of loading, should be viewed as first approximations and not as a replacement of detailed dynamic numerical models. For rock masses or discontinuous media, no close form solution exists. In such cases discrete element methods should be used. These methods are discussed in Chapter 5. It has been proposed, as a first approximation and for tunnels placed in the far field, that the stability of wedges around the opening can be estimated through a static analysis subjecting the wedge to the free-field vertical and horizontal accelerations induced by the earthquake (Hendron and Fernández, 1983). To the authors' knowledge, wedge stability analysis following this approach has not been validated. Given that the presence of the tunnel may significantly affect the response of the ground surrounding the opening, it appears that this procedure could be used for tunnels stiffer than the surrounding ground. For tunnels with a relative stiffness smaller than the ground that they replace, the seismic-induced motions close to the tunnel may be amplified relative to the free field, and thus the approach may give unsafe results.

Another note of caution must be given regarding underground structures placed in saturated soft soils. During an earthquake, large excess pore pressures may be accumulated due to the cycles of loading induced in the soil by the earthquake (excess pore pressures, in soils, may accumulate with cycles of loading). Excess pore pressures reduce the effective stresses in the soil, which may lead to soil liquefaction. Also, as the effective stresses in the soil decrease, the stiffness of the soil also decreases, and so the relative stiffness of the structure with respect to that of the ground increases with the number of cycles. That is, the structure becomes (relatively) stiffer. This effect is not included in the analytical solutions presented. However, numerical analyses by Sandoval and Bobet (2020b) suggest that a good approximation to the actual, dynamic response of the structure with pore pressure accumulation of the ground, can be obtained using a static analysis, with a soil model with appropriate non-linear response and pore pressure accumulation, by imposing statically the loading and unloading cycles in the far field obtained with 1D site response, as shown in Figure 6.25. The same authors, in Sandoval and Bobet (2020a), show that a quasi-static analysis, for undrained conditions, is appropriate when the ratio of the wavelength of the seismic motions to the size of the tunnel (e.g. diameter) is larger than about 8 to 9.

Given the level of uncertainty associated with estimation of seismic-induced deformations imposed on underground structures, their design must include a significant level of ductility such that brittle failures are avoided. This is particularly important in cut and cover structures, which have to resist significant bending moments, and thus their behavior is not different from that of aboveground structures. Collapse of a shallow rectangular underground structure may be induced, similar to aboveground structures, in critical elements where the moment capacity and/or shear capacity is exhausted. This is quite different from deep circular tunnels where the support works mostly in compression. Creation of plastic hinges in the support during earthquake-induced motions may not necessarily cause the collapse of the structure, as the compression capacity of the liner may not be significantly reduced. This provides a level of redundancy that is not available in (shallow) rectangular structures.

The approach recommended assumes that the axial dimension of the underground structure is much larger than the other two dimensions and that the tunnel support and ground are relatively homogeneous. There are cases, however, where a significant or abrupt change of stiffness along the axis of the tunnel may occur; e.g. connections between tunnels and buildings and transit stations, junctions, crossing of geologic media with different stiffness, or local restraints (Hashash et al., 2001). Rigid connections at the transitions may induce extremely large, generally unacceptable loads. Often the best solution is to place joints to accommodate the relative movements between the two sections connected. The magnitude of the relative movements can be quantified by numerical methods or, using the analytical methods discussed, estimated by treating each section independently from each other and as infinitely long.

6.3 PRESSURE TUNNELS

6.3.1 Introduction

The purpose of pressure tunnels is to convey water under pressure with acceptable losses and without causing instabilities. Pressure tunnels are mostly built in hydroelectric power plant projects. They are also used, albeit with smaller pressures, for water supply and wastewater conveyance.

Figure 6.26 illustrates different possibilities (Broch, 1984a). Figure 6.26a shows a design where the water from the intake flows to the powerhouse through a low-pressure (headrace) tunnel and a steel penstock (pressure shaft) placed on top of the natural slope. A surge shaft or surge tank is located at the end of the low-pressure tunnel. In Figure 6.26b, the pressure shaft is moved inside the rock mass and a tailrace tunnel discharges the water from the powerhouse to the lower reservoir or to another body of water. In Figure 6.26c, the surge shaft is replaced with a surge chamber with air cushion. The pressure heads in tunnels can be

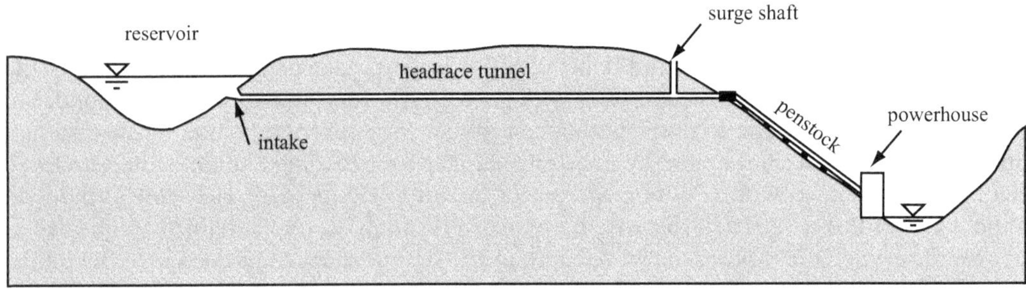

(a)

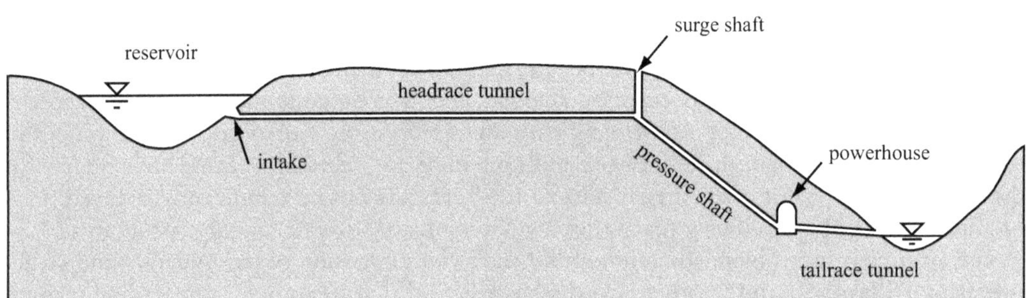

(b)

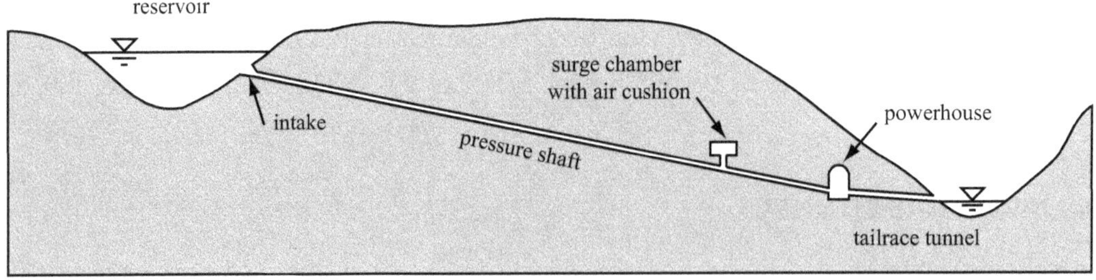

(c)

Figure 6.26 Development of hydroelectric power plants through time (After Broch, 1984a). (a) Design with headrace, surge shaft and penstock. (b) Design with headrace, surge shaft and tailrace tunnel. (c) Design with pressure shaft, surge chamber and tailrace.

substantial, and have increased steadily with time, with values larger than 1,000 m in today's designs.

Tunnel dimensions are typically between 4 and 6 m and several km in length (Deere, 1983), with a minimum economically feasible diameter estimated in 2.5 m (Brekke and Ripley, 1993). There are, however, many cases with dimensions outside these boundaries (e.g. the Háslslón headrace tunnel in Iceland has a diameter of 7.2–7.6 m and a length of 39.7 km). The longitudinal slope of headrace and tailrace tunnels is often constrained by the maximum practical grade of the equipment used during construction, which is about 1–2% for rail and 10–15% for rubber tire machinery. Inclined shafts have often angles steeper than 40° to 50° to ensure that the shaft is self-mucking (Brekke and Ripley, 1993).

The cross section of pressure tunnels is usually circular. A circular cross section is preferred because it has structural advantages given the large internal pressures, and also for hydraulic reasons. The cross-section area and the surface roughness (i.e. unlined tunnel or concrete or steel lining) depend on the head losses accepted over the length of the tunnel, provided that the opening is stable. Frictional head losses can be estimated using Darcy-Weisbach theory or Manning's empirical equation. The head loss is given by:

$$\begin{aligned} \Delta h &= \frac{f\,L\,v^2}{4\,r_o\,g} \quad \text{for Darcy-Weisbach} \\ \Delta h &= \frac{n^2\,L\,v^2}{R_h^{4/3}} \quad \text{for Mannig} \end{aligned} \tag{6.42}$$

where Δh is the head loss between two points separated by a distance L; v is the water velocity inside the tunnel; r_o is the tunnel radius; g is the acceleration of gravity; R_h is the hydraulic radius of the tunnel; and f and n are the Darcy-Weisbach roughness coefficient and the Manning roughness coefficient, respectively. The coefficients are given in Table 6.1 for some typical tunnel finishes (Brekke and Ripley, 1993).

The table shows that as the roughness of the surface of the tunnel decreases, the roughness coefficients decrease. Thus a smaller tunnel with a smooth surface may be hydraulically equivalent to a larger tunnel with a rough surface. Hendron et al. (1989) pointed out that the hydraulic characteristics of an unlined tunnel in rock are similar to those of a tunnel made smaller with a 0.3–0.6 m thick concrete liner, and thus the decision of placing a liner in a tunnel could be deferred after the tunnel is excavated, which allows for accurate determination of critical properties of the rock mass during construction. Flow velocities in unlined tunnels are normally limited to prevent erosion of the excavated ground. They are typically within

Table 6.1 Manning and Darcy-Weisbach roughness coefficients

Lining	*Manning's coefficient, n*	*Darcy-Weisbach's coefficient, f*			
		r_o = 1.25 m	*r_o = 2.5 m*	*r_o = 3.75 m*	*r_o = 5 m*
Unlined (drill and blast)	0.025–0.040	0.057–0.147	0.046–0.117	0.040–0.102	0.036–0.093
Unlined (tunnel boring machine)	0.016–0.022	0.023–0.044	0.019–0.035	0.016–0.031	0.015–0.028
Shotcrete (drill and blast)	0.018–0.025	0.030–0.057	0.024–0.046	0.021–0.040	0.019–0.036
Concrete	0.012–0.016	0.013–0.023	0.010–0.019	0.0092–0.016	0.0083–0.015
Steel	0.010–0.014	0.0092–0.018	0.0073–0.014	0.0064–0.012	0.0058–0.011

Source: (after Brekke and Ripley, 1993)

the range 1.0 to 1.5 m/s for unlined tunnels with unpaved inverts and 1.0 to 2.75 m/s for tunnels with paved inverts (Brekke and Ripley, 1993; USBR, 2014); however, larger velocities have been reported (e.g. Dann et al., 1964). A paved invert may slightly increase the cost of the project but provides a good surface during construction and maintenance, decreases head loss, and allows larger water velocities.

The modes of failure in pressure tunnels have been identified by Hendron et al. (1989) and by Fernandez (1994) as: (1) excessive leakage; (2) excessive pore water pressures; (3) failure of the liner; and (4) collapse of the opening. Excessive leakage may result in unacceptable economic losses for the project or flooding on the ground surface and at the limit can cause hydraulic fracturing of the surrounding rock mass (hydraulic fracturing occurs when pore pressures are large enough to produce a tensile crack in the rock; hydrojacking or hydraulic jacking are terms also used to describe the phenomenon and in this text they are used interchangeably). Also, tunnels excavated in erodible materials may experience loss of contact between support and ground causing collapse of the tunnel. Excessive pore pressures surrounding the excavation can reach the ground above the tunnel, and if not properly controlled, cause slope failures. Large pore pressures can build up behind impermeable soils accumulated on the surface above the tunnel, which may trigger landslides. Pore pressures can also induce excessive loads on nearby tunnels or on other adjacent structures compromising their stability. Failure of the lining is possible when the liner is subjected to loads or load magnitudes different than those assumed during design. For example, the steel liner of a pressure tunnel may buckle under external pressures much larger than those initially estimated or because of poor contact between the steel and the concrete. Collapse of the opening may be induced by loss of strength of rock materials exposed to water. The first three modes of failure, namely excessive leakage, excessive pore water pressures, and failure of the liner are related to the leakage through the liner with the consequent buildup of pore pressures in the ground immediately surrounding the tunnel. An undesirable effect of the pore pressure buildup is a decrease of the effective compressive stresses in the ground immediately behind the liner.

The relation between leakage from pressure tunnels and pore pressure buildup was already recognized by Terzaghi (1962) when discussing the sources of failures of steep slopes on hard unweathered rock. Terzaghi, based on the rock slides that occurred on a pressure tunnel in North-eastern Austria (based on data from Stini, 1956), suggested the following relation to prevent rock slides due to leakage from pressure tunnels:

$$H'_c = \frac{\gamma_w}{\gamma_r} H \approx 0.385H \qquad (6.43)$$

where H'_c is the minimum depth of rock above the tunnel, measured perpendicular to the ground surface, γ_r and γ_w are the unit weights of the rock and water, respectively, and H is the hydrostatic head of the water in the tunnel. Assuming relative unit weights for the water and rock of 1 and 2.6, the value of 0.385 on the right-hand side of the equation is obtained. Terzaghi suggests using as "rule of thumb," $H'_c = 0.5$ H, which carries a factor of safety of 1.3. Further discussion on minimum rock cover in tunnels is provided in Section 6.3.3.1. It is informative to quote Terzaghi's assessment on the importance of rock cover in pressure tunnels (Terzaghi, 1962):

"None of the rock slides which have so far come to the author's attention were caused by leakage from the tunnel at points where the depth of overburden was greater than Hc', Equation (6.43) (equation (4) in the original paper). Hence it appears that the rule expressed by $H'_c = 0.5$ H (equation (5) in the original paper) is reliable, provided the lining of the tunnel is designed and constructed in such a manner that it remains permanently intact. In order to satisfy this condition the reinforcement must be strong enough to withstand the full internal water pressure without being overstressed even if the rock support were entirely removed."

It is clear that an estimate of the magnitude of the pore pressures and an analysis of the poro-mechanical effects on the ground and liner are required. The following section provides the conceptual framework to understand and evaluate the interaction between ground, liner, and pore pressures.

6.3.2 Interaction between ground, liner, and water pressure

The liner of a pressure tunnel needs to be designed such that it can withstand the loads from the ground, external water pressure, if any, the internal pressure, and minimize the development of significant pore pressures at the liner–ground interface. The internal pressure results in an expansion of the liner which then transfers part of the load to the surrounding ground because of compatibility of stresses and deformations between the liner and the ground, assuming a good connection at the contact between the liner and the excavation. If there is seepage through the liner, pore pressures increase behind the liner. The pore pressures have a double effect: first they counteract the expansion of the liner caused by the inside pressure, and second they change the stress field in the ground around the tunnel by decreasing effective stresses (Seeber, 1985). Additional loads to the liner and surrounding ground may be the result of other operations, e.g. of prestressing the liner by high-pressure (> 5 bar) grout injected behind the liner (structural grouting; see Section 6.3.3). These loads are not considered in the following discussion.

An analytical solution has been developed by Fernández and Alvarez (1994) and Bobet and Nam (2007) that includes the interaction between ground, liner, and pore pressures in the medium (see also Schleiss, 1986, 1997). The solution is based on the following assumptions (Figure 6.27a): (1) circular cross section; (2) plane strain conditions in a direction perpendicular to the cross section of the tunnel; (3) the ground is fully saturated, homogeneous, and isotropic; (4) deformations of ground and liner remain within their respective elastic regimes; (5) the principle of effective stresses applies; (6) no slip between ground and support; and

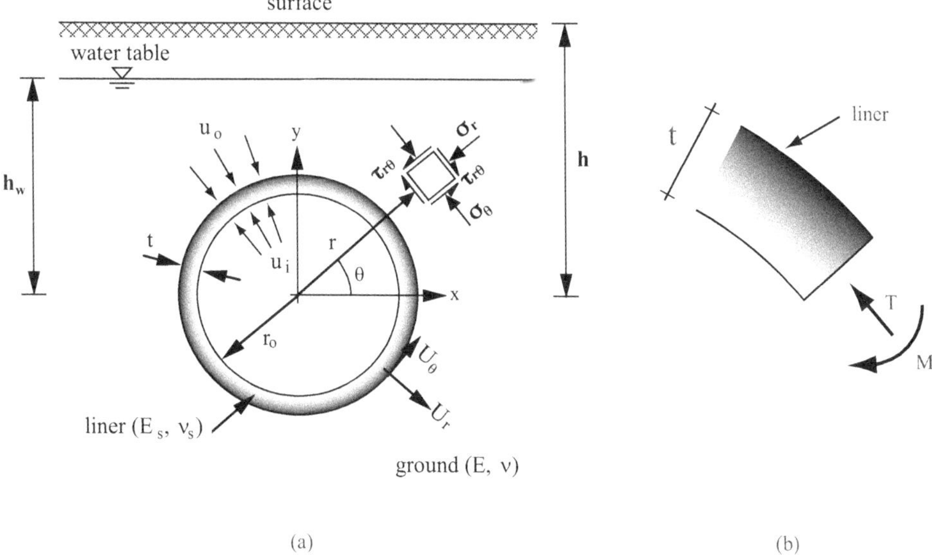

Figure 6.27 Pressure tunnel. (a) Notation. (b) Thrust and moment in liner.

(7) depth to radius ratio larger than 2.5. For smaller depth to radius values, the solution is not valid due to the effect of the ground surface (Mindlin, 1940).

The solution focuses on the excess pressures induced by the tunnel, and so the input pressures and stresses and displacements obtained are those due to the water pressures in excess to the ambient pore pressures induced by the existing water table. The excess water pressure inside the tunnel is u_i, and is assumed constant along the perimeter of the tunnel; $u_i = \gamma_w(h_i - h_w)$ where γ_w is the unit weight of water, h_i is the internal head and h_w is the head from the groundwater table. Because the liner may leak (i.e. flow through the thickness of a concrete liner, through cracks in the concrete, etc.) a pore pressure buildup, in excess of the external hydrostatic pressures, u_o behind the liner, at the contact between the liner and the ground, may occur (Figure 6.27a). Note that u_o is the difference between the water pressure behind the liner and the hydrostatic pressure $\gamma_w h_w$. It is also assumed that u_o is constant along the perimeter of the liner–ground interface. Note that the approach is correct because flow and associated stresses depends on the difference between the inside and outside fluid head/pressure in the tunnel. The liner, of course, will be subjected to the loading due to the water leaking through the tunnel and to a radial pressure of magnitude $\gamma_w h_w$.

The flow, Q^s, per unit length of tunnel, from the pressure tunnel into the ground through an uncracked liner is:

$$Q^s = 2\pi \frac{r_o}{t} K^s \frac{(u_i - u_o)}{\gamma_w} \tag{6.44}$$

where r_o is the radius of the tunnel, t the thickness of the liner, and K^s the permeability of the liner.

For a cracked liner, assuming laminar flow through the cracks and validity of the cubic law:

$$Q^s = 2\pi \frac{r_o}{t} K^s \frac{(u_i - u_o)}{\gamma_w} + \frac{N}{12\eta} \frac{W^3}{t}(u_i - u_o) \tag{6.45}$$

where N is the number of longitudinal cracks, W is the aperture of the cracks, t is the thickness of the liner, and η is the dynamic viscosity of water, equal to 10^{-3} Pa·s at 20°C. The first term in (6.45), which estimates the flow through the liner itself, i.e. Equation (6.44), is usually neglected because it is generally much smaller than the flow through the cracks.

The excess pore pressure distribution in the ground for the leaky tunnel problem is (Harr, 1962; Fernández and Alvarez, 1994; Bobet, 2001):

$$u = u_o \frac{\ln\left[1 + \frac{4h_w}{r}\left(\frac{h_w}{r} - \sin\theta\right)\right]}{\ln\left[1 + \frac{4h_w}{r_o}\left(\frac{h_w}{r_o} - \sin\theta\right)\right]} \tag{6.46}$$

where h_w is the depth of the tunnel below the water table, and r and θ are the polar coordinates.

Note that (6.46), as mentioned, represents the pore pressures in excess of the ambient pore pressures (note that as $r \to \infty$, $u = 0$), and so the total pore pressures will be the sum of those from (6.46) and from the hydrostatic pressures due to the water table. The excess pore pressures behind the liner u_o can be estimated by equating the flow through the liner, Q^s, with the flow into the ground, Q^g (i.e. $Q^s = Q^g$). The flow through the liner is given by

Equations (6.44) or (6.45), while the flow into the ground, Q^g, at the ground–liner interface is given by (Bobet and Nam, 2007):

$$Q^g = \frac{\left(\frac{h_w}{r_o}\right)^2}{\left(1+4\frac{h_w^2}{r_o^2}\right)\ln\left(1+4\frac{h_w^2}{r_o^2}\right)} 2\pi\alpha K^g \frac{u_o}{\gamma_w} \tag{6.47}$$

where K^g is the ground permeability. The parameter α is given in Table 6.2 (Bobet and Nam, 2007). It depends on the depth to radius ratio of the tunnel and reaches a theoretical value of eight for deep tunnels.

Effective stresses and pore pressures of the ground are given by:

$$\begin{aligned}
\sigma'_r &= \frac{a_o}{r^2} + 2b_o + c_o(1+2\ln r) - (6a'_2 r^{-4} + 4b'_2 r^{-2})\cos 2\theta - u_o \frac{\ln\left[1+\frac{4h_w}{r}\left(\frac{h_w}{r} - \sin\theta\right)\right]}{\ln\left[1+\frac{4h_w}{r_o}\left(\frac{h_w}{r_o} - \sin\theta\right)\right]} \\
\sigma'_\theta &= -\frac{a_o}{r^2} + 2b_o + c_o(3+2\ln r) + 6a'_2 r^{-4}\cos 2\theta - u_o \frac{\ln\left[1+\frac{4h_w}{r}\left(\frac{h_w}{r} - \sin\theta\right)\right]}{\ln\left[1+\frac{4h_w}{r_o}\left(\frac{h_w}{r_o} - \sin\theta\right)\right]} \\
\tau &= -(6a'_2 r^{-4} + 2b'_2 r^{-2})\sin 2\theta \\
u &= u_o \frac{\ln\left[1+\frac{4h_w}{r}\left(\frac{h_w}{r} - \sin\theta\right)\right]}{\ln\left[1+\frac{4h_w}{r_o}\left(\frac{h_w}{r_o} - \sin\theta\right)\right]}
\end{aligned} \tag{6.48}$$

Table 6.2 Parameter α with tunnel depth

h_w/r_o	α
2.5	8.602
3.0	8.391
3.5	8.274
4.0	8.204
4.5	8.157
5.0	8.125
7.5	8.053
10	8.030
20	8.009
25	8.006
50	8.003

Source: (after Bobet and Nam, 2007)

and displacements:

$$U_r = -\frac{1+\nu}{E}\left\{-\frac{a_o}{r} + 2(1-2\nu)b_o r + (1-4\nu)c_o r + 2(1-2\nu)(\ln r - 1)c_o r \right.$$
$$+ [2a'_2 r^{-3} + 4(1-\nu)b'_2 r^{-1}]\cos 2\theta$$
$$\left. - \frac{(1-2\nu)u_o}{\ln\left[1+\frac{4h_w}{r_o}\left(\frac{h_w}{r_o}-\sin\theta\right)\right]}\left[\begin{array}{l}(r-2h_w\sin\theta)\ln(r^2+4h_w^2-4h_w r\sin\theta)\\ -r\ln r^2 + 4h_w\cos\theta\tan^{-1}\left(\frac{r\cos\theta}{2h_w - r\sin\theta}\right)\end{array}\right]\right\}$$

$$U_\theta = -\frac{1+\nu}{E}\left\{4(1-\nu)c_o r\theta + [2a'_2 r^{-3} - 2(1-2\nu)b'_2 r^{-1}]\sin 2\theta \right. \tag{6.49}$$
$$\left. - \frac{(1-2\nu)u_o}{\ln\left[1+\frac{4h_w}{r_o}\left(\frac{h_w}{r_o}-\sin\theta\right)\right]}\left[\begin{array}{l}-2h_w\cos\theta\ln\left(r^2+4h_w^2-4h_w r\sin\theta\right)-2r\theta\\ +r\tan^{-1}\left(\frac{4h_w r\cos\theta}{4h_w^2-r^2}\right)+r\tan^{-1}\left(\frac{r^2\sin 2\theta}{4h_w^2+r^2\cos 2\theta}\right)\\ -4h_w\sin\theta\tan^{-1}\left(\frac{r\cos\theta}{2h-r\sin\theta}\right)\end{array}\right]\right\}$$

with the following values:

$$a_o = -\frac{\left\{\begin{array}{c}(1-2\nu)(C+F)u_o - (1-\nu)CF u_i\\ -2c_o\left[(C+F)\left(1-\nu+(1-2\nu)\ln\frac{r_o}{h_w}\right)-(1-\nu)CF\ln\frac{r_o}{h_w}\right]\end{array}\right\}}{(C+F)[r_o^{-2}+(1-2\nu)h_w^{-2}]+(1-\nu)(r_o^{-2}-h_w^{-2})CF}$$
$$b_o = -\frac{1}{2}[a_o h_w^{-2} + c_o(1+2\ln h_w)]$$
$$c_o = -\frac{1}{2}\frac{1-2\nu}{1-\nu}\frac{u_o}{\ln\left(1+\frac{4h_w^2}{r_o^2}\right)} \tag{6.50}$$
$$a'_2 = \frac{(1-\nu)[4\nu+(1-\nu)C]}{(1-\nu)[3-2\nu+(1-\nu)C]F+3[4(3-4\nu)+(1-\nu)(5-6\nu)C]}\frac{1}{3}\frac{r_o^6}{h_w^2}c_o$$
$$b'_2 = -\frac{(1-\nu)[2+(1-\nu)C]}{(1-\nu)[3-2\nu+(1-\nu)C]F+3[4(3-4\nu)+(1-\nu)(5-6\nu)C]}\frac{r_o^4}{h_w^2}c_o$$

where C and F are the compressibility and flexibility factors, defined as (Einstein and Schwartz, 1979):

$$C = \frac{E r_o(1-\nu_s^2)}{E_s A_s(1-\nu^2)}$$
$$F = \frac{E r_o^3(1-\nu_s^2)}{E_s I_s(1-\nu^2)} \tag{6.51}$$

A_s is the cross-section area of the liner per unit length of tunnel (i.e. $A_s = t$); I_s is the moment of inertia of the liner per unit length of tunnel (i.e. $I_s = 1/12\ t^3$; Figure 6.27b); E and ν are the Young's modulus and Poisson's ratio of the rock and E_s and ν_s those of the liner.

The axial force T and moment M in the liner are:

$$\begin{aligned} T &= -u_i r_o + a_o r_o^{-1} + 2b_o r_o + c_o r_o(1+2\ln r_o) - 2a'_2 r_o^{-3}\cos 2\theta \\ M &= (a'_2 r_o^{-2} + b'_2)\cos 2\theta \end{aligned} \tag{6.52}$$

The axial force is positive in compression and the moment is positive clockwise (Figure 6.27b). Equations (6.48) to (6.52) are valid as long as $\sigma'_r \geq 0$ at $r = r_o$ (with the assumption of small lining thickness, $t << r_o$). Otherwise the liner would pull from the ground, which is not possible. When $\boldsymbol{\sigma}'_r = 0$ a gap forms between the liner and the ground. Along the length of the gap, the liner is subjected to the full excess pore pressure u_o, and the ground to a total stress equal to u_o, to a pore pressure u_o, and thus to zero radial effective stresses. Note also that the solution provided in Equations (6.48) to (6.52) is for the problem of water leaking from the tunnel. That is, it does not include stresses from e.g. tunnel construction and so the actual radial effective stresses at the ground–liner interface should be those from Equation (6.48) plus those resulting from construction, etc. That is, the gap may form when the combined radial effective stresses is zero or negative.

Equations (6.48) to (6.52) can also be used for tunnels where there is no leakage. The solution is obtained taking $u_o = 0$, so the tunnel is only subjected to the internal pressure u_i. The equations can also be used to obtain stresses in the ground for an unsupported (unlined) tunnel using $C = F = \infty$. In this case the limitation of no tensile effective radial stresses at the tunnel perimeter does not apply since there is no liner. The following are the expressions for the parameters, for the unlined tunnel:

$$\begin{aligned} a_o &= \frac{2c_o \ln\frac{h_w}{r_o} + u_i}{r_o^{-2} - h_w^{-2}} \\ b_o &= -\frac{1}{2}\left[a_o h_w^{-2} + c_o(1+2\ln h_w)\right] \\ c_o &= -\frac{1}{2}\frac{1-2\nu}{1-\nu}\frac{u_i}{\ln\left(1+\frac{4h_w^2}{r_o^2}\right)} \\ a'_2 &= b'_2 = 0 \end{aligned} \tag{6.53}$$

At the tunnel perimeter, stresses are:

$$\begin{aligned} \sigma'_r &= 0 \\ \sigma'_\theta &= -\left[\frac{1-2\nu}{1-\nu}\left(1-\frac{2h_w^2}{h_w^2-r_o^2}\ln\frac{h_w}{r_o}\right)\frac{1}{\ln\left(1+\frac{4h_w^2}{r_o^2}\right)} + \frac{2h_w^2}{h_w^2-r_o^2}\right]u_i \\ \tau &= 0 \\ u &= u_i \end{aligned} \tag{6.54}$$

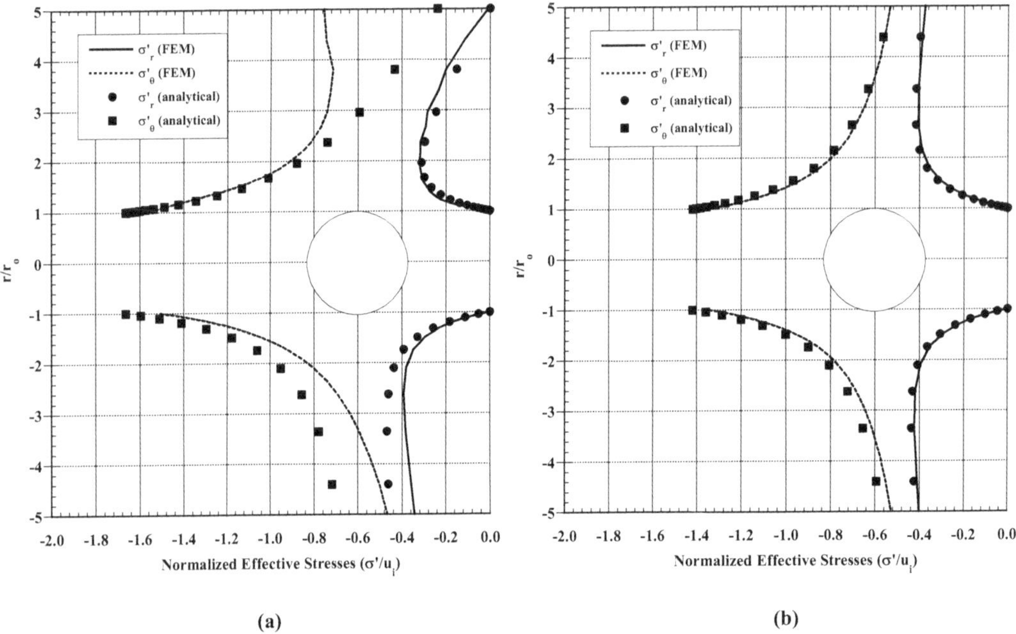

Figure 6.28 Unlined tunnel. Effective stresses along a vertical plane through the center of the tunnel. (a) $h_w/r_o = 5$. (b) $h_w/r_o = 30$.

Figure 6.28 presents effective radial and tangential stresses around an unlined shallow tunnel ($h_w/r_o = 5$) and a deep tunnel ($h_w/r_o = 30$), subjected to an excess internal pressure u_i. The input parameters are: $r_o = 4$ m, E = 500 MPa, and $\nu = 0.15$. The solution is compared with results from the Finite Element ABAQUS using the same geometry and material properties. The effective stresses are computed along a vertical plane through the center of the tunnel, with positive radial distances moving away from the tunnel. In the plots, stresses are normalized with respect to the pore pressure u_i and the radial distances are measured as multiples of the tunnel radius. The plots show that very close to the tunnel the analytical and the FEM results compare very well. As the distance from the tunnel increases, the differences increase very quickly. The discrepancy is produced because the analytical solution does not satisfy boundary conditions far from the tunnel (Bobet and Nam, 2007). Figure 6.28b indicates that the differences decrease as the depth of the tunnel increases. The comparisons shown in the figure indicate that the analytical solution gives acceptable results close to the tunnel, but it should be used with caution beyond the tunnel perimeter, in particular for shallow tunnels. The water flowing from the tunnel into the ground induces tensile effective radial and tangential stresses, with maximum magnitudes close to the tunnel substantially larger than the applied excess pore pressure. Another important observation is that the induced effective stresses decay very slowly as the radial distance from the tunnel increases. This is the result of the terms with the logarithm of radial distance present in the equations. The influence of the excess water pressure in the tunnel can reach radial distances of the order of twenty times the tunnel radius (Bobet, 2003).

Inspection of stress and displacement Equations (6.48) and (6.49) indicate that the following dimensionless factors determine the solution: relative stiffness between the ground and liner given by the compressibility and flexibility ratios, C and F; ground Poisson's ratio, ν; relative tunnel depth, h_w/r_o; and relative pressure magnitude u_i/u_o. The effects of the relative stiffness and Poisson's ratio have been extensively discussed by Einstein and Schwartz (1979) for deep tunnels, and their observations are also applicable to the present case. In essence, a very stiff liner will limit tunnel deformations at the expense of large liner stresses with a small change of the ground's stresses. A soft liner will have the opposite effect: small liner stresses, large tunnel deformations, and a large change in ground stresses. The effect of the ground Poisson's ratio is small, and in general can be neglected within the range of typical values. The effects of relative tunnel depth, h_w/r_o, have been explored by Bobet and Nam (2007) in incompressible tunnels, i.e. C = 0, where it was found that this factor has a modest effect, with effective stresses decreasing with depth by 10–15% over the range $5 \leq h_w/r_o \leq 100$.

The consequences of increasing the pore pressure, i.e. leakage, behind the liner are illustrated in Figure 6.29, where liner and ground stresses at the interface between ground and liner are plotted for different values of u_o/u_i (related to magnitude of the water leak through the liner). Only stresses due to the pore pressures are shown (the contribution of the ground effects is not included). The results correspond to a tunnel with the following properties: E = 500 MPa, $\nu = 0.15$, $E_s = 24000$ MPa, $\nu_s = 0.25$, $t/r_o = 0.05$, and $h/r_o = 5$. As the pore pressures u_o behind the liner increase, the tangential stresses of the ground become more tensile. For the particular case analyzed, the increase of tensile tangential stress of the ground is about threefold as u_o increases from zero (no leak) to the same magnitude as the internal pressure u_i. The change in the ground's effective radial stresses is more dramatic; σ'_r is reduced from a compressive $\sigma'_r/u_i \approx 0.2$, when there is no leak, to a tensile $\sigma'_r/u_i \approx -1.0$ when $u_o = u_i$. In contrast, the effect on the liner tangential stresses is modest, with an increase in tensile tangential stresses of only 30%. This is an interesting result that highlights the interplay that exists between the ground and the liner. Figure 6.29 also plots the total radial stresses at the interface, which show, as they should, the same trend as the liner tangential stresses. While

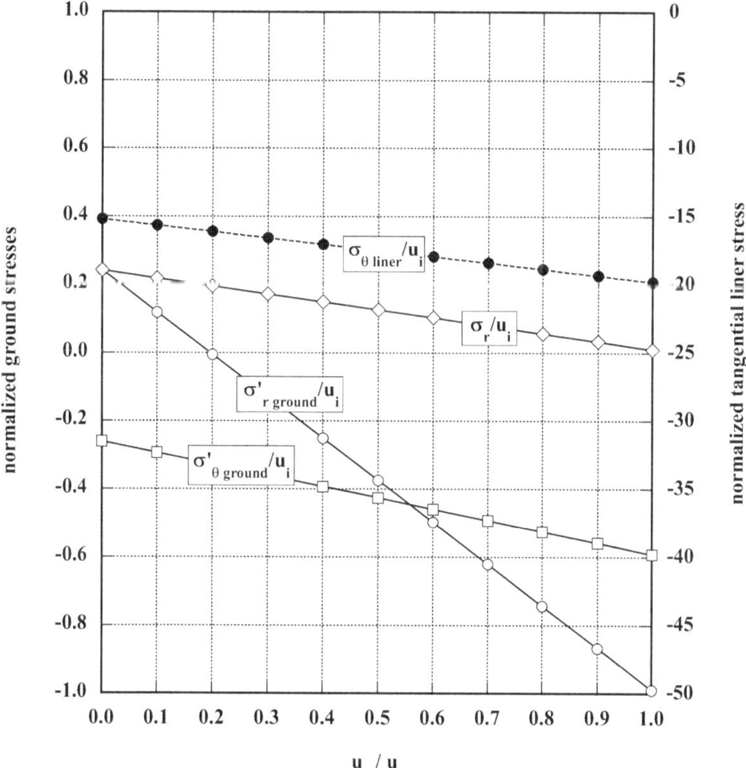

Figure 6.29 Normalized liner and ground stresses at interface due to water seepage through the support for $h/r_o = 5$, E = 500 MPa, $\nu = 0.15$, $E_s = 24000$ MPa, $\nu_s = 0.25$, $t/r_o = 0.05$ (Bobet and Nam, 2007). Republished with permission of Springer Nature.

the two materials, ground and liner, deform based on effective or total stresses, respectively, equilibrium at the interface must be satisfied in terms of total stresses. Thus while the ground experiences a significant reduction of effective stresses as the pore pressure u_0 increases, the total stresses do not change much, and hence the liner stresses (and its deformations) also do not change. This is a key observation because measurements of liner stresses or deformations may not be effective to estimate the magnitude of leakage through the liner or the effective stresses in the ground.

When a rapid loss of internal pressure occurs, i.e. rapid tunnel dewatering, the internal excess pressure u_i is zero, but the excess pore pressure at the liner-rock interface u_0 may remain. The analytical solution predicts tensile radial effective stresses at the contact between the liner and the rock (see e.g. Figure 6.29 that shows a decrease of effective radial stresses with an increase of u_0/u_i). If the effective radial stresses from the ground, tunnel excavation, etc. are not large enough, a net effective tensile radial stress may be produced at the interface. This is not possible, and so a gap forms. This indicates that the liner and the support detach; the ground is then subjected to a state of stress similar to that shown in Figure 6.28, and the support is subjected to the full excess pore pressure u_0.

6.3.3 Guiding principles for pressure tunnels

The final stresses in the liner and ground for a pressure tunnel depend on the interaction between the liner and the ground, on the stress change induced by pressurizing the tunnel and on the magnitude of the water flow, if any, through the liner. In the following discussion, it is assumed that the liner supports only the loads induced by the excess internal and excess external water pressures. This applies to tunnels where the rock needs no support or in tunnels where an initial support carries the loads from the ground and so the final support takes only the loads from the internal pressure.

Figure 6.30 shows three different zones along a pressure tunnel alignment. The zones are determined according to the position of the groundwater table with respect to the tunnel hydraulic pressure (Merritt, 1999). Zone C is defined as the tunnel sections where the hydraulic head inside the tunnel lies below the groundwater table. Flow is from the rock toward the tunnel, unless the groundwater table has been permanently lowered during construction, which is possible since the tunnel acts as a drain. Depending on the permeability of the rock mass, the drawdown may reach the invert of the tunnel. With time, it has been observed that the original groundwater table is recovered. A liner in zone C is not required for water control. A liner may be required for long-term support of the rock, to prevent erosion or swelling, to improve hydraulic properties of the opening, etc. In zones A and B

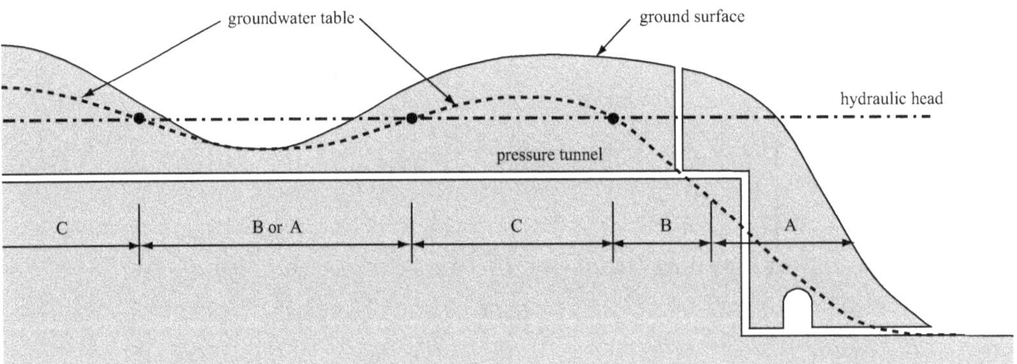

Figure 6.30 Tunnel lining distribution (after Merritt, 1999).

the groundwater table lies below the tunnel hydraulic head. In these zones flow is possible from the tunnel to the rock. Zone B applies to those sections of the tunnel where field testing has proven that the minimum principal stress in the rock is larger than the internal water pressure and so hydrojacking (or hydrofracturing) of the rock cannot occur. Zone B may be unsupported if the quality of the rock is good and there are no long-term detrimental effects and the permeability is low enough such that the water losses due to leakage are acceptable. If these conditions are not met, a liner must be provided. The liner may be unreinforced concrete if the permeability of the rock is low and its stiffness high, and so the liner is used as a protection for the rock around the opening. An unreinforced liner may crack under internal pressure, particularly if the stiffness of the rock is low, in which case substantial leakage may occur. If the permeability of the rock is high or its stiffness is low, a reinforced liner may be required. The decision of reinforced or unreinforced liner depends on the field measurements of the permeability of the rock and on its modulus of deformation. Zone A is defined as the section of the tunnel where the minimum principal stress of the rock is smaller than the internal water pressure. In Zone A, a steel liner may be needed. Other types of liners such as prestressed concrete (Douglass and Sundberg, 1987) and sandwich-type liners (Marulanda and Gutierrez, 1999) have been also used.

6.3.3.1 Unlined pressure tunnels

Given that the cost of an unreinforced concrete liner can be a substantial portion of the total cost of the tunnel (Merritt, 1999), it is desirable to have as many sections of the tunnel unlined as possible. Unlined pressure tunnels can be used provided that (Fernandez, 1994): (1) the opening is self-supporting and the rock is not susceptible to erosion, dissolution, deterioration or reduction of strength with time; (2) the minimum principal stresses are large enough with respect to the internal water pressure such that hydrojacking or hydrofracturing of the rock will not occur; (3) the permeability of the rock mass around the tunnel is low; and (4) localized zones of poor quality rock are treated with shotcrete, rockbolts, or dental concrete.

A final decision on what sections of a pressure tunnel can be left unlined can be made after excavation of the tunnel is completed. Empirical recommendations exist for a preliminary estimate of what sections can be left unlined (Brekke and Ripley, 1985). These recommendations are based on the premise that enough confinement should exist in the rock at the location of the tunnel such that hydraulic jacking or hydrofracturing of the rock would not occur (Benson, 1989; Hartmaier et al., 1998). An assumption is made in these criteria that the in situ stresses in the rock mass are related to the weight of the overlying rock. Four criteria are reviewed: (1) Terzaghi's criterion; (2) Norwegian criterion; (3) Snowy Mountains criterion; and (4) design charts criterion based on Finite Element Models.

The Terzaghi criterion (Terzaghi, 1962) has been introduced in Section 6.3.1 and provides the minimum rock cover following the recommendation that $H'_c \geq 0.5$ H, where H'_c is the minimum depth of rock above the tunnel, measured perpendicular to the ground surface and H is the hydrostatic head of the water in the tunnel. The minimum H'_c carries a factor of safety of 1.3.

The Norwegian criterion is based on the experience accumulated on both successes and failures from Norwegian hydraulic power plants. The criterion was first introduced by Bergh-Christensen and Dannevig in 1971 after the failure at Åskora in 1970 (Bergh-Christensen, 1982; Broch, 1984a; Selmer-Olsen, 1985). The criterion states that (Figure 6.31a):

$$L > \frac{F\gamma_w H}{\gamma_r \cos\beta} \quad \text{or} \quad C_{RV} > \frac{F\gamma_w H}{\gamma_r \cos^2\beta} \tag{6.55}$$

where L is the shortest distance between the surface and the tunnel; C_{RV} is the minimum vertical rock cover; γ_w and γ_r are the unit weight of water and rock, respectively; H is the static hydraulic head at the point in question; β is the average inclination of the valley side; and F is a factor of safety, typically 1.3. It is recommended to exclude in (6.55) surface soil deposits, and only consider in the computation of L bedrock. The criterion is based on experience on tunnels excavated in massive igneous and metamorphic rocks with low permeability, in a topography strongly influenced by recent glaciations, which have produced smooth valley slopes. Broch (1984b) points out that irregular noses and ridges on the sides of the valley could be stress-relieved and should be neglected when considering the rock overburden in (6.55). He suggests using simplified topographic maps with smooth contour lines that eliminate noses and ridges.

The Snowy Mountains criterion (Dann et al., 1964), developed for the Snowy Mountains hydroelectric project in Australia, establishes the minimum horizontal and vertical rock cover from the pressure tunnel. The minimum vertical rock cover C_{RV} is:

$$C_{RV} > \frac{\gamma_w H}{\gamma_r} \tag{6.56}$$

The minimum horizontal rock cover is two times the vertical rock cover (Figure 6.31b).

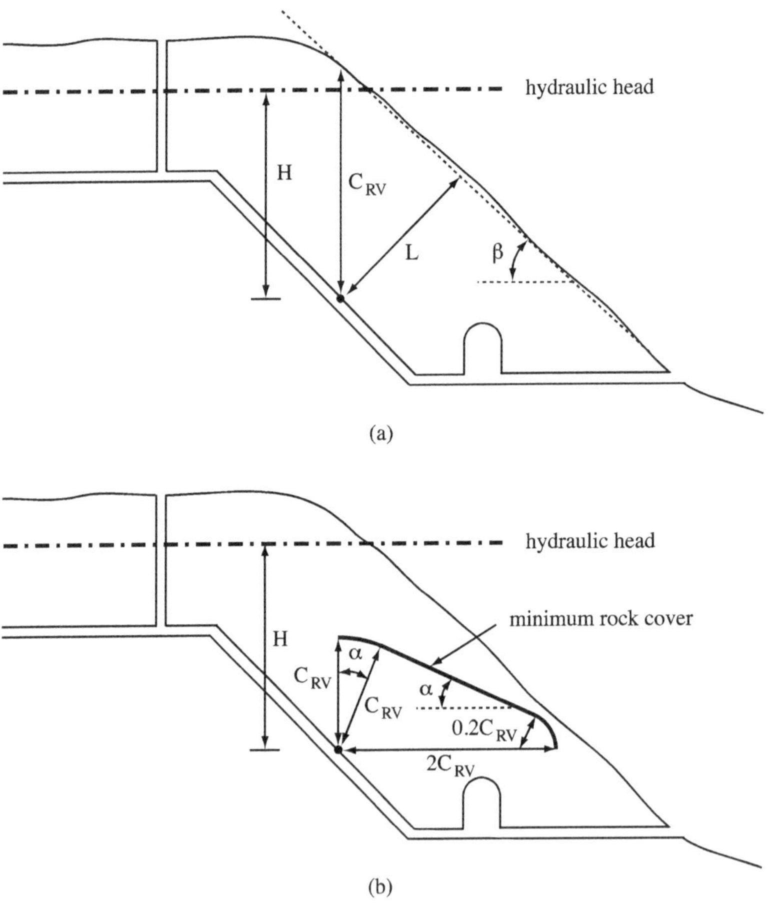

Figure 6.31 Minimum rock cover criteria. (a) Norwegian Criterion. (b) Snowy Mountains Criterion.

Both methods use the total (not effective or buoyant) unit weight of rock. Brekke and Ripley (1989) and (1993) have shown that both methods result in similar rock covers for valley slope angles β (Figure 6.31a) ranging from 0 to 70°.

Finite Element Methods have been used to evaluate the minimum principal stresses inside the valley slope where the pressure tunnel is placed (Broch, 1984a, 1984b). A cross section perpendicular to the valley is discretized with plane strain elements. Appropriate boundary conditions are imposed to the limits of the discretization; e.g. horizontal stresses, both tectonic and gravitational, are imposed to the lateral boundaries. The alignment of the tunnel is decided such that the internal water pressure, increased by the factor of safety, is smaller than the minor principal stress surrounding the tunnel obtained from the numerical analysis. A series of dimensionless design charts have been developed for different values of rock properties and valley inclination angles (Broch, 1984a). Amadei and Pan (1995) have expanded the criterion to anisotropic rock masses. The drawback of this method, compared to the Norwegian and Snowy Mountains methods, is that it requires a more detailed knowledge of rock properties. For rock conditions and topography commonly encountered in Norway, the alignment resulting from the design charts criterion is similar to that from the Norwegian criterion (Chang et al., 1988).

Fernández and Alvarez (1994) suggested to shift the location predicted using the design charts criterion (Broch, 1984a) toward inside the rock mass, by moving horizontally the tunnel a distance 0.25 L, where L is the minimum rock cover. The reason for the modification is that the influence of the pressure tunnel in the surrounding ground could extend up to 0.25 times the depth of the tunnel.

Broch (1984a, 1984b) mentioned that the Norwegian and the design charts criteria should be used when the rock mass is homogeneous and continuous and when the existence of natural joints has a minor influence on the distribution of virgin stresses. These are characteristics usually met in the projects in Norway. For other tectonic stresses (the ratio of horizontal to vertical stress in Norway is typically between 0.5 and 1.3, exceptionally 1.5), he recommended to rely on field measurements, e.g. hydraulic fracturing, to determine the magnitude of the minimum principal stress at the location of the tunnel.

Even though the Norwegian method has been successfully used in many projects, a number of failures have occurred using the criterion (Deere, 1983). Figure 6.32 shows a comparison between predictions made with the Norwegian criterion (similar predictions would result with the other criteria) and performance of pressure tunnels around the world. The database is taken from Alvarez (1997), who expanded the original database compiled by Brekke and Ripley (1986) and included cases from Deere (1983). Figure 6.32a shows tunnels excavated in generally massive plutonic and metamorphic rocks. In general, the Norwegian criterion results in acceptable predictions. Figure 6.32b shows a similar comparison with tunnels excavated in sedimentary and volcanic rocks. In the second case, the predictions from the Norwegian method are very poor. The key difference between the two scenarios is the role that discontinuities play on the behavior of pressure tunnels (Alvarez, 1997; Alvarez et al., 1999). Massive, isotropic igneous and metamorphic rocks tend to have the major principal stress parallel to the slope and the minor principal stress perpendicular to the slope. The resulting orientation of principal stresses is consistent with the three empirical criteria. Such state of stress may create a set of joints, superimposed to the regional joint system, parallel to the slope (sheeting joints), with joint spacing increasing toward the interior of the slope. Sheeting rarely occurs at distances larger than 15 m from the rock surface (Terzaghi, 1962). Volcanic and sedimentary rocks typically have very well developed discontinuity sets. For example, due to cooling vertical lava flows tend to have vertical discontinuities and sedimentary rocks with horizontal bedding are characterized by vertical discontinuities, perpendicular to bedding. Given that along discontinuities the maximum shear stress that can be

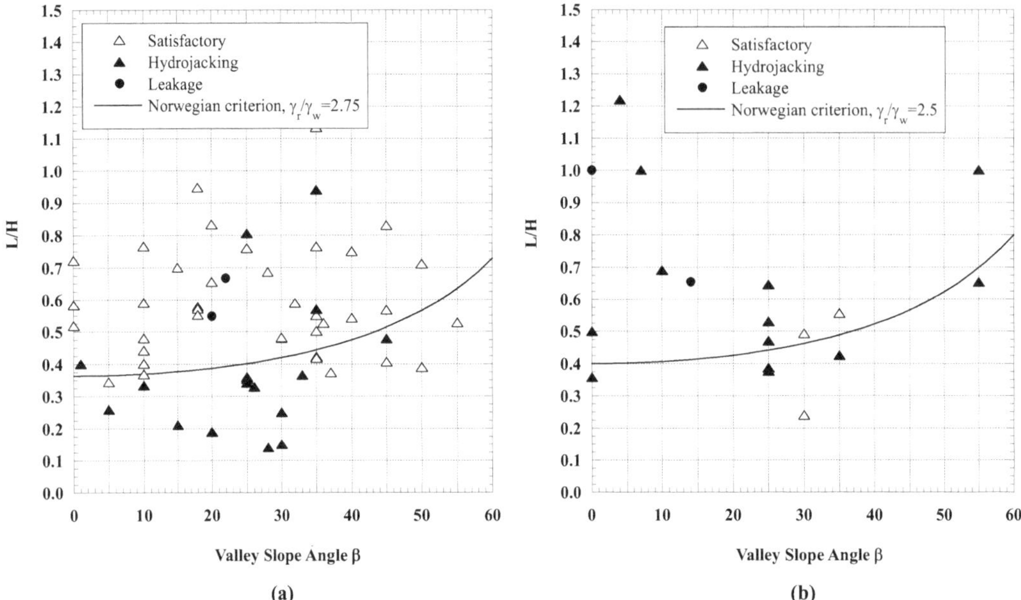

Figure 6.32 Comparison between Norwegian criterion and pressure tunnel data (data from Alvarez et al., 1999). (a) Plutonic and metamorphic rocks. (b) Volcanic and sedimentary rocks.

transmitted is limited to their frictional strength, the direction of principal stresses in rock masses next to natural slopes may not be related to the topography as with massive rocks. In fact, because of the lateral stress relief produced in the rock mass due to the excavation of the valley, vertical joints tend to open or at least unload which may result in magnitudes of the far-field minor principal stresses much smaller than those assumed by the empirical criteria (including continuum numerical methods) and in orientations that tend to be perpendicular to the vertical joints. This is apparently the cause for some of the failures displayed in Figure 6.32b (Alvarez, 1997; Alvarez et al., 1999). In light of some of the failures produced, Deere (1983) recommended a more conservative empirical rule of thumb for preliminary design: (1) extend the steel lining from the powerhouse into the mountain until the rock cover C_{RV} equals 0.8 H, where H is the hydraulic head at the point of interest due to the surge tank pressure, and the horizontal rock cover is 2.0 H (see Figure 6.31 for notation; note that this requirement is similar to the Snowy Mountains criterion); (2) a reinforced concrete transition section would extend from the steel liner an additional distance to the point where C_{RV} equals 1.3 H; (3) verify the end location of the steel and reinforced concrete liners by checking the in situ stresses, preferably during design, by conducting hydraulic fracturing tests. At the end of the steel lining, the hydraulic fracturing pressure should be at least 1.2 H and at the end of the reinforced concrete 1.4 H (again, H is the hydraulic head due to the tank surge pressure). If the hydraulic fracturing tests give higher values, the lengths of the linings could be reduced, but not below the vertical cover C_{RV} of about 0.6 H for the steel lining and 1.0 H for the reinforced concrete lining transition.

Alvarez (1997) and Alvarez et al. (1999) investigated numerically the development of pore pressures and effective stresses around pressure tunnels excavated in discontinuous rock masses. A Discrete Element Method (Chapter 5) coupled with a hydromechanical model for discontinuities was used to conduct a parametric study to determine the influence of joint orientation and valley topography on the local and global stability of pressure tunnels. It was found that the magnitude of the water pressures at which hydrojacking in the vicinity of the tunnel (local failure) and failure of the valley slope under which the tunnel was located (global failure) occurred were very sensitive to the orientation of the discontinuities, which in turn strongly influenced the magnitude and orientation of principal stresses in the rock under the valley. A new cover criterion, the Illinois criterion, was proposed for global failure, based on limit equilibrium analysis of a potential rock wedge limited by the tunnel, two discontinuities, and the valley slope (Figure 6.33). The criterion states that:

$$\frac{L}{H} > \frac{\gamma_w}{\gamma_r}\frac{\cos\beta}{\tan\alpha_1 - \tan\alpha_2}\left\{\frac{U_1}{U_1^{linear}}(\tan\beta - \tan\alpha_2)\left[1 + \frac{\tan\alpha_1}{\tan(\phi_2 - \alpha_2)}\right] - \frac{U_2}{U_2^{linear}}(\tan\beta - \tan\alpha_1)\left[1 + \frac{\tan\alpha_2}{\tan(\phi_2 - \alpha_2)}\right]\right\} \tag{6.57}$$

where L and H, as before, are the minimum distance from the tunnel to the valley slope and the hydraulic head in the tunnel, respectively, at the point of interest (Figure 6.33); γ_w and γ_r are the unit weight of water and rock; β is the valley slope angle; α_1 and α_2 are the dips of joints 1 and 2; ϕ_2 is the friction angle along joint 2; U_1 and U_2 are the resultant forces from the water pressure along joints 1 and 2, respectively; and U_1^{linear} and U_2^{linear} are the resultant forces of the water pressure along the joints assuming a linear distribution from the tunnel to the valley surface (Figure 6.33). The actual distribution of the water pressure needs to be estimated to obtain U_1 and U_2. The distribution ranges from a logarithmic function in massive rock or rock with closed joints, e.g. Equation (6.46), to a linear function for open joints. A linear function is of course more conservative. In a rock mass with vertical joints, $\alpha_1 = 90°$,

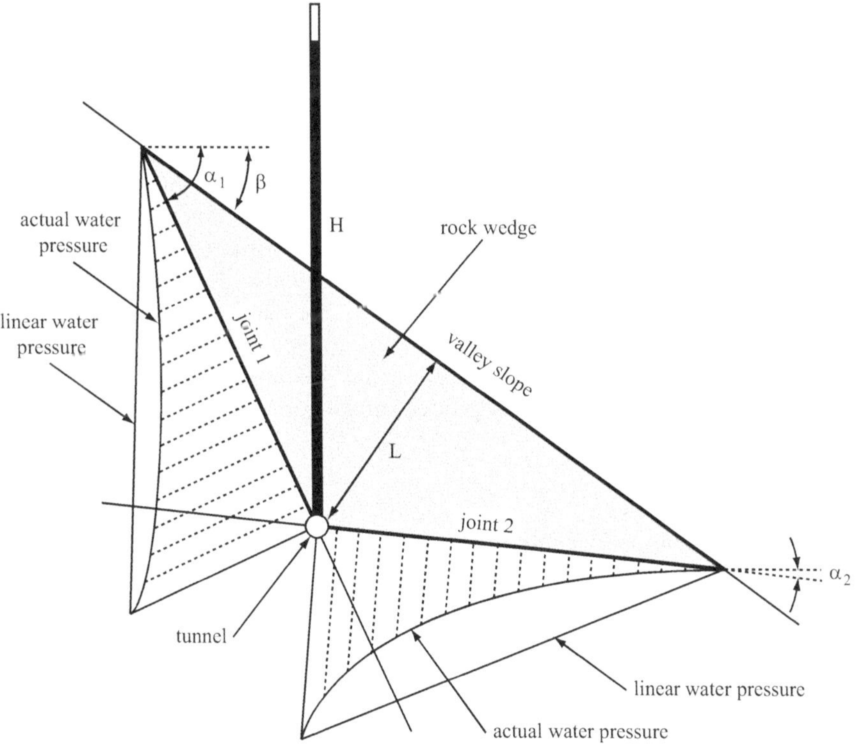

Figure 6.33 Limit equilibrium of a rock wedge with water pressure (adapted from Alvarez et al., 1999).

which would be representative of the cases shown in Figure 6.32b, Equation (6.57) simplifies into:

$$\frac{L}{H} > \frac{\gamma_w}{\gamma_r}\cos\beta\left[1+\frac{\tan\beta}{\tan(\phi_2-\alpha_2)}\right] \tag{6.58}$$

with the assumption of linear pore pressure distribution along the joints.

The Illinois criterion recovers the Norwegian criterion for the case of joints parallel and perpendicular to the slope, e.g. $\alpha_1 = \beta$ and $\alpha_2 = \beta - 90°$, assuming a linear pore pressure distribution and frictionless joints.

For horizontal ground with vertical and horizontal joints, the results of the parametric study showed that the empirical criteria are unconservative if the coefficient of earth pressure at rest, K_o, is less than 0.8. For near horizontal ground, $\beta < 20°$, failure through shallow discontinuities is not the critical mode of failure; instead hydrojacking may occur, and so failure depends on the ratio of in situ horizontal to vertical stresses, K_o. The rule of thumb proposed for $\beta < 20°$ is:

$$\begin{aligned} &\frac{L}{H} > \frac{\gamma_w}{\gamma_r} \quad \text{for } K_o \geq 1 \\ &\frac{L}{H} > \frac{\gamma_w}{\gamma_r}\frac{1}{K_o} \quad \text{for } K_o < 1 \end{aligned} \tag{6.59}$$

Note that the equation for $K_o \geq 1$ coincides with the Norwegian criterion, given in (6.55).

Equations (6.57) to (6.59) are derived without considering a factor of safety. Due to the important consequences of failure, a factor of safety of at least 1.3 is recommended when considering static head (Benson, 1989; Fernández and Alvarez, 1994) and 1.1 for normal surge operations (Ming and Brown, 1988; Benson, 1989).

When stability conditions are appropriate and there is no danger of long-term problems (erosion, dissolution, swelling, etc.), an unlined tunnel is an economic option, and the final decision is based on the leakage allowed. Historical records from Norwegian tunnels indicate that losses within the range of 0.5–5.0 l/s/km of tunnel are acceptable (Merritt, 1999; USACE, 1997). The flow per unit length of tunnel can be estimated with Equation (6.47). In addition, the hydraulic gradient within the rock mass may also be limited to prevent erosion. Maximum values should be established in a case-by-case basis. Benson (1989) proposed, as acceptable hydraulic gradients: 10–15 for massive hard rock, widely jointed; 8–10 for hard to moderately hard and moderately jointed rock; 5–7 for moderate to weak rock and moderately jointed; 3–5 for weak and closely jointed or sheared rock; and < 3 plus filters for very weak, possibly erodible rock.

6.3.3.2 Shotcrete or unreinforced concrete lining

Shotcrete is an effective means of providing stability to tunnels and to improve the hydraulic characteristics of the tunnel, as shown in Table 6.1. Shotcrete must be treated as a permeable material due to shrinkage cracking and crack formation during application of the internal pressure. The cracks, however, appear well distributed unless significant movements of the rock underneath occur. Shotcrete performs well during dewatering as it allows dissipation of external water pressures through the cracks without spalling. The shotcrete is often applied as an initial support in areas of poor rock, with or without rock bolts and mesh. Additional layers of shotcrete may be placed to complete the final lining, with total thicknesses ranging from 10 cm to 20 cm (additional information is provided in Section 7.5).

Cast in place unreinforced concrete linings are also an effective means of providing support to the rock while improving the hydraulic characteristics of the tunnel (see Section 7.2). Unreinforced concrete linings have a thickness of the order of 10–20% the radius of the opening, with a typical range between 20 cm and 40 cm. The lining is not impermeable because of the cracking of the concrete. There are two sources for cracking: (1) thermal shrinkage due to hydration of the cement; and (2) radial displacement due to the internal pressure.

Thermal shrinkage may be reduced by using low heat of hydration, Type 2, cement, with cool aggregates and cool water, and by limiting the amount of overbreak, thus reducing the volume of concrete per unit length of tunnel. Application of the internal water pressure induces outward movements of the concrete lining, the magnitude of which depends on the stiffness of the rock, the pressure applied, and the diameter of the tunnel. If the maximum tensile strain of the concrete, about 1.5×10^{-4}, is not reached the concrete will behave as an impermeable barrier with permeability of the order of 10^{-7} to 10^{-8} cm/s (Fernandez, 1994). For larger strains in the concrete, cracking will occur, typically in the form of one to two cracks perpendicular to the minimum principal stress. Large flow, even through cracks with small aperture, occurs with flow rates much larger than those through the uncracked concrete.

Prestressing the concrete liner is a technique extensively used because it introduces large compressive stresses in the liner, thus preventing tensile stresses and tensile cracking. This is accomplished by injecting grout at the liner–ground interface at high pressures, usually larger than 5 bars. Injection of the grout is done through the concrete liner, already in place through boreholes drilled through the concrete or through sleeve hoses embedded in the concrete prior to casting (Wannenmacher et al., 2015). The grouting pressure should be, on the one hand, small enough such that it does not induce failure on the concrete of the liner and also smaller than the minimum principal stress in the ground adjacent to the liner to prevent hydrojacking. On the other hand, it should be large enough such that the concrete will be in compression during tunnel operations, i.e. the water pressure in the tunnel should not induce tensile stresses in the concrete. The volume of grout to be injected depends, to a large extent, on the stiffness of the ground and the liner, as well as on the injection pressure. A number of methods exist for calculation of injection pressures, to estimate the volume of injected ground, etc. Those are beyond the scope of the book and can be found elsewhere (Seeber, 1975, 1984, 1985, 1999; Simanjuntak, 2015). It is important to mention that injection grout should be performed in all concrete-supported tunnels (unreinforced and reinforced). The injection serves two purposes (Wannenmacher et al., 2015): fill the gaps between the liner and the ground, and preliminary compression/prestressing of the support to ensure a tight bond between the ground and the liner. This is particularly important to ensure that good contact between the support and the ground remains after the first filling (when the tunnel is filled, the ground is loaded, but when emptied, it is unloaded. Since the ground usually has a steeper unloading than loading response – it is stiffer in unloading than in loading – a gap could form between the liner and the ground).

Computation of flow, stresses, and displacement of the lining and ground can be done using the theoretical framework presented in Section 6.3.2. An iterative process can be followed to obtain the final solution. This is so because stresses and displacements depend on the flow through the liner, which in turn depends on whether it is cracked or not, and if so, on the aperture of the cracks. In the first iteration one can assume that the concrete is not cracked and Equation (6.44) applies. Equation (6.46) gives the pore pressure distribution in the ground (note that the pore pressures in the equations should be the excess pore pressures, or the difference between internal or external and hydrostatic pore pressures from the groundwater table). Equating Equation (6.44) with (6.47), which gives the flow through the ground, imposes continuity of flow through the liner and ground and provides the magnitude of the excess pore pressure u_o at the liner-ground contact. Equations (6.48) and (6.49) with (6.50) give the stresses and displacements in the ground, and Equations (6.52) with (6.50) the loads of the liner. The stress distribution in the liner is not uniform and depends on the angular coordinate θ, as Equations (6.52) indicate. However, the stresses do not change much, with differences of about 5% or less around the tunnel. A conservative estimate of the tangential strain in the lining ε_θ^s is given by:

$$\varepsilon_\theta^s = \frac{1}{t}[-u_i r_o + a_o r_o^{-1} + 2b_o r_o + c_o r_o(1 + 2\ln r_o) + 2a'_2 r_o^{-3}] \qquad (6.60)$$

where t is the lining thickness, u_i is the internal excess pressure, u_o the excess pore pressure at the ground–liner interface, r_o the tunnel radius, and a_o, b_o, c_o are coefficients given by (6.50). If the tangential strain from (6.60) is larger than the maximum tensile strain of the concrete, the concrete cracks and the assumption of flow only through the concrete material is no longer valid. Instead, flow occurs mostly through the cracks (the cubic law, in the absence of specific information regarding seepage and crack aperture, can be used as a preliminary estimate). Thus Equation (6.45) should be used for cracked concrete. The aperture of the cracks is unknown, but to start the iteration process it can be assumed that there are two cracks, each with aperture:

$$W = \pi r_o \varepsilon_\theta^s \qquad (6.61)$$

where ε_θ^s is obtained from (6.60). Since the concrete liner is cracked, Equation (6.50) should be used with C = F = ∞, since the stiffness of the liner is lost. As before, stresses and displacements of the ground are computed from (6.48), (6.49), and (6.50), again with C = F = ∞. The liner loads are given by (6.52) and strains by (6.60). If the crack aperture now computed by (6.61) is the same as the aperture assumed at the beginning of the iteration, the final solution is reached; the crack aperture, stresses, water flow, etc. are those obtained in the last iteration. If not, a new iteration starts with the new crack aperture from (6.61).

It has to be mentioned that the formulation suggested does not consider any non-homogeneity or anisotropic characteristics of the rock mass. Even in fairly uniform rock masses, excavation of the tunnel can induce damage around the opening, in particular if drill and blast techniques are employed. The intensity and extent of damage, which can range from few centimeters to several meters, significantly increase the permeability of the rock and decrease its stiffness. No analytical solution exists for this case and numerical methods should be employed.

The unreinforced concrete lining should also be designed to resist the external hydrostatic pressure when the tunnel is emptied, if no provisions for drainage exist. If the internal pressure of the tunnel is larger than the external pressure from groundwater, then an external pressure equal to the internal pressure should be considered; if it is smaller, the external pressure should be the hydrostatic pressure from the groundwater. The liner is subjected to a compressive stress of magnitude:

$$\sigma_\theta = \frac{r_o}{t} u \qquad (6.62)$$

where u is the external pressure, of magnitude equal to either the internal or the external water pressure, whichever is the largest.

6.3.3.3 Reinforced concrete lining

Reinforced concrete lining is used to limit the amount of leakage from the tunnel and as a transition from sections of the tunnel with steel lining to sections with unreinforced concrete or unlined (Deere, 1983). Reinforced concrete cracks when the internal pressure is applied and may experience shrinkage cracking during hydration, so it is considered as a semi-permeable barrier.

Reinforced concrete liners may be placed in sections where cracking of unreinforced concrete would be unacceptable, and in sections where the rock is erodible or susceptible to deterioration. Concrete thicknesses are similar to those of unreinforced concrete liners, between 20 cm to 40 cm (Deere, 1983). Reinforced concrete liners are an effective means of controlling or reducing the potential of hydrofracturing. When the pore pressure behind the liner is large enough to open the fractures in the rock, i.e. when hydrojacking occurs, the hydraulic conductivity of the rock increases very rapidly by several orders of magnitude. However, the leakage is controlled by the amount of flow through the liner, which is determined by the aperture of the cracks in the concrete, which in turn is limited by the steel reinforcement. Thus the hydrojacking phenomenon, even though it may occur, is controlled by the limited cracking allowed in the liner by the steel reinforcement. As a result, reinforced concrete liners are beneficial in tunnel sections where in situ stresses are marginal to prevent hydrojacking.

An estimate of the loads and deformations of reinforced concrete liners, using an analytical solution, follows a similar process to that of unreinforced liners described in Section 6.3.3.2. In the first iteration it is assumed that the concrete is not cracked, so Equation (6.44) applies. The pore pressure behind the liner is obtained by equating the flow through the liner from (6.44) with the flow through the ground given by (6.47). Stresses and displacements in the ground are obtained from (6.48) and (6.49) given the parameters of (6.50). The liner loads are estimated from (6.52). The tangential strains are those from (6.60). If the strains computed are smaller than the maximum tensile strain of the concrete, the above equations provide the solution. However, if the strains are larger, the liner cracks. The crack aperture W and the number of cracks N can be estimated as (Hendron et al., 1989; Fernandez, 1994):

$$\begin{aligned} W &= \varepsilon_0^s S \\ N &= \frac{2\pi r_o}{S} \\ S &= 5(d-7.1)+33.8+0.08\frac{d}{\rho} \approx 0.1\frac{d}{\rho} \end{aligned} \tag{6.63}$$

In the third equation, S is the average crack spacing (in mm), which depends on the diameter of the reinforcing bars d (in mm) and on the ratio of the area of steel to area of concrete, $\rho = A_{steel}/t$. For typical values of d and ρ, the spacing can be approximated as $S = 0.1\ d/\rho$.

The flow through the cracked liner is given by (6.45), which depends on the aperture and number of cracks. An iterative process starts with the number and aperture of cracks from the previous iteration. The iteration process is completed when the input values compare well with the values obtained at the end of the iteration. In each iteration, the pore pressure behind the liner u_o is computed by equating (6.45) with (6.47). Stresses and displacements in the ground are given by (6.48) and (6.49) and, in the liner, by (6.52). Since the liner is cracked, the compressibility and flexibility factors are:

$$\begin{aligned} C &= \frac{E\, r_o(1-\nu_s^2)}{E_{steel}\, A_{steel}(1-\nu^2)} \\ F &= 0 \end{aligned} \tag{6.64}$$

where the Young's modulus and area of steel, E_{steel} and A_{steel}, are used instead of those of the entire support as in (6.51).

In each iteration, the effective radial stress at the liner–ground interface must be checked. If the result is positive, compressive stresses exist between the ground and the liner and part of the load from the internal pressure is carried by the rock. Negative effective radial stresses would imply that the liner pulls from the ground, which is not possible. Instead, a gap forms between the liner and the ground, the effective radial stresses are zero and the radial total stresses are equal to the water pressure, so there is no load transfer from the liner to the ground. In this case stresses, displacements of ground and loads and strains of the liner are computed with the coefficients given by:

$$\begin{aligned} a_o &= \frac{2c_o \ln \dfrac{h_w}{r_o} + u_o}{r_o^{-2} - h_w^{-2}} \\ b_o &= -\frac{1}{2}\left[a_o h_w^{-2} + c_o(1 + 2\ln h_w)\right] \\ c_o &= -\frac{1}{2}\frac{1-2\nu}{1-\nu}\frac{u_o}{\ln\left(1 + \dfrac{4h_w^2}{r_o^2}\right)} \\ a'_2 &= b'_2 = 0 \end{aligned} \tag{6.65}$$

The maximum crack aperture is often limited, typically between 0.25 and 0.5 mm (Hendron et al., 1989). Erosion in erodible rock may occur for crack apertures as small as 0.3 mm. Fernandez (1994) suggested to design the reinforcement such that the strain in the liner is below 6×10^{-4} to preclude the propagation of tension cracks into the rock and to maintain the permeability of the liner compatible with that of the rock mass. Larger strain levels may be acceptable if measures are implemented to reduce the permeability of the liner, e.g. with an impermeable membrane embedded in the concrete.

The same note of caution as that mentioned for unreinforced concrete liners applies. The equations have been derived with the assumption of homogeneous and isotropic ground. If this is not the case, e.g. because of damage to the surrounding rock during excavation, a numerical method must be used for the computations. Similar to unreinforced concrete liners, reinforced concrete liners need also be designed for external pressures. The compressive stresses are given by (6.62).

6.3.3.4 Steel liners

A steel liner is the industry standard for those sections of the tunnel where the minimum principal stress of the rock around the excavation is smaller than the internal tunnel pressure with a suitable factor of safety (Fernandez, 1994). Steel liners are also needed when the internal pressure is large and the surrounding rock has low modulus, so leakage control with reinforced concrete is not possible. A concrete-encased steel cylinder is the most common type of liner (Eskilsson, 1999); see Figure 6.34. A practical maximum thickness of the steel is 1.5 inches; for thicker plates proper welding would necessitate access to the back of the steel, which may not be practical. The steel liner needs to be designed to adequately sustain the internal pressures during operation, external hydraulic pressures during grouting and dewatering, and handling loads during transportation and erection (Brekke and Ripley, 1993).

Figure 6.34 shows how the internal pressure is distributed between the steel, concrete, and rock mass. As the internal pressure increases, the steel liner may initially take all the stress (this may occur if there is a gap between the steel and the concrete, e.g. due to differential

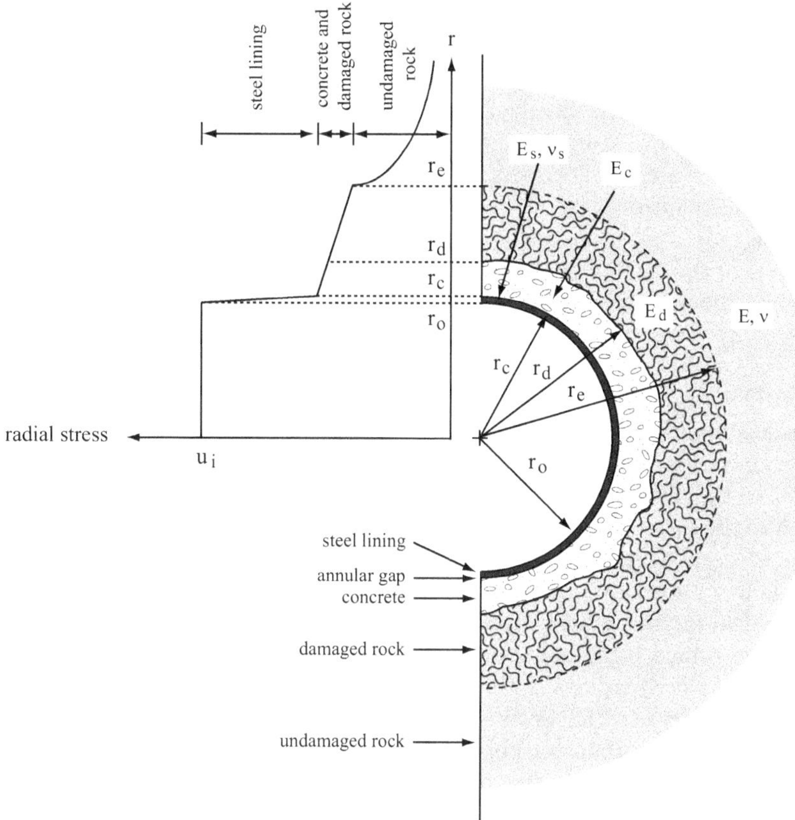

Figure 6.34 Distribution of radial stresses in a concrete-encased steel lining.

thermal contraction between the steel and the concrete, creep of the rock, cycles of loading and unloading as the tunnel is pressurized and emptied, or skin grouting of the steel-concrete interface is not done or is not effective). With further increase of internal pressure, as the steel liner expands, part of the loading is transferred to the concrete, to the rock damaged during excavation and to the undamaged rock. It is generally assumed that the concrete and the damaged rock can only transmit radial deformations as the damage prevents transfer of tangential stresses. This results in radial stresses that decrease with 1/r within the concrete and damaged rock zones, and $1/r^2$ in the undamaged rock zone. The magnitude of the radial stresses and corresponding radial displacements can be obtained by imposing equilibrium along the radial direction and compatibility of radial displacements at the damaged-undamaged rock boundary (Moore, 1989; USACE, 1997). Thus:

$$\sigma_r \big|_{r=r_d} = \frac{r_c}{r_d} \sigma_r \big|_{r=r_c}$$

$$\sigma_r \big|_{r=r_e} = \frac{r_c}{r_e} \sigma_r \big|_{r=r_c} \tag{6.66}$$

$$\sigma_r \big|_{r=r_c} = \frac{\dfrac{u_i\, r_o^2(1-\nu_s^2)}{E_s\, t} - C_s\, r_c\, \Delta T}{\dfrac{r_o^2(1-\nu_s^2)}{E_s\, t} + \dfrac{r_c}{E_c} \ln\dfrac{r_d}{r_c} + \dfrac{r_c}{E_d} \ln\dfrac{r_e}{r_d} + \dfrac{r_c}{E}(1+\nu)}$$

where u_i is the internal pressure, usually the dynamic pressure from normal surge operations; r_o, r_c, r_d, and r_e are the internal radius, radial distance to the steel-concrete boundary, radial distance to the concrete-damaged rock boundary, and radial distance to the damaged-undamaged rock contact, respectively; E_s and t are the Young's modulus and thickness of the steel; E_c, E_d, and E are the Young's moduli of the concrete, damaged and undamaged rock, respectively; ν_s and ν are the Poisson's ratios of steel and undamaged rock; C_s is the coefficient of thermal expansion of steel, $C_s = 12 \times 10^{-6}/°C$; and ΔT is the change in temperature in degrees centigrade.

The thickness of the steel is determined such that the maximum stress is smaller than the allowable stress. Diverse organizations recommend different allowable stress criteria (Brekke and Ripley, 1993). Two calculations are usually performed. One where no load transfer between the steel and the concrete is assumed; the allowable steel stress is either a high percentage, 75–100% of the yield stress of the steel, or a moderate percentage, 50–67%, of the ultimate tensile strength. The second calculation assumes load transfer, i.e. Equation (6.66), and the allowable stress is a moderate percentage of the yield stress, 50–67%, or a low percentage, 33%, of the ultimate tensile strength.

In addition to the internal pressure, the steel lining has to be designed to support, without buckling, the external pressure in the event of dewatering of the tunnel. The external pressure can range from a minimum of hydrostatic pressure from groundwater conditions to a maximum equal to the static hydraulic pressure inside the tunnel, depending on how much leakage from the tunnel has reached the rock mass behind sections with steel lining. It is prudent to take pressure readings of the rock behind the liner before dewatering. A factor of safety of 1.5 is typically used for design (ASCE, 1993).

Experience has shown that buckling usually occurs in a single lobe (Figure 6.35a). Several formulations exist to predict the critical pressure at which buckling occurs; typically, Amstutz's (1970) and Jacobsen's (1974), the latter being used the most (ASCE, 1993; Eskilsson, 1999). The critical external buckling pressure for an unstiffened steel liner depends on the gap between the steel and the concrete. The gap can realistically vary from 0 to 0.001 times the tunnel radius (ASCE, 1993; USACE, 1997). Following Jacobsen's

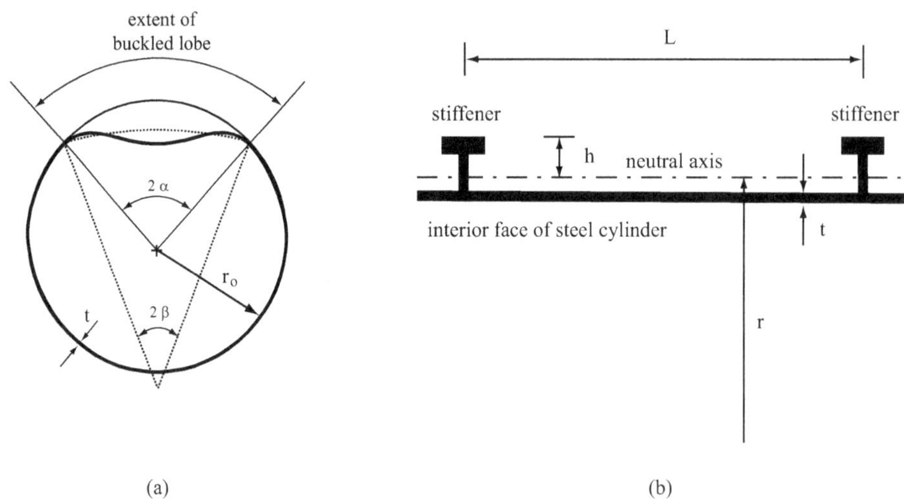

Figure 6.35 Buckling of thin cylindrical shells. (a) Single-lobe buckling. (b) Plan view of steel cyliner with stiffeners.

formulation, the three equations in (6.67) must be solved simultaneously for the three unknowns α, β, p_{crit}:

$$\frac{r_o}{t}=\sqrt{\frac{\left[\frac{9\pi^2}{4\beta^2}-1\right]\left[\pi-\alpha+\beta\left(\frac{\sin\alpha}{\sin\beta}\right)^2\right]}{12\left(\frac{\sin\alpha}{\sin\beta}\right)^3\left[\alpha-\pi\frac{\Delta}{r_o}-\beta\left(\frac{\sin\alpha}{\sin\beta}\right)[1+0.25\tan^2(\alpha-\beta)]\right]}}$$

$$\frac{p_{crit}}{E^*}=\frac{\frac{9}{4}\left(\frac{\pi}{\beta}\right)^2-1}{12\left(\frac{r_o}{t}\right)^3\left(\frac{\sin\alpha}{\sin\beta}\right)^3} \tag{6.67}$$

$$\frac{\sigma_y}{E^*}=\frac{t}{2r_o}\left[1-\frac{\sin\beta}{\sin\alpha}\right]+\left(\frac{p_{crit}\,r_o}{E^*\,t}\frac{\sin\alpha}{\sin\beta}\right)\left[1+\frac{4\beta r_o\sin\alpha\tan(\alpha-\beta)}{\pi t\sin\beta}\right]$$

where α is one-half the angle subtended to the center of the cylindrical shell by the buckled lobe (Figure 6.35a); β is one-half of the angle subtended by the new radius through the half waves of the buckled lobe; p_{crit} is the critical external buckling pressure; Δ/r_o is the ratio of the gap between steel and concrete and the radius of the steel lining; r_o is the interior tunnel radius; σ_y is the yield stress of the steel; t is the steel thickness; $E^*=E_s/(1-\nu_s^2)$, where E_s and ν_s are the Young's modulus and Poisson's ratio of the steel.

If the thickness of the steel liner becomes too large, using the above Equation (6.67), a possible alternative is to place stiffeners embedded in the concrete to limit the length of the steel cylinder that can buckle (Figure 6.35b). Alternatively, steel bolts/rods can be welded to the back of the steel liner and embedded in the concrete behind. The critical buckling pressure of the shell between stiffeners (Figure 6.35b) is computed by following the assumption of either rotary-symmetric buckling or Jacobsen (1974) single-lobe buckling. The critical pressure for rotary-symmetric buckling is obtained from Von-Mises equation, which involves the formation of multiple waves, and it is given by:

$$p_{crit}=\frac{E_s}{1-\nu_s^2}\frac{t}{r_s}\left[\frac{1-\nu_s^2}{(n^2-1)\left(\frac{n^2L^2}{\pi^2r_s^2}+1\right)^2}\right]+\frac{E_s}{12(1-\nu_s^2)}\left(\frac{t}{r_s}\right)^3\left(n^2-1+\frac{2n^2-1-\nu_s}{\frac{n^2L^2}{\pi^2r_s^2}-1}\right) \tag{6.68}$$

The variables in the equation have the same meaning as in (6.67). In addition n is the number of lobes or waves involved in buckling, L is the length of the cylinder between stiffeners, and r_s is the radius of the outside of the shell. The value of n is the integer number for which the critical pressure p_{crit} is minimum. For most practical cases, n is found between 6 and 12 (ASCE, 1993).

Jacobsen's critical pressure is obtained by solving simultaneously the following three equations:

$$\frac{r}{\sqrt{12\frac{J}{A_{ST}}}}=\sqrt{\frac{\left[\frac{9\pi^2}{4\beta^2}-1\right]\left[\pi-\alpha+\beta\left(\frac{\sin\alpha}{\sin\beta}\right)^2\right]}{12\left(\frac{\sin\alpha}{\sin\beta}\right)^3\left[\alpha-\pi\frac{\Delta}{r_o}-\beta\left(\frac{\sin\alpha}{\sin\beta}\right)[1+0.25\tan^2(\alpha-\beta)]\right]}}$$

$$\frac{p_{crit}L}{EA_{ST}}=\frac{\frac{9}{4}\left(\frac{\pi}{\beta}\right)^2-1}{\left(\frac{r\sin\alpha}{\sin\beta}\right)^3}\frac{J}{A_{ST}} \tag{6.69}$$

$$\frac{\sigma_y}{E}=\frac{h}{r}\left[1-\frac{\sin\beta}{\sin\alpha}\right]+\left(\frac{p_{crit}L\,r}{EA_{ST}}\frac{\sin\alpha}{\sin\beta}\right)\left[1+\frac{8\beta h r\sin\alpha\tan(\alpha-\beta)}{\pi 12\frac{J}{A_{ST}}\sin\beta}\right]$$

The variables are the same as in (6.67). In addition r is the radius to the neutral axis of the ring stiffener (Figure 6.35b); A_{ST} is the cross-section area of the stiffener and the pipe shell between the stiffeners; h is the distance from the neutral axis of the stiffener to the outer edge of the stiffener; and J is the moment of inertia of the stiffener and contributing width of the shell. The contributing width is $\left(1.57\sqrt{r_o t}+t_{SF}\right)$ where t_{SF} is the thickness at the base of the stiffener.

As an alternative, numerical analyses, generally using Finite Element Methods, can be used to compute the critical pressure (Ahlgren and Mattson, 1997; Frey et al., 2002). The advantage of numerical methods over analytical methods is that complex geometries and non-linear material models can be considered. In addition, there may be design solutions that can prevent or at least minimize the buildup of pore pressures behind the liner, e.g. by the construction of relief valves (Schleiss and Manso, 2012).

In general, dewatering of the tunnel must be done at a slow rate such that water pressures behind the steel liner can dissipate and so the liner is not subjected to unacceptable pressures that may damage or even fail the steel shell by buckling (Benson, 1989; Brekke and Ripley, 1993; Merritt, 1999). It is also prudent to fill the tunnel at a similar rate; this is particular important for the first fill because the water leaks would re-build the depressed water table or slowly saturate the rock behind the tunnel. Both operations should be carefully monitored through a network of piezometers to determine that actual and expected conditions do not differ significantly.

6.3.4 Summary

Pressure tunnels transport water under high pressure from the intake to the powerhouse. The leaks through the length of the tunnel change the water pressures in the rock mass, which, together with the stresses induced by excavation and support, determine the state of stress, and thus behavior, of the surrounding rock mass.

The following geological and geotechnical information is vital for the design of pressure tunnels: (1) hydraulic head; (2) location of the water table; (3) rock characteristics; (4) in situ rock stresses; (5) rock modulus; and (6) rock permeability.

The location of the ground water table, together with the hydraulic head in the tunnel, determines whether water flow will be from the tunnel to the rock or vice versa. If the hydraulic head inside the tunnel is larger than that from the water table, water will flow out from the tunnel. Rock characteristics such as its potential for erosion, swelling, and long-term degradation are needed to decide the type of support along the tunnel. Rock stresses around the tunnel need to be larger than the water pressures to prevent hydrofracturing or hydrojacking. The stiffness of the rock is needed to determine the contribution of the rock mass to support the internal pressure, as the load transfer from the liner to the rock depends largely on the stiffness of the liner relative to the surrounding rock. Finally, the permeability of the rock mass is required to estimate the leakage flow and thus the pore pressures behind the support.

The layout of the pressure tunnel can be decided based on empirical methods and the information obtained during field exploration, if available. Such empirical methods can provide initial estimates as to where a watertight liner is needed. Following this, calculations regarding flow rate, stresses in the liner and in the ground should be made to assess whether the leakage is acceptable and that the stress state in the liner and ground is adequate, as well as to provide an estimate of the factor of safety. It is imperative that hydraulic fracturing tests are conducted at critical locations along the tunnel to ensure that the water pressure in the rock mass induced by the tunnel is smaller than the minimum in situ stress, with an adequate factor of safety.

It is generally accepted in the tunneling profession that the geotechnical investigation must continue during construction and that the design cannot be finalized until the project is completed.

> Thus, at some time during the excavation of the tunnel and well before the decisions of final lining are made, the Engineer needs to inspect the tunnel, assess the conditions encountered, verify that the assumptions made in the Design Memorandum are still valid and make those in-situ tests judged necessary.
>
> (Merritt, 1999)

Continuous performance monitoring during normal operations of the tunnel may provide an early warning for potentially dangerous situations or distress of the lining or the rock around the excavation. At least numerous piezometers should be installed along the tunnel and routinely monitored. Measurements of flow leaks through the tunnel and flow from water springs in the vicinity of the tunnel should be systematically made.

6.4 ROCKBURSTS IN TUNNELS

6.4.1 Introduction

The term "rockburst" is applied to the damage that occurs in a tunnel as a result of, or directly associated with, a seismic event (Ortlepp and Stacey, 1994). Almost all rockbursts have been reported in highly stressed, brittle rock. A characteristic of any rockburst is the sudden or violent damage of the rock around the tunnel. The damage usually results in deformation and increase of tunnel convergence, but it may not necessarily result in ejection of the rock into the excavation (CAMIRO, 1995). Because of the volume of the rock involved in the phenomenon and the velocities of the pieces of rock that may be ejected, rockbursts pose a significant risk to tunnelers, it damages the rock surrounding the tunnel, and it may have a detrimental effect on the rate of advance of tunnel excavation.

Seismicity and associated rockburst induced by mining operations were recognized as a problem in early 20th century in deep mines in South Africa, India, Canada, and the United States (Ortlepp, 2000; Hoek, 2007). The relation between rockbursts and seismicity was first

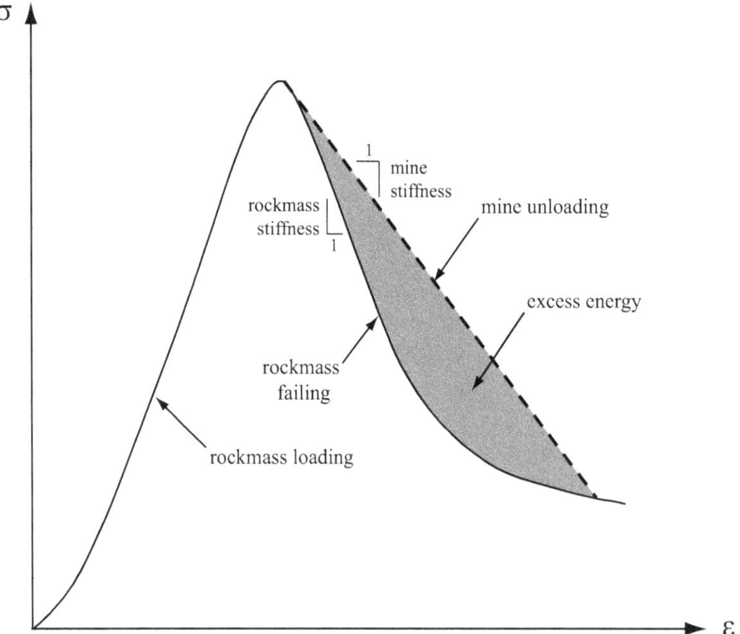

Figure 6.36 Conceptual rockburst mechanism in mines.

established by Cook (1964, 1965) after deploying and monitoring a sensor network on the East Rand Proprietary Mines in South Africa. The seismic sources were located in the intact rock ahead of the excavation, and it was postulated that the energy of the seismic events was the elastic strain energy stored in the rock mass, which was released when the stiffness of the loading mechanism at work in the mine was smaller than that of the failing rock; see Figure 6.36. The conclusions regarding the location of the seismic events were supported later on by observations made by Joughin and Jager (1983) at the Doornfontein gold mine in South Africa. The direct link between seismicity and rockburst was confirmed by McGarr and Wiebols (1977), who proposed a relation between the number of seismic events in a time period, N, larger than a given magnitude M, that follows the Gutenberg-Richter law:

$$\log N = a - bM \tag{6.70}$$

where a is a function on the strain rate of the rock, and b is a constant that depends on the particular region of interest.

Ortlepp (1984) identified, as additional sources for rockburst, slip along preexisting faults or along newly created faults due to mining operations, which were found at some distance from the excavation. The sudden release of energy caused by slip propagates in the rock mass as P- and S-waves, which, upon reaching the tunnel, may induce very high strain rates in the rock surrounding the excavation and cause damage.

It is important to distinguish between seismicity and rockbursts since the source for the two may not be the same. Also, a seismic event may not cause a rockburst, but a rockburst will always generate seismicity. In the following, a distinction is made between source of seismicity and result of associated damage. The main source mechanisms that produce seismic events in underground excavations are, ordered with increasing energy released (Stacey and Ortlepp, 1994; Ortlepp and Stacey, 1994): (1) strain bursting; (2) buckling; (3) face crushing; (4) virgin shear in the rock mass; and (5) slip along preexisting discontinuities.

Strain bursting is caused by surficial spalling with violent ejection of rock fragments. The source of the energy required for the phenomenon is the strain energy stored in the rock. The seismic motions induced are often too small to be detected and the energy released is generally quite modest, with Richter Magnitudes within the range of –0.2 to 0. Strainbursting is most likely to occur from half a diameter to three diameters behind the face or perimeter (Ortlepp and Stacey, 1994; Martin, 1997), but it has been also reported at the face (Saito et al., 1983). The source of buckling is large slabs of rock parallel to the tunnel perimeter or face that are ejected into the tunnel. The Richter Magnitude associated with buckling ranges from 0 to 1.5. Face crush consists of the violent expulsion of rock fragments from the face of the tunnel. The Richter Magnitude is from 1.0 to 2.5. Shear rupture is caused by the violent propagation of a shear fracture through the intact rock. The seismicity produced by shear rupture corresponds to Richter magnitudes from 2.0 to 3.5. Fault slip is the result of a sudden movement along an existing fault, and has the largest energy released, with Richter Magnitudes in the range of 2.5 to 5.0. For strain bursting, buckling, and face crushing phenomena, the location of damage in the tunnel and the source of seismicity coincide. The energy required for the phenomenon to occur is the energy stored in the rock mass, which is released in an unstable manner as the rock fails because the stiffness of the loading mechanism (the mine) is smaller than that of the rock mass (see Figure 6.36). Virgin shear and fault slip may occur hundreds of meters away from the tunnel and damage is caused by the P- and S-waves traveling through the rock mass and impinging the excavation. Due to the relative proximity between the underground excavation and the source, the seismic waves are characterized by high frequency content (see Section 6.2.2), which, upon reaching the tunnel, may create tensile stresses in the vicinity of the excavation.

For virgin shear and fault slip to occur, a large change in stresses in the rock mass is generally required. For this reason, these two seismic sources are associated almost exclusively with mining and not with civil engineering tunnels, where the stress change induced by the tunnel excavation and support tends to be moderate. Strain bursting, buckling, and face crushing are the seismic sources common to civil engineering projects, which, because they require smaller changes in stresses, release much less energy.

The damage associated with the seismic source mechanisms described above can be categorized as (Ortlepp and Stacey, 1994): (1) Strainbursts; (2) Buckling; (3) Ejection; (4) Arch collapse; and (5) Other. See Figure 6.37.

Strain bursting is characterized by surficial spalling with violent ejection of rock fragments. See Figure 6.37a. The location of strainbursts coincides with the location of the maximum tangential stress around the opening, and so they are directly related with the in situ stresses of the rock and the change of stresses induced by the excavation. Strainbursts are typically associated with massive rocks and may be produced at stress levels well below the unconfined compression strength of the intact rock. Stresses as small as 15–20% of the unconfined compression strength have been reported (Saito et al., 1983; Ortlepp and Stacey, 1994) to cause strainbursts, but typical values fall within the range of 30–50% (Dowding and Andersson, 1986; Martin et al., 1999). Strainbursting conditions pose a severe hazard to the personnel working in the tunnel and may lead to a reduction in tunneling progress rates.

Buckling damage is due to large slabs of rock parallel to the tunnel perimeter or face that are ejected into the tunnel (Figure 6.37b). Buckling is produced around excavations in jointed, laminated, or transversely anisotropic rock. The location of buckling is related to the stress field around the opening and to the orientation of the discontinuities with the excavation. The energy source for buckling is the energy stored in the rock, even though additional energy may be supplied by compressive waves originated from other seismic sources.

Ejection of the rock around the tunnel is caused by a transient seismic wave impinging on the excavation, which source may be located at some distance from the tunnel (Figure 6.37c).

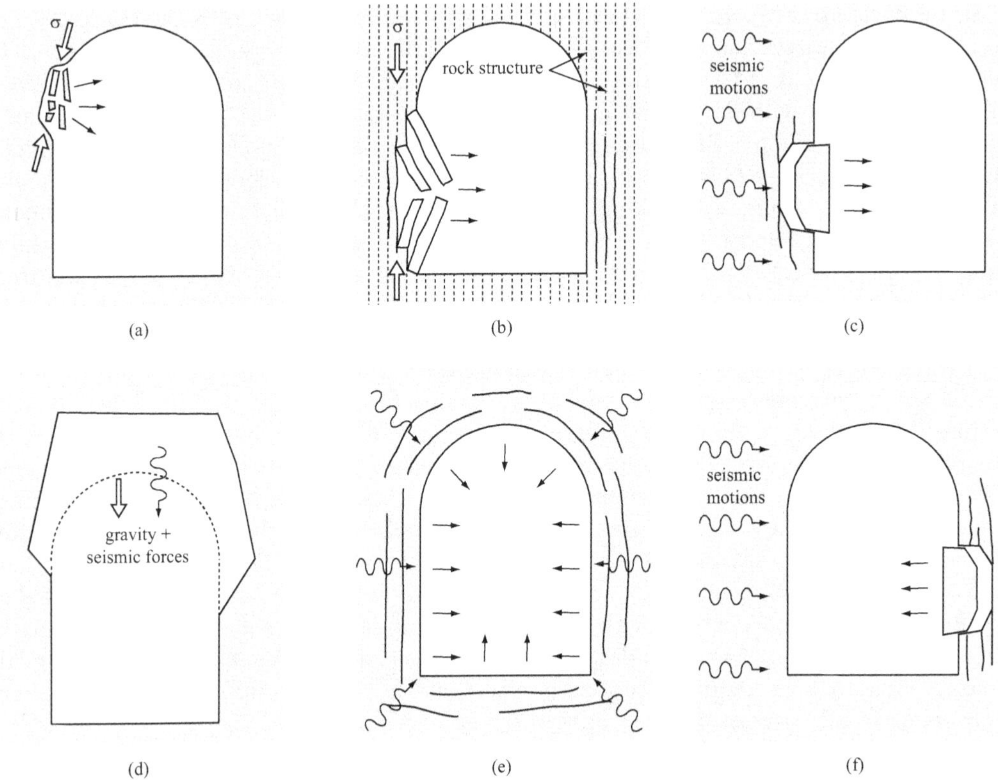

Figure 6.37 Rockburst damage mechanisms (adapted from Ortlepp and Stacey, 1994). (a) Strain bursting. (b) Buckling. (c) Ejection. (d) Arch collapse. (e) Implosive damage. (f) Inertial displacement.

The size and shape of the ejected pieces is normally determined by the discontinuities present in the rock mass. Large volumes of rock may be ejected at velocities of up to 10 m/s. Because this mechanism requires a seismic source, it is not generally encountered in civil engineering tunnels.

Arch collapse (Figure 6.37d), similar to falling rock, as a rockburst damage mechanism, is produced by the kinematic movement of large rock wedges, which are delimited by the structure of the rock mass. As with ejection, arch collapse requires an external seismic source, which augments the effects of gravity. Because of the need for a seismic source, this mechanism is observed only in mines.

Other mechanisms include implosive damage (Figure 6.37e) and inertial displacement of rock slabs (Figure 6.37f), both possible in mines, but unlikely in civil engineering tunnels.

The severity of rockbursts in mines is, given the previous discussion, much larger than in civil engineering tunnels because the seismic sources that generate rockbursts can release several orders of magnitude more energy than those associated with civil engineering tunnels. Discussion of rockburst mechanisms, damage, and mitigation in mining are beyond the scope of this book. The following is intended to direct the reader to a number of publications and research work that, albeit applicable to mines, characterize the same fundamental rockbursts phenomena as in tunnels, and may provide insight and guidelines to civil engineers when faced with particularly intense rockbursts.

6.4.2 Rockburst in mining

The impact that rockbursts have had and still have in mines is reflected in the large number of cases found in the technical literature (e.g. Ortlepp, 1984; Whyatt et al., 2000; Andrieux et al., 2013). Rockbursts in mining are still an active topic of research regarding their origin, triggering mechanisms, and required support (Salamon, 1983; Milev et al., 2000; Hagan et al., 2001). This is so because of the potential considerable damage associated with large seismic events, which may cause closure of large sections of the mine and pose unacceptable risks to mining personnel. The approach followed in deep mines in hard rock is a combination of monitoring of seismic events in the mine through a dense network of accelerometers, prevention of rockbursts, and mine support that can resist a rockburst without collapse of the opening (CAMIRO, 1995).

Monitoring consists of the deployment of an array of sensors, usually uniaxial and triaxial accelerometers. The sensors signals are amplified in underground stations and are sent via cable or wirelessly to the surface for data acquisition, storage, and analysis. The network density and sensor sensitivity are such that a seismic event can be located with an accuracy of few meters. Typically, the following information is collected: location of the seismic event, date and time, magnitude, frequency content of P- and S-waves, seismic moment, seismic energy, stress drop, apparent volume, and others (Srinivasan et al., 1997). The idea behind monitoring is that the source of the seismic event can be found quickly and so it is relatively easy to inspect the site for potential damage; also correlations (usually ad-hoc or mine-specific) can be established between mining operations, source, and intensity of rockbursts, which can then be used to optimize mine operations (Delgado and Mercer, 2006; Simser, 2006; Sweby et al., 2006; Wondrad and Chen, 2006). Intense research is currently being done to determine precursors to rockbursts such as event count rate, strain energy release, and frequency content from microseismic emissions, as indicators of damage of the rock around the mine (Srinivasan et al., 1999). Modern monitoring systems were first installed in deep mines in early 1990s and nowadays constitute a necessary requirement for mining operations.

Prevention or mitigation of large rockburst events in mining is accomplished by optimization of the excavation process, such that stress levels within the rock are kept below critical levels. Decisions are made based on the auscultation of the mine response to the different excavation processes obtained from the monitoring system in place. A successfully used alternative is preconditioning of the rock before excavation. Preconditioning consists of destressing the rock in advance of mining. This is done by pre-fracturing the rock ahead of the advancing mining faces by controlled blasting, with the idea that the high stresses resulting from the excavation are relieved by the yielding of the material in the pre-conditioned zone (Blake, 1984; Tang et al., 2000).

Support of the ground in burst-prone zones is decided following a three-step process (CAMIRO, 1995): (1) identification of the rockburst damage mechanism; (2) determination of the severity of the resulting rock damage; and (3) selection of the support.

The rockburst damage mechanisms are classified as: (i) rock bulking due to fracturing; (ii) rock ejection due to seismic energy transfer; and (iii) rockfalls induced by seismic shaking. The term rock bulking refers to the increase of volume of the rock as it fractures. Rock bulking includes the damage mechanisms shown in Figures 6.37a and b, strain bursting and buckling, even though a distinction is made between bulking without ejection and bulking with ejection. In the first case, the rock fails in a relatively stable manner, with displacement velocities smaller than about 1.5 m/s. This occurs when the tunnel is properly supported or the stored strain energy in the rock is absorbed by the formation of new fractures. If the energy in the rock is larger than the energy that can be consumed during fracturing, then the excess energy is transformed into kinetic energy and the mechanism corresponds to that of bulking with ejection. In this case, the ejection velocities are typically between 1.5 and 3 m/s. Mechanism (ii), rock ejection due to seismic energy transfer corresponds to the rockburst damage mechanisms shown in Figures 6.37c, e and f: ejection, implosive damage, and inertial displacement, and it generally induces rock ejection velocities in excess of 3 m/s. As described in Section 6.4.1, mechanism (ii) requires energy transmitted from a seismic source. Mechanism (iii), rockfalls induced by seismic shaking, corresponds to the arch collapse mechanism shown in Figure 6.37d, and it also requires a seismic energy source, even though gravity is the dominant driving force for failure.

The severity of the rockburst damage is classified as: (i) minor damage; (ii) moderate damage; and (iii) major damage. The classification is based on the depth and lateral extent of the damage to the rock around the tunnel if left unsupported. Minor damage involves a shallow skin of fractured or loose rock less than 0.25 m thick. The support shows signs of loading and the mesh, if any, signs of bagging with few broken wires. Moderate damage affects a depth of rock between 0.25 m and 0.75 m. With this level of severity, many support elements fail; e.g. shotcrete will be heavily fractured and the mesh will be bagged at capacity. Major damage involves fracturing of the rock for a depth larger than 0.75 m around the opening; most ground support elements fail and the opening is impassable due to the displaced rock.

Table 6.3 is a summary of the rockburst damage mechanism and damage severity. The data in the Table is taken from CAMIRO (1995). The table includes, for a given damage mechanism and severity of damage, the cause of the damage, the thickness of the rock fractured around the opening, the weight of the rock associated with the damage per square meter of excavation surface, the rock displacements (closure) at the perimeter of the opening, typical ejection velocity of rock fragments, and the kinetic energy that may be transmitted to the support system during the rockburst event.

The damage depth, or damage severity, is decided from observations from previous rockbursts in the mine, or it can be approximated from the stress-to-strength ratio of the rock in the vicinity of the opening. The depth of failure, $d_f = R_f - r_{eq}$, in massive or good quality rock can be estimated as:

$$\frac{d_f}{r_{eq}} = 1.25\frac{\sigma_{max}}{\sigma_c} - 0.51 \pm 0.1 \tag{6.71}$$

Table 6.3 Rockburst damage mechanisms and damage

Damage Mechanism	*Damage Severity*	*Cause of Damage*	*Thickness (m)*	*Weight (kN/m²)*	*Closure (mm)*	*Ejection Velocity (m/s)*	*Energy (kJ/m²)*
Bulking without ejection	Minor	highly stressed rock with little excess stored strain energy	< 0.25	< 7	15	< 1.5	not critical
	Moderate		< 0.75	< 20	30	< 1.5	not critical
	Major		< 1.5	< 50	60	< 1.5	not critical
Bulking with ejection	Minor	highly stressed rock with significant excess stored strain energy	< 0.25	< 7	50	1.5–3.0	not critical
	Moderate		< 0.75	< 20	150	1.5–3.0	2–10
	Major		< 1.5	< 50	300	1.5–3.0	5–25
Ejection	Minor	seismic energy transfer to jointed or broken rock	< 0.25	< 7	< 150	> 3.0	3–10
	Moderate		< 0.75	< 20	< 300	> 3.0	10–20
	Major		< 1.5	< 50	> 300	> 3.0	20–50
Rockfall	Minor	inadequate strength, forces increased by seismic acceleration	< 0.25	< 7g/(a + g)[1]	n/a[2]	n/a[2]	n/a[2]
	Moderate		< 0.75	< 20g/(a + g)[1]	n/a[2]	n/a[2]	n/a[2]
	Major		< 1.5	< 50g/(a + g)[1]	n/a[2]	n/a[2]	n/a[2]

Source: (After CAMIRO, 1995)
[1] a and g are the seismic and gravitational accelerations
[2] Not applicable

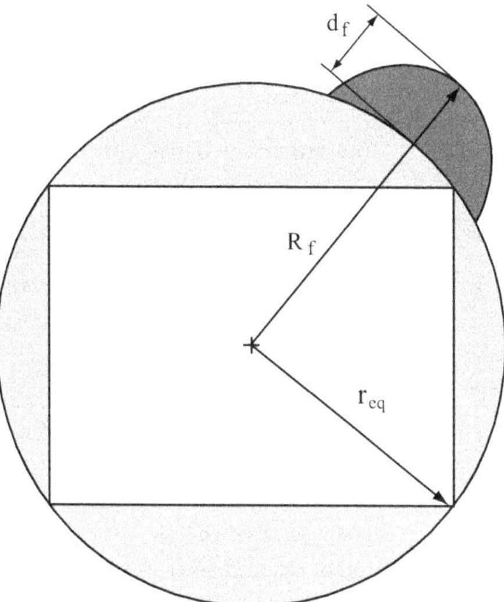

Figure 6.38 Equivalent radius r_{eq} of a rectangular opening.

d_f is the depth of failure, R_f is the radial distance to failure, r_{eq} is the equivalent opening radius (see Figure 6.38); σ_{max} is the maximum tangential stress along the perimeter of the opening, and σ_c is the unconfined compression strength of the intact rock obtained in the laboratory. Equation (6.71) is based on observations of damage in case studies, and will be discussed in Section 6.4.3. The equation should be limited to values of $d_f/r_{eq} < 0.6$. The maximum tangential stress is (CAMIRO, 1995):

$$\sigma_{max} = \sigma_{static} + 4\rho V_s V_{max}^s \tag{6.72}$$

The first term in (6.72), σ_{static}, is the maximum tangential stress around the opening due to regular working conditions; e.g. for a single deep circular opening in an elastic medium and subjected to free-field stresses σ_v and σ_h, it is $\sigma_{static} = 3\ \sigma_v - \sigma_h$ with the assumption $\sigma_v > \sigma_h$. The second term in Equation (6.72) is the additional stress contribution due the shear waves produced by a seismic event; ρ is the mass density of the rock; V_s is the shear wave velocity of the rock, and V_{max}^s is the peak particle velocity at the perimeter of the tunnel due to the shear wave produced by the event.

Finally, support of the mine in rockburst zones is designed based on the function that the support provides. There are three primary functions: to reinforce the rock mass, to retain broken rock, and to hold or tie back the retaining elements. Reinforcement strengthens the rock mass and prevents the loss of strength of the rock, thus restricting bulking. The support also needs to retain the rock even if it has failed to prevent progressive failure and unraveling. Finally the support has to tie the retaining elements to stable ground to prevent gravity-driven rockfalls. The fundamental principle behind the design of the support is that it must be able to survive the displacements imposed by the rockburst and remain functional after the event to provide stability to the opening. Hence the support design is based on the concept of stiffness rather than strength to accommodate the imposed displacements and, if necessary, absorb the kinetic energy of the ejected rock. Table 6.4 describes the role of the support for the different mechanisms and severity of damage.

Table 6.5 includes the peak load, displacement limit and energy absorption of typical support elements used in Canada, ordered in increasing displacement capacity. The results

Table 6.4 Role of support

Damage Mechanism	*Damage Severity*	*Role of Support*
Bulking without ejection	Minor	Tolerate minor damage or reinforce rock mass to prevent initiation of fracturing
	Moderate	Reinforce rock mass to limit rock mass bulking and control rock displacements with support pressure
	Major	Control rock mass bulking and survive large rock displacements
Bulking with ejection	Minor	Retain small volumes of ejected rock and limit rock displacements
	Moderate	Retain ejected rock with a tough retaining system and survive rock displacements
	Major	Retain ejected rock, survive large rock displacements, and absorb energy
Ejection	Minor	Retain small volumes of ejected rock with a retaining system that absorbs energy
	Moderate	Retain ejected rock, absorb energy with holding elements and survive rock displacements
	Major	Retain ejected rock, absorb energy with holding and retaining elements, and survive large rock displacements
Rockfall	Minor	Reinforce rock mass to prevent failure or unraveling
	Moderate	Reinforce rock mass, retain and hold unstable rock
	Major	Provide maximum holding capacity, maintain rock mass integrity with strong reinforcement and retaining system

Source: (After CAMIRO, 1995)

Table 6.5 Typical load-displacement data of support elements

Description	*Peak Load (kN)*	*Displacement Limit (mm)*	*Energy Absorption (kJ)*
19 mm resin-grouted rebar	120–170	10–30	1–4
16 mm cablebolt	160–240	20–40	2–6
16 mm, 2m mechanical bolt	70–120	20–50	2–4
16 mm, 4 m debonded cablebolt	160–240	30–50	4–8
16 mm grouted smooth bar	70–120	50–100	4–10
Split Set bolt	50–100	80–200	5–15
Yielding Swellex bolt	80–90	100–150	8–12
Yielding Super Swellex bolt	180–190	100–150	18–25
16 mm cone bolt	90–140	100–200	10–25
#6 gauge welded wire mesh	24–28	125–200	2–4/m^2
#4 gauge welded wire mesh	34–42	150–225	3–6/m^2
#9 gauge chain-link mesh	32–38	350–450	3–10/m^2
Shotcrete and welded wire mesh	2 × mesh	< mesh	3–5 × mesh[1]

Source: (After CAMIRO, 1995)
[1] at displacements below 100–150 mm

strongly depend on the rock mass and on testing conditions and should be taken as indicative. Actual values should be obtained from site-specific tests, when possible, and should be representative of the underground working environment.

The support system must provide one or more of the reinforcing, holding, or retaining functions. All the elements of the support system must be well connected such that they can

Table 6.6 Support systems in rockburst zones

Damage Mechanism	*Damage Severity*	*Load (kN/m²)*	*Displacement (mm)*	*Energy (kJ/m²)*	*Suggested Support*[1]
Bulking without ejection	Minor	50	30	not critical	Mesh with rockbolts or grouted rebars (and shotcrete)
	Moderate	50	75	not critical	Mesh with rockbolts and grouted rebars (and shotcrete)
	Major	100	150	not critical	Mesh and shotcrete panels with yielding bolts and grouted rebars
Bulking with ejection	Minor	50	100	not critical	Mesh with rockbolts and Split Set bolts (and shotcrete)
	Moderate	100	200	20	Mesh and shotcrete panels with rebars and yielding bolts
	Major	150	> 300	50	Mesh and shotcrete panels with strong yielding bolts and rebars (and lacing)
Ejection	Minor	100	150	10	Reinforced shotcrete with rockbolts or Split Set bolts
	Moderate	150	300	30	Reinforced shotcrete panels with rockbolts and yielding bolts (and lacing)
	Major	150	> 300	> 50	Reinforced shotcrete panels with strong yielding bolts and rebars and lacing
Rockfall	Minor	100	n/a	n/a	Grouted rebars and shotcrete
	Moderate	150	n/a	n/a	Grouted rebars and plated cablebolts with mesh and straps, or mesh reinforced shotcrete
	Major	200	n/a	n/a	As above plus higher-density cablebolting
Limit (MPSL)		200	300	50	Maximum Practical Support Limit

Source: (After CAMIRO, 1995)
[1] Items in parenthesis are beneficial but optional

work effectively together and there are no sources of weakness. Table 6.6 summarizes the support recommendations provided by CAMIRO (1995) as a function of the damage mechanism and the severity of the damage. The data in the table are ordered with increasing degree of displacement and energy absorption. When the severity of the damage is minor, the function of the support is to reinforce the rock mass and prevent unraveling. Most standard bolt and mesh support systems will be adequate. As the severity of the rockburst increases, yielding elements must be included with displacement capacities from 30 mm to 300 mm. These elements are not required to dissipate energy, but to deform sufficiently such that the energy is dissipated by fracturing of the rock; cablebolts are an example. For major severity damage, elements that can absorb significant amounts of energy must be added. These include Swellex or Cone bolts. Shotcrete and lacing are added in the Table as optional elements because of their high capacity of energy absorption. See Chapter 7 for further discussion on rockbolts and shotcrete. The support suggested in Table 6.6 assumes standard bolting of one bolt per square meter of excavation surface; where more than one bolt type is mentioned, it is assumed that the bolting patterns overlap resulting in two bolts per square meter. The support is intended for temporary mining drifts of approximately 4 m span. For larger openings and permanent openings, additional support may be required. The reader is directed to

CAMIRO (1995) and other more recent specialized publications for additional information (e.g. Kaiser et al., 1996; Kaiser and Cai, 2018; Simser, 2019), and in particular for the engineering design of the support in rockburst zones when the mining conditions fall outside the assumption made for Table 6.6.

When the rockburst conditions are so severe that the thickness of the damage zone exceeds 1.5 to 2 m, the wall movement is larger than about 300 mm, or the energy absorption required is larger than 50 kJ/m^2, there is no practical support that is able to maintain the integrity of the opening. The Maximum Practical Support Limit (MPSL) is reached. Mitigation of rockburst damage calls for a combination of support and changes in mining operations such as modified extraction geometry, sequencing, and distressing.

In the mines in South Africa, the preferred support is a combination of bolts, lacing and wire-mesh (Ortlepp, 1983; Roberts and Brummer, 1988; Simser et al., 2002a, 2002b). The bolts are fully grouted, have a diameter from 12 mm to 20 mm, are 2.5 m to 3 m long (for typical main mine tunnels 3 to 3.5 m wide), on a grid with 1m to 2 m spacing. The length should be compatible with expected depth of rockburst, which based on observations from past rockbursts, can be of the order of one third to one half the height of the tunnel walls (Ortlepp, 1983; Wagner, 1984). The most critical requirement for the bolts is that failure should occur by slip along the grout rather than by failure of the bolt itself, which may call for reduced bond strength. This is needed to ensure that the bolt yields and thus can displace relative to the rock. A particularly effective bolt is the cone-bolt, developed in South Africa, for support in rockburst areas. The conebolt is similar to a conventional mechanically anchored bolt except that it has a conical shape at is farther end and that the entire length of the bolt is coated with a debonding agent (Figure 7.12b). As the rock deforms, the cone at the end of the bolt is drawn through the surrounding grout producing work while energy is also absorbed by the grout as it deforms and fractures (resin has been also used as the bonding agent between the bolt and the rock). Some of the bolts may be instrumented to determine their loading under normal working conditions or after a rockburst event (Bawden and Jones, 2002).

Wagner (1984) provided the following general recommendations for a yielding support: (1) the support elements must be capable of yielding at closure rates in excess of 3 m/s; (2) the support system must be able to accommodate displacements of no less than 400 mm for a stope support and no less than 60 mm for a tunnel support; (3) the support must provide a minimum stress of 200 kN/m^2 for stope support and 100 kN/m^2 for tunnel support; and (4) the support system must be able to maintain the integrity of the rock mass surrounding the excavation during a rockburst event.

While the support designs described for both Canada and South Africa are based on experience, attempts have been made to develop a general approach to decide the type and characteristics of the support based on a number of factors, including stress conditions, excavation span, and geological structure (Heal et al., 2006). While the proposal is preliminary and much more work is needed, it is a step in the right direction, which ultimately will provide the means to quantify design stability based on factors of safety or through reliability analyses.

6.4.3 Rockbursts in civil engineering tunnels

The stress changes induced in the surrounding mass by the excavation of civil engineering tunnels are much smaller than those induced by mining operations. Because of that (see Section 6.4.1) strain damage, buckling and face crushing are usually the only possible seismic sources. The associated damage mechanisms are strainbursts and buckling (Section 6.4.1).

Strainbursts occur in highly massive brittle rocks subjected to high stresses. The phenomenon is commonly observed in deep tunnels excavated in massive igneous or metamorphic rocks. If the rock has structure (e.g. foliation or jointing) oriented parallel to the tunnel or perpendicular to the face, the phenomenon can be compounded by a combination of strainbursting and buckling. Figure 6.39 shows the evidence of strainbursting at the Lötschberg tunnel in Switzerland, excavated in massive gneiss. The tunnel has a diameter of 9.4 m and reaches a maximum depth of 2,000 m.

At the perimeter of the opening, or at the tunnel face, the rock has no confinement in the direction of the opening. The rock is subjected to a major principal stress σ_1 which results from the stress concentration around the excavation, and to a minor principal stress σ_3, which is zero. Thus the loading conditions of the rock at the perimeter of the tunnel are those of the uniaxial compression test. Hoek and Brown (1980a) compiled observations on the failure of rock surrounding square tunnels in massive quartzite in deep level gold mines in South Africa. Figure 6.40 shows the vertical in situ stress at which sidewall slabbing of the tunnels occurred, together with the uniaxial compressive stress of the intact rock. The ratio of vertical stress to unconfined compression strength σ_v/σ_c is used in the figure to divide the stability of the tunnel in six regions. For ratios less or equal than 0.1 the unsupported tunnel is stable and the behavior of the rock is mostly within its elastic regime. For σ_v/σ_c ratios between 0.1 and 0.2, minor sidewall spalling may occur, requiring light support. For ratios between 0.2 and 0.3, severe sidewall spalling is possible and moderate support may be necessary; between 0.3 and 0.4, heavy support is required to stabilize the opening. For ratios larger than 0.4 the opening may be very difficult to support.

Perhaps the best observations regarding strain bursting are those made at the Mine-by-Experiment at the Underground Research Laboratory of Atomic Energy of Canada Ltd., in southeastern Manitoba, Canada, and reported by Martin (1997) and Martin et al. (1997). The Mine-by-Experiment was performed at a depth of 420 m in massive porphyritic granite-granodiorite. At the depth of the tunnel, the in situ stresses measured were $\sigma_1 = 55 \pm 5$ MPa,

Figure 6.39 Rockburst damage at the Lötschberg tunnel, Switzerland.

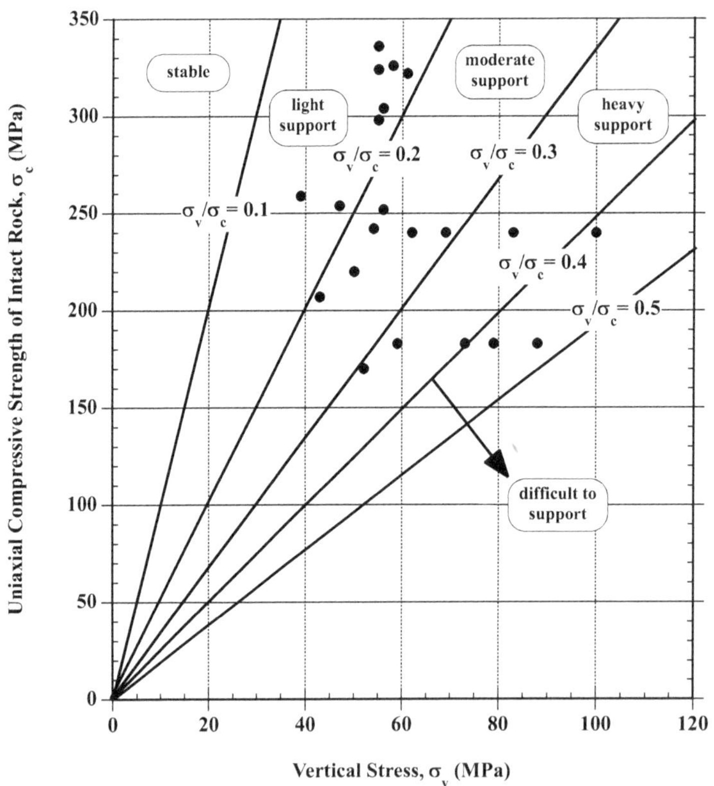

Figure 6.40 Instability of square tunnels in very good quartzite. Data from Hoek and Brown (1980).

$\sigma_2 = 48 \pm 5$ MPa, and $\sigma_3 = 14 \pm 1$ MPa. The unconfined compression strength σ_c of the granite was 213 ± 20 MPa and of the granodiorite 228 ± 20 MPa. The tunnel alignment was approximately parallel to the intermediate principal stress σ_2, and it was excavated using perimeter drilling and mechanical breaking of the central rock stub to prevent damage to the rock around the excavation. The tunnel was 46 m long and had a diameter of 3.5 m (radius, $r_o = 1.75$ m). During excavation, slabbing of the roof and floor was observed at about 0.5 to 1 m from the face, immediately after excavation. The failure progressed until a final V-shaped notch was formed, as shown in Figure 6.41, with a depth of 1.3 to 1.5 the tunnel radius. The thickness of the spalling slabs was very small, between few millimeters and 10 cm. The location of the notch coincided with the point of maximum tangential stress around the opening which, given the orientation of the tunnel with respect to the in situ principal stress, was along a radial direction parallel to σ_3 and perpendicular to σ_1 (Figure 6.41). The stages involved in the formation of the notch were (Figure 6.42): (1) Initiation (Figure 6.42a). This stage consists of the formation of microcracks in the rock, which are concentrated in a narrow region near the tunnel face; (2) Process Zone (Figure 6.42b). Crushing of the rock about 0.5 to 1 m behind the face. It is characterized by a very narrow process zone, 5 to 10 cm wide, which occurs at the location where the tangential stress exceeds the strength of the rock. In this zone, intense microcracking occurs, with all microseismic events having a constant stress drop between 200 Pa and 200 kPa, which can be associated with fractures only a few centimeters in size. Dilation within the process zone results in the formation of thin slabs, 2–5 mm thick, of the same size as the grain of the rock; (3) Slabbing and Spalling (Figure 6.42c). In this step larger slabs detach from the sides of the notch, typically smaller

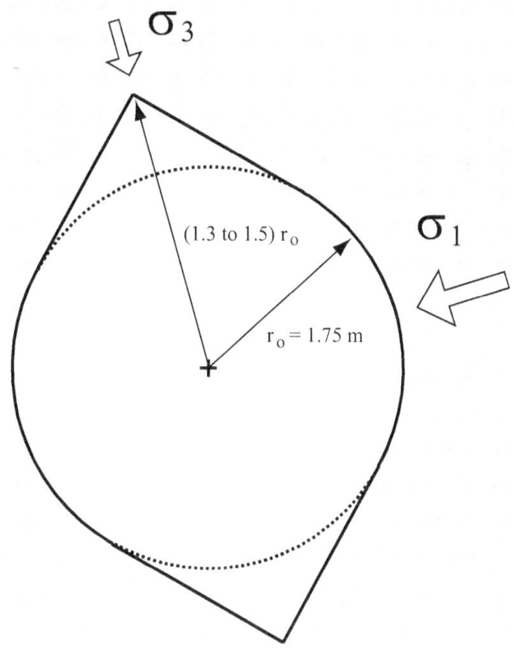

Figure 6.41 Mine-by-experiment tunnel with notch due to rockbursting. Adapted from Martin (1997).

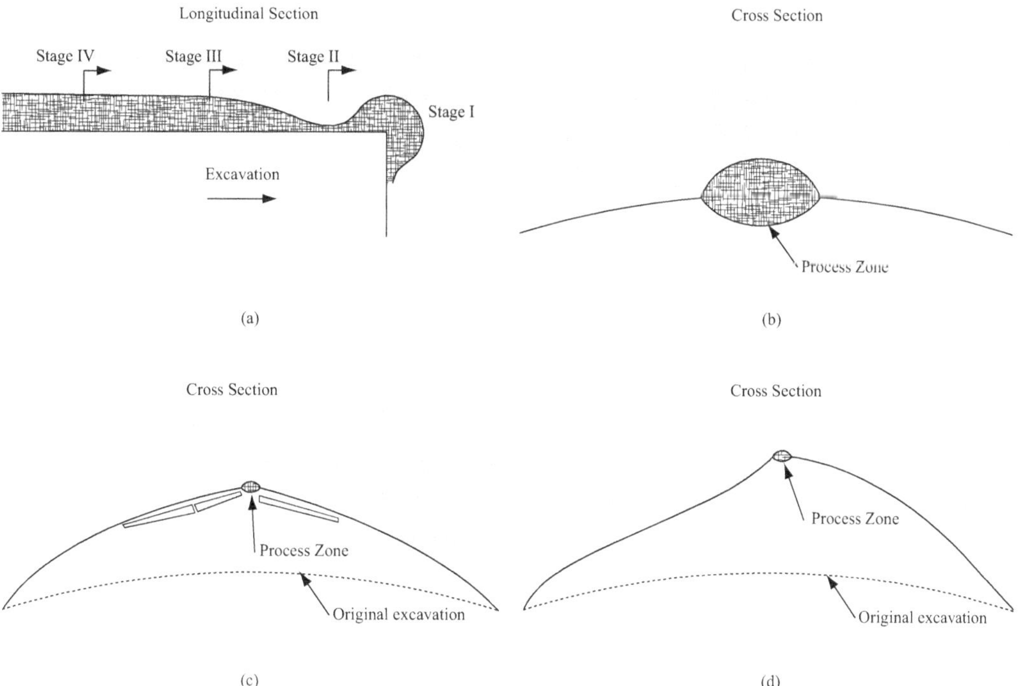

Figure 6.42 Progressive development of V-notch at mine-by-experiment. Adapted from Martin et al. (1997).

than few centimeters thick; and (4) Stabilization (Figure 6.42d). In this stage the notch is stable. It may take several months until the final depth of the notch is reached, and it appears that the final shape of the notch is such that a small confinement is induced in the rock at the apex of the notch thus preventing further progress of the failure. A very interesting observation was that the notch at the floor of the tunnel was smaller than at the crown, suggesting that even small confinement (the floor was covered by muck) can prevent the occurrence of strain bursts.

Data from the Mine-by-Experiment and from other tunnels, where strainbursting was also reported, was compiled by Martin et al. (1999). The data showed that a V-shaped notch was formed in tunnels subjected to strainburst regardless of the original shape of the tunnel, e.g. rectangular, circular, horseshoe. The data also indicated that the formation of the notch initiated at a tangential stress of a fraction of the unconfined compression strength of the intact rock, of the order of 0.4 σ_c. The authors associated the stress at which rockbursts started to occur with the stress at which crack initiation started in laboratory intact rock specimens subjected to uniaxial compression. In all cases, the notch progressed into the rock by slabbing and spalling of thin sheets of rock detaching from the surface of the excavation, as described in the Mine-by-Experiment. Figure 6.43 shows a correlation developed by Martin et al. (1999) between the depth of the notch relative to the size of the tunnel and the maximum tangential stress σ_{max} at the perimeter of the excavation (for circular tunnels, $\sigma_{max} = 3\,\sigma_1 - \sigma_3$). The relation is:

$$\frac{R_f}{r_o} = 1.25\frac{\sigma_{max}}{\sigma_c} + 0.49 \pm 0.1 \tag{6.73}$$

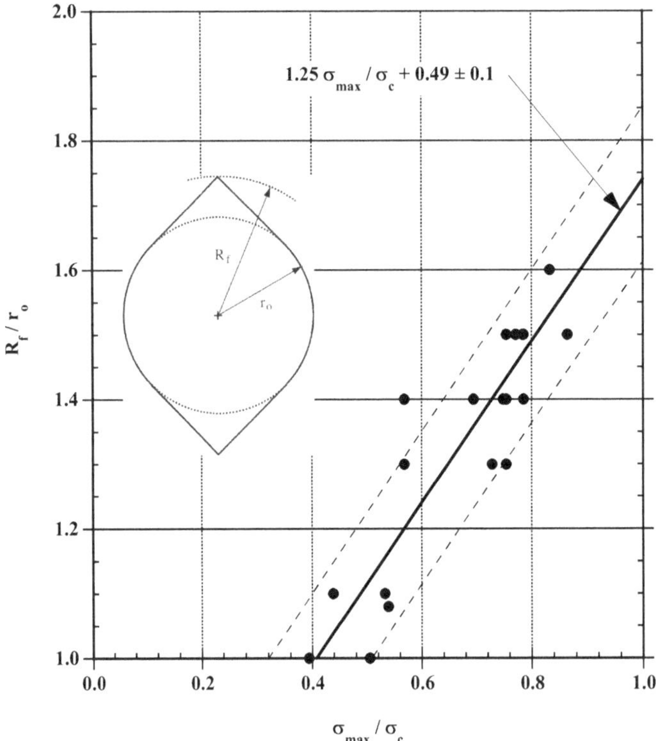

Figure 6.43 Damage initiation in tunnels in brittle rock. Adopted from Martin et al. (1999).

where R_f is the radial depth of the notch, r_o is the tunnel radius, σ_{max} is the maximum tangential stress along the perimeter of the opening, and σ_c is the unconfined compression strength of the intact rock obtained in the laboratory. The relation should not be used for $R_f/r_o > 1.6$ because this is the limit of the data available. Initiation of failure, according to Equation (6.73), can be estimated making $R_f = r_o$, which results in a ratio $\sigma_{max}/\sigma_c = 0.4$, again of the order of the stress required for the onset of microcracking of the intact rock in uniaxial compression. Equation (6.71) is derived from (6.73) since $d_f = R_f - r_{eq}$ (for circular tunnels, $r_{eq} = r_o$); see Figure 6.38.

Observations from the Mine-by-Experiment as well as those used to develop Equation (6.73) were made in brittle massive to moderately jointed rock. In these cases, failure occurred in the intact rock behind the face, and the structure of the rock, if any, played no significant role. Rockbursts, however, occur at the location of maximum tangential stresses, which may or may not be at the tunnel perimeter behind the face. In fact rockbursts at the face of the tunnel rather than in the tunnel itself, as observed in the Mine-by-Experiment, have been reported by Saito et al. (1983) on a deep tunnel excavated in massive quartz diorite and by Hirata et al. (2007) on a tunnel excavated in quartz diorite. Foliation and other closely spaced discontinuities present in the rock may exacerbate the problem because of buckling of the thin slabs created by the discontinuities (Ortlepp and Stacey, 1994), as they occurred at the face of the Lötschberg tunnel (Einstein, 2005). The location of rockbursts depends on the magnitude and orientation of the in situ principal stresses with respect to the tunnel, on the tunnel excavation procedure, and on the structure of the rock mass. A very typical case is the Mont Blanc tunnel with severe rockbursts at the face.

The conditions for rockbursting to occur, namely brittle rock and tangential stresses larger than 0.4 times the unconfined compression strength of the intact rock, are necessary conditions, but not sufficient. In other words, rockbursting, when observed, is associated with massive brittle rock under high stresses, but rockbursting may not occur even though these conditions are met. For example, rockbursts are more likely to occur in machine-excavated tunnels than in drill-and-blast tunnels. This is so because rockbursts are surface phenomena. If the rock on the surface is damaged during excavation, e.g. by blasting, it will yield rather than fail in a brittle matter; farther inside the rock, the radial stresses may have increased enough such that the small confinement prevents the rockburst.

There are no well-established procedures to assess whether a rockburst will occur or not. The Strain Energy Storage Index, W_{ET}, criterion was proposed by Kidybinski (1981) for coal, based on the rock's ability to store and release strain energy, which is the mechanism thought responsible for rockburst phenomena. Later on Jienan et al. (2005), Singh (1988), Wang and Park (2001), and Lee et al. (2004b) extended the criterion to other rocks. The Index is based on results of uniaxial compression tests on an intact rock specimen. The specimen is first loaded to a stress 80–90% of the unconfined compression strength. The specimen is then unloaded at the same loading rate used for first loading (about 0.04 MPa/s in the original publication). The Index is defined as (Figure 6.44):

$$W_{ET} = \frac{\int_{\varepsilon p}^{\varepsilon T} \sigma^U(\varepsilon)\,d\varepsilon}{\int_0^{\varepsilon T} \sigma_L(\varepsilon)\,d\varepsilon - \int_{\varepsilon p}^{\varepsilon T} \sigma^U(\varepsilon)\,d\varepsilon} \tag{6.74}$$

The numerator represents the elastic strain energy retained in the specimen, and the denominator the dissipated strain energy. The criterion is defined as: (I) $W_{ET} \geq 5.0$, high bursting liability; (II) $2 \leq W_{ET} \leq 5.0$, low bursting liability; and (III) $W_{ET} < 2.0$, no bursting liability.

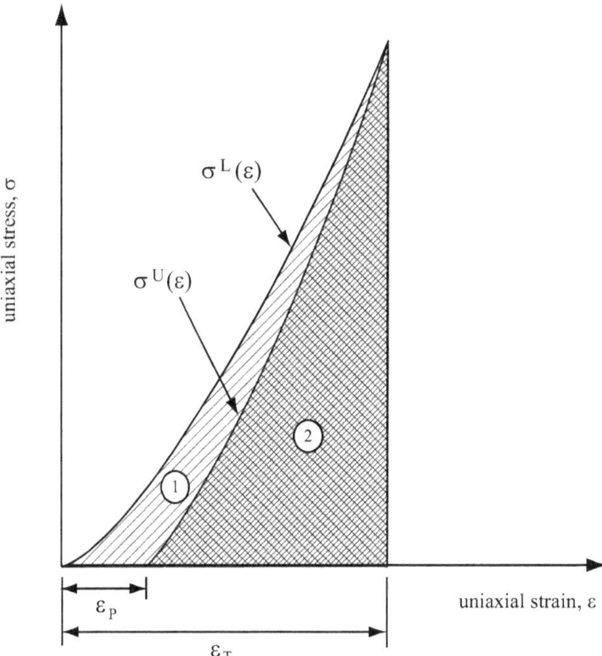

Figure 6.44 Strain energy storage index.

Support of rockburst areas in civil engineering tunnels has been successfully accomplished with systematic bolting in a grid with 1 to 2 m spacing, with lengths between 2 and 3 m (Flaate, 1983; Saito et al., 1983; Broch and Sørheim, 1984; Myrvang and Grimstad, 1984). The rockbolts are typically mechanically anchored or fully grouted and untensioned. They are installed as close to the face as possible and perpendicular to the tunnel surface. A combination of rockbolts and shotcrete has also proved successful, in particular shotcrete with steel fibers. Shotcrete, even if applied at some distance from the face, introduces some confinement to the rock and prevents further spalling (see Chapter 7 for further discussion on shotcrete and rockbolts).

6.4.4 Recommendations

The mechanisms for rockbursts in civil engineering tunnels are strain bursting and buckling. These mechanisms release small quantities of energy, as compared with rockbursts in mining. The result is the slabbing and spalling of rock at the tunnel perimeter and/or face, which may pose a safety risk to tunnel personnel as small pieces of rock are ejected from the tunnel. An important consequence of rockbursting is a decrease of production in tunnel advance as the cutters of the TBM encounter loose rock or the lateral grips of the machine may not have an adequate support.

Prediction of rockburst is difficult since there are no well-established and verified criteria which, given the nature of the rock, the in situ stress field, and the tunneling excavation method, can be used to determine whether the phenomenon will occur and if so at what intensity. Rockbursts are likely only in massive brittle rocks subjected to high in situ stresses. If the maximum stress in the perimeter or face of the tunnel exceeds about 0.4 times the unconfined compression strength of the rock, rockbursts may occur.

Rockburst phenomena decrease dramatically with even small confinement. For this reason, rockbolts and shotcrete have been very successful to prevent or mitigate the damage associated with rockbursts. Rockbolts and shotcrete should be placed as close to the face as possible to induce confinement in the early stages of the phenomenon.

The section has provided a review of criteria that can be used to estimate the onset and eventual extent of damage due to rockbursting in civil engineering tunnels. One needs to be cautious while using the information since these criteria are based on limited information.

6.5 TUNNELS IN SWELLING GROUND

6.5.1 Introduction

Problems when tunneling in swelling and squeezing ground are as old as major tunneling, i.e. railroad tunneling in the 19th century, as illustrated in Figure 6.45. The preceding sentence also points out a problem of definition in that swelling and squeezing, which describe time-dependent deformation of the ground, often occur together. This section is devoted to tunnels in ground that has time-dependent behavior because of its mineralogy. The time-dependent processes, as will be seen, are often a mix of swelling and squeezing in the narrow sense. Practically speaking and as will become evident in the review of case histories in Section 6.5.4, swelling, particularly in form of invert-heave in tunnels can range from a few cm to 1 m or more.

Section 6.5.2 defines or characterizes swelling ground, while the corresponding tests are described in Section 6.5.3. Section 6.5.4 summarizes and analyzes case histories, which will form the basis for describing the swelling/squeezing mechanics affecting tunnels. The corresponding analyses and design procedures are discussed in Section 6.5.5. Much of what is presented in the following is summarized from what has been published by the second author who has been working in the "tunnels in swelling rock" area for nearly 40 years. This includes also ISRM suggested methods. The summarized material will be designated as such.

6.5.2 Characterization of swelling ground and basic mechanisms

Phenomenologically, swelling of rock and soil can be defined as the time-dependent volume increase involving physicochemical reaction with water. Mineralogies favoring such reactions are clay minerals, anhydrite, and pyrite (marcasite) alone or in combination. It is important to note that swelling can occur in rock or soil. Defining a boundary between rock and soil in general and swelling rock and expansive soil, in particular, is often not possible. For this reason, it is acceptable, from a practical point of view, to consider e.g. an over-consolidated clay to be a rock or a soft clay shale to be a soil. The difference between swelling rock and expansive soil problems can be the fact that the latter often involve unsaturated soils with suction pressure changes as a major issue. This chapter will concentrate on swelling rock with a variety of mineralogies and swelling of clay-soils. Within rock, further distinctions are possible in that swelling of rock masses can occur as swelling of intact rock, as swelling of gouge and fillers in the discontinuous rock mass, or as swelling of a combination of the above. It is also of interest that rocks that are susceptible to swelling often weather (slake) strongly if subject to water content changes. Weathering (slaking) will not be considered here. Weathering (or slaking) of swell susceptible rocks can occur if they are subject to swelling-drying cycles or intensive air-drying by which the durability of rock can be significantly affected. The phenomenon of slaking, although often consequent upon swelling, is complex and may involve other causes and mechanisms. It is not treated here since so far no convincing empirical or analytical methods exist to consider it in tunnel design.

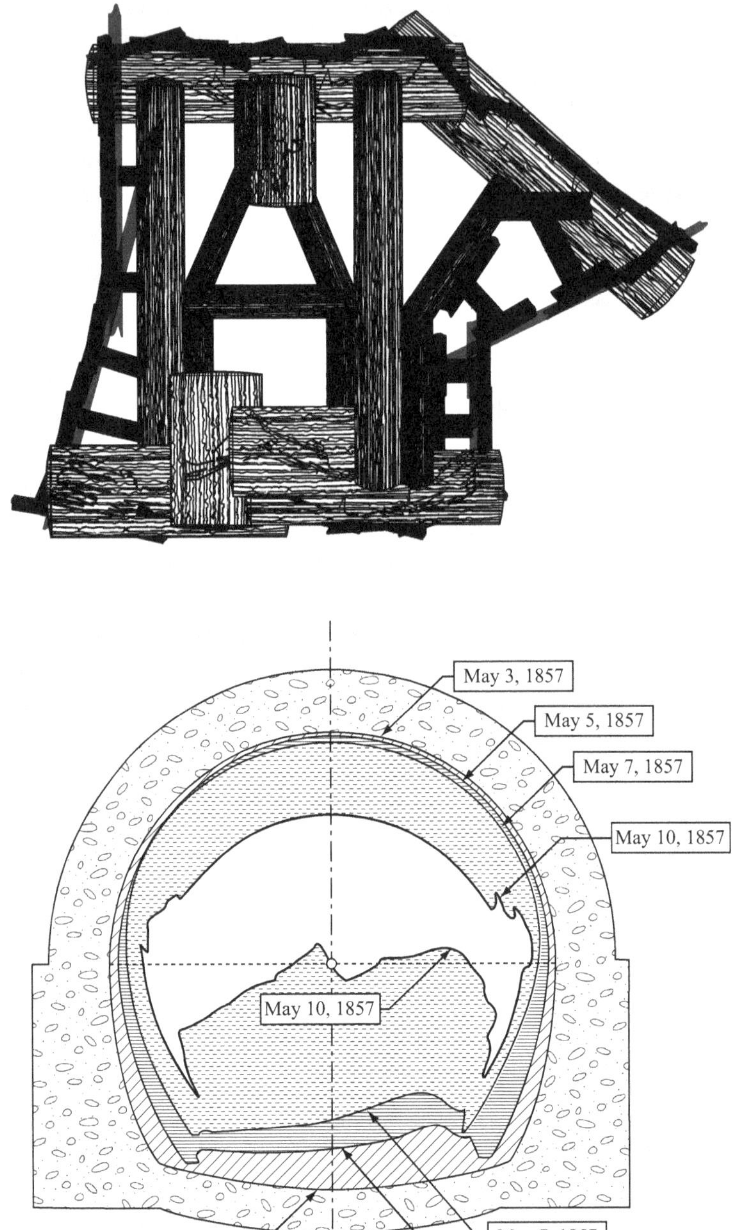

Figure 6.45 Deformations and destruction of initial support in swelling/squeezing rock. After Rziha 1867/1872 (1987).

It is important to note that swelling involves a combination of physicochemical reaction with water and stress relief. The physicochemical reaction with water is usually the major contributor to swelling but can only take place associated with stress reduction/stress change. This will be discussed in more detail in Section 6.5.3. It is now necessary to describe the basic physicochemical mechanisms. The basis for this description is ISRM (1983) expanded by work conducted by Madsen and Nüesch (1991). The major physicochemical mechanisms involving water are:

i. *Physicochemical mechanisms involving clay minerals*: Water is adsorbed at the exterior surfaces of clay minerals and is taken up at internal surfaces of clay minerals having expandable layers. Swell pressure depends on the interparticle distance of clay particles and intraparticle distance between expandable layers. Swell heave depends on the amount of clay particles and expandable layers. Rocks with expandable layer type particles (e.g. Smectite) swell more than those with particles that adsorb water only externally (e.g. Kaolinite).
ii. *Clay swelling is osmotic swelling*: The difference in ion-concentration between the double layer surrounding the clay minerals and the pore water is the underlying cause for water moving between and into the clay minerals. The swell potential depends on the surface area that is available to water, the distance between these surfaces and the volume fraction of the cations. This explains the difference between "non-swelling" / "slightly swelling" (e.g. Kaolinite, Illite) and "swelling" (smectite, mixed layer) clays. Clay swelling can lead to a volume increase of up to 20%. Observed swelling pressures in tunnels can reach up to 3 MPa as will be discussed in Section 6.5.4. The knowledge about these mechanisms, specifically about the role of the double layer led to the double layer theory (also called DLVO theory) by Madsen and Müller-VonMoos (1985, 1989) that states:

$$\sigma = \frac{\pi^2 \varepsilon \varepsilon_0 k^2 T^2}{2e^2 v^2 d^2} \qquad (6.75)$$

where σ is the swell pressure, in N/m^2; ε is the dielectric constant (for water = 80); ε_0 is the influence constant for vacuum, in C^2/Nm^2; k is the Boltzman Constant in J/°K; T is the absolute temperature, in °K; e is the elementary charge, in C; ν is the cation valence (for $Na^+ = 1$); and d is half the distance between clay particles, in m. Equation (6.75) results, for T = 293°K in:

$$\sigma = \frac{2.2710^{-12}}{v^2 d^2}\left({}^{N}\!/_{m^2}\right) \qquad (6.75)$$

The cation valence ν can be obtained through a determination of exchangeable cations and the half distance d can be derived from the water content. Additional details on the double layer theory can be obtained from Madsen and Müller-Vonmoos (1985, 1989) and Hauber et al. (2005). The latter includes also an example calculation and comparisons to laboratory measurements, which are very satisfactory. Actual cases involving clay swelling will be discussed in Section 6.5.4. In that section, the complete mechanism of clay swelling around a tunnel will also be discussed. What has been described above is the physicochemical mechanism, which is a part of the overall mechanism.
iii. *Physicochemical mechanisms of sulfatic rocks*: Here one needs to distinguish between pure sulfate rocks and clay-sulfate rocks (note that much of what is discussed below is taken from Hauber et al., 2005 and the underlying literature.)
iv. *Pure sulfates*: This involves the hydration of Anhydrite into Gypsum, namely Ca SO_4 + $2H_20$ → Ca SO_4 $2H_20$. Very important is the fact that under atmospheric pressure and up to 42°C (possibly 57°C), Anhydrite is more soluble than gypsum (e.g. at 25°C, gypsum solubility is 2.4 g/l, Anhydrite 2.7 g/l). This means that under these conditions,

anhydrite gets dissolved and then precipitates as gypsum if the concentration of the solution is sufficiently high. In an open system, where water has free access to the anhydrite and assuming that the anhydrite has zero porosity and is initially completely dry, a volume increase of 60% will occur when anhydrite is transformed into gypsum. In natural rock, which has greater porosity, this volume increase is smaller. As indicated above (open system) anhydrite to gypsum swelling also requires access to or existence of water. Two conditions are essential: (1) Water has to be able to access the anhydrite mineral surfaces. This can occur through microcracks or, as will be discussed below, through surrounding clay (evidence for the latter is that massive anhydrite swells much less compared to anhydrite embedded in clay); (2) The amount of water has to be limited since too much water will lead to lower concentration thus preventing gypsum precipitation and possibly transport of the dissolved sulfates away. As a matter of fact, the water–sulfate solution should be oversaturated for gypsum precipitation to occur. As will be discussed later, all this can possibly lead to gypsum deposits at other locations.

Physicochemical mechanisms in clay sulfate rocks: It is widely recognized that clay sulfate rocks, specifically anhydritic shales, show a much more rapid and intensive hydration of anhydrite into gypsum than pure anhydrite. Although the results are well known, the causes are still not entirely clear. The following four possibilities were hypothesized by Hauber et al. (2005) (see also Section 6.5.3. for newer investigations):

Clay swelling leads to cracks and thus facilitates water access to anhydrite: This is possible and such cracking has been observed, possibly including a feedback cycle "clay cracking – more water moves to anhydrite-anhydrite to gypsum transformation-causes further cracking," etc. On the other hand, it is well known that clay swelling acts in a self-healing manner as observed and relied upon in waste repositories.

Clayey rocks are 10^4 times more permeable than pure anhydrite and thus serve as "water reservoirs." Also, water adhering to clay minerals is used in the anhydrite-gypsum transformation. This is unlikely. Clay minerals when unloading/destressed (remember that stress relief is a necessary component in swelling) attract additional water in the double layers, which is thus not available for anhydrite to gypsum swelling.

The swelling clay minerals take up water and thus increase the gypsum concentration in the water leading to precipitation of gypsum. As mentioned in the preceding point, clay minerals attract water and may lead to a concentration increase. In addition, clay swelling can only occur with a porosity increase. It has been observed that this is possible through an increase in the number of small pores and a decrease of the larger ones with a related decrease in permeability. The pore size decrease favors capillary action. These two facts, i.e. lower permeability and capillary action, can explain the aforementioned self-healing and can also explain gypsum precipitation in areas at a distance from the original anhydrite deposit (Madsen and Nüesch, 1990).

Clay minerals are chemical catalysts. Possible and occasionally observed.

Overall, and as also evident from the preceding, it is likely that there is an interaction of these mechanisms, which can simply lead to a filling of the pores produced through clay swelling (passive mechanism) or the crystallization pressure of gypsum produces additional swelling (active mechanism). For the passive mechanisms, an increase of the sulfate (gypsum) concentration in the water is sufficient; for the active mechanism, this concentration has to be generally available and, in addition, in situ critical stresses must be less than a limiting pressure at which anhydrite becomes less soluble than gypsum (at atmospheric pressure, anhydrite is more soluble than gypsum, up to temperatures of 42°C; this temperature limit

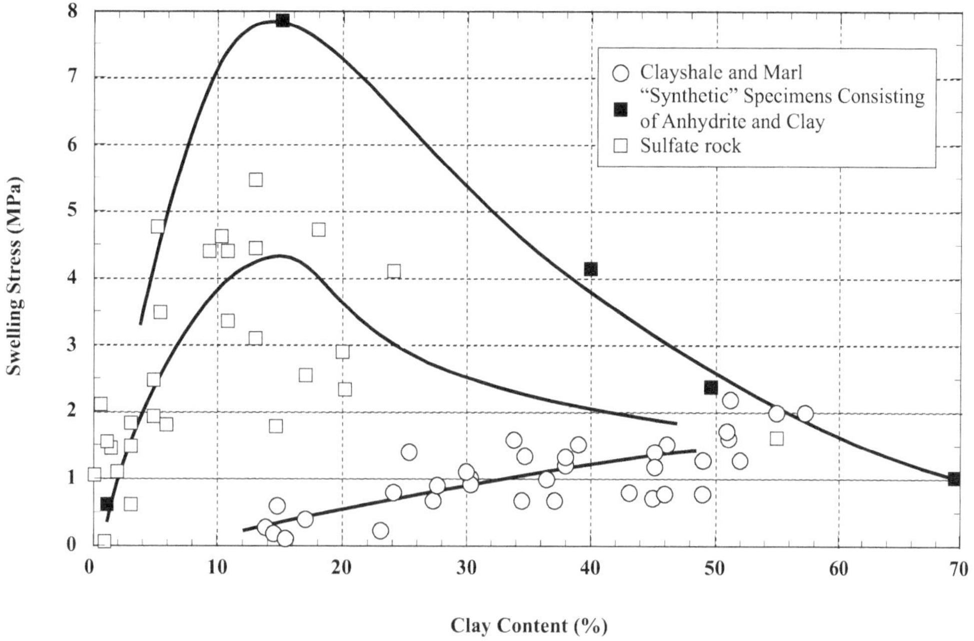

Figure 6.46 Swelling of sulfate containing and sulfate free clayey rocks (Hauber et al., 2005).

is not firmly established, see Hauber et al., 2005). The consequence of all this is that clay-sulfate rocks at clay contents between 5 and 40% swell more than pure clay rocks, as shown in Figure 6.46. On the other hand, the sulfate content in clay-sulfate rocks (shaley anhydrites) is what governs swelling stress and strain of such rocks, as shown in Figures 6.46 and 6.47, which represent results from a number of natural rock samples from Northern Switzerland and Southern Germany (the total test duration was 6 years). Figures 6.47a and 6.47b report the swelling strain and stress in relation to sulfate and clay content and compare reasonably well to what is shown in Figure 6.46. Figures 6.47c and d report the percentage of gypsum/anhydrite after completion of the test. In the free strain test, most of the sulfate minerals were gypsum, while in the stress tests they were anhydrite. As will be seen in the next section, stress

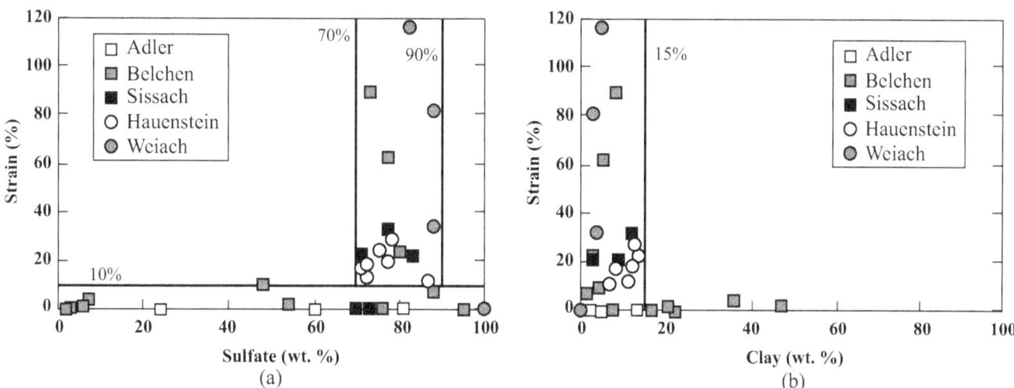

Figure 6.47a Strain observed in "free strain" swell tests at the end of test (Nüesch and Ko, 2000), versus. (a) Sulfate content. (b) Clay content.

(Continued)

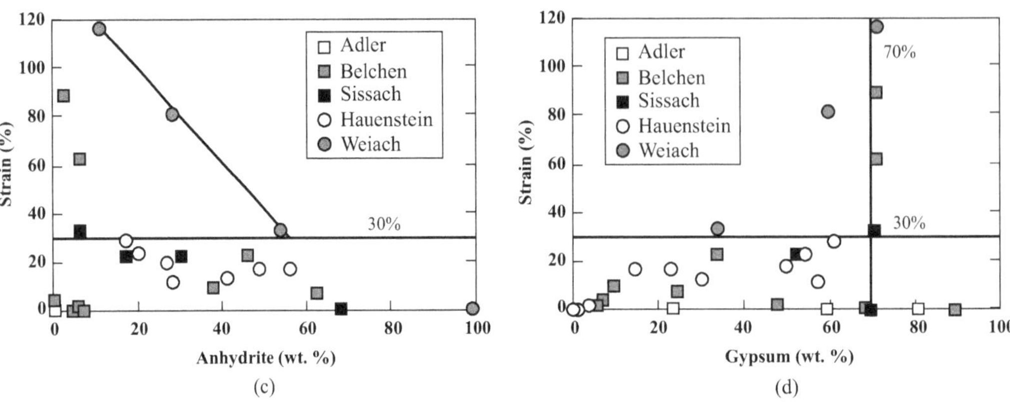

Figure 6.47a Strain observed in "free strain" swell tests at the end of test (Nüesch and Ko, 2000), versus. (c) Anhydrite content. (d) Gypsum content. (The names in the boxes identify the tunnels/quarries from which the samples were taken.)

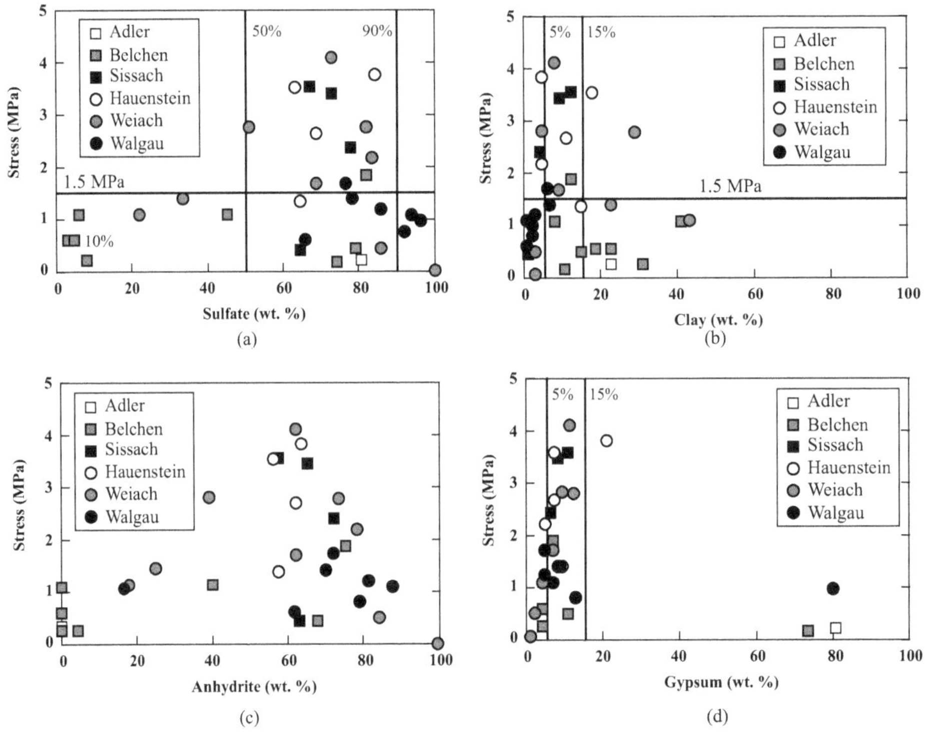

Figure 6.47b Stress observed in "stress" swell tests at the end of the test (Nüesch and Ko, 2000) versus. (a) Sulfate content. (b) Clay content. (c) Anhydrite content. (d) Gypsum content.

tests lead to a buildup of stress, which limits the hydration as discussed earlier (solution of anhydrite depends on temperature and pressure). In contrast, the free strain tests have essentially zero-stress on the specimen.

More recent work by Anagnostou and co-authors encompasses detailed models of gypsum crystallization (Serafeimidis and Anagnostou, 2014a, 2014b) as well as extensive thermodynamic/kinetic modeling of chemical reactions and associated swelling in clay sulfate rocks, Serafeimidis and Anagnostou (2012, 2013). This theoretical work has been most recently supplemented by extensive experimentation by Wanninger (2020). This research extends significantly what has been discussed above and the reader is referred to the just mentioned publications. Summarizing what this work so far shows: Crystallization pressure of gypsum strongly depends on ion concentration (activity) of the solution, and anhydrite and gypsum crystallization pressures interact with each other. This is then incorporated into the context of clay sulfate rocks where the clay chemistry affects the ion concentration (clay lowers the activity and intensifies surface energy effects because of small pore size, all of which increase the solubility of gypsum). The two authors also address the issue of sealing, i.e. the development of a gypsum skin on anhydrite that can effectively stop further anhydrite to gypsum transformation.

6.5.3 Swell testing

Qualitative tests to observe swelling by immersing specimens in water and observing/measuring volume increase have existed for a long time. More formal tests producing quantitative results are, e.g. those by Lambe (1960), for expansive soils and the relation between index properties of fault gauge and swell processes (measured in the laboratory) applied to tunnel problems by Brekke and Howard (1973).

The ISRM has made two efforts to define swelling tests. The first one, Suggested Methods for Laboratory Testing of Argillaceous Swelling Rocks (1989) includes tests for determining the maximum swell pressure, swell stress-strain tests and free-swell testing, but will not be discussed further since they have been superseded by the tests discussed below. In the same suggested methods, there is also a slaking test with which the mass loss due to slaking is measured by having 3 mm size particles rotated in a cage underwater. This is mentioned here since slaking often occurs together with swelling in clay rock.

Formal quantitative measurements of swelling originate from soil mechanics where odometers can be used to observe heave as a specimen is unloaded with water access (Lambe and Whitman, 1969). This concept has led to the first formally described testing procedure for swelling of clayey and clay sulfate rocks by Huder and Amberg (1970). In this, a specimen of the rock is placed in an odometer and subjected to a load-unload process under dry conditions. As will be discussed later, the testing of clay sulfate rocks may have to be different. Then, water is supplied and the specimen is unloaded, observing the displacement at each unloading step. A stress-strain swell curve (essentially a reverse consolidation curve) is observed. In essence, the test duplicates what occurs in a tunnel invert and it was actually developed for this purpose and then accordingly used (Grob, 1972). The odometer principle has been used as a basis for the ISRM tests, which are now described; specifically they are the test for determining the maximum swelling stress and the test for determining swelling strain as a function of actual swelling stress. These tests and the free swell strain test are described in ISRM (1989) and will be briefly summarized below. The tests can be used for argillaceous (clayey) rocks and rocks containing sulfates; differences will be commented upon later.

i. *Determining Axial Swelling Stress*
   The apparatus is shown in Figure 6.48. The specimens are circular cylinders with diameters between 50 and 100 mm (the dimensions of the apparatus vary correspondingly) and a thickness (height) of 10 mm. Specimens should, after suitable watertight storage, which should be as short as possible after removal from the ground, be prepared following the procedures described in ISRM (1989). This notably involves cutting without water (air cooling or anti-swelling medium). The swell test is conducted by placing the specimen in the apparatus, slightly preloading it (axial stress approximately 25 kPa) and then filling the container (No. 4 in Figure 6.48) with water. The water permeates into the specimen through the porous plates, which causes the specimen to swell in the axial direction (the stainless steel ring, No. 1 in Figure 6.48, prevents radial expansion).

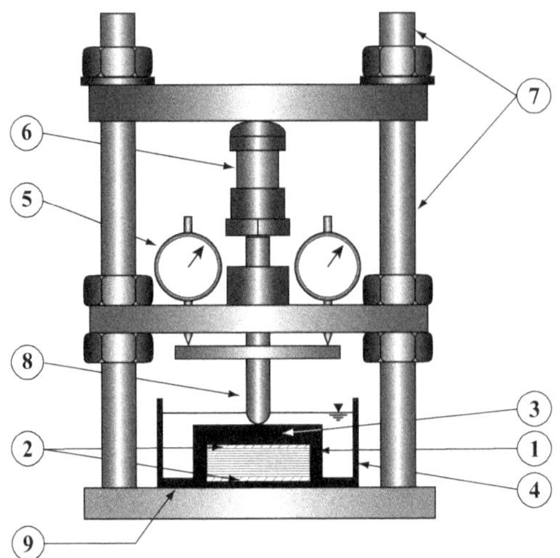

Figure 6.48 Apparatus for determining axial swelling stress. (1) Stainless steel ring. (2) Porous metal plates. (3) Stainless steel loading plate. (4) Container (filled with water). (5) Dial Gauges (attached to bottom of container – attachment not shown). (6) Load frame and load cell. (7) Rigid frame. (8) Loading piston. (9) Stainless steel plate. Adapted from ISRM (1999).

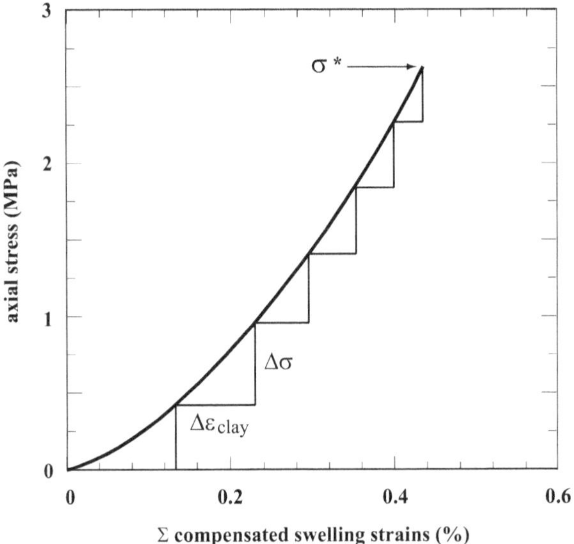

Figure 6.49 Axial stress versus compensated axial strain for argillaceous rock, σ^* = Maximum axial stress; (after ISRM, 1999).

In essence, given the rigid setup of the apparatus, the axial swelling will cause the axial stresses to increase till a maximum swelling stress has been reached. However, since the apparatus is not completely rigid, some axial strain will occur. If argillaceous rocks are tested, these strains can be compensated (axial dimension kept constant) in as many small steps as possible (Figure 6.49). This is allowable since swelling and consolidation in clayey material are reversible. If anhydrite exists (clay sulfate rocks), ISRM (1989) recommends that such compensation of strains must not be performed since it would lead to "unreasonably" high swelling stresses and/or exceed the capacity of

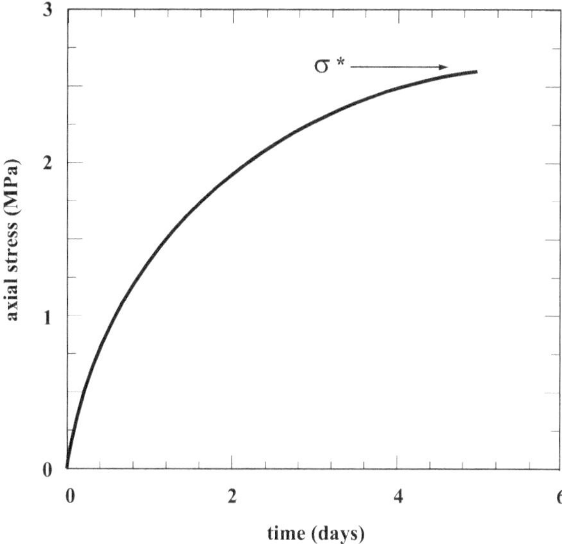

Figure 6.50 Axial stress versus time, σ^* = Maximum axial stress for argillaceous rock; (after ISRM, 1999).

the apparatus. Pimentel and Anagnostou (2013) have developed a testing apparatus, which is stiff enough to constrain deformation. As a matter of fact, this apparatus, which so far is used for research purposes, makes it possible to run also load controlled and stress-swell strain tests (see below). The result of the axial swelling stress tests are usually given in form of stress-time plots (Figure 6.50). The specimen has to be massed prior to and after testing and after subsequent drying; special procedures are necessary if the specimen contains anhydrite (see ISRM, 1989). Clearly, the mineralogy has also to be determined preferably both on the tested and on neighboring specimens from the source sample.

ii. *Determining Free Axial and Radial Swelling Strains*
In this test, a specimen is immersed in water in an apparatus as shown in Figure 6.51. The specimen dimensions are not specified but can be similar to those used in the axial stress test (cylindrical specimen between 50 and 100 mm diameter, 10 mm high). The axial displacement is measured with a "dial gage" or comparable electronic device. The circumferential expansion can be measured with a 0 to 1 mm thick flexible steel band which is calibrated at 0.1 mm intervals and fixed with a rubber band to the specimen. Alternatively, lateral expansion devices can be used. The measured displacements are transformed into axial and radial strain and plotted as shown in Figure 6.52.

iii. *Determining Axial Swelling Stress as a Function of Axial Swelling Strain*
This test is conducted in an apparatus as shown in Figure 6.53. The specimen, steel ring, and water container correspond to the stress test equipment shown in Figure 6.48. Different from that test is the loading which is done with deadloads (Figure 6.53). The specimen preparation is as before. The test is performed by preloading the specimen with a stress roughly comparable to the original overburden stress (Curve 1 in Figure 6.54) and then applying water and measuring the heave (displacement) at this instance (2 and 3 in Figure 6.54). Following this is the unloading in stepwise increments. These can be chosen by the investigator, but it is recommended that identical displacements are used for the entire test. ISRM (1999) recommends a 50% load reduction in each step. The specimen should not be completely unloaded, but one should leave a load producing roughly 25 kPa on the specimen (prevents uneven displacements). Very

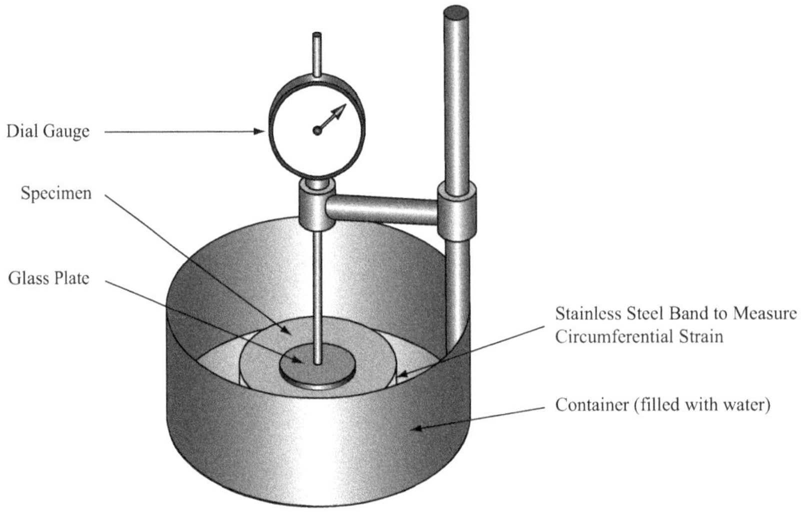

Dial Gauge

Specimen

Glass Plate

Stainless Steel Band to Measure Circumferential Strain

Container (filled with water)

Figure 6.51 Apparatus for measuring swelling. Adapted from ISRM (1999).

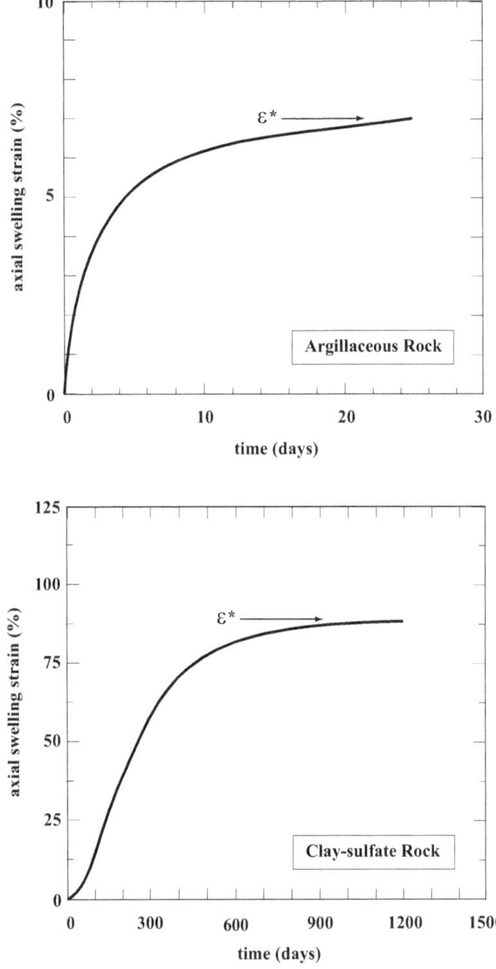

Figure 6.52 Axial swelling strain versus time curves; ε^* Maximum swelling strain (after ISRM, 1999).

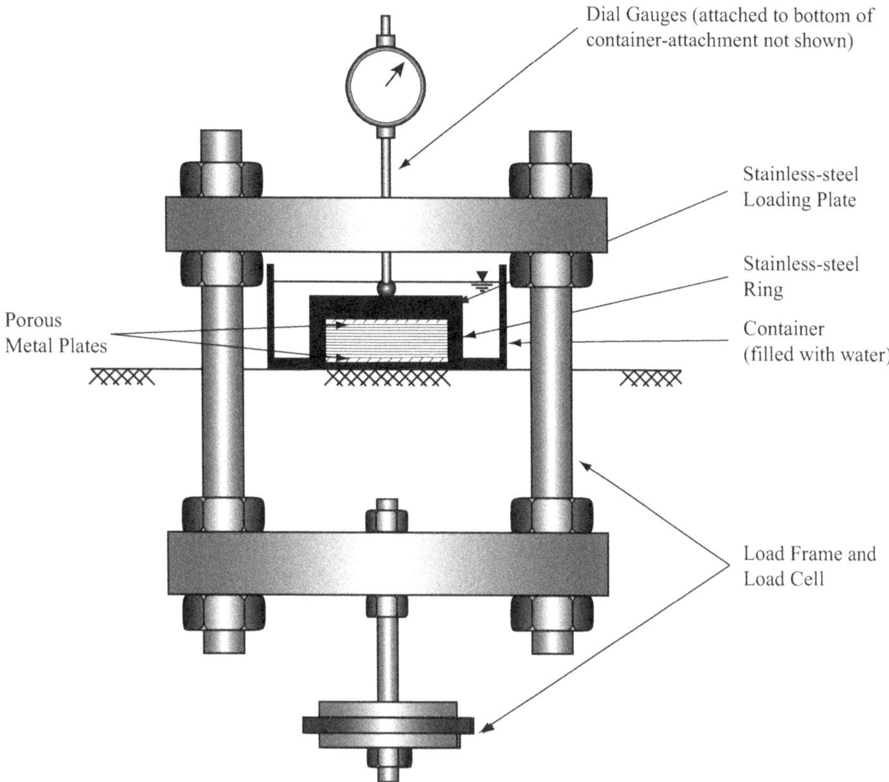

Figure 6.53 Apparatus for measuring axial swelling stress as a function of axial swelling strain. Adapted from ISRM (1999).

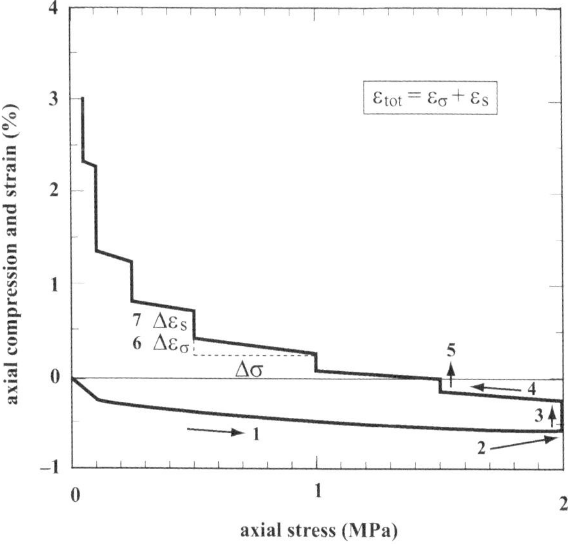

Figure 6.54 Axial stress versus total axial strain curve (after ISRM, 1999).

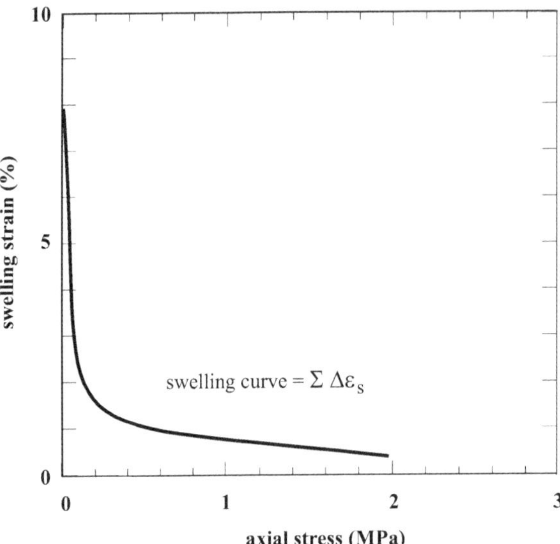

Figure 6.55 Axial stress versus swelling strain (after ISRM, 1999).

importantly, for each load-decrease $\Delta\sigma$, the initial (elastic or matrix strains) $\Delta\ \varepsilon_\sigma$ in Figure 6.54 are observed and then the swelling strain $\Delta\ \varepsilon_s$. The latter occurs over a time period and should be determined when no further displacement occurs for a particular load decrement. The axial stress–swelling strain curve is then obtained by plotting the stress at each load decrement and the strain of the $\Delta\varepsilon_s$ up to this point. This produces the curve shown in Figure 6.55.

The results of the three ISRM tests can be used in the analyses and design of tunnels in swelling rock as will be discussed in Section 6.5.5.

6.5.4 Cases and swelling mechanisms around tunnels

It is both interesting and useful for the understanding of these mechanisms to review what happened in a number of tunnels. Again, most of this material is taken from previous publications. The cases involving clayey/argillaceous rocks are reported first, followed by those in clay sulfate rock. Note, however, that some of the tunnels have stretches in both rock types.

6.5.4.1 Tunnels in swelling argillaceous rocks

The lithology involves clay shales and marls. Historically and, as mentioned in the introduction to this section, it is interesting to observe that damage due to swelling is as old as modern tunneling, i.e. the tunnels for railroads in the mid-19th century. Figure 6.56, which is taken from Rziha (1867/1872), shows failed timber support due to combined swelling and squeezing. This case is not only mentioned because it is one of the earliest ones reported but because the mechanism of combined swelling and squeezing is actually very basic. Einstein (2000) summarized a number of cases in Opalinus Clay shale. One of these cases is the Bözberg railroad tunnel (Swiss Jura Mountains) built in 1871–1875 and where invert heave and inward movement of the abutments (Figure 6.57) occurred shortly after construction. Evidently (Bridel and Beck, 1970), part of the drainage channels had to be replaced already in 1885 because they were so deformed. Figure 6.58a is a record of the continuous measurements

Figure 6.56 Deformation and destruction of initial (timber) support (after Rizha, 1867/1872 [1987]).

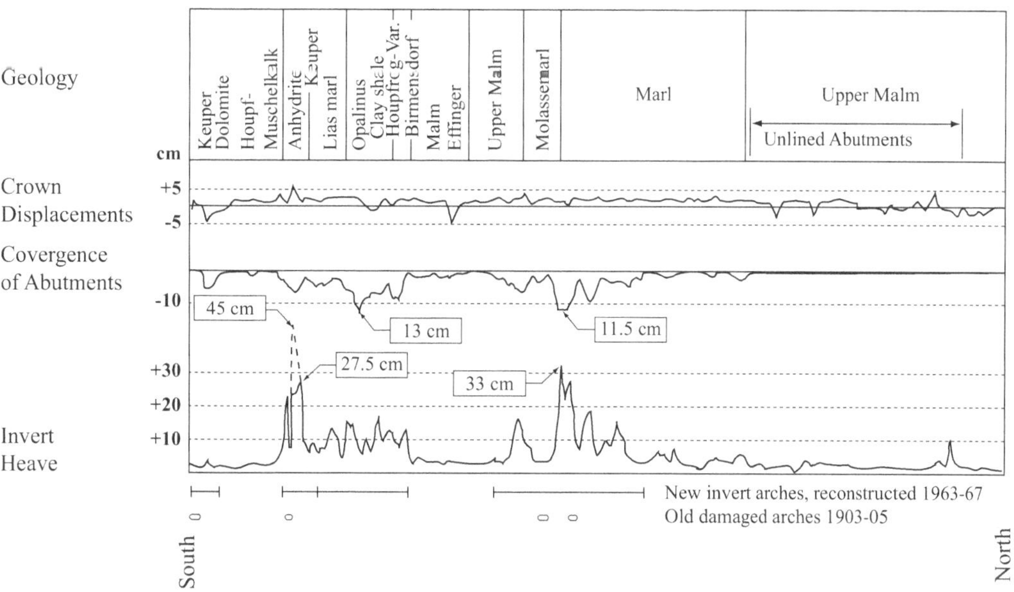

Figure 6.57 Bözberg tunnel deformation measurements 1923–1954 (after Beck and Golta, 1972).

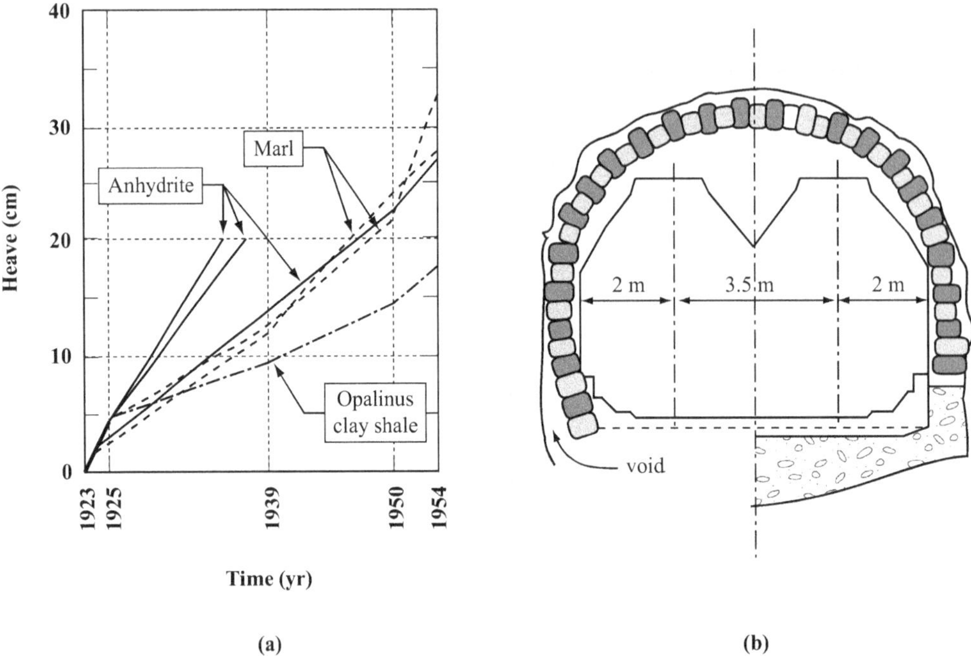

Figure 6.58 Bözberg tunnel (after Grob, 1976). (a) Invert Heave. (b) Abutment Convergence (left) Reconstructed Abutment (right).

taken in the period 1923–1954. The figure shows large heave and abutment movements in the Opalinus clay shale but also in the Molasse-Marl (another argillaceous rock). Note also the displacements in the "anhydrite," which will be discussed below. Two comments by Beck and Golta (1972) who studied this tunnel in detail are important: (1) The abutment movement is not caused by the rock behind it, pushing the abutment out but an indirect effect of invert heave (see Figure 6.58b); (2) Much of the swelling was caused by an ineffective drainage channel, which due to an irregular profile did not move the tunnel water away. The vibration induced by the railroad worsened the rock conditions. An invert arch, built in 1963–1967, mitigated the situation.

The effect of wetting, because of non-existing or non-effective invert protection, together with train vibration effects were also observed in other tunnels in the Opalinus Clay Shale in the Swiss Jura Mountains, namely, in the Grenchenberg tunnel (see e.g. Einstein, 2000; Steiner and Metzger, 1988) and in the Hauenstein Base Tunnel (Etterlin, 1988). The latter case is particularly interesting in that during reconstruction in 1980–1986, an 80 cm thick invert arch was placed (Figure 6.59). Stress measurements in 1988 showed that swelling stresses between 0.09 and 0.36 MPa acted on the tunnel liner. This indicates that swelling continues.

Einstein (2000) discussed some other cases in Opalinus-Clay shale. A general overview of swelling in tunnels is given in Einstein (1979). In that publication, a summary table is presented in which also cases in marl are listed.

Summarizing the observations in these clayey rocks, one can state that displacement (usually invert heave) rates between 4 and 20 mm/year were observed. The reported total displacements in these rocks ranged up to 400 mm; this does not necessarily mean that such total final displacement was observed since in a number of cases the heaving rock was removed (shaved off) at intervals.

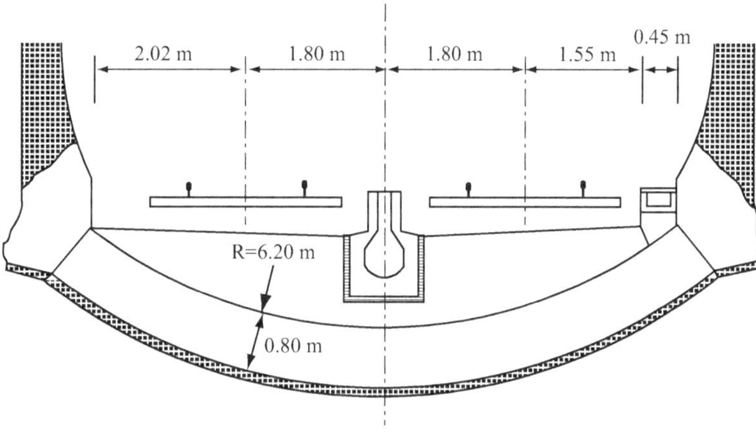

Figure 6.59 Hauenstein base tunnel – reconstructed invert arch (after Etterlin, 1988).

6.5.4.2 Tunnels in clay sulfate rocks

The (displacement) heave rates and total displacements are much more dramatic in these rocks compared to the argillaceous ones. Displacement rates up to 100 mm/yr have been observed but there are also statements in construction reports such as "in a few weeks" for the Hauenstein Base tunnel. Regarding total displacements (again this usually involved shaving off of rock at intervals), they reached the order of meters. Also, in some instances, swelling still seems to be going on decades after construction. Three cases will be discussed in more detail:

i. The Kappelisberg Tunnel: The tunnel, located in Southern Germany, is a two-track railroad tunnel and is the often quoted extreme case with heave rates between 20 and 50 mm per year and, most importantly, a total invert heave of 4,700 mm. A crown heave of 640 mm was also observed.
ii. *The Belchen Tunnel*: This is a 3.2-km-long highway tunnel through the Swiss Jura Mountains (two, two-lane tunnels with a horseshoe shape, 12 m wide, 10 m high) was completed in 1970. It runs through, among other formations, Opalinus Clay Shale and the Keuper. The latter consists of clay sulfate rocks. Figure 6.60 is a longitudinal cross section and also shows the invert heave, which is greatest in the clay-sulfate rock.

   The tunnel was constructed by excavating, in each tube, two invert drifts and then enlarging them to the full cross section. After the enlargement to the full cross section in the Keuper formation, the invert level rose 0.9 m within a few months. As a consequence, an invert arch was built with a radius of r = 10.4 m and a thickness of t = 0.45 m; see Figure 6.61. However, this was sheared off shortly after construction, and a further 0.6 m of invert heave occurred. This required the construction of a new invert arch of r = 8.12 m and t = 0.85 m (Figure 6.61). In 1974, stress cells at the rock-concrete interface were placed to measure the contact stresses, while the fiber-stresses in the concrete support were backfigured from deformation observations and a few overcoring tests. The maximum contact stresses were 3.5 MPa and the maximum fiber stresses were 27 MPa.

   The tunnel incurred higher stresses and also damages in the subsequent years. As a matter of fact, the problems essentially continue and, in addition to measuring stress

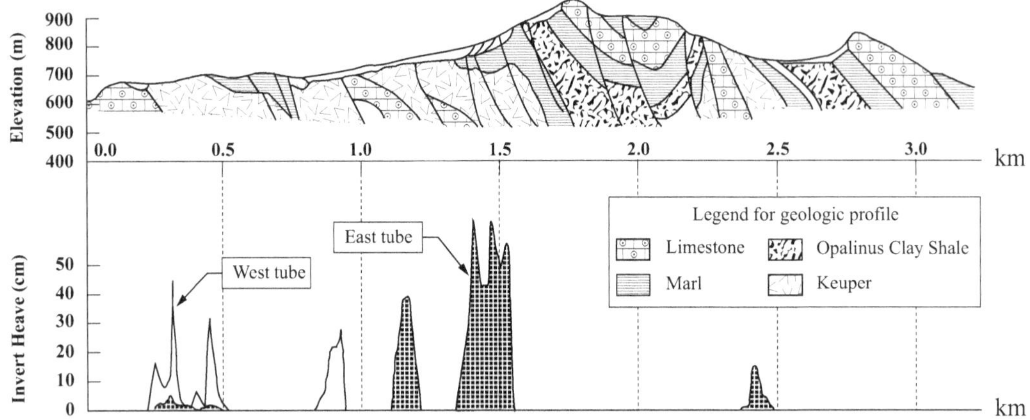

Figure 6.60 Belchentunnel: Longitudinal profile and invert heave (after Grob, 1972).

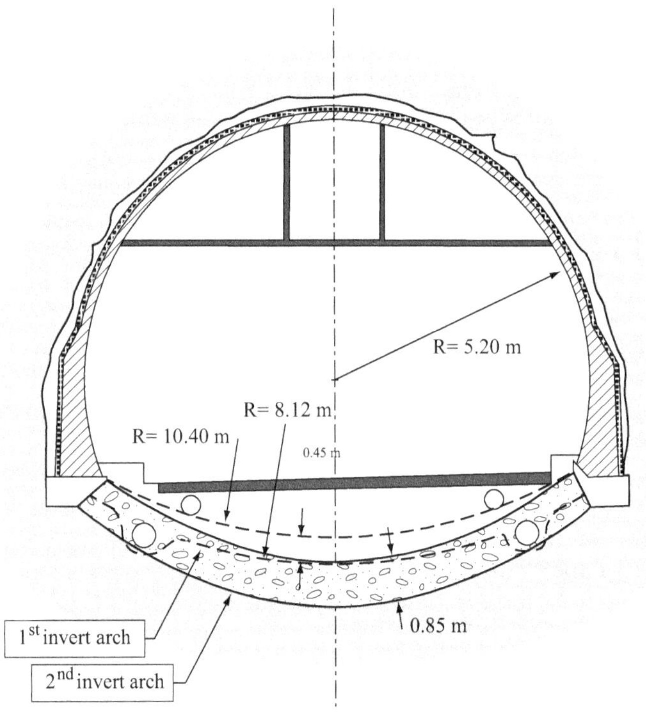

Figure 6.61 Belchentunel – Reconstruction of invert arch (after Grob, 1972).

and deformation, some minor remedial measures were taken and plans for major remedial measures were prepared and are under way at present (Chiaverio et al., 2017). A number of survey borings were made in the clay-sulfate rock section in the period 1986–1996 (Hauber et al., 2005). These borings were placed in zones of tunnel damage and included, in addition to core recovery, also displacement measurement with a sliding micrometer. A typical profile obtained in these borings is shown in Figure 6.62. As will be discussed later, this profile is very typical for sulfate (anhydrite) or clay sulfate rocks undergoing swelling. The active zone, in which gypsification and thus swelling occurs, is located between F and A. Movements in this zone were actually observed.

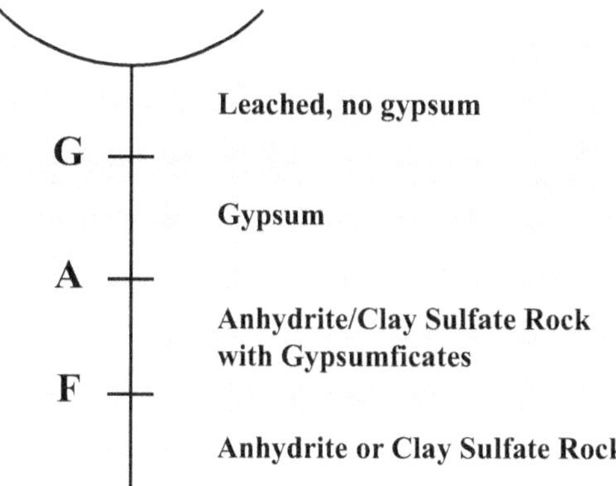

Figure 6.62 Typical rock profile – under the invert of the Belchen tunnel (after Hauber et al., 2005). F = Gypsification front where hydration of anhydrite into gypsum starts; A = Anhydrite level above which only gypsum exists; G = Upper boundary of gypsum.

Also interesting is the comparison of the rock conditions observed during construction and with those in the 1986–1996 borings. In three cases, there was no gypsum during construction but was observed in the 1986–1996 borings. From this it was possible to derive the rate of gypsification, which was above 20 mm/year. Similar rates were found in other tunnels in this formation.

Additional interesting observations were made during construction. It was, for instance, relatively conclusively shown that suction of water through pores and small cracks had the greatest effect, while joints seemed to play a much smaller role. Grout injection into the joints stopped the flow of water but not the swelling. Also, boreholes in the open invert were completely dry; however, after placement of the invert arch (i.e. once free swelling was inhibited) they filled with water. This is a strong indication that the pores are filled with water and, when taken together with the remark on joint grouting, confirms the existence of suction through pores. How water interacts with the tunnel and swelling is, however, not entirely clear: A 400-m-long drainage tunnel was built between 1998 and 2001, approximately 6 m below the invert, and drainage drill holes were placed between the drainage tunnel and the invert. The intent was that this would drain the zone under the invert and reduce/stop swelling. This did not happen. A hypothesis (Anagnostou et al., 2010) is that draining of water actually led to oversaturation with sulfates and thus led to continued swelling. Another important result is the fact that an invert arch per se does not necessarily stop invert heave, but that the proper combination of curvature and thickness is necessary to provide the required resistance, approaching that of a circular shaped opening.

iii. *Chienberg Tunnel*: This tunnel is also in the Swiss Jura Mountains near the town of Sissach; 1.4 km of a total of roughly 2.3 km is a mined tunnel. The geological profile of the mined section is shown in Figure 6.63. A large portion on the West side is in the "Gypsum-Keuper," the clay-sulfate rock formation already mentioned for the Belchen tunnel. As shown in Figure 6.64, the first 350 m on the western side were excavated first with a so-called crown tunnel (essentially a pilot tunnel) by drilling and blasting. This was joined by a TBM-driven pilot tunnel from the East (Figure 6.64). Following this, the excavation of the entire tunnel was done sequentially as shown in Figures 6.64 and 6.65 (widening of crown section/heading, bench, invert) from West to East. Figure 6.65 shows that quite strict limitations on the distances/time intervals for the sequential steps were imposed. The reason for this was to quickly achieve ring closure (Figure 6.65). The tunnel was designed following the "resistance principle" i.e. constructing a liner

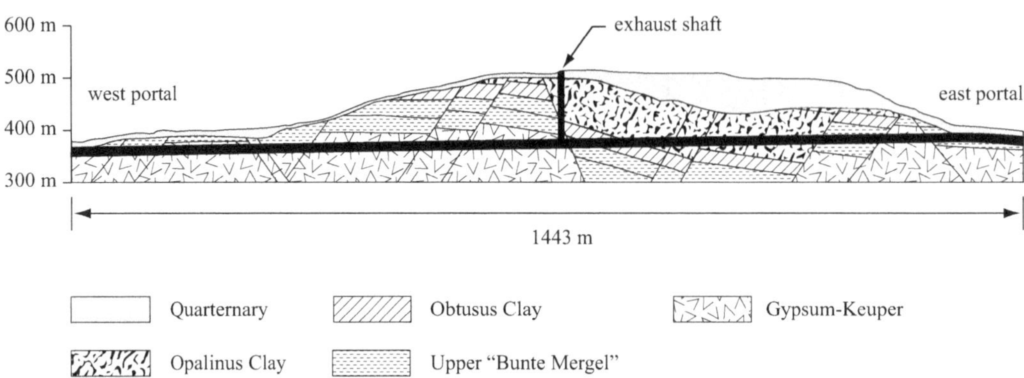

Figure 6.63 Chienbergtunnel – Geological profile of mined section (after Chiaverio, 2002). Reproduced with permission of F. Chiaverio.

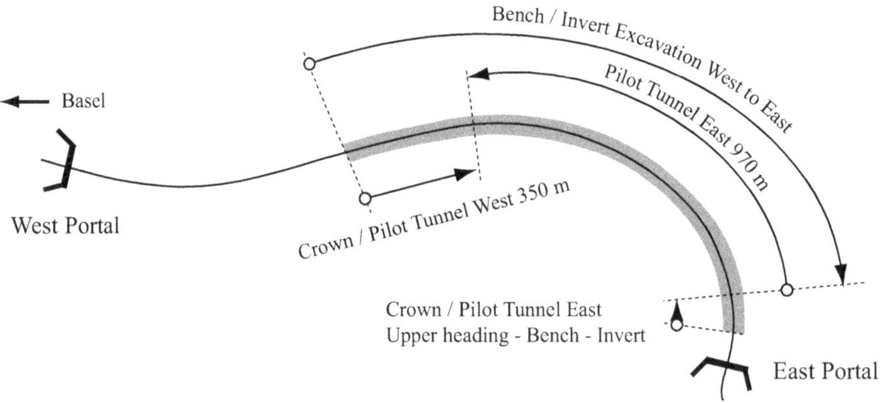

Figure 6.64 Chienbergtunnel – Mined section plan view and excavation schematic (after Chiaverio, 2002). See also Figure 6.65. Reproduced with permission of F. Chiaverio.

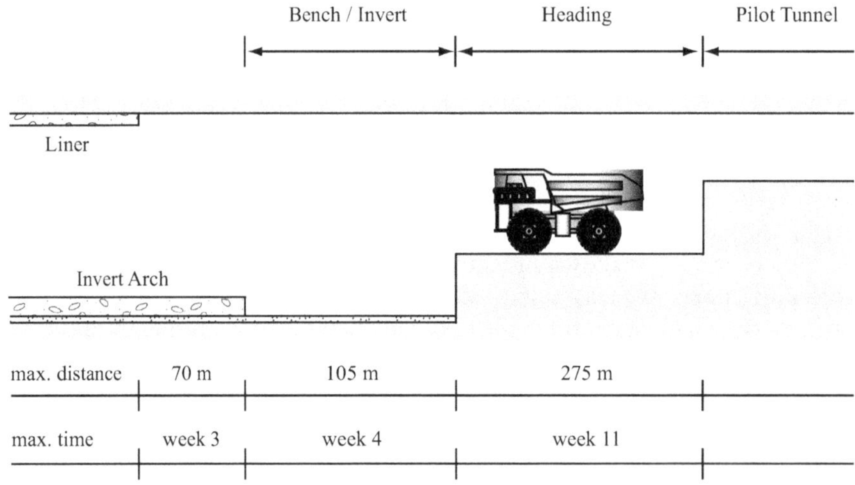

Figure 6.65 Chienbergtunnel – Schematic of excavation of mined section (after Chiaverio, 2002). Reproduced with permission of F. Chiaverio.

that will resist swelling pressure. The cross section with the up to 1.05 m thick liner is shown in Figure 6.66.

Construction of the mined section started in 2000. On February 1, 2002, at Sta. 1100, a daylight collapse occurred as the bench and invert were excavated in this section. This is roughly the location where the original crown tunnel and the TBM pilot tunnel met (Figure 6.64). The collapse is shown in Figures 6.67a, b. Note that the crown section (heading) widening had advanced beyond this collapsed section. The situation was remedied by constructing a bypass tunnel and then excavating the caved-in material from both sides, which lasted till September 2002. Already four weeks after the collapse, very strong heave (1.5 m) of the crown invert, on the east side of the collapsed zone, was observed. This is shown in Figure 6.68.

The excavated and partially completed section on the west side was regularly monitored. This showed invert heaves starting in fall 2002 and values of tunnel heave of 101 mm in March 2004 at Sta. 880 m. Interestingly, there was also a heave of 45 mm at the surface. Note that the overburden is relatively low (~25 m) in this part of the tunnel. Such values are clearly too high to be tolerated. Figure 6.69 shows that there are two different heave sections. Comments about this will be made later in this section.

Initially, the countermeasures considered tiebacks to provide a counterpressure in the order of 1 MPa in the endangered tunnel section. Since the tie-down may lead to overstressing of the tunnel liner in the sidewalls, this was not considered to be a viable permanent solution. Such a solution was found in the use of a so-called deformable or yielding invert as shown in Figure 6.70. The yielding invert consists of yielding elements in the liner which can deform to a certain extent before transmitting a load, thus limiting the forces on the tunnel liner. Deformations of up to 50% of a 1-m-long element were observed in the laboratory before the final load increased. Also, the horizontal

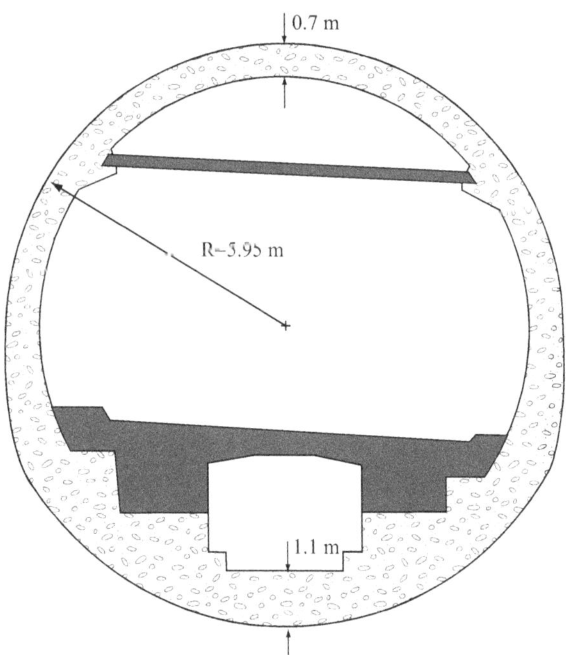

Figure 6.66 Chienbergtunnel – Original cross section of mined tunnel (after Chiaverio and Thut, 2010). Reproduced with permission of F. Chiaverio.

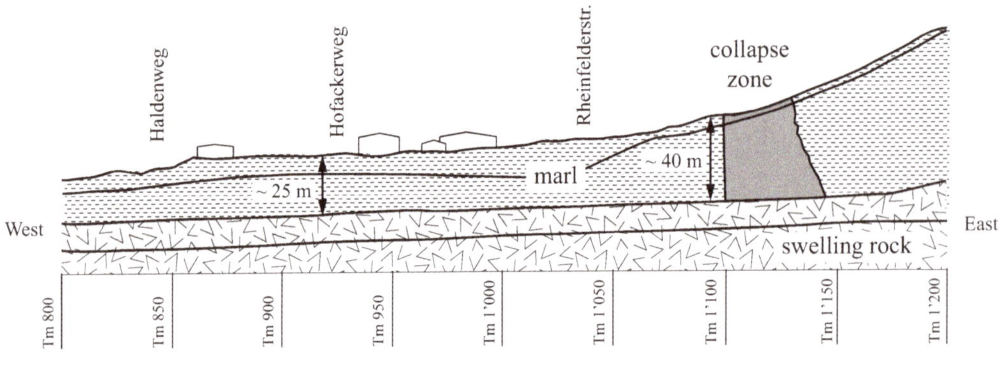

(a)

(b)

Figure 6.67 Chienbergtunnel – Daylight collapse (after Chiaverio, 2002). (a) Vertical cross section. (b) Photograph of the surface (made available by Chiaverio, 2022). Reproduced with permission of F. Chiaverio.

invert (Figure 6.70) is being held in place by vertical tiebacks (anchors), which also have yielding elements in the anchor heads. All this required removal of the lower part (invert arch) of the existing liner and involved extra costs on the order of 35 million sfr. (about $36.5 million).

The case, therefore, shows nicely the application of the two basic design principles: resistance (stiff) support versus deformable (flexible) support. This will be extensively discussed in Section 6.5.5.

Most recent research on the Chienbergtunnel (Butscher et al., 2011a, 2011b, 2011c) made it possible to suggest an explanation for the occurrence of swelling in the two separate zones. This can be done through consideration of regional hydrogeology. Including hydrology in the swelling mechanisms of rock around tunnels was always known to be important. What was found at the Chienbergtunnel extends the scale from local to regional. Figure 6.71 is a detailed longitudinal cross section of the Chienbergtunnel showing all relevant formations as well as the two swelling sections

Figure 6.68 Chienbergtunnel – Heave in top heading (after Chiaverio, 2002). Reproduced with permission of F. Chiaverio.

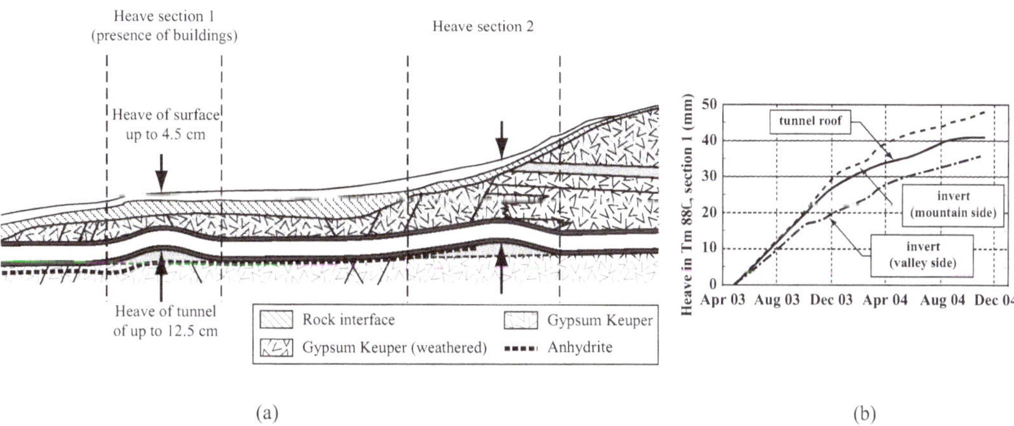

Figure 6.69 Chienbergtunnel – Two sections with heavy swelling phenomena (after Chiaverio and Thut, 2010). (a) Longitudinal cross-section. (b) Time heave diagram at section 1 at Sta. 880. Reproduced with permission of F. Chiaverio.

and the location of four transverse sections (A–D), which were hydraulically modeled (Butscher et al., 2011b, 2011c). What needs to be noted is that the tunnel in the relevant sections is located in the so-called Gypsum-Keuper (GK) just touching or somewhat below the weathered Gypsum-Keuper (GKw) (see Figure 6.72). The Gypsum-Keuper consists of the clay-sulfate rock mentioned earlier. When weathered, GKw is strongly jointed to crumbled and much more permeable than GK. In the upper part of GK, the

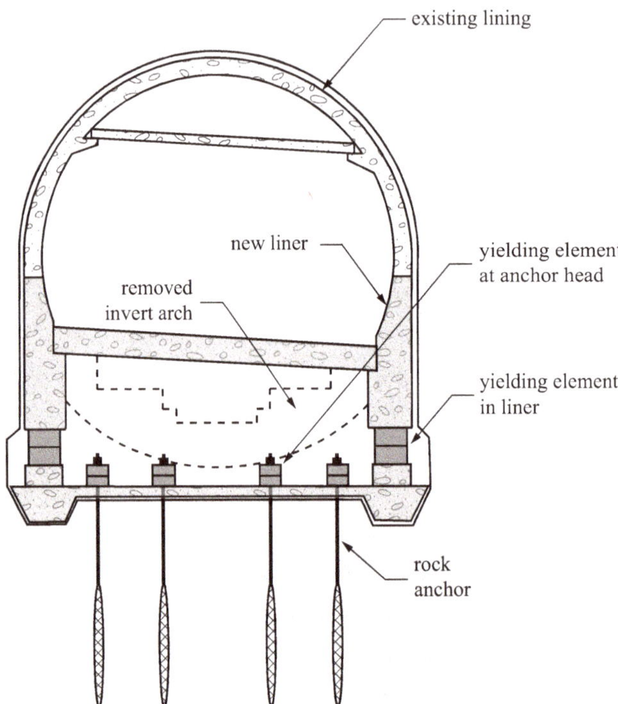

Figure 6.70 Chienbergtunnel – Deformable invert (after Chiaverio and Thut, 2010). Reproduced with permission of F. Chiaverio.

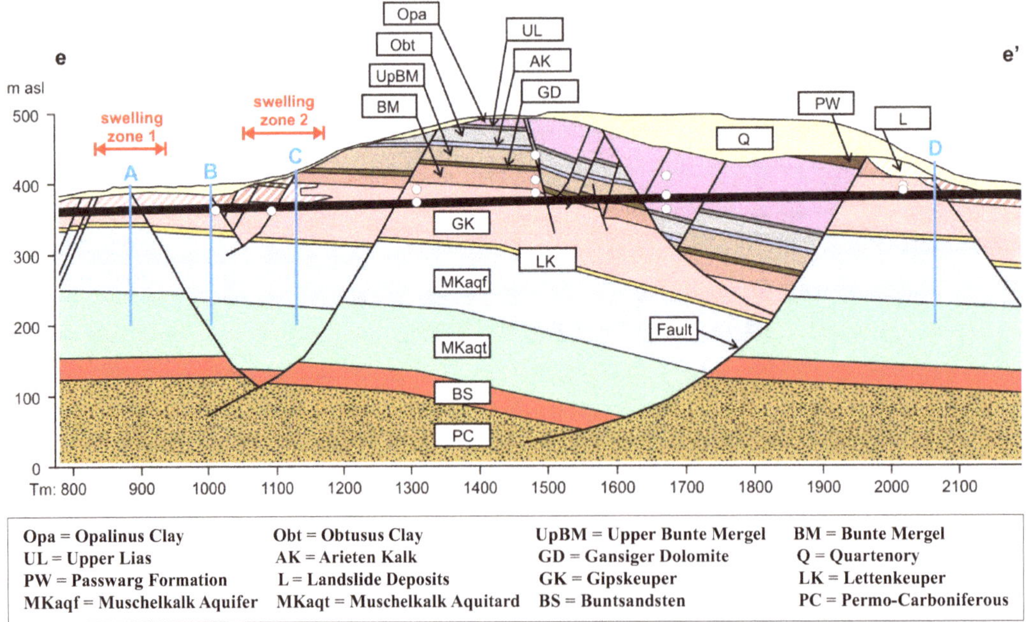

Figure 6.71 Longitudinal cross section. Weathered geological units are shaded. Blue lines indicate position of transverse cross sections and white dots of pore pressure probes (after Butscher et al., 2011b). Reproduced with permission of Elsevier.

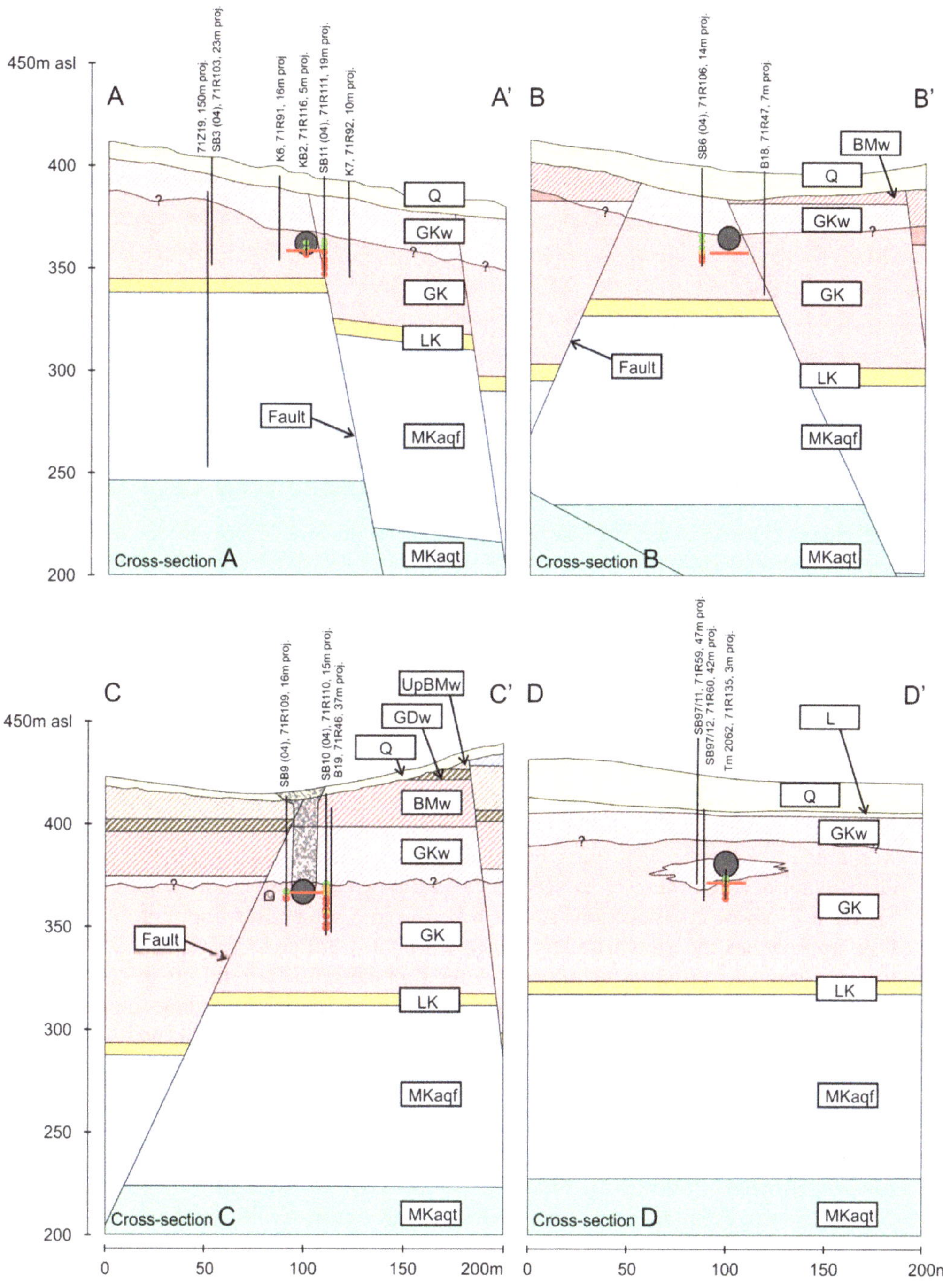

Figure 6.72 Transverse cross sections (abbreviations of geological units are explained in Figure 6.71). Black lines indicate exploration drill holes. Dots indicate sulfate analysis samples (green: sulfate present as gypsum; red: sulfate present as anhydrite; orange: sulfate present as gypsum and anhydrite). Red line indicates anhydrite level. Dotted area in Section C represents a cave-in area. Question marks indicate the uncertain position of the boundary layer Weathered/Unweathered Gipskeuper away from tunnel (after Butscher et al., 2011b). Reproduced with permission of Elsevier.

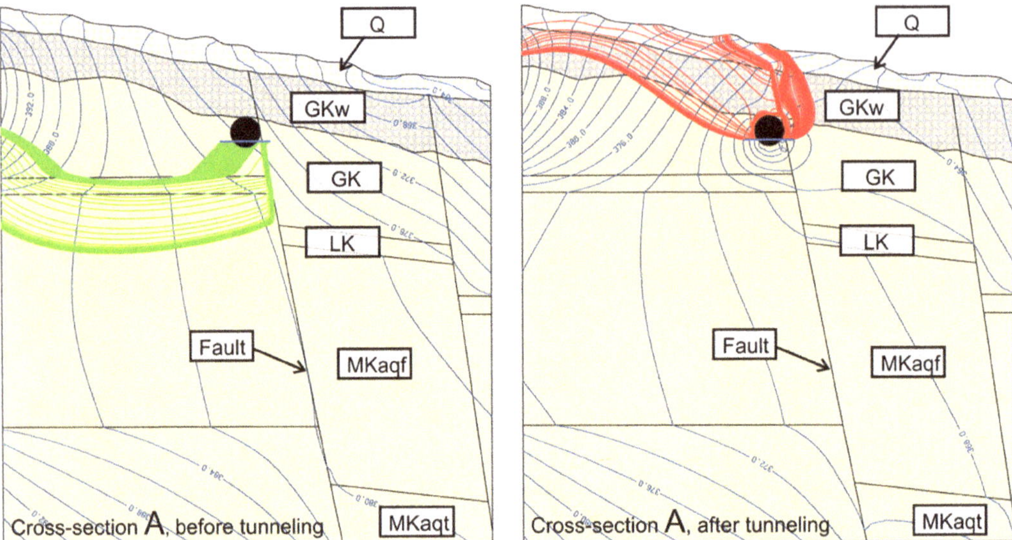

Figure 6.73 Modeled hydraulic head and flow paths toward anhydrite level in cross section A before (left) and after tunneling (right). The Weathered Gipskeuper is indicated in gray with a dotted signature; other weathered bedrock and quaternary colluvium in white with a dotted signature. After tunnel excavation: the anhydrite level is hydraulically connected to the weathered Gipskeuper by the tunnel and the surrounding EDZ (after Butscher et al., 2011b). Reproduced with permission of Elsevier.

sulfates are gypsum; in the lower part, they are anhydrite. Figure 6.72 shows the transverse cross sections A–D. The boundary between the gypsum and the anhydrite, the so-called anhydrite level, is located within the tunnel in Sections A and C and below the tunnel in sections B and D. This is already a strong indication of why swelling occurs in A and C and not in B and D. However, as discussed before, anhydrite needs access to water in order to swell. Butscher et al. (2011b) developed two-dimensional finite element models for these cross sections to study groundwater flow. Figures 6.73 and Figure 6.74 show cross sections A and B before and after tunneling. Before tunneling, flow approaches the anhydrite level from below in both cross sections, i.e. the flow is through low permeability formations. After tunneling, the conditions in cross section B are still the same. However, in cross section A, the tunnel, including the excavation damaged zone (EDZ) around it, introduces a radical change in groundwater flow from the weathered Gipskeuper, which is permeable, to the anhydrite level. Detailed calculations of flow rate in Butscher et al. (2011c) show that the flow in sections A and C leads to magnitudes of swell heave that correspond well with what has been observed in the field. The study in Butscher et al. (2011c) also shows that the conductivity at the tunnel perimeter ("outflow"), which can be affected by waterstops (sealing), and the conductivity of the EDZ have a major effect on groundwater flow and thus on swelling.

To summarize what has been observed in these cases:

i. Swelling in tunnels in argillaceous rock and clay-sulfate rocks occurs mostly in the invert but can, in the case of clay sulfates, also lead to a "lifting" of the entire tunnel and possibly also to heave on the overlying surface.
ii. Access of water is essential. It appears that most of the water involved in swelling comes from the rock mass. The regional and local hydrogeology, specifically permeable and

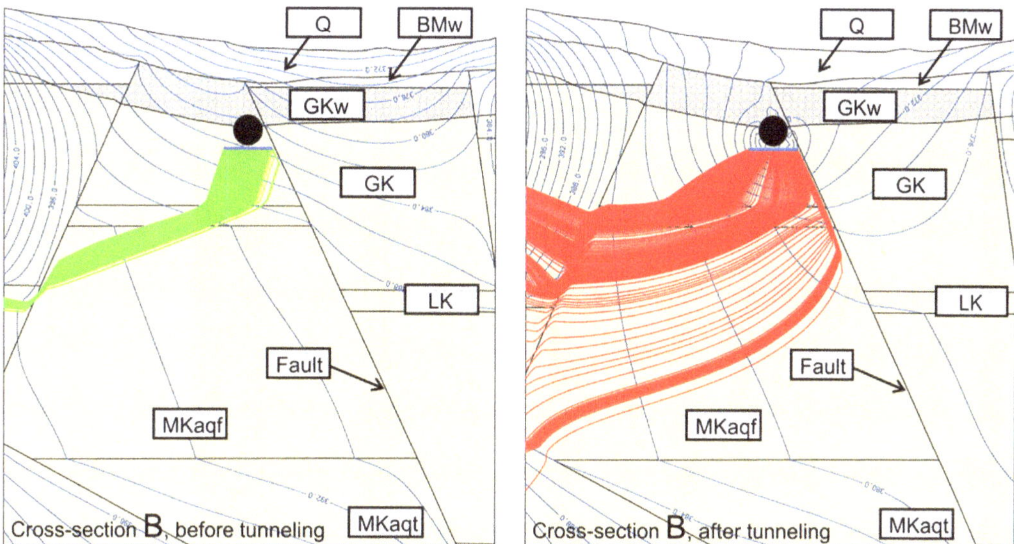

Figure 6.74 Modeled hydraulic head and flow paths toward anhydrite level in cross section B before (left) and after tunneling (right). The Weathered Gipskeuper is indicated in gray with a dotted signature; other weathered bedrock and quaternary colluvium in white with a dotted signature. Groundwater flow approached the anhydrite level from before and after tunneling (from Butscher et al., 2011b). Reproduced with permission of Elsevier.

less permeable formations play an important role. However, the hydraulic conditions of the tunnel (sealed, not sealed) and of the excavation damaged zone around it significantly affect groundwater flow. Tunnel construction water that exists if the tunnel is not well drained can produce additional swelling.

iii. Swell heave in argillaceous rock can reach up to a total of 400 mm and in sulfate/clay sulfate rock, up to 4,000 mm. The larger values occur when swell heave is not restricted by the tunnel liner.

iv. Swell pressure observed behind rigid liners (and below rigid invert arches) can reach values of up to 0.5 MPa for Opalinus rock and of several MPa for clay sulfate rock.

v. The zone under the invert affected by swelling can be in the order of 3–5 m for argillaceous rock and 6–10 m for clay sulfate rock. Note that the affected depths depend on the size of the opening where larger openings dimensions produce deeper affected zones.

From this and from the comments on the physicochemical mechanisms made in Section 6.5.2, one can make some general statements about the swelling rock in tunnels.

Water, which is the essential ingredient in argillaceous rock-, sulfate rock-, and clay sulfate rock–swelling, flows into the critical zone mostly under the invert. This is caused by the fact that the creation of an opening will often lead to a lowering of the octahedral (volumetric) stresses under the invert and above the crown under the usual in situ stress conditions (see Chapter 1). A reduction of mean stresses leads to negative excess pore pressures and thus suction of pore water into the distressed zones. Equally important is the fact that with a permeable liner (or no liner) the pore pressures at the tunnel perimeter are zero, which introduces a gradient. Considering the pore water gradient alone, there will be more groundwater (GW) flow toward the invert than toward the rest of the tunnel because of head differences (Figure 6.75).

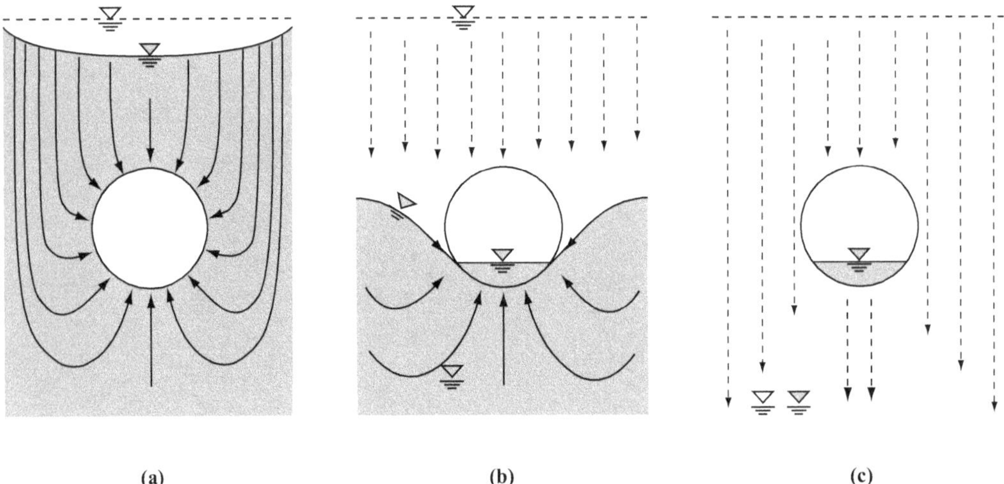

Figure 6.75 Groundwater – Tunnel water flow (after Hauber et al., 2005). (a) Original GW level above tunnel - Low permeability. (b) Original GW level above tunnel - High permeability. (c) original GW level below tunnel

While Figure 6.75 is conceptual, real conditions are more complex, as discussed in the Chienberg case (Figures 6.72 to 6.74) and the combination of regional and local geohydrological conditions together with the effect of the tunnel will govern groundwater flow and thus affect swelling.

In addition to water from the rock mass, so called tunnel water may also play a role. This is water that flows from the rock mass into the tunnel and then possibly back into the rock mass. In order for this "tunnel parallel" flow to happen, conditions such as those shown in Figure 6.72c (permeable zone around tunnel) have to exist. Tunnel water is an issue as water may penetrate the tunnel from non-swelling rock and flow back into the rock mass in rock subject to swelling. It also can play a role in transporting dissolved sulfates over distance and precipitating them at different locations.

To summarize, one can state that stress changes due to tunnel excavation and hydraulic gradients lead water into the space near tunnel zones. Some evidence exists that the invert is favored, but at least conceptually the crown, and, depending on the stress conditions, the sidewalls could be also affected.

The stress changes around a tunnel require some further investigation and comments. Figure 6.76 shows the stress paths for two elements, one on the invert and one in the springline of a tunnel. The excavation of the tunnel, essentially an undrained unloading or loading, produces negative excess pore pressures. The corresponding stress states are I' and S' in Figure 6.76. The negative excess pore pressure will induce suction, i.e. pore water flow and swelling follows. Note that in the geotechnical testing framework the I-specimen undergoes extension, while the S-specimen undergoes compression. What is now of great importance is the stress path during swelling, i.e. from I' to I and from S' to S. The I'-I stress path touches or comes close to the failure envelope (depending on the stress state, this may also happen for the S' to S path). This means shear failure can occur together with swelling. This behavior was postulated by Bellwald (1991) and was actually observed in tests on shale by Sun Jun et al. (1984). This means that the rock at a particular location near the tunnel perimeter undergoes both swelling and shear failure (squeezing). Even if no outright failure occurs, the stress path near the failure surface may lead to time-dependent shear deformation (creep). All this was investigated by Aristorenas (1992) through an extensive testing series on Opalinus Clay Shales. The schematic behavior shown in Figure 6.76 was observed, specifically dilation

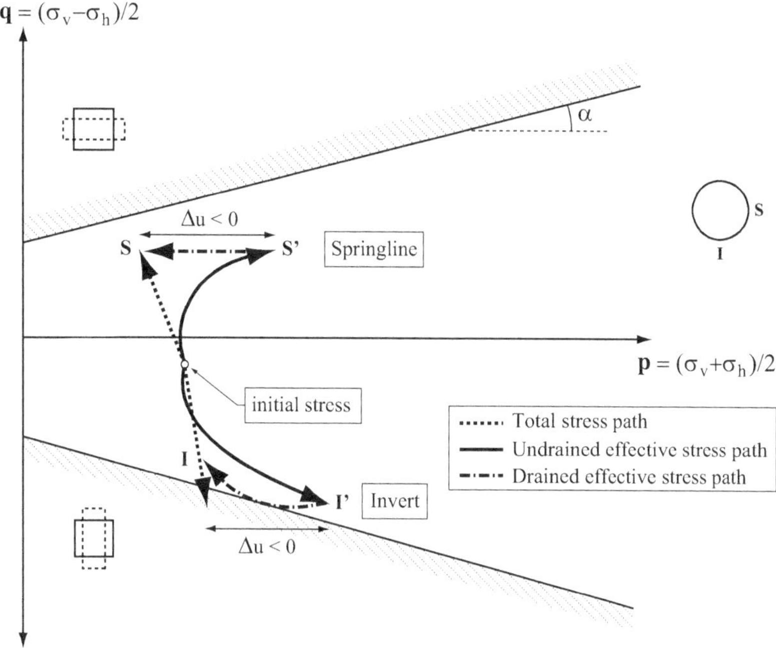

Figure 6.76 Combined swelling and creep after Bellwald (1991) and Einstein (2000). S, S': Total and Effective Stress State in Springline. I, I': Total and Effective Stress State in Invert.

leading to negative excess pore pressure, then suction and then swelling. Specific to Opalinus Clay Shale were the following observations:

i. It is anisotropic and dilatant i.e. shearing occurs under isotropic loading and volume changes occur under pure shear.
ii. Both shear and volumetric creep occur. After consolidation, no time-dependent volume reduction should occur. However, if shearing-induced volume changes through dilation occur, volumetric creep is possible.
iii. The creep rate increases with shear stress levels.

While specific to Opalinus Clay Shale, it is likely that similar behavior occurs in other argillaceous rocks since the depositional regime is usually similar, i.e. producing anisotropic materials. As will be seen in Section 6.5.5, the results of Aristorenas' (1992) tests were incorporated in an analytical model.

Regarding sulfate or clay sulfate rocks, it is necessary, similar to what was done for argillaceous rocks, to look at two scales: the tunnel scale and the somewhat smaller scale of the rock. Figures 6.77a and b present the anhydrite, gypsum, leached rock (from bottom to top) sequence that was observed in tunnels (and mentioned briefly in the Belchen and Chienberg cases). Essentially, the figures show the so-called Gypsification Front (F) where the hydration of anhydrite into gypsum starts, the anhydrite level (A) above which only gypsum exists, and the gypsum level (G) above which no solid sulfates exist (see also Figure 6.62). The difference between Figures 6.77a and b is that Figure 6.77a shows the process involving tunnel water penetration into the rock, while Figure 6.77b shows the process when rock mass water flows toward the tunnel. Although the movement of water and dissolved sulfates is different, the end result is the same. From what was explained earlier, the major swelling takes place between F and A, while above G it is actually possible that volume loss might occur. In

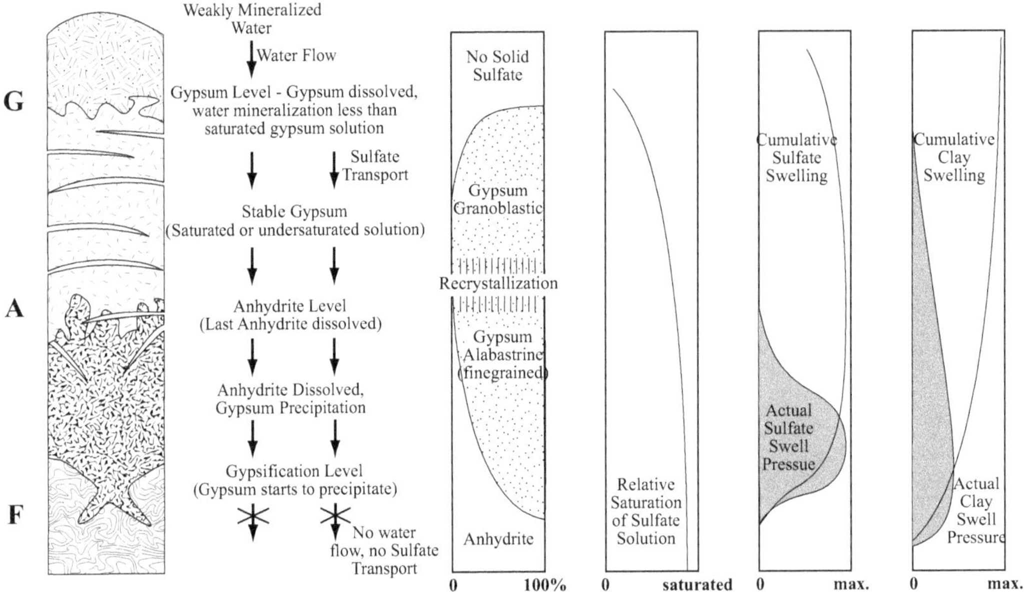

Figure 6.77a Gypsification caused by tunnel water (after Hauber et al., 2005).

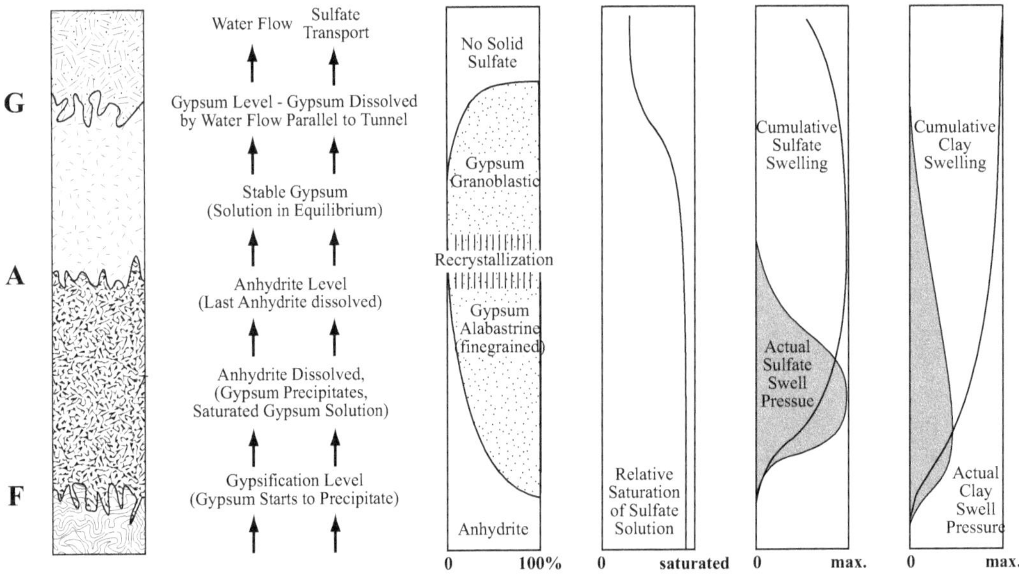

Figure 6.77b Gypsification caused by water flowing toward tunnel (after Hauber et al., 2005).

essence, this large scale process takes also place in clay sulfate rocks, with swelling affected by the relative proportion of anhydrite (sulfates) and clay, as discussed in Section 6.5.2 in which also the most recent theoretical developments in the context (e.g. Serafeimidis and Anagnostou, 2012, 2013, 2014a, 2014b; Wanninger, 2020) were mentioned.

Very interesting is what might happen on the smaller scale in clay sulfate rocks and how clay might contribute to anhydrite swelling and to the combined shearing - swelling mechanism

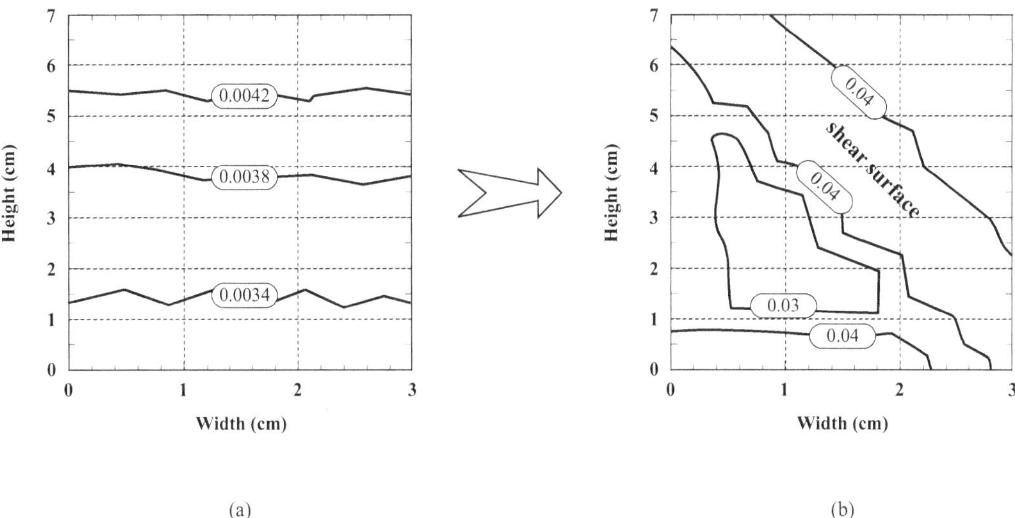

Figure 6.78 Tenfold increase of gypsum content near shear surface (after Einstein and Mayer, 1999). Numbers Indicate Gypsum Content. (a) Specimen before shear testing. (b) Specimen after shear testing. Note increase in gypsum content.

mentioned in the context of clayey rocks. Experiments at MIT (Figure 6.78) on artificially prepared specimens consisting of 15% bentonite and 85% anhydrite show an enrichment of gypsum in the sheared zone after they have been subjected to a triaxial testing. This can be explained by the stress path model of Figure 6.76. Shearing during the triaxial test causes dilation with water flow (suction) toward the dilated zone. This water dissolves the anhydrite and re-precipitates as gypsum. While so far only shown in a few experiments (Chen, 1996) this process offers, nevertheless, a sensible explanation of anhydrite-gypsum swelling in clay-sulfate and possibly in pure sulfate rocks. It also indicates that a similar combination of swelling and squeezing, as is observed in clayey rocks, might occur in sulfate rocks.

6.5.5 Analysis and design of tunnels in swelling rock

As shown in the preceding sections, much work has been done to develop testing methods for swelling rock and to understand what goes on as swelling occurs in and around tunnels. As has also become clear, much is still not satisfactorily known, particularly regarding swelling of sulfate and clay sulfate rock. What is presented here are, therefore, some basic principles, a short review of few existing analytical approaches and a summary of design/construction concepts.

6.5.5.1 Basic principles

Figure 6.79 shows the characteristic curves as they were already shown in Chapter 1. As was mentioned earlier and shown here again, there are two basically different principles: (1) Use a support that can deform and limit the support load at the cost of additional deformation (it has to be pointed out that the structural deformation is practically limited); or (2) Have a rigid structure (similarly, the structure may not be able to sustain stresses beyond a limit). All of this will be discussed at the end of this section when discussing design concepts and details.

The discussion here follows the write-up in ISRM Commission on Swelling Rock (1994). The discussion here is a summary of what is described in that publication with some additional comments reflecting more recent work. As usual, there are besides analytical approaches, also

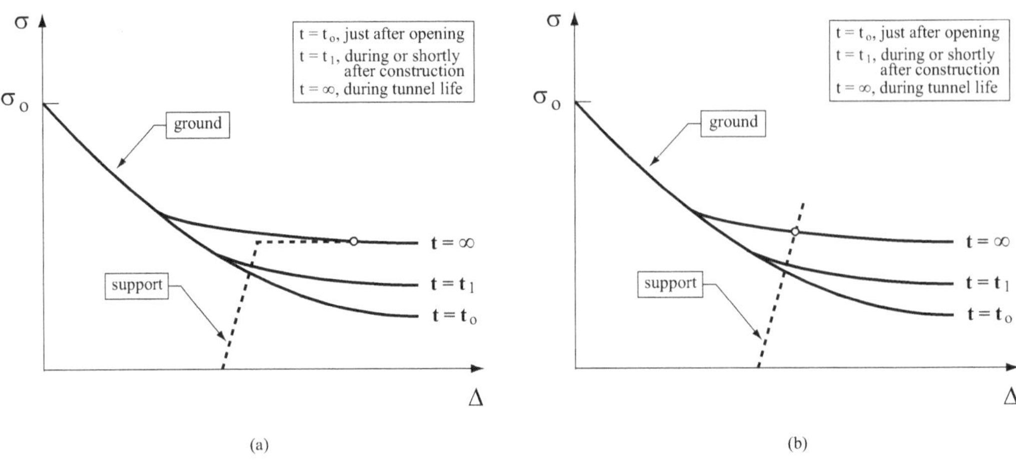

Figure 6.79 Tunnels in swelling ground with different support types. (a) Swelling rock and deformable support. (b) Swelling rock and rigid support.

empirical relationships. These have been mentioned earlier in this book (e.g. Terzaghi's classification and Barton's Q Method, both include swelling rock, see Chapter 2, as well as the approaches by Lambe [1960] and Brekke and Howard [1973] mentioned earlier in this section). The set of analytical methods can be characterized as: (1) Time-independent methods; (2) Rheological models; (3) Derived rheological models; and (4) Mechanistic models.

"Time-independent" or "swelling law" methods are based on odometer swelling tests and, in their simplified form, e.g. Grob (1972), propose relations between swelling stress and swelling strain assuming that the odometer test is a model for what happens at the tunnel invert. The concept and details as proposed by Grob are shown in Figure 6.80. Note also that the swell curve in Figure 6.80 is similar, but not identical, to what is obtained in the ISRM test shown in Figures 6.54 and 6.55.

At this point, it is very important to mention that testing of clay-sulfate rock in an odometer setup does often not produce the curve shown in Figure 6.80 (this curve is often referred to as the semi-logarithmic curve or semi-logarithmic law) but swelling strains that are largely independent of swelling stresses (see Pimentel and Anagnostou, 2013). Newer work (Wanninger, 2020) further addresses the applicability of the semi-logarithmic law. In the context of this book this has been discussed with regard to the cases in Section 6.5.4 and will be discussed in 6.5.5.2 (design), and indicates that much is not known regarding the basic behavior and, particularly, the engineering implications of clay-sulfate rock swelling.

Simultaneously, Einstein et al. (1972) and also later Grob's essentially one-dimensional approach was extended to three dimensions, in which the first stress invariant of total stresses was related to volumetric swelling (see also Wittke and Rissler, 1976; Gysel, 1977, 1987; Schwesig and Duddeck, 1985, Frölich, 1986). One has to be aware of the fact that all these approaches provide good first estimates but are limited by: (1) Assumption of linear elasticity; (2) Only the strain state at final equilibrium is calculated; (3) If the initial stress state is isotropic, no swelling is predicted; and (4) Generalization of the odometer conditions to reality. The latter limitation can be eliminated by conducting swelling tests in triaxial equipment, as was done by Pregl et al. (1980), whose results were then used by Kiehl (1990) in an extended swelling law. Kiehl's approach in a way represents a transition between the swelling law and mechanistic models.

"Rheological models" use different combinations of the three basic elements: (1) Hooke-Spring element; (2) Newton-Dashpot element; and (3) St. Venant Slider-element. These

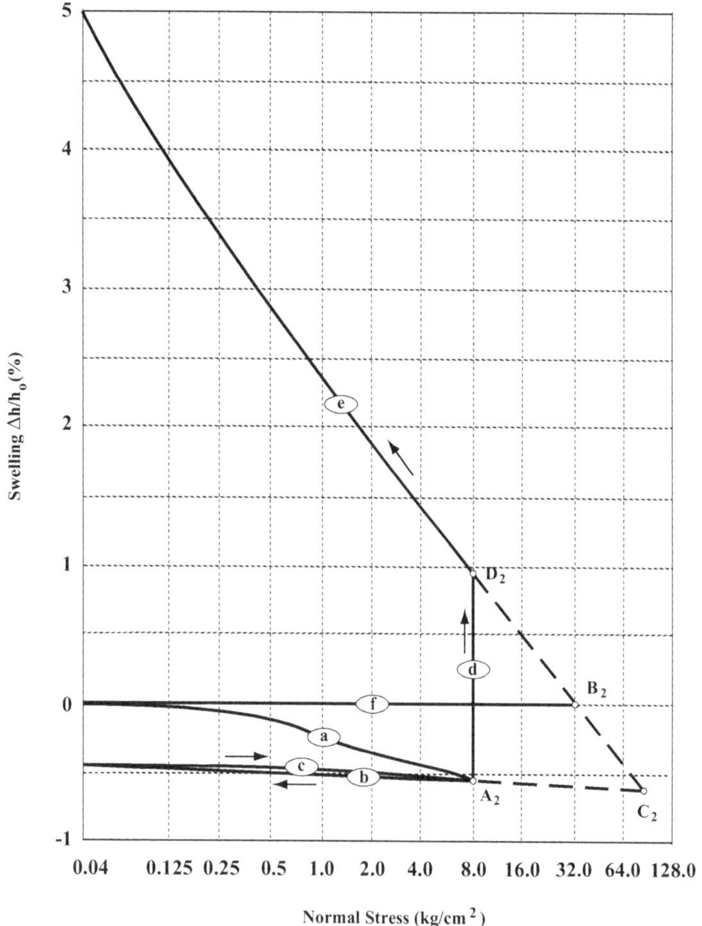

Figure 6.80 Swelling of marl in an oedometer test under variable load (after Huder/Amberg; from Grob, 1972). a-b-c: Load-unload-reload cycle. Water under (no water added) natural conditions; d: Addition of water, constant load; e: Load reduction with simultaneous swell; f: Zero swell level.

rheological models represent a phenomenological approach to model the time-dependent swelling and creep processes. The reason for choosing this approach is the complexity of the underlying mechanisms. They are scale independent and are stress-shear-time models with which a variety of creep/swell behavior can be represented. Many of the early rheologic models could also handle the deviatoric (creep) but not the volumetric (swelling) component. Later models expanded on this: The reader is referred to ISRM (1994) for a more detailed discussion.

"*Derived rheological models*" are obtained from curve fitting to results from in situ observation or laboratory tests; see e.g. Semple et al. (1973) and Sulem et al. (1987).

"*Mechanistic models*" attempt to represent the details of the mechanics underlying time-dependent behavior of rocks and soils. Note that the following (indented) text is a verbatim duplication of what is said in ISRM (1994).

> The models presented so far (swelling laws, rheological and derived rheological models) model swelling largely without specifically considering the essential element of swelling: water. The mechanistic models directly incorporate the effect of water and thus provide a more rational approach. The mechanical response of a fluid-saturated porous

material is characterized by deformation-diffusion processes, specifically consolidation when there is a volume decrease or swelling for a volume increase. Note that only mechanical swelling, as opposed to physicochemical swelling, is considered in argillaceous rock. Mechanical swelling can be modeled by a diffusion law and obeys only the laws of mechanics. Physicochemical swelling, which is due to electro-chemical interactions between the water and the minerals, is not considered here.

Mechanistic models for consolidation or swelling can be uncoupled or coupled. The uncoupled models were first developed by Terzaghi (1923) in one dimension and extended to three dimensions by Rendulic (1935). In these models, the solutions are obtained by solving the solid and matrix deformation independently of the (hydraulic) diffusion. In coupled models, however, deformation and diffusion are solved interactively. The simplest theory, which accounts for such coupled processes, is Biot's (1941, 1956) poroelasticity theory. It differs from the uncoupled theories in the following respects: (1) A mechanism for the generation of pore water pressure (the pore water pressure generation mechanism can be related to Skempton's (1954) B-coefficient); (2) The effective stress governing the deformation of the porous solid is characterized by $\sigma'_{ij} = \sigma_{ij} - \alpha u \delta_{ij}$ where σ'_{ij} is the effective stress, σ_{ij} is the total stress, α is the Biot coefficient ($0 < \alpha < 1$; Terzaghi and Rendulic implicitly assume $\alpha = 1$ in their theories), u is the pore pressure and δ_{ij} is the Kronecker delta; (3) The diffusion of the pore water pressure is coupled to the rate of the volumetric strain. Coupled models have been applied to the tunneling problem by Carter and Booker (1982), Detournay and Cheng (1988) for elastic conditions and Carter (1988) for elasto-plastic conditions.

A further step in the use of coupled models is made by Anagnostou (1991). He interprets the time-dependent development of swelling strain as a consequence of the dissipation of negative pore pressures. In addition, the flow of water within the rock mass is taken into consideration. This and the modeling of swelling rock as an anisotropic non-linear-elastic, perfectly plastic material, produces realistic predictions of swelling strain, specifically in the form of invert heave in tunnels.

The newest development described here is based on what has been presented in Figure 6.76 and which is partly repeated in Figure 6.81, namely, that two types of time-dependent behavior, volume increase, and shearing (creep) can occur simultaneously. Otherwise expressed and as proven through extensive testing on Opalinus clay shale by Aristorenas (1992) one can state: (1) Shearing occurs under isotropic loading and volume change occurs under pure shear; (2) Dilation related to isotropic volume increase or related to the above-mentioned shear-related dilation induces negative pore pressures and suction of water into the increased pore volume and is time-dependent, i.e. swelling occurs; and (3) Shear and volumetric creep occur, and both are linearly related to time.

While these rigorous observations are related to a variety of Opalinus Clay Shales, Bellwald (1991) observed similar behavior in other clayshales. It is a reasonable, but not fully substantiated, assumption that the characteristic phenomena mentioned do occur with most anisotropic clayshales. Aristorenas accordingly developed an analytical model for (Opalinus) clay shale, which is now described (for details, see Aristorenas, 1992; Bobet et al., 1999).

The model has nine parameters to describe deformability (5 parameters), strength (2 parameters), permeability (1 parameter), and creep (1 parameter) characteristics.

i. *Deformation parameters*: The compressibility C_{Be} is the slope of the $\sigma'_{oct} - \varepsilon_{vol}$ (effective octahedral stress-volume change) curve measured in isotropic consolidation tests. The "shear stress-shear strain" behavior measured in triaxial tests can be expressed in the hyperbolic form:

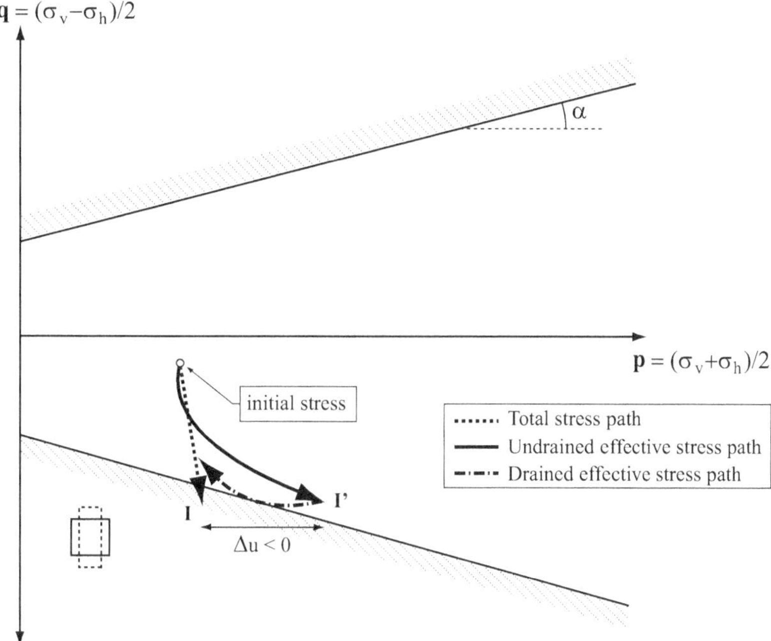

Figure 6.81 Stress path during unloading and swelling of a point of the tunnel invert for a circular tunnel with initial stresses $\sigma_h/\sigma_v = 1.5$.

$$\frac{q}{\sigma'_{co}} = \frac{\gamma}{a + b\gamma} \tag{6.77}$$

Equation (6.77) follows the concept developed by Ladd and Foott (1974), where $q = \frac{1}{2}(\sigma_1 - \sigma_3)$, $\gamma = |\varepsilon_1 - \varepsilon_3|$, and σ'_{co} is the effective octahedral consolidation stress. Hence, parameters a and b characterize the shear deformation, with "a" defining the compliance and "b" the non-linearity. The type of triaxial test (undrained extension or compression, or drained compression; see Einstein, 2000) has to be chosen to represent the appropriate stress path of the ground element being considered.

In an anisotropic material such as shale, the shear strains and volumetric strains are coupled. Specifically, in the pure shear experiments conducted on Opalinus Clayshale, shear-induced volume changes express themselves as positive (caused by contraction) or negative (caused by dilation) excess pore pressures in undrained tests. In drained tests, one obtains, $q/\sigma'_{co} - \varepsilon_{vol}$ curves, which are expressed as:

$$\varepsilon_{vol} = m_1\left(\frac{q}{\sigma'_{co}}\right) + m_2\left(\frac{q}{\sigma'_{co}}\right)^2 \tag{6.78}$$

where the first and second terms, respectively, describe the linear and non-linear anisotropic behavior.

ii. *Strength parameters*: These are simply the cohesion c and friction angle ϕ, determined in triaxial tests.
iii. *Permeability*: In essence, permeability can be determined from consolidation tests or from the consolidation phase prior to triaxial shear testing. It is also possible – and

this was done in the experiments by Aristorenas (1992) on Opalinus Clayshale – to derive permeability in drained triaxial testing in which shear stress/effective octahedral stress increments were applied and the volume change over time was measured after each increment; however, determining permeability with this procedure is difficult if the increment is shear-dominated.

iv. *Creep*: Information on creep, the time-dependent deformation under constant effective stresses, can be obtained by measuring the time-dependent deformation after the end of primary consolidation in drained tests. The volumetric and shear creep increase linearly with time and the slopes of these curves, the parameters $m_\varepsilon = de_{vol}/d$ (log t) and $m_\gamma = d\gamma/d$ (log t) are the volumetric and shear creep parameters. The slopes increase with the magnitude of the applied normalized shear stress q/σ'_{oct}. This confirms what has been said in conjunction with Figures 6.76 and 6.81, namely, that creep increases with the proximity to failure. Interestingly, in the tests in Opalinus Clayshale, m_ε and m_γ are linearly related to each other (Aristorenas, 1992). The dependence or independence of shear – and volumetric creep is a major open question in soil and rock mechanics and cannot be covered here. For the purposes of the clayshale model, the linear relationship between m_ε and m_γ allows one to work with only one of the creep parameters.

The clayshale analytical model was incorporated in a finite element model and used in several investigations (Aristorenas, 1992; Einstein et al., 1995a, 1995b; Bobet et al., 1999). It was possible to model different tunnel sizes and shapes as well as drained and undrained conditions. The short- and long-term behavior could be clearly shown (see above mentioned literature. See also Section 4.3.6 and Figures 4.37 to 4.42). At the present time, the Aristorenas clayshale model still represents the state-of-the-art and should/can be used. Work on more sophisticated models has been performed (Pei and Einstein, 2011; Pei et al., 2018) but requires additional confirmation through testing.

Regarding analytical methods for sulfate containing rocks, this is an area where much is unknown or known to a limited extent only. In principle, one can again use swelling laws, rheologic models, and mechanistic models. For the reasons just mentioned, really representative mechanistic models do not yet exist. Work on coupled hydro-chemo-mechanical models on such rocks is underway at ETH (see e.g. Anagnostou et al., 2010 and later).

The preceding discussion on analytical approaches assumed that water in sufficient amounts is available. If this is indeed the case, it needs to be checked with hydrologic analyses. This is best done with numerical models that can represent the different conductivities of geologic formations, of the Excavation Damage Zone (EDZ) around the tunnel and considers if the tunnel perimeter is sealed or not. The reader is referred to the article by Butscher et al. (2011c) in which such a modeling procedure is explained in detail.

6.5.5.2 Design of tunnels in swelling rock

This section is again strongly based on ISRM (1994) but includes newer developments as they were already discussed in the description of case studies (Section 6.5.4) and as provided in the special session at the Salzburg Geomechanik Kolloquium in 2010.

In essence, a structure in or on swelling rock, as any other structure, must be built such that it remains safe and serviceable throughout its lifetime. In swelling rock, this means that stresses and deformations in the structure and as much as possible also in the geological environment have to remain below material dependent limits (the latter is often not possible).

As illustrated, in principle, in Figure 6.79, the basic idea underlying any design in swelling rock is to either accommodate swelling or prevent it. Swelling can be accommodated by passive (flexible or deformable) design, which, in essence, allows all or most of the swelling displacements to take place without distress of the structure. Active (rigid, also called "resistance principle") design prevents or limits swelling and will lead to a buildup of stresses acting on the structure. Since swelling is a consequence of a change in the original stress state and water inflow, it can be reduced by counterstresses and/or by limiting access to water. With increasing frequency, one finds also design approaches that are intermediate between active (rigid) and passive (flexible).

The following descriptions illustrate the three design concepts with some practical solutions, the latter including also design details. The solutions represent only a sampling of many possible approaches.

i. *Passive design*: In the most extreme passive design, the rock is allowed to swell freely and is regularly removed (shaved off) such that the facility continues to be useable. Examples are road tunnels in which the rock surface is left uncovered and where the heaved rock is regularly removed. Many old railroad tunnels in swelling rock unintentionally followed this design concept in that the ballast was placed on the unprotected invert, which heaved and had to be regularly removed.

   Analogous to this concept, but practically much more satisfactory, is the one in which a void is left between the rock surface and an internal rigid structure (Figure 6.82). This solution requires careful design of the rigid part of the structure such that it can support the swelling stresses and does not undergo excessive inward movement of the abutments.

   Probably the most important aspect of passive design is the shaping of the excavated opening. As was mentioned in the preceding section, swelling is a result of the change from the original in situ state of stress and of water inflow. Swelling can be prevented or significantly reduced if the original stress state, under proper consideration of the effects of anisotropy, can be maintained or reproduced. Passive designs attempt to achieve this through proper shaping of the opening but without the action of an artificial structure.

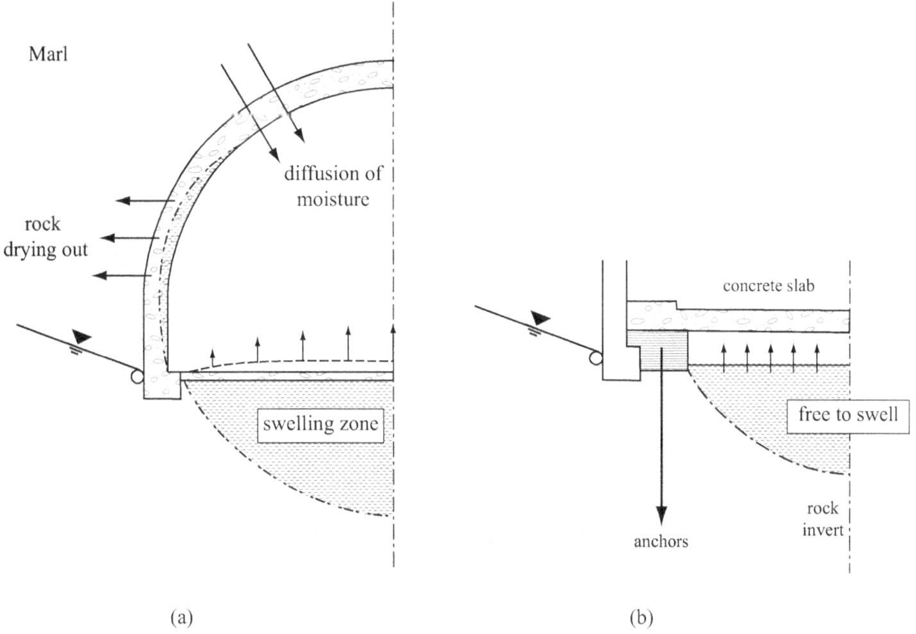

Figure 6.82 Schematic illustration of void between swelling rock and structure (after Lombardi, 1984). (a) During Construction. (b) Final Situation.

Circular and other curved tunnel surfaces with relatively small radii can produce a stress state after excavation, which is close to the original one. The curved shape also provides constraints if the near surface rock expands and swells. This will produce counterstresses, even without an artificial tunnel support, which limit swelling provided that the yield limit of the rock mass has not been reached.

Examples of passive design are: (1) No rock support – Felsenau gypsum mines, Switzerland; (2) Void between rock and support – Seelisberg Tunnel and Neuchâtel West Tunnel, Switzerland.

ii. *Active design*: Active design involves artificial means to reduce water inflow, artificial application of counterstresses or combinations, thereof. Drainage (removal) of water flowing into the tunnel is one of the most effective measures against swelling. Channeling water into lined drainage channels or pipes reduces the contact of most of the exposed rock surface with water. If these drainage facilities have to serve long-term purposes, their upkeep is important and may also involve considerable cost. If swelling cannot be completely prevented with drainage, the drainage facilities may be damaged. This will then, due to contact of large amounts of water with the rock at the location of the damage, accelerate swelling and further destruction.

Sealing of all exposed surfaces is also a very effective measure to reduce flow of water from the tunnel into the rock. This can also relate to moisture in the tunnel air. The air in tunnels, particularly before they are holed through, is usually highly saturated and stress relieved rock often provides the necessary pressure gradient to attract some of the water carried in the air. Temperature differences between rock and air may increase these effects, which can be prevented by sealing. Effective sealing requires fast application of low permeability seals that are compatible with the other support and lining methods used in the particular case.

Most water flows to the stress relieved zones from the interior of the rock masses. Particularly in material with small pore sizes, the suction effect is very significant. Very importantly the outflow of groundwater into the tunnel and/or into the EDZ moves significant quantities of water into zones that can swell. Both the suction and outflow effect can be controlled by making the rock mass less permeable. However, it is practically impossible to stop this flow by artificially blocking small pores. The situation is different if the water supply through larger openings like joints and permeable fault material is considered. In such cases, grouting can provide a barrier against further water inflow.

The application of so-called chemical inhibitors has not seen much application in tunneling but is common in the oil industry where additives in the drilling slurry of oil production wells cause a change in electrolyte content, which in turn reduces swelling.

Rigid structures, which produce significant counterstresses as the rock starts to swell such as thick walled, curved liners, represent he best-known active design (Figure 6.83). The curved (usually circular) shape not only increases the rigidity of the structure but, through the corresponding shape of the rock surface, it also leads to the production of natural counterstresses as discussed earlier. Another possibility involves bolting or anchoring where the swell displacement mobilizes the bolt resistance and/or where counterstresses are applied through prestressing (Figure 6.84). While Figure 6.84 shows an actual case in slightly swelling rock where the number and lengths of bolts were sufficient, this may not be the case in other rocks. Bolt number and length may be such as to make this solution uneconomical.

Examples of active design are the earlier mentioned Belchen Tunnel (with limited success!) and the Bözberg Highway Tunnel (Figure 6.83c) in Switzerland.

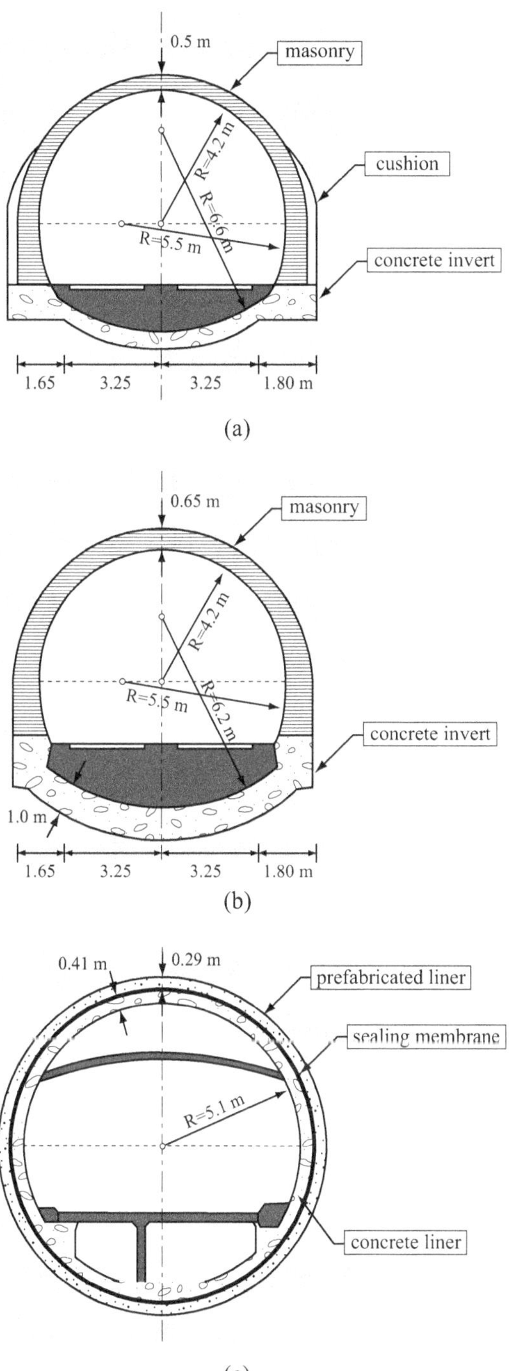

Figure 6.83 Examples of tunnels with invert arch. (a) Hauenstein tunnel, Switzerland. Original design by Berger (Steiner 2020). (b) Hauenstein tunnel, Switzerland. Tunnel section without lateral cushions and modified later (Steiner 2020). (c) Circular lining design. Bözberg Road Tunnel, Switzerland.

(iii) *Intermediate design*: Frangible backpacking between liner and rock allows some deformation to take place until it is completely compressed upon which the structure starts to act (Figure 6.85). In concept, analogous but different in detail, is the use of a

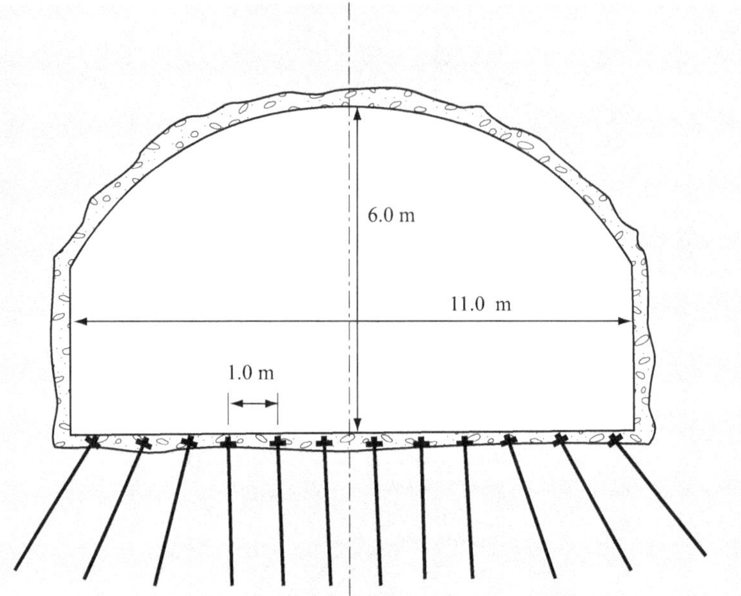

Figure 6.84 Storage cavern in Switzerland. Example of bolted invert.

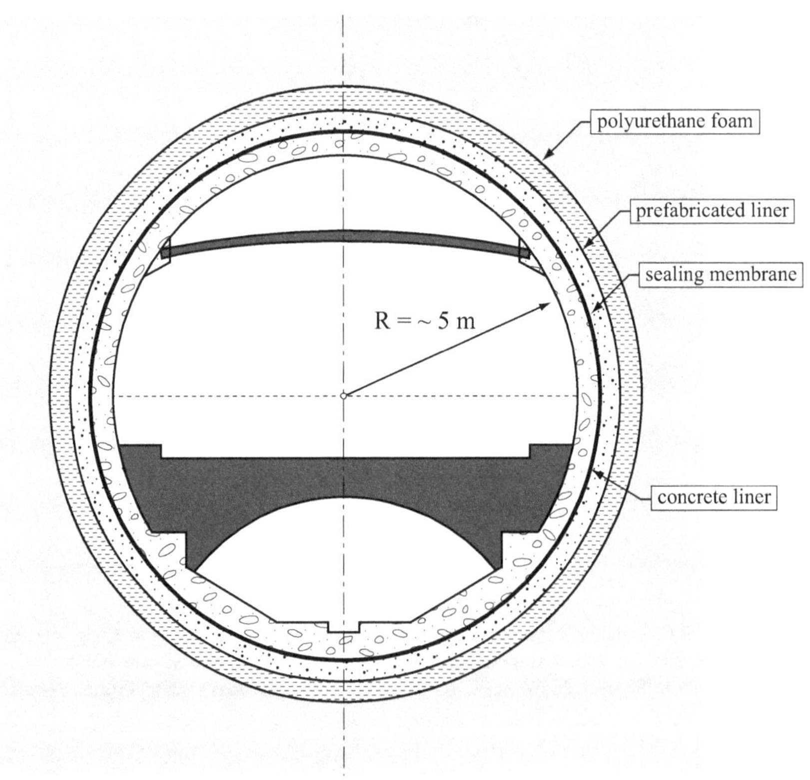

Figure 6.85 Proposed tunnel support with frangible backpacking (after Lombardi, 1981).

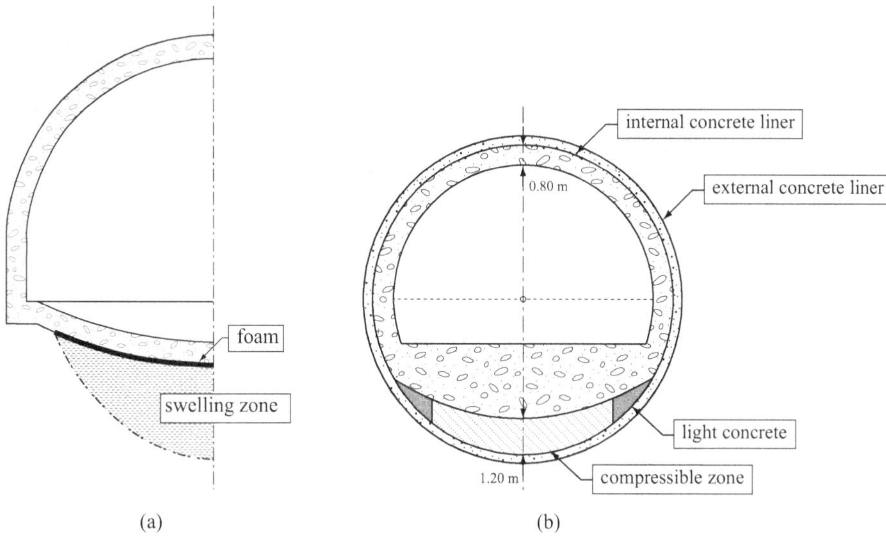

Figure 6.86 Compressible layer between swelling rock and structure. (a) Buechberg Tunnel, Switzerland (after Lombardi, 1984). (b) Freudenstein Tunnel, Germany (after Kirschke, 2010).

compressible support layer between the swelling rock and the structure as used in the Buechberg Tunnel (Figure 6.86a) or the use of a compressible zone between the external and inner lining in the case of the Freudenstein Tunnel (Figure 6.86b). Alternatives are compression elements between prefabricated liners (Figure 6.87). The so far most complex and sophisticated intermediate design is that for the Chienberg tunnel, which was described in detail in Section 6.5.4.2, (see also Chiaverio and Thut, 2010) where an "empty zone" under the tunnel invert, consisting of deformable tendons in the invert and

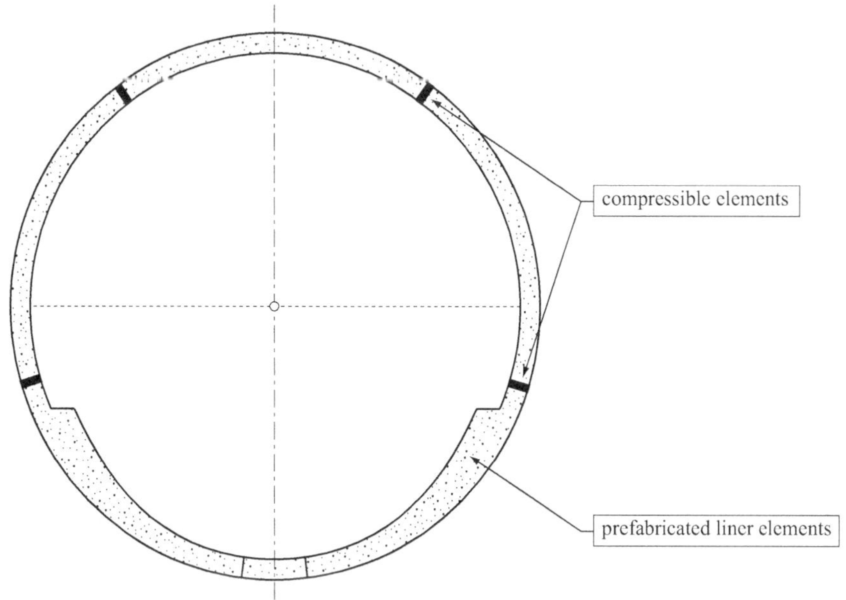

Figure 6.87 Tunnel support with compressible elements between rigid prefabricated liner elements.

compressible elements in the liner were used. One has to be aware of the fact that such intermediate designs implicitly assume that there is a limit to swell displacement, which is the case for most argillaceous rocks but may not be so for many sulfate-argillaceous rocks, which may swell to an extent as to make intermediate designs inapplicable.

Construction is an integral part of the design concept. Passive, active, or intermediate designs have thus to be realized, i.e. constructed such that they achieve the intended purpose. Specifically, passive or intermediate structures have to be built such as to allow the swell displacements to take place while active structures should be rigid from the start. In any case, it is very important to prevent access of construction water to exposed rock surfaces independent of the design concept. The rock should, therefore, be sealed immediately upon excavation.

Rock characteristics cannot be established with certainty; hence, accurate analytical predictions of performance are thus not possible and the design may not necessarily be adequate. Observation of the encountered conditions during construction and particularly observation of displacements of the rock and of the structure as well as of the stresses in the structure indicate to what extent assumptions and associated predictions are correct. If the performance is not as predicted, rock characteristics can be back-calculated and additional tests can be run. Most importantly, it is on this basis that other design alternatives can be chosen. This is particularly important for the intermediate design concepts, which may be sensitive to differences in swelling in that slightly larger than anticipated deformations can produce unacceptable loads in the rigid part of the structure. It is important to emphasize that simply being able to monitor is insufficient. Following the philosophy of the observational method (see Section 1.4), a feedback cycle is required with easy and quick adaptation of design and construction in order to be successful.

6.5.6 Final comments on tunneling in swelling ground

The preceding sections went into quite a bit of detail about swelling behavior, its causes, and how to test, analyze, and design. As has become clear, what has been presented is largely based on the Swiss experience including work at ETH-Zurich (formerly Madsen and coworkers, then Kovari and now Anagnostou and coworkers), as well as the work by the second author of this book. The Swiss work also includes applications by authors such as Hauber, Steiner, and Chiaverio and their coworkers. This emphasis on the Swiss work is acceptable because the problems and their solutions are widely representative.

Nevertheless, we would like to suggest to the readers that they also consult work by the groups around Alonso in Spain (e.g. Alonso et al., 2013; Alonso and Olivella, 2008; Alonso and Ramon, 2013), Sulem in France (e.g. Mohajerani et al., 2011; Braun et al., 2021) and Wittke (e.g. Wittke and Rissler, 1976; Wittke, 1990; Wittke et al., 2017) and Kirschke (2010) in Germany.

Chapter 7

Analysis of structural components

7.1 INTRODUCTION

The tunnel liner is a structural system that provides support to the ground, limits the inflow of water, supports appurtenances, and provides the final finished surface of the tunnel (FHWA, 2009). The support of the tunnel, since it is a structural element, must be designed to perform adequately during the life of the facility. Specifically, the support has to withstand all the loads that the structure can experience. These include (ITA, 2000):

Loads that must always be considered: ground pressure; water pressure; dead load; surcharge; subgrade reaction.
Loads that, if necessary, should be considered: loads from inside; loads during construction; earthquake.
Special loads: effects of adjacent tunnels; effects of settlement; other loads.

Other loads may include (DAUB, 2001; FHWA, 2009; BTS and ICE, 2004) temperature, shrinkage, creep, chemical attack, fire and explosions; see Table 7.1.

Clearly, the structural elements of the tunnel will be subjected to load combinations that have different probabilities of occurrence, and thus calculations using adequate load combinations must be conducted to check that all the elements will perform adequately. For road tunnels, FHWA (2009) suggests using the LRFD (Load and Resistance Factor Design) and AASHTO specifications. Table 7.1 (Table 10-1 in FHWA, 2009) contains the load factors, loads acting on the tunnel and the load combinations that should be considered for the design of the tunnel liner.

In addition, the structure must perform within the limits of serviceability, which include deformations, settlements, limited cracking and water inflow, and any other performance-based conditions.

ITA (2000) suggests to perform calculations on the cross sections for the following critical sections: (1) section with the deepest overburden; (2) section with the shallowest overburden; (3) sections with the highest groundwater table; (4) section with the lowest groundwater table; (5) section with the largest surcharge; (6) section with eccentric loads; (7) section with unleveled surface; and (8) section with adjacent tunnel at present or planned one in the future. These recommendations are for shield tunnels, but seem to be applicable to other general cases. They are not exclusive, as there may be other critical sections, depending on the project considered, e.g. tunnels in urban areas close to existing facilities, tunnels crossing a fault, pressure tunnels, etc.

Tunnel linings (the terms "lining" and "liner" are used interchangeably in this book) are used as initial support, permanent support, or a combination. The lining can be built as a one-pass or as a two-pass (multi-pass) system. A one-pass system, as the name indicates,

DOI: 10.1201/9781003328940-7

Table 7.1 Load Factors and Load Combination Table for Road Tunnels (from FHWA, 2009)

Load Combination/ Limit State	*DC*		*DW*		*EH[1] EV[2]*		*ES*		*LL, IM, LS, CT, PL*	*WA*	*TU, CR, SH*		*TG*
	Max.	*Min*	*Max.*	*Min*	*Max.*	*Min*	*Max.*	*Min*			*Max.*	*Min*	
Strength I	1.25	0.90	1.50	0.65	1.35	0.90	1.50	0.75	1.75	1.00	1.20	0.50	0.00
Strength II	1.25	0.90	1.50	0.65	1.35	0.90	1.50	0.75	1.35	1.00	1.20	0.50	0.00
Strength III	1.25	0.90	1.50	0.65	1.35	0.90	1.50	0.75	0.00	1.00	1.20	0.50	0.00
Service I	1.00		1.00		1.00		1.00		1.00	1.00	1.20	1.00	0.50
Service IV	1.00		1.00		1.00		1.00		0.00	1.00	1.20	1.00	1.00
Extreme Event I	1.25	0.90	1.50	0.65	1.35	0.90	1.50	0.75	EQ[3]	1.00	N/A	N/A	N/A

1 The load factors shown are for at-rest earth pressure. At-rest pressure should be used for all conditions of design of cut and cover tunnel structures.

2 The load factors shown are for rigid frames. All cut and cover structures are considered rigid frames.

3 This load factor is determined on a project specific basis (see Section 6.2). When developing the loads to be applied to the structure, each possible combination of load factors should be developed.

Load definitions:

DC: Dead Load. It includes the self weight of the structural components and the loads associated with nonstructural attachments

DW: Dead Load. This is the self weight of wearing surfaces and utilities

EH: Horizontal Earth Pressure Load

ES: Earth Surcharge Load. This is the vertical earth load due to fill over the structure that was placed above the original ground line. It is recommended that a minimum 400 psf (19.2 kPa) load be used for the design of tunnels. If there is the potential for future development adjacent to the tunnel structure, the actual surcharge should be used, or a minimum of 1,000 psf (47.9 kPa)

EV: Vertical Earth Pressure

CR: Creep

CT: Vehicular Collision Force

EQ: Earthquake. This load should be applied to the tunnel lining as appropriate for the seismic zone of the tunnel. Other extreme events such as explosive blast should be considered.

IM: Vehicle Dynamic Load Allowance. The load is applied to the roadway slabs. It can also be transmitted to a tunnel lining through the ground surface when the tunnel is under a highway, railroad, or runway.

LL: Vehicular Live Load. The load is applied to the roadway slabs. It can also be transmitted to a tunnel lining through the ground surface when the tunnel is under a highway, railroad, or runway.

LS: Live Load Surcharge. The load is applied to the lining of tunnels that are constructed under other roadways, rail lines, or other facilities that carry moving vehicles.

PL: Pedestrian Live Load.

SH: Shrinkage.

TG: Temperature Gradient. This is due to local and seasonal changes of temperature across the lining.

TU: Uniform Temperature. The load is used primarily to design expansion joints in the structure.

WA: Water Load. The load represents the hydraulic pressure outside the tunnel structure.

consists of a liner that is used as the initial and final support. These include shotcrete and rockbolts, steel sets, segmental lining, and, in some cases, cast in place concrete. A two-pass system consists of two linings, placed at different times. The second, final lining is typically, but not necessarily, cast in place concrete. It is generally placed much later than the initial support, and thus most of the ground deformations have taken place. If the tunnel is below the groundwater table, the initial lining may not be fully impermeable, in which case, the final liner may need to be designed considering the effectiveness of the drainage if the tunnel is watertight. Typically, only normal stresses may be transmitted between the two liners. If a full connection between the liners is desired, i.e. such that both normal and shear stresses are transmitted, a special design is required, which may need placement of steel reinforcement or dowels crossing the contact between the two liners. Very importantly, and as indicated above, the final lining may serve as the structure carrying the entire water pressure if a waterstop is

Table 7.2 Typical support systems for initial support (after FHWA, 2009)

Ground	*Rockbolts*	*Rockbolts with wire mesh*	*Rockbolts with shotcrete*	*Steel ribs and lattice girder with shotcrete*	*Cast-in-place concrete*	*Concrete segments*
Strong Rock						
Medium Rock						
Soft Rock						
Soil						

placed between the two liners. It is also important to note that there is some controversy if in a two-pass system the initial liner should be considered as a load bearing component in the final structure or not.

The design of tunnel supports is an iterative process that has to consider a number of factors such as performance, cost, availability of materials, and local expertise (FHWA, 2009). Thus, there are no clear or quantitative guidelines to decide on the optimum support. Table 7.2 lists typical support systems used for the initial support (after FHWA, 2009):

The chapter provides a description of the following types of support: Cast-in-place concrete and Segmental liners (Section 7.2); Steel Sets (Section 7.3); Rock Reinforcement Systems (Section 7.4); Shotcrete (Section 7.5); and Composite Liners (Section 7.6). These are, arguably, the most commonly used types of tunnel support. The objective of the chapter is to provide an overview of the types of support, their advantages and disadvantages, construction issues, as well as guidelines for their structural analysis. All this is supported by data taken from a number of codes and standards, with the objective of providing the reader with information about expected or typical properties or performance of the materials, construction procedures, etc. The intent is not to give all design details, an impossible task since each country has its own codes and regulations, which are updated on a regular basis and must be followed (some, if not most, of the codes or regulations included in the chapter may have been replaced or updated at the time of publication of the book and so the reader is cautioned to take the data provided as informative, and is encouraged to consult the current codes and regulations for the most recent information). Rather, the goal is to discuss the working principles for each type of support and the types of analyses that must be performed. Finally, Section 7.7 includes a short overview of water handling (waterstop and drainage).

7.2 CAST-IN-PLACE CONCRETE AND SEGMENTAL LINERS

The loads that cast-in-place concrete and segmental liners will experience due to the ground and/or water can be obtained by any of the methods discussed, namely empirical (Chapters 2 and 3), analytical (Chapter 4), or numerical (Chapter 5). One of the key aspects of all liners is their stiffness. Proper consideration of the relative stiffness of the support and the ground is key to determine the loading of the support, as discussed in Chapters 1 and 4. Cast-in-place concrete produces a monolithic structure and thus its moment of inertia can be computed using the cross section of the liner. Segmental liners can also be considered as a monolithic structure if the elements are bolted together to provide complete moment and shear load transmission. Non-bolted segmental liners produce a support that is not continuous, as it is made by the assembly of pieces or segments, connected to one-another through joints. Modeling of the liner requires proper consideration of the rotational and shear stiffness of the joints. Muir Wood (1975) provides a first approximation of the moment of inertia of a liner with joints, provided that there are more than four joints.

$$I_e = I_j + I\left(\frac{4}{n}\right)^2 \text{ for } I_e < I, n > 4 \tag{7.1}$$

where I_e is the effective moment of inertia, i.e. used for calculations, I_j is the moment of inertia of the joint (conservatively taken as zero), I is the moment of inertia of the gross cross section of the liner, and n is the number of joints in the liner. Note that as the number of joints increases, the moment of inertia of the liner decreases, thus its stiffness decreases. In other words, the liner becomes more flexible in bending but remains quite stiff in compression. Note also that this is the concept that Peck used to provide recommendations for the stress distribution on the liner (see Section 1.3 and Figure 1.11; further discussion can be found in Section 4.2.1).

The characteristics of each type of liner, and construction method associated with it are discussed in the following sections.

7.2.1 Cast-in-place concrete

Cast-in-place concrete liners are typically used in two-pass liner systems and as the second and final liner. They may also be used as a single liner in good-quality ground where initial support is not needed, but there are cases where a cast-in-place concrete liner is also used in disadvantageous ground conditions (see Section 2.3.6.5). The advantages of this type of support are (FHWA, 2009): suitable for use with any excavation and initial ground support method; corrects irregularities in the excavation; can be constructed to any shape; provides a regular sound foundation for tunnel finishes; and provides a durable, low maintenance structure.

Second pass cast-in-place concrete liners are usually placed much later than the initial support and thus the ground loads they have to withstand are usually different than those taken by the initial support. If the tunnel is waterproofed, they may need to be designed to resist the full hydrostatic load. Figure 7.1 shows a sketch of a two-pass liner with cast-in-place concrete with a waterproofing system. Waterproofing will be discussed and defined in Section 7.7.

The cast-in-place concrete liner may be unreinforced or reinforced. Steel reinforcement is not often done due to the high construction costs associated with the placement of the reinforcement and concrete, as well as the increased time for construction. Because of this, when reinforcement is needed, the use of steel fibers in the concrete is a practical solution that has been used in the past several years.

Two different types of formwork are used: (1) Continuous casting. In this case, the formwork is generally 10 to 20 m long and continually advances as the concrete is poured (the length depends on the angle at which the fresh concrete slopes, the pumping rate and the hardening rate of the concrete). Construction joints may have to be included, however. (2) Sectioned casting. Here, the concrete is placed using a formwork in pour lengths of about 8 to 12 m. Each block is connected to the next by a construction joint. The joint has to be able to allow the deformations of each block due to shrinkage of the concrete while curing and hardening, as well as due to temperature fluctuations during the life of the tunnel. If the liner is impermeable, special joints will need to be placed to keep the liner watertight.

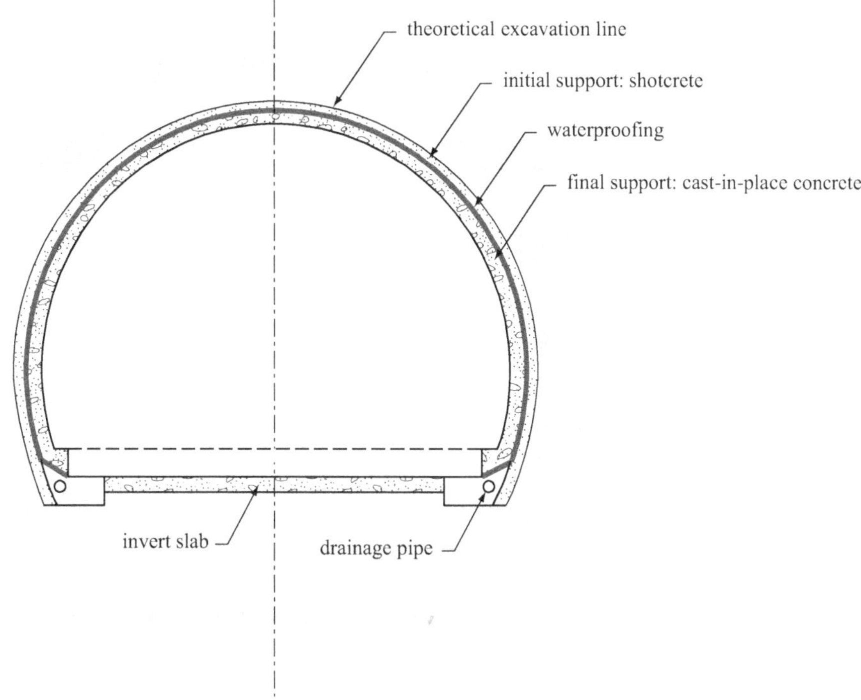

Figure 7.1 Two-pass liner system with cast-in-place concrete for final support.

The concrete used needs to satisfy a number of requirements. It needs to be fluid enough such that it can be pumped behind the form and fill the gaps between the form and the ground or initial liner, while it should have enough early strength to facilitate stripping as soon as possible. In addition, it has to have enough resistance to cracking, as well as resistance to any chemical attacks. A slump of 5″ (12.7 cm) is recommended (FHWA, 2009), which is accomplished with the addition of plasticizers and/or fly ash, as well as entrained air (3–5%). Typical cement content used is of the order of 280 to 320 kg/m^3, with 28 day compressive strengths in the range of 25–30 MPa. As mentioned, the concrete is placed by pumping and it is vibrated, most commonly with vibrators attached to the form. Usually, the concrete is poured in a form covering the entire cross section. In very large openings, the formwork and pours are subdivided, for instance, one for each flank, and one for the crown. This creates construction joints that need to be treated carefully. The most problematic location is the crown, where a void may form between the concrete and the ground/initial liner. Grouting is normally done at the crown, once the concrete has gained its full strength, at about 28 days after pouring.

7.2.2 Segmental liners

Segmental liners are used as one-pass or two-pass liners, even though the most common application is as one-pass, final liner. They are mostly used in circular tunnels excavated with a shield and are assembled under the protection of the shield to form a complete ring. There are several types of segmental liners: precast reinforced concrete, cast iron, and steel elements as well as steel plates. Today, by far, most commonly used as support are precast segmental liners. Steel plate liners are also described in this section for completeness.

Table 7.3 Distortion for circular tunnels in soft ground (BTS and ICE, 2004)

Soil type	*Distortion* $\Delta r_o/r_o$ *(%)*
Stiff to hard clays, overconsolidation ratio OCR > 2.5–3.0	0.15–0.40
Soft clays or silts, overconsolidation ratio OCR < 2.5–3.0	0.25–0.75
Dense or cohesive sands, most residual soils	0.05–0.25
Loose sands	0.10–0.30

The BTS and ICE (2004) suggest that segmental linings can be represented as confined flexible rings, and recommend a maximum acceptable distortion of the liner of 2% of the difference between the maximum and minimum diameters of the lining. They also suggest using the distortion values in Table 7.3, in terms of the change in radius over the radius of the tunnel, $\Delta r_0/r_0$, for design, to check bending moments in the liner and to assess performance during construction.

7.2.2.1 *Precast segmental liners*

Precast segmental liners are made of reinforced concrete. Segments of unreinforced concrete may be used for tunnels with small cross section and low quality requirements (DAUB, 2001). They can also be post-tensioned, but there are only few cases (e.g. sewage tunnels in Osaka, Japan, with outside diameter of 3.55 m, as reported by Nishikawa, 2003).

Individual segments are produced in factories outside the tunnel, either existing ones, or exclusively setup for the job, if the volume warrants the initial investment. This means that the quality of construction can be very high, with very tight tolerances. The segments are transported into the tunnel and assembled under the tail of the shield or tunnel boring machine. Figure 7.2 is an idealization of a one-pass segmental liner. The construction process follows the excavation and support cycle discussed in Section 3.3, which is illustrated in Figure 3.3. After a round of excavation, equal to the length of a ring (Figure 7.2), the hydraulic jacks are retracted, the shield is advanced, and the ring is constructed by placing the segments, usually with an erector at the end of the shield. The number of segments per ring depends on the designer and on the diameter of the tunnel. There are two different design philosophies: (1) Few long elements – usually one on the invert, two each in the lower and upper sidewalls and one "key" in the crown (the "key" can also be at the bottom, but then two invert elements are needed.); (2) Short elements – usually 2 to 3 m long. The segments can be straight or with skewed joints, with dimensions often between 20 and 30 cm thick and 1.0 to 1.5 m long (along the direction of the tunnel, i.e. the length of the ring). Construction of the ring creates two types of joints: ring or circumferential joints and radial joints (Figure 7.2). The segments are sometimes rotated from ring to ring, particularly for short elements, so the radial joints are not aligned. The curvature of the tunnel is accommodated using tapered segments and so, by rotating the elements, changes in vertical or horizontal alignment can be accomplished.

If the segmental liner constitutes the final support of the tunnel, gaskets are placed around the segments to prevent water leakage (Figure 7.3a). Neoprene or other materials are embedded into a groove around the perimeter of the segment which, when properly compressed, makes the joint watertight (Figure 7.3b). Radial compression is obtained by bolting the segments in the circumferential direction and or by pushing on the keyblock. Bolts may also be placed in the longitudinal direction, but usually the force to advance the shield or TBM is enough to provide sufficient compression. If large water pressures exist, a two-line gasket may be required: one can be made by a hydrophilic material. The joints between segments

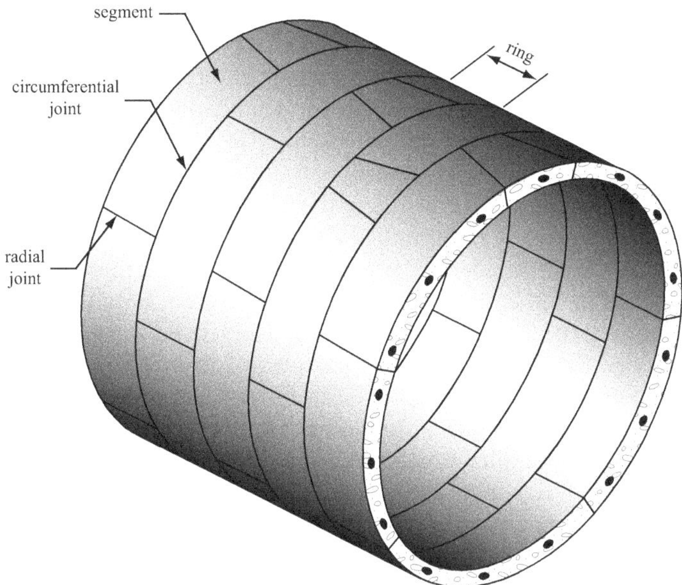

Figure 7.2 Segmental liner.

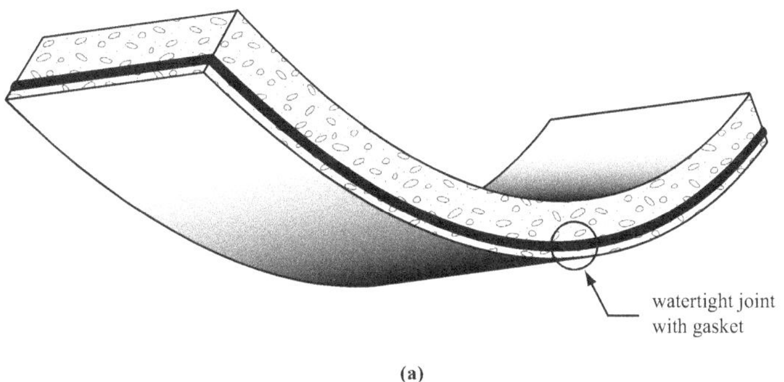

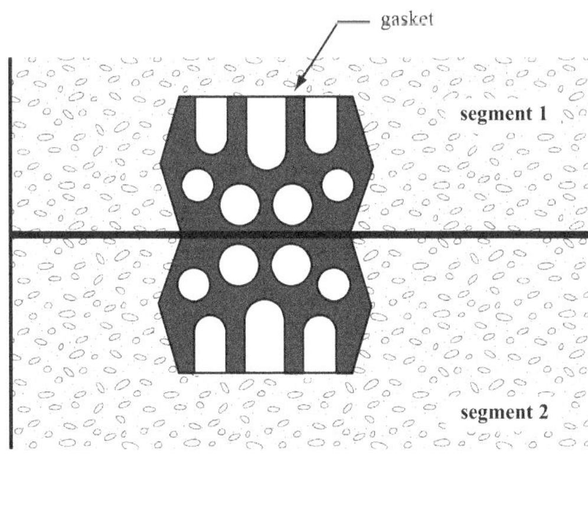

Figure 7.3 Segment and watertight joint. (a) Segment. (b) Detail of watertight joint with gasket.

may be flat, concave-concave or concave-convex. It seems that the practice is going toward flat joints, as they may require less end reinforcement (FHWA, 2009).

If the segmental liner is used for the initial support and the tunnel needs to be watertight, an impermeable membrane is placed between the segmental liner and the final support. In this case, the segments do not need to have gaskets.

Grouting between the segments and the ground is done in two stages. The first stage is done immediately after ring erection and consists of the grouting of the annular space between the segment and the ground (see e.g. Figure 3.8) and usually consists of pea gravel (no binding agent, see Section 3.3). This is needed to prevent the ground from moving toward the annulus left between the shield and the ground and thus reducing undesirable deformations, which are particularly important in soft ground tunneling. The second stage is completed at a much later time and is done through grout ports installed in the segments during construction. Grout pressures must be controlled so no undesirable deformations and/or loads are induced in the ground or support. The pressure should be larger than the hydrostatic pressure of the groundwater, but should be small enough to not damage the liner.

The segments must be designed to withstand all loads, both during construction, as well as during the service of the tunnel. An important load condition, some times more critical than those occurring during tunnel operation, is construction and erection of the liner, including temporary storage, transport of the segments to the tunnel face, erection and jacking from the shield or TBM. While each case may be different and thus a case-specific analysis is required, as a rule of thumb, a liner thickness to tunnel radius of about ten seems to work well, particularly for shallow tunnels (Bakker and Blom, 2009).

Segments are manufactured under very strict control conditions, with concrete mixes not requiring special design. Typical values of cement content are in the range of 340–380 kg/m^3, with water to cement ratios of 0.43–0.48 (DAUB, 2001). Concrete strengths in the range of 27–35 MPa are often sufficient for initial lining, and between 34 to 48 MPa for one-pass lining. Special cements may be needed if the concrete is exposed to chemical attack (this applies to cast in place also). Air entrainment may be needed if the segments are going to be stored outside for extended periods of time and may be subjected to cycles of thawing and freezing (FHWA, 2009). Reinforcement is included in the form of steel bars or welded wire mesh. However, very good results have been achieved using precast concrete reinforced with steel fibers (Hudoba, 1997).

While there are significant advantages in using precast concrete lining such as integration between shield tunneling and support, good and durable final lining, there are also a number of disadvantages (FHWA, 2009): (1) segments must be fabricated to very high tolerances; (2) on-site storage is required; (3) segments can be damaged if mishandled or due to jacking of the shield; (3) gasketed segments must be installed with low tolerance to ensure proper performance; (4) cracking may allow water seepage through the segment and reduce the life of the lining; and (5) chemical attack in certain ground may compromise the integrity of the liner.

7.2.2.2 Steel plate liners

Steel plate liners, as the name indicates, are formed by the assembly of steel segments, typically under a shield similar to what is done with the precast concrete liner (Figure 7.4a). The segments have flanges along their perimeter that are used to bolt the segments together (Figures 7.4b, c). Rubber gaskets may be placed between the flanges to provide watertightness to the tunnel. As with precast concrete segments, these are also rotated from ring to ring so no continuous longitudinal joints are produced. Tapered segments can be used to negotiate changes of vertical or horizontal alignment of the tunnel.

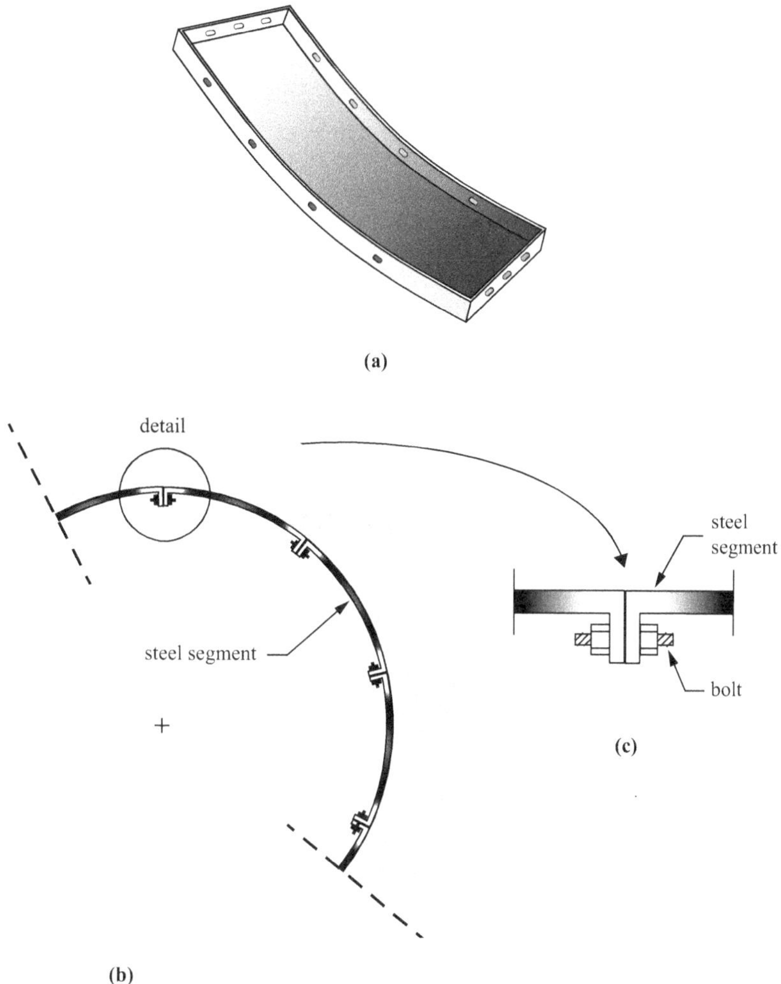

Figure 7.4 Steel plate liners. (a) Steel segment. (b) Transverse cross section with steel panels. (c) Segment connection.

Unlike the concrete segments, steel plate segments come in standard sizes of either 12″ (25.4 cm) or 24″ (50.8 cm), according to FHWA (2009). The loading demands are similar to those of precast concrete segments, except that corrosion protection needs to be provided to the steel plates. In addition, the design should be checked against general buckling, and attention should be paid to resisting the shield or TBM jacking forces. Given the dimensions of the plates, a jacking ring may be needed to distribute the loads from the jacks to prevent damage to the plates (see Section 3.3).

The advantages of using steel plate liners are (FHWA, 2009): (1) they provide complete stable ground support; (2) materials are easily transported and handled inside the tunnel; and (3) no forming or curing is required. The disadvantages are: (1) the thrust applied from the jacks may need to be limited; (2) corrosion may be a problem in tunnels because of their damped environment; (3) fire may cause the plate to buckle and fail, so cast-in-place concrete may be needed for fire protection.

Steel plate liners are not used as much as in the past and they have been largely replaced by precast concrete segments (see Section 7.2.2.1)

7.2.3 Design considerations

There are three methods that can be used for the design of the structural elements of a tunnel: the Allowable Stress Design (ASD); the Limit State Design (LSD); and the Load and Resistant Factor Design.

In the ASD method, the safety of the cross section of the structural element under design is evaluated by comparing the maximum stress demand on the concrete and reinforcement with the allowable stress. That is (ITA, 2000):

$$\sigma_c \le \sigma_{ca} \le \frac{f_{ck}}{F_c}$$
$$\sigma_s \le \sigma_{sa} \le \frac{f_{yd}}{F_s} \tag{7.2}$$

where σ_c is the maximum stress in the concrete, σ_{ca} is the allowable stress of the concrete; f_{ck} is the characteristic (nominal) strength of concrete; F_c is the factor of safety of the concrete; σ_s is the stress in the reinforcement, σ_{sa} is the allowable stress of the reinforcement; f_{yd} is the yield stress of the reinforcement; and F_s is the factor of safety of the reinforcement. Figure 7.5a shows an example of a possible stress distribution in a reinforced concrete cross section. Actual strain/stress distributions depend on appropriate stress-strain relations for concrete and reinforcement, generally prescribed by design codes.

The LSD method is based on the notion that the axial force and moment demand on the liner of the tunnel are below the capacity of the section of the liner under design. The axial force, N_D in Figure 7.5b, and moment demand, M_D in the figure, can be computed using any of the methods described in the preceding chapters. The capacity of the section is given by the combination of the ultimate axial load, N_C, and moment, M_C, that the section can withstand and is usually represented graphically by a curve similar to that shown in Figure 7.5b, after ITA (2000). The ultimate loads are computed following appropriate codes, e.g. ACI, AASHTO, and assuming that the strength capacity of the section of interest is exhausted either by axial load, moment or a combination of axial load and moment. Appropriate stress-strain relations for concrete and steel are used, reducing the strength of the concrete and/or steel by appropriate factors. The safety of the design is evaluated by confirming that the load demand falls inside the ultimate capacity of the section, and the ratio between the axial or moment capacity (N_C, M_C) and the axial or moment demand (N_C, M_C), N_C/N_D and M_C/M_D provide a measure of the factor of safety.

The designer can include additional safety considerations in this procedure. It should be noted that while the analogy "eccentrically loaded column" – "tunnel liner" makes sense, failure and thus safety against failure are different since the tunnel liner when deforming outward is supported by the ground. Related to this consideration is the fact that nonlinear analysis (e.g. representing cracked concrete) will usually lead to a higher critical thrust as described in detail by Paul et al. (1983).

Most modern design codes follow the Load and Resistance Factor Design (LRFD) method (e.g. AASHTO Bridge Design Specifications). The following is the fundamental equation:

$$\Sigma \eta_i \gamma_i Q_i \le \phi R_n \tag{7.3}$$

where η_i are factors that take into account the ductility, redundancy and importance of the structural element. AASHTO, for road tunnels, decomposes η_i into three components: (1) a factor related to ductility, $\eta_D = 1$ for tunnel linings designed following AASHTO; (2) a factor related to redundancy, $\eta_R = 1$ for mined tunnel linings; and (3) a factor related to the importance of the structure, $\eta_I = 1.05$.

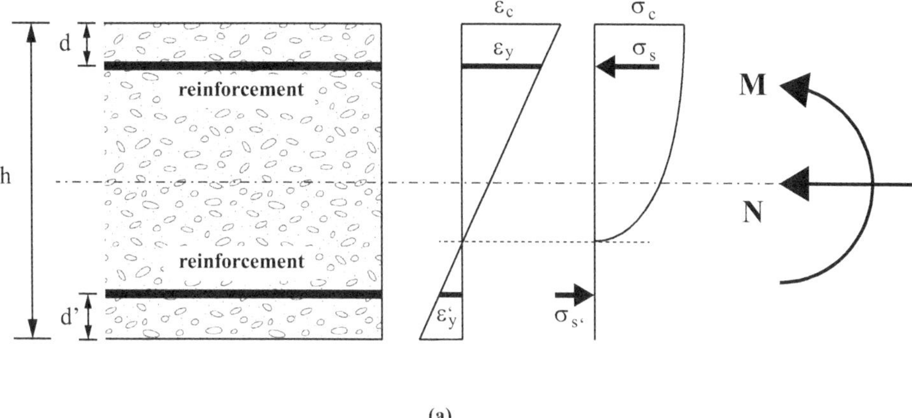

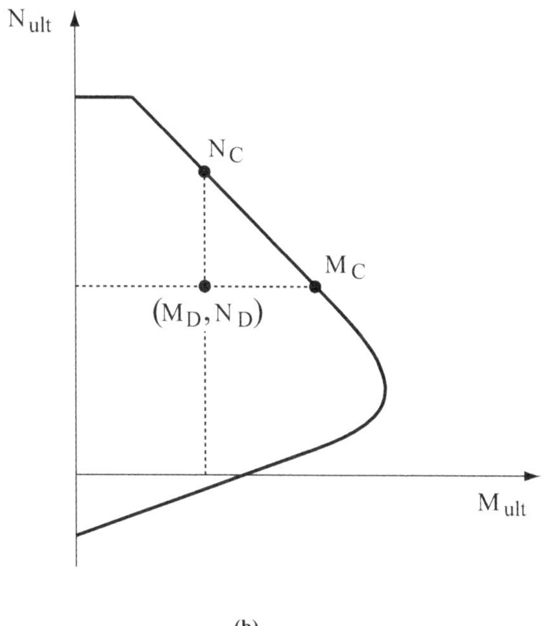

Figure 7.5 Allowable Stress Design (ASD) and Limit State Design (LSD) methods (after ITA, 2000). (a) ASD method: Stress-strain distribution. (b) LSD Method: Thrust-moment interaction diagram. N_C, M_C: Capacity; N_D, M_D: Demand.

γ_i is a factor applied to the load Q_i acting on the structural member under design. The factor depends on the particular load combination used for the calculation. Table 7.4 lists the factors and load combinations recommended by FHWA (2009), which are based on AASHTO specifications, as they pertain to road tunnels (also Table 7.1).

The loads in the Table correspond to the following description: DC: Dead load of structural components and nonstructural attachments; DW: Dead load of wearing surfaces and utilities; EH: Horizontal earth pressure load; ES: Earth surcharge load due to fill above the structure placed above the original ground line; EV: Vertical earth pressure; EQ: Earthquake load; LL: Vehicle live load; IM: Vehicle dynamic load; LS: Live load surcharge; CT: Vehicle

Table 7.4 Load factors γ_i and load combinations (after FHWA, 2009)

Load combination & limit state	*DC*		*DW*		*EH EV*		*ES*		*LL, IM, LS, CT, PL*	*WA*	*TU, CR, SH*		*TG*
	Max.	*Min.*	*Max.*	*Min.*	*Max.*	*Min.*	*Max.*	*Min.*			*Max.*	*Min.*	
Strength I	1.25	0.90	1.50	0.65	1.35	0.90	1.50	0.75	1.75	1.00	1.20	0.50	0.00
Strength II	1.25	0.90	1.50	0.65	1.35	0.90	1.50	0.75	1.35	1.00	1.20	0.50	0.00
Strength III	1.25	0.90	1.50	0.65	1.35	0.90	1.50	0.75	0.00	1.00	1.20	0.50	0.00
Service I	1.00		1.00		1.00		1.00		1.00	1.00	1.20	1.00	0.50
Service IV	1.00		1.00		1.00		1.00		0.00	1.00	1.20	1.00	1.00
Extreme Event I	1.25	0.90	1.50	0.65	1.35	0.90	1.50	0.75	γ_iEQ[1]	1.00	N/A	N/A	N/A

[1] Based on the specific project

collision load; PL: Pedestrian live load; WA: Water load; TU: Uniform temperature; CR: creep; SH: Shrinkage; TG: temperature gradient. Some of these loads, namely earth, water and earthquake can be computed using any of the methods discussed in the preceding chapters. Clearly, for structures other than road tunnels, e.g. railroad, pressure tunnels, load combinations other than those listed in Table 7.4 need to be considered that account for the specific use of the tunnel.

ϕ in equation (7.3) is the resistance factor that affects the nominal resistance R_n of the structural element considered. The resistance factors depend on the material, being plain concrete, reinforced concrete, or structural steel, and are given by the corresponding codes. For example, AASHTO LRFD specifications for reinforced concrete list the following (taken from FHWA, 2009):

For reinforced concrete linings: $\phi = 0.90$ for flexure; $\phi = 0.70$ for shear; $\phi = 0.75$ for axial compression; and $\phi = 0.70$ for bearing on concrete.

For structural steel members: $\phi = 1.00$ for flexure; $\phi = 1.00$ for shear; $\phi = 0.90$ for axial compression for plain steel and composite members; $\phi = 1.00$ for minimum wall area and buckling; and $\phi = 0.90$ minimum longitudinal seam strength.

As already noted, elements need to be designed for other loads that can occur during construction, notably bending during transportation, placement and shield thrust. Other considerations are the concentrated loads at the joints that may require intense reinforcement.

7.3 STEEL SETS

The use of steel ribs (steel sets) for support revolutionized tunneling. Before the 1920s, tunnel support was done mostly with timber (see Chapter 3). Tunnel support with steel ribs has a number of advantages such as (Rivas Vargas, 1997): (1) high strength in both compression and tension; (2) high stiffness and ductility; (3) relatively easy to fabricate and adapt to the tunnel cross section; (4) more durable than timber; (5) homogeneous material of good quality; (6) reusable (in principle) if the elastic limit has not been reached.

The first use of steel for tunnel support was as steel plates for the liner of tunnels excavated in soft ground, and as steel ribs for tunnels in rock. Steel plates are no longer used, except perhaps for tunnels excavated with a shield (see Section 7.2.2) or possibly as lagging when steel ribs are used for support.

Steel ribs are still used for support of tunnels excavated in rock when (although some applications in soil tunnels also occur): the rock is highly fractured, rockbolts do not work well given the type of rock, the in situ rock stresses are very high, or for tunnel portals, tunnel intersections, or short tunnels (e.g. Rivas Vargas, 1997). They are commonly employed as the initial support, with cast-in-place concrete for the final or permanent support.

The most traditional method of support with steel ribs is the so-called rib-and-blocking method. The construction process using the rib-and-blocking method consists of cycles of excavation as shown in Figures 7.6 for a full-face tunnel construction. The tunnel is excavated such that the unsupported length at the face usually equals the spacing of the steel ribs (Figure 7.6a). Once excavation is completed, a new steel rib is placed and new blocking is installed to support the ground (Figure 7.6b).

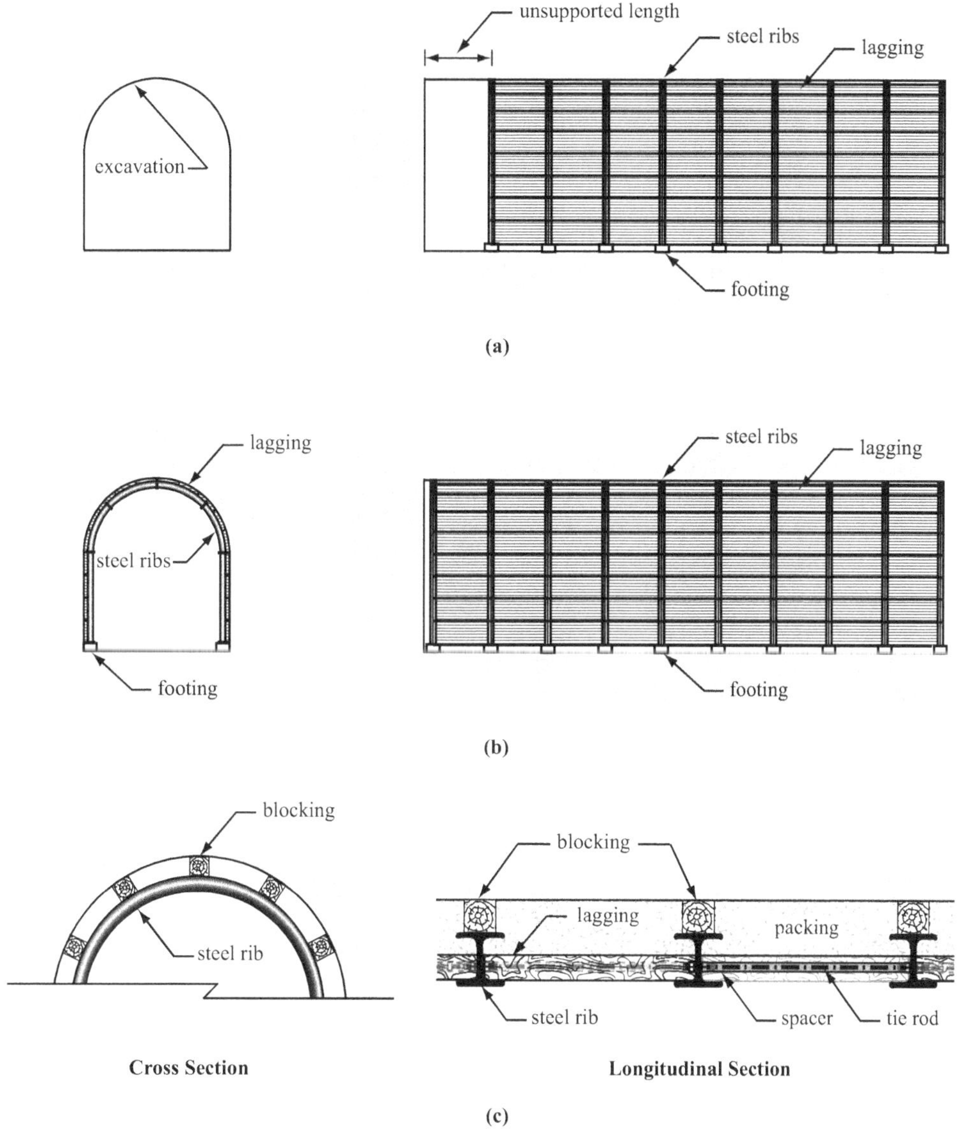

Figure 7.6 Support with steel ribs and lagging. (a) Excavation round. (b) Support round. (c) Support details.

The support consists of the following components (Proctor and White, 1977):

Rib sets: These are structural elements made of structural steel. The common profile is a flanged profile, typically H-beam, I-beam, W-beam and channels (Figures 7.6b, c). The rib is divided into two or more segments (Figure 7.7) to be able to install the rib inside the tunnel. The number and length of the segments depends on the method of excavation, tunnel cross-section, and erection process.

Blocking: These are structural elements that transmit the ground loads to the rib. They are placed between the ribs and the ground. Originally and still very often, blocks consist of wood (Figure 7.6c); today, shotcrete is often used. Blocking is crucial since it transfers the load from the ground to the ribs or vice versa. The wood blocking system, therefore, involves wedges to make certain that the load transfer takes place. Also the distance between blocks governs the moments in the ribs since one essentially considers this as a beam on multiple supports (see also Proctor and White, 1968). It is important to realize that even with the blocks, the steel ribs will not provide a stable system in the longitudinal direction. For this reason, the ribs are connected by spacer bars and tie rods as shown in Figure 7.6c.

Lagging: While steel ribs with blocking, together with the longitudinal stabilization, provide a structural support system, they do not (or only to a limited extent) prevent individual pieces of rock from falling and endangering the crew. For this reason, lagging is placed usually on the lower flange of the structural steel as shown in Figures 7.6c and 7.8. This is done in the crown and possibly on the sidewalls. Lagging consists of timber boards but is now often done with steel plates and, possibly, with shotcrete.

Footing: The load from the ground is transmitted to the blocking, then to the ribs, and ultimately back to the ground. While some of this load from the rib is transferred back by some of the blocks (depending on the loading and deformation of the rib) most of it is transferred to the footing. The footing consists of structural elements, such as wood (rarely today) or concrete, that are wide enough to transmit a load that is smaller than the bearing capacity of the ground (Figures 7.6 and 7.7).

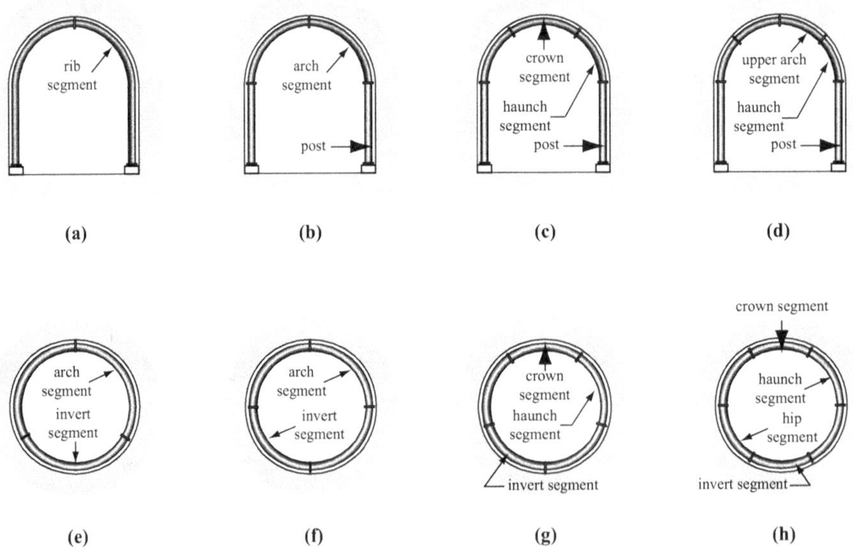

Figure 7.7 Typical tunnel ribs (after Proctor and White, 1977). (a) 2 piece set. (b) 4 piece set. (c) 5 piece set. (d) 6 piece set. (e) 3 piece set. (f) Inertial displacement. (f) 4 piece set. (g) 5 piece set. (h) 6 piece set

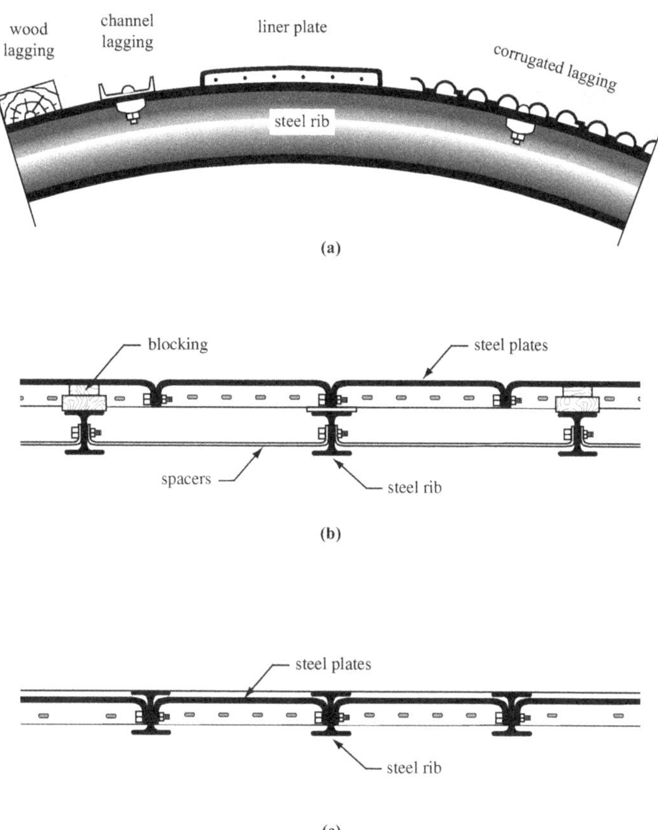

Figure 7.8 Steel rib support and lagging (after Proctor and White, 1977). (a) Types of lagging. (b) Ribs under liner plates. (c) Ribs between liner plates.

The loads on the ribs can be estimated using the rock load method developed by Terzaghi (described in detail in Section 2.3.2.3 or 3.2). The method was established for this particular support and is based on the assumption of the development of full or partial arching of the ground above the tunnel. This requires sufficient deformations of the support, which are possible due to the flexibility of the ribs. Transfer of the rock loads to the support can be assumed as a uniform load, when the connection between the blocking and the ribs is continuous, or as discrete, concentrated loads at the blocking points. The blocking points can transfer load, but not moment, according to Proctor and White (1968), and because the shear stiffness of the blocks is generally small, the blocks can only transmit a radial load to the ribs. Rib support with blocking is not used as much as in earlier times. Proctor and White (1968) proposed a graphical, iterative approach to obtain the thrust and moment acting on the ribs. These calculations can also be done using numerical methods, as those described in Chapter 5, using appropriate boundary conditions to represent the blocking points, e.g. roller connections.

Nowadays, ribs are used in addition to other methods of support such as rockbolts and shotcrete (Figure 7.9). In this case, the support is treated as a composite structural element with stiffness and strength according to the properties and geometry of each of the components, as well as their layout within the support (see Section 7.6 for additional discussion).

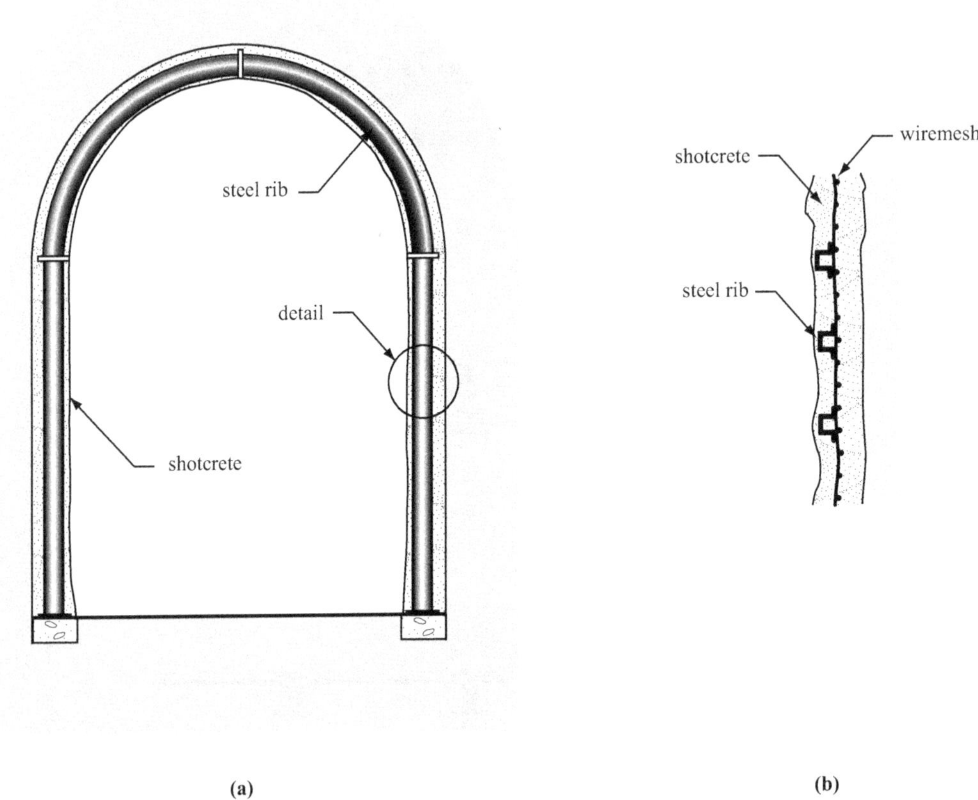

Figure 7.9 Tunnel with steel rib support. (a) Steel ribs and shotcrete. (b) Support detail.

7.4 ROCK REINFORCEMENT SYSTEMS

The use of rockbolts for tunnel support dates as far back as 1913 (Kovári, 2003). In the United States rock bolts for roof support were introduced in 1927 in non-coal mining and in 1946 in coal mining operations (Bolstad et al., 1983). Rockbolt use experienced a rapid increase in the 1970s, and nowadays they are extensively used as support in both civil and mining tunnels.

Rockbolts are part of a larger family of support methods, denoted in the following as "rock reinforcement systems." A distinction is made between support and reinforcement. Following Windsor and Thompson (1993), tunnel support provides a surface restraint to the ground by installation of a structural element on the tunnel perimeter, while reinforcement improves the behavior of the ground by installing structural elements within the ground. Another way to reinforce the rock mass is through grouting. This is done by injecting grout into discontinuities (fractures) and in essence cement them. This produces a rock mass acting as a less deformable and higher strength continuum compared to the non-grouted rock mass. Grouting is often used together with bolting and possibly in the form of the umbrella method (see Figure 1.20b).

The focus of this section is on rock reinforcement systems that include structural elements, i.e. rockbolts. The following provides a classification of the systems and recommendations for their design. The emphasis of the discussion is on support methods typical of civil engineering projects, but details regarding other types of support used preferentially in mining are also included since, after all, the principles underlying the support methods are the same.

7.4.1 Classification of rock reinforcement systems

A reinforcement system comprises the following components (Windsor, 1997): (1) the rock mass; (2) the rock reinforcement element; (3) the element internal fixture; and (4) the element external fixture. See Figure 7.10. The rock mass is in general an integral part of the reinforcement system and has to be included as a component. The rock reinforcement element is a steel, fiberglass, resin bar, or cable. While steel is the material most used, fiber reinforced polymer composites are beginning to be attractive alternatives due to their good engineering characteristics and durability (Peterson et al., 1992). The internal fixture connects the rock reinforcement element to the rock mass and can be limited to the end of the element along an anchor zone length or can extend over the entire length of the element. It is usually a resin or a cement-based grout filling the space between the element and the rock. The external fixture may or may not be always present. If present, it ties the rock reinforcement element to the rock surface. It typically comprises a steel plate to distribute the load of the rockbolt over its surface and a nut to tie the external fixture to the reinforcement element. There is usually a spherical seat between the nut and the plate, or the nut itself has a spherical shape, to prevent the generation of large moments in the element when loaded. The fundamental idea behind the rock reinforcement concept is that as the rock moves toward the excavation, the reinforcement element, which is attached to the rock, elongates and so it creates tension in the element (thus the need for element materials with good tensile capacity) that is transferred to the rock as compression, thus helping the rock support itself by increasing confinement. In essence, the load transfer from the reinforcement to the rock is based on three basic mechanisms (Windsor and Thompson, 1993; Windsor, 1997): (1) rock movement and load transfer from an unstable, *displacing* zone to the reinforcing

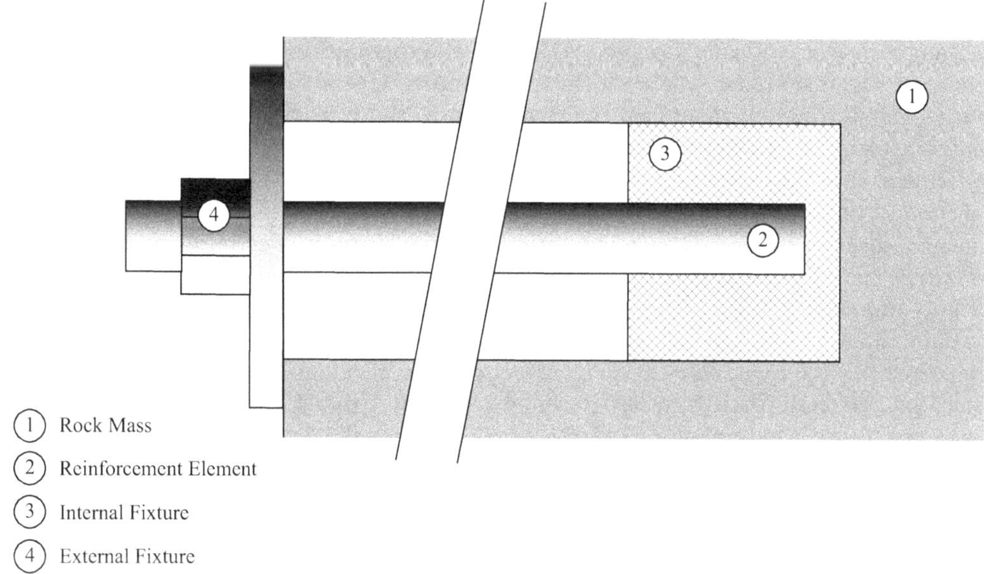

Figure 7.10 Components of a reinforcement system (after Windsor, 1997).

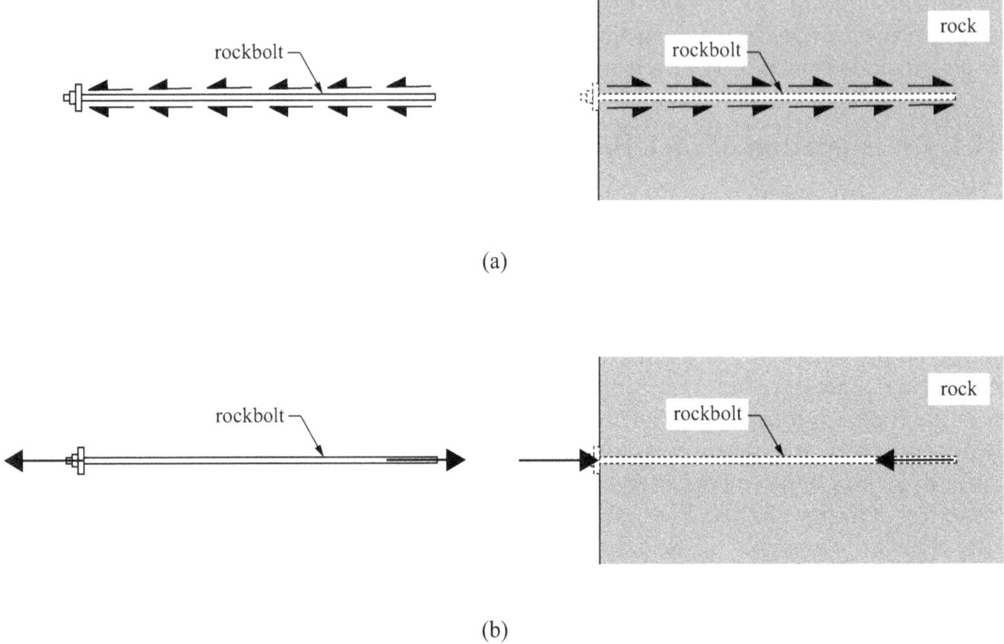

Figure 7.11 Schematic of load transfer between rock and reinforcement device. (a) Load transfer in CMC and CFC systems. (b) Load transfer in DMFC systems.

element; (2) transfer of load from the unstable, *displacing* rock region to a stable interior region via the reinforcement element; and (3) transfer of the reinforcement element load to the stable rock mass.

The types of reinforcement devices can be classified as: (1) Continuously Mechanically Coupled (CMC); (2) Continuously Frictionally Coupled (CFC); and (3) Discretely Mechanically or Frictionally Coupled (DMFC). The classification is based on how the element load is transferred to the rock. In CMC and CFC systems, the load transfer occurs continuously through the length of the element; in CMC systems, through a bond (usually grout) between the element and the rock, or through an intimate frictional contact between the element and the rock in CFC elements (Figure 7.11a). The load transfer between the element and the rock in DMFC systems occurs over a short length at the distal end of the element, either by bonding the element with the rock with resin or grout or by a mechanical device that provides frictional capacity (Figure 7.11b).

The reinforcement systems can be installed as pattern reinforcement, spot reinforcement or a combination of the two. Pattern reinforcement consists of a regular uniform geometric arrangement of the reinforcement elements, either as a rectangular or oblique pattern. The pattern has dimension a × b, where a denotes the row spacing and b the column spacing between elements. A rectangular pattern is obtained when the directions of the columns and rows are perpendicular to each other, e.g. along and transverse to the tunnel, and oblique when the directions meet at an angle other than 90°. A square or diamond pattern is obtained when a = b. A spot reinforcement consists of a discretionary arrangement to support localized zones; e.g. a particular rock block.

7.4.2 Types of rock reinforcement system devices

There are three types of rock reinforcement system devices (Windsor, 1997): (1) rockbolts and dowels; they are generally few meters in length; (2) cablebolts, which are typically in the range from 3 m to 15–20 m; and (3) ground anchors or tiebacks, which are generally longer than 10 m.

7.4.2.1 Rockbolts

There are four different types of rockbolts: (1) mechanically anchored rockbolts; (2) resin-anchored rockbolts; (3) grouted rockbolts or grouted dowels; and (4) friction rockbolts or friction dowels.

Mechanically anchored rockbolts are a DMFC reinforcement system. The rockbolts are anchored to the rock mass at their distal end through a mechanical device, which typically consists of a pair of wedges and a tapered cone. See Figure 7.12a. As the rockbolt is tensioned, the tapered cone forces the wedges against the wall of the drilled hole providing a mechanical anchor. Installation of a rockbolt starts by drilling the hole, which typically needs to be about 100 mm longer than the bolt to prevent damage to the mechanical anchor. The bolt is then placed inside the hole. A sharp pull at the end of the bolt provides an initial seating of the anchor. The anchor is connected to the rock at the perimeter of the opening by a faceplate, a nut, and a tapered or domed washer to produce a spherical seat. By tightening the nut, the wedges expand further into the rock. Mechanically anchored rockbolts are appropriate for

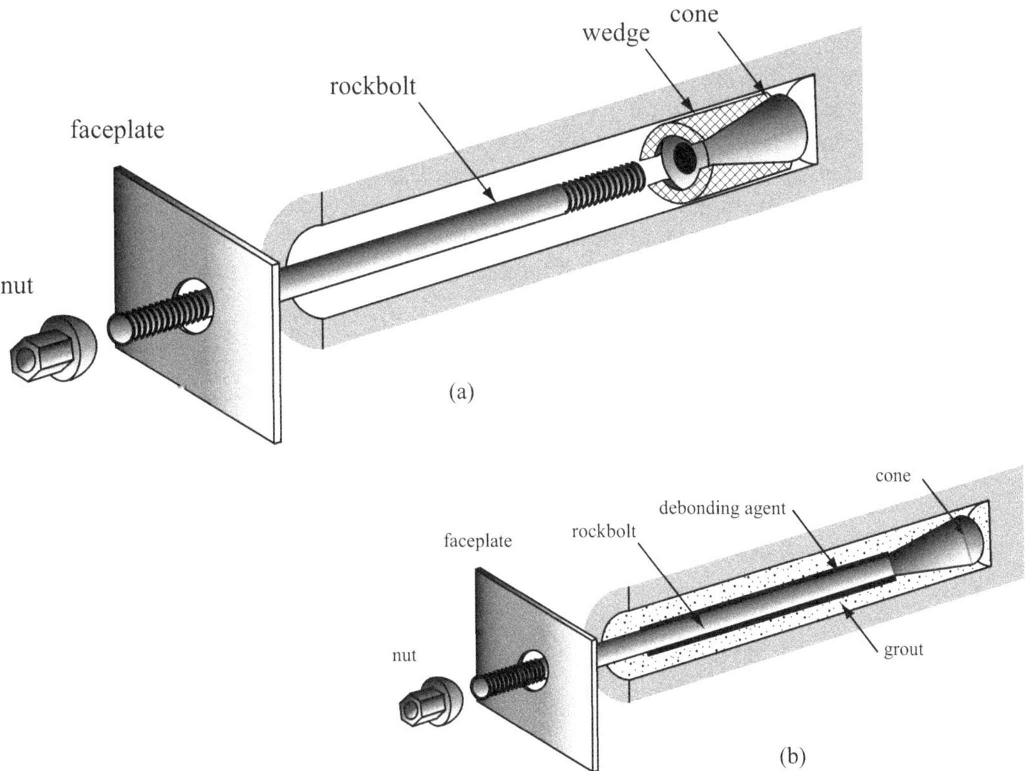

Figure 7.12 Mechanically anchored rockbolt. (a) Wedge-anchored rockbolt. (b) Conebolt.

hard rock but may not work well in soft rocks, as the required anchor force at the end of the rockbolt may not be attained. Installation of the rockbolts is very fast, but it is very important that the bolt remains in tension to ensure that the load is transferred through its different components to the rock. Figure 7.12b shows the conebolt, which is used in mining in areas prone to rockburst (Cai et al., 2010). Under normal, static, working conditions, the bolt works as a mechanically anchored rockbolt. However, when subjected to dynamic loading, i.e. during rockbursts, the bolt yields or the cone at the distal end ploughs through the grout, thus dissipating energy (see Section 6.4.2 for further discussion). When the rockbolt is used as permanent support, it is good practice to provide protection against corrosion. This is usually accomplished by grouting. Grouting is done through a short tube extending into the hole from the face plate and displacing the air through a hole in the bolt or vice versa (depending on the inclination of the bolt). The requirement for the grout is not to provide strength but protection against corrosion, and so the water/cement ratio is designed such that the grout is fluid enough so that it is easily pumpable and fills the space between the bolt and the hole. Water/cement ratios are generally within the range 0.4 to 0.5 (Hoek et al., 2000).

Resin-anchored rockbolts are also a DMFC reinforcement system. They usually provide a better anchor support than mechanically anchored rockbolts and so they are appropriate to be used in soft rocks or near blast areas where a mechanical anchor may be dislodged by the vibrations (Hoek et al., 2000). After drilling the hole, fast-setting resin cartridges are pushed to the end of the hole. Afterward the rockbolt, typically a steel rebar, is inserted and given a few rotations. Insertion of the rebar breaks the resin cartridge and spinning mixes the components of the cartridge: the resin and the catalyst. The mixing time of the cartridge components must follow the manufacturer's specifications to accomplish a thorough mixture, but also to prevent weakening of the resin if the rebar is rotated after the resin has set. Setting of the resin is very fast and the bolt can be tensioned a few minutes after installation. Corrosion protection can be quickly obtained by inserting after the fast-setting resin cartridge slow-setting resin cartridges, as shown in Figure 7.13. The slow-setting resin sets after about 30 minutes, which provides enough time to tension the rockbolt. Installation of resin-anchored rockbolts is fast, and these rockbolts, since they can be tensioned, can be used to provide an active load to the rock. The cost of these rockbolts is higher than other rockbolts because of the resin, but in some operations the cost may be offset because of the speed of installation. One has to be aware of the fact that the setting time is temperature-dependent and great care needs to be taken to have the proper setting time.

Figure 7.14 shows a grouted rockbolt, also called grouted dowel. These are Continuously Mechanically Coupled (CMC) rock reinforcement systems. Installation, as with other

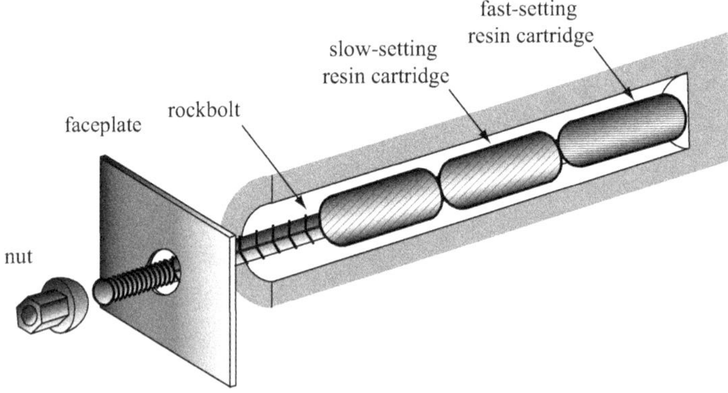

Figure 7.13 Resin-anchored rockbolt.

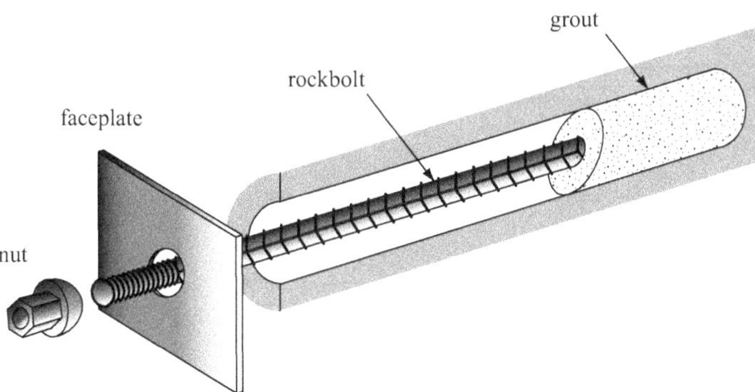

Figure 7.14 Grouted rockbolt or grouted dowel.

rockbolts, starts by drilling the hole where the dowel is placed. Afterward a cement grout is pumped into the hole by inserting the grout tube till the end of the hole and slowly withdrawing it as the grout is pumped. The dowel is afterward pushed in the hole. A plate and a nut connect the end of the dowel to the tunnel perimeter. Note that the grouts can also be resin grouts applied analogously to what is shown in Figure 7.13. A small load is typically applied by tightening the nut. The grout is designed for two purposes: the first one is to intimately connect the dowel to the rock and the second to provide corrosion protection. In contrast to the DMFC rockbolts that impart an active support to the rock, grouted dowels provide a passive support. The dowel takes load in tension as the rock deforms toward the tunnel. The load is transmitted from the rock to the steel through the grout, and so the grout must be designed to have good strength capabilities. Typical water/cement ratios range from 0.3 to 0.35 (Hoek et al., 2000). Because the rock has to deform, for the dowel to start providing support, it is good practice to place grouted rockbolts as close to the face of the tunnel as possible.

Friction dowels transmit load from the rock to the dowel through friction between the shaft of the dowel and the rock. There are two types of friction dowels: Split Set stabilizers and Swellex dowels. Both types are classified as Continuously Frictionally Coupled (CFC) reinforcement systems.

Split Set stabilizers were developed by Scott (1976, 1983). The dowel consists of a slotted steel tube with one end tapered for easy insertion into the drill hole. The other end has a ring flange to hold a dome plate. See Figure 7.15a. The dowel is inserted into a hole that is slightly smaller than the outside diameter of the steel tube. As the tube is pushed into the hole, its diameter is compressed and the slot partially closes. This effect creates radial compression at the tube-rock contact, which provides a frictional strength at the dowel-rock interface. As the rock deforms, the shear stresses at the contact with the dowel increase. This generates tension in the stabilizer and compression in the rock, which provides support. Once the shear stress at the interface reaches the frictional strength, slip between the dowel and the rock occurs. Table 7.5 provides specifications for the Split Set stabilizers.

Installation of Split Set stabilizers is very fast. Corrosion, however, may be an important issue if the system is used as permanent support. The dowels can be galvanized, which can slow the corrosion phenomenon.

Swellex dowels (Figure 7.15b) were developed by Atlas Copco. They consist of a steel tube that is folded during manufacturing. The tube is inserted into the drill hole that has a diameter larger than the folded tube. A high-pressure water injection applied to the tube unfolds the steel and pushes it into an intimate contact with the rock in the hole. Interlocking between the dowel and the rock and friction between the rock-and the dowel are the mechanisms that

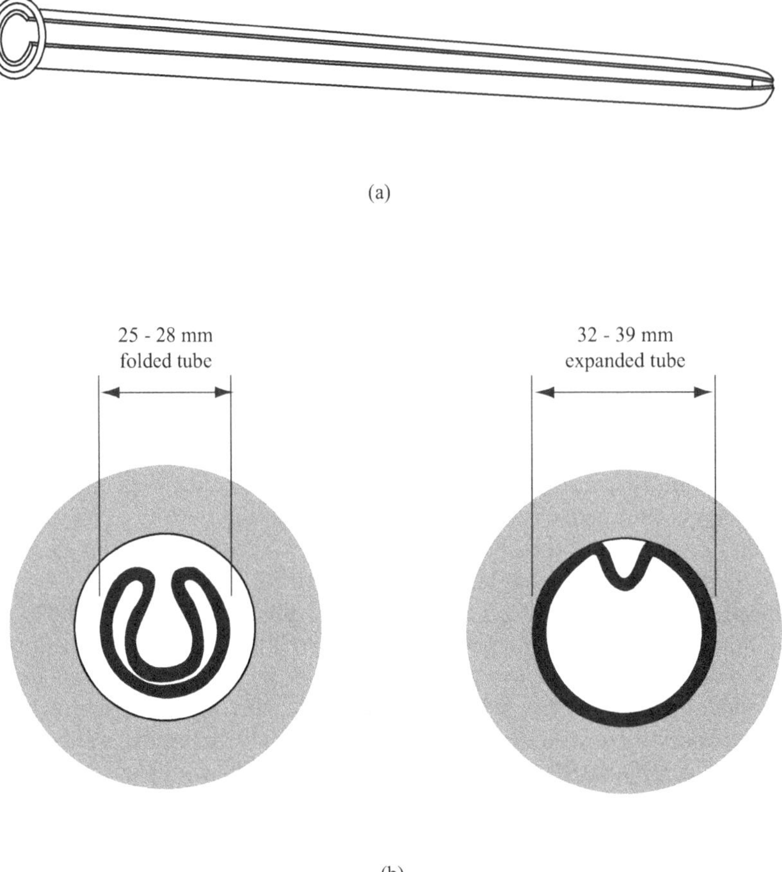

Figure 7.15 Friction dowels. (a) Split set dowel. (b) Standard swellex dowel.

Table 7.5 Split set stabilizer specifications (from International Rollforms, Inc.)

Type	*SS - 33*	*SS - 39*	*SS – 46*
Outside Diameter (mm)	33	39	46
Length (mm)	762–2438	762–3048	914–3658
Drill Diameter (mm)	30–33	35–38	41–45
Initial Load (tons)	3–6	3–6	6–9
Domed Plate (mm)	150 × 150 × 4	150 × 150 × 4	150 × 150 × 4

generate the shear strength along the Swellex-rock interface. As the rock displaces toward the tunnel opening, friction develops along the interface, which induces tension in the dowel and compression in the rock. Once the shear strength along the interface is exhausted, there is slip between the dowel and the rock. There are three types of Swellex dowels: Standard, Manganese (Swellex Mn), and Premium (Swellex Pm). Table 7.6 contains a summary of the technical specifications of the dowels. The difference between Swellex Premium and Manganese is the lower elongation at failure of the Swellex Premium.

Table 7.6 Swellex dowels specifications (from Atlas Copco)

Type	*Swellex Mn12*	*Swellex Mn16*	*Swellex Mn24*	*Swellex standard*
Breaking Load (kN)	120	160	240	100
Typical Elongation	30% (10%)[1]	30% (10%)[1]	30% (10%)[1]	15%
Inflation Pressure (MPa)	30	24	30	30
Recommended Hole Diameter (mm)	32–39	43–52	43–52	32–39
Optimal Hole Diameter (mm)	35–38	45–51	45–51	35–38
Length (mm)	600–6500	1800–6500	1800–7000	600–12000

[1] Values in parenthesis are for Swellex Premium

The Swellex dowels can be coated (Coated Swellex Rockbolts) for corrosion protection in case the dowels need to provide permanent support. As with the Split Set stabilizers, the advantage of these dowels is the speed of installation, which can be done using automated rockbolters.

Figure 7.16 shows the load-deformation characteristics from laboratory tests carried out on rockbolts. The tests were conducted by Stillborg (1994) by pulling apart, using displacement control, two concrete blocks linked together by a rockbolt. The rockbolts were installed in the concrete blocks using techniques similar to those used in mines. The figure shows that the frictional rockbolts, i.e. EXL Swellex dowel (the EXL Swellex dowel was introduced in the market in 1993 and was made with high-strength, ductile steel) and SS 39 Split Set stabilizer,

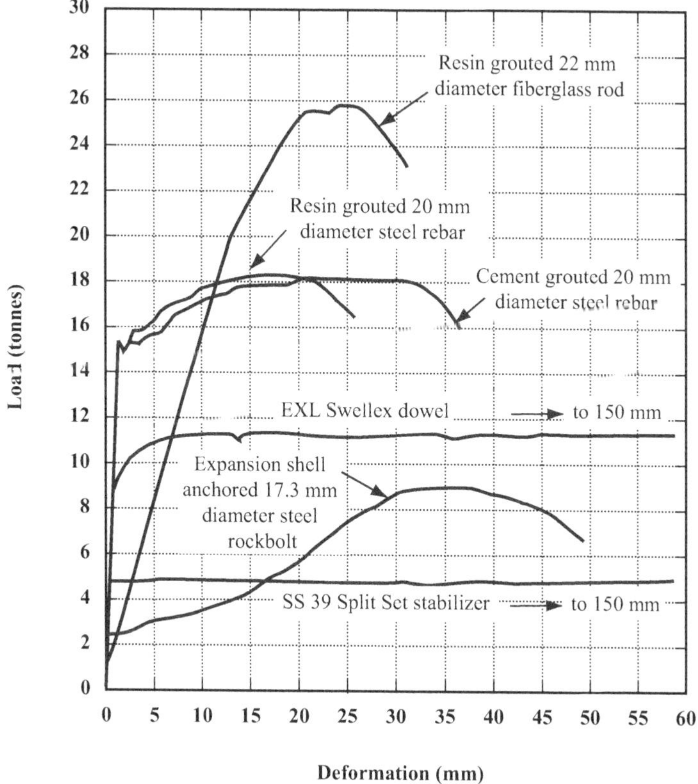

Figure 7.16 Load-deformation laboratory test results for different rockbolts (after Stillborg, 1994).

have a marked rigid-plastic type of behavior, with very small initial deformations until failure. Failure was produced by slip of the rockbolt. The rockbolt, however, was able to maintain a constant load during slip until a fairly large deformation, up to 150 mm. The expansion shell-anchored rockbolt (analogous to the mechanically anchored rockbolt in Figure 7.12a) begins to show a non-linear response at about 4 tons. At this load, the face plate has significant plastic deformations. Maximum load was reached at 9 tons with a deformation of 35 mm. Once failure occurred, the rockbolt load capacity decreased very quickly. The results for the cement- or resin-grouted rockbolts are similar. The linear response ended at a load of about 15 tons. The load increased with deformation until a maximum load of 18 tons. Both rockbolts lost capacity after failure. It appears that the cement-grouted rockbolt can sustain larger deformations that the resin-grouted rockbolt. The largest maximum load was obtained with a resin-grouted fiberglass rockbolt. In this case, the non-linear response started at a fairly small load and was induced by slip between the rockbolt and the resin. Slip progressed with larger loads until the maximum load of 26 ton, at 25 mm deformation, was reached.

7.4.2.2 Cablebolts

Cablebolts are long, grouted, steel cables with high tensile strength. Cablebolts are often used in mining. Their application in civil engineering projects is usually for the support of large caverns in weak rock. Cablebolts were first used in mining at the Willroy mine in Canada and at the Free State Geduld Mines Ltd. in South Africa in the early 1960s (Fuller, 1983; Hoek et al., 2000). Given the similarities between the purpose of cablebolts and pretensioning/posttensioning of bridges, some of these technologies are also being used in cablebolts (e.g. VSL, 1982).

Figure 7.17 provides a summary of the different types of cablebolts used by the mining industry. The summary was compiled by Windsor (1992, 2001) and has been adopted by a number of authors (e.g. Hoek et al., 2000; Brady and Brown, 2004). Early applications consisted of the Multiwire tendon (Figure 7.17a), which consists of seven 7 mm diameter single steel wires, typically used for prestressed concrete, arranged with plastic spacers (Clifford, 1974). Their performance was rather poor due to the low frictional characteristics between the wire and the grout and because the Poisson's effect that, when the wire is loaded in tension, tends to reduce its diameter and thus facilitate separation between the wire and the grout. The cablebolt performance was improved with the Birdcage Multiwire tendon (Jirovec, 1978), which is shown in Figure 7.17b; however, its installation was rather difficult. An important improvement in cablebolt technology came in the early 1970s with the use of single strand cables, which contained seven wires: one straight central wire, and six other wires around the central one forming a spiral. Different manufacturing options include normal, indented, and drawn strand. Their systematic use in cut-and-fill stopes is described by Hunt and Askew (1977). The cables can be sheathed, coated, or encapsulated for corrosion protection (VSL Systems, 1982; Dorsten et al., 1984). The performance improvement with single strand cables is due to a better load-transfer between the cable and the grout. As the strand is tensioned, the spiral shape of the wires induces dilation at the interface with the grout, and so the radial stresses increase. With the increase of the stresses normal to the grout-cable interface, the shear strength of the interface also increases. Cablebolt failure almost exclusively occurs by slip between the strand and the grout (Fuller, 1983). Further developments in the early 1980s comprise the addition of anchors spaced along the strand (Matthews et al., 1983; Schmuck, 1979). The birdcage strand was first used at Mt. Isa in Australia (Windsor, 2001). It consists of a rewound of a strand that results in an open cross section, which significantly increases the load-transfer between the strand and the grout (Hutchins et al., 1990). Additional enhancements of the cross section included the bulbed

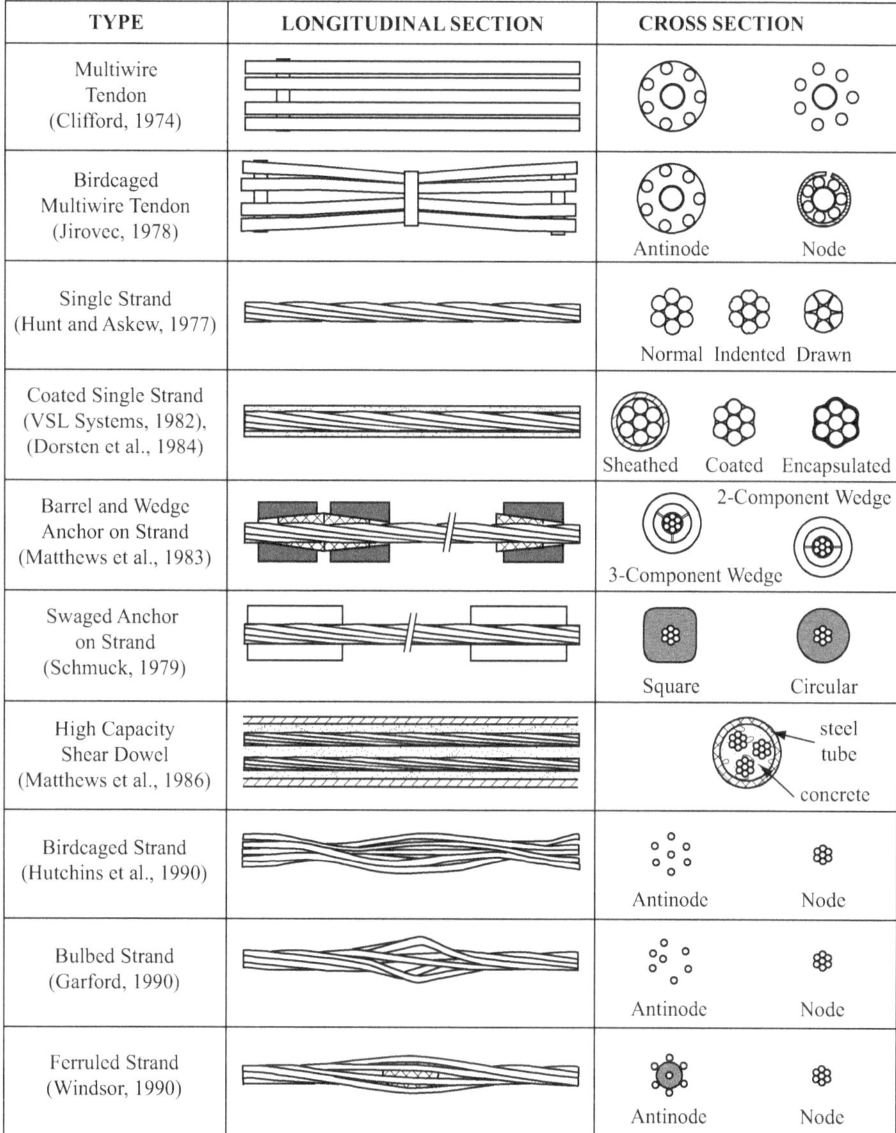

Figure 7.17 Summary of the development of cablebolt configurations (after Windsor, 1992).

strand (Garford, 1990) and the ferruled strand (Windsor, 1990). In hard rock mining applications, the most common cablebolts used are the plain seven-wire strand or the modified cablebolts birdcaged, bulbed, or ferruled strands. Because the cablebolt-rock load transfer is very sensitive to confinement, modified cablebolts are preferred in areas where a decrease of confinement may occur after cablebolt installation; e.g. due to the unloading associated with excavation. When used for permanent support, cablebolts are grouted into a corrugated plastic sleeve for corrosion protection, and the whole assembly is grouted to the rock. In civil engineering projects, it is common to grout first a few meters at the end of the cablebolt to form an anchor. Once the grout has set, the cablebolt is tensioned and the rest of the system is grouted (Hoek et al., 2000).

Figure 7.18 shows results of pullout tests of a 15.2 mm seven-wire strand cable embedded in 250 mm cement grout (Hyett et al., 1995). The tests were performed under different

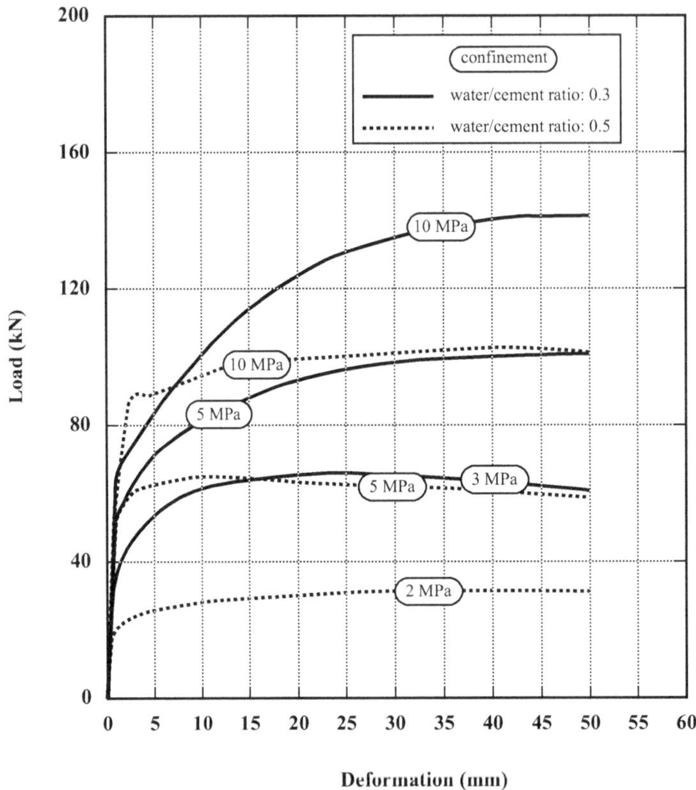

Figure 7.18 Load-deformation laboratory test results for seven-wire strand cables embedded 250 mm in cement grout. Adapted from Hyett et al. (1995).

confinement pressures and for different water/cement ratios. Failure always occurred by slip along the cable-grout interface, which is typical of what happens with cablebolts in the field. The results in the figure clearly illustrate the importance of the water/cement ratio. As the ratio was increased from 0.3 to 0.5 the load capacity of the cablebolts decreased dramatically. Also, the pullout capacity increased with confinement. All the tests showed similar behavior: an initial linear elastic response, followed by a non-linear load-displacement behavior until failure, which was reached at 40–50 mm displacement. The linear portion is controlled by the stiffness of the grout and of the cable, and the properties of the interface. As the load increases, slip between the cable and the grout occurs, which is the cause of the loss of linear behavior. Slip is associated with radial fracturing of the grout annulus and/or with shear through the grout flutes. Radial fracturing starts at the cable/grout contact and is caused by dilation as the cable stretches and it is originated by the geometric mismatch at the cable/grout interface and by the tendency of the cable to unravel with tension. Both mechanisms are pressure-sensitive. With further load, radial cracking propagates toward the outer boundary of the grout, which is progressively divided into interlocking wedges. Once the radial cracks reach the outer grout boundary, the tangential stresses in the grout disappear, the radial stresses decrease and failure occurs. In the field, confinement is provided by the stiffness of the rock mass, as dilation of the cable under tension tries to expand the drill hole, which is resisted by the rock, thus increasing the normal stresses at the grout/rock contact. Pullout laboratory and field tests performed on different rocks confirm this mechanism, as higher pullout loads are attained in rocks with higher stiffness (Hassani et al., 1992; Hyett et al., 1992b; Yazici and Kaiser, 1992).

An important reduction of confinement can occur because of de-stressing of the rock mass where the cables are installed during excavation and other mining operations. Pullout tests at the Winston Lake Mine in Minnova, Ontario (Maloney et al., 1992) show a considerable reduction of cable pullout capacity as the excavation front passed the cable location, which was associated with rock mass stress reduction. Failure occurred by slip between the cable and the grout, which, as discussed in the previous paragraph, is a pressure-sensitive phenomenon and thus a reduction of confinement due to rock mass unloading carries a reduction of cable pullout capacity. These observations are supported by numerical simulations of the unloading caused by mine excavation (Kaiser et al., 1992) and are also consistent with the laboratory results shown in Figure 7.16, after Stillborg (1994). This observation is relevant as the modeling of the different reinforcement systems is discussed in the following sections.

7.4.2.3 Ground anchors or tiebacks

Ground anchors are devices that are able to transmit to the ground a high tensile load. They are constituted by three distinct parts, as illustrated in Figure 7.19: anchor head, unbonded length, and bonded length. The tensile load applied to the anchor is transferred to the ground surface by the anchor head and is resisted by the bond between the anchor and the surrounding ground at the distal end. The unbonded length is required to stretch the anchor until the desired load is attained. The tendon can be a steel bar or cable (typically) that has from one to several wire strands, which can be several tens of meters long and can reach loads of several mega-Newtons. Anchors are typically installed across potentially unstable discontinuities or soil masses and are anchored in stable ground. In mining, there are reports of anchor use for rock support as early as 1918 in the Mir Mine in Poland, and in civil engineering since 1934 in the Cheurfas dam in Algeria (Littlejohn, 1993).

Anchor installation consists of the following steps (Wyllie and Mah, 2004): drilling the hole, anchor installation, bond construction, tensioning, anchor head protection, and testing. The diameter of the hole should be large enough to allow for the insertion of the tendon and provide space for the grout between the anchor and the walls of the hole. After drilling is completed, the tendon has to be installed as soon as possible to prevent cave-ins or deterioration of the hole surfaces that may reduce bond strength. The tendons are typically supplied to the site by the manufacturer with the required length, and their installation down the drill hole is commonly done mechanically because of the large weight of long tendons. Centralizers are normally placed along the tendon to ensure a minimum grout cover between the tendon and

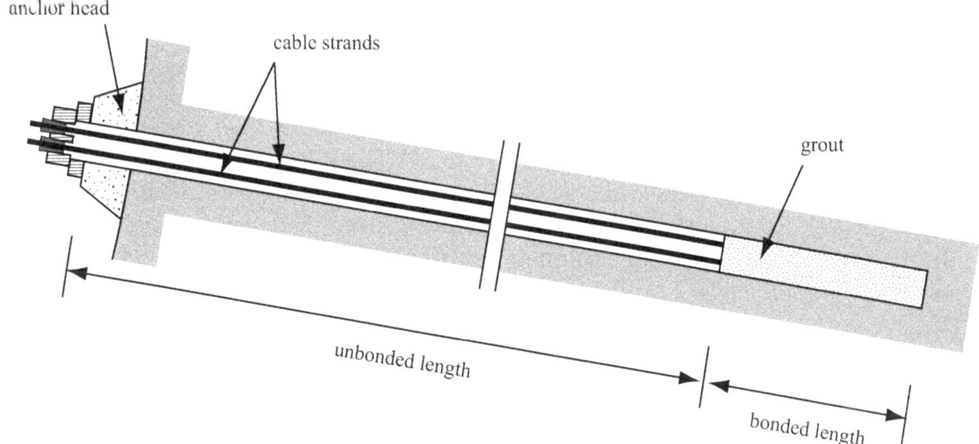

Figure 7.19 Multi-strand cable rock anchor.

the ground. The tendons are protected against corrosion to prevent loss of cross-section area over time in aggressive environments. In corrosive environments grout alone may not be sufficient to provide an adequate level of protection and additional measures such as protective coatings or coverings may be required. The bond is attained by inserting a bonding agent at the distal end of the anchor. The agent can be a resin or a cement grout, the latter being the industry standard for long anchors. The grout is pumped from the tunnel surface through a plastic tube usually attached to the tendons. As with cablebolts, the water to cement ratio (w/c) is the most critical factor that determines the quality of the bond. Low ratios make the grout very viscous and increase the difficulty of pumping into the hole. In addition, below a w/c = 0.3, the scatter of the strength of the grout increases significantly in laboratory tests (Hyett et al., 1992a). Water/cement ratios larger than 0.5 make the grout very fluid but at the expense of low engineering properties. Typical ratios are between 0.4 and 0.45 (Littlejohn, 1992, 1993; Wyllie, 1999; Wyllie and Mah, 2004). The total bond length is generally between 3 and 10 m. Even though in hard rock smaller bond lengths may be adequate to successfully transfer the tension load to the surrounding rock, a longer bond length allows for the possibility of encountering along the bond length areas of low-quality rock. After the grout has set and has reached the desired strength, which may take up to several days, the tendons are tensioned from the tunnel. By controlling the load and displacement of the tendon, an estimate of the free anchor length can be made. Anchors are typically designed for working loads from 50% of their ultimate load for permanent support, to 62.5% for temporary support, which correspond to factors of safety of 2.0 and 1.6, respectively. They should not be stressed beyond 80% of their strength (Littlejohn, 1992, 1993). Steel relaxation may be an important factor to consider in those cases where the anchor is used as permanent support and is subjected to a substantial tensile load. For example in ordinary stress-relieved steel a load loss of 5–10% of the applied stress occurs at stress levels of 75% of the ultimate strength at 20°C. For loads smaller than 55% of the yield load of the tendon, relaxation is negligible (Wyllie, 1999). Once the tensile load is applied to the anchor, it is transferred to the perimeter of the tunnel excavation through the anchor head. The anchor head consists of a bearing plate and a stressing head where the tendons are held in place, usually through a system of barrel and wedges (Thompson, 1992). The anchor head, as with the unbonded tendon, need to be protected against corrosion. This can be done in different ways, e.g. by encasing the tendon in a grease filled pipe. A fundamental part of the installation process is testing. A procedure must be put in place to ensure that each anchor is working as designed. Recommendations for acceptance procedures and guidelines are provided by Littlejohn (1993) and Wyllie and Mah (2004).

As was briefly mentioned in the preceding discussion, two problems will usually affect the performance of bolts: unloading/loading of the rock mass; and effects on bolt-grout-rock interface.

The load transfer at the bolt-grout-rock interface is affected by the surface roughness of the rock and the bolt. Bolts consisting of ribbed reinforcing steel or cables with irregular surfaces are advantageous. On the other hand, cracking of the rock and, particularly, of the grout may lead to a reduction of transmittable loads. The load transfer is also affected by the "Poisson effect" of the element under tension, i.e. usually tension on the bolt will reduce its transverse cross section and may cause detachment from the grout or rock. This can be avoided if the grout is placed after tensioning or with the arrangement shown in Figure 7.20. The arrangement in the figure can be used for cablebolts or ground anchors/tiebacks.

7.4.3 Support and reinforcement design philosophy

The design philosophy is to provide enough support such that the rock mass can withstand its own weight. The advantage of rock reinforcement systems is that they are very flexible

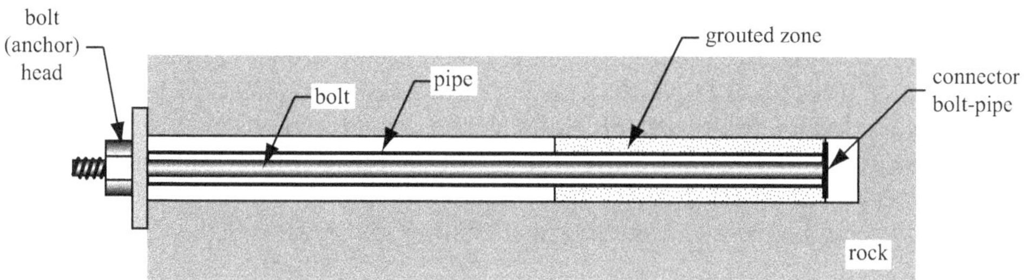

Figure 7.20 Grouted zone in compression for a tension bolt.

and can be easily adapted to changing rock conditions during tunnel excavation. For example, to control excessive convergence, the number of rockbolts and/or their length can be systematically increased until the deformations are acceptable. No other support system is so versatile, which is one of the reasons why rockbolts are used in tunnels constructed e.g. following the NATM.

A distinction is made between local support and general or systematic support. Local support is required when unstable blocks or wedges can occur along the perimeter of the excavation. General or systematic support is needed when excessive deformations occur in the tunnel, which are generally associated with non-linear behavior of the rock mass. The discussion focuses on static loading. The reader is directed to Chapters 5 and 6.2 for seismic design and to Section 6.4 for the support in rockburst zones.

Failure of a reinforcing element can occur in one or more of the following modes (Littlejohn, 1993): (1) failure within the ground; (2) failure at the ground/grout bond, or at the ground/element bond for those reinforcement devices where the element is in contact with the ground; (3) failure at the grout/element bond; and (4) failure of the element or element head.

The load that causes failure at an interface, T^f, is approximated by the following expression:

$$T^f = \pi d L \tau^f \tag{7.4}$$

where d is the diameter of the interface (drill hole diameter for the ground/grout interface, or diameter of the reinforcement element for the grout/element interface); L is the length of the bond; and τ^f is the shear strength at the interface. Equation (7.4) is based on the assumption that the shear stress is uniformly distributed along the bond length. This is not correct. As it will be discussed in more detail in Section 7.4.5 (see also Figure 7.38), the shear stress is the largest close to where the load is applied, at the proximal end of the bond, and decreases quickly with increasing distance along the bond, i.e. toward the distal end of the bond. The shear strength is reached first near the load application point and, with increasing load, failure progresses along the bond until the capacity of the system is attained when the entire interface has reached its shear strength. Nevertheless, experience shows that Equation (7.4) gives adequate results, at least for routine analyses. The value of τ^f can be obtained from field or laboratory tests (Benmokrane et al., 1992), or in the absence of reliable experiments, a first estimate is provided by the following empirical relation (Ballivy and Martin, 1983; Littlejohn, 1993):

At the rock-grout contact:

$$\tau^f = 1/30\sigma_c^{g,r} \quad (\text{max } 1.4\text{MPa}) \tag{7.5}$$

where $\sigma_c^{g,r}$ is the smallest unconfined compression strength of the rock mass or the grout. A factor of safety of three has been introduced in the above expression (that is, the equation gives values of τ^f three times larger than the actual strength). For weak rocks, equation (7.5) may yield values that are too small. Littlejohn (1993) provides a collection of strength values for cement-grout anchorages for a variety of igneous, metamorphic, and sedimentary rocks. Additional data can be found in Wyllie and Mah (2004).

At the grout/reinforcement element interface (Ballivy and Martin, 1983):

$$\begin{aligned} \tau^f &= 0.2\sigma_c \quad (\text{max. } 1.1\,\text{MPa}) \quad \text{for smooth steel bar} \\ \tau^f &= 0.1\sigma_c \quad (\text{max. } 2.4\,\text{MPa}) \quad \text{for deformed steel bar} \end{aligned} \tag{7.6}$$

where σ_c is the unconfined compression strength of the grout. The values of τ^f in (7.6) already incorporate a factor of safety between 2 and 2.5. Littlejohn (1993) has similar recommendations, with maximum τ^f values of 1.0 MPa for clean plain wire or plain bar; 1.5 MPa for clean crimped wire; 2.0 MPa for clean strand or deformed bar; and 3.0 MPa for locally nodded strands.

Analytical work on the pullout strength of rock anchors by Serrano and Olalla (1999) provide support for the empirical expressions in (7.4) to (7.6). The maximum pullout load of a long anchor is:

$$T^f = \pi d_g L \left(\frac{s^3}{8m^2} \right)^{\frac{1}{4}} \sigma_c \tag{7.7}$$

where T^f is the ultimate pullout load, d_g is the hole diameter, L is the anchor length, s and m are rock strength parameters, from the Hoek-Brown failure criterion, and σ_c is the unconfined compression strength of the intact rock. The equation is based on the hypothesis that failure occurs through the rock mass. Note that the product of $\pi\ d_g\ L$ is the area over which failure occurs, and thus the remaining term $(s^3/(8m^2))^{1/4}$ provides the shear stress at failure of the bolt-rock contact. For sound to very sound rocks ($80 \le \text{RMR} \le 100$) expression (7.7) provides a result very close to that of Equation (7.4), which gives a shear strength of about 10% of the unconfined compression strength of the rock. If Equation (7.7) is used to estimate τ^f, the results should be divided by an appropriate factor of safety, e.g. three. As mentioned, previous expressions, i.e. (7.5) and (7.6), already include a factor of safety.

Equations (7.4) to (7.7) rely on the assumption that the drill hole is large enough to accommodate the reinforcement elements and to provide a good embedment of the grout. For resin-bonded anchored rockbolts, the ratio of area of drill hole to area of reinforcement element should be based on resin-manufacturer specifications. For cement grouted rockbolts, the ratio of drill hole diameter d_g to element reinforcement diameter d_b should be within the range, d_g/d_b = 1.5–2.5 (Wyllie, 1999).

7.4.3.1 Local support

Section 4.2.3 contains a summary of the methods that can be used to identify unstable blocks or wedges and to quantify the load that needs to be applied to make the rock block stable with a satisfactory factor of safety. Ideally, the support should be rigid and installed as soon as possible. This is because small slip along the discontinuities that constitute the wedge can reduce the shear resistance from peak to residual. Mechanical or resin-anchored rockbolts are well suited for this task because, in addition to a fast installation, they can be tensioned, thus limiting deformations as long as the load in the rockbolts does not exceed its pre-load.

Grouted dowels and frictional rockbolts can also be used. In this case, however, the reinforcement system needs to deform with the rock to develop support. Such deformations can occur through rigid-body motions originated by translation or rotation of the wedge as it slips or detaches from its faces. In the case that deformations are allowed along discontinuities, the support design must satisfy equilibrium and kinematic conditions. In other words, the support must be able to provide the required load for a given factor of safety, but also the magnitude of the load must be compatible with the displacements that will occur. As the rock deforms, the rockbolt will experience both axial and shear loads, as depicted in Figure 7.21a. The magnitude of the loads depends, in addition to the engineering properties of the bolt, grout, and rock, on the slip along the discontinuity and on the angle of the rockbolt with respect to the direction of slip. When the rockbolt is perpendicular to the plane of the discontinuity, the failure of the bolt usually requires significant slip displacements. If, however, the bolt axis has a small angle with the plane of the discontinuity, the amount of slip needed to fail the bolt is very small (Pellet and Egger, 1996). It is thus required, for proper dimensioning of the rockbolts, to compute both the loads and displacements. The rockbolt load-response can be obtained experimentally (Ludvig, 1983; Spang and Egger, 1990; Grasselli, 2005) analytically (Pellet and Egger, 1996) or numerically (Spang and Egger, 1990; Brady and Lorig, 1988; Brady and Brown, 2004; Grasselli, 2005).

An analytical closed-form solution has been provided by Pellet and Egger (1996). The bolt is idealized as a semi-infinite elastic beam with an axial load N and a shear load Q at the location where the bolt intersects the plane of the discontinuity. See Figure 7.21b. The forces that produce first yield in the rockbolt, Q^e and N^e, are as follows:

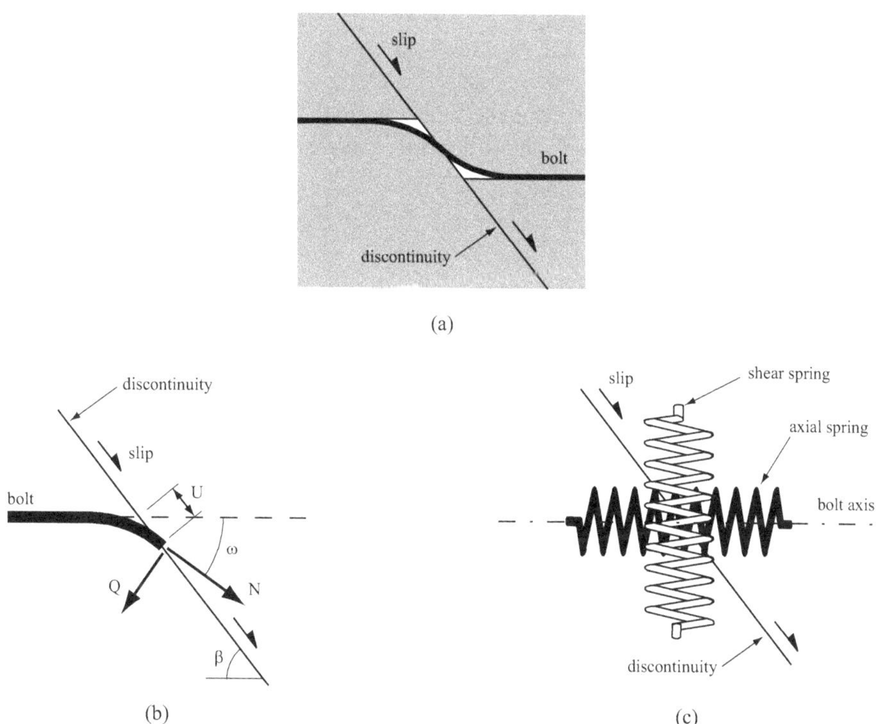

Figure 7.21 Models for rockbolts subjected to axial and shear loads. (a) Shearing of bolted discontinuity. (b) Forces and displacements of bolt (Pellet and Egger, 1996). (c) Axial and shear spring (Brady and Brown, 2004).

$$Q^e = \frac{1}{2}\sqrt{p_u d_b\left(\frac{1}{4}\pi d_b^2 \sigma_y - N^e\right)} \tag{7.8}$$

where p_u is the yield pressure of the grout filling the space between the bolt and the rock in the drill hole, d_b is the diameter of the bolt, and σ_y is the yield stress of the bolt material. The value of p_u can be estimated as:

$$p_u = K\sigma_c d_b \tag{7.9}$$

where K is a load factor ($K \geq 1$) and σ_c the unconfined compression strength of the grout.

The magnitude of N^e is obtained from Equation (7.8), after the solution of the following third order equation that gives Q^e:

$$(Q^e)^3 + (Q^e)^2 \frac{3p_u \pi^3 d_b \tan\beta}{256b} - \frac{3p_u^2 \pi^4 d_b^4 \sigma_y \tan\beta}{4096b} = 0 \tag{7.10}$$

where β is the angle between the plane of the discontinuity and the rockbolt axis (Figure 7.21b), and b is a constant, with approximate value b = 0.27.

The rockbolt displacement, U^e, and rotation, ω^e, at first yield are (Figure 7.21b):

$$\begin{aligned} U^e &= \frac{8192(Q^e)^4 b}{E_b \pi^4 d_b^4 p_u^3 \sin\beta} \\ \omega^e &= \frac{2048(Q^e)^3 b}{E_b \pi^3 d_b^4 p_u^2} \end{aligned} \tag{7.11}$$

where E_b is the Young's modulus of the rockbolt material.

Failure of the rockbolt is given by the following expression:

$$Q^f = \frac{1}{8}\pi d_b^2 \sigma_f \sqrt{1 - 16\left(\frac{N^f}{\pi d_b^2 \sigma_f}\right)^2} \tag{7.12}$$

where Q^f and N^f are the shear and axial load that induce failure, and σ_f is the failure stress of the bolt material. The loads and displacements at failure are:

$$\begin{aligned} Q^f &= Q^e \\ N^f &= \frac{1}{4}\pi d_b^2 \sigma_f \sqrt{1 - 64\left(\frac{Q^e}{\pi d_b^2 \sigma_f}\right)^2} \\ U^f &= U^e + \frac{Q^e \sin(\Delta\omega)}{p_u \sin(\beta - \Delta\omega)} \\ \omega^f &= \omega^e + \Delta\omega \\ \Delta\omega &= \cos^{-1}\left[\frac{1}{1+\varepsilon_f}\sin^2\beta \pm \cos\beta\sqrt{1 - \frac{\sin^2\beta}{(1+\varepsilon_f)^2}}\right] \end{aligned} \tag{7.13}$$

where ε_f is the bolt strain at failure.

An approach suitable for implementation into a numerical code has been suggested by Brady and Lorig (1988) and Brady and Brown (2004). The rockbolt is replaced by two springs, as shown in Figure 7.21c. One of the springs is oriented along the direction of the rockbolt axis (axial spring) and the other in the direction perpendicular to the rockbolt axis (shear spring). The springs have the following stiffness:

$$K_a = \pi d_b \sqrt{\frac{1}{2} \frac{G_g E_b}{(d_g/d_b - 1)}}$$
$$K_s = E_b I \left[\frac{1}{2} \frac{E_g}{E_b I} \frac{1}{(d_g/d_b - 1)} \right]^{3/2} \tag{7.14}$$

where K_a and K_s are, respectively, the stiffness of the axial and shear springs; d_b and E_b are the diameter and Young's modulus of the rockbolt; d_g, E_g, and G_g are the diameter, Young's modulus and shear stiffness of the grout; and I is the moment of inertia of the rockbolt-grout composite. The load-displacement relation is approximated with the following expression:

$$F_{a,s}^{t+\Delta t} = F_{a,s}^{t} + K_{a,s} \left| F_{a,s}^{f} - F_{a,s}^{t} \right| \frac{F_{a,s}^{f} - F_{a,s}^{t}}{(F_{a,s}^{f})^2} \left| U_{a,s}^{\Delta t} \right| \tag{7.15}$$

where F and U are force and displacement, and the indices a, s indicate the axial or shear spring. In the equation, $t + \Delta t$ is used to specify the value, in this case of the force, at the end of the current load increment, t is used for the values at the end of the previous increment, and Δt for the incremental value from the end of the previous to the current load increment. $F_{a,s}^{f}$ is the ultimate force of the axial or shear spring; it can be obtained from experiments, from equation (7.13), or from Brady and Lorig (1988):

$$F_a^f = 0.67 d_b^2 \sqrt{\sigma_y \sigma_c}$$
$$F_s^f = \pi d L \tau^f \tag{7.16}$$

where σ_y is the yield strength of the bolt; σ_c the unconfined compression strength of the rock; L is the bond length, τ^f is the peak shear strength of either the bolt-grout or the grout-rock interface, whichever is the most critical, and d the corresponding diameter of the cylinder where failure occurs: bolt-grout contact, d_b, or grout-rock contact, d_g. The value of τ^f can be obtained from field experiments or, in the absence of reliable experiments, from the relations given in (7.4) to (7.7).

7.4.3.2 General or systematic support

Systematic support is required to limit generalized deformations of the tunnel excavation and to help the rock support itself as it deforms and yields. A common feature of all rock reinforcement systems is that they contain elements that have a high capacity in tension and are usually stiffer than the supported rock (as mentioned earlier, the focus of the discussion is on reinforcement elements, i.e. rockbolts, and not on grouting of the ground). These two qualities are required because the reinforcement is embedded into the rock and attached to it such that it follows the displacements of the rock as the tunnel is excavated. As the ground moves toward the tunnel opening, the reinforcement elements elongate and are loaded in tension.

The tensile load is transmitted to the rock as a compression load, which increases the rock confinement, thus helping the rock support itself.

There are in essence three methods for the design of the support: empirical, analytical, and numerical. Empirical methods are presented in Chapter 2, which includes quantitative guidelines to estimate the spacing, size, length, etc. of the rockbolts needed for adequate support. The following sections provide the analytical, when available, and numerical methods that can be used for the design of rock reinforcement support. A distinction is made based on how the load is transferred from the rock mass to the support element, and thus mechanically and resin-anchored rockbolts, grouted dowels, split set and Swellex rockbolts, and cablebolts are each discussed separately.

In the following, it is assumed that the axial stiffness of a rockbolt is much larger than its shear stiffness and so a bolt can be represented as idealized one-dimensional element. This assumption is generally appropriate provided that the rock displacement field is continuous, but it may not be entirely appropriate when the rockbolts cross a discontinuity where localized slip occurs. In this case, the shear stiffness of the reinforcing element should be included or at least its design should consider the shear loads induced by the relative displacements between the two faces of the discontinuity.

7.4.4 Mechanically and resin-anchored rockbolts

Figure 7.22a shows an idealization of anchored rockbolts distributed around a tunnel (in the following, "anchored rockbolt" will be used to refer to mechanically and resin-anchored rockbolts). The bolts can be modeled as springs attached on one end to the rockbolt head and on the other end to the rockbolt bond. The rockbolt force F can be expressed as follows:

$$F = F_o + K_{anchor}\,\Delta U + F_{head} + F_{bond} \tag{7.17}$$

where F_o is the initial, pre-tension load if any, K_{anchor} is the stiffness of the spring; ΔU the relative displacements between the two ends of the bolt; and F_{head} and F_{bond} are the load losses at the rock–rockbolt connections. The spring constant, K_{anchor}, can be obtained from the material properties of the rockbolt and the free length of the bar, as follows:

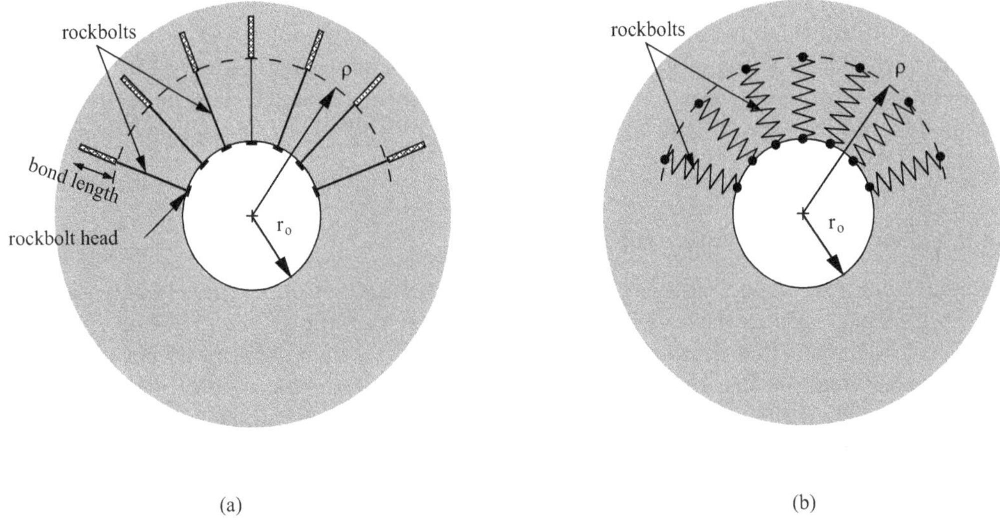

Figure 7.22 Anchored rockbolts. (a) Rockbolts. (b) Rockbolt model.

$$K_{anchor} = \frac{\pi E_b d_b^2}{4(\rho - r_o)S_z} \tag{7.18}$$

where E_b is the Young's modulus of the bar, d_b is the bar diameter, ρ the radial distance from the center of the tunnel to the end of the bolt, r_o is the radial distance from the center of the tunnel to the rockbolt head (i.e. the free length of the bolt, L, is $L = \rho - r_o$), and S_z the spacing of the anchored rockbolts along the tunnel axis. It is important to realize that the source of displacements may not be only the elongation of the rockbolt, but deformations may also be produced at the bond (F_{bond}), where the rockbolt may slip with respect to the grout or the grout may slip at the contact with the rock, or at the rockbolt head (F_{head}) where, e.g. the rock under the bolt plate may be damaged and the plate may lose support. The displacements associated with these other two sources are typically dependent on the magnitude of the load and may be significant, to the point that any tension that the rockbolt may have can be lost. It only takes a few millimeters to lose a load of magnitude equal to the capacity of a steel rebar 2 m long with $d_b = 20$ mm. Hence care should be taken during installation to accomplish very stiff connections between the rockbolt and the rock.

Figure 7.23 shows the results of a two-dimensional analysis of a deep circular tunnel with and without support. The results are obtained with the Finite Element Method ABAQUS. The

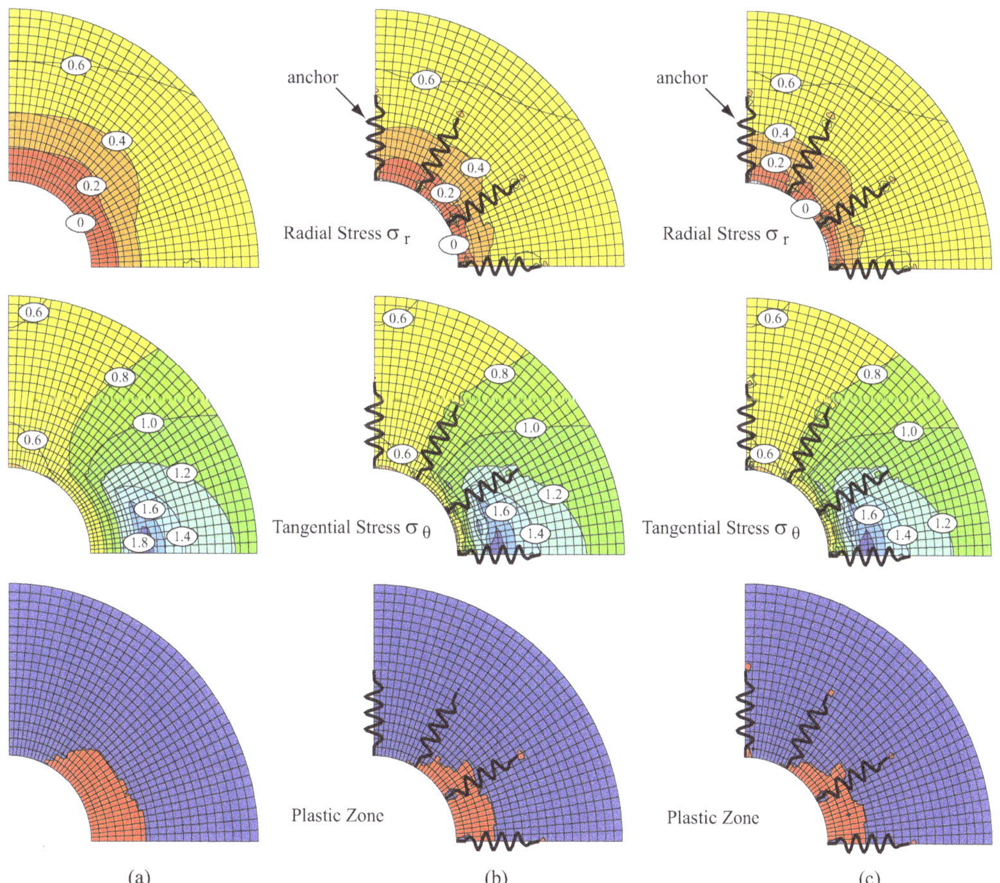

Figure 7.23 Tunnel support with anchored rockbolts; $r_o = 2$ m, $\sigma_v = 1.0$ MPa, $K_o = 0.5$, anchor length, L = 2 m, $d_b = 20$ mm. (a) Unsupported tunnel. (b) Untensioned anchors. (c) Tensioned anchors, F=100 kN.

tunnel is circular with radius, r_0 = 2 m, and subjected to far-field stresses σ_v = 1 MPa, K_o = 0.5. The ground is modeled as elasto-plastic with Coulomb failure with the following properties: Young's modulus E = 500 MPa, Poisson's ratio ν = 0.25; internal friction angle ϕ = 30°, dilation angle ψ = 20°, cohesion c = 0.15 MPa (the values adopted are too small for rock, even for weak rock; they are taken to facilitate discussion, as they provide a large contrast between the stiffness of the medium and the rockbolt and induce yielding). Two support cases are analyzed, both with steel-anchored rockbolts distributed uniformly around the perimeter of the tunnel at 30° angles (at about 1 m spacing) and at 1 m spacing along the axis of the tunnel. The anchored rockbolts have a free length of 2 m and have a diameter of 20 mm. In one case, the rockbolts are not tensioned and, in the other case, a pre-tension load of 100 kN (about 320 MPa stress) is applied. The analysis is done assuming a simultaneous excavation of the tunnel and support installation. As discussed in Chapter 4, such an analysis does not consider the deformations of the ground prior to the installation of the anchored rockbolts, and so the results underestimate ground deformations and overestimate rockbolt loads. The objective of the analysis is to illustrate the ground-anchored rockbolt interaction. For actual design, a three-dimensional analysis would be recommended. Also, no attempts have been made to optimize the discretization of the model. This is particularly important in problems such as this one where the magnitude of a concentrated load depends on displacements (Bobet, 2006). The results appear to be particularly sensitive to the discretization adopted when the ratio between the spring stiffness and the rock's Young modulus is high; i.e. for relatively soft rocks when K/E > 0.1, where K is the spring constant, as defined in equation (7.18), and E is the rock's Young modulus. For these cases, a fine mesh is generally needed, which needs to be designed and optimized on a case-by-case basis (in the case of Figure 7.23, the ratio is K/E = 0.07).

Figure 7.23a includes contour plots of the radial stresses, tangential stresses, and yield area of the ground around the tunnel when no support is provided. The figure shows yielding around the springline. At the springline, the tangential stresses are the largest and they occur at some distance from the excavation. As discussed in Chapter 4 this is because the ground can support larger deviatoric stresses as the confinement increases. Near the excavation, the confinement, provided by the radial stresses, is zero. With increasing radial distance, the radial stresses increase and so the tangential stresses also increase. With further distance, the effects of the excavation diminish and both the radial and tangential stresses slowly approach the far-field stresses. The results in Figure 7.23a can be compared with those of Figure 4.23. The difference between the two figures is that the analysis presented in Figure 4.23 is three-dimensional and in Figure 7.23a is two-dimensional. Figure 7.23b shows analogous plots but when non-tensioned anchored rockbolts are placed. The result is a slight increase of the radial stresses, a decrease of the tangential stresses and a reduction of the volume of the material that yields. Note also that the location of maximum tangential stress moves toward the excavation, as a result of the additional support provided by the rockbolts in the form of increased radial stress (the magnitude of the tangential stress is limited when the material yields; with the additional confinement provided by the anchored rockbolts, the extent of the plastic area around the tunnel decreases, as shown in the figure, and so the rock can attain higher tangential stresses). Figure 7.23c illustrates the effects of pre-tensioning the anchored rockbolts at a load of 100 kN. The radial stresses slightly increase, and the tangential stresses and the size of the ground yielding decrease. The benefits of tensioning the rockbolts are, however, not significant. This is because, in the case analyzed, the magnitude of the pre-load is too small to prevent deformations of the ground, and so the final rockbolt loads are larger than the initial pre-load. This can be observed better in Table 7.7 and Figure 7.24, which contain the rockbolt stresses and radial displacements of the ground along the perimeter of the excavation, respectively.

Table 7.7 Anchored rockbolt stresses. Results from tunnel of Figure 7.23

Rockbolt location θ^1	*Untensioned*	*Tensioned*
0°	361 MPa	346 MPa
30°	270 MPa	296 MPa
60°	213 MPa	335 MPa
90°	236 MPa	353 MPa

[1] Angle measured from horizontal; see inset in Figure 7.24

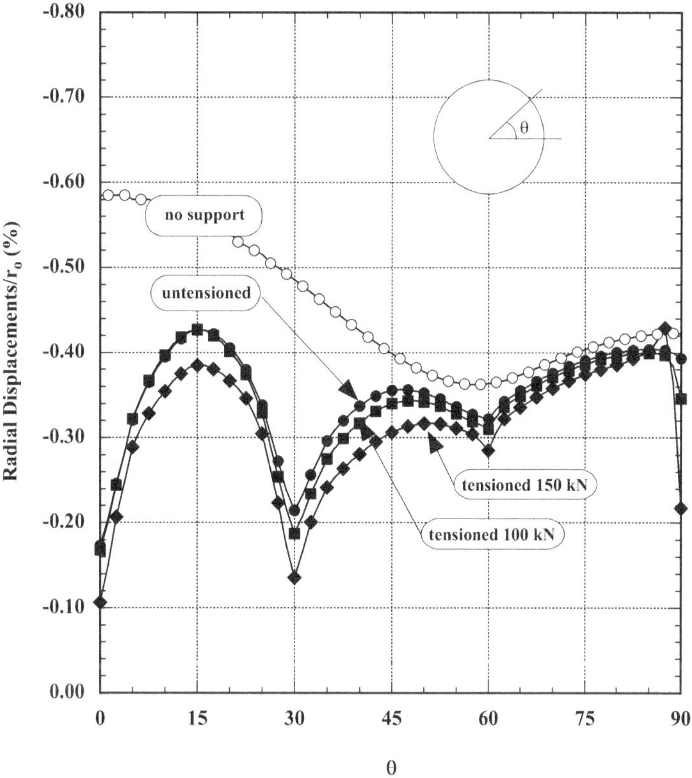

Figure 7.24 Radial displacement at tunnel perimeter, with r_o = 2 m, σ_v = 1.0 MPa, K_o = 0.5, anchor length = 2 m, d_b = 20 mm. Untensioned and Tensioned (100 kN and 150 kN) Anchored rockbolts. (Bobet and Einstein, 2011). Reproduced with permission of Elsevier.

Pre-tensioning of the anchored rockbolts better distributes the rockbolt loads around the tunnel, and even though the maximum load is approximately the same with and without pre-load, the radial displacements are much smaller, by a factor larger than two at the crown. This is, of course, at the expense of transferring loads to the rockbolts around the crown. Better results could have been obtained by increasing the magnitude of the pre-tension load, which in this case would have required a larger bolt diameter. Indeed when the anchor diameter is increased to 25 mm and the rockbolts are tensioned to 120 kN, no further displacements of the ground occur and the rockbolts keep the pre-tension load. This results in smaller tunnel convergence and smaller size of the plastic zone (not shown in the figures). Figures 7.23 and

7.24 show that, even with a small number of anchored rockbolts, the support benefits in terms of reduced plastic zone size and smaller tunnel convergence are significant.

Analytical solutions exist for elastic and elasto-plastic rock when the problem is axisymmetric, which requires a circular cross section, homogeneous, and isotropic ground and uniform far-field loading. A solution has been found by treating the anchored rockbolts forces as a two uniformly compressive distributed loads applied at both ends of the rockbolt (Labiouse, 1992, 1996; Bobet, 2002). Also, an analytical solution has been provided for any number of anchored rockbolts, with any distribution around the tunnel and for non-uniform far-field stresses (Bobet, 2006), but with the assumption of elastic ground response. In the solution, the anchored rockbolts are simulated as two collinear and opposite concentrated forces, one at the rockbolt head and the other at the rockbolt bond. The magnitude of the forces is equal to the anchored rockbolt spring constant times the relative displacement between the points of application of the forces, or equal, if the rockbolt is pre-tensioned, to the pre-load plus any additional load due to the relative displacements of the rockbolt end points for loads beyond the initial load.

The advantage of a closed-form solution is that the key variables that determine the ground-anchored rockbolt response can be readily identified and their relative importance quickly estimated. From results of a parametric study, Bobet (2006) determined that the most important factor that controls the load on an anchored rockbolt in an elastic ground is its relative stiffness with respect to that of the ground, defined in terms of the ratio E/E_b, or the ratio between the Young's modulus of the ground and that of the rockbolt material.

Given the limitations inherent in the analytical solutions, a three-dimensional numerical method will be generally required to compute the loads of the anchored rockbolts, and the stresses and displacements of the ground. An iterative scheme will be typically needed to optimize the number, diameter, length and pre-tension (if any) of rockbolts required to make the excavation stable and limit the radial convergence. The maximum rockbolt load should be smaller than the yield load of the bolt material, the yield load of the rockbolt head, and smaller than the pullout load of the rockbolt bond. The length of the rockbolt bond can be estimated using equations (7.4) to (7.7). Also, deformations that occur with time need to be evaluated since they may contribute to reduce the rockbolt loads. These include creep of the rock under the anchored rockbolt head plate and around the rockbolt bond.

It is informative to discuss the interaction between mechanically and resin-anchored rockbolts (anchored rockbolts), the tunnel, and the ground using the analytical solution obtained by Bobet and Einstein (2011). The solution was obtained given the following assumptions (see Figure 7.25): (a) circular cross section with radius r_o; (b) the tunnel is deep, and so far-field stresses can be approximated with a uniform distribution with magnitude σ_o equal to the unit weight of the rock times the depth of the tunnel center; (c) the ground is homogeneous and isotropic, and remains elastic until reaching failure, which is defined by a Coulomb failure envelope; (d) post-failure behavior follows a non-associated flow rule defined by the dilation angle of the ground; (e) the anchored rockbolts are uniformly distributed around the tunnel cross section with spacing S°_{θ} and along the tunnel axis with spacing S_z, and their load is assumed as a uniformly distributed pressure; and (f) construction effects, i.e. three-dimensional effects, can be approximated using the β-method (see below). As shown in Figure 7.25, the rockbolts have a length of ρ-r_o and the tunnel has an internal support pressure with magnitude σ_i. Under the conditions described, the problem is axisymmetric. In other words, stresses and displacements are independent of the polar coordinate θ.

The β-method consists of two steps (Section 4.3.3). In the first step, the tunnel is excavated and a pressure equal to $\beta\ \sigma_o$ is applied to the perimeter of the opening; σ_o is the far-field stress and $0 < \beta < 1$ is the stress reduction factor. In the second step the "support," i.e. the reinforcement is placed and the stress $\beta\ \sigma_o$ is shared by the ground and the support. The magnitude

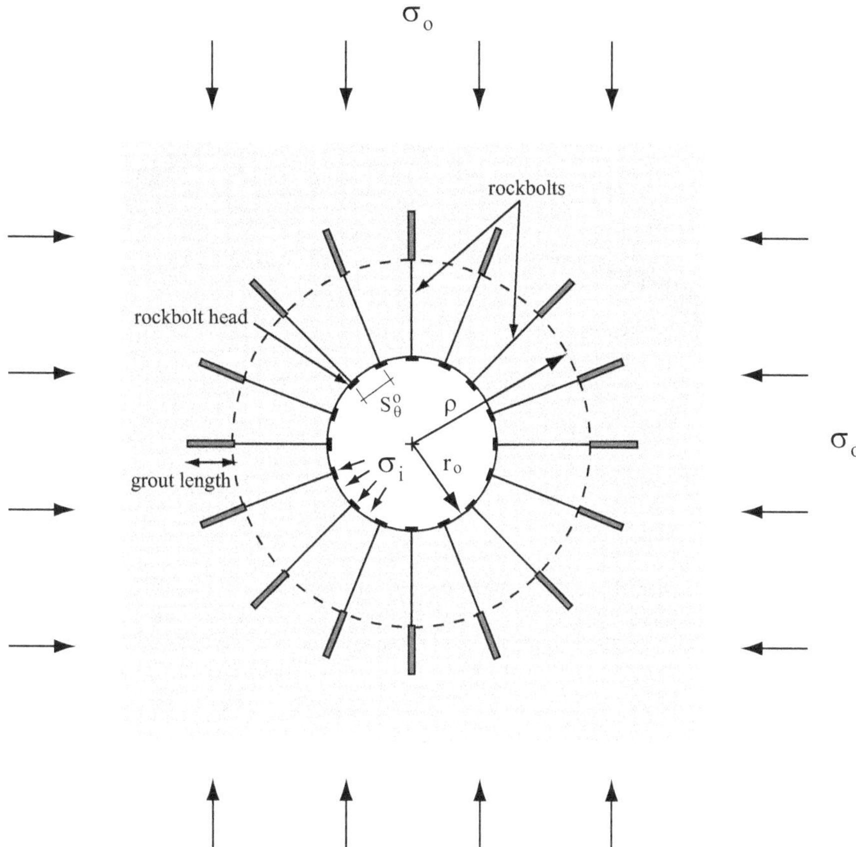

Figure 7.25 Tunnel with anchored rockbolts (Bobet and Einstein, 2011). Reproduced with permission of Elsevier.

of β is related to the "delay" in placing the support. Thus a large unsupported length of the tunnel is associated with a small β; and vice versa, a short unsupported length has a large β. A large factor β results in small displacements of the ground, small radial displacements at the perimeter of the excavation and large loads of the support. In contrast, a small β factor results in large ground displacements and small support loads. In the method, the magnitude of β must be known a priori. A typical value of β is 0.5, but it can range from 0.2 to 0.8. In addition, appropriate values of β strongly depend on the construction method, ground stiffness, plasticity model, K_o, unsupported tunnel length, etc. (Möller, 2006).

Figures 7.26 and 7.27, for untensioned anchored rockbolts, and 7.28 and 7.29, for tensioned anchored rockbolts, show a comparison between results obtained with the analytical solution and the Finite Element Method (FEM) software ABAQUS (ABAQUS, 2005). Figures 7.26a and 7.28a are plots of the rock radial stresses with radial distance; Figures 7.26b and 7.28b show rock tangential stresses, and Figures 7.27 and 7.29 show the radial displacements also as a function of radial distance from the tunnel center (positive radial displacements denote outward movement). The results apply to a tunnel with radius r_o = 3 m, subjected to a far-field stress σ_o = 1 MPa and internal pressure σ_i = 0. The properties of the ground are: E = 500 MPa, ν = 0.2, peak and residual strength $\phi^p = \phi^r$ = 30°, $c^p = c^r$ = 0.1 MPa. The rockbolts have a length of 3 m, i.e. ρ = 6 m, diameter d_b = 25 mm, and elastic properties E_b = 210 GPa, ν_b = 0.3, and T_o = 0.177 MN for the load applied to the tensioned rockbolts. The initial

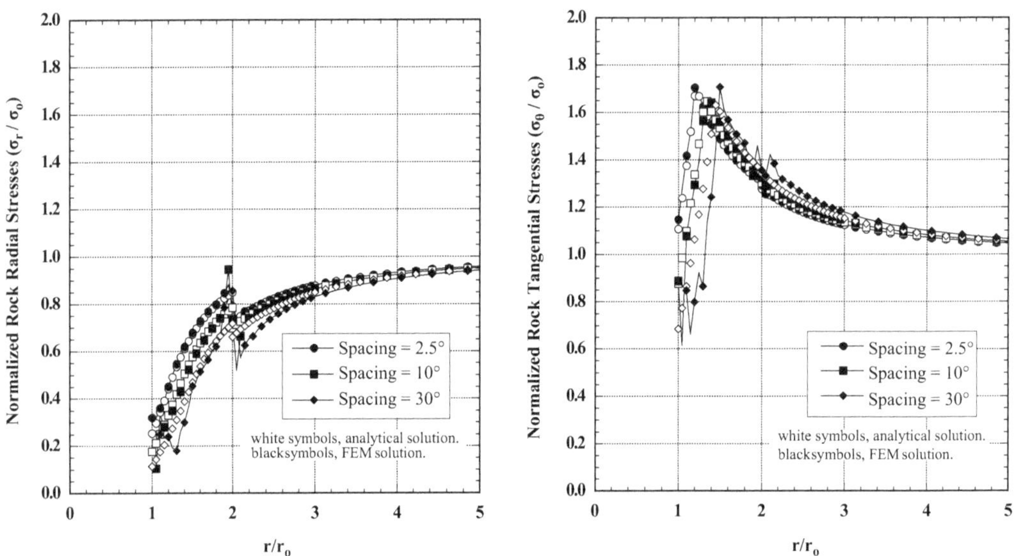

Figure 7.26 Untensioned anchored rockbolts. Normalized rock stresses for r_o = 3 m, ρ = 6 m, σ_o= 1 MPa, σ_i = 0, E = 500 MPa, ν = 0.2, $\phi^p = \phi^r$ = 30°, $c^p = c^r$ = 0.1 MPa, d_b = 25 mm, E_b = 210 GPa, ν_b = 0.3, S_z = 1 m, β = 0.3. (Bobet and Einstein, 2011). Reproduced with permission of Elsevier.

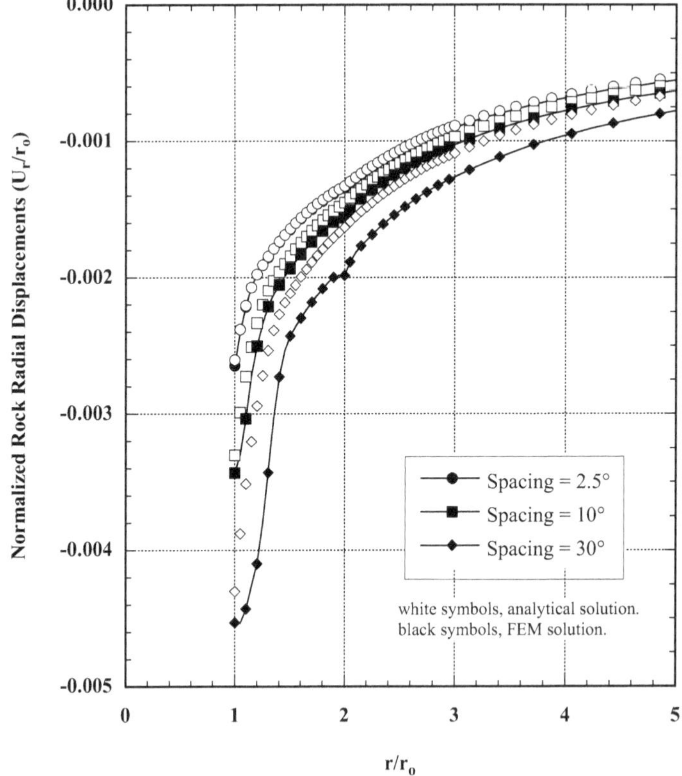

Figure 7.27 Untensioned anchored rockbots. Normalized radial displacements for r = r_o = 3 m, ρ = 6 m, σ_o= 1 MPa, σ_i = 0, E = 500 MPa, ν = 0.2, $\phi^p = \phi^r$ = 30°, $c^p = c^r$ = 0.1 MPa, d_b = 25 mm, E_b = 210 GPa, ν_b = 0.3, S_z = 1 m, β = 0.3. (Bobet and Einstein, 2011). Reproduced with permission of Elsevier.

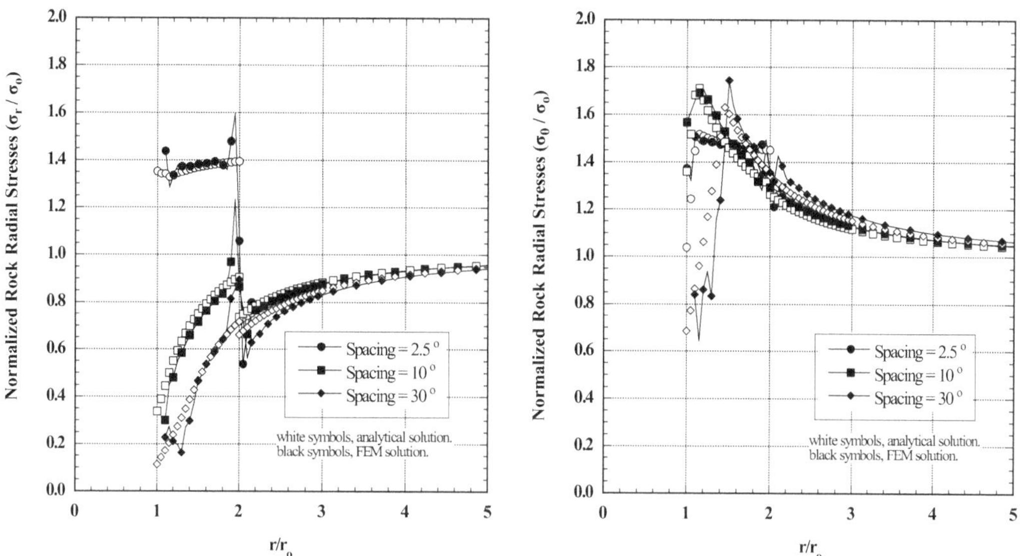

Figure 7.28 Tensioned anchored rockbolts. Normalized rock stresses for r_o = 3 m, ρ = 6 m, σ_o= 1 MPa, σ_i = 0, E = 500 MPa, ν = 0.2, $\phi^p = \phi^r$ = 30°, $c^p = c^r$ = 0.1 MPa, d_b = 25 mm, E_b = 210 GPa, ν_b = 0.3, S_z = 1 m, T_o = 0.177 MN, β = 0.3. (Bobet and Einstein, 2011). Reproduced with permission of Elsevier.

unloading of the excavation is represented with β = 0.3. The results are plotted for three different spacings $S°_\theta$ (S_z = 1 m in all cases), given by the angle between the anchored rockbolts, 2.5°, 10° and 30°, which correspond for r_o = 3m, to 0.13 m, 0.52 m, and 1.57 m spacing. The first spacing is impractically small, while the other two are more reasonable. The small spacing, however, aligns better with the assumption of a distributed rockbolt load; the other two cases with larger spacing are used to evaluate the sensitivity of the analytical solution to the smeared approach (with an spacing of 1.57 m, there are only 12 anchored rockbolts around the tunnel perimeter). The figures show a good agreement between the numerical and the analytical solutions, with small errors, albeit increasing as the rockbolt spacing increases. The radial stresses, as expected, increase with radial distance from the tunnel perimeter; the magnitude is not zero at the perimeter of the excavation due to the support provided by the anchored rockbolts heads. The jumps in radial stresses in Figures 7.26a and 7.28a occur at the distal end of the rockbolt, where the bolt load creates a discontinuity in the rock radial stress. The tangential stresses, in contrast, increase from the opening until they reach a peak and decrease afterward with radial distance until they reach the far-field stress magnitude. The peaks in Figures 7.26b and 7.28b occur at the transition between the plastic and the elastic zones of the rock. Note that the size of the plastic zone, determined by the position of the peak in the tangential stress plots, increases with rockbolt spacing. Close to the tunnel, the radial stresses are small and so the tangential stresses must also be small as their magnitude is limited by the failure envelope of the rock. As the radial distance increases, the radial stresses increase, and so the tangential stresses also increase. As mentioned, the peak tangential stress is reached at the limit of the plastic zone. With increasing radial distance the response is elastic, which is characterized by a slow increase in radial stresses and a decrease in tangential stresses; both stresses slowly converge to the magnitude of the far-field stress. Radial displacements, which are depicted in Figures 7.27 and 7.29, are larger near the opening and decrease in magnitude as the distance from the opening increases. The slope of the displacement plots, or in other words, the rate of displacement change, is much larger near the opening than far from the opening, which is due to the yielding of the rock. The case for

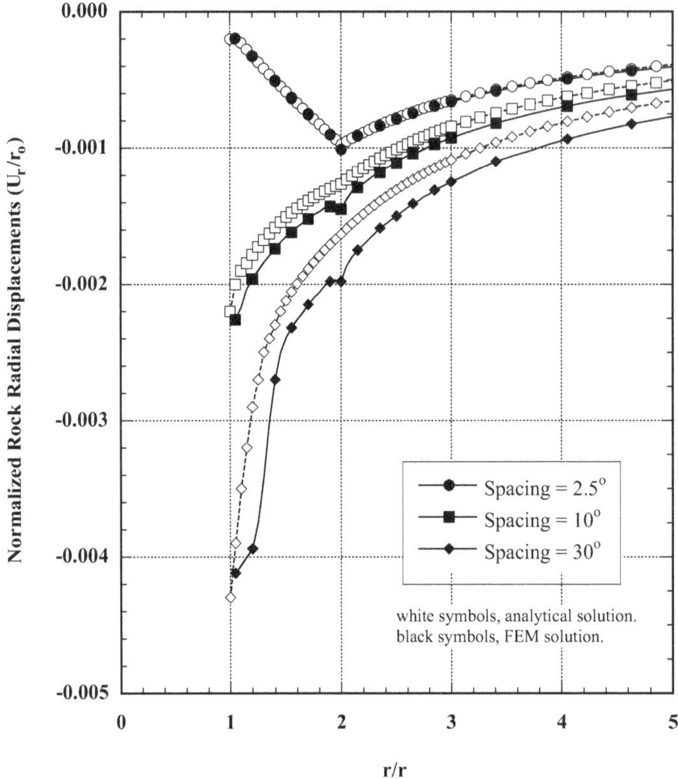

Figure 7.29 Tensioned anchored rockbolts. Normalized rock stresses for r_o = 3 m, ρ = 6 m, σ_o= 1 MPa, σ_i = 0, E = 500 MPa, ν = 0.2, $\phi^p = \phi^r$ = 30°, $c^p = c^r$ = 0.1 MPa, d_b = 25 mm, E_b = 210 GPa, ν_b = 0.3, S_z = 1 m, T_o = 0.177 MN, β = 0.3. (Bobet and Einstein, 2011). Reproduced with permission of Elsevier.

Table 7.8 Errors between analytical solution and FEM

	Untensioned anchored rockbolts		*Tensioned anchored rockbolts*
Spacing (degrees)	*Tension load (MN)*	*Plastic zone size (m)*	*Plastic zone size (m)*
2.5°	2.3%	1.0%	1.2%
5°	3.4%	0.8%	1.2%
10°	1.8%	0.0%	2.8%
15°	7.6%	5.6%	3.9%
30°	8.3%	0.2%	0.2%

the tensioned anchored rockbolts with spacing 2.5° in Figure 7.29 shows a complete reversal of slope compared to the other cases; this is due to the large load induced in the ground by pre-tensioning; the presence of such a large load is supported by the large radial stresses induced in the rock near the excavation, as shown in Figure 7.28a.

Table 7.8 lists the errors between the analytical and numerical solutions of the plastic zone sizes for the cases shown in the figures, as well as the errors of the rockbolt loads for the case of untensioned anchored rockbolts. The small magnitude of the errors supports the previous conclusion that the analytical solution provides acceptable results.

Figure 7.30 plots the ground characteristic curves for the same cases as those in Figures 7.26 to 7.29. All anchored rockbolts are untensioned. Figure 7.30a shows the characteristic curves for the unsupported ground and for anchored rockbolts at spacings 2.5°, 10°, and 30°. Even with a relatively small number of rockbolts, the benefits, in terms of radial displacements are significant; there is a reduction by about a factor of two with respect to the displacements of the excavation without reinforcement. As expected, decreasing the rockbolt spacing decreases the radial deformations while simultaneously decreasing the rockbolt loads, as shown in Figure 7.30b. It is interesting to note that the magnitude of the load, as the spacing decreases, decreases much faster than the reduction in radial deformations, which indicates that, with appropriate ground conditions, a smaller number of anchored rockbolts with larger loads may be a more economical solution that placing anchors at a smaller spacing with smaller loads. Figure 7.31 explores the effects of spacing, rockbolt length, rockbolt cross-section area, and placement of rockbolts with respect to the tunnel face. A base case is analyzed first, with the following input parameters: $r_o = 3$ m, $\rho = 6$ m, $\sigma_o = 1$ MPa, E = 500 MPa, $\nu = 0.2$, $\phi^p = \phi^r = 30°$, $c^p = c^r = 0.1$ MPa, $d_b = 25$ mm, $E_b = 210$ GPa, $\nu_b = 0.3$, spacing $S_\theta° = 10°$, $S_z = 1$ m, and $\beta = 0.3$. The tunnel radial deformations and rockbolt loads are plotted in Figures 7.31a and b, respectively, and are compared with results obtained with the same input parameters when the rockbolt spacing is reduced by half, i.e. spacing = 5°, the rockbolt length is doubled, i.e. $\rho = 9.0$ m, the rockbolt cross section is doubled, i.e. $d_b = 35$ mm, and when β is increased to 0.6 representing a construction process where the anchored rockbolts are placed closer to the tunnel face (i.e. a decrease in support delay). Figure 7.31a shows that the effects on tunnel convergence are modest, even though the changes imply doubling the material (e.g. steel) used for reinforcement or decreasing the support delay. The changes on the rockbolt load are most significant when the spacing is reduced by half. Increasing the length of the anchored rockbolts, the rockbolts' cross section and the support delay have

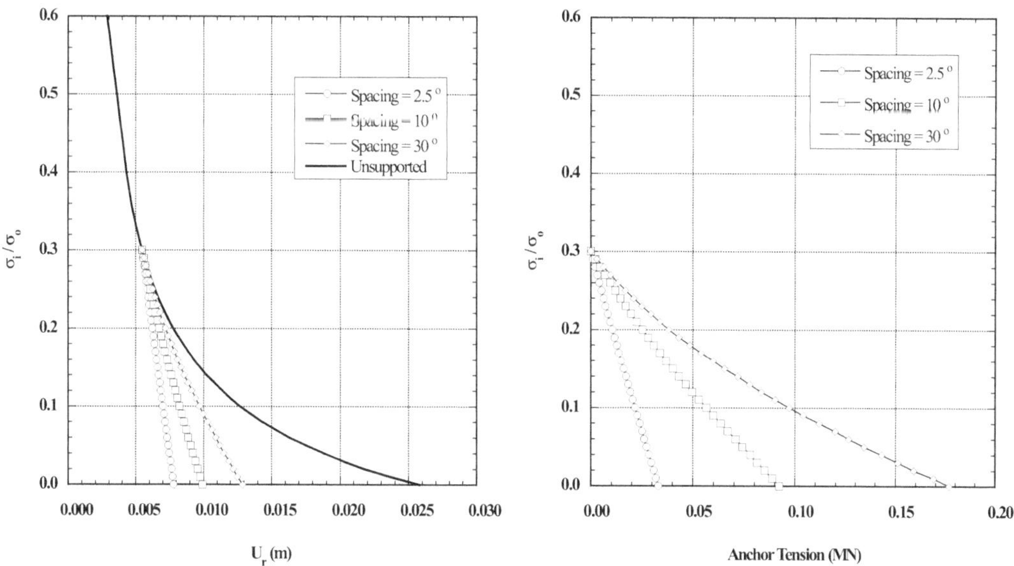

Figure 7.30 Untensioned anchored rockbolts. Radial displacements and anchor tension for r_o= 3 m, ρ = 6 m, σ_o = 1 MPa, E = 500 MPa, ν = 0.2, $\phi^p = \phi^r$ = 30°, $c^p = c^r$ = 0.1 MPa, d_b = 25 mm, E_b = 210 GPa, ν_b = 0.3, S_z = 1 m, β = 0.3. (Bobet and Einstein, 2011). Reproduced with permission of Elsevier.

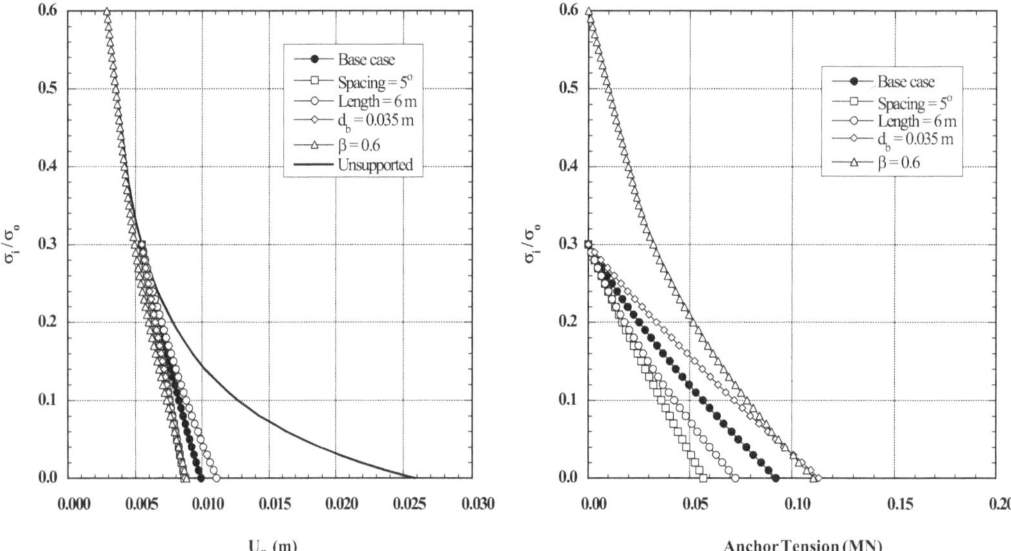

Figure 7.31 Untensioned anchored rockbolts. Effect of anchor spacing, length and cross section. Radial displacements and anchor tension. Base case: r_o= 3 m, ρ = 6 m, σ_o = 1 MPa, E = 500 MPa, ν = 0.2, $\phi^p = \phi^r$ = 30°, $c^p = c^r$ = 0.1 MPa, E_b = 210 GPa, d_b = 25 mm, S_z = 1 m, S_θ = 10°, β = 0.3. (Bobet and Einstein, 2011). Reproduced with permission of Elsevier.

modest effects. The results from the support delay seem counterintuitive, as one expects that the closer the support is placed to the tunnel face, the smaller the tunnel deformations and the larger the support loads. This is still the case, as shown in Figure 7.31, except that the effects are small. This is caused by the fact that the anchored rockbolts have been placed while the ground is still undergoing elastic deformations. Plastic deformations start to occur quite late, at σ_i/σ_o = 0.3. Within the interval $0.3 \le \sigma_i/\sigma_o \le 0.6$ the ground deformations are thus small and so the contribution of the anchored rockbolts is modest. This is an interesting observation that indicates that the installation of anchored rockbolts is most effective while the ground undergoes plastic deformations. Installation of reinforcement within the elastic regime of the ground does not provide significant benefits. This is a most important consideration, and has larger implications than the length of the rockbolts, spacing, and cross section.

There are two important issues not addressed by the analytical solution: (1) non-uniform far-field loading; and (2) non-uniform distribution of rockbolts around the tunnel. These issues are numerically investigated in the following. The FEM code ABAQUS is used for the simulations, following the steps described previously for the verification of the analytical solution. The ground properties and bolt characteristics are the same as the base case (i.e. r_o = 3 m, ρ = 6 m, σ_o = 1 MPa, E = 500 MPa, ν = 0.2, $\phi^p = \phi^r$ = 30°, $c^p = c^r$ = 0.1 MPa, d_b = 25 mm, E_b = 210 GPa, ν_b = 0.3, spacing $S_\theta^o = 10^\circ$, and β = 0.3), except for the far-field loading and bolt distribution. Four cases are analyzed. In the first case the far-field stresses are given by a vertical stress σ_v = 1 MPa, and a horizontal stress σ_h = 0.5 MPa; i.e. K_o = 0.5 with rockbolts uniformly distributed at a spacing of 10°. In the second case, the far-field stresses are uniform, $\sigma_v = \sigma_h = \sigma_o$ = 1 MPa but only seven rockbolts are placed symmetrically at the crown with a spacing of 10° (i.e. the first rockbolt is placed at 60° from the springline and the last at 120°). The third and fourth cases combine the previous two: K_o = 0.5 and seven rockbolts either at the crown (case 3) or at the springline (case 4). See Figure 7.32. The two first cases are intended to explore the effects of K_o and non-uniform bolt distribution independently, while the last two cases explore the effects of the two issues combined.

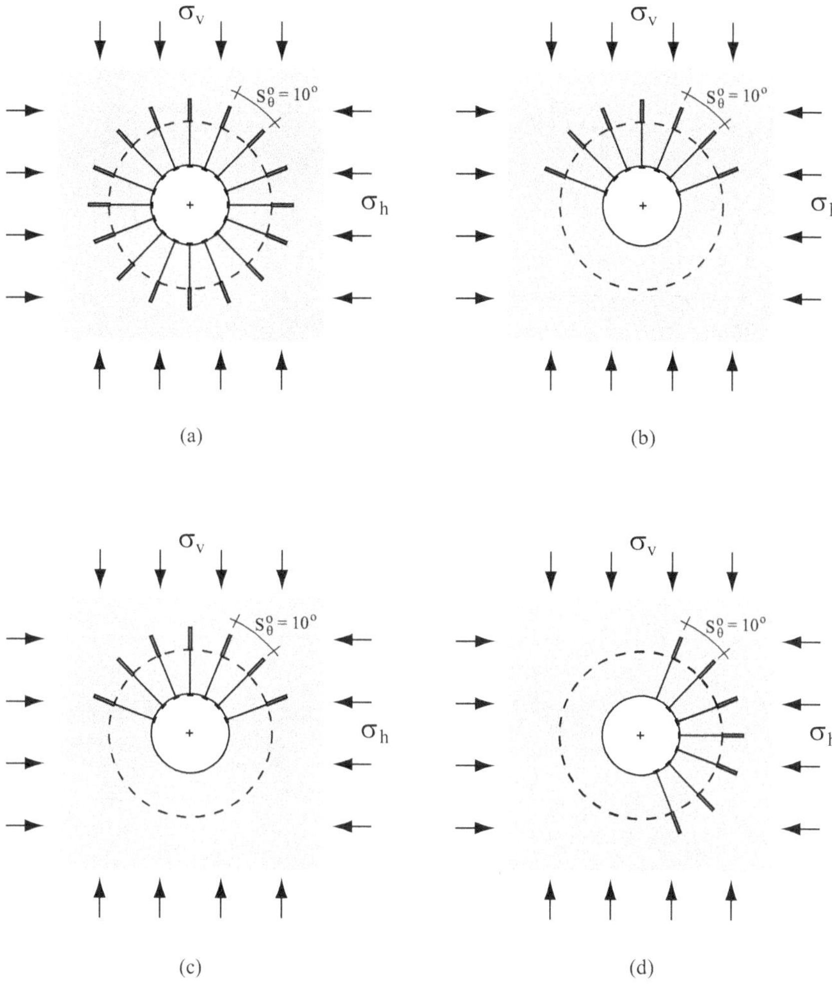

Figure 7.32 Cases investigated numerically (Bobet and Einstein, 2011). (a) Case 1. σ_v = 1 MPa, σ_h = 0.5 MPa, $S_0°$ = 10°. (b) Case 2. $\sigma_v = \sigma_h$ = 1 MPa, $S_0°$ = 10°. (c) Case 3. σ_v = 1 MPa, σ_h = 0.5 MPa, $S_0°$ = 10°. (d) Case 4. σ_v = 1 MPa, σ_h = 0.5 MPa, $S_0°$ = 10°. Reproduced with permission of Elsevier.

The results are presented in Figure 7.33. Figure 7.33a is a plot of the radial displacements at the tunnel perimeter with internal pressure σ_i for cases 1 and 2, and Figure 7.33c is an analogous plot, but for cases 3 and 4. Both figures include the characteristic curves of the unreinforced ground for K_o = 1 and 0.5. Figures 7.33b and 7.33d show, respectively for cases 1–2 and 3–4, the bolts' tension for different values of the internal support stress σ_i.

Effect of K_o: A non-uniform far-field stress changes the response of the unreinforced tunnel. Figure 7.33a includes the radial displacements of a point at the springline and a point at the crown for K_o = 0.5, and of a point at the tunnel perimeter for K_o = 1. The non-uniform loading induces at the springline smaller deformations than the uniform far-field loading. At the crown, the radial displacements are larger with K_o = 0.5 than K_o = 1 for internal support pressures larger than about 14% of the vertical far-field stress. A uniform distribution of the anchored rockbolts (case 1 and Figure 7.33a) produces the largest decrease of radial convergence at the springline. A comparison with results

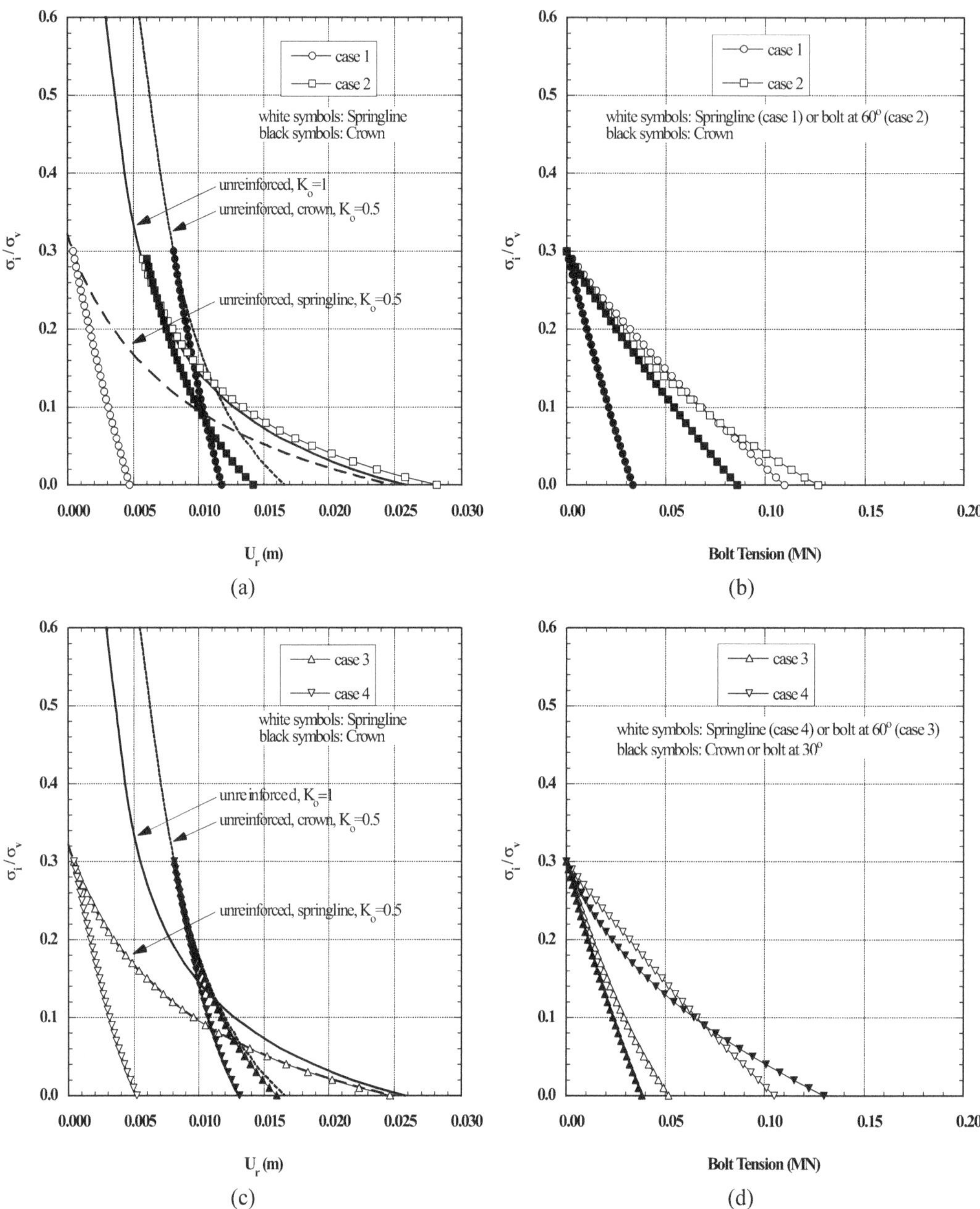

Figure 7.33 Untensioned anchored rockbolts. FEM results for radial displacements and bolt tension. Effects of K_o and non-uniform distribution of rockbolts. See Figure 7.32 for description of cases. r_o= 3 m, ρ = 6 m, σ_v = 1 MPa, E = 500 MPa, ν = 0.2, $\phi^p = \phi^r$ = 30°, $c^p = c^r$ = 0.1 MPa, E_b = 210 GPa, ν_b = 0.3, d_b = 25 mm, S_z = 1 m, S_θ = 10°, β = 0.3. (Bobet and Einstein, 2011). Reproduced with permission of Elsevier.

shown in Figure 7.30, which are obtained for K_o = 1 and also uniform rockbolt distribution, indicates that the reinforcement with non-uniform loading results in a substantial decrease of deformations at the springline, compared with the case with K_o = 1, and in somewhat larger deformations at the crown. This is associated with a smaller rockbolt load at the crown and a larger load at the springline (Figures 7.33b and 7.30b).

Effect of rockbolt distribution. Case 2 in Figures 7.33a and 7.33b can be used to observe the effect of concentrated anchored rockbolts with uniform far-field distribution. The results can be compared with those of Figure 7.30, which apply to the same case but with uniform rockbolt distribution. Figure 7.33a shows that the reinforcement is ineffective at the springline, where radial displacements are similar to those of the unreinforced tunnel. A larger reduction is found at the crown, but the radial displacements are somewhat larger than those of the tunnel with uniform rockbolt distribution (Figure 7.30). The largest rockbolt load however (case 2 in Figure 7.33b, bolt at 60°) is larger than with uniform reinforcement (Figure 7.30). The effects of different rockbolt distributions and $K_o = 0.5$ can be determined by inspection of results from cases 1, 3 and 4 in Figures 7.33a and 7.33c for radial displacements and 7.33b and 7.33d for rockbolt loads. Figure 7.33c shows that concentrating the anchored rockbolts at the crown (case 3) has negligible effects in terms of tunnel convergence since the radial displacements plotted in the figure are very similar to those of the unreinforced tunnel. The rockbolt loads, however, are substantial, albeit much smaller than those found in cases 1 and 4. Placement of the rockbolts around the springline (case 4) results in tunnel deformations at the springline similar to those of case 1, tunnel with uniform bolt distribution, and in slightly larger deformations at the crown. The rockbolt loads are larger for case 4 than for case 1. This observation, together with the results from case 3, highlights the importance of proper placement of the reinforcement given the far-field stress conditions, which determine what areas of the rock, if unreinforced, will deform more than others and thus will require concentrated support. This can be estimated based on the characteristic curves of the unreinforced ground that show that the ground at the springline would deform much more than at the crown, and thus the springline should be the location where rockbolts are more effective.

7.4.5 Grouted dowels, split set stabilizers, and Swellex rockbolts

What distinguishes grouted dowels/bolts, Continuously Mechanically Coupled (CMC) rockbolts, and Split Set Stabilizers and Swellex, Continuously Frictionally Coupled (CFC), rockbolts from the Mechanically and Resin Anchored, DMFC, rockbolts discussed in Section 7.4.4, is that the former become part of the rock as they are intimately connected to it, while the latter can be considered as an external support. In the following, the discussion focuses on grouted dowels, even though many of the observations made and conclusions reached are also applicable to CMC, CFC, Split Set Stabilizers, and Swellex rockbolts, unless noted otherwise.

Figure 7.34 shows contour plots of radial, tangential stresses and plastic zone size for a tunnel with the same geometry, ground properties, and loading as the one discussed in Section 7.4.4. That is, $r_o = 2$ m, $\sigma_v = 1$ MPa, $K_o = 0.5$, $E = 500$ MPa, $\nu = 0.25$, $\phi = 30°$, $\psi = 20°$, $c = 0.15$ MPa. Two support cases are analyzed, both with grouted dowels with diameter $d_b = 20$ mm, drill hole diameter $d_g = 32$ mm, at 1 m spacing along the axis of the tunnel. In the first case, Figure 7.34a, the dowels are spaced at 30° (at about 1 m spacing) while in the second case, Figure 7.34b, the spacing is at 15° (at about 0.5 m between rockbolt axes). A comparison between Figure 7.23a, which plots contours for the unsupported tunnel, and Figure 7.34a shows that the addition of the grouted dowels increases the radial stresses, thus increasing confinement, the maximum tangential stresses move toward the opening, and the size of the plastic zone decreases. As the number of dowels increases, Figure 7.34b, the support effects increase; as the radial stresses increase, the largest tangential stresses move further toward the opening and the volume of rock that yields decreases. This is because there is a transfer of load from the rock to the dowels, which work in tension, as indicated in Figure 7.35, a plot

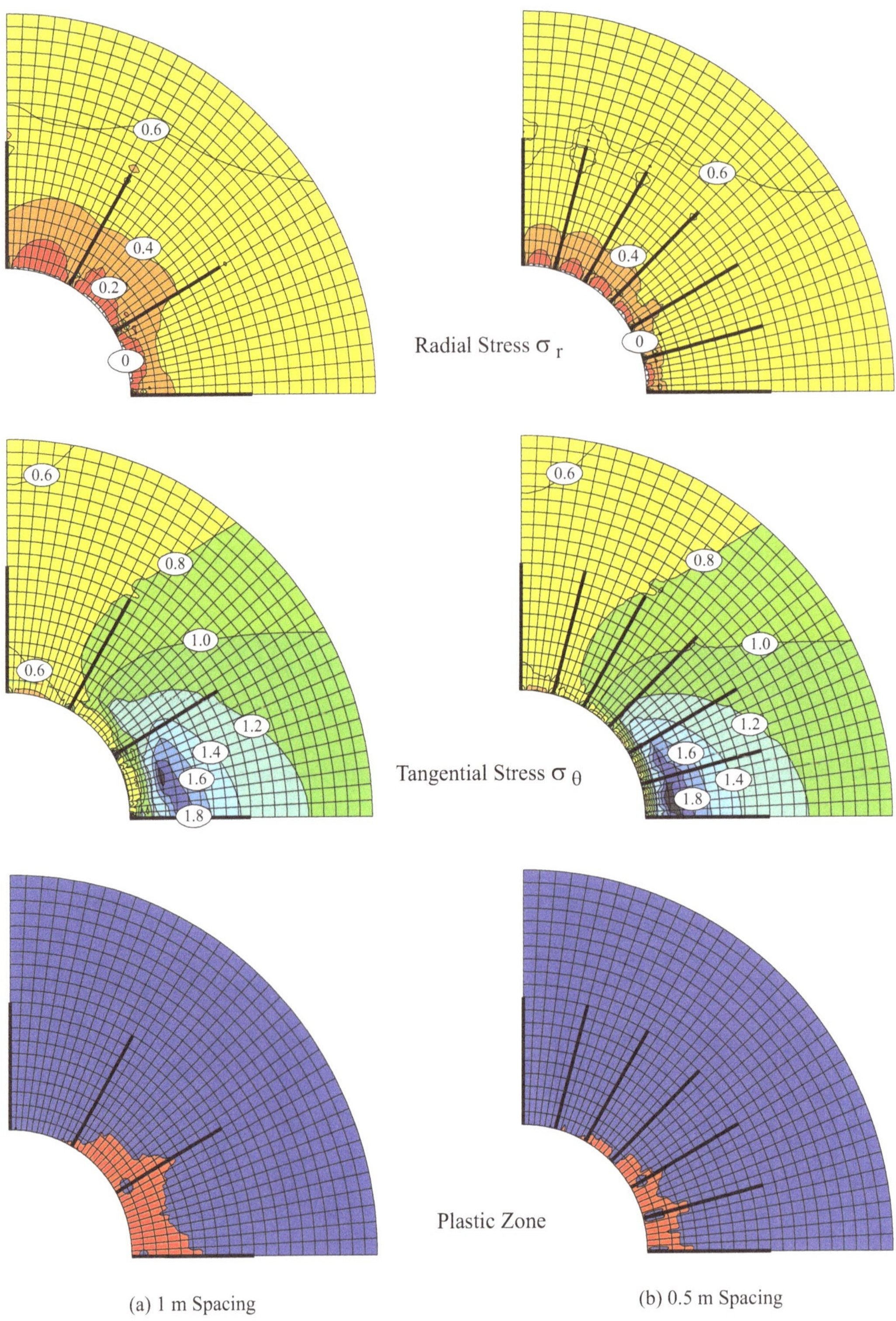

Figure 7.34 Effect of dowels on tunnel support; r_o = 2 m, σ_v = 1.0 MPa, K_o = 0.5, Rockbolt length, L = 2 m, d_b = 20 mm, d_g = 32 mm. (a) 1 m Spacing. (b) 0.5 m Spacing.

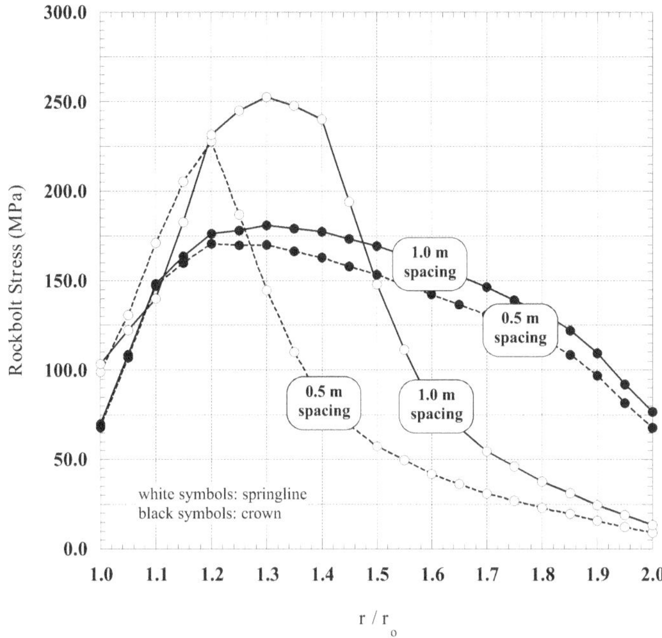

Figure 7.35 Radial stresses of dowels in tunnel with r_o= 2 m, σ_v = 1 MPa, K_o = 0.5, Rockbolt length = 2 m, d_b = 20 mm, d_g = 32 mm.

of the stresses along the dowels placed at the springline and at the crown (in the figure the radial distance from the center of the tunnel, r, is normalized by the radius of the tunnel, r_0). The largest stresses are found at the springline, which is where the rock yields and where the displacements are thus the largest (Figure 7.36). The dowels decrease the tunnel convergence with respect to the unsupported tunnel. The reduction is most significant at the springline, which is where the rock yields and thus where the addition of the dowels is more effective. A comparison between Figures 7.34a, 7.35 and 7.36, which provide results for grouted dowels, with Figures 7.23b, Table 7.7 and Figure 7.24 are useful to evaluate the different performance of anchored rockbolts and dowels (grouted rockbolts). The first observation is that the rock stresses are not much different, with a slight increase of radial stresses (confinement) with the dowels, which may be the reason for a smaller plastic zone size also with the dowels. The radial displacements at the perimeter of the excavation are much smaller with the dowels (grouted rockbolts) than with the anchored rockbolts, while the stresses in the dowels are also smaller. This points to a somewhat better performance of the dowels than the anchored rockbolts. The source of the benefits may be found in the continuous distribution of support along the dowel rather than at the two ends of the anchored rockbolts that provides a more efficient load transfer from the rock to the bolt.

As the dowel spacing in Figures 7.34 to 7.36 is decreased from 1 m to 0.5 m, the radial displacements at the tunnel perimeter decrease and the dowel stresses also decrease. Close to the tunnel, the dowel stresses are not zero (Figure 7.35) due to the connection of the bolt to the rock by the faceplate. With increasing radial distance, the stresses increase until they reach a maximum at a distance of 1.2 to 1.3 times the tunnel radius, depending on the location of the dowel, and afterward they decrease with increasing distance from the tunnel. Such behavior was first recognized by Freeman (1978) from field measurements of instrumented grouted rockbolts (dowels) at the Kielder experimental tunnel.

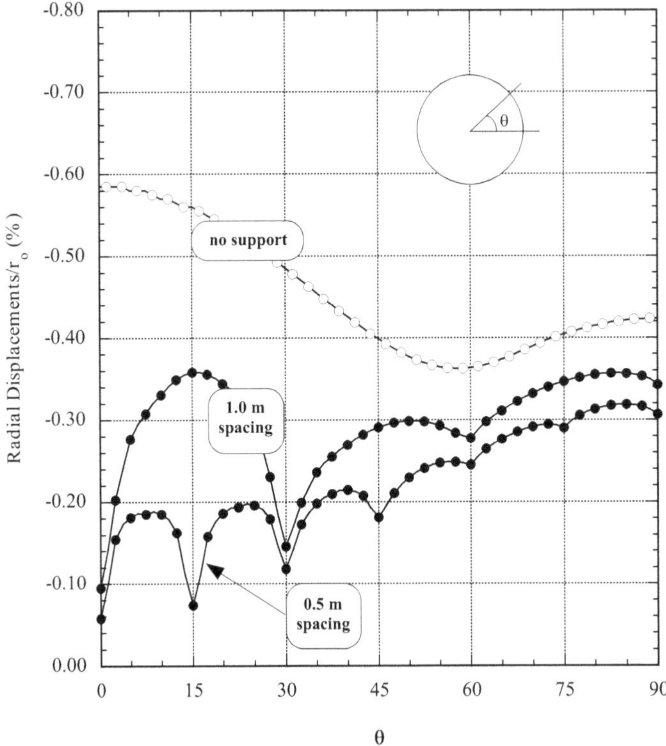

Figure 7.36 Radial displacement at tunnel perimeter with dowels, $r_o = 2$ m, $\sigma_v = 1$ MPa, $K_o = 0.5$, Rockbolt length = 2 m, $d_b = 20$ mm, $d_g = 32$ mm.

Figure 7.37 shows a conceptual model for the load transfer from the rock to the dowel. Figure 7.37a shows the rock immediately after installation of the dowels. With further tunnel advance, the rock near the excavation moves toward the opening more than the rock further inside. The rockbolt displacements can be decomposed into a rigid body motion where the entire length of the bolt moves toward the tunnel and a relative displacement between the bolt and the rock. Near the excavation (pickup length) the rock tends to move more than the dowel and so shear stresses are produced at the rockbolt-rock contact that induce tension in the bolt. With further distance along the bolt toward the inside of the rock, a neutral point is reached where there are no relative displacements between the dowel and the rock and so the shear stresses are zero. Beyond the neutral point, the shear stresses at the bolt-rock contact reverse sign because the rock tends to have relative displacements smaller than the rockbolt ("anchor length"). See Figure 7.37b. The shear stresses reduce the tensile load in the dowel, which then decreases as the distance from the tunnel increases. See Figure 7.37c.

The radial distance from the center of the tunnel to the neutral point is often referred to using the symbol ρ; see Figure 7.37b. Yu and Xian (1983) provided the following relation to estimate the location of the neutral point:

$$\rho = \frac{L}{\ln\left(\frac{L + r_o}{r_o}\right)} \tag{7.19}$$

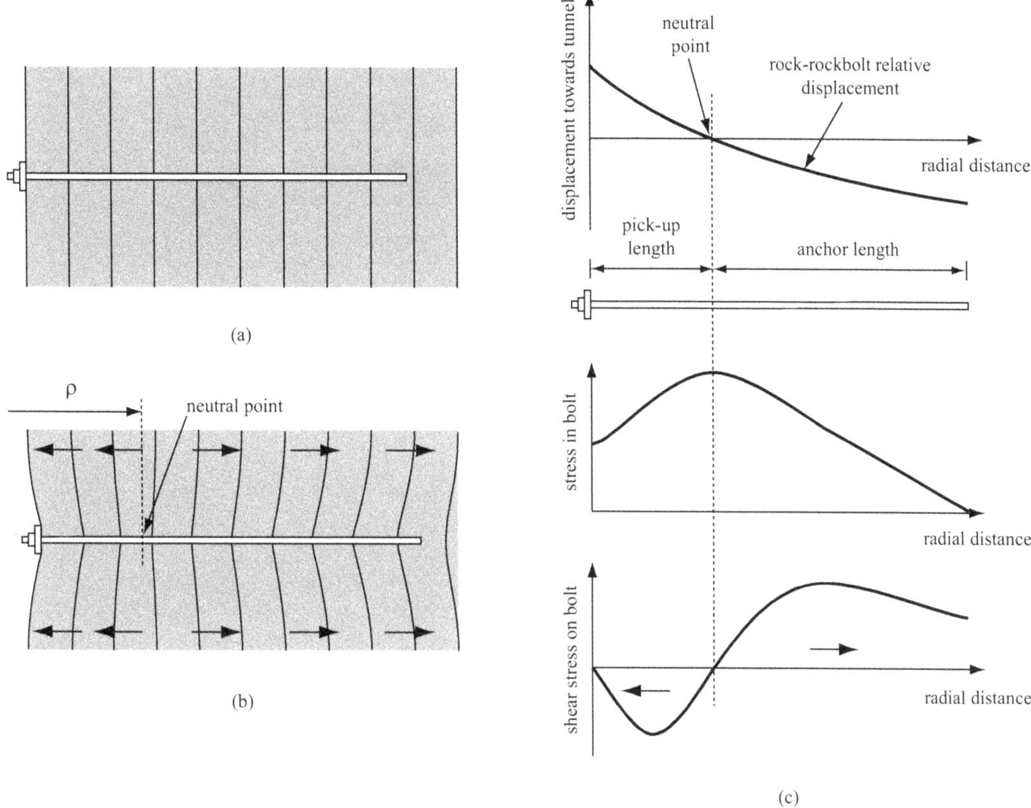

Figure 7.37 Interaction between rock and dowel. (a) Undeformed ground. (b) Deformed ground. (c) Rockbolt stresses and deformations. Adapted from Freeman (1978).

where L is the length of the dowel and r_0 is the radius of the tunnel. The expression is independent of the properties of the ground or the rockbolt, and only depends on geometric parameters. This is so, as discussed by Hyett et al. (1996), because of the assumption of a rigid bolt in the derivation of equation (7.19). In general the location of the neutral point will depend on the geometric conditions but also on the engineering properties of the rock and the bolt. Consideration of the bolt as rigid may lead to erroneous distributions of the shear stresses along the rockbolt (Hyett et al., 1996).

Two types of problems concerning dowels/grouted rockbolts have been approached analytically: the pullout of an isolated rockbolt embedded in an infinite elastic medium, and the support of a deep circular tunnel with rockbolts uniformly distributed around the excavation and subjected to a hydrostatic far-field stress.

The pullout load problem is a particular example of the more general case of a stiff inclusion in an elastic medium. Figure 7.38a is an idealization of the rockbolt, the grout and the ground. The transfer of the rockbolt stress σ_b to the surrounding ground is a difficult problem and no exact solution exists. An approximate solution can be found using shear-lag analysis (Nairn, 1997; Nairn and Mendels, 2001). This approach has been often followed to solve problems concerning materials with inclusions (Abramento and Whittle, 1993, 1995a, 1995b; Hsueh, 1988, 1990a; Ochiai et al., 1999), such as the problem of a rockbolt inside a rock mass (Farmer, 1975; Hsueh, 1990b; Li and Stillborg, 1999; Li, 2000). With the shear-lag analysis, simplifications are usually made to decouple the x and y directions, and thus

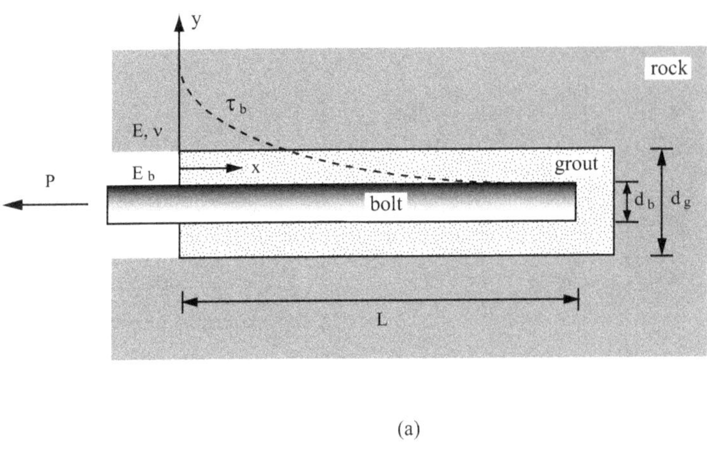

(a)

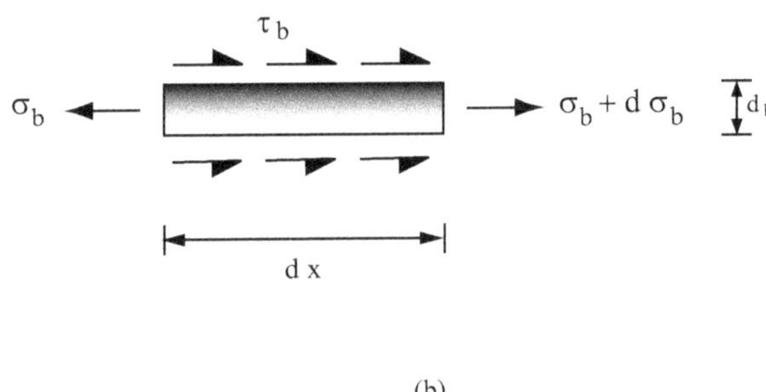

(b)

Figure 7.38 Pullout test of dowel/grouted rockbolt. (a) Dowel/Grouted rockbolt. (b) Stress components.

reduce the 2D problem to a 1D problem. It is often assumed that displacements along the y-axis are functions only of 'y' (the 'y' axis is perpendicular to the dowel; see Figure 7.38a). In other words:

$$\frac{\partial U_y}{\partial x} = 0, \text{ or} \left|\frac{\partial U_y}{\partial x}\right| << \left|\frac{\partial U_x}{\partial y}\right| \tag{7.20}$$

where U_x and U_y are the displacements along the "x" and "y" axes, respectively. As a consequence,

$$\tau = G\gamma = G\left(\frac{\partial U_x}{\partial y} + \frac{\partial U_y}{\partial x}\right) \approx G\frac{\partial U_x}{\partial y} \tag{7.21}$$

where τ is the shear stress, G is the shear modulus, and γ is the shear strain. A common approach is to assume a particular function for the shear stress and use equation (7.21) to determine displacements and then stresses. Despite the approximate underlying assumptions, this formulation has been proven very effective to obtain accurate solutions in composites (Hsueh et al., 1997; Nairn, 1997; Nairn and Mendels, 2001), where the problem of most interest is that of a single inclusion in a homogeneous matrix, i.e. an axisymmetric problem (Cox, 1952; Hsueh, 1988, 1990a, b, c, d). A function for the shear stress commonly adopted is of the form:

$$\tau = \frac{A}{y} \tag{7.22}$$

where A is a constant that depends on the boundary conditions, e.g. on the magnitude of the shear stress at the interface (τ_b, at $y = d_b/2$). τ_b can be obtained from (7.22) and depends on the axial stress of the rockbolt, σ_b, as follows (see Figure 7.38b):

$$\tau_b = -\frac{1}{4} d_b \frac{d\sigma_b}{dx} \tag{7.23}$$

where d_b is the diameter of the bolt. The following is the solution reached by Li and Stillborg (1999) for dowels/fully grouted rockbolts, which is somewhat more general than the solution proposed by Farmer (1975):

$$\begin{aligned} \tau_b &= \frac{2P\alpha}{\pi d_b^2} e^{-2\alpha \frac{x}{d_b}} \\ \sigma_b &= \frac{4P}{\pi d_b^2} e^{-2\alpha \frac{x}{d_b}} \\ \alpha^2 &= \frac{2GG_g}{E_b \left[G \ln \frac{d_g}{d_b} + G_g \ln \frac{d_o}{d_g} \right]} \end{aligned} \tag{7.24}$$

where P is the force at the free end, x = 0 in Figure 7.38a; d_g is the external diameter of the grout (bonding material); d_o is the size of the area of influence of the dowel, which can be estimated as $d_o = 10\ d_g$ (Li, 2000); G is the rock shear modulus; G_g is the grout (bonding material) shear modulus; and E_b is the Young's modulus of the bolt. The shape of the shear stress τ_b distribution is shown in Figure 7.38a. The shear stress is maximum at the free end and quickly decreases along the bond length, due to the negative exponent in equation (7.24).

As the pullout load increases, the shear stress at the free end may reach the shear strength of the rockbolt-grout interface (in the following an assumption is made that failure occurs at the bolt-grout contact, but analogous equations can be obtained by using the peak and residual strength of the grout-rock contact, and substituting the diameter d_b by d_g). With further loading, a fully decoupled zone is formed near the pullout end. In this zone, $0 \leq x \leq x_0$, as denoted in Figure 7.39a, the shear stress at the interface is negligible. A partially decoupled zone appears afterward, $x_0 \leq x \leq x_1$, where the shear strength at the interface has dropped to its residual value s_r. A transition zone follows, $x_1 \leq x \leq x_2$, where the shear strength increases from residual to peak s_p; an assumption is made that the increase is linear with axial distance. Beyond the x_2 coordinate, the interface is fully coupled and an exponential decrease of shear stress from the peak value is assumed, according to equation (7.24).

For dowels/grouted rockbolts, the stresses in each of the sections can be estimated as follows (Li and Stillborg, 1999):

For $0 \leq x \leq x_0$,

$$\begin{aligned} \tau_b &= 0 \\ \sigma_b &= \frac{4P}{\pi d_b^2} \end{aligned} \tag{7.25a}$$

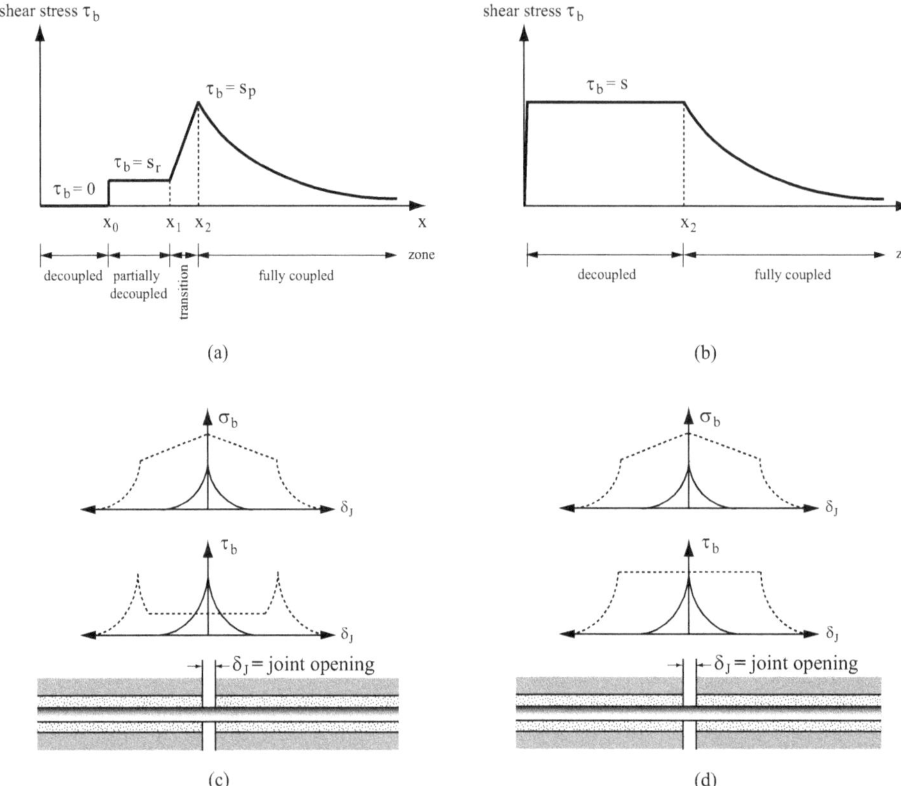

Figure 7.39 Shear and axial stresses in dowels and frictional rockbolts. (a) Pullout dowel. (b) Pullot frictional rockbolt. (c) Joint opening. Dowel. (d) Joint opening. Frictional rockbolt. Adapted from Li and Stillborg (1999).

For $x_0 \leq x \leq x_1$,

$$\begin{aligned} \tau_b &= s_r \\ \sigma_b &= \frac{4P}{\pi d_b^2} - \frac{4s_r}{d_b}(x - x_0) \end{aligned} \tag{7.25b}$$

For $x_1 \leq x \leq x_2$,

$$\begin{aligned} \tau_b &= s_r + \frac{x - x_1}{x_2 - x_1}(s_p - s_r) \\ \sigma_b &= \frac{4P}{\pi d_b^2} - \frac{2}{d_b}\left[2s_r(x - x_0) + (s_p - s_r)\frac{(x - x_1)^2}{x_2 - x_1}\right] \end{aligned} \tag{7.25c}$$

And for $x \geq x_2$,

$$\begin{aligned} \tau_b &= s_p e^{-2\alpha\frac{x - x_2}{d_b}} \\ \sigma_b &= \frac{2s_p}{\alpha} e^{-2\alpha\frac{x - x_2}{d_b}} \end{aligned} \tag{7.25d}$$

where α is given by (7.24). The position of the decoupling front and the maximum pullout force P_{max} are:

$$x_2 = x_0 + \frac{1}{2}\left[\frac{2P}{\pi d_b s_r} - \frac{d_b}{\alpha}\frac{s_p}{s_r} - \frac{(s_p - s_r)}{s_r}(x_2 - x_1)\right]$$
$$P_{max} = \pi d_b\left[s_r\left(L + \frac{d_b}{2\alpha}\ln\frac{s_r}{s_p} - x_2 + x_1 - x_0\right) + \frac{1}{2}(x_2 - x_1)(s_p - s_r) + \frac{d_b}{2\alpha}(s_p - s_r)\right] \quad (7.26)$$

For frictional rockbolts (split sets and Swellex), for which the bolt is in contact with the rock, equation (7.24) still applies if the Young's modulus and Poisson's ratio of the grout are made equal to those of the rock (i.e. $\alpha^2 = 2\ G/[E_b \ln(d_o/d_b)]$. Since the rockbolts are frictional, the residual and peak shear strength values can be taken as equal, $s = s_r = s_p$, and so the shear stress at the interface and the axial stress of the rockbolt are (Figure 7.38b):

For $0 \le x \le x_2$,

$$\tau_b = s$$
$$\sigma_b = \frac{P}{A_b} - \frac{\pi d_b s}{A_b}x \quad (7.27a)$$

And for $x \ge x_2$,

$$\tau_b = s e^{-2\alpha\frac{x - x_2}{d_b}}$$
$$\sigma_b = \frac{2s}{\alpha}e^{-2\alpha\frac{x - x_2}{d_b}} \quad (7.27b)$$

The position of the decoupling front and the maximum pullout force P_{max} are:

$$x_2 = \left[\frac{P}{\pi d_b s} - \frac{2A_b}{\pi d_b \alpha}\right]$$
$$P_{max} = \pi d_b s L \quad (7.28)$$

where A_b is the cross-section area of the rockbolt.

The peak and residual strength values s_p and s_r for dowels and s for frictional rockbolts can be determined from laboratory or field experiments (e.g. Kilic et al., 2002, 2003; Karanam and Dasyapu, 2005) or from analytical and empirical relations such as those given in (7.4) to (7.7).

Equations (7.25) to (7.28) provide an acceptable approximation for pullout tests (Farmer, 1975; Li and Stillborg, 1999; Cai et al., 2004a) but cannot be considered for the support design of a tunnel because the rock displacements are not fully included in the derivation. A good approximation, however, has been found in some field cases relating the shear stresses in the rockbolt to the rock displacements with the following equation (Li and Stillborg, 1999):

$$\tau = \frac{EE_b S_z S_\theta}{1/4\pi d_b^2 E_b + ES_z S_\theta}\left(\frac{d_b}{4}\frac{d^2U_r}{dr^2} - \frac{\alpha}{2}\int_{r_0}^{r}\frac{d^2U_r}{dt^2}e^{-2\alpha\frac{r-t}{d_b}}dt\right) \quad (7.29)$$

where S_z and S_θ are, respectively, the axial and transverse spacing of the rockbolts, and U_r is the radial displacements of the rock around the tunnel. All other parameters are as previously defined. The limitation of equation (7.29) is that the values of U_r must be known.

The shear-lag analysis has also been used to estimate the stress distribution in a rockbolt that crosses a joint and is subjected to an opening of magnitude δ_J, as denoted in Figure 7.39c for dowels and Figure 7.39d for frictional rockbolts (note the differences in model behavior in Figures 7.39a and 7.39b). The axial stresses in the vicinity of the discontinuity, due to the opening δ_J are (Li and Stillborg, 1999):

For dowels (it is assumed that $x_0 = 0$),

$$\begin{aligned}
\delta_J &\approx \frac{\sigma_b d_b}{\alpha E_b}, \quad \text{if} \sigma_b \le \frac{2s_p}{\alpha} \\
\delta_J &= \frac{2}{E_b}\left[\sigma_b x_2 - 2s_r \frac{x_1^2}{d_b} - (2s_r + s_p)\frac{2(x_2 - x_1)^2}{3d_b} + \frac{s_p}{\alpha^2} d_b\right] \quad \text{otherwise, and} \\
x_2 &= \frac{d_b}{2}\left[\frac{\sigma_b}{2s_r} - \frac{s_p}{\alpha s_r} - \frac{(x_2 - x_1)}{d_b}\left(\frac{s_p}{s_r} - 1\right)\right]
\end{aligned} \tag{7.30}$$

For fully frictional rockbolts,

$$\begin{aligned}
\delta_J &\approx \frac{\sigma_b d_b}{\alpha E_b}, \quad \text{if} \sigma_b \le \frac{2s_p}{\alpha} \\
\delta_J &= \frac{2}{E_b}\left[\sigma_b x_2 - s d_b\left(\frac{\pi x_2^2}{2A_b} - \frac{1}{\alpha^2}\right)\right] \quad \text{otherwise, and} \\
x_2 &= \frac{A_b}{\pi s d_b}\left[\sigma_b - \frac{2s}{\alpha}\right]
\end{aligned} \tag{7.31}$$

where σ_b is the axial stress in the rockbolt at the center of the discontinuity (i.e. the maximum stress). It is important to note that equations (7.30) and (7.31) indicate that small values of the joint aperture δ_J may induce very large stresses in the rockbolt, and thus yielding may occur even with small joint movements. This phenomenon is particularly important in hard rocks, where large forces have been measured in rockbolts at locations where they intersect major discontinuities (Björnfot and Stephansson, 1983).

7.4.5.1 Rock–rockbolt interaction: analytical methods

The problem of the design of tunnel support with rockbolts, which requires rock-support interaction analysis, is complex and there is no general analytical solution. Over the years, a number of semi-empirical recommendations have been provided, each applicable to a specific case (Hoek and Brown, 1980; Bergman and Bjurström, 1983; Sinha and Schoeman, 1983, who provide a review of some of the methods in the early 1980s; Singh et al., 1995; Carranza-Torres and Fairhurst, 2000; Brady and Brown, 2004). A shortcoming of such semi-empirical recommendations is that it is not clear what the factor of safety is and that they cannot be systematically and safely applied for design. Analytical solutions based on fundamental mechanics and behavior of rock and support offer the advantage that the key parameters can be identified and that they provide a conceptual framework to understand the coupled response of rock and support.

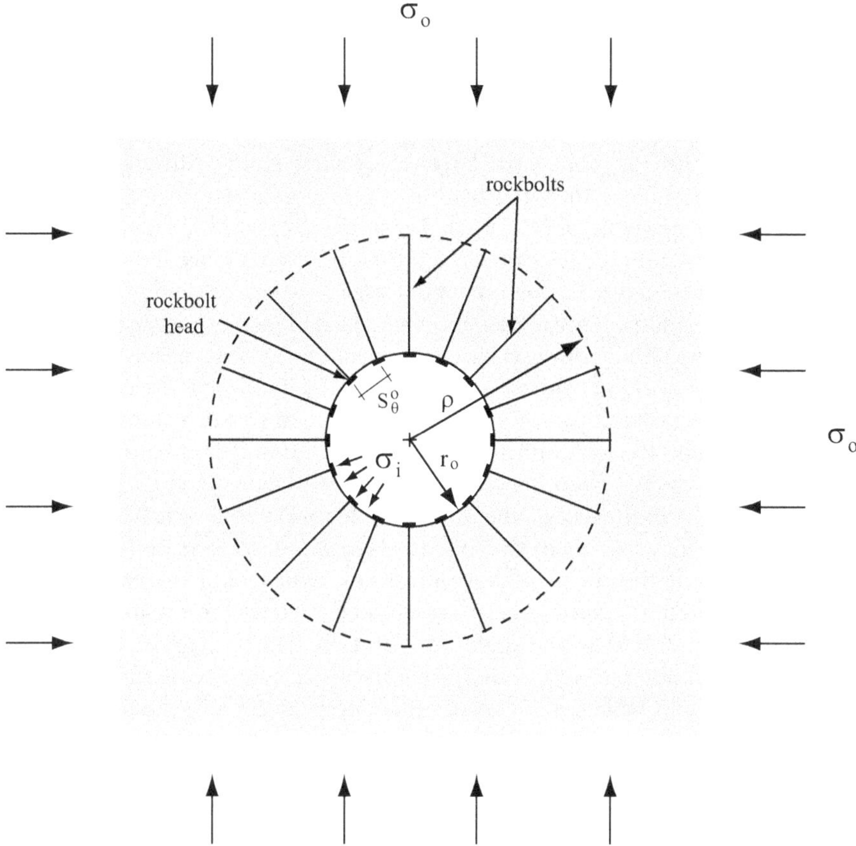

Figure 7.40 Tunnel with dowels/grouted rockbolts.

The analytical solutions that have been proposed for tunnel design can be classified in two groups: (1) solutions that are derived based on the assumption of an "equivalent" material or homogenization method; and (2) solutions obtained with a "smeared" approach. In both cases the following assumptions are made: (1) deep circular tunnel; and (2) plane strain conditions (i.e. two-dimensional approach). See Figure 7.40.

In the first group, the properties of the medium surrounding the tunnel are those of a composite that includes both the dowels and the rock. The solution is found based on the engineering properties of the equivalent material. It is assumed that the dowels are placed at regular intervals with spacing much smaller than a characteristic dimension of the problem such as the tunnel radius or the rockbolt length. This approach has been used to solve the cases numerically (Sharma and Pande, 1988), or analytically (Indraratna and Kaiser, 1990a, b; Pelizza et al., 1994; Peila and Oreste, 1995; Cai et al., 2004b; Osgoui and Oreste, 2007). Simple analytical solutions have been found with the following additional assumptions: (1) uniform far-field stresses (i.e. axisymmetric problem); (2) the stiffness of the medium is not changed by the addition of the rockbolts; (3) the mechanical contribution of the rockbolts can be approximated by a force $\Delta T = \pi\ d_b \sigma_\theta\ \mu\ dr$, where d_b is the bolt diameter, σ_θ is the tangential stress, μ is the coefficient of friction between the rockbolt and the rock and dr is the differential radial coordinate; and (4) the neutral point is given by equation (7.19).

While this approach is very attractive because it is simple to use, the assumptions raise questions. As already mentioned, equation (7.19) leads to an erroneous distribution of the shear stresses along the rockbolt (Hyett et al., 1996). Perhaps a more serious concern is the hypothesis that the shear stress along the rockbolt is proportional to the tangential stress through a coefficient of friction, i.e. assumption (3) above. The implications are twofold. The first one is that the rockbolt slips. Laboratory tests from Stillborg (1994) and Kilic et al. (2003) appear to indicate that the rockbolt stress-elongation response can be approximated better as elastic-perfectly plastic, with a residual shear stress similar to the peak; see Figure 7.16. The results from the frictional rockbolts shown in Figure 7.16 indicate that once the peak strength is reached the reinforcement deformations increase without further increasing the load, i.e. the bolt slips. For the dowels/grouted rockbolts, however, an increase in load with slip is observed. As will be discussed in more detail in the next section, when discussing cablebolts, the increase in load with slip may be due to dilation at the bolt-grout contact and due to the cracking of the grout. The magnitude of the load increase seems, however, not to be significant. The second implication is that the full tangential stress is acting on the rockbolt. This ignores the installation process of the rockbolts where, after drilling the hole, the radial stresses at the perimeter of the hole are effectively zero. Excavation of the tunnel does change the tangential stresses in the rock (the tangential stresses at the tunnel perimeter are the radial stresses at the rock/rockbolt interface), which tend to increase with further excavation; this benefit, if any, should be considered by including in the formulation only the increase of tangential stress after the rockbolt is installed and the grout has hardened. The effect on frictional rockbolts is even more questionable since the post-installation tangential stresses in the rock would be larger than the radial stress applied to the hole by the split set or Swellex rockbolt.

The equivalent material or homogenization method was used by Bobet (2009) to obtain the elastic properties of the composite or equivalent rock–rockbolt material. The properties are estimated using the shear-lag method, and are as follows:

$$\begin{aligned} E_r &= E + \frac{1}{4}\frac{\pi d_b^2}{S_\theta S_z} E_b \\ E_\theta &= E_z = \frac{E}{1 - \nu^2 (1 - E/E_r)} \\ \nu_{r\theta} &= \nu_{z\theta} = \nu_{rz} = \nu_{\theta z} = \nu \\ \nu_{\theta r} &= \nu_{zr} = \nu \frac{E}{E_r} \end{aligned} \qquad (7.32)$$

The shear modulus of the composite can be taken as the shear modulus of the rock. In (7.32), r, θ and z represent the polar coordinates and the longitudinal axis of the tunnel, respectively; E and ν are the Young's modulus and Poisson's ratio of the rock; E_b is the Young's modulus of the rockbolt material; d_b is the rockbolt diameter; and S_θ and S_z are the tangential and axial rockbolt spacings. The relations in (7.32) are obtained with the assumption that $d_b << S_\theta$. The presence of the rockbolts increases the radial stiffness of the rock mass while the tangential and axial stiffness is not much changed. This makes the reinforced area to have cylindrical anisotropy. A complete analytical solution for a deep tunnel in an elastic medium subjected to a general far-field loading was obtained considering the anisotropy introduced by the dowels (Bobet, 2009). Stresses and displacements are obtained then from either the analytical solution or from a numerical method where the elastic properties of the reinforced rock are those given in equation (7.32). The results are those of the composite.

Using shear-lag analysis, the radial stresses in the rock, σ_r^m, and the stresses in the dowels, σ_b, are (Bobet, 2009):

$$\sigma_r^m = \frac{\nu(1+\nu)(\sigma_\theta - \sigma_\theta^o) + S_\theta^o \frac{r}{r_o} S_z \frac{E}{A_b E_b}(\sigma_r - \sigma_r^o)}{1 - \nu^2 + S_\theta^o \frac{r}{r_o} S_z \frac{E}{A_b E_b}} + \sigma_r^o \tag{7.33}$$

$$\sigma_b = (1+\nu)\frac{E_b}{E}\left[(\sigma_r^m - \sigma_r^o) - \nu(\sigma_r^m + \sigma_\theta - \sigma_r^o - \sigma_\theta^o)\right]$$

In (7.33) σ_r and σ_θ are the radial and tangential stresses obtained in the analysis and σ_r^o and σ_θ^o are the radial and tangential stresses at the point where the stresses are computed prior to the installation of the dowels. For example, with the assumption of simultaneous tunnel excavation and rockbolt installation, often made with the relative stiffness method, σ_r^o and σ_θ^o would be the radial and tangential stresses due to the initial far-field stresses, i.e. equation (4.37), and σ_r and σ_θ would be given by equation (4.41). S_θ^o is the tangential spacing of the dowels measured at the perimeter of the excavation (see e.g. Figure 7.40), S_z is the axial spacing of the rockbolts, r is the radial distance where the stress is computed, r_o is the tunnel radius, and A_b is the cross-section area of the rockbolt. Displacements and tangential and shear stresses in the composite and in the rock are similar and do not need to be modified.

The elastic solution shows that in addition to the rockbolt length, the contribution of the dowels depends on the additional stiffness they introduce, which can be estimated from (7.32) as:

$$\frac{E_r}{E} = 1 + \frac{1}{4}\frac{\pi d_b^2}{S_\theta S_z}\frac{E_b}{E} = 1 + \frac{A_b}{S_\theta S_z}\frac{E_b}{E} \tag{7.34}$$

where E_r is the modulus of the composite rock/rock bolt material. The second term on the right-hand side gives the increase of the radial stiffness provided by the reinforcement. It is only for relatively soft rocks with very close rockbolt spacing that the reinforcement will significantly change the stresses and displacements around the excavation. For E_r/E ratios smaller than, say 10%, the elastic properties of the reinforced rock can be taken equal as those of the unreinforced rock. This observation strengthens the concept that rockbolts are effective in limiting or at least reducing yielding of the rock since the plastic deformations can be thought of as associated with a material with reduced stiffness. If excavation of the tunnel does not induce plastic deformations in the rock, then the use of rockbolts for support may not be effective, unless the stiffness of the rock is relatively small and the increase of stiffness provided by the reinforcement, according to equation (7.34), is substantial.

Rock-support interaction using the "smeared" underlying approach is based on the assumption that the reinforcement contribution is in the form of a radial load smeared within the zone of influence of the rockbolt. In other words, the tensile load of the rockbolt T introduces a radial compression in the rock with magnitude $T/(S_\theta S_z)$. See Figure 7.41. With this hypothesis, together with the assumptions of axial symmetry (i.e. tunnel with circular cross section and uniform far-field loading), and very close spacing of the dowels, the equations of equilibrium can be written as:

$$\frac{d\sigma_r}{dr} + \frac{\sigma_r - \sigma_\theta}{r} + \frac{1}{S_\theta S_z}\frac{dT}{dr} = 0 \tag{7.35}$$

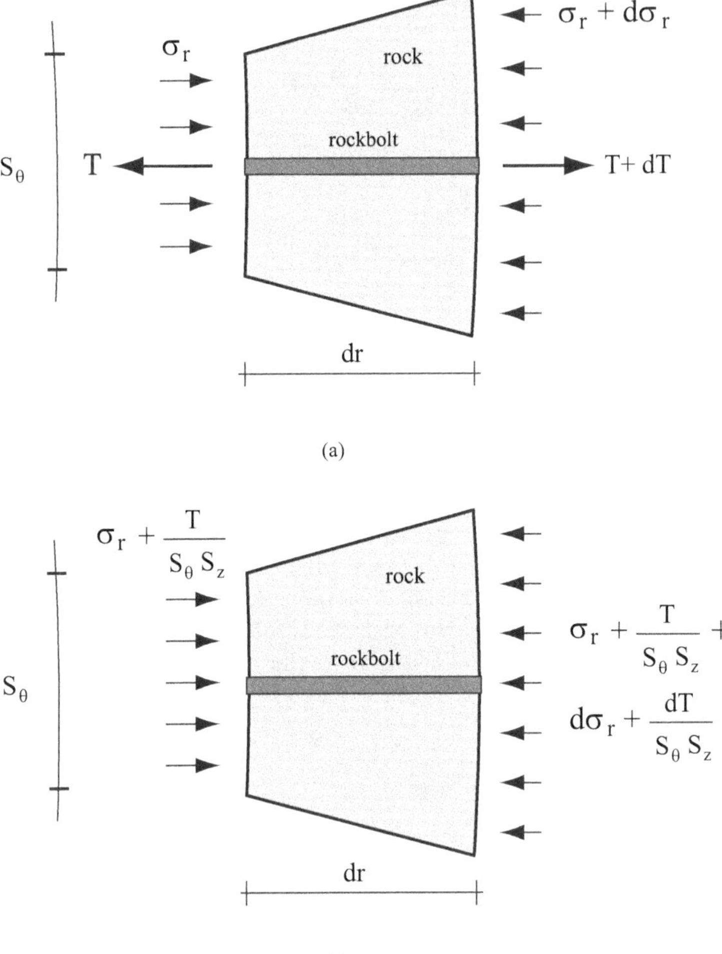

Figure 7.41 Smeared approach (Bobet and Einstein, 2011). (a) Radial stresses in rock and bolt tension. (b) Equivalent stresses. Reproduced with permission of Elsevier.

where σ_r and σ_θ are the radial and tangential stresses in the rock, r is the radial coordinate, S_θ and S_z are the tangential and axial spacing of the dowels, and T is the compression force provided by the dowel. Note that, because of the axial symmetry, the shear stresses are zero and the results do not depend on the angular coordinate. This approach with some variations has been followed by a number of authors (Stille, 1983; Stille et al., 1989; Oreste and Peila, 1996; Fahimifar and Soroush, 2005; Guan et al., 2007).

The force T depends on the strains induced in the rockbolt after installation. Hence equation (7.35) can be rewritten as:

$$\frac{d\sigma_r}{dr} + \frac{\sigma_r - \sigma_\theta}{r} + \frac{A_b E_b}{S_\theta S_z} \frac{d(\varepsilon_r - \varepsilon_r^o)}{dr} = 0 \tag{7.36}$$

where A_b and E_b are the cross-section area of the rockbolt and the Young's modulus of the rockbolt material; ε_r is the radial strain of the rock and ε_r^o is the strain in the ground before rockbolt installation. This assumes that the strains in the dowels are those of the rock; i.e. there is no relative slip. Note that in (7.35) the tangential spacing S_θ depends on the radial distance, i.e. as the radial coordinate increases the distance between rockbolts increases. To render the problem tractable analytically, Bobet and Einstein (2011) assumed that the dependency of the tangential spacing on the radial distance could be neglected, and it could be represented by an "average" spacing $\bar{S}_\theta$.

Following the approach described, Bobet and Einstein (2011) found a closed-form solution for the stresses and displacements of the rock and dowels. Assumptions similar to those used for anchored rockbolts were taken, i.e. circular cross section with radius r_o; deep tunnel with uniform far-field stresses σ_o; homogeneous and isotropic ground, that remains elastic until reaching failure, which is defined by a Coulomb failure envelope; non-associated flow rule for post-failure behavior, defined by the dilation angle of the material; uniform distribution of dowels with spacing S^o_θ and S_z; and the use of the β-method to capture the three-dimensional effects. Figure 7.40 shows that the dowels have a length ρ-r_o and the tunnel may have an internal support pressure with magnitude σ_i.

Similar to what has been done in the preceding section on mechanically and resin-anchored rockbolts (Section 7.4.4), the analytical solution can be used to provide insight into the interaction between tunnel excavation, rockbolt support and ground.

Figures 7.42 and 7.43 provide a comparison between the results obtained with the analytical solution and those from the Finite Element Method code ABAQUS. The problem solved is analogous to that used in Section 7.4.4: a tunnel with radius r_o = 3 m, subjected to a far-field stress σ_o = 1 MPa and internal pressure σ_i = 0; the properties of the ground are: E = 500 MPa, ν = 0.2, peak and residual strength $\phi^p = \phi^r$ = 30°, $c^p = c^r$ = 0.1 MPa (as mentioned, the values are too low and may not be representative of rock, but they are taken to facilitate discussion); the dowels have a length of 3 m, i.e. ρ = 6 m, and have properties d_b = 25 mm, E_b = 210 GPa, ν_b = 0.3. The initial unloading of the excavation is represented with β = 0.3. In the analytical solution, the average spacing $\bar{S}_\theta$ is taken as the distance between dowels measured

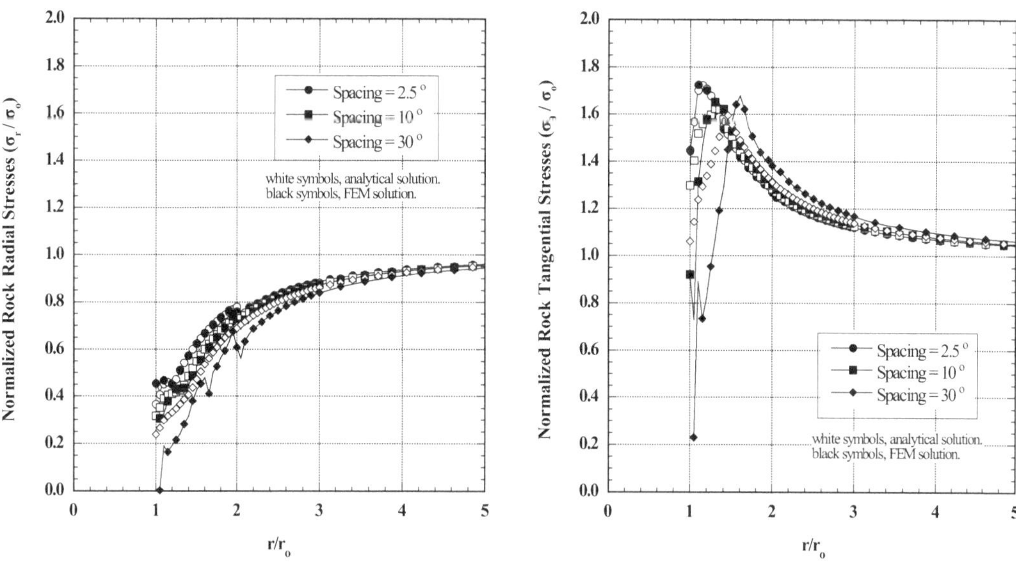

Figure 7.42 Dowels. Normalized rock stresses for r_o= 3 m, ρ = 6 m, σ_o = 1 MPa, σ_i = 0, E = 500 MPa, ν = 0.2, $\phi^p = \phi^r$ = 30°, $c^p = c^r$ = 0.1 MPa, d_b = 25 mm, E_b = 210 GPa, ν_b = 0.3, S_z = 1 m, β = 0.3. (Bobet and Einstein, 2011). Reproduced with permission of Elsevier.

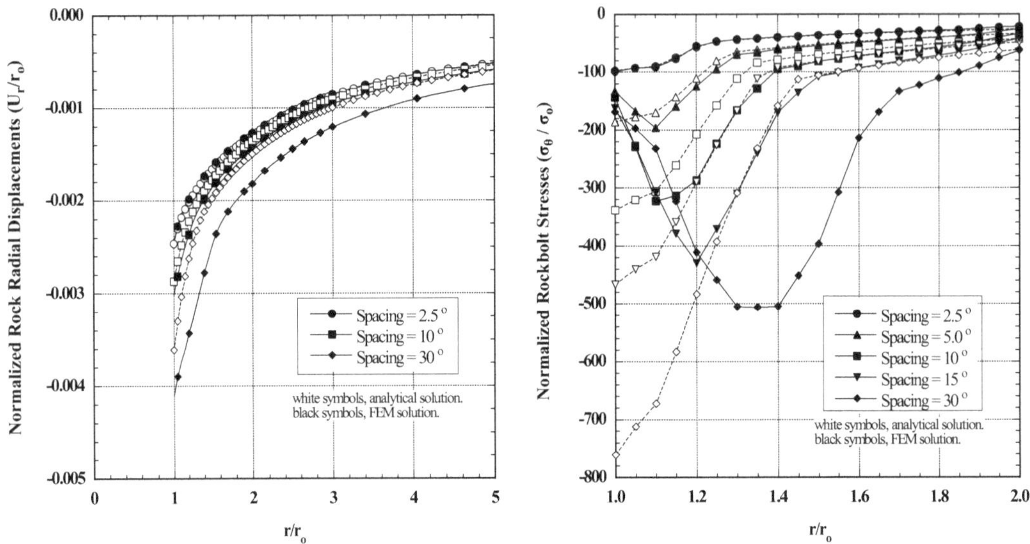

Figure 7.43 Dowels. Normalized radial displacements and rockbolt stresses for r_o= 3 m, ρ = 6 m, σ_o = 1 MPa, σ_i = 0, E = 500 MPa, ν = 0.2, $\phi^p = \phi^r$ = 30°, $c^p = c^r$ = 0.1 MPa, d_b = 25 mm, E_b = 210 GPa, ν_b = 0.3, S_z = 1 m, β = 0.3. (Bobet and Einstein, 2011). Reproduced with permission of Elsevier.

at 0.25 times the length of the dowels; i.e. at a radial distance $r_o + 0.25(\rho\text{-}r_o)$. The results are plotted for three different spacings, which are given by the angle between dowels: 2.5°, 10° and 30°. The angles correspond, for r_o = 3m, to 0.13 m, 0.52 m, and 1.57 m distance between dowels along the tunnel perimeter. Figure 7.42a is a comparison of the radial stresses in the rock between the analytical solution and the numerical model and Figure 7.42b of the tangential stresses of the rock mass. Figures 7.43a and 7.43b, respectively, compare the radial displacements at the tunnel opening and the stresses along the dowels. The comparisons for stresses and displacements of the rock mass show good agreement between the two solutions up to spacings of 15° (not shown in Figure 7.43a). For larger spacings, i.e. 30°, the errors are large, around 20–30%. While the rock response is fairly well captured by the analytical solution, the bolt stresses are not. The analytical solution yields acceptable results up to a spacing of about 5°. For spacings of 10° and 15° the maximum stress predicted by the solution compares well with that obtained by ABAQUS. However, the shape of the stress distribution is quite different. While the analytical solution shows a monotonic decrease in stress with increasing radial distance, the numerical solution shows an increase first up to a certain distance inside the rock where the bolt load attains a maximum and a decrease afterward. The results obtained with ABAQUS correspond better to observations made in the field (e.g. Freeman, 1978; Li and Stillborg, 1999). The reverse in load distribution is due, as shown in ABAQUS, to two factors. The first one, and perhaps the most important one, is the absence of support of the rock between rockbolts at the opening; the second one is the development of shear stresses in the rock mass between the rockbolts in the vicinity of the tunnel opening. Because of the first factor, the rock has to have relatively large deformations until there is a transfer of stress to the dowels, where the stresses slowly increase toward the inside of the ground. This zone corresponds to the "pick-up length" in Figures 7.37b and 7.37c. The second factor, the presence of shear stresses, produces a rotation of the principal stresses; that is, the radial and tangential stresses are not the principal stresses as assumed in the analytical derivation. The end result is that the rock can carry more load than predicted by the analytical solution (Figure 7.42 shows reduced radial and tangential stresses; note, however, that

Table 7.9 Errors between analytical solution and FEM

	Dowels/grouted rockbolts	
Spacing (degrees)	*Maximum tension load (MN)*	*Plastic zone size (m)*
2.5°	1.5%	0%
5°	5.4%	1.9%
10°	4.8%	1.5%
15°	8.6%	6.0%
30°	50.4%	15.9%

these are no longer principal stresses), it deforms more (see Figure 7.43a), and there is less load transferred to the dowels (Figure 7.43b). It is interesting to note that such a phenomenon is not observed at very small spacings, i.e. 2.5°, where the analytical solution perfectly matches the numerical results. Table 7.9 summarizes the differences between the analytical solution and the FEM in terms of size of the plastic zone and maximum load in the dowels. Again, the differences may be considered acceptable up to spacings of 15°. For larger spacings the errors increase very quickly and the analytical solution overestimates stresses in the rock and rockbolts and underestimates the size of the plastic zone and the rock deformations.

Based on these results, it can be concluded that the analytical solution is capable of producing reasonable results for small spacings and should not be used when the spacing between dowels is large. For this reason, the following results from the analytical solution are limited to spacings smaller than 15°.

It is interesting to compare results obtained when the tunnel is supported with anchored rockbolts and with dowels. This can be done by inspection of Figures 7.26 and 7.27, for the anchored rockbolts, and Figures 7.42 and 7.43 for the dowels. The figures show that the dowels result in slightly larger radial and tangential stresses for the ground and somewhat smaller radial deformations. The differences are not significant, of the order of 10% to 20%. What is significant is the stresses induced in the reinforcing elements. Table 7.10 provides a comparison of the maximum loads in the reinforcement elements and the size of the plastic zone in the rock. The comparison is done in terms of the ratio between results from dowels/grouted rockbolts and anchored rockbolts.

The table shows that the size of the plastic zone is not much affected by the type of rockbolt, which is consistent with the observation of similar stresses and deformations for the rock. The maximum load in the dowels, however, is significantly larger than in the anchors. This is due to the interaction between each type of reinforcing element and the rock. Figure 7.44 is a conceptual sketch that may be used to illustrate the behavior of the two types of rockbolts. Figure 7.44b shows stress or strain in an anchored rockbolt. Because the bolt is only

Table 7.10 Ratio between results with dowels/grouted rockbolts and anchored rockbolts

Spacing (degrees)	*Maximum tension load ratio*	*Plastic zone size ratio*
2.5°	1.4	1.0
5°	1.6	1.0
10°	1.7	1.0
15°	1.9	1.0
30°	1.5	1.2

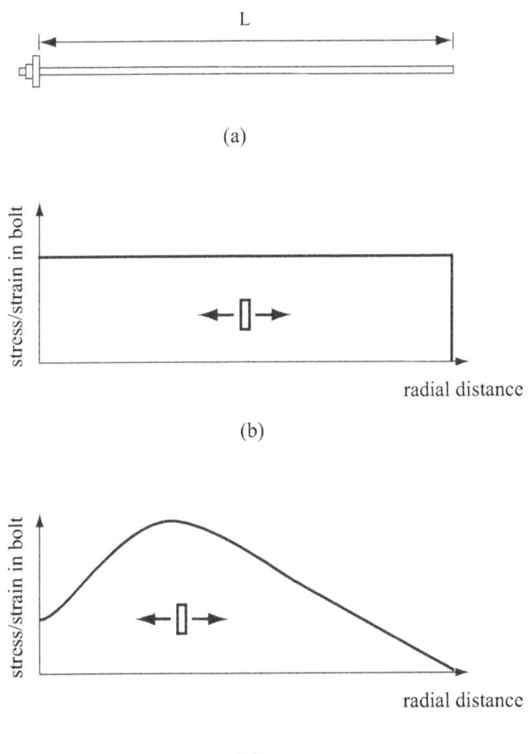

Figure 7.44 Stress/strain distribution along anchored rockbolts and dowels. (Bobet and Einstein, 2011). (a) Rockbolt. (b) Anchored rockbolt. (c) Dowel. Reproduced with permission of Elsevier.

connected to the rock at its ends, the stress/strain is constant. The magnitude of the rockbolt load can then be viewed as the result of the cumulative deformation of the rock between the two ends of the bolt, and so it is the result of an "average" behavior of the supported rock. Dowels (grouted rockbolts), however, are in intimate contact with the rock over their entire length and so stress/strain distributions are not constant (Figure 7.44c). The magnitude and location of the point of maximum load is associated with the magnitude and location of the largest strain in the rock.

Figure 7.45 is a plot of the characteristic curves of the reinforced rock with dowels and the maximum tension in the dowels. As expected, the figure shows that as the dowel spacing increases, the rock deformations increase and the maximum tension increases. A comparison with Figure 7.30, which is a similar plot for reinforcement with anchored rockbolts, shows that the dowels decrease the deformations more significantly but at the expense of larger loads. This is in agreement with the results shown in Table 7.10.

Figure 7.46 explores the effects of spacing, dowel length, and diameter and delay of support installation (given by the parameter β). The figure shows plots of the ground characteristic curves and bolt tension load obtained by reducing the dowel spacing from 10° to 5°, increasing the length of the dowels from 3 m to 6 m, and increasing the bolt diameter from 0.025 m to 0.035 m. These changes correspond to doubling the amount of material (e.g. steel) in the reinforcement with respect to the base case. The base case is defined as: r_o = 3 m, ρ = 6 m, σ_o = 1 MPa, E = 500 MPa, ν = 0.2, $\phi^p = \phi^r$ = 30°, $c^p = c^r$ = 0.1 MPa, d_b = 25 mm, E_b = 210 GPa, ν_b = 0.3, spacing S_θ° = 10°, and β = 0.3. Figure 7.46 also shows results by changing the parameter β from 0.3 to 0.6; the latter value representing an "earlier" installation of the

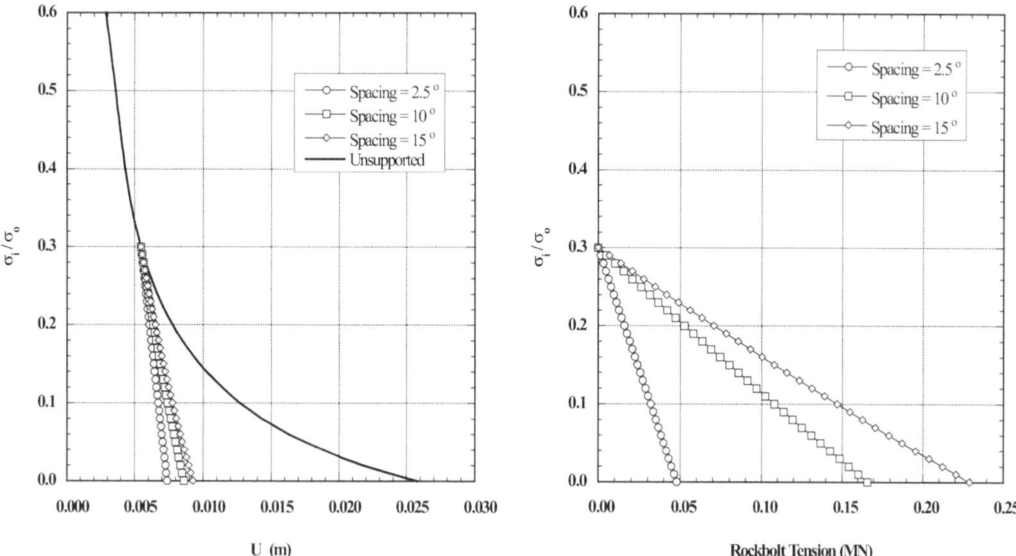

Figure 7.45 Dowels. Radial displacements and maximum rockbolt tension for r_o = 3 m, ρ = 6 m, σ_o = 1 MPa, E = 500 MPa, ν = 0.2, $\phi^p = \phi^r$ = 30°, $c^p = c^r$ = 0.1 MPa, d_b = 25 mm, E_b = 210 GPa, ν_b = 0.3, S_z = 1 m, β = 0.3. (Bobet and Einstein, 2011). Reproduced with permission of Elsevier.

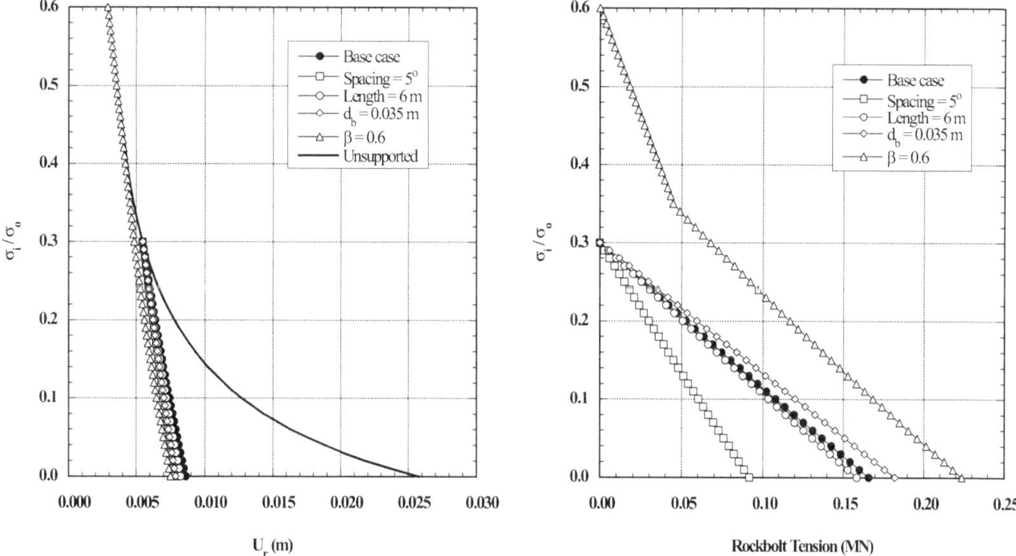

Figure 7.46 Dowels. Radial displacements and rockbolt tension. Base case: r_o = 3 m, ρ = 6 m, σ_o = 1 MPa, E = 500 MPa, ν = 0.2, $\phi^p = \phi^r$ = 30°, $c^p = c^r$ = 0.1 MPa, E_b = 210 GPa, d_b = 25 mm, S_z = 1 m, S_θ = 10°, β = 0.3. Effect of rockbolt spacing, length and cross section. (Bobet and Einstein, 2011). Reproduced with permission of Elsevier.

dowels. The results indicate a very small improvement of ground deformations by changing any of the parameters (Figure 7.46a). The effects on the rockbolt loads, in contrast, are significant (Figure 7.46b). An increase in length or diameter does not change substantially the maximum tension in the dowels. A reduction of the spacing, however, is associated with a significant reduction in the maximum load; in this case by a factor of about two. An early installation of the dowels (approximated by increasing the value of β to 0.6) has minimal effect on the ground deformations but causes a very large increase in the dowels load. This result is due, as shown in Figure 7.46a, to the installation of the dowels within the elastic regime of the ground, which is defined by the steep portion of the unsupported ground characteristic curve. A change in β within the plastic regime would have had a larger effect. This last case is useful to illustrate the observation that the installation of dowels is more effective when the ground undergoes plastic deformations. A similar conclusion was reached for support with anchored rockbolts. A comparison between Figures 7.46, for dowels, and 7.31, for anchored rockbolts, indicates a relatively more beneficial effect of doubling the parameters for anchored rockbolts than for dowels. This beneficial effect produces a comparatively smaller increase in loads for the anchored rockbolts.

Figure 7.47 is a plot of numerical results using ABAQUS to explore four cases when $K_o \neq 1$ and/or the dowels are not uniformly distributed around the opening; see Figure 7.32 for case definition. These problems need to be analyzed numerically because the assumption of axial symmetry does not apply. The following cases are solved: spacing $S_\theta^o = 10°$ and $K_o = 0.5$ (case 1); $K_o = 1$ with only seven dowels around the crown, i.e. $\theta = 60°, 70°, 80°, 90°, 100°, 110°, 120°$ (case 2); $K_o = 0.5$ and seven dowels around the crown (case 3); and $K_o = 0.5$ with seven dowels around the springline, i.e. $\theta = -30°, -20°, -10°, 0°, 10°, 20°, 30°$, (case 4). The observations are similar to those reached with analogous cases for anchored rockbolts, presented in Figure 7.33, and can be summarized as follows:

Effect of K_o. The results shown in Figures 7.47a and 7.47b for case 1 can be compared with those in Figure 7.46. The comparison indicates that with $K_o = 0.5$ the largest reduction of tunnel deformations with uniform reinforcement occurs at the springline. The maximum bolt load for the case with $K_o = 0.5$ occurs at the springline and is larger than for $K_o = 1$.

Effect of dowel distribution. Results from case 2, shown in Figures 7.47a and 7.47b, illustrate the effects of concentrated dowels with uniform far-field stress distribution. The results can be compared with those of Figure 7.46, which apply to a similar tunnel but with uniform reinforcement. Figure 7.47a shows that the reinforcement is ineffective at the springline. The displacements at the crown are smaller than those of the unreinforced tunnel, but are somewhat larger than those of the tunnel with uniform dowel distribution (Figure 7.46). The maximum rockbolt load, however, is similar to that with uniform reinforcement (Figure 7.46).

Cases 1, 3, and 4 in Figure 7.32 can be used to investigate the effects of K_o and dowel distribution. As with anchored rockbolts, clustering the reinforcement at the crown (case 3) has negligible effects in reducing tunnel convergence (Figure 7.47c). The rockbolt loads are smaller than those found in cases 1 and 4, but still substantial. Placement of the dowels around the springline (case 4) results in tunnel deformations at the springline similar to those of case 1, tunnel with uniform reinforcement, and in larger deformations at the crown. The maximum rockbolt load is larger for case 4 than for case 1. Similar to what was mentioned for anchored rockbolts, it is advantageous to distribute the reinforcement around the tunnel according to the far-field loading, and thus in the areas where largest deformations of the unreinforced tunnel will occur.

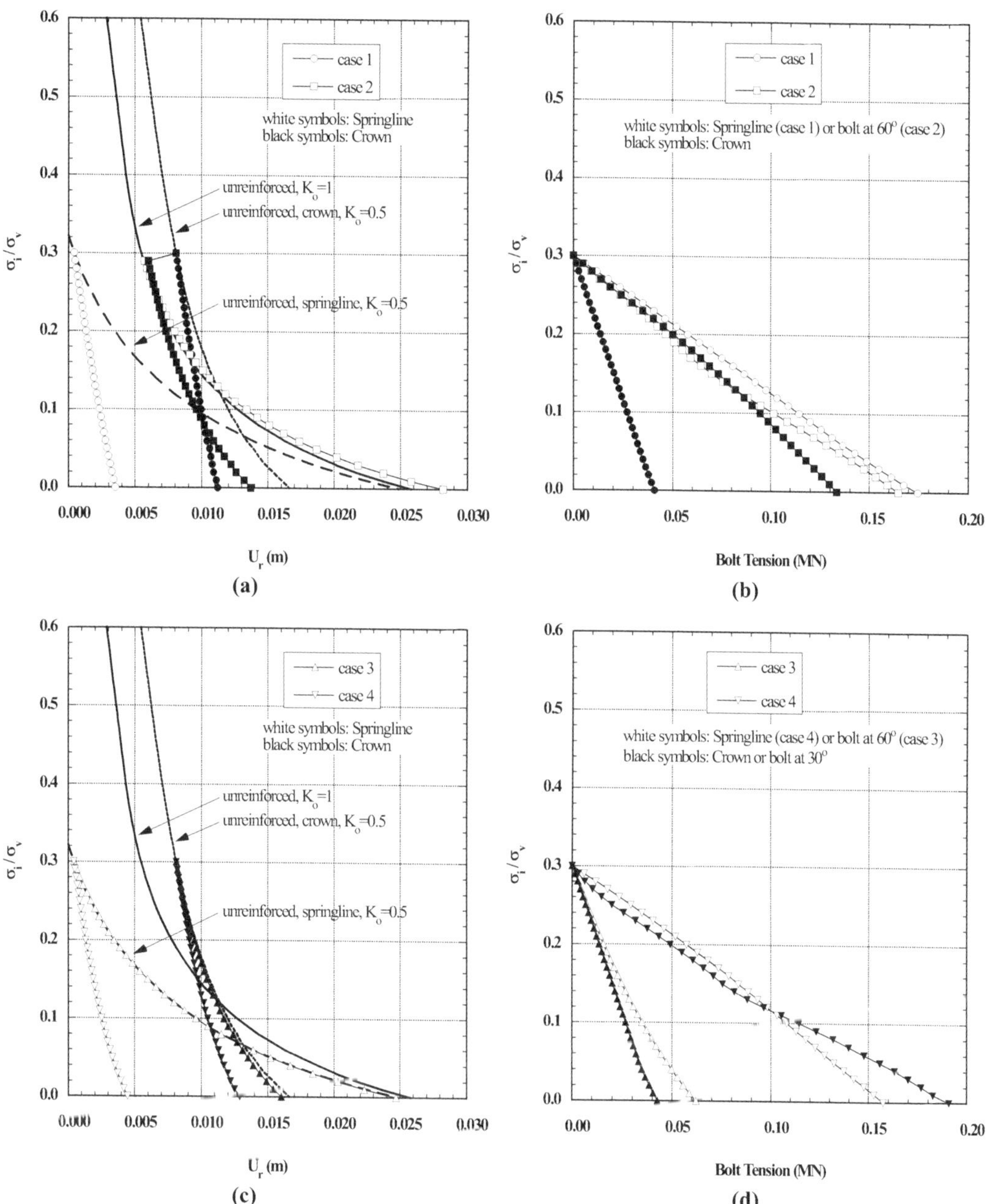

Figure 7.47 Dowels. FEM results for radial displacements and maximum bolt tension. Effects of K_o and Non-uniform distribution of rockbolts. See Figure 7.32 for description of cases. r_o = 3 m, ρ = 6 m, σ_v = 1 MPa, E = 500 MPa, ν = 0.2, $\phi^p = \phi^r$ = 30°, $c^p = c^r$ = 0.1 MPa, E_b = 210 GPa, ν_b = 0.3, d_b = 25 mm, S_z = 1 m, S_θ = 10°, β = 0.3. (Bobet and Einstein, 2011). Reproduced with permission of Elsevier.

As shown in Figures 7.42 to 7.47, reinforcement with dowels may induce large tensile loads in the rockbolts. In few of the cases however, the results are not realistic because the rockbolts would yield, given the particular input parameters, or at least debonding and slip would occur between the bolt and the grout or between the grout and the rock. To investigate the effects of slip, an additional case has been analyzed numerically with the properties of the base case, except that the maximum tension allowed in the rockbolt is 195 MPa. This is the maximum stress carried by the dowels for the base case. This allows one to further compare the tunnel performance when the ground is reinforced with anchored or with dowels. Figure 7.48a shows the radial and tangential stresses of the ground when the dowel is allowed to yield and when it is not. Figure 7.48b plots the radial displacements of the ground and the tensile stress of the dowel. The figure indicates that when the dowel yields, the radial stresses in the ground decrease, and as a consequence the tangential stresses in the ground also decrease, albeit by not much in this case given the input parameters chosen for the analysis. This is because, as the ground yields, a reduction of confinement carries a reduction of strength. Because the ground carries less load, the deformations are larger. As expected, the load carried by the dowel decreases. Interestingly, there is a load transfer within the dowel from the area of maximum stress demand to areas deeper into the rock (Figure 7.48b). This is associated with an increase, with respect to the case with no yield, of tangential stresses carried by the ground. The effects of yield are more noticeable in the ground characteristic curves, shown in Figure 7.49a, and in the rockbolt maximum tension, in Figure 7.49b. Yielding of the dowels is associated, as expected, with an increase of ground deformations and a decrease of the maximum tension carried by the reinforcement.

7.4.5.2 Rock–rockbolt interaction: numerical methods

The analytical solutions described in the preceding section have the critical limitation that most of them only apply to axisymmetric problems. For all other cases, e.g. a tunnel with a cross section other than circular, non-uniform far-field stresses (with limited exceptions), gravity forces, or when the rockbolts are not uniformly distributed around the tunnel, or for three-dimensional analysis, the approach discussed does not apply. Numerical methods must be used instead (Brady and Lorig, 1988; Tannant et al., 1995; Hyett et al., 1996; Oreste and Peila, 1996; Chen et al., 2004; Goel et al., 2007; Guan et al., 2007; Malmgren and Nordlund, 2008).

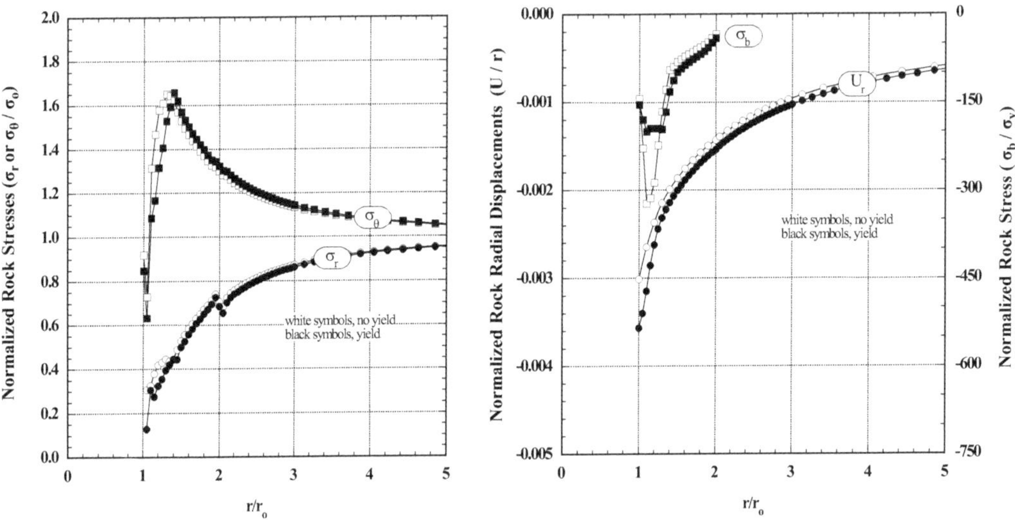

Figure 7.48 Dowels. Effect of rockbolt yield. r_o = 3 m, ρ = 6 m, σ_v = 1 MPa, K_o = 1, E = 500 MPa, ν = 0.2, $\phi^p = \phi^r$ = 30°, $c^p = c^r$ = 0.1 MPa, E_b = 210 GPa, ν_b = 0.3, d_b = 25 mm, S_z = 1 m, S_θ = 10°, β = 0.3. (Bobet and Einstein, 2011). Reproduced with permission of Elsevier.

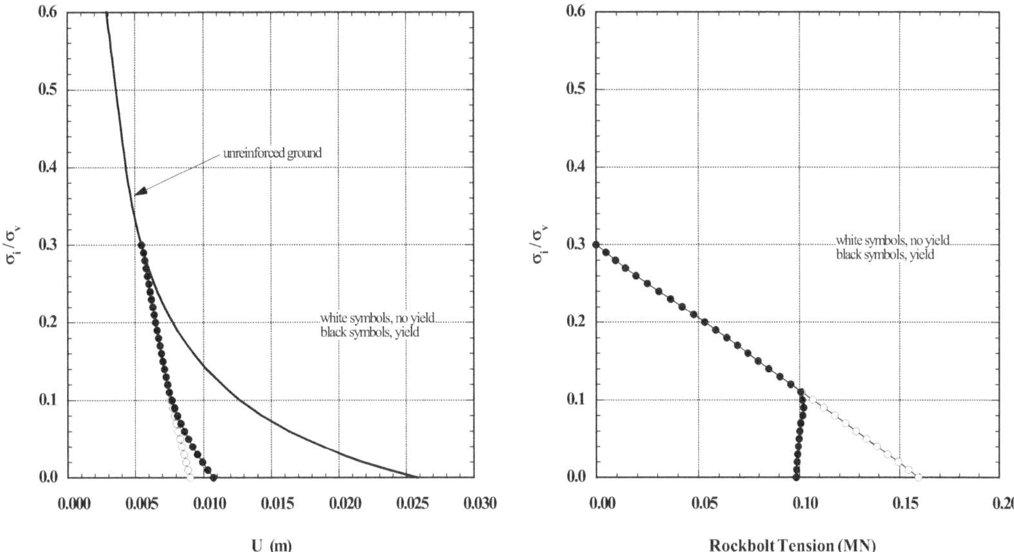

Figure 7.49 Dowels. Effect of rockbolt yield. r_o = 3 m, ρ = 6 m, σ_v = 1 MPa, K_o = 1, E = 500 MPa, ν = 0.2, $\phi^p = \phi^r$ = 30°, $c^p = c^r$ = 0.1 MPa, E_b = 210 GPa, ν_b = 0.3, d_b = 25 mm, S_z = 1 m, S_θ = 10°, β = 0.3. (Bobet and Einstein, 2011). Reproduced with permission of Elsevier.

In numerical methods, there are two models used to simulate dowels: (1) the tied model; and (2) the relative slip model. See Figure 7.50. The two models have similar characteristics. The rock and dowel in Figure 7.50a are discretized as shown in Figure 7.50b following e.g. a Finite Element Method or Finite Different Method scheme.

In the tied method (Wang and Garga, 1992; Tannant et al., 1995; Oreste and Peila, 1996; Guan et al., 2007) the dowel between two nodes is replaced by a spring with a tied connection with the rock (i.e. there is no relative displacement between the dowel and the rock at the nodes). The axial force F_a in the spring per unit length is (Figure 7.50c):

$$\begin{aligned} &F_a = K_a \Delta U < F_a^{max}, \quad \text{otherwise} \\ &F_a = F_a^{res} \\ &K_a = E_b A_b \left(1 + \frac{E_g A_g}{E_b A_b}\right) \end{aligned} \tag{7.37}$$

where K_a is the axial stiffness; E_b and A_b are the Young's modulus and cross-section area of the bolt; E_g and A_g are the Young's modulus and cross-section area of the grout (i.e. total area of the hole minus the area of the bolt); ΔU is the relative displacement between nodes i and i + 1; F_a^{max} is the axial yield force and F_a^{res} is the residual axial force when yield has been reached. The values of F_a^{max} can be determined experimentally, or can be estimated following semi-empirical relations such as those of equations (7.4) to (7.7) with $F_a^{max} = \pi\, d_b \tau^f$. Brady and Brown (2004) suggest a perfect-plastic behavior with $F_a^{res} = F_a^{max}$. Guan et al. (2007) proposed values based on Coulomb's law between the bolt and the grout of the following form: $F_a^{max} = \pi\, d_b\, (\sigma_\theta \tan\phi + c)$ and $F_a^{res} = \pi\, d_b \sigma_\theta \tan\phi$, where σ_θ is the tangential stress acting on the bolt (note that this stress should be the incremental stress after installation of the dowels, since drilling of the hole eliminates any existing stresses), c and ϕ are the cohesion and friction angle at the bolt-grout interface.

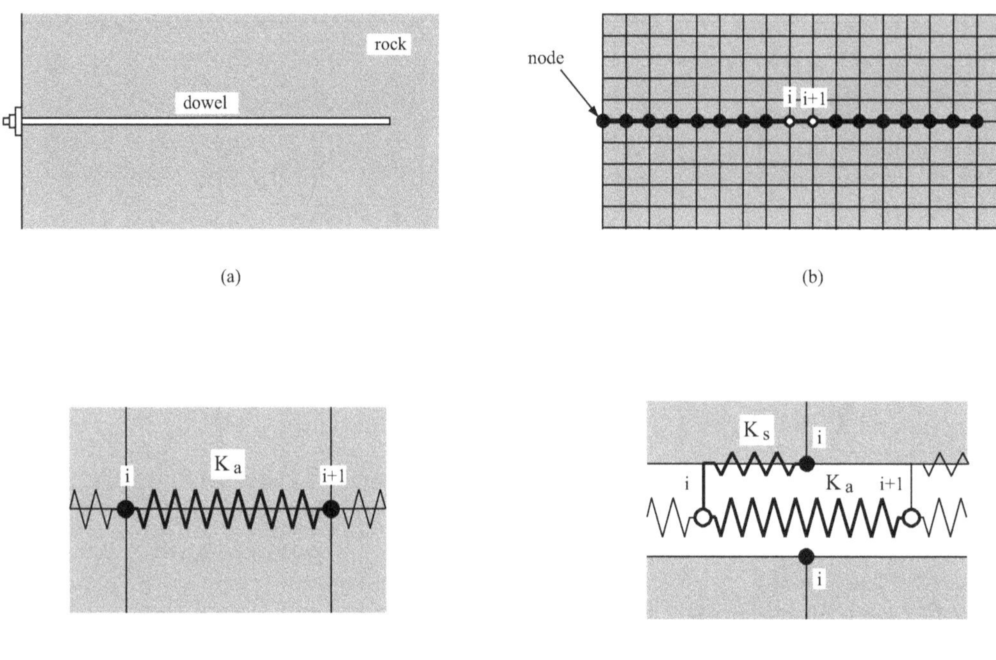

Figure 7.50 Dowel models. (a) Rock and dowel. (b) Rock and dowel model. (c) Tied model. (d) Relative slip model.

The relative displacement method is based on the assumption that the force in the dowel is the result of the relative displacement between the bolt and the rock (Brady and Lorig, 1988; Hassani et al., 1992; Hyett et al., 1996; Brady and Brown, 2004). See Figure 7.50d. The dowel induces a shear force in the rock F_s that has a magnitude per unit length:

$$F_s = K_s(U_{rock} - U_{rockbolt}) \leq F_s^{max}$$
$$K_s = \frac{2\pi G_g}{\ln\left(1 + \frac{d_g - d_b}{d_b}\right)} \qquad (7.38)$$

where K_s is the stiffness of the bond between the dowel and the rock (also denoted as K_{bond} in the technical literature); $(U_{rock} - U_{rockbolt})$ is the relative displacement between the rock and the rockbolt; and F_s^{max} the maximum bond force. K_s can be determined from field experiments, or approximated with the second expression in (7.38). Typical values for K_s fall within the range 1×10^3 to 2×10^3 MN/m/m (Zipf, 2007). F_s^{max} can be obtained also from field experiments or approximated using equations (7.4) and (7.5) with $F_a^{max} = \pi\delta_b\tau^f$, or from (7.7). Zipf (2007) provides typical values for different rocks. In (7.38) the dowel displacements are related to the dowel force as follows:

$$F_s = F_a = E_b A_b \Delta U \leq F_a^y \qquad (7.39)$$

where F_s or F_a are forces per rockbolt unit length; ΔU is the dowel elongation and F_a^y is the yield force of the rockbolt. The assumption in (7.38) and (7.39) is that after debonding or yield the maximum or yield force is maintained; i.e. a perfectly plastic response is assumed.

While the approach discussed is applicable when rock displacements are continuously distributed around the opening, additional loads in the dowel may be introduced due to opening of joints as the result of the change of stresses induced by tunnel construction. The additional stresses can be computed using Equations (7.30) and (7.31); see also Figures 7.39c and 7.39d.

7.4.6 Cablebolts

Cablebolt performance is influenced by the following factors (Bawden et al., 1992): (1) effect of the rock mass, which is defined in terms of fracture spacing and confinement related to rock mass stiffness and in-situ stresses; (2) excavation-induced stress changes; (3) cablebolt geometry, spacing and orientation, and surface fixtures; (4) grout water/cement ratio; and (5) materials and construction quality. These factors are relevant because failure of cablebolts occurs at the cable-grout interface and strength mobilization is due to frictional rather than adhesional resistance between the steel and the grout, as discussed in Section 7.4.2.2.

Empirical and numerical methods can be used for the design of cablebolt support of an underground opening. The empirical method is based on the work by Potvin (1988), Potvin and Milne (1992), and Nickson (1992), who developed the Stability Graph method for the design of open stope backs in mines. The method is based on the experience collected in more than 350 Canadian mines (Hoek et al., 2000) and requires the calculation of two factors: the stability number, N, and the hydraulic radius, S. The stability number is a measure of the ability of the rock to stand up under a given stress condition, and the hydraulic radius is a measure of the shape of the opening.

The stability number N is defined as:

$$N = Q'ABC \tag{7.40}$$

where Q' is based on the NGI rock mass classification, as defined in Chapter 2; A is the rock stress factor; B is the joint orientation adjustment factor; and C is the gravity adjustment factor.

The factor Q' is the same as the standard NGI rock index Q, except that the stress reduction factor, SRF, is set equal to 1.0. Since the method has not been applied to underground openings with significant groundwater, the joint water reduction factor is set to 1.0.

The rock stress factor A is determined based on the ratio between the unconfined compressive strength of the intact rock, σ_c, and the stress acting parallel to the face of the stope, σ_1 (major principal stress; see inset in Figure 7.51a). The unconfined compressive strength can be obtained from laboratory experiments and the induced stress due to the stope can be estimated from numerical methods (Hoek et al., 2000). The factor is:

$$\begin{aligned}
A &= 0.1 && \text{for } \frac{\sigma_c}{\sigma_1} < 2 \\
A &= 0.1125\frac{\sigma_c}{\sigma_1} - 0.125 && \text{for } 2 < \frac{\sigma_c}{\sigma_1} < 10 \\
A &= 1.0 && \text{for } \frac{\sigma_c}{\sigma_1} > 10
\end{aligned} \tag{7.41}$$

The joint orientation adjustment factor B takes into account the influence of the critical joint on the stability of structurally controlled failures. It is a function of the relative dip difference between the critical joint and the stope surface, θ, as defined in the inset in Figure 7.51a, and on the difference between the strike of the critical joint and the slope of the surface under consideration. The magnitude of the factor B is found from Figure 7.51a.

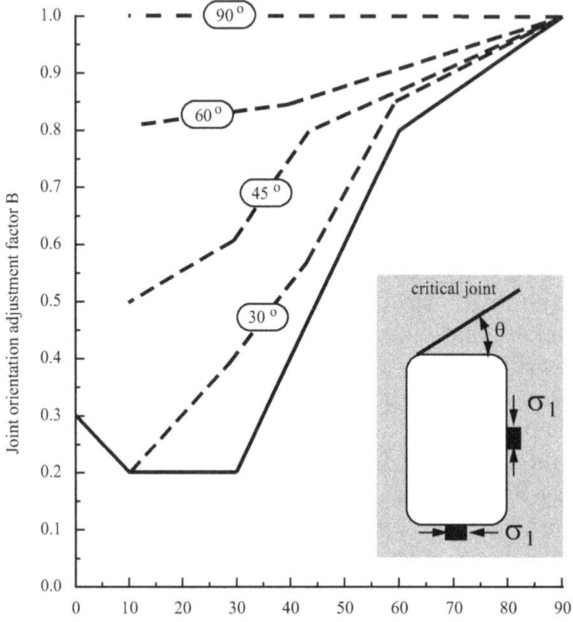

(a)

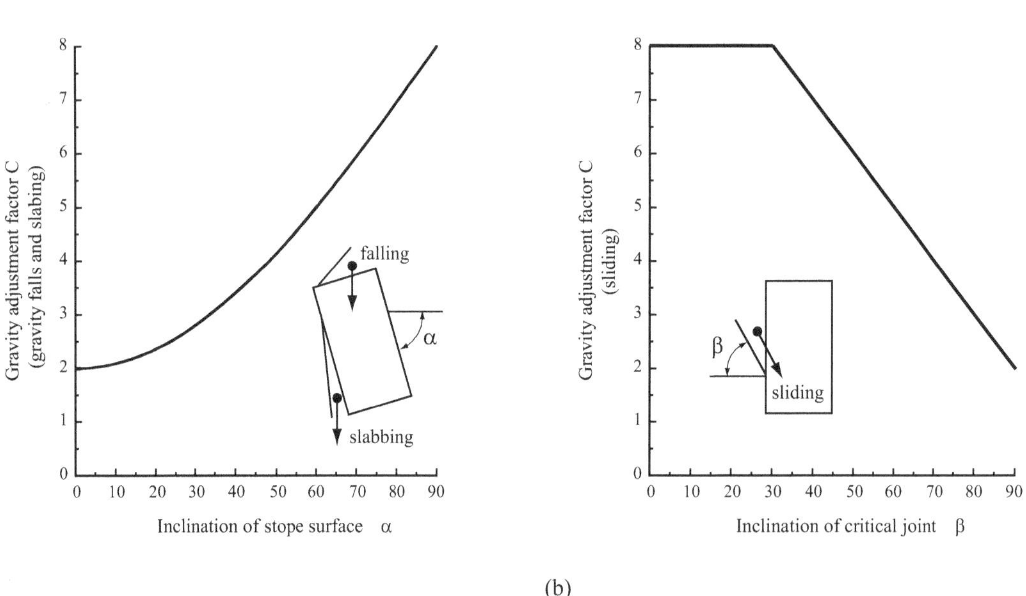

(b)

Figure 7.51 The stability graph method. Factors B and C (after Potvin, 1988). (a) Factor B. (b) Factor C.

The gravity adjustment factor C depends on the type of failure induced by the critical joint: gravity falls or sliding, Figure 7.51b. Gravity-induced failure depends on the inclination of the stope surface α, and can be calculated from $C = 8 - 6 \cos \alpha$. The factor has a minimum of 2 for horizontal stope backs and a maximum of 8 for vertical walls. The values of C for sliding failures depend on the inclination of the critical joint β, as shown in Figure 7.51b.

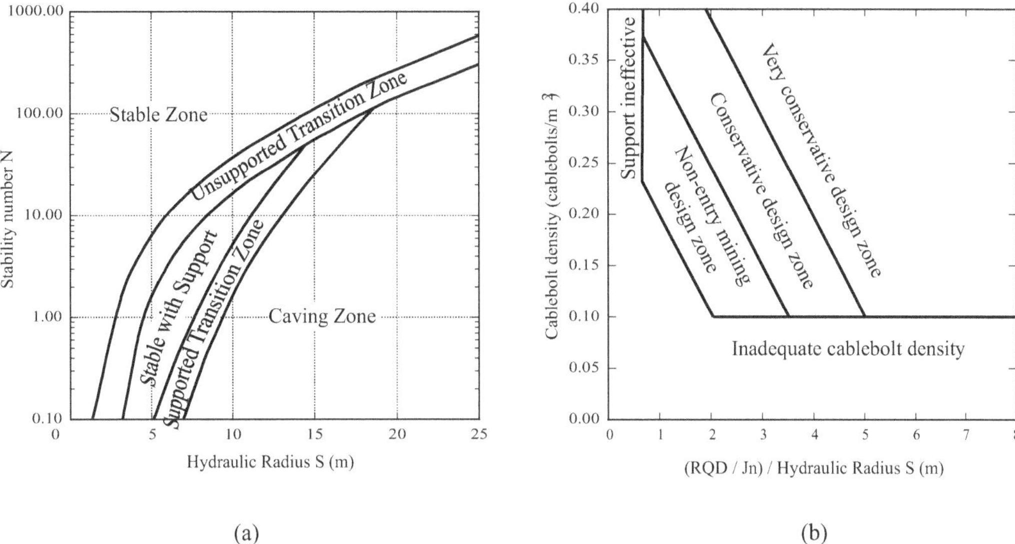

Figure 7.52 Stability graph method. Stability graph and cablebolt design chart (after Potvin, 1988). (a) Stability graph. (b) Cablebolt density chart.

Figure 7.52a provides the stability of the stope given the values of the stability number N, computed using equation (7.40) and the hydraulic radius S, which is the ratio between the area and the perimeter of the stope. The stability graph is divided into a stable zone, a transition zone, an unstable but supportable zone and an unstable, or caving zone, even with support. If the stope plots between the stable and the caving zone, cablebolts can be used to provide sufficient stability. Figure 7.52b provides an estimate of the density of cablebolts given the relative block size factor (RQD/J_n)/ hydraulic radius; J_n is the factor for number of joint sets of the Q rock mass classification system. The graph of Figure 7.52b is divided into four areas: support ineffective or inadequate cablebolt density, non-entry mining design zone, conservative design zone, and very conservative design zone. Cablebolt support is generally ineffective if the relative block size factor is less than 0.75 or the cablebolt density is less than 0.1. The zone labeled as "non-entry mining design zone" is the preferred design guideline for non-entry open stope backs. The conservative design zone is recommended for entry mining. As one can see in Figure 7.52b, the stability of the stope increases as the stability number increases or as the size of the opening decreases.

The use of the stability graph method must be limited to cases similar to those of the mines used in the database. Geologic conditions such as faults, shear zones, dykes, or waste inclusions, the creation of a slot, or brow in the stope may cause instability and under these circumstances the method is unreliable. In addition, the stope surface must be entirely covered by the cablebolting pattern and the cablebolts must extend to areas that are stable. Figure 7.52b gives the cablebolt density but not the pattern. Cablebolt spacing of about 2 m × 2m is common (Bawden et al., 1992) and appears to be sufficient to contain the larger rock blocks. Additional rockbolts between cablebolts can provide support for smaller blocks.

Cablebolts can be modeled numerically following a formulation similar to that used for dowels/grouted rockbolts, and given by equations (7.37) and (7.38). It has to be noted that this approach may be suitable to capture the initial or elastic response of the cablebolt. With large loads, the response of the cable and grout is not linear, as shown in Figure 7.18. The non-linearity is due to the interaction between the cable, the grout, and the rock, which is pressure/confinement sensitive. Better models have been proposed to capture the strain-hardening behavior of cablebolts (Yazici and Kaiser, 1992; Kaiser et al., 1992; Hyett et al., 1995; Hyett et al., 1996; Moosavi and Grayeli, 2006). They take a general form:

$$\begin{aligned} F_a^{t+\Delta t} &= F_a^t + K_1 \sigma_r^{\Delta t} + K_2 U_a^{\Delta t} \\ U_r^{\Delta t} &= U_r^t + K_3 \sigma_r^{\Delta t} + K_4 U_a^{\Delta t} \end{aligned} \tag{7.42}$$

where F_a and U_a are the axial force and axial displacement, respectively, of the cablebolt; σ_r and U_r are the radial stress and displacement at the grout-rock contact interface; and K_1, K_2, K_3, and K_4 are parameters that depend on the geometry and material properties of the cable, grout, drill hole and rock, and are pressure-dependent (Hyett et al., 1995, 1996). In the equation, the indices t + Δt correspond to the end of the current load increment; t denotes values at the end of the previous increment; and Δt indicates incremental values. If the rock mass can be approximated as an elastic material, expression (7.42) simplifies into:

$$F_a^{t+\Delta t} = F_a^t + \left(\frac{K_4 K_1 K_r}{1 - K_3 K_r} + K_2 \right) U^{\Delta t} + \frac{K_1}{(1 - K_3 K_r)(1 + \nu)} (2\sigma_r^{\Delta t} - \nu \sigma_z^{\Delta t}) \tag{7.43}$$

where K_1 to K_4 are the parameters defined in Equation (7.42), ν is the Poisson's ratio of the rock mass, $K_r = 4G/d$, with G the shear modulus of the rock mass, and d the diameter of the cablebolt hole; $U^{\Delta t} = (U_{rock}\text{-}U_{cable})$, which is the relative displacement between the rock and the cable; and σ_z is the stress in the rock along the axis of the cable. Equation (7.43) indicates that the axial force can be decomposed into two components: the first one due to incremental slip, which is similar to rockbolts, as denoted in equations (7.37) and (7.38); and the second one, due to stress-induced changes, e.g. due to mining operations.

7.5 SHOTCRETE

Shotcrete (sprayed concrete in Europe) can be described as a mortar or concrete projected pneumatically at high velocity onto a surface (Wood, 1992). The origins of shotcrete can be traced back to early 20th century and started with the invention and patent of a cement-gun in 1911 by the taxidermist C.E. Akeley. In its early use, the sprayed mortar was called "gunite" and "torcrete," a mixture of cement, sand, and water, which was mostly employed as a sealant to prevent rock mass degradation (Barrett and McCreath, 1995; Kovári, 2003). In the early 1950s, equipment was developed in Switzerland in the context of the Maggia Powerplant that made it possible to spray material with large size aggregate, i.e. concrete. This material and the technology was used to provide ground support. The American Concrete Institute adopted the term shotcrete to describe the dry mix process (Barrett and McCreath, 1995). Since then, the term has been expanded and now it also includes the wet mix process. Improvements on quality and strength of the mixture followed by adding coarser aggregate, lowering the water/cement ratio, and introducing additives. The wet mix began to be used in the late 1970s as a means of reducing rebound, and more recently Steel Fiber Reinforced Shotcrete (SFRS) started to replace steel mesh reinforced shotcrete. Today shotcrete is used worldwide as the sole support or as a component of the final lining in civil engineering tunnels and in mines (Franzén et al., 2001).

As alluded to above, there are two methods for application of shotcrete: dry mix and wet mix (Spearing and Naismith, 1999; Hoek et al., 2000). The main difference between the two methods resides on where the water is added. In the dry mix method, aggregates and cement and solid additives are placed in a hopper and mixed under continuous agitation (Figure 7.53a). Compressed air propels the mixture through a hose to the point of application and water and liquid additives are supplied at the nozzle. In the wet mix method, the water and additives are also placed in the mixer, prior to pumping; see Figure 7.53b.

Each method has advantages and disadvantages. For example the advantages of the wet mix method are (Melbye, 1999; Melbye and Dimmock, 2001): low rebound, low dust, thicker layers, better control of water/cement ratio, and high production capacity (12 m^3/hr plus, compared to 4–5 m^3/hr for dry mix); and the disadvantages: limiting conveying distance (about 300 m), higher equipment cleaning cost, and bulkier equipment. The final shotcrete quality is similar in both methods, thus the use of one method or the other depends on the time and economic constraints for each particular job and so the final decision is usually left to the contractor. In applications with difficult access, long pumping distances, or local use only, the dry mix method may be more advantageous, while the wet mix method may be preferable in instances with easy access and high production demands. In today's tunneling operation, with relatively large cross sections and high advance rates, the wet mix method is becoming standard (Franzén, 1992; Melbye, 1999).

There are three requirements to obtain a good shotcrete product (Wood, 1992; Melbye and Dimmock, 2001): appropriate design, flexible technical specifications to refine and adapt shotcrete mix to local site conditions to meet minimum performance specifications, and experienced nozzle and pump operators to minimize rebound and maximize compaction. Shotcrete mix and application are reviewed in the next section, while shotcrete design is discussed in Section 7.5.2.

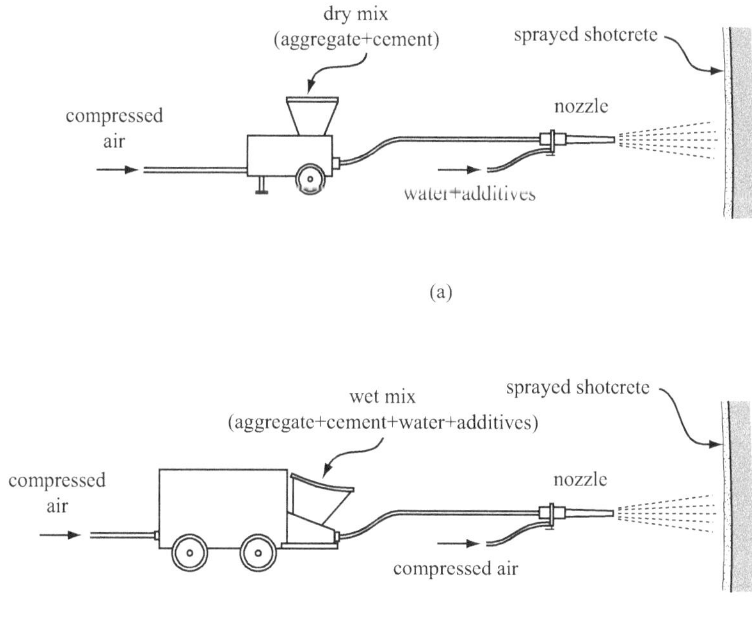

Figure 7.53 Shotcrete systems. (a) Dry-mix system. (b) Wet-mix system.

7.5.1 Materials and application

Shotcrete can be viewed as one of several ways to construct with concrete (Melbye, 1999), and thus its fabrication must comply with all the usual requirements for concrete. The following is a succinct summary of the characteristics and properties of the materials used for the manufacturing of shotcrete as well as of its application. The interested reader is directed to other publications for more detailed information (e.g. Franzén, 1992; Wood, 1992; AFTES, 1996; López, 1997; Melbye, 1999; Melbye and Dimmock, 2001; ACI 506R-05 and EM 1110-2-1005 for North America, and EFNARC, 1996 for Europe, among others). Note in particular that there are regular "shotcrete for underground support" conferences.

Shotcrete is made with some or all of the following components: cement, aggregates, water, additives, (admixtures), curing agents, and fibers (ACI 506.2-95). Aggregate ranges up to maximum grain sizes of 25 to 30 mm; the water/cement ratio ranges between 0.35 and 0.55 (upper range for larger aggregates and wet shotcrete); cement (usually Portland cement) at 300 to 450 kg/m^3; accelerators are usually 6% but can go up to 20%. Table 7.11 shows an example mix for steel fiber reinforced shotcrete (ACI 506.1R-98).

The relatively large cement content as well as accelerators (see additional comments later) are needed to obtain fast early strength, which is necessary for shotcrete to adhere to surfaces and produce early support (see below). Mason and Lorig (1981) mention strengths with 4% accelerators: after 2 hours, 1.4 to 1.7 MPa (200 to 250 psi), after 12 hours, 5.5 MPa (800 psi). As with regular concrete, the water/cement ratio is one of the most important factors, as the strength of the shotcrete decreases with added water. For wet-mix shotcrete, it is important, however, to ensure pumpability of the mixture and so a higher water content is required to obtain slumps within the range of 40–75 mm.

Figure 7.54 shows recommendations for the gradation of the aggregate based on ASTM C 1436 (also ACI 506R-05) and EFNARC (1996). Note that the recommendations are quite similar. ACI 506-05 recommends Grading No. 1 for fine-aggregate shotcrete and Grading No. 2 for all other shotcrete.

Pozzolanic additives are typically included in the shotcrete mixture as a cement replacement to enhance workability or pumpability of the concrete, increase the resistance to sulfate attacks, increase strength, and enhance adhesion. The most common additions are fly ash, ground granulated blastfurnace slag (GGBS), and microsilica or silica fume. EFNARC (1996) limits the amount of additives that can be added to the concrete, as shown in Table 7.12.

Admixtures may be used to enhance desirable properties of the shotcrete to accelerate or retard the hydration process, reduce water, etc. Admixtures are classified as accelerators, water reducers, and air entrainment. Accelerators increase the hydration rate of the cement and are commonly used for wet-mix shotcrete to reduce the slump at the moment of spraying, so the shotcrete can rapidly adhere to the ground. Because the hydration time is reduced, the shotcrete can develop very quickly high early strength and so it can be used for initial

Table 7.11 Typical steel fiber reinforced shotcrete mix (ACI 506.1R-98)

Material	*Fine aggregate mix (kg/m^3)*	*Aggregate – 9 mm mix (kg/m^3)*
Cement	446 to 558	445
Sand – 6 mm maximum	1,679 to 1,483	880 to 697
Aggregate – 9 mm maximum	-	700 to 875
Steel fiber	39 to 157	39 to 150
Accelerator	Varies	Varies
Water-cement by weight	0.4 to 0.45	0.4 to 0.45

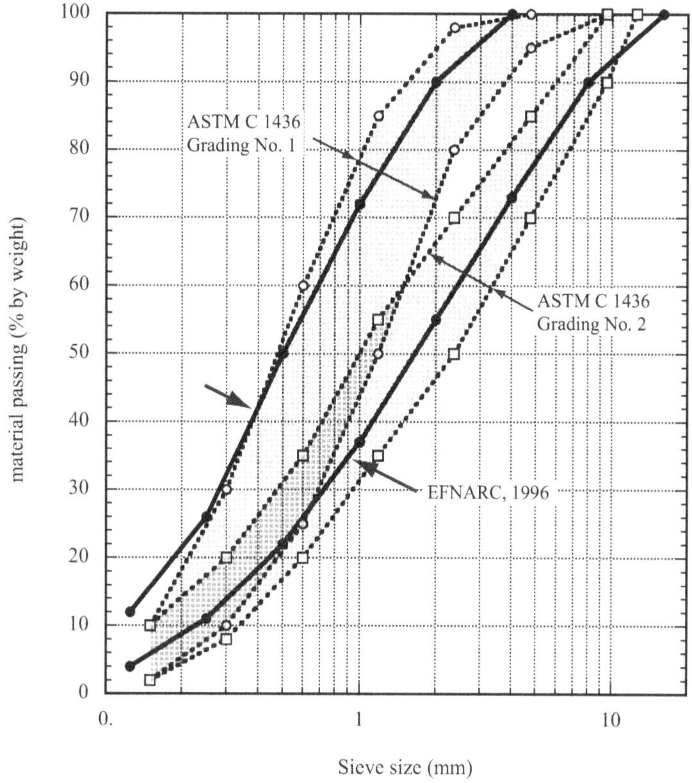

Figure 7.54 Aggregate gradation for shotcrete.

Table 7.12 Maximum levels of additions by weight to shotcrete (EFNARC, 1996)

Cementitious material	*Maximum addition*
Silica Fume	15% of Portland cement
Fly Ash	30% of Portland cement
	15% of Portland/Fly Ash cement
	20% of Portland blastfurnace slag cement
GGBS	30% of Portland cement

support and for soft ground tunneling. An undesirable effect of accelerators is loss of ultimate strength and durability of the concrete. Typical dosages are of the order of 3–6% by weight of the cementitious material (Melbye and Dimmock, 2001), which with good quality control should not produce a strength loss of more than 20%, with typical values in the range of 10–15% loss. Water reducing admixtures (superplastizers and hyperplastizers) increase the slump of the concrete without the need of increasing the water/cement ratio. They are used for wet-mix shotcrete with normal dosages of 4 to 10 kg/m^3. Retarders act as a hydration modifier and are also typically used in wet-mix shotcrete when the time from mixing to spraying may be significant. They are added in two components. The first component is the stabilizer admixture and is supplied at the batching plant to prevent hydration of the cement. Depending on the dosage, hydration can be delayed from 3 to 72 hours. The second component is added at the nozzle and is an activator that disperses the first component allowing for the hydration of the cement. Air entrainment is also added to wet-mix shotcrete to improve pumpability and to improve durability when the shotcrete is subjected to cycles of freeze and thaw under saturated conditions. Curing agents may be included in the mixture to ensure that the cement remains fully hydrated as it gains strength over time.

Finally, fibers may be added to the shotcrete to increase its ductility, toughness, impact resistance, and reduce crack propagation (e.g. Ding and Kusterle, 2000). The overall performance of fiber reinforced shotcrete is comparable to the traditional shotcrete and mesh reinforcement (Holmgren, 1987; Morgan et al., 1999), and yet the productivity is often more than doubled (Melbye and Dimmock, 2001). Steel fibers are commonly used, with typical lengths of 13 to 40 mm and dosage ranging from 20 to 60 kg/m^3. A dose of 40 kg/m^3 is shown to limit the crack width of the shotcrete to less than 0.2 mm (Melbye and Dimmock, 2001). The length of the fibers is sometimes limited by the size of the hoses to prevent blockage; e.g. EFNARC (1996) limits the length of steel fibers to 0.7 times the internal diameter of the pipes or hoses used. Synthetic fibers, such as glass or polypropylene are also employed and in fact are gaining acceptance. Polypropylene fibers are quite common and are 25 to 50 mm long, with dosages of the order of 1 to 3 kg/m^3.

Even though steel fibers are increasingly used, one has to be aware of the fact that they often produce a surface with the fibers sticking out like needles. This affects the safety of personnel and can also affect the waterproofing membranes. Polypropylene fibers, which may be used in addition to steel fibers, serve a special purpose: fire proofing of shotcrete (or concrete). These fibers melt at temperatures around 130–170°C and leave open pores into which steam produced from heating the water in the concrete can expand. This reduces the spalling of concrete/shotcrete under effects of fire (clearly such fibers can also be added to cast-in-place concrete).

While adequate quality of materials, dosage, additives, and admixtures determines the overall engineering properties of the material, the actual quality of the applied shotcrete relies heavily on the craftsmanship of the operators, and in particular on the skills of the nozzleman. In fact, some codes require certification of the nozzle operator (ACI 506R-05). One of the major requirements is that the shotcrete jet be as perpendicular as possible to the application surface. It should be noted that shotcrete application in major tunnel projects is today largely mechanized and automated. Preconstruction testing is generally advisable. This is needed both to familiarize the nozzle operator with the process and, most importantly, to check the particular combination of aggregate, cement and additives. These material components vary and checking the particular combination is of utmost importance. This is done by shooting test panels that simulate job conditions. Before shotcrete application, it is important to ensure that the application surface is adequately prepared. This usually requires removal of all loose material such as debris, chips, mud, and dirt that may prevent a good bond between the shotcrete and the ground. Additional surface preparation may include milling and sandblasting to increase the surface roughness. The receiving surface should be pre-wet to the point that the material is saturated but the surface is dry; this is done to prevent water from the shotcrete to migrate to the ground. The shotcrete should be sprayed against the surface at a distance between 0.6 to 1.8 m and at an angle perpendicular to the receiving surface. If the angle of application deviates from the perpendicular, the shotcrete rolls or folds over and creates an uneven surface. Each layer of shotcrete is built through several passes over the work area. The thickness of the layer placed in one pass should not exceed 5 cm. When the shotcrete is first applied to the perimeter of the tunnel, significant rebound may be produced. Rebound is aggregate and cement paste that bounces off the surface during shotcrete application. Once a thin layer of cement-rich, plastic, fine shotcrete is obtained against the surface, the rebound decreases. The amount of rebound depends on the shotcrete mix method and on the final thickness of the layer (Parker, 1999). Table 7.13 provides indications on the

Table 7.13 Approximate ranges of rebound losses (ACI 506R-05)

Surface	*Percent of rebound (%)*
Floor or slabs	0 to 10
Sloping and vertical walls	10 to 30
Overhead work	10 to 30

rebound expected for typical jobs (ACI 506R-05). Note that rebound cannot be reused and also requires special disposal management since it includes cement and chemicals in addition to the aggregate.

Once the shotcrete is in place, it must cure for over a period of several days so that it can develop its full strength. This is generally accomplished by keeping the surface of the shotcrete moist. Ideally, the temperature of the shotcrete from application to curing should be maintained within the range of 10–30°C. Shotcrete application should be discontinued if the ambient temperature drops below 5°C (EFNARC, 1996).

Specifications for quality control such as sampling and testing depend on the size and complexity of the project. They may just call for minimum unconfined compressive strength at 28 days, but normally additional requirements need to be satisfied such as flexural strength, impact resistance, toughness, pullout strength, tensile strength, and bond strength (e.g. ACI 506.1R-98). ACI 506R-05 suggests testing once a day or every 40 m^3 or shotcrete, whichever is greater. Table 7.14 includes the frequency of testing suggested by EFNARC (1996).

The tests can be performed from specimens sawed or cored from the liner or from test panels. With a typical mix such as that shown in Table 7.11, the following properties are expected (ACI 506.1R-98 and ACI 506R-05): (1) unconfined compression strength at 28 days may range from 29 to 52 MPa; (2) flexural strength between 5.5 and 7.6 MPa; (3) flexural toughness measured from small beams 100 x 100 x 350 mm (ASTM C 78) in the range of 10–20 times that of plain concrete; (4) pullout strength (ASTM C 78) from 6.9 to 12.4 MPa; (5) tensile bond strength between 0.7 and 1 MPa; (6) drying shrinkage within the range of 0.06%–0.1% after three months, following ASTM C 157; (7) density of good-quality shotcrete within 2,230 and 2,390 kg/m^3; and (8) modulus of elasticity within the range 17 to 50 GPa, similar to that of concrete.

The engineering property of shotcrete that is perhaps of most interest to tunnel construction is its strength at early age. Because of longitudinal arching, significant ground deformations occur within one to two diameters behind the tunnel face. As a result, significant loads are expected in the liner placed at this location. Given modern construction operations, these loads are applied to shotcrete that is from about one day of age, to few days old. Thus there

Table 7.14 Frequency of control testing (EFNARC, 1996). Values in m^2 between tests

Type of control	*Minor*	*Normal*	*Extended*
Compressive strength	500	250	100
Flexural strength		500	250
Residual strength value		1000	500
Energy absorption		1000	500
Bond		500	250
Fiber content		250	100
Thickness	50	25	10

is a need of attaining a material that has substantial strength shortly after it is placed. This is generally accomplished by introducing admixtures into the shotcrete mix, as discussed above.

7.5.2 Design

Shotcrete is used as temporary or permanent support of tunnels. It has been proven effective as a support system in excavations in blocky rock where failure is structurally controlled (local-type failure) and in soft and fractured rocks, and in soils where failure is ductile (general-type failure).

Shotcrete design for tunnel-support is difficult for two reasons: (1) the material is not well defined, as has been discussed; and (2) the shotcrete support mechanisms of the rock are complex. Figure 7.55 shows possible mechanisms as introduced in the shotcrete research by the University of Illinois. Only considering the shell action with the hoop stress formula will produce shotcrete support that is too thin. It is, therefore, necessary to design against bending and thrust as done for cast-in-place concrete but, in addition, against shearing because of the punch- through mechanisms shown in Figure 7.55. More specifically, Figure 7.55a describes the shotcrete shell action, i.e. a thin-walled cylinder. Figure 7.55b depicts how shotcrete is forced into a discontinuity and, thus, reinforces the rock mass. Figure 7.55c illustrates shotcrete-groundwater interaction; it has a positive effect in that it prevents flow and thus piping and a negative effect because of the buildup of water pressure. Figure 7.55d is a schematic of: (i) how shotcrete adheres to rock (denoted by dotted lines) and prevents movement of rock blocks; and also (ii) of the shearing resistance to resist punch-through. The effects of nonplanar surfaces in rock excavated by drill and blast have multiple effects, namely: waviness of the shotcrete that creates a "corrugated support," producing greater resistance to moments (Figure 7.55e); greater thickness of shotcrete filling concave parts (Figure 7.55f); and smaller thickness in the convex parts (Figure 7.55g).

To prevent structurally-controlled instabilities, shotcrete needs to be applied immediately after excavation to decrease initial deformations of the rock, since it only takes very small relative displacements for the strength of the discontinuities to decrease from peak to residual. The function of the shotcrete is then to preserve the integrity of the rock and prevent raveling mechanisms (Barrett and McCreath, 1995; Ratcliffe, 1999). The shotcrete supports the weight of the rock, which requires good adhesion between the shotcrete and the rock. Shotcrete, as the sole means of stabilization of structurally-controlled failures, is not a cost-effective mechanism as it can only provide support where it intersects the block discontinuities. Instead, it is used in conjunction with rockbolts.

Design of the shotcrete must consider all potential modes of failure. For structurally-controlled instabilities, the most likely failure mode is punching failure. The thickness and strength of the shotcrete must be such that, upon application, it can safely carry the rock block load. The load can be determined using the methods discussed in Section 4.2.3. The capacity of early age fiber reinforced shotcrete can be estimated based on the following equations (Bernard, 2008):

$$\begin{aligned} V &= vpt \\ v &= 0.28(f_c)^{0.26} - 0.11 \end{aligned} \tag{7.44}$$

where V is the shearing resistance of a cross section of shotcrete with thickness t; v is the shear strength of the shotcrete in MPa; p is the length of the intersection of the rockblock

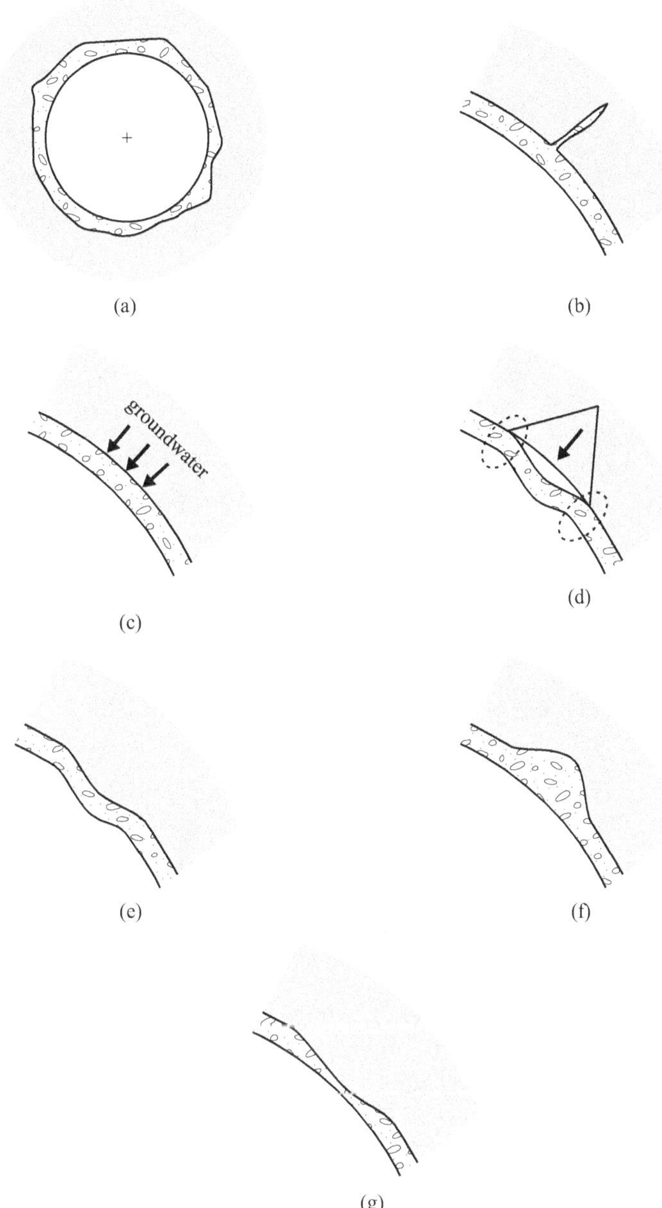

Figure 7.55 Shotcrete mechanisms. (a) Shell action. (b) Shotcrete into discontinuities. (c) Shotcrete-ground water interaction. (d) Shotcrete adhesion Shotcrete prevents punch through. (e) Shotcrete waviness. (f) Shotcrete concave - Thicker. (g) Shotcrete convex - Thinner.

with the shotcrete liner; and f_c is the unconfined compression strength of the shotcrete (MPa) at a given age. Equation (7.44) appears to be insensitive to the type of fiber used.

For general-type failures, shotcrete can be treated for design as a continuous thin membrane that deforms with the ground. This is the approach that has been taken by a number of researchers (e.g. Pöttler, 1990; Oreste and Peila, 1997; Chryssanthakis et al., 1997; Lackner

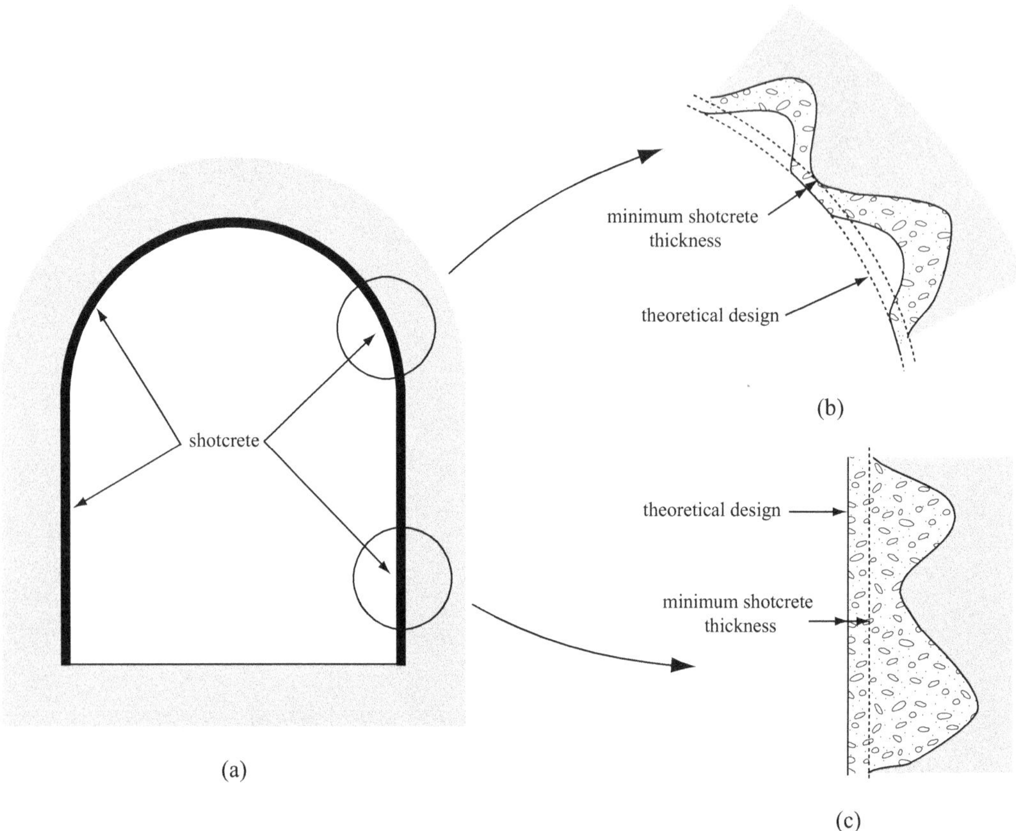

Figure 7.56 Continuous or discontinuous shotcrete liner. (a) Tunnel with shotcrete lining. (b) Discontinuous support. (c) Continuous support.

et al., 2002; Oreste, 2003a; Graziani et al., 2005; Boldini et al., 2005). It has to be noted, however, that for this assumption to be applicable, the shotcrete must have flexural capacity. This is possible only if a smooth, continuous section of shotcrete is obtained during construction. Figure 7.56 illustrates the concept. Figure 7.56a depicts a tunnel where a thin layer of shotcrete is designed as the sole means of support. If the finish of the excavation is rough and a minimum thickness of shotcrete is effectively applied on the cross section, there are two limit scenarios that can be identified (Windsor and Thompson, 1999). In the first scenario, illustrated in Figure 7.56b, a uniform smooth shotcrete liner is not obtained. The shotcrete coats the ground and locally may provide support, but it is doubtful that the assumption of a continuous, smooth membrane over the entire cross section is realistic. In the second scenario, Figure 7.56c, the shotcrete fills completely the over-excavation and produces a continuous smooth liner with the thickness required. In this scenario the assumption of a continuous membrane appears to be valid.

The load carried by the shotcrete, with the assumption of a continuous membrane, can be estimated using the analytical methods presented in Chapter 4 or the numerical methods in Chapter 5. It is also interesting and informative to discuss early proposals to approximate the loads that must be carried by the shotcrete. One of the earlier analytical methods was that by

Deere et al. (1969), who suggested to compute the thrust acting on the shotcrete by extending the hoop stress formula, such that:

$$\sigma_{shotcrete} = \sigma_{hoop} + \sigma_{distortion} = \frac{\sigma_{ext} r_o}{t} + E\frac{t\Delta r_o}{r_{o,i} r_{av}} \tag{7.45}$$

where $\sigma_{shotcrete}$ is the stress on the shotcrete, which is obtained by adding to the hoop stress, the stresses induced in the shotcrete by the distortion of the tunnel; σ_{ext} is the external pressure, r_o is the radius of the tunnel (and also the exterior radius of the shotcrete), t is the shotcrete thickness, E is the Young's modulus of the shotcrete, r_o and r_i are the interior and exterior radii of the inner and outer fiber of the shotcrete, r_{av} is the average radii, equal to $(r_o + r_c)/2$, and Δr_o is the change in radius due to the distortion. Deere et al. (1969) suggest a change of radius/distortion of 1% for normal conditions and 2% for extreme conditions. It should be emphasized that this approach assumes a flexible liner with only small moment development. Refer to Chapter 1 for a detailed discussion on the assumptions of flexible and rigid liners.

Heuer (1974) was perhaps one of the first authors who proposed a design approach based on load and resistance factors. He suggested computing the ultimate required thrust, $T_{ult, req}$, as:

$$T_{ult,req} = LF\sigma_{ext} r_o \tag{7.46}$$

where σ_{ext} is the external pressure, r_o is the radius of the tunnel, and LF the load factor, with recommended values of 1.4 for reinforced concrete and 1.9 for earthwork. The ultimate thrust capacity recommended by Heuer, $T_{ult,cap}$ was given by:

$$T_{ult,cap} = \phi 0.85 f'_c t \tag{7.47}$$

where ϕ is the capacity reduction factor (for reinforced concrete equal to 0.9 and for flexural members in plain concrete, 0.65), f'_c is the unconfined compression strength of concrete and t, the thickness of the shotcrete. Equating equations (7.46) and (7.47) provides the thickness of the shotcrete. That is,

$$t = \frac{LF}{0.85\phi}\frac{\sigma_{ext}}{f_c'} r_o \tag{7.48}$$

where, after Heuer, LF/(0.85 ϕ) should be 2.0 for temporary support and 2.5–3.0 for permanent support. Heuer (1974) then recommends to add to the theoretical thickness in Equation (7.48) 2″ for shotcrete applied on smooth surfaces, in machine-bored tunnels, and 4″ for shotcrete applied on rough surfaces, in drill and blast tunnels. The added shotcrete is to account for local variations in the shotcrete thickness and irregularities of the excavation surface.

Because shotcrete sets and hardens with time, its engineering properties change with time. It is not clear then if the most critical condition occurs when the shotcrete is fully hardened (Oreste and Peila, 1997). The deformations of the shotcrete are produced by the following mechanisms: elasto-plastic, viscous (short-term creep), flow (long-term creep) and chemical shrinkage. Advanced numerical models have been proposed that account for the thermo-chemo-mechanical, time-dependent, behavior of shotcrete (Hellmich et al., 2001; Lackner et al., 2002; Boldini et al., 2005). Such models are still research tools and have been used with three-dimensional FE methods to investigate the evolution of stresses in the shotcrete with time. The analyses show that, for time-independent ground deformations, the chemo-mechanical properties of shotcrete tend to reduce the stresses in the liner. This is so because the ground deformations are the largest close to the tunnel face, which is where the shotcrete is applied. At this early age, the shotcrete material is compliant and so it can absorb significant deformations without taking much load; viscous creep and flow tend to reduce even more the loads. The result is a liner with reduced loading with larger tunnel convergence. For tunnels excavated in ground that undergoes deformations with time, e.g. consolidation due to pore pressure dissipation, creep, swelling, etc., the numerical analyses show that in those cases where ground deformations occur at a rate much slower than the hydration processes in the shotcrete, a simplified elasto-plastic analysis is sufficient. For those cases where the rate of ground deformations is comparable to the hydration rate of the shotcrete, time-effects need to be considered to proper model the response of the liner (Boldini et al., 2005). As already mentioned, this can be done with three-dimensional analyses with advanced material models for the shotcrete. As an alternative, simplified approximations have been proposed to capture the time-increase of the stiffness and strength of the shotcrete. There are two approaches. In the first approach the stress-strain and time-strength of the shotcrete are approximated with a hyperbolic relationship (Kuwajima, 1999; Hisatake, 2003). In the second approach the shotcrete is assumed elastic with Young's modulus and strength dependent on time through an exponential function (Oreste and Peila, 1997; Oreste, 2003a; Graziani et al., 2005). The following relations approximate the stiffness and strength dependence of the shotcrete with time (Graziani et al., 2005):

$$
\begin{aligned}
\frac{E(t)}{E_{28}} &= c_1 \exp\left(\frac{c_2}{t^{c_3}}\right) \\
\frac{\sigma_c(t)}{\sigma_{c,28}} &= d_1 \exp\left(\frac{d_2}{t^{d_3}}\right)
\end{aligned}
\tag{7.49}
$$

where $E(t)$ and $\sigma_c(t)$ are the Young's modulus and strength of the shotcrete at time t, expressed in days; E_{28} and $\sigma_{c,28}$ are the Young's modulus and strength at 28 days; and c_1, c_2, c_3, d_1, d_2, and d_3 are material constants, with the following average values: $c_1 = 1.062$, $c_2 = -0.446$, $c_3 = 0.6$, $d_1 = 1.105$, $d_2 = -0.743$, $d_3 = 0.7$, when time is expressed in days (Graziani et al., 2005).

While the use of analytical and numerical procedures, as discussed in Chapters 4 and 5, is attractive because all relevant parameters are incorporated into a rational analysis, the material models presented still require in-depth comparisons with field cases to bring confidence in their predictions. In addition, shotcrete support is seldom used alone. It is typically combined with other support methods such as rock reinforcement and steel sets, which complicates the use of rational methods. Given the complex mechanisms and, particularly, the variability of ground conditions, shotcrete is very often designed based on empirical methods. These have been discussed in detail in Chapter 2. Particularly for initial support, it may be most appropriate to use these approaches possibly running a quick check with one of the methods described above.

Table 7.15 lists shotcrete recommendations for underground mining, after Hoek et al. (2000). When combined with rockbolts, it is desirable to apply the shotcrete first to stabilize the ground, and then place the faceplate of the rockbolts on top of the shotcrete. This allows for an effective transfer of the rockbolt load to the underlying rock and a reduction of loss of load at the head of the rockbolt due to creep or weathering of the rock.

Table 7.15 Summary of recommended shotcrete applications in underground mining, for different rock mass conditions: (After Hoek et al. 2000)

Rock mass description	*Rock mass behavior*	*Support requirements*	*Shotcrete application*
Massive metamorphic or igneous rock. Low stress conditions.	No spalling, slabbing or failure.	None.	None.
Massive sedimentary rock. Low stress conditions.	Surfaces of some shales, siltstones, or claystones may slake as a result of moisture content change.	Sealing surface to prevent slaking.	Apply 25 mm thickness of plain shotcrete to permanent surfaces as soon as possible after excavation. Repair shotcrete damage due to blasting.
Massive rock with single wide fault or shear zone.	Fault gouge may be weak and erodible and may cause stability problems in adjacent jointed rock.	Provision of support and surface sealing in vicinity of weak fault of shear zone.	Remove weak material to a depth equal to width of fault or shear zone and grout rebar into adjacent sound rock. Weldmesh can be used if required to provide temporary rock-fall support. Fill void with plain shotcrete. Extend steel fiber reinforced shotcrete laterally for at least the width of gouge zone.
Massive metamorphic or igneous rock. High stress conditions.	Surface slabbing, spalling and possible rock-burst damage.	Retention of broken rock and control of rock mass dilation.	Apply 50 mm shotcrete over weldmesh anchored behind bolt faceplates, or apply 50 mm of steel fiber reinforced shotcrete on rock and install rockbolts with faceplates; then apply second 25 mm shotcrete layer. Extend shotcrete application down sidewalls where required.
Massive sedimentary rock. High stress conditions.	Surface slabbing, spalling and possible squeezing in shales and soft rocks.	Retention of broken rock and control of squeezing.	Apply 75 mm layer of fiber reinforced shotcrete directly on clean rock. Rockbolts or dowels are also needed for additional support.
Metamorphic or igneous rock with a few widely spaced joints. Low stress conditions.	Potential for wedges or blocks to fall or slide due to gravity loading.	Provision of support in addition to that available from rockbolts or cables.	Apply 50 mm of steel fiber reinforced shotcrete to rock surfaces on which joint traces are exposed.
Sedimentary rock with a few widely spaced bedding planes and joints. Low stress conditions.	Potential for wedges or blocks to fall or slide due to gravity loading. Bedding plane exposures may deteriorate in time.	Provision of support in addition to that available from rockbolts or cables. Sealing of weak bedding plane exposures.	Apply 50 mm of steel fiber reinforced shotcrete on rock surface on which discontinuity traces are exposed, with particular attention to bedding plane traces.
Jointed metamorphic or igneous rock. High stress conditions.	Combined structural and stress controlled failures around opening boundary.	Retention of broken rock and control of rock mass dilation.	Apply 75 mm plain shotcrete over weldmesh anchored behind bolt faceplates or apply 75 mm of steel fiber reinforced shotcrete on rock, install rockbolts with faceplates and then apply second 25 mm shotcrete layer. Thicker shotcrete layers may be required at high stress concentrations.
Bedded and jointed weak sedimentary rock. High stress conditions.	Slabbing, spalling and possibly squeezing.	Control of rock mass failure and squeezing.	Apply 75 mm of steel fiber reinforced shotcrete to clean rock surfaces as soon as possible, install rockbolts, with faceplates, through shotcrete, apply second 75 mm shotcrete layer.
Highly jointed metamorphic or igneous rock. Low stress conditions.	Raveling of small wedges and blocks defined by intersecting joints.	Prevention of progressive raveling.	Apply 50 mm of steel fiber reinforced shotcrete on clean rock surface in roof of excavation. Rockbolts or dowels may be needed for additional support for large blocks.
Highly jointed and bedded sedimentary rock. Low stress conditions.	Bed separation in wide span excavations and raveling of bedding traces in inclined faces.	Control of bed separation and raveling.	Rockbolts or dowels required to control bed separation. Apply 75 mm of fiber reinforced shotcrete to bedding plane traces before bolting.
Heavily jointed igneous or metamorphic rock, conglomerates or cemented rockfill. High stress conditions.	Squeezing and "plastic" flow of rock mass around opening.	Control of rock mass failure and dilation.	Apply 100 mm of steel fiber reinforced shotcrete as soon as possible and install rock-bolts, with faceplates, through shotcrete. Apply additional 50 mm of shotcrete if required. Extend support down sidewalls if necessary.
Heavily jointed sedimentary rock with clay coated surfaces. High stress conditions.	Squeezing and "plastic" flow of rock mass around opening. Clay rich rocks may swell.	Control of rock mass failure and dilation.	Apply 50 mm of steel fiber reinforced shotcrete as soon as possible, install lattice girders or light steel sets, with invert struts where required, then more steel fiber reinforced shotcrete to cover sets or girders. Forepoling or spiling may be required to stabilize face ahead of excavation. Gaps may be left in final shotcrete to allow for movement resulting from squeezing or swelling. Gap should be closed once opening is stable.
Mild rockburst conditions in massive rock subjected to high stress conditions.	Spalling, slabbing and mild rockbursts.	Retention of broken rock and control of failure propagation.	Apply 50 to 100 mm of shotcrete over mesh or cable lacing which is firmly attached to the rock surface by means of yielding rock-bolts or cablebolts.

7.6 COMPOSITE LINERS

Typical tunnel liners are built with more than one type of structural component, perhaps with the sole exception of segmental concrete liners placed behind a shield or TBM. Steel set supports are combined with shotcrete or with a secondary concrete liner, and shotcrete and rockbolts are regularly used together to provide permanent support, etc. The end result is a composite liner that is made of two or more structural components. The following classification is proposed to catalog composite liners: primary and secondary composite liners. Primary composite liners are produced when all structural components are placed at the same time. A secondary composite is obtained when at least one structural component is placed much later than the other components.

Figure 7.57a shows the support reaction curves for two structural components and for their composite support. Those are denoted as support 1, e.g. shotcrete, and support 2, e.g.

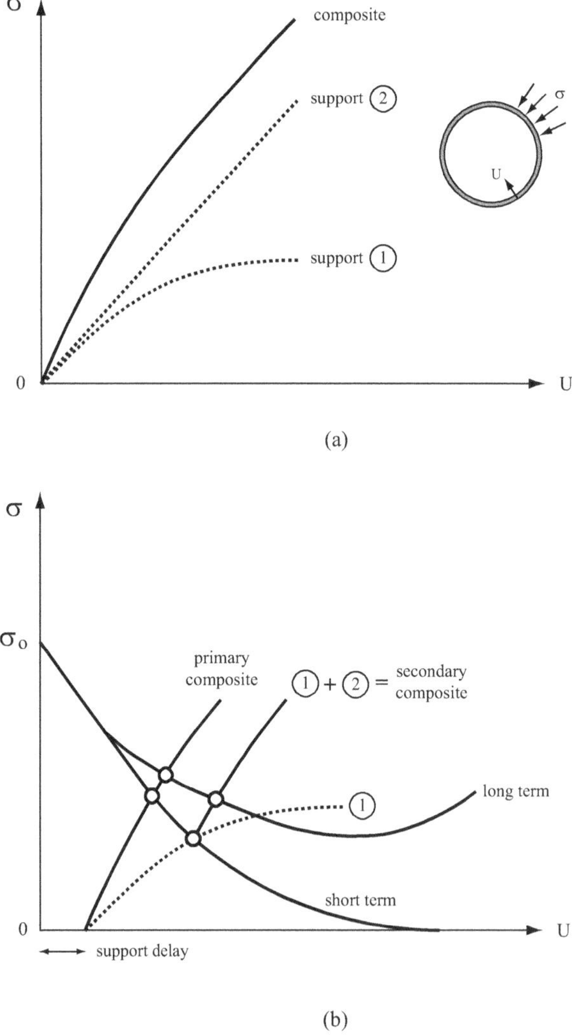

Figure 7.57 Reactions curves for composite support. (a) Support reaction curve. (b) Composite support and ground reaction curves.

steel sets or rockbolts. Figure 7.57b shows the interaction between the composite support and the ground reaction curve. In the figure, two scenarios are illustrated. In the first scenario (primary composite) the composite reaction curve from Figure 7.57b is used for permanent support for both short-term and long-term conditions. Short-term and long-term conditions may be different in a tunnel where, e.g. excess pore pressures are produced during excavation. As expected, Figure 7.57b shows that the support stresses and displacements increase from short-term to long-term conditions. In the second scenario, support 1 is placed first and is used for short-term conditions, while the long-term stresses are resisted by both support 1 and support 2, which is added at a later time. The figure shows that even though the same composite liner is placed, the overall response is quite different, with smaller stresses and larger displacements obtained with the secondary composite (support 1 and later support 2).

In general, the response of the composite behavior depends on the support stiffness and on the construction delay that either the composite liner or each structural component has. The final load of the composite liner must be shared by each of the structural components. If none of the components yields, the final load in each component depends on its relative stiffness with respect to the stiffness of the composite. Simplified relations can be obtained with the assumption that the displacements of each structural component must be the same as those of the composite after the structural component is added. For example for a composite with two elastic components placed at the same time, the following relations apply:

$$
\begin{aligned}
dU_r &= \frac{r_o}{k_1 + k_2} d\sigma = \frac{r_o}{k_{eq}} d\sigma \\
d\sigma_1 &= \frac{k_1}{k_{eq}} d\sigma \\
d\sigma_2 &= \frac{k_2}{k_{eq}} d\sigma
\end{aligned}
\tag{7.50}
$$

where dU_r is the incremental radial displacement of the composite; r_o the tunnel radius; k_1 and k_2 are the stiffness of the structural components 1 and 2, respectively, which can depend on displacement; $d\sigma_1$ and $d\sigma_2$ are the incremental radial stresses of the structural components 1 and 2, respectively (note that because the components may have different stiffnesses, the stresses they carry may be different); $d\sigma$ is the incremental radial stress of the composite; and $k_{eq} = k_1 + k_2$ is the equivalent stiffness of the composite. For a linear elastic composite, the relations are:

$$
\begin{aligned}
U_r &= \frac{r_o}{k_1 + k_2} \sigma = \frac{r_o}{k_{eq}} \sigma \\
\sigma_1 &= \frac{k_1}{k_{eq}} \sigma \\
\sigma_2 &= \frac{k_2}{k_{eq}} \sigma
\end{aligned}
\tag{7.51}
$$

7.7 WATER MANAGEMENT

Groundwater has a number of effects that need to be considered: (1) it changes the stress conditions in the ground and thus the mechanical behavior of the ground. This, and how it can be handled, has been discussed in Chapters 2 and 4; (2) the groundwater flows into the tunnel and needs to be removed – water flow problem; (3) if there is an impermeable or partially permeable liner, the groundwater exerts pressure on the liner, and this needs to be considered in the design – water pressure problem.

As will be seen, much of what needs to be done in this regard is related to construction- and design details that are beyond the intent of this book. However, important principles and required procedures will be addressed. Although the two problems are interrelated (stopping water flow will likely end up increasing the water pressure), this will be treated separately.

7.7.1 Water flow

This can occur as sudden influx of groundwater that stops construction or as continuous flow during the lifetime of the tunnel or both. During construction, drainage ditches prevent water from covering the tunnel invert. In horse-shoe or similarly shaped tunnels, they are placed at one or both sides while in circular tunnels they are usually placed in the middle (Figure 7.58).

The cross section of these trenches needs to be derived from the anticipated flow, usually not the maximum flow but the flow that one wants to accommodate without lengthy construction disruption. Inclination of the drainage trenches ranges from 0.5 to 0.1%. It is important to note that the same trenches will eventually be used to place the final drainage pipe (see later), and the contractor needs to decide how the necessary cross sections need to be created. As a matter of fact, in long tunnels, water during construction will flow both through the trenches and the final drainage pipes.

One of the specific problems with water inflow during construction is concentrated flow from joints, which makes e.g. shotcrete applications difficult. This can be handled by inserting hoses into the joints during the shotcreting operation, as shown in Figure 7.59.

Overall, water inflow in tunnels is handled in the "classic" way of dealing with water inflow (not only in tunneling) through a combination of drainage and sealing. There are many ways to do this but the design used for the new Gotthard Base Tunnel corresponds to what is done in many if not most tunnel designs (Figure 7.60), and consists of: (1) A drainage layer placed against the initial support or the rock. This is usually a felt-like geotextile. In the past, when the final liner was a structure created at a distance from the rock, the drainage layer consisted of gravel filled in the space between rock and support; recent developments also involve such gravel based drainage layers. (2) Watertight membranes, i.e. the "seal" consisting of a flexible membrane. It is important to realize that these watertight membranes

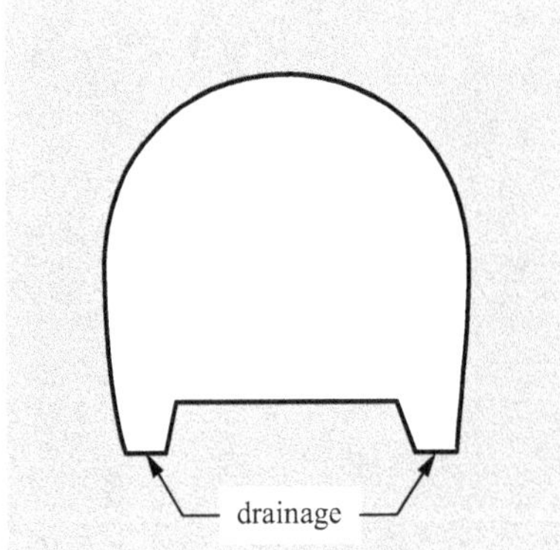

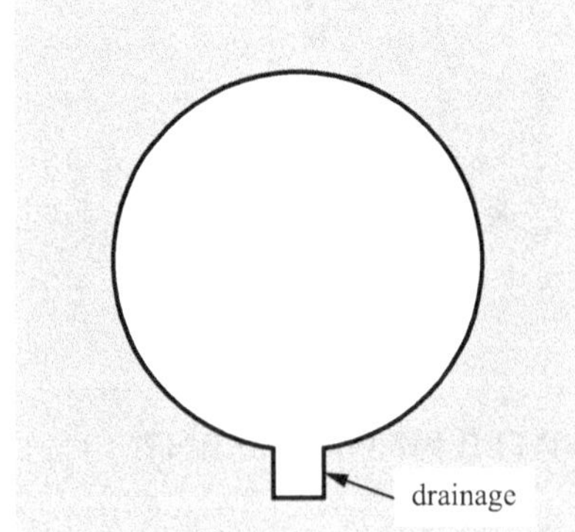

Figure 7.58 Drainage (channels/ditches).

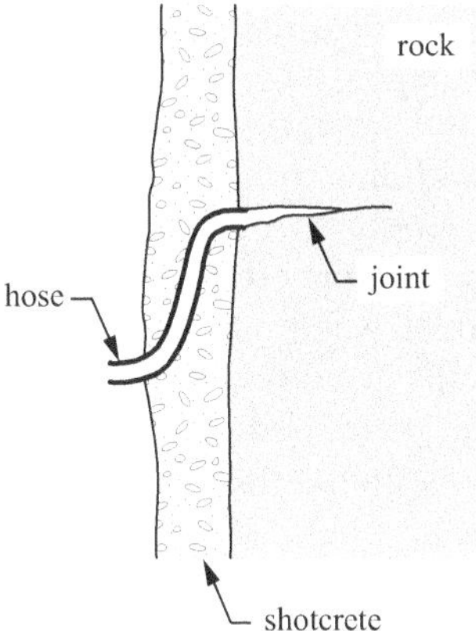

Figure 7.59 Capture of concentrated flow from a joint.

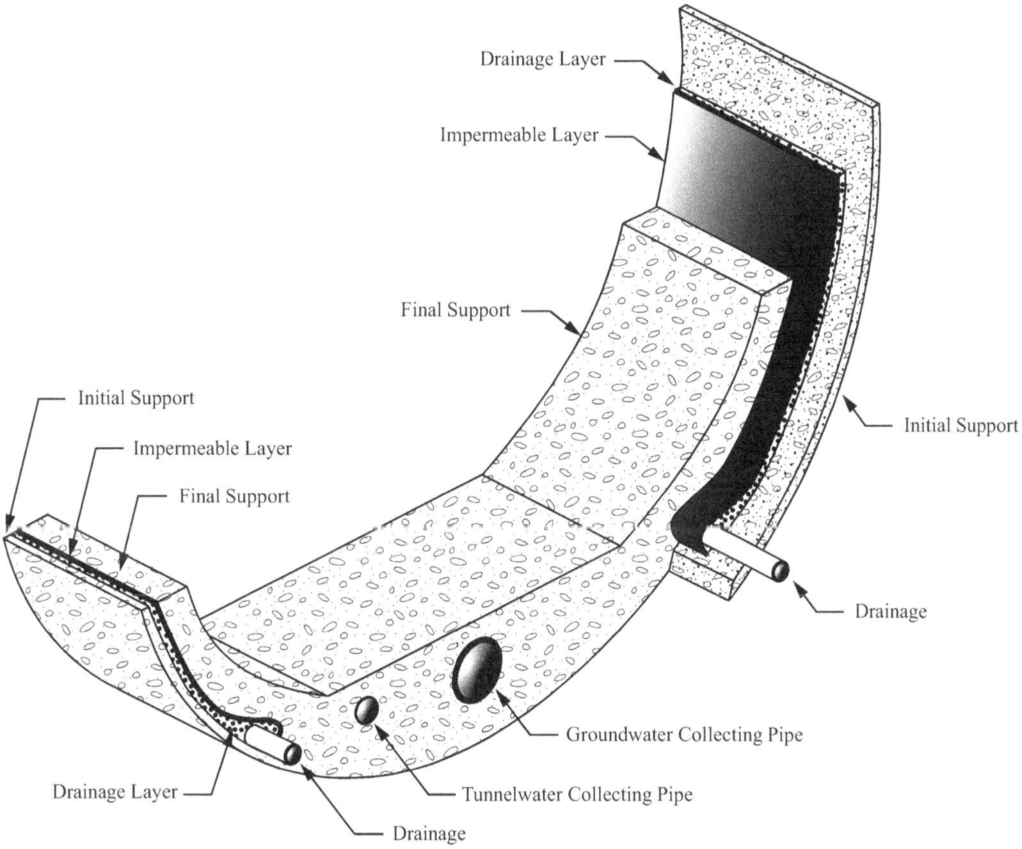

Figure 7.60 Schematic of tunnel support, drainage, and other conduits Bodio section, Gotthard Base Tunnel.

need to be eventually supported by the final liner. If they are simply applied to the face of the initial liner or the barerock, water will push them in and may break them. (3) As shown in Figure 7.60, the water then flows through the drainage layer into drainage pipes. These pipes usually run at the edge of the invert, and they often are installed during initial construction as discussed above. (4) The drain pipes are connected at regular intervals to a collection pipe which can, but does not have to, be below the center as shown in Figure 7.60. Note that the connection drainage layer-drainage pipe and drainage-pipe to collection pipe require very careful design and construction to make certain that no leaking occurs. (5) There can be (as shown in Figure 7.60) an additional pipe for the "tunnel water." This is water flowing in the finished tunnel. Since it is often contaminated, it is conveyed separately from the groundwater, which is essentially "clean" (see comments about clogging below).

Keeping groundwater and tunnel water separate may also have other reasons. In the Transalpine tunnels, for instance, the groundwater has high temperatures up to 50° and cannot be directly drained into surface water bodies, e.g. rivers, lakes, but needs to be cooled (as a matter of fact the warm ground water is used to heat a greenhouse and sturgeon farm at the Loetschberg Base tunnel).

It is important to note that what has just been described is one possibility of a water control system. There are others, such as watertight concrete of the final liner serving as a seal and the above-mentioned gravel drainage layer. Also, this is a domain in which much development is going on.

The effectiveness of drainage systems is affected by mineral deposition/precipitation. Most important in this regard is calcite precipitation although other minerals can also be precipitated. Garmisch and Girmscheid (2005, 2007) provide a detailed explanation of this process and some example calculations. The calcite dissolution/precipitation process essentially involves CO_2 in the groundwater interacting with $CaCO_3$ in the rock. The dissolved Ca carbonate is transported in the water and when the CO_2 concentration in the water changes, calcite precipitates. Expressed in a simplified way, lower pressures, higher temperatures, and higher pH content favor precipitation. Most important, given that the dissolved CO_2 plays an essential role, is to prevent it from coming out of solution, which is related to open water surfaces, i.e. water–air contact. In other words, drainage water should not flow with an open surface. This means that pipes have to be designed such that there is full pipe flow (which has the added advantage of higher pressures). While this is relatively easy to fulfill in the standard section shown in Figure 7.60, it is difficult to fulfill at connections. Very important in all this is the possibility to clean the drain pipes at regular intervals, which requires access installations.

The question arises why this drainage is so important, and this is related to the buildup of water pressure, which will now be discussed.

7.7.2 Water pressure

There are two extremes: (1) The tunnel acts as an open drain with water pressure at its boundary equaling the atmosphere pressure; there are, however, seepage pressures in the ground around the tunnel; and (2) fully sealed tunnel with no drainage, where the water pressure at the rock/seal interface is equal to the hydrostatic groundwater pressure. In both cases, water pressure through pore pressure, affects the ground behavior. In case 2 and cases between 1 and 2, the liner needs to be designed for water pressure.

With regard to ground behavior, it is probably best to quote Deere et al. (1970):

> If a tunnel lining below the ground water table is perfectly watertight, it must withstand the combined loading from the hydrostatic water pressure and the pressure of the soil in terms of effective stresses. If, on the other hand, the lining is perfectly permeable, it must withstand just the effective pressure from the soil. However, if the gradient near the tunnel wall is appreciable, then the effective stresses in the soil are increased by the seepage pressure. It is not possible to predict in general the increase in the effective stress but it may be estimated fairly accurately, e.g. by means of flownets. Trial computations indicate

that the increase is of the same order as the effective stress computed while disregarding the seepage pressure. Whereas the loading of a tunnel in a frictionless medium is nearly the same whether the lining is permeable or not, the loading in a frictional medium may be appreciably reduced by making the lining permeable.

Ground–structure interaction including the effect of pore pressures has been discussed in detail in Chapters 4–6 and is the basis for design discussed in this section. Complications arise when the support consists of an initial and a final liner.

Different philosophies and, for that matter, code requirements, exist in which the initial liner is considered to provide no support or partial support. Looking at the comments on ground–structure interaction and support delay related to the place and time where the final support is installed, the latter may be only subject to time-dependent behavior of the ground and the water pressure of case 2. This may be unconservative, and, as just indicated, some design approaches, therefore, assume that all of the support loads on the initial support have to be carried by the final support.

The design against water pressure (case 2) seems to be more straightforward but is not. Assuming full water pressure is very conservative and would make tunneling at great depth impossible. This is one of the reasons for drainage layers and weep holes. As indicated in the Deere et al. comments, there is uncertainty regarding the pore pressures in the ground, and this is directly related to the uncertainty regarding water pressure acting on the liner. In shallow tunnels in permeable ground, it is advisable to consider full water pressure; there are also some cases of deep lying tunnels where this needs to be done (e.g. the Engelberg tunnel, Switzerland, in Karst Terrain).

As mentioned at the beginning of this section, water management is essential. It needs to be considered in design but is strongly related to construction details and procedures.

References

ABAQUS (1998, 2005). Finite Element Program, from Hibbit, Karlson & Sorensen, Inc. 1080 Main Street, Pawtucket, RI, USA.

Abdel Salam, M.E. and Abdel Fattah, M. (1994). Seismic hazards for Greater Cairo underground metro lines. *Tunnelling and Underground Conditions. Proceedings of the International Congress on Tunnelling and Ground Conditions.* M.E. Abdel Salam, Editor. A.A. Balkema, Rotterdam, Holland, pp. 293–301.

Abdel-Meguid, M., Rowe, R.K., and Lo, K.Y. (2002). 3D Effects of Surface Construction Over Existing Subway Tunnels. *The International Journal of Geomechanics*, Vol. 2, No. 4, pp. 447–469.

Abramento, M. and Whittle, A.J. (1993). Shear-lag Analysis of a Planar Soil Reinforcement in Plane Strain Compression. *Journal of Engineering Mechanics*, ASCE, Vol. 119, No. 2, pp. 270–291.

Abramento, M. and Whittle, A.J. (1995a). Analysis of Pullout Tests for Planar Reinforcements in Soil. *Journal of Geotechnical Engineering*, ASCE, Vol. 121, No. 6, pp. 476–485.

Abramento, M. and Whittle, A.J. (1995b). Experimental Evaluation of Pullout Analyses for Planar Reinforcements. *Journal of Geotechnical Engineering*, ASCE, Vol. 121, No. 6, pp. 486–492.

ACI 506.1R-98 (1998). *Committee Report on Fiber Reinforced Shotcrete.* Reported by ACI Committee 506. American Concrete Institute.

ACI 506.2-95 (1995). *Specification for Shotcrete.* Reported by ACI Committee 506. American Concrete Institute.

ACI 506R-05 (2005). *Guide to Shotcrete.* Reported by ACI Committee 506. American Concrete Institute.

Adachi, T., Kimura, M., and Kishida, K. (2003). Experimental Study on the Distribution of Earth Pressure and Surface Settlement through Three-Dimensional Trapdoor Tests. *Tunneling and Underground Space Technology*, Vol. 18, No. 2, pp. 171–183.

Addenbrooke, T.I. and Potts, D.M. (2001). Twin Tunnel Interaction: Surface and Subsurface Effects. *The International Journal of Geomechanics*, Vol. 1, No. 2, pp. 249–271.

AFTES, Working Group No. 6 (1996). AFTES Recommendations on Fibre-Reinforced Sprayed Concrete Technology and Practice. *Tunnelling and Underground Space Technology*, Vol. 11, No. 2, pp. 205–214.

Ahlgren, C.S. and Mattson, R.A. (1997). Buckling Capacity of Tunnel and Penstock. *Proceedings of the International Conference on Hydropower, Waterpower '97*, D.J. Mahoney, Editor. ASCE, New York, NY, pp. 95–102.

Alber, M. (1996). Prediction of Penetration and Utilization of Hardrock TBMs. *Proceedings Rock Characterization ISRM Symposium: Eurock '96, Turin, Italy.* G. Barla, Editor. Balkema, Rotterdam, The Netherlands, pp. 721–725.

Alejano, L.R., Ramirez-Oyanguren, P., and Taboada, J. (1999). FDM Predictive Methodology for Subsidence Due to Flat and Inclined Coal Seam Mining. *International Journal of Rock Mechanics and Mining Sciences*, Vol. 36, pp. 475–491.

Alonso, E. and Olivella, S. (2008). Modelling Tunnel Performance in Expansive Gypsum Claystone. *12th International Conference on Computer Methods and Advances in Geomechanics*, Goa 2008, pp. 891–910.

Alonso, E. and Ramon, A. (2013). Heave of a Railway Bridge Induced by Gypsum Crystal Growth: Field Observations. *Géotechnique*, Vol. 63, No. 9, pp. 707–719.

Alonso, E., Berdugo, I., and Ramon, A. (2013). Extreme Expansive Phenomena in Anhydritic-gypsiferous Claystone: The Case of Lilla Tunnel. *Géotechnique*, Vol. 63, pp. 584–612.

Alvarez, T.R. (1997). *A Study of the Coupled Hydromechanical Behavior of Jointed Rock Masses around Pressure Tunnels*. Ph.D. Thesis, University of Illinois at Urbana-Champaign, Urbana, IL.

Alvarez, T.R., Cording, E.J., and Fernandez, G.G. (1999). Pressure Tunnels in Fractured Rock: Minimum Cover Criteria from the Stability of Rock Wedges. *Geo-Engineering for Underground Facilities*. Geotechnical Special Publication No. 90. G. Fernandez and R.A. Bauer, Editors. ASCE, Reston, VA, pp. 459–470.

Amadei, B. and Pan, E. (1995). Role of Topography and Anisotropy When Selecting Unlined Pressure-Tunnel Alignment. *Journal of Geotechnical Engineering*, Vol. 121, No. 12, pp. 879–885.

Amstutz, E. (1970). Buckling of Pressure Shafts and Tunnel Linings. *Water Power*, Vol. 22, No. 11, pp. 391–399.

Anagnostou, G. (1991). *Untersuchungen zur Sttik des Tunnel-baus im quellfähigen Gebirge*. ETH-Dissertation No. 9553.

Anagnostou, G. (2007a). Continuous tunnel excavation in a poro-elastoplastic medium. *Proceedings of the Tenth International Symposium on Numerical Models in Geomechanics (NUMOG X)*. G.N. Pande and S. Pietruszczak, Editors. Taylor & Francis Group.

Anagnostou, G. (2007b). The One-Step Solution of the Advancing Tunnel Heading Problem. *Proceedings of the ECCOMAS Thematic Conference on Computational Methods in Tunneling (EURO:TUN 2007)*. J. Eberhardsteiner, G. Beer, C. Hellmich, H.A. Mang, G. Meschke, and W. Schubert, Editors. CD-ROM.

Anagnostou, G. (2008). The Effect of Tunnel Advance on the Surface Settlements. *Proceedings of the 12th International Conference of International Association for Computer Methods and Advances in Geomechanics (IACMAG)*, 1-6 October, Goa, India. CD-ROM.

Anagnostou, G., Pimentel, E., and Serafeimidis, K. (2010). Swelling of Sulphatic Claystones – Some Fundamental Questions and Their Practical Relevance. *Geomechanics and Tunnelling*, Vol. 3, No. 5, pp. 567–572.

Anagnostou, G. and Kovari, K. (1993). Significant Parameters in Elastoplastic Analysis of Underground Opening. *Journal of Geotechnical Engineering*, Vol. 199, No. 3, pp. 401–419.

Anagnostou, G. and Kovari, K. (1994). The Face Stability of Slurry-shield-driven Tunnels. *Tunnelling and Underground Space Technology*, Vol. 9, No. 2, pp. 165–174.

Anagnostou, G. and Kovari, K. (1996a). Face Stability Conditions with Earth-Pressure-Balanced Shields. *Tunnelling and Underground Space Technology*, Vol. 11, No. 2, pp. 165–173.

Anagnostou, G. and Kovari, K. (1996b). Face stability in slurry and EPB shield tunneling. *Proceedings of the International Symposium on Geotechnical Aspects of Underground Construction in Soft Ground, London*. R.J. Mair and R.N. Taylor, Editors. A.A. Balked, Rotterdam, pp. 453–458.

Anderson, D.G., Martin, G.R., Lam, I., and Wang, J.N. (2008). *Seismic Analysis and Design of Retaining Walls, Buried Structures, Slopes and Embankments. NCHRP Report 611*. Transportation Research Board, Washington, DC.

Andrieux, P., Blake, W., Hedley, D., Nordlund, E., Phipps, D., Simser, B., and Swan, G. (2013). *Rockburst case histories 1985, 1990, 2001 & 2013*. CAMIRO Mining Division for the Deep Mining Research Consortium, Sudbury, Canada (2013).

Argyris, J.H. (1955). Energy Theorems and Structural Analysis. *Aircraft Engineering*, Vol. 27, No. 315, pp. 145–158.

Aristorenas, G. (1992). *Time Dependent Behavior of Tunnels Excavated in Shale*. Ph.D. Thesis, MIT, Camdrige, MA.

Asakura, T. and Sato, Y. (1996). Damage to Mountain Tunnels in Hazard Area. *Soils and Foundations*, Vol. 36, Supplement, pp. 301–310.

ASCE Task Committee on Manual of Practice for Steel Penstocks (1993). *Steel Penstocks*. ASCE Manuals and Reports on Engineering Practice No. 79. ASCE, New York, NY.

Atkinson, J.H. and Potts, D.M. (1977). Stability of a Shallow Circular Tunnel in Cohesionless Soil. *Géotechnique*, Vol. 27, No. 2, pp. 203–215.

Atkinson, J.H., Brown, E.T., and Potts, D.M. (1977). Ground Movements near Shallow Model Tunnels in Sand. *Large Ground Movements and Structures*, John Wiley and Sons, New York, pp. 373–386.

Attewell, P.B. (1977). Ground Movements Caused by Tunneling in Soil. *Proceedings of the Conference on Large Ground Movements and Structures.* John Wiley & Sons, New York, pp. 812–948.

Attewell, P.B. (1981). Engineering contract, site investigation and surface movements in tunneling works. *Soft-Ground Tunneling: Failures and Displacements.* D. Reséndiz and M.P. Romo, Editors. A.A. Balkema, Rotterdam, pp. 5–12.

Attewell, P.B. and Farmer, I.W. (1974). Ground Deformations Resulting from Shield Tunnelling in London Clay. *Canadian Geotechnical Journal*, Vol. 11, pp. 380–395.

Attewell, P.B., Yeates, J., and Selby, A.R. (1986). *Soil Movements Induced by Tunnelling and Their Effects on Pipelines and Structures.* Blackie, Glasgow, UK.

Austrian Concrete Society (1999). *Shotcrete Specification.* Austrian Shotcrete Society, Vienna, Austria.

Austrian Society of Geomechanics (2010). *Guideline for the Geotechnical Design of Underground Structures with Conventional Excavation.* Austrian Society of Geomechanics, Salzburg, 37 pp.

Babuska, I. and Melenk, J.M. (1996). The Partition of Unity Finite Element Method: Basic Theory and Applications. *Computer Methods in Applied Mechanics and Engineering*, Vol. 139, pp. 289–315.

Bakker, K.J. and Blom, C.B.M. (2009). Ultimate Limit State Design for Linings of Bored Tunnels. *Geomechanics and Tunnelling*, Vol. 2, No. 4, pp. 345–358.

Ballivy, G. and Martin, A. (1983). The dimensioning of grouted anchors. *Rock Bolting. Theory and Application in Mining and Underground Construction. Proceedings of the International Symposium on Rock Bolting.* O. Stephansson, Editor. A.A. Balkema, Rotterdam, The Netherlands, pp. 353–365.

Banerjee, P.K. and Cathie, D.N. (1980). A Direct Formulation and Numerical Implementation of the Boundary Element Method for Two-dimensional Problems of Elasto-plasticity. *International Journal of Mechanical Sciences*, Vol. 22, pp. 233–245.

Banerjee, P.K., Cathie, D.N., and Davies, T.G. (1979). Two- and three-dimensional problems of elasto-plasticity. *Developments in Boundary Element Methods -1.* P.K. Banerjee and R. Butterfield, Editors. Applied Science Publishers, London, pp. 65–95.

Bardet, J.P. and Davis, C.A. (1995). Lower San Fernando Corrugated Metal Pipe Failure. *Proceedings of the Fourth US Conference on Lifeline Earthquake Engineering*, Technical Council on Lifeline Earthquake Engineering, ASCE, San Francisco, California, pp. 644–651.

Barrett, S.V.L. and McCreath, D.R. (1995). Shotcrete Support Design in Blocky Ground: Towards A Deterministic Approach. *Tunnelling and Underground Space Technology*, Vol. 10, No. 1, pp. 79–89.

Barton, N. (1976). Recent Experiences with the Q-System of Tunnel Support Design. *Proceedings of the Symposium on Exploration for Rock Engineering*, Johannesburg, South Africa. Z.T. Bieniawski, Editor. A.A. Balkema, Cape Town, South Africa, pp. 107–115.

Barton, N. (1991). Geotechnical Design. *World Tunnelling*, Vol. 4, No. 7, pp. 410–416.

Barton, N. (1995). The Influence of Joint Properties in Modelling Jointed Rock Masses. *Proceedings of the 8th ISRM Congress*, Tokyo, Japan, Vol. 3. T. Fujii, Editor. A.A. Balkema, Rotterdam, The Netherlands, pp. 1023–1032.

Barton, N. (1999). TBM Performance Estimation in Rock Using Q(TBM). *Tunnels and Tunnelling International*, Vol. 31, No. 9, pp. 30–34.

Barton, N. (2000). *TBM Tunneling in Jointed and Faulted Rock.* A.A. Balkema, Rotterdam, The Netherlands, 173 pp.

Barton, N. (2001). Are Long Tunnels Faster by TBM? *Proceedings of the Rapid Excavation and Tunneling Conference*, San Diego, CA. W.H. Hansmire and M.I. Gowring, Editors. Society for Mining, Metallurgy, and Exploration, Colorado, pp. 819–828.

Barton, N. (2002). Some New Q-Relations to Assist in Site Characterization and Tunnel Design. *International Journal of Rock Mechanics and Mining Sciences*, Vol. 39, No. 2, pp. 185–216.

Barton, N. (2005). Comments on 'A Critique of QTBM.' *Tunnels and Tunnelling International*, Vol. 37, No. 7, pp. 16–19.

Barton, N. (2006). Rock Mass Characterization and Modelling Aspects of Mining and Civil Engineering. *XI Congreso Colombiano de Geotecnia - VI Congreso Suramericano de Mecánica de Rocas*, 8–13 October, Cartagena, Colombia, pp. 45–75.

Barton, N. (2017). Minimizing the Use of Concrete in Tunnels and Caverns: Comparing NATM and NMT. *Innovative Infrastructure Solutions*, Vol. 2, article No. 52, 37 pp. Springer. doi: 10.1007/s41062-017-0071-x.

Barton, N. and Grimstad, E. (2014). Forty Years with the Q-System in Norway and Abroad. *Proceedings of Fjellsprengningsteknikk; Bergmekanikk/Geoteknikk*. Oslo, pp. 4.1–4.25.

Barton, N.R., Lunde, J., and Lien, R. (1974a). *Analysis of Rock Mass Quality and Support Practice in Tunneling, and a Guide for Estimating Support Requirements*. Internal Report No. 54206, Norwegian Geotechnical Institute, Oslo, 74 pp.

Barton, N.R., Lien, R., and Lunde, J. (1974b). Engineering Classification of Rock Masses for the Design of Tunnel Supports. *Rock Mechanics*, Vol. 6, No. 4, pp. 189–236.

Barton, N.R., Lien, R., and Lunde, J. (1975). Estimation of Support Requirements for Underground Excavations. *Design Methods in Rock Mechanics: Proceedings of the 16th Symposium on Rock Mechanics*. University of Minnesota, MN. C. Fairhurst and S.L. Crouch, Editors. American Society of Civil Engineers (ASCE), New York, pp. 165–177.

Barton, N., Løset, N., Lien, R., and Lunde, J. (1980). Application of the Q-System in Design Decisions Concerning Dimensions and Appropriate Support of Underground Installations. *International Conference on Underground Space, Rockstore*, Stockholm, Sweden, pp. 553–561.

Bathe, K.J. (1982). *Finite Element Procedures in Engineering Analysis*. Prentice-Hall, Inc., Englewood Cliffs, NJ.

Bawden, W.F. and Jones, S. (2002). Ground support design and performance under strong rockburst conditions. *Proceedings of the North American Rock Mechanics Symposium: NARMS-TAC 2002*, Toronto, Canada, pp. 923–931.

Bawden, W.F., Hyett, A.J., and Cortolezzis, D. (1992). Towards a methodology for performance assessment in cable bolt design. *Rock Support in Mining and Underground Construction. Proceedings of the International Symposium on Rock Support*. P.K. Kaiser and D.R. McCreath, Editors. A.A. Balkema, Rotterdam, The Netherlands, pp. 277–284.

Beck, A. and Golta, A. (1972). Tunnelsanierungen der Schweiz. *Bundesbahnen, Schweiz Bauzeitung*, Vol. 90, No. 36, pp. 857–863.

Beer, G. (1993). Rock mechanics. *Boundary Element Techniques in Geomechanics*. G.D. Manolis and T.G. Davies, Editors. Elsevier Applied Science, London, pp. 295–326.

Bellwald, P. (1991). *A Contribution to the Design of Tunnels in Argillaceous Rocks*. Ph.D. Thesis, MIT, Cambridge, MA.

Belytschko, T., Krongauz, Y., Organ, D., Fleming, M., and Krysl, P. (1996). Meshless Methods: An Overview and Recent Developments. *Computer Methods in Applied Mechanics and Engineering*, Vol. 139, pp. 3–47.

Belytschko, T., Lu, Y.Y., and Gu, L. (1994). Element-free Galerkin Methods. *International Journal for Numerical Methods in Engineering*, Vol. 37, pp. 229–256.

Benito, J.J., Ureña, F., and Gavete, L. (2001). Influence of Several Factor in the Generalized Finite Difference Method. *Applied Mathematical Modelling*, Vol. 25, pp. 1039–1053.

Benmokrane, B., Chennouf, A., and Ballivy, G. (1992). Study of bond strength behaviour of steel cables and bars anchored with different cement grouts. *Rock Support in Mining and Underground Construction. Proceedings of the International Symposium on Rock Support*. P.K. Kaiser and D.R. McCreath, Editors. A.A. Balkema, Rotterdam, The Netherlands, pp. 293–301.

Benson, R.P. (1989). Design of Unlined and Lined Pressure Tunnels. *Tunnelling and Underground Space Technology*, Vol. 4, No. 2, pp. 155–170.

Bergh-Christensen, J. (1982). Design of Unlined Pressure Shaft at Mauranger Power Plant, Norway. *Proceedings of the ISRM Symposium Rock Mechanics: Caverns and Pressure Shafts*. W. Wittke, Editor. A.A. Balkema, Rotterdam, pp. 531–536.

Bergman, S.G.A. and Bjurström, S. (1983). Swedish experience of rock bolting. A keynote lecture. *Rock Bolting. Theory and Application in Mining and Underground Construction. Proceedings of the International Symposium on Rock Bolting*. O. Stephansson, Editor. A.A. Balkema, Rotterdam, The Netherlands, pp. 243–255.

Bernard, E.K. (2008). Early-age Load Resistance of Fibre Reinforced Shotcrete Linings. *Tunnelling and Underground Space Technology*, Vol. 23, pp. 451–460.

Bernaud, D. and Rousset, G. (1996). The 'New Implicit Method' for Tunnel Analysis. *International Journal for Numerical and Analytical Methods in Geomechanics*, Vol. 20, pp. 673–690.

Berry, D.S. and Sales, T.W. (1962). An Elastic Treatment of Ground Movement Due to Mining, III, Three-dimensional Problem, Transversely Isotropic Ground. *Journal of Mechanics and Physics of Solids*, Vol. 10, pp. 73–83.

Beskos, D.E. (1993). Wave propagation through ground. *Boundary Element Techniques in Geomechanics*. G.D. Manolis and T.G. Davies, Editors. Elsevier Applied Science, London, pp. 359–406.

Betti, E. (1872). Teori Dell Elasticita. *Il Nuovo Cimento*, Vol. 7, No. 1, pp. 5–21.

Bhasin, R. and Hoeg, K. (1998). Parametric Study for a Large Cavern in Jointed Rock Using a Distinct Element Model (UDEC-BB). *International Journal of Rock Mechanics and Mining Sciences*, Vol. 35, No. 1, pp. 17–29.

Bhasin, R.K., Barton, N., Grimstad, E., Chryssanthakis, P., and Shende, F.P. (1996). Comparison of Predicted and Measured Performance of a Large Cavern in the Himalayas. *International Journal of Rock Mechanics and Mining Sciences*, Vol. 33, No. 6, pp. 607–626.

Bieniawski, Z.T. (1972). *Engineering Classification for Rock Masses*. Rep. South Afr. Council. Sci. and Res. ME 327. Rock Mechanics, Pretoria, South Africa.

Bieniawski, Z.T. (1973). Engineering Classification of Jointed Rock Masses. *The Civil Engineer in South Africa*, Vol. 15, No. 12, pp. 335–344.

Bieniawski, Z.T. (1974). Geomechanics Classification of Rock Masses and Its Application in Tunneling. *Proceedings of the 3rd International Congress on Rock Mechanics*, International Society for Rock Mechanics (ISRM), Denver, CO, Vol. IIA, pp. 27–32.

Bieniawski, Z.T. (1976). Rock Mass Classifications in Rock Engineering. *Proceedings of the Symposium on Exploration for Rock Engineering*, Johannesburg, South Africa. Z.T. Bieniawski, Editor. A.A Balkema, Rotterdam, The Netherlands, pp. 97–106.

Bieniawski, Z.T. (1978). Determining Rock Mass Deformability: Experience from Case Histories. *International Journal of Rock Mechanics and Mining Sciences*, Vol. 15, No. 5, pp. 237–247.

Bieniawski, Z.T. (1979a). The Geomechanics Classification in Rock Engineering Applications. *Proceedings of the 4th International Congress on Rock Mechanics*, ISRM, Montreux, Switzerland, Vol. 2. A.A. Balkema, Rotterdam, The Netherlands, pp. 41–48.

Bieniawski, Z.T. (1979b). *Tunnel Design by the Rock Mass Classifications*. Technical Report GL 799-19, U.S. Army Engineer Waterways Experiment Station, Vicksburg, MS, pp. 50–62.

Bieniawski, Z.T. (1979c). *Tunnel Design by Rock Mass Classifications*. Technical Report GL-79-19, Report to Office, Chief of Engineers, U.S. Army, Washington, DC, 131 pp.

Bieniawski, Z.T. (1989). *Engineering Rock Mass Classifications: A Complete Manual*. John Wiley & Sons, New York, 272 pp.

Bieniawski, Z.T. (1997). Quo-Vadis Rock Mass Classification. *Felsbau*, Vol. 15, No. 3, pp. 177–178.

Bieniawski, Z.T. (2004). Aspectos clave en la elección del método constructivo de túneles. *Proceedings of Jornada Técnica*, Madrid, Spain, pp. 1–37.

Bieniawski, Z.T. and Celada, B. (2008). Mechanized Excavability Rating for Hard-Rock Masses. *Proceedings of the International Workshop on Rock Mass Classification in Underground Mining*, Information Circular 9498. C. Mark, R. Pakalnis, and R.J. Tuchman, Editors. National Institute for Occupational Safety and Health (NIOSH), Pittsburgh, PA, pp. 15–23.

Bieniawski, Z.T. and Orr, C.M. (1976). Rapid Site Appraisal for Dam Foundations by the Geomechanics Classification. *Proceedings of the 12th International Commission on Large Dams (ICOLD) Congress*, Mexico City, pp. 483–501.

Bieniawski, Z.T., Celada Tanames, B., Galeva Fernandez, J.M., and Hernandez, Alvarez (2006). Rock Mass Excavability Indicator: New Way to Selecting the Optimum Tunnel Construction Method. *Tunneling and Underground Space Technology*, Vol. 21, No. 3–4, pp. 237.

Bierbaumer, A. (1913). *Die Dimensionerung des Tunnelmauerwerks*. W. Engelmann, Leipzig, 101 pp.

Biot, M.A. (1941). General Theory of Three-Dimensional Consolidation. *Journal of Applied Physics*, Vol. 12, pp. 155–164.

Biot, M.A. (1956). General Solutions of the Equations of Elasticity and Consolidation for a Porous Material. *Transactions of the ASME. Journal of Applied Mechanics*, Vol. 78, pp. 91–96.

Bittnar, Z. and Sejnoha, J. (1996). *Numerical Methods in Structural Mechanics*. ASCE Press, New York, NY.

Bjerrum, L. (1963). Discussion on: *Proceedings of the European Conference on Soil Mechanics and Foundation Engineering, Vol. III. Publication No 98*, Norwegian Geotechnical Institute, Oslo, Norway, pp. 1–3.

Björnfot, F. and Stephansson, O. (1983). Interaction of Grouted Rock Bolts and Hard Rock Masses at Variable Loading in a Test Drift of the Kiirunavaara Mine, Sweden. *Theory and Application in Mining and Underground Construction. Proceedings of the International Symposium on Rock Bolting*. O. Stephansson, Editor. A.A. Balkema, Rotterdam, The Netherlands, pp. 377–395.

Blake, W. (1984). Rock Preconditioning as a Seismic Control Measure. *Proceedings of the 1st International Congress on Rockbursts and Seismicity in Mines, The South African Institute of Mining and Metallurgy Symposium Series, No. 6*. N.C. Gay and E.H. Wainwright, Editors. The South African Institute of Mining and Metallurgy, Johannesburg, South Africa, pp. 225–229.

Blindheim, O.T. (2005). A Critique of QTBM. *Tunnels and Tunnelling International*, Vol. 37, No. 6, pp. 32–35.

Board, M.P. and Fairhurst, C. (1983). Rockburst control through distressing – a case example. *Rockbursts: Prediction and Control*. The Institution of Mining and Metallurgy, London, England, pp. 91–101.

Bobet, A. (2000). The Initiation of Secondary Cracks in Compression. *Engineering Fracture Mechanics*, Vol. 66, No. 2, pp. 187–219.

Bobet, A. (2001a). Analytical Solutions for Shallow Tunnels in Saturated Ground. *ASCE Journal of Engineering Mechanics*, Vol. 127, No. 12, pp. 1258–1266.

Bobet, A. (2001b). A Hybridized Displacement Discontinuity Method for Mixed Mode I-II-III Loading. *International Journal of Rock Mechanics and Mining Sciences*, Vol. 38, pp. 1121–1134.

Bobet, A. (2001c). Numerical Solution of Initiation of Tensile and Shear Cracks. *Proceedings of the 38th U.S. Rock Mechanics Symposium*, Washington, DC, USA, pp. 731–738.

Bobet, A. (2002). Mechanically Anchored Rockbolts in Tunnels in Saturated Ground. *Proceedings of the North American Rock Mechanics Symposium: NARMS-TAC 2002*, pp. 797–804.

Bobet, A. (2003). Effect of Pore Water Pressure on Tunnel Support during Static and Seismic Loading. *Tunnelling and Underground Space Technology*, Vol. 18, pp. 377–393.

Bobet, A. (2006). A Simple Method for Analysis of Point Anchored Rockbolts in Circular Tunnels in Elastic Ground. *Rock Mechanics and Rock Engineering*, Vol. 39, No. 4, pp. 315–338.

Bobet, A. (2009). Elastic Solution for Deep Tunnels. Application to Excavation Damage Zone and Rockbolt Support. *Rock Mechanics and Rock Engineering*, Special Issue on "Deep Tunnel: Issues in Rock Engineering", Vol. 42, No. 2, pp. 147–174.

Bobet, A. (2010a). Characteristic Curves for Deep Circular Tunnels in Poroplastic Rock. *Rock Mechanics and Rock Engineering*, Vol. 43, pp. 185–200.

Bobet, A. (2010b). Numerical Methods in Geomechanics. *The Arabian Journal for Science and Engineering*, Vol. 35, No. 1B, pp. 27–48.

Bobet, A. (2010c). Response by the Author to: Comments on "Characteristic Curves for Deep Circular Tunnels in Poroplastic Rock" by G. Anagnostou and R. Schürch. *Rock Mechanics and Rock Engineering*, Vol. 43, pp. 235–239.

Bobet, A. (2011). Lined Circular Tunnels in Transversely Anisotropic Rock at Depth. *Rock Mechanics and Rock Engineering*, Vol. 44, pp. 149–167.

Bobet, A. (2016a). Lined Circular Tunnels in Transversely Anisotropic Rock at Depth: Complementary Solutions. *Rock Mechanics and Rock Engineering*, Vol. 49, No. 9, pp. 3817–3822.

Bobet, A. (2016b). Deep Tunnel in Transversely Anisotropic Rock with Groundwater Flow. *Rock Mechanics and Rock Engineering*, Vol. 49, No. 12, pp. 4817–4832.

Bobet, A. and Einstein, H.H. (1998). Numerical Modeling of Fracture Coalescence in Rock Materials. *International Journal of Fracture*, Vol. 92, No. 3, pp. 221–252.

Bobet, A. and Einstein, H.H. (2008). Deep tunnels in clay shales: Evaluation of key properties for short and long term support. *GeoCongress 2008: Characterization, Monitoring, and Modeling of GeoSystems, New Orleans, LO*. A.N. Alshawabkeh, K. Reddy, and M.V. Khire, Editors. ASCE Press, Reston, VA.

Bobet, A. and Einstein, H.H. (2011). Tunnel Reinforcement with Rockbolts. *Tunnelling and Underground Space Technology*, Vol. 26, pp. 100–123.

Bobet, A. and Mutlu, O. (2005). Stress and Displacement Discontinuity Element Method for Undrained Analysis. *Engineering Fracture Mechanics Journal*, Vol. 72, pp. 1411–1437.

Bobet, A. and Nam, S. (2007). Stresses around Pressure Tunnels with Semi-Permeable Liners. *Rock Mechanics and Rock Engineering*, Vol. 40, No. 3, pp. 287–315.

Bobet, A., Aristorenas, G., and Einstein, H.H. (1999). Feasibility Analysis for a Radioactive Waste Repository Tunnel. *Tunnelling and Underground Space Technology*, Vol. 13, No. 4, pp. 409–426.

Bobet, A., Fakhimi, A., Johnson, S., Morris, J., Tonon, F., and Yeung, M. (2009). Numerical Models in Discontinuous Media: A Review of Advances for Rock Mechanics Applications. *ASCE Journal of Geotechnical and Geoenvironmental Engineering*, Vol. 135, No. 11, pp. 1547–1561.

Bobet, A., Fernandez, G., Huo, H., and Ramirez, J. (2008). A Practical Procedure to Estimate Seismic-Induced Deformations of Shallow Rectangular Structures. *Canadian Geotechnical Journal*, Vol. 45, No. 7, pp. 923–938.

Boidy, E., Bouvard, A., and Pellet, F. (2002). Back Analysis of Time-Dependent Behaviour of a Test Gallery in Claystone. *Tunnelling and Underground Space Technology*, Vol. 17, pp. 415–424.

Boldini, D., Lackner, R., and Mang, H.A. (2005). Ground-Shotcrete Interaction of NATM Tunnels with High Overburden. *Journal of Geotechnical and Geoenvironmental Engineering*, Vol. 131, No. 7, pp. 886–897.

Bolstad, D.D., Hill, J.R.M., and Karhnak, J.M. (1983). US Bureau of Mines rock bolting research. *Rock Bolting. Theory and Application in Mining and Underground Construction. Proceedings of the International Symposium on Rock Bolting*. O. Stephansson, Editor. A.A. Balkema, Rotterdam, The Netherlands, pp. 313–320.

Boone, S.J. (1996). Ground-Movement-Related Building Damage. *ASCE Journal of Geotechnical Engineering*, Vol. 122, No. 1, pp. 886–896.

Boone, S.J., Westland, J., and Nusik, R. (1999). Ground-Movement-Related Building Damage. *Canadian Geotechnical Journal*, Vol. 36, pp. 210–223.

Boresi, A.P., Chong, K.P., and Saigal, S. (2003). *Approximate Solution Methods in Engineering Mechanics*. John Wiley and Sons, Inc., New Jersey.

Boscardin, M.D. and Cording, E.J. (1989). Building Response to Excavation-Induced Settlement. *ASCE Journal of Geotechnical and Geoenvironmental Engineering*, Vol. 115, No. 1, pp. 1–21.

Bottou, L. (2010). Large-Scale Machine Learning with Stochastic Gradient Descent. *Proceedings of COMPSTAT'2010*. Y. Lechevallier and G. Saporta, Editors. Physica-Verlag HD. https://doi.org/10.1007/978-3-7908-2604-3_16

Boutt, D. and McPherson, B. (2002). The Role of Particle Packing in Modeling Rock Mechanical Behavior Using Discrete Elements. *Discrete Element Methods. Numerical Modeling of Discontinua*. Geotechnical Special Publication No. 117. B.K. Cook and R.P. Jensen, Editors. ASCE, Reston, VA, pp. 86–92.

Brady, B.H.G. (1987). Boundary element and linked methods for underground excavation. *Analytical and Computational Methods in Engineering Rock Mechanics*. E.T. Brown, Editor. Allen & Unwin, London, pp. 164–204.

Brady, B.H.G. and Brown, E.T. (2004). *Rock Mechanics for Underground Mining*. 3rd Edition. Kluwer Academy Publishers, Dordrecht, The Netherlands.

Brady, B.H.G. and Lorig, L. (1988). Analysis of Rock Reinforcement using Finite Difference Methods. *Computers and Geotechnics*, Vol. 5, pp. 123–149.

Braun, Ph., Ghabezloo, S., Delage, P., Sulem, J., and Conil, N. (2021). Transversely Isotropic Poroelastic Behaviour of the Callovo-Oxfordian Claystone: A Set of Stress-Dependent Parameters. *Rock Mechanics and Rock Engineering*, Vol. 1, pp. 377–395

Brebbia, C.A., Telles, J.C.F., and Wrobel, L.C. (1984). *Boundary Element Techniques*. Springer-Verlag, Berlin.

Brekke, T.L. and Howard, T.R. (1973). *Functional Classification of Gouge Material from Seams and Faults in Relation to Stability Problems in Underground Openings*. Final Technical Report to ARPA, NTIS AD 766 046, 195 pp.

Brekke, T.L. and Ripley, B.D. (1985). Leakage from Pressure Tunnels and Shafts. *Proceedings of an International Conference on Hydropower, Waterpower '85*. M.J. Roluti, Editor. ASCE, New York, NY, pp. 1228–1237.

Brekke, T.L. and Ripley, B.D. (1986). *Strategies for Pressure Tunnels and Shafts*. Report EPRI, contract No. Rp-1747-17.

Brekke, T.L. and Ripley, B.D. (1989). Geotechnical engineering challenges in the design of pressure tunnels and shafts. *The Art of Science of Geotechnical Engineering*. E.J. Cording, W.J. Hall, J.D. Haltiwanger, A.J. Hendron, and G. Mesri, Editors. Prentice Hall, Urbana-Champaign, IL, pp. 521–535.

Brekke, T.L. and Ripley, B.D. (1993). Design of pressure tunnels and shafts. *Comprehensive Rock Engineering. Principles, Practice and Projects*. J.A. Hudson and C. Fairhurst, Editors. Pergamon Press, New York, NY, Vol. 2, pp. 349–369.

Bridel, G. and Beck, A. (1970). Zwei Tunnelsanierungen bei den SBB. *Eisenbahntechnische Rundschau*, Heft 1, 2.

British Tunnelling Society (BTS) and The Institution of Civil Engineers (ICE) (2004). *Tunnel Lining Design Guide*. Thomas Telford Publishing, London, England.

Broch, E. (1984a). Development of Unlined Pressure Shafts and Tunnels in Norway. *Tunnels and Deep Space*, Vol. 8, pp. 177–184.

Broch, E. (1984b). Unlined High Pressure Tunnels in Areas of Complex Topography. *Water Power and Dam Construction*, Vol. 36, No. 11, pp. 21–23.

Broch, E. and Sørheim, S. (1984). Experiences from the Planning, Construction and Supporting of a Road Tunnel Subjected to Heavy Rockbursting. *Rock Mechanics and Rock Engineering*, Vol. 17, pp. 15–35.

Broms, B.B. and Bennermark, H. (1967). Stability of Clay at Vertical Openings. *Journal of Soil Mechanics and Foundation Division*, ASCE, Vol. 93, No. (SM1), pp. 71–94.

Brown, E.T. (1981). *Rock Characterization, Testing and Monitoring - ISRM Suggested Methods*. Brown, Ed., Oxford, Pergamon, pp. 171–183.

Brown, E.T., Bray, J.W., Ladanyi, B., and Hoek, E. (1983). Ground Response Curves for Rock Tunnels. *ASCE Journal of Geotechnical Engineering*, Vol. 109, No. 1, pp. 15–39.

Bruland, A. (1998). *Hard Rock Tunnel Boring*. Ph.D. Thesis, Norwegian Science and Technology University, Trondheim.

Bruneau, G., Hudyma, M.R., Hadjigeorgiou, J., and Potvin, Y. (2003). Influence of Faulting on a Mine Shaft – A Case Study: Part II – Numerical Modeling. *International Journal of Rock Mechanics and Mining Sciences*, Vol. 40, pp. 113–125.

Bucky, P.B. (1931). *The Use of Models for the Study of Mining Problems*. Technical Publication 425. American Institute of Mining and Metallurgical Engineers, New York, NY.

Bucky, P.B. and Fentress, A.L. (1934). Application of Principles of Similitude to Design of Mine Workings. *Transactions of the American Institute of Mining and Metallurgical Engineers*, Vol. 1, pp. 25–50.

Burland, J.B. (1995). Assessment of Risk of Damage to Buildings due to Tunnelling and Excavation. *Proceedings of the First International Conference on Earthquake Geotechnical Engineering, IS Tokyo 95*. Tokyo, Japan. K. Ishihara, Editor. A.A. Balkema, Rotterdam, pp. 1189–1201.

Burland, J.B., Broms, B.B., and de Mello, V.F.B. (1977). Behavior of Foundations and Structures. State-of-the-Art Report. *Proceedings of the 9th International Conference on Soil Mechanics and Foundation Engineering, II*, Tokyo, Japan, pp. 495–546.

Burland, J.B. and Wroth, C.P. (1974). Settlement of Buildings and Associated Damage. *Proceedings of the Conference on the Settlement of Structures*, Cambridge, UK. Pentech Press, London, England, pp. 611–654.

Burns, J.Q. and Richard, R.M. (1964). Attenuation of Stresses for Buried Cylinders. *Proceedings of the Symposium on Soil-Structure Interaction*, Tucson, Arizona, pp. 378–392.

Butscher, C., Huggenberger, P., and Zechner, E. (2011a). Impact of Tunneling on Regional Groundwater Flow and Implications for Swelling of Clay-Sulfate Rocks. *Engineering Geology*, Vol. 117, pp. 198–206.

Butscher, C., Huggenberger, P., Zechner, E., and Einstein, H.H. (2011b). Relation between Hydrogeological Setting and Swelling Potential of Clay-Sulfate Rocks in Tunneling. *Engineering Geology*, Vol. 122, No. 3–4, pp. 204–214.

Butscher, Ch., Einstein, H.H., and Huggenberger, P. (2011c). Effects of Tunneling on Groundwater Flow and Swelling of Clay Sulfate Rocks. *Water Resources Research*, Vol. 47. doi: 10.1029/2011WR011023.

Bwalya, M.M. and Moys, M.H. (2002). The use of PFC^{2D} to simulate milling. *Numerical Modeling in Micromechanics via Particle Methods*. H. Konietzky, Editor. Balkema, Netherlands, pp. 73–77.

Cai, M., Champaigne, D., and Kaiser, P.K. (2010). Development of a fully debonded cone for rockburst support. *Deep Mining, 2010: Proceedings of the 5th International Seminar on Deep and High Stress Mining*, M. Van Sint Jan and Y. Potvin, Editors. Santiago, Chile, pp. 329–342.

Cai, Y., Esaki, T., and Jiang, Y. (2004a). An Analytical Model to Predict Axial Load in Grouted Rock Bolt for Soft Rock Tunneling. *Tunnelling and Underground Space Technology*, Vol. 19, pp. 607–618.

Cai, Y., Jiang, Y.J., and Esaki, T. (2004b). A Study of Rock Bolting Design in Soft Rock. *International Journal of Rock Mechanics and Mining Sciences*, Vol. 41, No. 3, Paper 2B 14, CD-ROM.

CAMIRO (Canadian Mining Industry Research Organization) (1995). *Canadian Rockburst Research Program 1990–1995*. CAMIRO Mining Division, Sudbury, Ontario, Canada.

Cantieni, L. and Anagnostou, G. (2009). The Effect of the Stress Path on Squeezing Behavior in Tunneling. *Rock Mechanics and Rock Engineering*, Vol. 42, pp. 289–318.

Cao, L.F., Teh, C.I., and Chang, M.-F. (2002). Analysis of Undrained Cavity Expansion in Elasto-plastic Soils with Non-linear Elasticity. *International Journal for Numerical and Analytical Methods in Geomechanics*, Vol. 26, pp. 25–52.

Carranza-Torres, C. (2003). Technical Note. Dimensionless Graphical Representation of the Exact Elasto-plastic Solution of a Circular Tunnel in a Mohr-Coulomb Material Subject to Uniform Far-field Stresses. *Rock Mechanics and Rock Engineering*, Supplement 1, pp. 629–639.

Carranza-Torres, C. (2004). Elasto-plastic Solution of Tunnel Problems using the Generalized Form of the Hoek-Brown Failure Criterion. *International Journal of Rock Mechanics and Mining Sciences*, Vol. 34, No. 3–4, paper No. 075.

Carranza-Torres, C. and Fairhurst, C. (1997). On the Stability of Tunnels under Gravity Loading, with Post-peak Softening of the Ground. *International Journal of Rock Mechanics and Mining Sciences*, Vol. 34, No. 3–4, paper No. 075.

Carranza-Torres, C. and Fairhurst, C. (1999). The Elasto-plastic Response of Underground Excavations in Rock Masses That Satisfy the Hoek-Brown Failure Criterion. *International Journal of Rock Mechanics and Mining Sciences*, Vol. 36, pp. 777–809.

Carranza-Torres, C. and Fairhurst, C. (2000). Application of the Convergence-Confinement Method Design to Rock Masses that Satisfy the Hoek-Brown Failure Criterion. *Tunnelling and Underground Space Technology*, Vol. 15, No. 2, pp. 187–213.

Carrer, J.A.M. and Telles, J.C.F. (1993). Static and dynamic nonlinear stress analysis. *Boundary Element Techniques in Geomechanics*. G.D. Manolis and Davies, Editors. Elsevier Applied Science, London, pp. 37 61.

Carter, J.P. (1988). A Semi-Analytical Solution for Swelling Around a Borehole. *International Journal of Rock Mechanics, Mining Sciences and Geomechanics Abstracts*, Vol. 12, pp. 197–212.

Carter, J.P. and Booker, J.R. (1982). Elastic Consolidation Around Deep Circular Tunnel. *International Journal of Solids and Structures*, Vol. 18, No. 12, pp. 537 – 544.

Casarin, C. and Mair, R.J. (1981). The assessment of tunnel stability in clay by model tests. *Soft-Ground Tunneling: Failures and Displacements*. D. Reséndiz and M.P. Romo, Editors. A.A. Balkema, Rotterdam, pp. 33–44.

Caulfield, R., Kieffer, D.S., Tsztoo, D.F., and Cain, B. (2005). Seismic Design Measures for the Retrofit of the Claremont Tunnel. *Chapter 89, RETC Proceedings*, pp. 1128–1139.

Cayol, V. and Cornet, F.H. (1997). 3D Mixed Boundary Elements for Elastostatic Deformation Field Analysis. *International Journal of Rock Mechanics and Mining Sciences*, Vol. 34, No. 2, pp. 275–287.

Cecil, O.S. (1970, 1975). *Correlations of Rock Bolt-Shotcrete Support and Rock Quality Parameters in Scandinavian Tunnels*. Ph.D. Thesis, University of Illinois, and *Swedish Geotechnical Institute Proceedings*, No. 27, Stockholm, 275 pp.

Celestino, T.B., Gomes, R.A.M.P., and Bortolucci, A.A. (2000). Errors in Ground Distortions due to Settlement Trough Adjustment. *Tunnelling and Underground Space Technology*, Vol. 15, No. 1, pp. 97–100.

Chambon, P. and Corté, J.F. (1994). Shallow Tunnels in Cohensionless Soil: Stability of Tunnel Face. *Journal of Geotechnical Engineering*, Vol. 120, No. 7, pp. 1148–1165.

Chan, H.C.M., Li, V., and Einstein, H.H. (1990). A Hybridized Displacement Discontinuity and Indirect Boundary Element Method to Model Fracture Propagation. *International Journal of Fracture*, Vol. 45, pp. 263–282.

Chang, L., Zhao-Qi, G., and Min-Zhong, C. (1988). Study on unlined pressure tunnels in stratified or jointed rock masses. *Rock Mechanics and Power Plants.* M. Romana, Editor. Balkema, Rotterdam, pp. 359–366.

Chen, A.G. (1996). *The Swelling Behavior of Clay Sulfate Rocks.* M.Sc. Thesis, MIT, Cambridge, MA.

Chen, S.-H., Qiang, S., Chen, S.-F., and Egger, P. (2004). Composite Element Model of the Fully Grouted Rock Bolt. *Rock Mechanics and Rock Engineering*, Vol. 37, No. 3, pp. 193–212.

Chen, S.G. and Zhao, J. (1998). A Study of UDEC Modelling for Blast Wave Propagation in Jointed Rock Masses. *International Journal of Rock Mechanics and Mining Sciences*, Vol. 35, No. 1, pp. 93–99.

Chen, W.W., Shih, B., Chen, Y.-C., Hung, J.-H., and Hwang, H.H. (2002). Seismic Response of Natural Gas and Water Pipelines in the Ji-Ji Earthquake. *Soil Dynamics and Earthquake Engineering*, Vol. 22, pp. 1209–1214.

Chen, Y., Lee, J., and Eskandarian, A. (2006). *Meshless Methods in Solid Mechanics.* Springer Science+Business Media Inc., New York, NY.

Cheng, Y.M. and Zhang, Y.H. (2002). Coupling of FEM and DDA Methods. The *International Journal of Geomechanics*, Vol. 2, No. 4, pp. 503–517.

Cheng, Y.P., Bolton, M.D., Nakata, Y. (2004). Crushing and Plastic Deformation of Soils Simulated Using DEM. *Geotechnique*, Vol. 54, No. 2, pp 131–141.

Chiaverio, F. (2002). Chienbergtunnel (Umfahrung Sissach) Tunnel in quellhaften Juragestein. *Mitteilungen der Schweizerischen Gesellschaft für Boden- und Felsmechanik*, Vol. 145, pp. 27–37.

Chiaverio, F., Straumann, U., Böheim, S., and Massignani, S. (2017). *Dritte Belchentunnelröhre: Neue Lösiungen für Vortrieb beim Tunnelbau in stark quellendem Gebirge.* Forschung + Praxis: U-Verkehr und unterirdisches Bauen, Issue No 49, Bauverlag GmbH pp. 226–231.

Chiaverio, F. and Thut, A. (2010). Chienbergtunnel: Rehabilitation Using Yielding Elements of the Section in Keuper Sediments Affected by Heave. *Proceedings of the 59th Geomechanics Colloquium*, Salzburg, pp. 573–582.

Chou, W. and Bobet, A. (2002). Predictions of Ground Deformations in Shallow Tunnels in Clay. *Tunnelling and Underground Space Technology*, Vol. 17, pp. 3–19.

Chryssanthakis, P., Barton, N., Lorig, L., and Christianson, M. (1997). Numerical Simulations of Fiber Reinforced Shotcrete in a Tunnel using the Discrete Element Method. *International Journal of Rock Mechanics and Mining Sciences*, Vol. 34, No. 3–4, paper No. 054, 14 pp.

Chua, C.G. and Goh, T.C. (2005). Estimating Wall Deflections in Deep Excavations Using Bayesian Neural Networks. *Tunnelling and Underground Space Technology*, Vol. 20, pp. 400–409.

Chuhan, Z., Yuntao, R., Pekau, O.A., and Feng, J. (2004). Time-Domain Boundary Element Method for Underground Structures in Orthotropic Media. *Journal of Engineering Mechanics*, Vol. 130. No. 1, pp. 105–116.

Churchill, R.V. (1960). *Complex Variables and Applications.* McGraw-Hill, New York, NY.

Clayton, C.R.I., Van der Berg, J.P., and Thomas, A.H. (2006). Monitoring and Displacements at Heathrow Express Terminal 4 Station Tunnels. *Géotechnique*, Vol. 56, No. 5, pp. 323–334.

Clifford, R.L. (1974). Long Rockbolt Support at New Broken Hill Consolidated Limited. *Proceedings of the Australian Institute of Mining and Metallurgy*, Vol. 251, pp. 21–26.

Clough, R.W. (1960). The Finite Element in Plane Stress Analysis. *Proceedings of the 2nd ASCE Conference on Electronic Computation*, Pittsburgh, PA, pp. 345–378.

Clough, G.W. and Schmidt, B. (1981). Design and performance of excavations and tunnels in soft clay. *Soft Clay Engineering.* E.W. Brand and R.P. Brenner, Editors. Elsevier, New York, NY, pp. 569–634.

Cocchi, G.M. (2000). The Finite Difference Method with Arbitrary Grids in the Elastic-static Analysis of Three-dimensional Continua. *Computers and Structures*, Vol. 75, pp. 187–208.

Cook, B.K. and Jensen, R.P. (2002). *Discrete Element Methods: Numerical Modeling of Discontinua.* Geotechnical Special Publication No. 117, ASCE.

Cook, N.G.W. (1964). The Application of Seismic Techniques to Problems in Rock Mechanics. *International Journal of Rock Mechanics and Mining Sciences*, Vol. 1, pp. 169–179.

Cook, N.G.W. (1965). A Note on Rockbursts Considered as a Problem of Stability. *Journal of the South African Institute of Mining and Metallurgy*, Vol. 65, No. 8, pp. 437–446.

Coon, R., and Merritt, A. (1970). Predicting In-Situ Modulus of Deformation Using Rock Quality, Indexes. *Determination of the In-Situ Modulus of Deformation of Rock Masses*, Special Technical Publication No. 477, American Society for Testing and Materials (ASTM), pp. 154–173.

Coon, R.F. (1968). *Correlation of Engineering Behavior with the Classification of the In-Situ Rock.* Ph.D. Thesis, University of Illinois, 236 pp.

Cooper, M.L., Chapman, D.N., Rogers, C.D.F., and Chan, A.H.C. (2002). Movements in the Piccadilly Line tunnels due to Heathrow Express construction. *Géotechnique*, Vol. 52, No. 4, pp. 243–257.

Corbetta, F., Bernaud, D., and Nguyen-Minh, D. (1991). Contribution à la Méthode Convergence-confinement par le Principe de la Similitude. *Revue Française de Géotechnique*, Vol. 54, pp. 5–11.

Cording, E.J. and Mahar, J.W. (1978). Index Properties and Observations for Design of Chambers in Rock. *Engineering Geology*, Vol. 12, pp. 113–142.

Cording, E.J., Long, J.H., Son, M., and Laefer, D.F. (2001). Modelling and Analysis of Excavation-induced Building Distortion and Damage Using a Strain-based Damage Criterion. *Proceedings of the London Conference for Responses of Buildings to Excavation-Induced Ground Movements in London*, London.

Cording, E.J., Long, J.L., Son, M., Laefer, D., and Ghahreman, B. (2010). Assessment of Excavation-induced Building Damage. *Proceedings of the Earth Retention Conference 3.* R.J. Finno, Y.M.A. Hashash, and P. Arduino, Editors. Geotechnical Special Publication GSP 208, pp. 101–120. ASCE, Reston, VA.

Cording, E.J. and Hansmire, W.H. (1975). Displacements around Soft Ground Tunnels. *General Report, Session IV: Tunnels in Soil, Proceedings 5th Pan American Conference in Soil Mechanics and Foundation Engineering*, ISRMFE, Buenos Aires, Vol. IV, pp. 571–632.

Cording, E.J., and Deere, D.U. (1972). Rock Tunnel Supports and Field Measurements. *Proceedings of the 1st North American Rapid Excavation and Tunneling Conference*, Chicago, IL. K.S. Lane and L.A. Garfield, Editors. Society of Mining Engineers of the American Institute of Mining, Metallurgical, and Petroleum Engineers (AIME), New York, pp. 601–622.

Cording, E.J., Brierley, G.S., Mahar, J.W., and Boscardin, M.D. (1989). Controlling ground movements during tunneling. *The Art and Science of Geotechnical Engineering.* E.J. Cording, W.J. Hall, J.D. Haltiwanger, A.J. Hendron, and G. Mesri, Editors. Prentice Hall, pp. 477–505.

Cording, E.J., Hansmire, W.H., MacPherson, H.H., Lenzini, P.A., and Vonderole, A.D. (1976). *Displacements around Tunnels in Soil.* Final Report by the University of Illinois on Contract No. DOT FR 30022 to Office of the Secretary and Federal Railroad Administration, Dept. of Transportation, Washington.

Cording, E.J., O'Rourke, T.D., and Boscardin, M. (1978). Ground movements and damage to structures. *Evaluation and Prediction of Subsidence.* S.K. Saxena, Editor. ASCE, Pensacola Beach, FL, USA, pp. 516–537.

Cox, H.L. (1952). The Elasticity and Strength of Paper and Other Fibrous Materials. *British Journal of Applied Physics*, Vol. 3, pp. 72–79.

Craig, R.N. and Muir Wood, A.M. (1978). *A Review of Tunnel Lining in the United Kingdom.* Supplement 335, Transport and Road Research Laboratory, Crowthorne, Berkshire, England.

Crouch, S.L. and Starfield, A.M. (1983). *Boundary Element Methods in Solid Mechanics.* Allen and Unwin, London.

Cruse, T.A. (1969). Numerical Solutions in Three-Dimensional Elastostatics. *International Journals of Solids and Structures*, Vol. 5, pp. 1259–1274.

Culmann, K. (1866). *Die Graphische Statik.* Meyer & Zeller, Zurich, 633 pp.

Cummings, R.A., Kendorski, F.G., and Bieniawski, Z.T. (1982). *Caving Mine Rock Mass Classification and Support Estimation.* Report to U.S. Bureau of Mines, 195 pp.

Cundall, P.A. (1971). A computer model for simulating progressive large scale movements in blocky rock systems. *Proceedings of the Symposium of the International Society of Rock Mechanics*, Vol. 1, Nancy, France, paper No. II-8.

Cundall, P.A. (1987). Distinct element models of rock and soil structure. *Analytical and Computational Methods in Engineering Rock Mechanics.* E.T. Brown, Editor. Allen & Unwin, London, pp. 129–163.

Cundall, P.A. (1988). Formulation of a Three-dimensional Distinct Element Model – Part I. A Scheme to detect and Represent Contacts in a System Composed of Many Polyhedral Blocks. *International Journal of Rock Mechanics and Mining Sciences*, Vol. 25, No. 3, pp. 107–116.

Cundall, P.A. and Hart, R.D. (1992). Numerical Modelling of Discontinua. *Engineering Computations*, Vol. 9, pp. 101–113.

Cundall, P.A. and Strack, O.D.L. (1979). A Discrete Numerical Model for Granular Assemblies. *Geotechnique*, Vol. 29, No. 1, pp. 47–65.

Cundall, P.A., Ruest, M.A., Guest, A.R., and Chitombo, G. (2002). Evaluation of schemes to improve the efficiency of a complete model of blasting and rock fracture. *Numerical Modeling in Micromechanics via Particle Methods*. H. Konietzky, Editor. Balkema, Netherlands, pp. 107–115.

Daemen, J.J.K. (1975). Tunnel Support Loading Caused by Rock Failure. *Technical Report MRD-3-75*, Missouri River Division, U.S. Corps of Engineers, Omaha, NE.

Daemen, J.J.K. (1977). Problems in Tunnel Support Mechanics. *Underground Space*, Vol. 1, No. 3, pp. 163–172.

Daemen, J.J.K. and Fairhurst, C. (1972). Rock Failure and Tunnel Support Loading. *Proceedings of the International Symposium on Underground Openings*, Lucerne, Switzerland. Swiss Society for Soil Mechanics and Foundation Engineering, pp 341–347

Dalgiç, S. (2002). Tunneling in Squeezing Rock, the Bolu Tunnel, Anatolian Motorway, Turkey. *Engineering Geology*, Vol. 67, pp. 73–96.

Dann, H.E., Hartwig, W.P., and Hunter, J.R. (1964). Unlined Tunnels of the Snowy Mountains Hydro-Electric Authority, Australia. *Journal of the Power Division*, ASCE, Vol. 90, pp. 47–79.

DAUB (2001). Concrete Linings for Tunnel Built by Underground Construction, Recommendations by DAUB (German Committee for Underground Construction, Inc). *Tunnel*, Vol. 5, pp. 50–66.

Davies, T.G. (1993). Numerical implementation. *Boundary Element Techniques in Geomechanics*. G.D. Manolis and T.G. Davies, Editors. Elsevier Applied Science, London, pp. 101–145.

Davis, C.A. (1999). Stress in buried pipes from shear distortions. *Geo-Engineering for Underground Facilities*. G. Fernandez and R.A. Bauer, Editors. Geotechnical Special Publication No. 90, ASCE, Reston, VA, pp. 483–494.

Davis, E.H., Gunn, M.J., Mair, R.J., and Seneviratnes, H.N. (1980). The Stability of Shallow Tunnels and Underground Openings in Cohesive Material. *Geotechnique*, Vol. 30, No. 4, pp. 397–416.

De Farias, M.M., Moraes, A.H., and de Assis, A.P. (2004). Displacement Control in Tunnels Excavated by the NATM: 3-D Numerical Simulations. *Tunnelling and Underground Space Technology*, Vol. 19, pp. 283–293.

De Wilde, P. (1996). *Neural Network Models: An Analysis*. Springer-Verlag, London.

Deb, D., Kumar, A., and Rosha, R.P.S. (2006). Forecasting Shield Pressures at a Longwall Face Using Artificial Neural Networks. *Geotechnical and Geological Engineering*, Vol. 24, pp. 1021–1037.

Deere, D.U. (1963). Technical Description of Rock Cores for Engineering Purposes. *Felsmechanik and Ingenieurgeologie*, Vol. 1, No. 1, pp. 17–22.

Deere, D.U. (1969). Personal Files.

Deere, D.U. (1983). Unique Geotechnical Problems at Some Hydroelectric Projects. *Proceedings of the Seventh Panamerican Conference on Soil Mechanics and Foundation Engineering. Canadian Geotechnical Society*, Montreal, Quebec, CA, pp. 865–888.

Deere, D.U., Hendron, A.J., Patton, D.F., and Cording, E.J. (1967). Design of Surface and Near Surface Construction in Rock. *Failure and Breakage of Rock*: *Proceedings of 8th Symposium on Rock Mechanics*, University of Minnesota, MN. C. Fairhurst, Editor. AIME, Littleton, Colorado, pp. 237–301.

Deere, D.U., Merritt, A.H., and Coon, R.F. (1968). *Engineering Classification of In-Situ Rocks*. Report by University of Illinois to Air Force Weapons Lab, Kirtland Air Force Base, New Mexico. Publication No. AFWL-TW-67-144, National Technical Information Service (NTIS) Publication No. AD 848798, 272 pp.

Deere, D.U., Merritt, A.H., and Cording, E.J. (1974). Engineering Geology and Underground Construction. General Report. *Session VII, 2nd International Congress of the International Association of Engineering Geology*, Sao Paulo, Brazil. pp. VII-GR, 1–26.

Deere, D.U., Peck, R.B., Monsees, J.E., and Schmidt, B. (1969). *Design of Tunnel Liners and Support Systems*. Report of University of Illinois to Office of High-Speed Ground Transportation, U.S. Department of Transportation (DOT), NTIS Publication No. PB-183799, 420 pp.

Deere, D.U., Peck, R.B., Parker, H.W., Monsees, J.E., and Schmidt, B. (1970). *Design of Tunnel Support Systems*. Highway Research Record, No. 339, pp. 26–33.

Delgado, J. and Mercer, R. (2006). Microseismic Monitoring and Rockbursting Activity at the Campbell Mine, Ontario, Canada. *41st U.S. Rock Mechanics Symposium*, Golden, CO. Paper 06-1160, 11 pages.

Deng, J., Yue, Z.Q., Tham, L.G., and Zhu, H.H. (2003). Pillar Design by Combining Finite Element Methods, Neural Networks and Reliability: A Case Study of the Feng Huangshan Copper Mine, China. *International Journal of Rock Mechanics and Mining Sciences*, Vol. 40, pp. 585–599.

Deng, J.H. and Lee, C.F. (2001). Displacement Back Analysis for a Steep Slope at the Three Gorges Project Site. *International Journal of Rock Mechanics and Mining Sciences*, Vol. 38, pp. 259–268.

Dershowitz, W.S. (1979). *Jointed Rock Mass Deformability, A Probabilistic Approach*. SMCE Thesis, Massachusetts Institute of Technology, Cambridge, 292 pp.

Desai, D.B., Merritt, J.L., and Chang, B. (1989). "Shake and Slip to Survive" – Tunnel Design. *Proceedings of the Rapid Excavation and Tunneling Conference (RETC)*. R.A. Pond and P.B. Kenny, Editors. Society of Mining, Metallurgy, and Exploration, Littleton, CO, pp. 13–30.

Detournay, E. and Cheng, A.H.D. (1988). Poroelastic Response of a Borehole in a Non-Hydrostatic Stress Field. *International Journal of Rock Mechanics and Mining Sciences*, Vol. 25, No. 3, pp. 171–182.

Detzlhofer, H. (1960). *Die Stollenarbeiten für das Innkraftwerk Prutz-Imst der Tiroler Wasserkraftwerke AG*. Montan Rundschau, Sonderhelft Stollen- und Tunnelbau, pp. 137–138.

Detzlhofer, H. (1974). Erfahrungen und Vorschläge für die Gebirgsklassifizierung beim Bau von Wasserstollen. *Strassenforschung*, Vol. 18, pp. 37–50.

Dhawan, K.R., Singh, D.N., and Gupta, I.D. (2002). 2D and3D Finite Element Analysis of Underground Openings in an Inhomogeneous Rock Mass. *International Journal of Rock Mechanics and Mining Sciences*, Vol. 39, pp. 217–227.

Dhawan, K.R., Singh, D.N., and Gupta, I.D. (2004). Dynamic Analysis of Underground Openings. *Rock Mechanics and Rock Engineering*, Vol. 37, No. 4, pp. 299–315.

Ding, Y. and Kusterle, W. (2000). Compressive Stress-Strain Relationship of Steel Fibre-Reinforced Concrete at Early Age. *Cement and Concrete Research*, Vol. 30, pp. 1573–1579.

Dorsten, V., Frederick, F.H., and Preston, H.K. (1984). Epoxy Coated Seven-Wire Strand for Prestressed Concrete. *Prestressed Concrete Institute Journal*, Vol. 29, No. 4, pp. 1–11.

Douglass, P.M. and Sundberg, C.C. (1987). Design of PCCP Pressure Tunnel Liners. *Proceedings of the 1987 Rapid Excavation and Tunneling Conference*, J.M. Jacobs and R.S. Hendricks, Editors. Society of Mining Engineers, Inc. Littleton, CO, pp. 63–86.

Dowding, C.H. (1985). Earthquake response of caverns: Empirical Correlations and Numerical Modeling. *Proceedings of the Rapid Excavation and Tunneling Conference (RETC)*. C.D. Mann and M.N. Kelley, Editors. Society of Mining Engineers of the American Institute of Mining, Metallurgical, and Petroleum Engineers, New York, NY, pp. 71–83.

Dowding, C.H. and Andersson, C.-A. (1986). Potential for Rock Bursting and Slabbing in Deep Caverns. *Engineering Geology*, Vol. 22, pp. 265–279.

Dowding, C.H. and Belytschko, T.B. (1983). Earthquake response of caverns in jointed rock: Effects of frequency and jointing. *Seismic Design of Embankments and Caverns*. T.R. Howard, Editor. ASCE, New York, pp. 142–156.

Dowding, C.H. and Rozen, A. (1978). Damage to Rock Tunnels from Earthquake Shaking. *Journal of Geotechnical Engineering Division*, ASCE 104 GT2, pp. 175–191.

Dowding, C.H., Belytschko, T.B., and Dmytryshyn, O. (2000). Dynamic Response of Million Block Cavern Models with Parallel Processing. *Rock Mechanics and Rock Engineering*, Vol. 33, No. 3, pp. 207–214.

Dowding, C.H., Belytschko, T.B., and Yen, H.J. (1983). A Coupled Finite Element-Rigid Block Method for Transient Analysis of Rock Caverns. *International Journal of Numerical and Analytical Methods in Geomechanics*, Vol. 7, pp. 117–127.

Du, Q., Gunzburger, M.D., and Ju, L. (2003). Voronoi-Based Finite Volume Methods, Optimal Voronoi Meshes, and PDEs on the Sphere. *Computer Methods in Applied Mechanics and Engineering*, Vol. 192, No. 35–36, pp. 3933–3957.

Duarte, C.A. and Oden, J.T.(1996). An h-p Adaptive Method Using Clouds. *Computer Methods in Applied Mechanics and Engineering*, Vol. 139, pp. 237–262.

Eberhardsteiner, J., Hofstetter, G., Kropik, Chr., and Mang, H.A. (1993). Elastoviscoplastic three-dimensional hybrid BE-FE analysis of the excavation of tunnels. *Boundary Element Techniques in Geomechanics*. G.D. Manolis and T.G. Davies, Editors. Elsevier Applied Science, London, pp. 327–358.

Eberhardt, E. (2001). Numerical Modeling of Three-dimension Stress Rotation Ahead of an Advancing Tunnel Face. *International Journal of Rock Mechanics and Mining Sciences*, Vol. 38, pp. 499–518.

Eberhardt, E., Stead, D., and Coggan, J.S. (2004). Numerical Analysis of Initiation and Progressive Failure in Natural Rock Slopes – the 1991 Randa Rockslide. *International Journal of Rock Mechanics and Mining Sciences*, Vol. 41, pp. 69–87.

Ebisu, S.P., Komarka, S., Aydan, Ö., and Kauramato, R. (1992). Characterization of Rock Masses for Rock Anchor Foundations. *Proceedings 1st Symposium on Fractured and Jointed Rock Masses*, Lake Tahoe, CA. L.R. Myer, Editor. A.A. Balkema, Rotterdam, The Netherlands, 770 pp.

EFNARC (1996). *European Specification for Sprayed Concrete*. EFNARC.

Einstein, H.H. (1979). Tunneling in Swelling Rocks. *Underground Space*, Vol. 4, No. 1, pp. 51–61.

Einstein, H.H. (2000). Tunnels in Opalinus Clayshale – A Review of Case Histories and New Developments. *Tunnelling an Underground Space Technology*, Vol. 15, No. 1, pp. 13–29.

Einstein, H.H. (2005). Personal communication.

Einstein, H.H. and Mayer, T. (1999). Puzzles in Rocks. Müller Lecture. *Proceedings of the 9th ISRM Congress*, 25–28 August, Paris, France, pp. 1707–1740.

Einstein, H.H. and Schwartz, C.W. (1979). Simplified Analysis for Tunnel Supports. *ASCE Journal of the Geotechnical Engineering Division*, Vol. 105, No. GT4, pp. 499–518.

Einstein, H.H. and Baecher, G.B. (1982). Probabilistic and Statistical Methods in Engineering Geology I. Problem Statement and Introduction to Solution. *Rock Mechanics*, Suppl. 12, pp. 47–61.

Einstein, H.H. and Bobet, A. (1997). Mechanized Tunneling in Squeezing Ground - From Basic Thoughts to Continuous Tunneling. *Proceedings of the World Tunnel Congress '97*. Vienna, Austria, pp. 619–632.

Einstein, H.H., Azzouz, A.G., McKown, A.F., and Thompson, D.E. (1983). Evaluation of design and performance – porter square transit chamber lining. *Proceedings RETC*. AIME, New York, pp. 597–620.

Einstein, H.H., Bischoff, N., and Hoffmann, E. (1972). Verhalten von Stollensohlen in quellendem Mergel. *International Symposium on Underground Openings*. Lucerne, pp. 296–319.

Einstein, H.H., Bobet, A., and Aristorenas, G. (1995a). *Feasibility Study Opalinuston*. NAGRA, Report Nib, 95-61, Volumes A through D.

Einstein, H.H., Bobet, A., and Aristorenas, G. (1995b). *Feasibility Study Opalinuston*, for NAGRA (Nationale Gesellschaft für die Lagerung radioaktiver Abfälle), Switzerland. M.I.T. Internal Report, MA, USA.

Einstein, H.H., Schwartz, C.W., Steiner, W., Baligh, M.M., and Levitt, R.E. (1977). *Improved Design for Tunnel Supports*, Final Report on Research to DOT, Contract No. DOT-OS-60-136, 504 pp.

Eisenstein, Z. and Branco, P. (1991). Convergence-Confinement Method in Shallow Tunnels. *Tunnelling and Underground Space Technology*, Vol. 6, No. 3, pp. 343–346.

Eisenstein, Z. and Ezzeldine, O. (1994). The role of face pressure for shields with positive ground control. *Proceedings of the International Congress on Tunnelling and Ground Conditions*, Cairo, Egipt. M.E. Abdel Salam, Editor. A.A. Balkema, Rotterdam, The Netherlands, pp. 557–571.

Eisenstein, Z.D. and Rossler, K. (1995). Geotechnical Criteria for Double Shield Tunnel Boring Machines. *Worldwide Innovations in Tunnelling, World Tunnel Congress*, pp. 192–201.

El Tani, M. (2003). Circular Tunnel in a Semi-infinite Aquifer. *Tunnelling and Underground Space Technology*, Vol. 18, pp. 49–55.

El-Nahhas, F., El-Kadi, F., and Ahmed, A. (1992). Interaction of Tunnel Linings and Soft Ground. *Tunnelling and Underground Space Technology*, Vol. 7, No. 1, pp. 33–43.

EM 1110-2-2005 (1993). *Standard Practice for Shotcrete.* Engineering Manual 1110-2-2005. Department of the Army, U.S. Corps of Engineering, Washington, DC.

Emmert-Streib, F., Yang, Z., Feng, H., Tripathi, S., and Dehmer, M. (2020). An Introductory Review of Deep Learning for Prediction Models with Big Data (2020). *Frontiers in Artifical Intelligence*, Vol. 28. doi: 10.3389/frai.2020.00004.

Engesser, F. (1882). Über den Erdduck gegen innere Stützwande (Tunnelwände). *Deutsche Bauzeitung*, No. 16, pp. 91–93. *Engineering*, Vol. 131, No. 7, pp. 886–897.

Eskilsson, J.N. (1999). Design of Pressure Tunnels. *Geo-Engineering for Underground Facilities.* Geotechnical Special Publication No. 90. G. Fernandez and R.A. Bauer, Editors. ASCE, Reston, VA, pp. 442–458.

Esmaeili, M., Vahdani, S., and Noorzad, A. (2006). Dynamic Response of Lined Circular Tunnel to Plane Harmonic Waves. *Tunnelling and Underground Space Technology*, Vol. 21, No.5, pp. 511–519.

Etterlin, A. Editor (1988). *Rekonstruktion Hauenstein-Basistunnel.* Hasler and Holtz, Olten, Switzerland.

Evans, C.H. (1983). *An Examination of Arching in Granular Soils.* S.M. Thesis, Department of Civil Engineering, MIT, Cambridge, MA.

Exadaktylos, G.E. and Stavropoulou, M.C. (2002). A Closed-Form Elastic Solution for Stresses and Displacements around Tunnels. *International Journal of Rock Mechanics and Mining Sciences*, Vol. 39, No. 7, pp. 905–916.

Exadaktylos, G.E., Liolios, P.A., and Stavropoulou, M.C. (2003). A Semi-Analytical Elastic Stress-Displacement Solution for Notched Circular Openings in Rocks. *International Journal of Solids and Structures*, Vol. 40, pp. 1165–1187.

Fahimifar, A. and Soroush, H. (2005). A Theoretical Approach for Analysis of the Interaction between Grouted Rockbolts and Rock Masses. *Tunnelling and Underground Space Technology*, Vol. 20, pp. 333–343.

Fakhimi, A., Carvalho, F. Ishida, T., and Labuz, J.F. (2002). Simulation of Failure Around a Circular Opening in Rock. *International Journal of Rock Mechanics and Mining Sciences*, Vol. 39, pp. 507–515.

Fakhimi, A.A. and Labuz, J.F. (2002). Modeling Rock Failure around a Circular Opening. *Discrete Element Methods: Numerical Modeling of Discontinua.* Geotechnical Special Publication No. 117, ASCE, pp. 323–328.

Fang, Y.S., Lin, J.S., and Su, C.S. (1994). An Estimation of Ground Settlement Due to Shield Tunneling by the Peck-Fujita Method. *Canadian Geotechnical Journal*, Vol. 31, pp. 431–443.

Farmer, I.W. (1975). Stress Distribution along Resin Grouted Rock Anchor. *International Journal of Rock Mechanics, Mining Sciences, and Geomechanics Abstracts*, Vol. 12, pp. 347–351.

Feng, X.-T. and An, H. (2004). Hybrid Intelligent Method Optimization of a Soft Rock Replacement Scheme for a Large Cavern Excavated in Alternate Hard and Soft Rock Strata. *International Journal of Rock Mechanics and Mining Sciences*, Vol. 41, pp. 655–667.

Feng, X.-T., Wang, Y.-J., and Yao, J.-G. (1996). A Neural Network Model for Real-time Roof Pressure Prediction in Coal Mines. *International Journal of Rock Mechanics and Mining Sciences*, Vol. 33, No. 6, pp. 647–653.

Fenner, R.T. (1986). *Engineering Elasticity. Application of Numerical and Analytical Techniques.* Ellis Horwood Limited, West Sussex, England.

Fernandez, G. (1994). Behavior of Pressure Tunnels and Guidelines for Liner Design. *Journal of Geotechnical Engineering*, Vol. 120, No. 10, pp. 1768–1791.

Fernández, G. and Alvarez Jr., T.A. (1994). Seepage-Induced Effective Stresses and water Pressures around Pressure Tunnels. *Journal of Geotechnical Engineering*, ASCE, Vol. 120, No.1, pp. 108–128.

Fernández, G., Tirso, A., and Alvarez Jr., A. (1994). Seepage-Induced Effective Stresses and water Pressures around Pressure Tunnels. *Journal of Geotechnical Engineering*, ASCE, Vol. 120, No. 1, pp. 108–128.

FHWA (2009). *Technical Manual for Design and Construction of Road Tunnels – Civil Elements.* U.S. Department of Transportation, Federal Highway Administration. Publication FHWA-NHI-10-034.

Finno, R.J. and Clough, G.W. (1985). Evaluation of Soil Response to EPB Shield Tunneling. *Journal of Geotechnical Engineering*, Vol. 111, No. 2, pp. 155–173.

Flaate, K. (1983). Fibre Reinforced Shotcrete Reinforces Rockbolts in Bid to Control Rockbursts. *Tunnels and Tunnelling*, Vol. 15, No. 4, pp. 40–41.

Flügge, W. (1966). *Stresses in Shells*, Springler-Verlag, Inc., New York, NY.

Franzén, T. (1992). Invited lecture: Shotcrete for underground support – A state of the art report with focus on steel fibre reinforcement. *Rock Support in Mining and Underground Construction, Proceedings of the International Symposium on Rock Support.* P.K. Kaiser and D.R. McCreath, Editors. A.A. Balkema, Rotterdam, pp. 91–104.

Franzén, T., Garshol, K.F., and Tomisawa, N. (2001). Sprayed Concrete for Final Linings: ITA Working Group Report. *Tunnelling and Underground Space Technology*, Vol. 16, pp. 295–309.

Franzius, J.N. and Potts, D.M. (2005). Influence of Mesh Geometry on Three-Dimensional Finite-Element Analysis of Tunnel Excavation. *International Journal of Geomechanics*, Vol. 5, No. 3, pp. 256–266.

Freeman, T.J. (1978). The Behaviour of Fully-bonded Rock Bolts in the Kielder Experimental Tunnel. *Tunnels and Tunnelling*, June 1978, Vol. 10, No. 5, pp. 37–40.

Frey, F., Rebora, B., Sarf, J.-L., and Truty, A. (2002). Buckling of Steel Linings for Hydropower Tunnels. *Hydropower and Dams*, Vol. 9, No. 1, pp. 79–86.

Frischmann, W.W., Hellings, J.E., Gittoes, G., and Snowden, C. (1994). Protection of the Mansion House against damage caused by ground movement due to the Docklands Light Railway Extension. *Proceedings of the Institution of Civil Engineers, Geotechnical Engineering*, Vol. 107, pp. 65–76

Frölich, B. (1986). Anisotropes Quellverhalten diagenetisch verfestigter Tonsteine. *Veröffentlichung des Inst. F. Boden- und Felsmechanik der Univ. Friedericiana*, Karlsruhe, Heft 99, 139 pp.

Fuller, P.G. (1983). Cable support in mining. A keynote lecture. *Rock Bolting. Theory and Application in Mining and Underground Construction. Proceedings of the International Symposium on Rock Bolting.* O. Stephansson, Editor. A.A. Balkema, Rotterdam, The Netherlands, pp. 511–522.

Gajewski, J. and Jonak, J. (2006). Utilization of Neural Networks to Identify the Status of the Cutting Tool Point. *Tunnelling and Underground Space Technology*, Vol. 21, pp. 180–184.

Gandhe, A., Venkateswarlu, V., and Gupta, R.N. (2005). Extraction of Coal Under a Surface Water Body – a Strata Control Investigation. *Rock Mechanics and Rock Engineering*, Vol. 38, No. 5, pp. 399–410.

Garford Pty Limited (1990). *An improved, economical method for rock stabilization.* Perth, 4 pages.

Garmisch, T. and Girmscheid, G. (2005). Lifecycle Oriented Construction and Maintenance of Traffic Tunnels – Strategy Assessment to Develop Drainage Systems with Low Calcification and Minimal Required Maintenances. *Advancing Engineering Management and Technology, Proc Intl. Conference on Construction, Athens, Greece.* S.M. Ahmed, I. Ahmed, and J.-P. Pantouvakis, Editors. pp. 115–122.

Garmisch, T. and Girmscheid, G. (2007). *Versinterungsprobleme in Bauentwässerungen.* Bauwerk, 665 pp.

Gavete, L., Gavete, M.L., and Benito, J.J. (2003). Improvements of Generalized Finite Difference Method and Comparison with Other Meshless Method. *Applied Mathematical Modelling*, Vol. 27, pp. 831–847.

Gazetas, G., Gerolymos, N., and Anastasopoulos, I. (2005). Response of Three Athens Metro Underground Structures in the 1999 Parnitha Earthquake. *Soil Dynamics and Earthquake Engineering*, Vol. 25, pp. 617–633.

Gerçek, H. (1997). An Elastic Solution for Stresses Around Tunnels with Conventional Shapes. *International Journal of Rock Mechanics and Mining Sciences*, Vol. 34, No. 3–4, paper No. 096.

Gerçek, H. (1991). Stresses around tunnels with arched roof. *7th International Congress on Rock Mechanics*, W. Wittke, Editor. A.A. Balkema, Rotterdam, Vol. 2, pp. 1297–1299.

Ghaboussi, J. and Barbosa, R. (1990). Three-dimensional Discrete Element Method for Granular Materials. *International Journal of Numerical and Analytical Methods in Geomechanics*, Vol. 14, pp. 451–472.

Gioda, G. (1984). A Simple Boundary Equation Technique for Elastic Stress Analysis on Underground Cavities. *Rock Mechanics and Rock Engineering*, Vol. 17, pp. 147–165.

Gioda, G. and Carini, A. (1985). A Combined Boundary Element-Finite Element Analysis of Lined Openings. *Rock Mechanics and Rock Engineering*, Vol. 18, pp. 293–302.

Glossop, N.H. and O'Reilly, M.P. (1982). Settlement Caused by Tunneling through Soft Marine Silty Clay. *Tunnels and Tunnelling*, Vol. 14, No. 9, pp. 13–16.

Goel, R.K., Swarup, A., and Sheorey, P.R. (2007). Bolt Length Requirement in Underground Openings. *International Journal of Rock Mechanics and Mining Sciences*, Vol. 44, pp. 802–811.

Goh, A.T.C. (1994). Seismic Liquefaction Potential Assessed by Neural Networks. *ASCE Journal of Geotechnical Engineering*, Vol. 120, No. 9, pp. 1467–1480.

Golser, H. and Schubert, W. (2003). Chapter 15: Application of numerical simulation at the tunnel site. *Numerical Simulation in Tunnelling*. G. Beer, Editor. Springer-Verlag, Wien, Austria, pp. 427–473.

Goodman, R.E. (1989). *Introduction to Rock Mechanics*. John Wiley and Sons, New York, NY.

Goodman, R.E. and Shi, G.-H. (1985). *Block Theory and Its Application to Rock Engineering*. Prentice Hall, Inc., New Jersey, US.

Goricki, A. (2003). *Classification of Rock Mass Behavior and an Hierarchical Rock Mass Characterization for the Design of Underground Structures*. Ph.D. Thesis, Graz University of Technology, 98 pp.

Grasselli, G. (2005). 3D Behaviour of Bolted Rock Joints: Experimental and Numerical Study. *International Journal of Rock Mechanics and Mining Sciences*, Vol. 42, pp. 13–24.

Graziani, A., Boldini, D., and Ribacchi, R. (2005). Practical Estimate of Deformations and Stress Relief Factors for Deep Tunnels Supported by Shotcrete. *Rock Mechanics and Rock Engineering*, Vol. 38, No. 5, pp. 345–372.

Grimstad, E. (1981). Engineering-geology at the Holmestrand Tunnel (in Norwegian). *Fjellsprengningsteknikk/Bergmekanikk/Geoteknikk*, pp. 30.1–30.8, Tapir Press.

Grimstad, E. (2007a). Må vi heretter støpe ut alle våre veitunneler? [Is cast concrete lining the only rock support in Norwegian road tunnels in the future?]. *Fjellsprengimngsteknikk/Bergmekanikk/Geoteknikk -2007* pp. 22.1–22.12.

Grimstad, E. (2007b). The Norwegian Method of Tunnelling – a Challenge for Support Design. *Proceedings of the XIV European Conference on Soli Mechanics and Geotechnical Engineering*, Madrid, Spain.

Grimstad, E. and Barton, N. (1988). Design Method of Rock Support. Norwegian Tunneling Today, Norwegian Soil and Rock Mechanics Association, Publication No. 5, pp. 59–64.

Grimstad, E. and Barton, N. (1993). Updating of the Q-System for NMT. *Proceedings of the International Symposium on Sprayed Concrete*, Fagernes, Norway. R. Kompen, O. Opsahl, and K. Berg, Editors. Norwegian Concrete Association, Oslo, 464 pp.

Grimstad, E., Barton, N., Lien, R., Lunde, J., and Løset, F. (1986). Classification of rock masses with respect to tunnel stability – new experiences with the Q-system. *Fjellsprengningsteknikk, Bergmekanikk, Geoteknikk*. A.M. Heltzen, R. Kjølberg, R. Lauritzen, and K.R. Berg, Editors. Taipir Press, Trondheim, pp. 30.1–30.18.

Grob, H. (1972). Schwelldruck im Belchentunnel. *International Symposium on Underground Openings*, Lucerne, pp. 99–119.

Grob, H. (1976). Swelling and Heave in Swiss Tunnels. *Bulletin of the International Association of Engineering Geology*, Vol. 13, pp. 55–60.

Guan, Z., Jiang, Y., Tanabasi, Y., and Huang, H. (2007). Reinforcement Mechanics of Passive Bolts in Conventional Tunnelling. *International Journal of Rock Mechanics and Mining Sciences*, Vol. 44, pp. 625–636.

Gysel, M. (1977). A Contribution to the Design of a Tunnel Lining in Swelling Rock. *Rock Mechanics*, Vol. 10, No. 1–2, pp. 55–77. Errata *Rock Mechanics*, Vol. 11, No. 3, pp. 132.

Gysel, M. (1987). Design Methods for Structures in Swelling Rock. *Proceedings of the 6th International Conference on Rock Mechanics*, Montreal, A.A. Balkema, Rotterdam, pp. 377–381

Hagan, T.O., Milev, A.M., Spottiswoode, S.M., Hildyard, M.W., Grodner, M., Rorke, A.J., Finnie, G.J., Reddy, N., Haile, A.T., Le Bron, K.B., and Grave, D.M. (2001). Simulated rockburst experiment – an overview. *The Journal of the South African Institute of Mining and Metallurgy*, August 2001, pp. 217–222.

Hansmire, W.H. and Cording, E.J. (1972). Performance of a Soft Ground Tunnel on the Washington Metro. *Proceedings of the 1st North American Rapid Excavation and Tunnelling Conference*, AIME, Vol. 1, pp. 371–389.

Hao, Y.H. and Azzam, R. (2005). The Plastic Zones and Displacements Around Underground Openings in Rock Masses Containing a Fault. *Tunnelling and Underground Space Technology*, Vol. 20, pp. 49–61.

Harr, M.E. (1962). *Groundwater and Seepage*. McGraw-Hill Book Co., Inc., New York, NY.

Harriss, J. (1975). *The Tallest Tower: Eiffel and the Belle Epoque*. Unlimited Publishing, Bloomington, IN, 212 pp.

Hart, R., Cundall, P.A., and Lemos, J. (1988). Formulation of a Three-dimensional Distinct Element Model – Part II. Mechanical Calculations for Motion and Interaction of a System Composed of Many Polyhedral Blocks. *International Journal of Rock Mechanics and Mining Sciences*, Vol. 25, No. 3, pp. 117–125.

Hartmaier, H.H., Doe, T.W., and Dixon, G. (1998). Evaluation of Hydrojacking Tests for an Unlined Pressure Tunnel. *Tunnelling and Underground Space Technology*, Vol. 13, No. 4, pp. 393–401.

Hashash, Y., Hook, J.J., Schmidt, B., and Yao, J. (2001). Seismic Design and Analysis of Underground Structures. *Tunnelling and Underground Space Technology*, Vol. 16, pp. 247–293.

Hashash, Y.M.A., Phillips, C., and Groholski, D. (2010). Recent advances in non-linear site response analysis. *5th International Conference on Recent Advances in Geotechnical Earthquake Engineering and Soil Dynamics*, paper No. OSP4.

Hassani, F.P., Mitri, H.S., Khan, U.H., and Rajaie, H. (1992). Experimental and numerical studies of the cable bolt support systems. *Rock Support in Mining and Underground Construction. Proceedings of the International Symposium on Rock Support*. P.K. Kaiser and D.R. McCreath, Editors. A.A. Balkema, Rotterdam, The Netherlands, pp. 411–417.

Hatzor, Y.H. and Feintuch, A. (2001). The Validity of Dynamic Block Displacement Prediction Using DDA. *International Journal of Rock Mechanics and Mining Sciences*, Vol. 38, pp. 599–606.

Hatzor, Y.H., Arzi, A.A., Zaslavsky, Y., and Shapira, A. (2004). Dynamic Stability Analysis of Jointed Rock Slopes Using the DDA Method: King Herod's Palace, Masada, Israel. *International Journal of Rock Mechanics and Mining Sciences*, Vol. 41, pp. 813–832.

Hauber, L., Jordan, P., Madsen, F., Nüesch, R., and Vögtli, B. (2005). *Clay Minerals and Sulfates as Source of Swelling of Sediments, Reasons and Effects of Swelling*. Final Report Res. Projects 55/92 and 52/96, Swiss Federal Roads Office (ASTRA), Bern, Switzerland.

Hazzard, J.F. and Young, R.P. (2004). Dynamic Modeling of Induced Seismicity. *International Journal of Rock Mechanics and Mining Sciences*, Vol. 41, pp. 1365–1376.

Heal, D., Potvin, Y., and Hudyma, M. (2006). Evaluating Rockburst damage Potential in Underground Mining. *41st U.S. Rock Mechanics Symposium*, Golden, CO. Paper 06-1020, 12 pages.

Hefny, A.M. and Lo, K.Y. (1999). Analytical Solutions for Stresses and Displacements around Tunnels Driven in Cross-Anisotropic Rocks. *International Journal for Analytical and Numerical Methods in Geomechanics*, Vol. 23, pp. 161–177.

Hellmich, C., Macht, J., Lackner, R., and Mang, H.A. (2001). Phase Transitions in Shotcrete: From Material Modeling to Structural Safety Assessment. *Shotcrete: Engineering Developments. Proceedings of the International Conference on Engineering Developments in Shotcrete*. E. Stefan Bernard, Editor. A.A. Balkema, Hobart, Tasmania, Australia, pp. 173–184.

Hellmich, C., Ulm, F.-J., Mang, H.A. (1999). Multisurface Chemoplasticity. I: Numerical Studies of NATM Tunneling. *Journal of Engineering Mechanics*, Vol. 125, No. 6, pp. 692–701.

Hellmich, C., Ulm, F.-J., Mang, H.A. (2000). Modeling of Early-age Creep of Shotcrete. II: Application to Tunneling. *Journal of Engineering Mechanics*, Vo. 126, No. 3, pp. 292–299.

Hendron, A.J. and Fernández, G. (1983). Dynamic and static design considerations for underground chambers. *Seismic Design of Embankments and Caverns*. T.R. Howard, Editor. ASCE, NewYork, pp. 157–197.

Hendron Jr., A.J., Fernandez, G., Lenzini, P.A., and Hendron, M.A. (1989). Design of pressure tunnels. *The Art and Science of Geotechnical Engineering*. E.J. Cording, W.J. Hall, J.D. Haltiwanger, A.J. Hendron, and G. Mesri, Editors. Prentice Hall, Urbana-Chapaign, IL, USA, pp. 161–192.

Heuer, R. (1974). Selection/Design of Shotcrete for Temporary Support. *Use of Shotcrete for Underground Support, Proceedings of the Engineering Foundation Conference*, Berwick Academy, also: ACI Publication SP-45.

Hewett, B.H.M. and Johammesson, S. (1922). *Shield and Compressed Air Tunneling*. McGraw Hill, New York, NY.

Hinton, G.E., Osindero, S., and Teh, Y.-W. (2006). A Fast Learning Algorithm for Deep Belief Nets. *Neural Computation*, Vol. 18, 1527–1554. doi: 10.1162/neco.2006.18.7.1527.

Hirata, A., Kameoka, Y., and Hirano, T. (2007). Safety Management Based on Detection of Possible Rock Bursts by AE Monitoring during Tunnel Excavation. *Rock Mechanics and Rock Engineering*, Vol. 40, No. 6, pp. 563–576.

Hisatake, M. (2003). Effects of Steel Fiber Reinforced High-Strength Shotcrete in a Squeezing Tunnel. *Tunnelling and Underground Space Technology*, Vol. 18, pp. 197–204.

Hocking, G. (1992). The Discrete Element Method for Analysis of Fragmentation of Discontinua. *Engineering Computations*, Vol. 9, pp. 145–155.

Hoek, E. (1994a). The Challenge of Input Data for Rock Engineering. *ISRM News Journal*, Vol. 2, pp. 23–24.

Hoek, E. (1994b). Strength of Rock and Rock Masses. *ISRM News Journal*, Vol. 2, pp. 4–16.

Hoek, E. (2007). *Practical Rock Engineering*. Electronic format, http://www.rocscience.com/hoek/pdf/Practical_Rock_Engineering.pdf

Hoek, E. and Brown, E.T. (1980a). *Underground Excavations in Rock*. Institution of Mining and Metallurgy, London, 527 pp.

Hoek, E. and Brown, E.T. (1980b). Empirical Strength Criterion for Rock Masses. *Journal of the Geotechnical Engineering Division*, Vol. 106, No. 9, pp. 1013–1035.

Hoek, E. and Brown, E.T. (1988). The Hoek-Brown Failure Criterion – a 1988 update. *Proceedings 15th Canadian Rock Mechanics Symposium*, University of Toronto, Canada. J.H. Curran, Editor. Toronto, pp. 31–38.

Hoek, E. and Brown, E.T. (1997). Practical Estimates of Rock Mass Strength. *International Journal of Rock Mechanics and Mining Sciences*, Vol. 34, No. 8, pp. 1165–1186.

Hoek, E. and Brown, E.T. (2019). The Hoek-Brown Failure Criterion an GSI -2018 Edition. *Journal of Rock Mechanics and Geotechnical Engineering*, Vol. 11, No. 3, pp. 445–463.

Hoek, E., Carranza-Torres, C.T., and Corkum, B. (2002). Hoek-Brown Failure Criterion - 2002 Edition. *NARMS-TAC 2002*, Toronto, Canada. University of Toronto Press, Vol. 1, pp. 267–273.

Hoek, E. and Diedrichs, M.S. (2006). Empirical Estimation of Rock Mass Modulus. *International Journal of Rock Mechanics and Mining Sciences*, Vol. 43, No. 2, pp. 203–215.

Hoek, E., Kaiser, P.K., and Bawden, W.F. (2000). *Support of Underground Excavations in Hard Rock*. A.A. Balkema, Rotterdam, The Netherlands.

Hoek, E., Wood, D., and Shah, S. (1992). A Modified Hoek-Brown Criterion for Jointed Rock Masses. *Proceedings Rock Characterization ISRM Symposium: Eurock '92*, Chester, UK. J.A. Hudson, Editor. British Geotechnical Society, London, pp. 209–213.

Holmgren, J. (1987). Bolt-Anchored, Steel-Fibre-Reinforced Shotcrete Linings. *Tunnelling and Underground Space Technology*, Vol. 2, No. 3, pp. 319–333.

Holt, R.M., Kjølaas, J., Li, L., Pilliteri, A.G., and Sønstebø, E.F. (2005). Comparison between Controlled Laboratory Experiments and Discrete Particle Simulations of the Mechanical Behaviour of Rock. *International Journal of Rock Mechanics and Mining Sciences*, Vol. 42, pp. 985–995.

Hsueh, C.H. (1988). Elastic Load Transfer from Partially Embedded Axially Loaded Fibre to Matrix. *Journal of Materials Science Letters*, Vol. 7, pp. 497–500.

Hsueh, C.H. (1990a). Interfacial Debonding and Fiber Pull-out Stresses of Fiber-reinforced Composites. *Materials Science and Engineering*, Vol. A123, pp. 1–11.

Hsueh, C.H. (1990b). Effects of Interfacial Bonding on Sliding Phenomena during Compressive Loading of an Embedded Fibre. *Journal of Materials Science*, Vol. 25, pp. 4080–4086.

Hsueh, C.H. (1990c). Fibre Pullout against Push-down for Fibre-reinforced Composites with Frictional Interfaces. *Journal of Materials Science*, Vol. 25, No. 2A, pp. 811–817.

Hsueh, C.H. (1990d). Interfacial Friction Analysis for Fibre-reinforced Composites during Fibre Push-down (Indentation). *Journal of Materials Science*, Vol. 25, No. 2A, pp. 18–828.

Hsueh, C.H., Young, R.J., Yang, X., and Becher, P.F. (1997). Stress transfer in a model composite containing a single embedded fiber. *Acta Materialia*, Vol. 45, No. 4, pp. 1469–1476.

Huder, J. and Amberg, G. (1970). Quellung im Mergel, Opalinuston und Anhydrit. *Scheizerische Bauzeitung*, Vol. 88, No. 43, pp. 975–980.

Hudoba, I. (1997). Contribution to Static Analysis of Load-bearing Concrete Tunnel Lining Built by Shield-driven Technology. *Tunnelling and Underground Space Technology*, Vol. 12, No. 1, pp. 55–58.

Hunt, R.E.B. and Askew, J.E. (1977). Installation and Design Guidelines for Cable Dowel Ground Support at ZC/NBHC. *Proceedings Underground Operators Conference*. Australian Institution of Mining and Metallurgy, Melbourne, Australia, pp. 113–122.

Huo, H. (2005). *Seismic Design and Analysis of Rectangular Underground Structures*. Ph. D. Thesis, Purdue University, West Lafayette, IN, USA.

Huo, H., Bobet, A., Fernández, G., and Ramírez, J. (2005). Load Transfer Mechanisms between Underground Structure and Surrounding Ground: Evaluation of the Failure of the Daikai Station. *ASCE Journal of Geotechnical and Geoenvironmental Engineering*, Vol. 131, No. 12, pp. 1522–1533.

Huo, H., Bobet, A., Fernández, G., and Ramírez, J. (2006). Analytical Solution for Deep Rectangular Structures Subjected to Far-Field Shear Stresses. *Tunnelling and Underground Space Technology*, Vol. 21, No. 6, pp. 613–625.

Hutchins, W.R., Bywater, S., Thompson, A.G., and Windsor, C.R. (1990). A Versatile Grouted Cable Dowel Reinforcing System for Rock. *Proceedings Australian Institute of Mining and Metallurgy Annual Conference*, Vol. 1, pp. 25–29.

Hwang, R.N. and Lysmer, J. (1981). Response of Buried Structures to Traveling Waves. *Journal of the Geotechnical Engineering Division*, ASCE, Vol. 107, No. GT2, pp. 183–200.

Hyett, A.J., Bawden, W.F., and Coulson, A.L. (1992a). Physical and mechanical properties of normal Portland cement pertaining to fully grouted cable bolts. *Rock Support in Mining and Underground Construction. Proceedings of the International Symposium on Rock Support*. P.K. Kaiser and D.R. McCreath, Editors. A.A. Balkema, Rotterdam, The Netherlands, pp. 341–348.

Hyett, A.J., Bawden, W.F., and Reichert, R.D. (1992b). The Effect of Rock Mass Confinement on the Bond Strength of Fully Grouted Cable Bolts. *International Journal of Rock Mechanics, Mining Sciences and Geomechanics Abstracts*, Vol. 29, No. 5, pp. 503–524.

Hyett, A.J., Bawden, W.F., Macsporran, G.R., and Moosavi, M. (1995). A Constitutive Law for Bond Failure of Fully-grouted Cable Bolts Using a Modified Hoek Cell. *International Journal of Rock Mechanics, Mining Sciences and Geomechanics Abstracts*, Vol. 32, No. 1, pp. 11–36.

Hyett, A.J., Moosavi, M., and Bawden, W.F. (1996). Load Distribution along Fully Grouted Bolts, with Emphasis on Cable Bolt Reinforcement. *International Journal for Numerical and Analytical Methods in Geomechanics*, Vol. 20, pp. 517–544.

Ichikawa, Y., Shimizu, H., and Kawanobe, M. (1982). Some dynamic behaviors of cavern walls during earthquakes. *Rock Mechanics: Caverns and Pressure Shafts, Proceedings of the International Society of Rock Mechanics Symposium*. W. Wittke, Editor. A.A. Balkema, Rotterdam, Holland, pp. 307–320.

Ieronymaki, E.S., Whittle, A.J., and Sureda, D.S. (2017). Interpretation of Free-Field Ground Movements Caused by Mechanized Tunnel Construction. *ASCE Journal of Geotechnical and Geoenvironmental Engineering*, Vol. 143, No. 4. doi: 10.1061/(ASCE)GT.1943-5606.0001632.

Iglesia, G.R. (1991). *Trapdoor Experiments on the Centrifuge: A Study of Arching in Geomaterials and Similitude in Geotechnical Models*. Ph.D. Thesis, Department of Civil Engineering, MIT, Cambridge, MA.

Iglesia, G.R., Einstein, H.H., and Whitman, R.V. (2011). Validation of Centrifuge Model Scaling for Soil Systems via Trapdoor Tests. *ASCE Journal of Geotechnical and Geoenvironmental Engineering*, Vol 137, No. 11, pp. 1075–1089

Iglesia, G.R., Einstein, H.H., and Whitman, R.V. (1999). Determination of vertical loading on underground structures based on arching evolution concept. *Geo-Engineering for Underground Facilities, Urbana-Champaign, IL*. G. Fernandez and R.A. Bauer, Editors. ASCE GSP No. 90, pp. 495–506.

Iglesia, G.R., Einstein, H.H., and Whitman, R.V. (2014). Investigation of Soil Arching with the Centrifuge Tests. *ASCE Journal of Geotechnical and Geoenvironmental Engineering*, Vol. 140, No. 2, pp. 040113005-1 to 13.

Iida, H., Hiroto, T., Yoshida, N., and Iwafuji, M. (1996). Damage to Daikai Subway Station. *Special Issue of Soils and Foundations*, JSCE. January, 1996. pp. 283–300.

Ikeda, K., Tanaka, T., and Hlguchi, I. (1966). *The Loosening of the Rock Around the Tunnel and Its Effect on the Steel Supports*. Quarterly Report Railway Technical Research Institute (Japan), Vol. 7, No. 4, pp. 26–30.

Indraratna, B. and Kaiser, P.K. (1990a). Design of Grouted Rock Bolts Based on the Convergence Control Method. *International Journal for Rock Mechanics, Mining Sciences and Geomechanics Abstracts*, Vol. 27, No. 4, pp. 269–281.

Indraratna, B. and Kaiser, P.K. (1990b). Analytical Model for the Design of Grouted Rock Bolts. *International Journal for Numerical and Analytical Methods in Geomechanics*, Vol. 14, pp. 227–251.

Ishida, T., Chigira, M., and Hibino, S. (1987). Application of the Distinct Element Method for Analysis of Toppling observed in a Fissured Rock Slope. *Rock Mechanics and Rock Engineering*, Vol. 20, pp. 277–283.

ISRM (1999). Suggested Methods for Laboratory Testing of Swelling Rocks. *International Journal of Rock Mechanics and Mining Sciences*, Vol. 36, pp. 291–306.

ISRM Commission on Swelling Rock (1983). *Characterization of Swelling Rock*. Pergamon Press, Oxford.

ISRM Commission on Swelling Rock (1989). Suggested Methods for Laboratory Testing of Argillaceous Rock. *International Journal of Rock Mechanics, Mining Sciences and Geomechanics Abstracts*, Vol. 26, pp. 415–426.

ISRM Commission on Swelling Rock (1994). *Comments and Recommendations on Design and Analysis Procedures for Structures in Argillaceous Swelling Rock*. Pergamon Press, Oxford.

ITA (2000). Guidelines for the Design of Shield Tunnel Lining. Working Group No. 2, International Tunnelling Association. *Tunnelling and Underground Space Technology*, Vol. 15, No. 3, pp. 303–331.

ITA/AITES Report 2006 (2007). Settlements Induced by Tunneling in Soft Ground. *Tunnelling and Underground Space Technology*, Vol. 22, pp. 119–149.

Jacobsen, S. (1974). Buckling of Circular Rings and Cylindrical Tubes under External Pressure. *Water Power and Dam Construction*, Vol. 26, pp. 400–407.

Janssen, H.A. (1895). Versuche über Getreidedruck in Silozellen. *Z.d. Vereins deutscher Ingenieure*, Vol. 39, pp. 1045 (partial English translation in Proceedings of the Institute of Civil Engineers, London, 1896, pp. 553).

Jasak, H. and Weller, H.G. (2000). Application of the Finite Volume Method and Unstructured Meshes to Linear Elasticity. *International Journal for Numerical Methods in Engineering*, Vol. 48, pp. 267–287.

Javadi, A.A. (2006). Estimation of Air Losses in Compressed Air Tunneling Using Neural Network. *Tunnelling and Underground Space Technology*, Vol. 21, pp. 9–20.

Javadi, A.A. and Snee, C.P.M. (2002). Numerical Modeling of Air Losses in Compressed Air Tunneling. *The International Journal of Geomechanics*, Vol. 2, No. 4, pp. 399–417.

Jayanthu, S., Singh, T.N., and Singh, D.P. (2004). Stress Distribution During Extraction of Pillars in a Thick Coal Seam. *Rock Mechanics and Rock Engineering*, Vol. 37, No. 3, pp. 171–192.

Jenny, R.J. (1983). Compressed Air Use in Soft Ground Tunneling. *ASCE Journal of Construction Engineering and Management*, Vol. 109, No. 2, pp. 206–213.

Jiang, Q.H. and Yeung, M.R. (2004). A Model of Point-to-Face Contact for Three-Dimensional Discontinuous Deformation Analysis. *Rock Mechanics and Rock Engineering*, Vol. 37, No. 2, pp. 95–116.

Jienan, P., Meng, Z., and Li, Y. (2005). Analysis of rock burst potential and its influencing factors based on composition & texture of sedimentary rock. *U.S. Rock Mechanics Symposium*, paper 05-684, 5 pp.

Jing, L. (1998). Formulation of Discontinuous Deformation Analysis (DDA) – An Implicit Discrete Element Model for Block Systems. *Engineering Geology*, Vol. 49, pp. 371–381.

Jing, L. (2003). A Review of Techniques, Advances and Outstanding Issues in Numerical Modeling for Rock Mechanics and Rock Engineering. *International Journal of Rock Mechanics and Mining Sciences*, Vol. 40, pp. 283–353.

Jing, L. and Hudson, J.A. (2002). Numerical Methods in Rock Mechanics. *International Journal of Rock Mechanics and Mining Sciences*, Vol. 39, pp. 409–427.

Jirovec, P. (1978). Wechselwirkung zwischen anker und gebirge. *Rock Mechanics*, Vol. 7, pp. 139–155.

Johannessen, G. and Askilsrud, O.G. (1993). Meraaker Hydro-Tunnelling the "Norwegian Way." *Proceedings RETC*, Boston, Massachusetts. L.D. Bowerman and J.E. Monsees, Editors. Society for Mining, Metallurgy, and Exploration, Littleton, CO, pp. 415–429.

John, M. (1976). Geotechnical Measurements in the Arlberg-Tunnel and their Consequences on Construction (in German), *Rock Mechanics*, Suppl. No. 5, Springer, New York, pp. 175–177.

John, M. (1978a). Personal Communication, Ground Classification for Dalaas Tunnel.

John, M. (1978b). Design of the Arlberg Expressway Tunnel and the Pfänder Tunnel. *Proceedings Shotcrete Conference*, St. Anton, Austria. 16 pp.

Jonak, J. and Gajewski, J. (2006). Identifying the Cutting Tool Type Used in Excavations Using Neural Networks. *Tunnelling and Underground Space Technology*, Vol. 21, pp. 185–189.

Jong, S.C., Ong, D.E.L., and Oh, E. (2021). State-of-the-art Review of Geotechnical-driven Artificial Intelligence Techniques in Underground Soil-structure Interaction. *Tunnelling and Underground Space Technology*, Vol. 113. doi: 10.1016/j.tust.2021.103946.

Joseph, P.G., Einstein, H.H., and Whitman, R.V. (1987). *A Literature Review of Geotechnical Centrifuge Modeling with Particular Emphasis on Rock Mechanics*. MIT Report to the Air Force Engineering and Services Center, Tyndall Air Force Base, Florida.

Joseph, P.G. and Einstein, H.H. (1987). *Rock Modelling Using the Centrifuge*. Report to AFESC, Contract DACA 88-86-D-0013.

Joughin, N.C. and Jager, A.J. (1983). Fracture of rock at stope in South African gold mines. *Rockbursts: Prediction and Control*. M. Board, C. Fairhurst, and E. Grimstadt, Editors. The Institution of Mining and Metallurgy, London, England, pp. 53–66.

Kabiesz, J. (2006). Effect of the Form of Data on the Quality of Mine Tremors Hazard Forecasting Using Neural Networks. *Geotechnical and Geological Engineering*, Vol. 24, pp. 1131–1147.

Kahraman, S., Altun, H., Tezekici, B.S., and Fener, M. (2006). Sawability Prediction of Carbonate Rocks from Shear Strength Parameters Using Artificial Neural Networks. *International Journal of Rock Mechanics and Mining Sciences*, Vol. 43, pp. 157–164.

Kaiser, P.K. and Cai, M. (2018). *Rockburst Support Volume 1 – Rockburst Phenomenon and Support Characteristics*. MIRARCO Mining Innovation, Laurentian University, Sudbury, Canada.

Kaiser, P.K., Diederichs, M., and Yazici, S. (1992). Cable bolt performance during mining induced stress change – Three case examples. *Rock Support in Mining and Underground Construction. Proceedings of the International Symposium on Rock Support*. P.K. Kaiser and D.R. McCreath, Editors. A.A. Balkema, Rotterdam, The Netherlands, pp. 377–384.

Kaiser, P.K., McCreath, D.R., and Tannant, D.D. (1996). *Canadian Rockburst Support Handbook*. Geomechanics Research Centre, Laurentian University, Sudbury, Canada.

Kamata, H. and Mashimo, H. (2003). Centrifuge model Test of Tunnel Face Reinforcement by Bolting. *Tunnelling and Underground Space Technology*, Vol. 18, pp. 205–212.

Karakus, M. (2007). Appraising the Methods Accounting for 3D Tunneling Effects in 2D Plane Strain FE Analysis. *Tunnelling and Underground Space Technology*, Vol. 22, pp. 47–56.

Karakus, M. and Fowell, R.J. (2003). Effects of Different Tunnel Face Advance Excavation on the Settlement by FEM. *Tunnelling and Underground Space Technology*, Vol. 18, pp. 513–523.

Karakus, M. and Fowell, R.J. (2005). Back Analysis for Tunneling Induced Ground Movements and Stress Redistribution. *Tunnelling and Underground Space Technology*, Vol. 20, pp. 514–524.

Karanam, U.M.R. and Dasyapu, S.K. (2005). Experimental and Numerical Investigations of Stresses in a Fully Grouted Rock Bolt. *Geotechnical and Geological Engineering*, Vol. 23, pp. 297–308.

Kasper, T. and Meschke, G. (2006). On the Influence of Face Pressure, Grouting Pressure and TBM Design in Soft Ground Tunneling. *Tunnelling and Underground Space Technology*, Vol. 21, pp. 160–171.

Kemeny, J. (2005). Time-dependent Drift Degradation Due to the Progressive Failure of Rock Bridges along Discontinuities. *International Journal of Rock Mechanics and Mining Sciences*, Vol. 42, pp. 35–46.

Kendorski, F.S., Cummings, R.A., Bieniawski, Z.T., and Skinner, E.H. (1983). Rock Mass Classification for Block Caving Mine Drift Support. *Proceedings of the 15th ISRM Congress*, Melbourne, Australia. A.A. Balkema, Rotterdam, The Netherlands, pp. B101–B113.

Kidybinski, A. (1981). Bursting Liability Indices of Coal. *International Journal of Rock Mechanics and Mining Sciences*, Vol. 18, pp. 295–304.

Kiehl, J.R. (1990). Ein drei dimensionales Quellgesetz und seine Anwendung auf den Felshohlraumbau. *Proceedings of the 9th Nationales Felsmechanik Sympium*, Aachen, Germany.

Kielbassa, S. and Duddeck, H. (1991). Stress-Strain Fields at the Tunnelling Face – Three-dimensional Analysis for Two-dimensional Technical Approach. *Rock Mechanics and Rock Engineering*, Vol. 24, pp. 115–132.

Kilic, A., Yasar, E., and Atis, C.D. (2003). Effect of Bar Shape on the Pull-out Capacity of Fully-grouted Rockbolts. *Tunnelling and Underground Space Technology*, Vol. 18, pp. 1–6.

Kilic, A., Yasar, E., and Celik, A.G. (2002). Effect of Grout Properties on the Pull-out Load Capacity of Fully Grouted Rock Bolt. *Tunnelling and Underground Space Technology*, Vol. 17, pp. 355–362.

Kim, Y.I., Amadei, B., and Pan, E. (1999). Modeling the Effect of Water, Excavation Sequence and Rock Reinforcement with Discontinuous Deformation Analysis. *International Journal of Rock Mechanics and Mining Sciences*, Vol. 36, pp. 949–970.

Kimura, T. and Mair, R.J. (1981). Centrifugal Testing of Model Tunnels in Soft Clay. *Proceedings of the 10th International Conference on Soil Mechanics and Foundation Engineering*, A.A. Balkema, Stockholm, Vol. 1, pp. 319–322.

Kirschke, D. (2010). Approaches to Technical Solutions for Tunneling in Swelling Ground. *Geomechanics and Tunnelling*, Vol. 3, No. 5, pp. 547–556.

Kirzhner, F. and Rosenhouse, G. (2000). Numerical Analysis of Tunnel Dynamic Response to Earth Motions. *Tunnelling and Underground Space Technology*, Vol. 15, No. 3, pp. 249–258.

Kitiyodom, P., Matsumoto, T., and Kawaguchi, K. (2005). A Simplified Analysis Method for Piled Raft Foundations Subjected to Ground Movements Induced by Tunneling. *International Journal for Numerical and Analytical Methods in Geomechanics*, Vol. 29, pp. 1485–1507.

Klar, A., Osman, A.S., and Bolton, M. (2007). 2D and 3D Upper Bound Solutions for Tunnel Excavation Using 'Elastic' Flow Fields. *International Journal for Numerical and Analytical Methods in Geomechanics*, Vol. 31, No. 12, pp. 1367–1374.

Klar, A., Vorster, T.E.B., Soga, K., and Mair, R.J. (2005). Soil-pipe Interaction Due to Tunneling: Comparison between Winkler and Elastic Continuum Solutions. *Géotechnique*, Vol. 55, No. 6, pp. 461–466.

Kleier, A.J. and Kleinschrodt, H.D. (2002). Discontinuous mechanical modeling of granular solids by means of PFC and LS-Dyna. *Numerical Modeling in Micromechanics via Particle Methods*. H. Konietzky, Editor. Balkema, Netherlands, pp. 37–43.

Kolymbas, D. (2005). *Tunnelling and Tunnel Mechanics. A Rational Approach to Tunnelling*. Springer-Verlag, Berlin Heidelberg, Germany.

Kommerell, O. (1940). *Statische Berechnung von Tunnelmauerwerk*. 2nd Edition, W. Ernst & Sohn, Berlin, 174 pp. This is the second edition of the book, originally published in 1912.

Konietzky, H. (2002). Numerical Modeling in Micromechanics via Particle Methods. *Proceedings of the 1st International PFC Symposium*, Gelsenkirchen, Germany, Balkema Publishers, The Netherlands.

Konietzky, H., te Kamp, L., and Bertrand, G. (2002). Modeling of cyclic fatigue under tension with PFC. *Numerical Modeling in Micromechanics via Particle Methods*. H. Konietzky, Editor. Balkema, Netherlands, pp. 37–43.

Koo, C.Y. and Chern, J.C. (1998). Modification of the DDA Method for Rigid Block Problems. *International Journal of Rock Mechanics and Mining Science and Geomechanics Abstracts*, Vol. 35, pp. 683–693.

Kooi, C.B. and Verruijt, A. (2001). Interaction of Circular Holes in an Infinite Elastic Medium. *Tunnelling and Underground Space Technology*, Vol. 16, pp. 59–62.

Kourepinis, D., Bicanic, N., and Pearce, C.J. (2003). A Higher-order Variational Numerical Manifold Method Formulation and Simplex Integration Strategy. *Development and Application of Discontinuum Modelling for Rock Engineering. Proceedings of the 6th International Conference on Analysis of Discontinuous Deformation*, M. Lu, Editor. A.A. Balkema Publishers, The Netherlands, pp. 145–151.

Kouretzis, G.P., Bouckovalas, G.D., and Gantes, C.J. (2006). 3-D Shell Analysis of Cylindrical Underground Structures Under Seismic Shear (S) Wave Action. *Soil Dynamics and Earthquake Engineering*, Vol. 26, pp. 909–921.

Kovári, K. (2003). History of the Sprayed Concrete Lining Method – Part I: Milestones up to the 1960s. *Tunnelling and Underground Space Technology*, Vol. 18, pp. 57–69.

Kramer, S.L. (1996). *Geotechnical Earthquake Engineering*. Prentice Hall, Upper Saddle River, NJ.

Kuesel, T.R. (1969). Earthquake Design Criteria for Subways. *ASCE Journal of the Structural Division*, Vol. 95, No. ST6, pp. 1213–1231

Kumar, P. (2000). Infinite Elements for Numerical Analysis of Underground Excavations. *Tunnelling and Underground Space Technology*, Vol. 15, No. 1, pp. 117–124.

Kuriyama, K., Mizuta, Y., Mozumi, H., and Watanabe, T. (1995). Three-dimensional Elastic Analysis by the Boundary Element Method with Analytical Integrations over Triangular Leaf Elements. *International Journal of Rock Mechanics and Mining Sciences*, Vol. 32, No. 1, pp. 77–83.

Kuwajima, F.M. (1999). Early Age Properties of Shotcrete. Shotcrete. *Shotcrete for Underground Support VIII. Proceedings of the Eighth International Conference.* T.B. Celestino and H.W. Parker, Editors. ASCE, Reston, VA, pp. 153–173.

Kveldsvik, V., Einstein, H.H.E., Nilsen, B., and Blikra, L.H. (2009). Numerical Analysis of the 650,000 m^2 Åknes Rock Slope based on Measured Displacements and Geotechnical Data. *Rock Mechanics and Rock Engineering*, Vol. 42, pp. 689–728.

Labiouse, V. (1992). The rock-support interaction analysis applied to ungrouted tensioned rock-bolts. *Rock Support in Mining and Underground Construction. Proceedings of the International Symposium on Rock Support.* P.K. Kaiser and D.R. McCreath, Editors. A.A. Balkema, Rotterdam, The Netherlands, pp. 75–82.

Labiouse, V. (1996). Ground Response Curves for Rock Excavations Supported by Ungrouted Tensioned Rockbolts. *Rock Mechanics and Rock Engineering*, Vol. 29, No. 1, pp. 19–38.

Labreche, D.A. (1983). Damage mechanisms in tunnels subjected to explosive loads. *Seismic Design of Embankments and Caverns.* T.R. Howard, Editor. ASCE, NewYork, pp. 128–141.

Lackner, R., Macht, J., Hellmich, C., and Mang, H.A. (2002). Hybrid Method for Analysis of Segmented Shotcrete Tunnel Linings. *Journal of Geotechnical and Geoenvironmental Engineering*, Vol. 128, No. 4, pp. 298–308.

Ladanyi, B. and Hoyaux, B. (1969). A Study of the Trap-Door Problem in a Granular Mass. *Canadian Geotechnical Journal*, Vol. 6, No. 1, pp. 1–15.

Ladd, C.C. and Foott, R. (1974). New Design Procedure for Stability of Soft Clays. *ASCE Journal of Geotechnical Engineering Division*, Vol 100, No. GT 7, pp 763–786.

Lahrmann, A. (1992). An Element Formulation for the Classical Finite Difference and Finite Volume Method Applied to Arbitrarily Shaped Domains. *International Journal for Numerical Methods in Engineering*, Vol. 35, pp. 893–913.

Laing, P.L. (1969). Private Communication to Monsees, J.E.

Lambe, T.W. (1960). *The Character and Identification of Expansive Soils – Soil PVC Meter*. Publication 701, Federal Housing Administration, Washington, DC.

Lambe, T.W. and Whitman, R.V. (1969). *Soil Mechanics*. John Wiley & Sons, Inc., New York, NY.

Lane, R.G.T. (1964). Rock Foundation Diagnosis of Mechanical Properties and Treatment. *Transactions 8th International Conference on Large Dams (ICOLD)*, Edinburgh, Scotland, pp. 141–165.

Lässer-Feizlmayr. (1978). Vorbereitung, Planung und Bauleitung für den Arlberg Strassentunnel. *Österreichische Ingenieurzeitschrift*, Vol. 21, No. 11, pp. 356–373.

Laubscher, D.H. (1977). Geomechanics Classification of Jointed Rock Masses – Mining Applications. *Transactions of the Institution of Mining and Metallurgy*, Vol. 93, pp. A70–A81.

Laubscher, D.H. and Taylor, H.W. (1976). The Importance of Geomechanics of Jointed Rock Masses in Mining Operations. *Proceedings of the Symposium on Exploration for Rock Engineering*, Johannesburg, South Africa. Z.T. Bieniawski, Editor. A.A. Balkema, Cape Town, South Africa, pp. 119–128.

Lauffer, H. (1958). Gebirgsklassifizierung für den Stollenbau. *Geologie und Bauwesen*, Vol. 24, No. 1, pp. 46–51.

Lauffer, H. (1960). Die neuere Entwicklung der Stollenbautechnik. *Österreichnische Ingenieurzeitschrift*, Vol. 3, No. 1, pp. 13–24.

Lauffer, H. (1988). Zur Gebirgsklassifizierung bei Fräsvortrieben. *Felsbsau*, Vol. 6, No. 3, pp. 137–149.

Law, H.K. and Lam, I.P. (1999). Seismic Performance of the Yerba Buena Island Tunnel. *Geo-Engineering for Underground Facilities. Geotechnical Special Publication No. 90.* G. Fernandez and R.A. Bauer, Editors. ASCE, Reston, VA, pp. 659–670.

Law, H.K. and Lam, I.P. (2003). Evaluation of Seismic Performance for Tunnel Retrofit Project. *Journal of Geotechnical and Geoenvironmental Engineering*, Vol. 129, No. 7, pp. 575–589.

Leca, E. and Dormieux, L. (1990). Upper and Lower Bound Solutions for the Face Stability of Shallow Circular Tunnels in Frictional Material. *Geotechnique*, Vol. 40, No. 4, pp. 581–606.

Lee, C.J., Wu, B.R., Chen, H.T., and Chiang, K.H. (2006a). Tunnel Stability and Arching Effects during Tunneling in Soft Clayey Soil. *Tunnelling and Underground Space Technology*, Vol. 21, pp. 119–132.

Lee, G.T.K. and Ng, C.W.W. (2003). Effects of Advancing Open Face Tunneling on an Existing Loaded Pile. *Journal of Geotechnical and Geoenvironmental Engineering*, Vol. 131, No. 2, pp. 193–201.

Lee, I.-M. and Nam, S.-W. (2001). The Study of Seepage Forces Acting on the Tunnel Lining and Tunnel Face in Shallow Tunnels. *Tunnelling and Underground Space Technology*, Vol. 16, pp. 31–40.

Lee, I.-M. and Nam, S.-W. (2004). Effect of Tunnel Advance Rate on Seepage Forces Acting on the Underwater Tunnel Face. *Tunnelling and Underground Space Technology*, Vol. 19, pp. 273–281.

Lee, I.-M., Lee, J.S., and Nam, S.-W. (2004a). Effect of Seepage Force on Tunnel Face Stability Reinforced with Multi-step Pipe Grouting. *Tunnelling and Underground Space Technology*, Vol. 19, pp. 551–565.

Lee, K.M. and Rowe, R.K. (1990). Finite Element Modelling of the Three-Dimensional Ground Deformations due to Tunnelling in Soft Cohesive Soils: Part 2 – Results. *Computers and Geotechnics*, Vol. 10, pp. 111–138.

Lee, K.M., Rowe, R.K., and Lo, K.Y. (1992). Subsidence Owing to Tunneling. I: Estimating the Gap Parameter. *Canadian Geotechnical Journal*, Vol. 29, pp. 929–940.

Lee, S.-W., Jung, J.-W., Nam, S.-W., and Lee, I.-M. (2006b). The Influence of Seepage Forces on Ground Reaction Curve of Circular Opening. *Tunnelling and Underground Space Technology*, Vol. 22, pp. 28–38.

Lee, S.M., Park, B.S., and Lee, S.W. (2004b). Analysis of Rockbursts That Have Occurred in a Waterway Tunnel in Korea. *International Journal of Rock Mechanics and Mining Sciences*, Vol. 41, No. 3, CD_ROM, 6 pp.

Lemos, J.V., Hart, R.D., and Cundall, P.A. (1985). A generalized distinct element program for modeling jointed rock mass. *Fundamentals of Rock Joints.* O. Stephansson, Editor. Centak Publishers, pp. 335–343.

Leontovich, V. (1959). *Frames and Arches: Condensed Solutions for Structural Analysis.* McGraw Hill Book Company, Inc., New York, NY.

Li, C. (2000). Analytical study of the behavior of rock bolts. *Pacific Rocks 2000.* J.M. Girard, M. Liebman, C. Breeds, and T. Doe, Editors. Balkema, The Netherlands.

Li, C. and Stillborg, B. (1999). Analytical Models for Rock Bolts. *International Journal of Rock Mechanics, Mining Sciences, and Geomechanics Abstracts*, Vol. 36, pp. 1013–1029.

Li, L. and Holt, R.M. (2002). Development of discrete particle modeling towards a numerical laboratory. *Numerical Modeling in Micromechanics via Particle Methods.* H. Konietzky, Editor. Balkema, Netherlands, pp. 19–27.

Li, S., Cheng, Y., and Wu, Y.-F. (2005a). Numerical Manifold Method Based on the Method of Weighted Residuals. *Computational Mechanics*, Vol. 35, pp. 470–480.

Li, S. and Wang, M. (2008). An Elastic Stress-displacement Solution for a Lined Tunnel at Great Depth. *International Journal of Rock Mechanics and Mining Sciences*, Vol. 45, No. 4, pp. 486–494.

Li, Z., Liu, H., Dai, R., and Su, X. (2005b). Application of Numerical Analysis Principles and Key Technology for High Fidelity Simulation to 3-D Physical Model Tests for Underground Caverns. *Tunnelling and Underground Space Technology*, Vol. 20, pp. 390–399.

Lin, C.T., Amadei, B., Jung, J., and Dwyer, J. (1996). Extensions of Discontinuous Deformation Analysis for Jointed Rock Masses. *International Journal of Rock Mechanics and Mining Sciences*, Vol. 33, No. 7, pp. 671–694.

Linder, R. (1963). Spritzbeton im Felshohlraumbau. *Bautechnik*, Vol. 40, No. 10, pp. 326–331, No. 11, pp. 383–388.

Link, H. (1964). Evaluation of Elasticity Moduli of Dam Foundation Rock Obtained Seismically in Comparison to those Arrived at Statically. *Proceedings 8th ICOLD Congress*, Edinburgh, Scotland. Vol. 1, Patis, pp. 853–858.

Littlejohn, G.S. (1992). Keynote lecture: Rock anchorage practice in civil engineering. *Rock Support in Mining and underground construction. Proceedings of the International Symposium on Rock Support.* P.K. Kaiser and D.R. McCreath, Editors. A.A. Balkema, Rotterdam, The Netherlands, pp. 257–268.

Littlejohn, S. (1993). Overview of rock anchorages. *Comprehensive Rock Engineering. Principles, Practice & Projects.* J.A. Hudson, Editor. Pergamon Press, Oxford, UK, Vol. 4, pp. 413–450.

Liu, W.K., Jun, S., and Zhang, Y.F. (1995). Reproducing Kernel Particle Methods. *International Journal of Numerical Methods in Engineering*, Vol. 20, pp. 1081–1106.

Lo, K.Y., Ng, R.M.C., and Rowe, R.K. (1984). Predicting settlement due to tunneling in clays. *Tunneling in Soil and Rock, Proceedings of two sessions at GEOTECH '84*, K.Y. Lo, Editor. American Society of Civil Engineers, pp. 48–76.

Loganathan, N. and Poulos, H.G. (1998). Analytical Prediction for Tunneling-induced Ground Movements in Clays. *Journal of Geotechnical and Geoenvironmental Engineering*, Vol. 124, No. 9, pp. 846–856.

Loganathan, N., Poulos, H.G., and Stewart, D.P. (2000). Centrifuge Model Testing of Tunneling-induced Ground and Pile Deformations. *Géotechnique*, Vol. 50, No. 3, pp. 283–294.

Lombardi, G. (1973). Dimensioning of Tunnel Rings with Regard to Construction Procedure. *Tunnels and Tunneling*, Vol. 5, No. 4, pp. 340–353

Lombardi, G. (1981). Bau von Tunneln bei grossen Verformungen des Gebirges/ Tunnel Construction at Great Rock Deformations. *Proceedings of the Congress on Tunnelling, Tunnel 1981*, pp. 353–383.

Lombardi, G. (1984). Underground Openings in Swelling Rock. *Proceedings of the 1st International Conference on Case Histories in Geotechnical Engineering*, Lahore, Pakistan.

López, C. (1997). *Manual de Túneles y Obras Subterráneas.* Entorno Gráfico, S.L., Madrid, Spain.

Løset, F. and Bhasin, R. (1991). Engineering Geology-Gjøvik Ice Hockey Cavern. *Research Project*: "*Publikumshall I Berg,*" SINTEF-NGI-Østlandsforskning, Norway.

Lu, M. (2003). Development and Application of Discontinuous Modelling for Rock Engineering. *Proceedings of the 6th International Conference on Analysis of Discontinuous Deformation.* A.A. Balkema, The Netherlands.

Lucy, L.N. (1977). A Numerical Approach to the Testing of the Fission Hypothesis. *The Astronomical Journal*, Vol. 8, No. 12, pp. 1013–1024.

Ludvig, B. (1983). Shear tests on rock bolts. *Rock Bolting. Theory and Application in Mining and Underground Construction. Proceedings of the International Symposium on Rock Bolting.* O. Stephansson, Editor. A.A. Balkema, Rottterdam, The Netherlands, pp. 113–123.

Lueke, J.S. and Ariaratnam, S.T. (2006). Numerical Characterization of Surface Heave Associated with Horizontal Directional Drilling. *Tunnelling and Underground Space Technology*, Vol. 21, pp. 106–117.

Luo, S., Zhang, X., and Cai, Y. (2001). The Variational Principle and Application of Numerical Manifold Method. *Applied Mathematics and Mechanics*, Vol. 22, No. 6, pp. 658–663.

Mack, M.G. (1993). The displacement discontinuity method. *Boundary Element Techniques in Geomechanics.* G.D. Manolis and T.G. Davies, Editors. Elsevier Applied Science, New York, NY.

Macklin, S.R. (1999). The Prediction of Volume Loss Due to Tunneling in Overconsolidated Clay Based on Heading Geometry and Stability Number. *Ground Engineering*, Vol. 32, No. 4, pp. 30–33.

MacLaughlin, M.M. and Berger, E.A. (2003). A Decade of DDA Validation. *Development and Application of Discontinuous Modelling for Rock Engineering. Proceedings of the 6th International Conference on Analysis of Discontinuous Deformation*, M. Lu, Editor. A.A. Balkema, The Netherlands, pp. 13–31.

MacLaughlin, M.M. and Clapp, K.K. (2002). Discrete Element Analysis of an Underground Opening in Blocky Rock: An Investigation of the Differences between UDEC and DDA Results. *Discrete Element Methods: Numerical Modeling of Discontinua*, B.K. Cook and R.P. Jensen, Editors. Geotechnical Special Publication No. 117, ASCE.

MacLaughlin, M.M. and Doolin, D.M. (2006). Review of Validation of the Discontinuous Deformation Analysis (DDA) Method. *International Journal of Numerical and Analytical Methods in Geomechanics*, Vol. 30, pp. 271–305.

MacLaughlin, M.M., Brady, T.M., and Pakalnis, R. (2006). Use of Distinct Element Models to Investigate Collapse and Support of an Underground Opening as a Function of varying Rock Mass Quality. *41st U.S. Rock Mechanics Symposium*, Golden, CO. Paper 06-967, 12 pages.

MacLaughlin, M.M., Sitar, N., Doolin, D.M., and Abbot, T. (2001). Investigation of Slope-stability Kinematics Using Discontinuous Deformation Analysis. *International Journal of Rock mechanics and Mining Sciences*, Vol. 38, pp. 753–762.

Madsen, F.T. and Nüesch, R. (1990). Langzeitquellverhalten von Tongesteinen und tonigen Sulfatgesteinen. Beiträge zur Geologic der Schweiz. *Teotechnische Serie*, 85 pp.

Madsen, F.T. and Müller-Vonmoos, M. (1985). Swelling Pressure Calculated from Mineralogical Properties of a Jurassic Opalinum Shale, Switzerland. *Clays and Clay Mineralogy*, Vol. 6, pp. 501–509.

Madsen, F.T. and Müller-Vonmoos, M. (1989). The Swelling Behavior of Clays. *Applied Clay Science*, Vol. 4, pp. 143–156.

Madsen, F.T. and Nüesch, R. (1991). The Swelling Behavior of Clay-Sulfate Rocks. *Proceedings of the 7th International Congress of the International Society for Rock Mechanics*. W. Wittke, Editor. Vol. 1, pp. 285–288. Aachen, Germany.

Maidl, B., Herrenknecht, M., and Anheuser, L. (1996). *Mechanised Shield Tunnelling*. Ernst & Sohn, Berlin, Germany.

Mair, R.J. (1979). *Centrifugal Modelling of Tunnel Construction in Soft Clay*. Ph.D. Thesis, Cambridge University Engineering Department, Cambridge, UK.

Mair, R.J. and Taylor, R.N. (1997). Bored Tunnelling in the Urban Environment. *Proceeding of the 14th International Conference on Soil Mechanics and Foundation Engineering*. A.A. Balkema, Rotterdam, Hamburg, Germany. pp. 2353–2385.

Mair, R.J., Gunn, M.J., and O'Reilly, M.P. (1981). Ground Movements around Shallow Tunnels in Soft Clay. *Proceedings of the 10th International Conference on Soil Mechanics and Foundation Engineering*, A.A. Balkema, Stockholm, Vol. 1, pp. 323–328.

Mair, R.J., Taylor, R.N., and Bracegirdle, A. (1993). Subsurface Settlement Profiles above Tunnels in Clays. *Géotechnique*, Vol. 43, No. 2, pp. 315–320.

Malmgren, L. and Nordlund, E. (2008). Interaction of Shotcrete with Rock and Rock Bolts – A Numerical Study. *International Journal of Rock Mechanics and Mining Sciences*, Vol. 45, No. 4, pp. 538–553.

Maloney, S., Fearon, R., Nose, J., and Kaiser, P.K. (1992). Investigations into the effect of stress change on support capacity. *Rock Support in Mining and Underground Construction. Proceedings of the International Symposium on Rock Support*. P.K. Kaiser and D.R. McCreath, Editors. A.A. Balkema, Rotterdam, The Netherlands, pp. 367–376.

Manolis, G.D., Davies, T.G., and Beskos, D.E. (1993). Overview of boundary element techniques in geomechanics. *Boundary Element Techniques in Geomechanics*. G.D. Manolis and T.G. Davies, Editors. Elsevier Applied Science, London, pp. 1–35.

Mansur, W.J. and Brebbia, C.A. (1983). Transient elastodynamics using a time-stepping technique. *Boundary Elements*. C.A. Brebbia, T. Futagami, and M. Tanaka, Editors. Springer-Verlag, Berlin, pp. 677–698.

Mansur, W.J. and Brebbia, C.A. (1985). Transient Elastodynamics. *Topics in Boundary Element Research*, Vol. 2, C.A. Brebbia, Editor. Springer-Verlag, Berlin, pp. 124–155.

Marinos, P. and Hoek, E. (2000). GSI – A Geologically Friendly Tool for Rock Mass Strength. *Proceedings GeoEng 2000: An International Conference on Geotechnical and Geological Engineering*, Melbourne, Australia. Technomic Publishing Co., Lancaster, PA, pp. 1422–1440.

Marinos, P. and Hoek, E. (2001). Estimating the Geotechnical Properties of Heterogeneous Rock Masses such as Flysch. *Bulletin of Engineering Geology and the Environment*, Vol. 60, No. 2, pp. 85–92.

Marinos, V. and Carter, T.G. (2018). Maintaining Geological Reality in Application of GSI for Design of Engineering Structures in Rock. *Engineering Geology*, Vol. 239, pp. 282–297.

Marston, A. and Anderson, A.O. (1913). *The Theory of Loads on Pipes in Ditches and Tests of Cement and Clay Drain Tile and Sewer Pipe*. Bulletin 31, Iowa Engineering Experiment Station, Iowa.

Martin, C.D. (1997). Seventeenth Canadian Geotechnical Colloquium: The Effect of Cohesion Loss and Stress Path on Brittle Rock Strength. *Canadian Geotechnical Journal*, Vol. 34, pp. 698–725.

Martin, C.D., Kaiser, P.K., and McCreath, D.R. (1999). Hoek-Brown Parameters for Predicting the Depth of Brittle Failure around Tunnels. *Canadian Geotechnical Journal*, Vol. 36, pp. 136–151.

Martin, C.D., Read, R.S., and Martino, J.B. (1997). Observations of Brittle Failure Around a Circular Test Tunnel. *International Journal of Rock Mechanics and Mining Sciences*, Vol. 34, No. 7, pp. 1065–1073.

Marulanda, A. and Gutierrez, R. (1999). Experience with Steel Seal Membranes for Liners in Pressure Shafts and Tunnels. *Geo-Engineering for Underground Facilities*. Geotechnical Special Publication No. 90. G. Fernandez and R.A. Bauer, Editors. ASCE, Reston, VA, pp. 634–646.

Mason, R.M. and Lorig, L. (1981). *Conventional Shotcrete*. In: The Atlanta Research Chamber, Applied Research Report and Monographs, Chapter VII. U.S. Department of Transportation, Washington, DC.

Matthews, S.M., Thompson, A.G., Windsor, C.R., and O'Bryan, P.R. (1986). A novel reinforcement system for large rock caverns in blocky rock masses. *Large Rock Caverns*. K.H.O. Saari, Editor. Pergamon Press, Oxford, England, Vol. 2, pp. 1541–1552.

Matthews, S.M., Tillman, V.H., and Worotnicki (1983). A Modified Cablebolt System for Support of Underground Openings. *Proceedings Australian Institute of Mining and Metallurgy Annual Conference*, Broken Hill, pp. 243–255.

Mauldon, M., Chou, K.C., and Wu, Y. (1997). Limit Analysis of 2-D Tunnel Keyblocks. *International Journal of Rock Mechanics and Mining Sciences*, Vol. 34, No. 3–4, paper No. 193.

Maynar, M.J.M. and Rodríguez, L.E.M. (2003). Discrete Numerical Model for Analysis of Earth Pressure Balance Tunnel Excavation. *Journal of Geotechnical and Geoenvironmental Engineering*, ASCE, Vol. Vol. 131, No. 10, pp. 1234–1242.

McCulloch, W. and Pitts, W. (1943). A Logical Calculus of the Ideas Immanent in Nervous Activity. *Bulletin of Mathematical Biophysics*, Vol. 5, 115–133.

McGarr, A. and Wiebols (1977). Influence of Mine Geometry and Closure Volume on Seismicity in a Deep-Level Mine. *International Journal of Rock Mechanics and Mining Sciences*, Vol. 14, pp. 139–145.

Meguid, M., Saada, O., Nines, M., and Mallar, J. (2008). Physical Modeling of Tunnels in Soft Ground: A Review. *Tunneling and Underground Space Technology*, Vol. 23, No. 5, pp. 258–267

Melbye, T.A. (1999). Keynote lecture: International practices and trends in sprayed concrete. *Rock Support and Reinforcement Practice in Mining, Proceedings of the International Symposium on Ground Support*. E. Villaescusa, C.R. Windsor, and A.G. Thompson, Editors. A.A. Balkema, Rotterdam, pp. 163–182.

Melbye, T.A. and Dimmock, R.H. (2001). Modern Advances and Applications of Sprayed Concrete. *Shotcrete: Engineering Developments. Proceedings of the International Conference on Engineering Developments in Shotcrete*. E. Stefan Bernard, Editor. A.A. Balkema, pp. 7–29.

Merritt, A.H. (1972). Geologic Predictions for Underground Excavations. *Proceedings 1st RETC*, Chicago, IL. AIME, New York, pp. 115–123.

Merritt, A.H. (1999). Geologic and Geotechnical Considerations for Pressure Tunnel Design. *Geo-Engineering for Underground Facilities*. Geotechnical Special Publication No. 90. G. Fernandez and R.A. Bauer, Editors. ASCE, Reston, VA, pp. 66–81.

Merritt, J.L., Monsees, J.E., and Hendron, Jr. A.J. (1985). Seismic Design of Underground Structures. *Proceedings of the Rapid Excavation and Tunneling Conference (RETC)*. C.D. Mann and M.N. Kelley, Editors. Society of Mining Engineers of the American Institute of Mining, Metallurgical, and Petroleum Engineers, New York, NY, pp. 104–131.

Meulenkamp, F. and Alvarez Grima, M. (1999). Application of Neural Networks for the Prediction of the Unconfined Compressive Strength (UCS) from Equotip Hardness. *International Journal of Rock Mechanics and Mining Sciences*, Vol. 36, pp. 29–39.

Milev, A.M., Sottiswoode, S.M., Hildyard, M.W., Rorke, A.J., and Finnie, G.J. (2000). Simulated rockburst – Source, design, seismic effect and damage. *Proceedings of the 4th North American Rock Mechanics Symposium, NARMS 2000*, J. Girard, M. Liebman, C. Breeds, and T. Doe, Editors. A.A. Balkema, Rotterdam, The Netherlands, pp. 327–334.

Mindlin, R.D. (1940). Stress Distribution around a Tunnel. *Transactions of the ASCE*, Vol. 105, No. 1, pp. 1117–1153.

Mindlin, R.D. (1948). Stress Distribution around a Hole Near the Edge of a Plate under Tension. *Proceedings of the Society of Experimental Stress Analysis*, Vol. 5, pp. 56–57.

Ming, L. and Brown, E.T. (1988). Designing Unlined Pressure Tunnels in Jointed Rock. *Water Power and Dam Construction*, November 1988, pp. 37–41.

Mitchell, R.J. (1983). *Earth Structure Engineering*. Allen and Unwin, Boston, MA.

Mohajerani, M., Delage, P., Monfared, M., Tang, A.M., Sulem, J., and Gatmiri, B. (2011). Oedometric Compression and Swelling Behaviour of the Callovo-Oxfordian Argillite. *International Journal of Rock Mechanics and Mining Sciences*, Vol. 48, pp. 606–615.

Mohammadi, S. (2003). *Discontinuum Mechanics: Using Finite and Discrete Elements*. WIT Press, Southhampton, UK.

Mohan, G.M., Sheorey, P.R., and Kushwaha, A. (2001). Numerical Estimation of Pillar Strength in Coal Mines. *International Journal of Rock Mechanics and Mining Sciences*, Vol. 38, pp. 1185–1192.

Möller, S. (2006). *Tunnel Induced Settlements and Structural Forces in Linings*. Ph.D. Thesis, Universität Stuttgart, Germany.

Mollon, G., Dias, D., and Soubra, A. (2010). Face Stability Analysis of Circular Tunnels Driven by a Pressurized Shield. *Journal of Geotechnical and Geoenvironmental Engineering*, Vol. 136, No. 1, pp. 215–229.

Monsees, J.E. (1970). *Design of Support Systems for Tunnels in Rock*. Ph.D. Thesis, University of Illinois, 253 pp.

Monsees, J.E. and Merritt, J.L. (1988). Seismic Modeling and Design of Underground Structures. *Numerical Methods in Geomechanics, Innsbruck 1988. Proceedings of the Sixth International Conference on Numerical Methods in Geomechanics*. G. Swoboda, Editor. Balkema, Rotterdam, Holland, pp. 1833–1842.

Moore, E.T. (1989). Design of Steel Tunnel Liners. *Proceedings of the International Conference on Hydropower, Hydropower '89*. A.J. Eberhardt, Editor. ASCE, New York, NY, pp. 384–394.

Moosavi, M. and Grayeli, R. (2006). A Model for Cable Bolt-rock Mass Interaction: Integration with Discontinuous Deformation Analysis (DDA) Algorithm. *International Journal of Rock Mechanics and Mining Sciences*, Vol. 43, pp. 661–670.

Morgan, D.R., Heere, R., McAskill, N., and Chan, C. (1999). Comparative Evaluation of System Ductility of Mesh and Fibre Reinforced Shotcrete. *Shotcrete for Underground Support VIII. Proceedings of the Eighth International Conference*. T.B. Celestino and H.W. Parker, Editors. ASCE, Reston, VA, pp. 216–239.

Morris, J.P. and Block, G.I. (2006). Simulations of Underground Structures Subjected to Dynamic Loading using Combined FEM/DEM/SPH Analysis. *41st U.S. Rock Mechanics Symposium*, Golden, CO. Paper 06-1078, 10 pages.

Morris, J.P., Glenn, L.A., and Blair, S.C. (2002a). The Distinct Element Method – Application to Structures in Jointed Rock. *Lecture Notes in Computational Science and Engineering*, Vol. 26, pp. 291–306.

Morris, J.P., Glenn, L.A., Heuze, F.E., and Blair, S.C. (2002b). Simulations of Underground Structures Subjected to Dynamic Loading using the Distinct Element Method. *Discrete Element Methods: Numerical Modeling of Discontinua*. Geotechnical Special Publication No. 117, ASCE, pp. 392–396.

Morris, J.P., Rubin, M.B., Blair, S.C., Glenn, L.A., and Heuze, F.E. (2004). Simulations of Underground Structures Subjected to Dynamic Loading Using the Distinct Element Method. *Engineering Computations*, Vol. 21, No. 2/3/4, pp. 384–408.

Motok, M.D. (1997). Stress Concentration on the Contour of a Plate Opening of an Arbitrary Corner Radius of Curvature. *Marine Structures*, Vol. 10, No. 1, pp. 1–12.

Mow, C.C. and Pao, Y.H. (1971). *The Diffraction of Elastic Waves and Dynamic Stress Concentrations*. Document No. R-482-PR. The Rand Corporation, Santa Monica, CA.

Muir Wood, A.M. (1975). The Circular Tunnel in Elastic Ground. *Géotechnique*, Vol. 25, No. 1, pp. 115–127.

Muskhelishvili, N.I. (1954). *Some Basic Problems of the Mathematical Theory of Elasticity*. Noordhoof Ltd, Groningen, the Netherlands.

Myrvang, A.M. and Grimstad, E. (1984). Coping with the Problem of Rockbursts in Hard Rock Tunneling. *Tunnels and Tunnelling*, Vol. 16, No. 7, pp. 13–15.

Nairn, J.A. (1997). On the Use of Shear-lag Methods for Analysis of Stress Transfer in Unidirectional Composites. *Mechanics of Materials*, Vol. 26, pp. 63–80.

Nairn, J.A. and Mendels, D.A. (2001). On the Use of Planar Shear-lag Methods for Stress-transfer Analysis of Multilayered Composites. *Mechanics of Materials*, Vol. 33, pp. 335–362.

Nam, S. and Bobet, A. (2006). Liner Stresses in Deep Tunnels below the Water Table. *Tunnelling and Underground Space Technology*, Vol. 21, No. 6, pp. 626–635.

Nam, S. and Bobet, A. (2007). Radial Deformations Induced by Groundwater Flow on Deep Circular Tunnels. *Rock Mechanics and Rock Engineering*, Vol. 40, No. 1, pp. 23–39.

Nayroles, B., Touzot, G., and Villon, P. (1992). Generalizing the Finite Element Method: Diffuse Approximation and Diffuse Elements. *Computational Mechanics*, Vol. 10, pp. 307–318.

Neaupane, K.M. and Adhikari, N.R. (2006). Prediction of Tunneling-Induced Ground Movement with the Multi-layer Perceptron. *Tunnelling and Underground Space Technology*, Vol. 21, pp. 151–159.

New, B.M. and Bowers, K.H. (1993). Ground Movement Model Validation at the Heathrow Express Trial Tunnel. *Tunnelling*, 94, IMM London, pp. 301–329.

New, B.M. and O'Reilly, M.P. (1991). Tunnelling Induced Ground Movements: Predicting Their Magnitude and Effects. *Ground Movements and Structures: Proceedings of the 4th International Conference*. J.D. Geddes, Editor. Pentech Press, London, pp. 671–697.

Ng, R.M.C. (1991). A Procedure for Prediction of Settlement Due to Tunnels in Clays. *Proceedings of the 9th Pan American Conference on Soil Mechanics and Foundation Engineering*, Vol. 3, pp. 1413–1430.

Nguyen-Minh, D. and Guo, C. (1993). Sur un principe d'interaction massif-soutenement des tunnels en avancement stationnaire. *Eurock '93*. L.M. Ribeiro e Sousa and N.F. Grosmann, Editors. Balkema, Rotterdam, pp. 171–177.

Nguyen-Minh, D. and Guo, C. (1996). Recent progress in convergence confinement method. *Eurock '96*. G. Barla, Editor. Balkema, Rotterdam, pp. 855–860.

Nickson, S.D. (1992). *Cable Support Guidelines for Underground Hard Rock Mine Operations*. M.S. Thesis, Department of Mining and Mineral Processing, University of British Columbia, Vancouver, Canada.

Nishikawa, K. (2003). Development of a Prestresses and Precast Concrete Segmental Lining. *Tunnelling and Underground Space Technology*, Vol. 18, pp. 243–251.

Nomikos, P.P., Yiouta-Mitra, P.V., and Sofianos, A.I. (2006). Stability of Asymmetric Roof Wedge Under Non-Symmetric Loading. *Rock Mechanics and Rock Engineering*, Vol. 39, No. 2, pp. 121–129.

Nüesch, R. and Ko, S.C. (2000). Influence of mineralogical composition to experimental swelling behaviour of shaly anhydrite rocks. *Applied Mineralogy*. D. Rammlmair, J. Mederer, T.H. Oberthür, R.B. Heimann, and H. Pentinghaus, Editors. Balkema, Rotterdam, Vol. 2, pp. 611–617.

O'Sullivan, C., Bray, J.D., and Cui, L. (2006). Experimental Validation of Particle-Based Discrete Element Methods. *GeoCongress 06*. CD ROM Proceedings, 18 pp.

O'Sullivan, C., Bray, J.D., and Riemer, M.F. (2002). The Influence of Particle Shape and Surface Friction Variability on Macroscopic Frictional Strength of Rod-Shaped Particulate Media. *Journal of Engineering Mechanics, ASCE*, Vol. 128, No. 11. pp 1182–1192.

O'Sullivan, C., Bray, J.D., and Riemer, M.F. (2004). An Examination of the Response of Regularly Packed Specimens of Spherical Particles Using Physical Tests and Discrete Element Simulations. *Journal of Engineering Mechanics, ASCE*, Vol. 130, No. 10, pp 1140–1150.

O'Connor, K.M. and Dowding, C.H. (1992). Hybrid Discrete Element Code for Simulation of Mining-Induced Strata Movements. *Engineering Computations*, Vol. 9, pp. 235–242.

O'Reilly, M.P. and New, B.M. (1982). Settlements above tunnels in the United Kingdom – their magnitude and prediction. *Tunnelling '82*. The Institution of Mining and Metallurgy, London, pp. 173–181.

Obert, L. and Duvall, W.I. (1967). *Rock Mechanics and the Design of Structures in Rock*. John Wiley & Sons, New York, NY.

Ochiai, S., Hojo, M., and Inoue, T. (1999). Shear-lag Simulation of the Progress of Interfacial Debonding in Unidirectional Composites. *Composites Science and Technology*, Vol. 59, pp. 77–88.

Olson, J., Narayanasamy, R., Holder, J., Rauch, A., and Comacho, B. (2002). DEM Study of Wave Propagation in Weak Sandstone. *Discrete Element Methods: Numerical Modeling of Discontinua.* Geotechnical Special Publication No. 117, ASCE, pp. 335–339.

Olson, J.E. and Pollard, D.D. (1988). Inferring stress states from detailed joint geometry. *Proceedings: 29th US Symposium on Rock Mechanics.* A.A. Balkema, Editor. Rotterdam, Minneapolis, MN, pp. 159–167.

Olson, J.E. and Pollard, D.D. (1989). Inferring Paleostresses from Natural Fracture Patterns: A New Method. *Geology*, Vol. 17, pp. 345–348.

Olson, J.E. and Pollard, D.D. (1991). The initiation of en échelon veins. *Journal of Structural Geology*, Vol. 13, No. 5, pp. 595–608.

Oñate, E., Cervera, M., and Zienkiewicz, O.C. (1994). A Finite Volume Formulation for Structural Mechanics. *International Journal for Numerical Methods in Engineering*, Vol. 37, pp. 181–201.

Oñate, E., Idelsohn, S., Zienkiewicz, O.C., and Taylor, R.L. (1996a). Finite Point Method in Computational Mechanics. Applications to Convective Transport and Fluid Flow. *International Journal for Numerical Methods in Engineering*, Vol. 39, pp. 3839–3866.

Oñate, E., Idelsohn, S., Zienkiewicz, O.C., Taylor, R.L., and Sacco, C. (1996b). Stabilized Finite Point Method for Analysis of Fluid Mechanics Problems. *Computer Methods in Applied Mechanics and Engineering*, Vol. 139, pp. 3–47.

Onodera, T.F. and Asoka-Kumara, H.M. (1980). Relation between Texture and Material Properties of Crystalline Rocks. *Bulletin of the International Association of Engineering Geology*, Vol. 22, pp. 173–177.

Oreste, P.P. (2003a). A Procedure for Determining the Reaction Curve of Shotcrete Lining Considering Transient Conditions. *Rock Mechanics and Rock Engineering*, Vol. 36, No. 3, pp. 209–236.

Oreste, P.P. (2003b). Analysis of Structural Interaction in Tunnels Using the Convergence-Confinement Approach. *Tunnelling and Underground Space Technology*, Vol. 18, pp. 347–363.

Oreste, P.P. and Peila, D. (1996). Radial Passive Rockbolting in Tunnelling Design with a New Convergence-confinement Model. *International Journal for Rock Mechanics, Mining Sciences and Geomechanics Abstracts*, Vol. 33, No. 5, pp. 443–454.

Oreste, P.P. and Peila, D. (1997). Modelling Progressive Hardening of Shotcrete in Convergence-Confinement Approach to Tunnel Design. *Tunnelling and Underground Space Technology*, Vol. 12, No. 3, pp. 425–431.

Ortlepp, W.D. (1983). Considerations in the Design of Support for Deep Hard-Rock Tunnels. *Proceedings of the 5th Congress of the International Society for Rock Mechanics*, Vol. 2, pp. D180-D187, A.A. Balkema, Rotterdam, The Netherlands.

Ortlepp, W.D. (1984). Rockbursts in South African Mines: A Phenomenological View. *Proceedings of the 1st International Congress on Rockbursts and Seismicity in Mines, The South African Institute of Mining and Metallurgy Symposium Series, No. 6.* N.C. Gay and E.H. Wainwright, Editors. The South African Institute of Mining and Metallurgy, Johannesburg, South Africa, pp. 165–178.

Ortlepp, W.D. (2000). Observation of Mining-Induced Faults in an Intact Rock Mass at Depth. *International Journal of Rock Mechanics and Mining Sciences*, Vol. 37, pp. 423–436.

Ortlepp, W.D. and Stacey, T.R. (1994). Rockburst Mechanisms in Tunnels and Shafts. *Tunnelling and Underground Space Technology*, Vol. 9, No. 1, pp. 59–65.

Osgoui, R.R. and Oreste, P. (2007). Convergence-Control Approach for Rock Tunnels Reinforced by Grouted Bolts, Using the Homogeneization Concept. *Geotechnical and Geological Engineering*, Vol. 25, pp. 431–440.

Öttl, G., Stark, R.F., Stelzer, R., and Hofstetter, G. (2003). A coupled FE-model for tunneling by means of compressed air. *Numerical Simulation in Tunnelling.* G. Beer, Editor. Springer-Verlag, New York, NY, pp. 303–351.

Owen, G.N. and Scholl, R.E. (1981). *Earthquake Engineering of Large Underground Structures.* Report No. FHWA/RD-80/195. Federal Highway Administration, U.S. Department of Transportation.

Owen, G.N. and Scholl, R.E. (1985). Earthquake Engineering of Tunnels – Revisited. *Proceedings of the Rapid Excavation and Tunneling Conference (RETC).* C.D. Mann and M.N. Kelley, Editors. Society of Mining Engineers of the American Institute of Mining, Metallurgical, and Petroleum Engineers, New York, NY, pp. 63–70.

Owen, G.N., Scholl, R.E., and Brekke, T.L. (1979). Earthquake Engineering of Tunnels. *Proceedings of the Rapid Excavation and Tunneling Conference (RETC)*. A.C. Maevis and W.A. Hustrulid, Editors. Society of Mining Engineers of the American Institute of Mining, Metallurgical, and Petroleum Engineers, New York, NY, pp. 709–721.

Pacher, F., Rabcewicz, L.V., and Golser, J. (1974). Zum derzeitigen Stand der Gebirgsklassifizierung im Stollen – und Tunnelbau. *Strassenforschung*, Vol. 18, pp. 51–58.

Pakbaz, M.C. and Yareevand, A. (2005). 2-D Analysis of Circular Tunnel against Earthquake Loading. *Tunnelling and Underground Space Technology*, Vol. 20, pp. 411–417.

Palmer, J.H. and Belshaw, D.J. (1980). Deformations and Pore Pressures in the Vicinity of a Precast, Segmented, Concrete-Lined Tunnel in Clay. *Canadian Geotechnical Journal*, Vol. 17, pp. 174–184.

Palmström, A. (1974). *Characterization of Jointing Density and the Quality of Rock Masses*. Internal Report, A.B. Berdal, Norway, 26 pp.

Palmström, A. (2005). Measurements and Correlations between Block Size and Rock Quality Designation (RQD). *Tunneling and Underground Space Technology*, Vol. 20, No. 4, pp. 362–377.

Palmström, A. and Broch, E. (2006). Use and Misuse of Rock Mass Classification Systems with Particular Reference to the Q-System. *Tunneling Underground Space Technology*, Vol. 21, No. 6, pp. 575–593.

Palmström, A. and Stille, H. (2007). Ground Behavior and Rock Engineering Tools for Underground Excavation. *Tunneling and Underground Space Technology*, Vol. 22, No. 4, pp. 363–376.

Palmström, A., Blindheim, O.T., and Broch, E. (2002). The Q-System Possibilities and Limitations. *Norwegian National Conference on Tunneling*, Norwegian Tunneling Association, pp. 41.1–41.3.

Pan, E., Amadei, B., and Kim, Y.I. (1998). 2-D BEM Analysis of Anisotropic Half-plane Problems application to Rock Mechanics. *International Journal of Rock Mechanics and Mining Sciences*, Vol. 35, No. 1, pp. 69–74.

Pan, X.F., Zhang, X., and Lu, M.W. (2005). Meshless Galerkin Least-Squares Method. *Computational Mechanics*, Vol. 35, pp. 182–189.

Pande, G.N., Beer, G., and Williams, J.R. (1990). *Numerical Methods in Rock Mechanics*. John Wiley and Sons, Ltd., West Sussex, England.

Panek, L.A. (1952). Centrifugal testing apparatus for mine-structure stress analysis. *Report of Investigations 4883, U.S. Bureau of Mines, College Park MD*.

Panet, M. (1976). Analyse de la stabilité d'un tunnel creusé dans un massif rocheux en tenant compte du comportement aprés la rupture. *Rock Mechanics*, Vol. 8, No. 6, pp. 209–233.

Panet, M. (1995). Calcul des tunnels par a la méthode de convergence-confinement. *Presse de l'Ecole Nationale des Ponts et Chaussées*. Paris, France.

Panet, M. and Guellec, P. (1974). Contribution à l'étude du soutènement d'un tunnel á l'arrière du front de taille. *Proceedings of the 3rd International Congress of the ISRM*. Denver. National Academy of Sciences. Washington, DC, pp. 1163–1168.

Panet, M. and Guenot, A. (1982). Analysis of convergence behind the face of a tunnel. *Proceedings International Symposium on Tunnelling '82*, IMM, London, pp. 197–204.

Park, K.-H. and Kim, Y.-J. (2006). Analytical Solution for a Circular Opening in an Elastic-brittle-plastic Rock. *International Journal of Rock Mechanics and Mining Sciences*, Vol. 43, pp. 616–622.

Parker, H.W. (1999). Early Developments of Shotcrete and Steel Fiber Shotcrete. *Shotcrete for Underground Support VIII. Proceedings of the Eighth International Conference*. T.B. Celestino and H.W. Parker, Editors. ASCE, Reston, VA, pp. 240–258.

Paul, S.L., Hendron, A.J., Cording, E.J., Sgouros, G.E., and Saha, P.K. (1983). *Design Recommendations for Concrete Tunnel Linings, Volume II Summary and Proposed Recommendations*. Final Report to US Dept of Transportation, Urban Mass Transportation Administration, 155 pp.

Peck, R.B. (1969a). Advantages and Limitations of the Observational Method in Applied Soil Mechanics (9th Rankine Lecture). *Géotechnique*, Vol. 19, No. 2, pp. 171–187.

Peck, R.B. (1969b). State of the Art Report. Deep Excavation and Tunneling in Soft Ground. *Proceedings 7th International Conference on Soil Mechanics and Foundation Engineering*, State-of-the-Art Volume, Mexico City, Mexico. Sociedad Mexicana de Mecánica de Suelos, pp. 225–290.

Peck, R.B., Deere, D.U., and Capacete, J.L. (1956). Discussion on "Allowable Settlements of Buildings." *Proceedings of the Institution of Civil Engineers, Part II*. The Institution of Civil Engineers, London, England.

Peck, R.B., Deere, D.U., Monsees, J.E., Parker, H.W., and Schmidt, B. (1969). *Some Design Considerations in the Selection of Underground Support Systems*. Contract 3-0152, Office of High-Speed Ground Transportation, DOT, Washington, DC.

Peck, R.B., Hendron, A.J., and Mohraz, B. (1972). State of the Art of Soft-Ground Tunneling. *Proceedings of the North American Rapid Excavation and Tunneling Conference*. K.S. Lane and L.A. Garfield, Editors. Society of Mining Engineers of the American Institute of Mining, Metallurgical, and Petroleum Engineers, pp. 259–286.

Peck, W. (2000). Determining the Stress Reduction Factor in Highly Stressed Jointed Rock. *Australian Geomechanics*, Vol. 35, No. 2, pp. 57–60.

Pei, J. and Einstein, H.H. (2011). A new failure criterion for transversely isotropic rocks and its validation against triaxial tests. *Chapter 16, in True Triaxial Testing of Rock*. M. Kaisniewski, X. Li, and M. Takahasti, Editors. CRC Press.

Pei, J., Einstein, H.H., and Whittle, A.J. (2018). The Normal Stress Space and Its Application to Constructing a New Failure Criterion for Cross-Anisotropic Geomaterials. *International Journal of Rock Mechanics and Mining Sciences*, Vol. 106, pp. 364–373.

Peila, D. and Oreste, P.P. (1995). Axisymmetric Analysis of Ground Reinforcing in Tunnelling Design. *Computers and Geotechnics*, Vol. 17, pp. 253–274.

Pelizza, S., Oreste, P.P., Peila, D., and Oggeri, C. (2000). Stability Analysis of a large Cavern in Italy for Quarrying Exploitation of a Pink Marble. *Tunnelling and Underground Space Technology*, Vol. 15, No. 4, pp. 421–435.

Pelizza, S., Peila, D., and Oreste, P.P. (1994). A new approach for ground reinforcing design in tunnelling. *Proceedings of the International Congress on Tunnelling and Ground Conditions*, Cairo, Egypt, pp. 517–522.

Pellet, F. and Egger, P. (1996). Analytical Model for the Mechanical Behaviour of Bolted Rock Joints Subjected to Shearing. *Rock Mechanics and Rock Engineering*, Vol. 29, No. 2, pp. 73–97.

Pender, M.J. (1980). Elastic Solutions for a Deep Circular Tunnel. *Geotechnique*, Vol. 30, pp. 216–222.

Penzien, J. (2000). Seismically Induced Raking of Tunnel Linings. *Earthquake Engineering and Structure Dynamics*, Vol. 29, pp. 683–691.

Perrone, N. and Kao, R. (1975). A General Finite Difference Method for Arbitrary Meshes. *Computers and Structures*, Vol. 5, pp. 45–58.

Peterson, D.A., Pakalnis, R., and Mah, G.P. (1992). Fibreglass cable bolts – An alternative. *Rock Support in Mining and Underground Construction. Proceedings of the International Symposium on Rock Support*. P.K. Kaiser and D.R. McCreath, Editors. A.A. Balkema, Rotterdam, The Netherlands, pp. 319–326.

Pichler, B., Lackner, R., and Mang, H.A. (2003). Chapter 9: Soft computing-based parameter identification as the basis for prognoses of the structural behaviour of tunnels. *Numerical Simulation in Tunnelling*. G. Beer, Editor. Springer-Verlag, Wien, Austria, pp. 201–223.

Pimentel, E. and Anagnostou, G. (2013). New Apparatus and Experimental Setup for Long-Term Swelling Tests in Sulphatic Claystone. *Rock Mechanics and Rock Engineering*, Vol. 46, No. 6, pp. 1271–1285.

Pinto, F. (1999). *Analytical Methods to Interpret Ground Deformations Due to Soft Ground Tunneling*. S.M. Thesis, Department of Civil and Environmental Engineering, Massachusetts Institute of Technology (MIT), Cambridge, MA.

Pinto, F. and Whittle, A.J. (2014). Ground Movements due to Shallow Tunnels in Soft Ground I. Analytical Solutions. *ASCE Journal of Geotechnical and Geoenvironmental Engineering*, Vol. 140, No. 4. doi: 10.1061/(ASCE)GT.1943-5606.0000948.

Pinto, F., Zymis, D.M., and Whittle, A.J. (2014). Ground Movements due to Shallow Tunnels in Soft Ground II. Analytical Interpretation and Prediction. *ASCE Journal of Geotechnical and Geoenvironmental Engineering*, Vol. 140, No. 4. doi: 10.1061/(ASCE)GT.1943-5606.0000947.

Pöchhacker, H. (1974). Tunneling in Squeezing Rock, Theory and Practice. Separate print from *Porr-Nachrichten* No. 57 and RETC, AIME, New York, pp. 464–512.

Pokrovsky, G.Y. and Fedorov, I.S. (1936). Studies of Soil Pressures and Soil Deformations by Means of a Centrifuge. *Proceedings of the 1st International Conference on Soil Mechanics and Foundation*

Engineering. A. Casagrande, P.C. Rutledge, and J.D. Watson, Editors. Harvard University, Cambridge, MA, Vol. 1, p. 70.

Pollard, D.D., Zeller, S., Olson, J., and Thomas, A. (1990). Understanding the process of jointing in brittle rock masses. *Proceedings: 31st US Symposium on Rock Mechanics, Golden, CO.* A.A. Balkema, Editor. Rotterdam, pp. 447–454.

Polshin, D.E. and Tokar, R.A. (1957). Maximum Allowable Nonuniform Settlement of Structures. *Proceedings of the 4th International Conference on Soil Mechanics and Foundation Engineering*, Vol. 1, pp. 402–405, Butterworth, England.

Pöttler, R. (1990). Time-dependent Rock-shotcrete Interaction: A Numerical Shortcut. *Computers and Geotechnics*, Vol. 9, pp. 149–169.

Potvin, Y. (1988). *Empirical Open Stope Design in Canada.* Ph.D. Thesis, Department of Mining and Mineral Processing, University of British Columbia, Vancouver, Canada.

Potvin, Y. and Milne, D. (1992). Empirical cable cable bolt support design. *Rock Support in Mining and Underground Construction. Proceedings of the International Symposium on Rock Support.* P.K. Kaiser and D.R. McCreath, Editors. A.A. Balkema, Rotterdam, The Netherlands, pp. 269–275.

Potyondy, D.O. and Cundall, P.A. (2004). A Bonded-particle Model for Rock. *International Journal of Rock Mechanics and Mining Sciences*, Vol. 41, pp. 1329–1364.

Pratt, H.R., Stephenson, D.E., Zandt, G., Bouchon, M., and Hustrulid, W.A. (1979). Earthquake Damage to Underground Facilities. *Proceedings of the Rapid Excavation and Tunneling Conference (RETC).* A.C. Maevis and W.A. Hustrulid, Editors. Society of Mining Engineers of the American Institute of Mining, Metallurgical, and Petroleum Engineers, New York, NY, pp. 19–51.

Pregl, O., Fuchs, M., Müller, H., Petschl, G., Riedmüller, G., and Schwaighofer, B. (1980). Dreiaxiale Schwellversuche an Tongestein. *Geotechnik*, Vol. 3, Heft 1, pp. 1–7.

Priest, S.D. and Hudson, J.A. (1976). Discontinuity Spacing in Rock. *International Journal of Rock Mechanics and Mining Sciences*, Vol. 13, No. 5, pp. 135–148.

Proctor, R.V. and White, T.L. (1946, 1968). *Rock Tunneling with Steel Supports.* Commercial Shearing and Stamping Company, Youngstown, Ohio, 95 pp.

Proctor, R.V. and White, T.L. (1977). *Earth Tunneling with Steel Supports.* Commercial Shearing Inc., Youngstown, Ohio, US.

Rabcewicz, L.V. (1957). Die Ankerung im Tunnelbau ersetzt bisher gebräuchliche Einbaumethoden. *Schweizerische Bauzeitung*, Vol. 79, No. 9, pp. 123–131.

Rabcewicz, L.V. (1963). Bemessung von Hohlraumbauten, die Neue Österreichische Bauweise und ihr Einfluss auf Gebirgsdruckwirkung und Dismensionierung. *Felsmechanik und Ingenieurgeologie*, Vol. I/3-4, pp. 224–244.

Rankin, W.J. (1988). Ground Movements Resulting from Urban Tunnelling; Predictions and Effects. *Conference on Engineering Geology of Underground Movement*, Nottingham. British Geological Society, Engineering Geology Special Publication No. 5, pp. 79–92.

Ratcliffe, R. (1999). The important properties of steel fibre reinforced shotcrete (SFRC) used in mining. *Rock Support and Reinforcement Practice in Mining, Proceedings of the International Symposium on Ground Support.* E. Villaescusa, C.R. Windsor, and A.G. Thompson, Editors. A.A. Balkema, Rotterdam, pp. 209–218.

Redlich, K.A., Terzaghi, K., and Kampe, R. (1929). *Ingenieurgeologie.* Springer, Wien, 708 pp.

Rendulic, L. (1935). Porenziffer und Porenwasserdruck in Tonen. *Der Bauingenieur.* Berlin, Vol. 17, No. 51/53, pp. 559–564.

Richtlinie für die Geomechanische Planung zu Untertagebauarbeiten mit zyklischem Vortrieb (2001). ÖGG– Austrian Society for Geomechanics, Salzburg, 56 pp.

Ritter, W. (1879). *Die Statik der Tunnelgewölbe.* Springer, Berlin, 66 pp.

Rivas Vargas, F. (1997). Sostenimineto con Cerchas Metálicas. *Manual de Túneles y Obras Subterráneas.* C. López Jimeno, Editor. Entorno Gráfico, S.L., Madrid, Spain, Chapter 15, pp. 515–545.

Rizzo, F.J. (1967). An Integral Equation Approach to Boundary Value Problems of Classical Elastostastics. *Quarterly of Applied Mathematics*, Vol. 25, pp. 83–95.

Roberts, M.K.C. and Brummer, R.K. (1988). Support Requirements in Rockburst Conditions. *Journal of the South African Institute of Mining and Metallurgy*, Vol. 88, No. 3, pp. 97–104.

Rodriguez-Roa, F. (2002). Ground Subsidence due to a Shallow Tunnel in Dense Sandy Gravel. *Journal of Geotechnical and Geoenvironmental Engineering*, Vol. 128, No. 5, pp. 426–434.

Romana, M. (1985). New Adjustment Ratings for Application of Bieniawski Classification to Slopes. *Proceedings of the International Symposium on the Role of Rock Mechanics in Excavations for Mining and Civil Works*. International Society of Rock Mechanics, Zacatecas, pp. 49–57.

Rowe, R.K. (1986). The Prediction of Deformations Caused by Soft Ground Tunneling – Recent Trends. *Canadian Geotechnical Journal*, Vol. 23, pp. 91–108.

Rowe, R.K. and Kack, G.J. (1983). A Theoretical Examination of the Settlements Induced by Tunneling: Four Case Histories. *Canadian Geotechnical Journal*, Vol. 20, pp. 299–314.

Rowe, R.K. and Lee, K.M. (1992). Subsidence Owing to Tunneling. II. Evaluation of a Prediction Technique. *Canadian Geotechnical Journal*, Vol. 29, pp. 941–954.

Rziha, F. (1867 Vol. 1/ 1872 Vol. 2). Reproduced by Glückauf (1986). Lehrbuch der gesammten Tunnelbaukunst. *Verl. Ernst u Korn*, Berlin, Vol. 1, 723 pp.; Vol. 2, 868 pp.

Sagaseta, C. (1987). Analysis of Undrained Soil Deformation due to Ground Loss. *Géotechnique*, Vol. 37, pp. 301–320.

Saito, T., Tsukada, K., Inami, E., Inoma, H., and Ito, Y. (1983). Study on Rockbursts at the Face of a Deep Tunnel, the Kanetsu Tunnel in Japan being an Example. *Proceedings of the 5th Congress of the International Society for Rock Mechanics*, Vol. 2, pp. D203–D206, A.A. Balkema, Rotterdam, The Netherlands.

Sakellariou, M.G. and Ferentinou, M.D. (2005). A Study of Slope Stability Prediction Using Neural Networks. *Geotechnical and Geological Engineering*, Vol. 23, pp. 419–445.

Salamon, M.D.G. (1983). Rockburst hazard and the fight for its alleviation in South African gold mines. *Rockbursts: Prediction and Control*. M. Board, C. Fairhurst, and E. Grimstadt, Editors. The Institution of Mining and Metallurgy, London, England, pp. 11–36.

Sandoval, E. and Bobet, A. (2020a). Seismic Response of Underground Structures under Undrained Loading with Excess Pore Pressures Accumulation. *Tunnelling and Underground Space Technology*, Vol. 99. doi: 10.1016/j.tust.2019.103255.

Sandoval, E. and Bobet, A. (2020b). Effect of Input Frequency on the Seismic Response of Deep Circular Tunnels. *Soil Dynamics and Earthquake Engineering*, Vol. 139. doi: 10.1016/j.soildyn.2020.106421.

Sapigni, M., Berti, M., Bethaz, E., Bustillo, A., and Cardone, G. (2002). TBM Performance Estimation Using Rock Mass Classifications. *International Journal of Rock Mechanics and Mining Sciences*, Vol. 39, No. 6, pp. 771–788.

Savigamin, Ch. and Bobet, A. (2021). Seismic Response of a Deep Circular Tunnel Subjected to Axial Shear and Axial Bending. *Tunnelling and Underground Space Technology*. doi: 10.1016/j.tust.2021.103863.

Savin, G.N. (1961). *Stress Concentration around Holes*. Pergamon Press, London, UK.

Scavia, C. (1995). A Method for the Study of Crack Propagation in Rock Structures. *Géotechnique*, Vol. 45, No. 3, 447–463.

Schalkoff, R.J. (1997). *Artificial Neural Networks*. The McGraw-Hill Companies, Inc., New York, NY.

Schleiss, A. (1986). Design of Pervious Pressure Tunnels. *International Water Power and Dam Construction*, Vol. 38, No. 5, pp. 21–26, Wilmington Publishing, Ltd., Church Hill, UK.

Schleiss, A.J. (1997). Design of Reinforced Concrete Linings of Pressure Tunnels and Shafts. *Hydropower and Dams*, Vol. 4, No. 3, pp. 88–94.

Schleiss, A.J. and Manso, P.A. (2012). Design of Pressure Relief Valves for Protection of Steel-Lined Pressure Shafts and Tunnels Against Buckling During Emptying. *Rock Mechanics and Rock Engineering*, Vol. 45, pp. 11–20.

Schmidt, B. (1974). Prediction of Settlements due to Tunneling in Soil: Three Case Histories. *Proceedings of the 2nd North American Rapid Excavation and Tunnelling Conference*, Vol. 2, pp. 1179–1199.

Schmuck, C.H. (1979). Cable Bolting at the Homestake Gold Mine. *Mining Engineering*, December, Vol. 31, No. 12, pp. 1677–1681.

Schnabel, P.B., Lysmer, J., and Seed, H.B. (1972). *SHAKE: A Computer Program for Earthquake Response Analysis of Horizontally Layered Sites*. Report No. UCB/EERC-72/12, Earthquake Engineering Research Center, University of California, Berkeley, 102 pp.

Schubert, W., Goricki, A. and Riedmüller G. (2003). The Guideline for the Geomechanical Design of Underground Structures with Conventional Excavation. *Felsbau*, Vol. 21, No. 4, pp. 13–18.

Schwartz, C.W. and Einstein, H.H. (1979). Improvement of Ground-Support Performance by Full Consideration of Ground Displacements. *Transportation Research Record*, Vol. 684, pp. 14–20.

Schwartz, C.W. and Einstein, H.H. (1980a). Simplified Analysis for Ground-Structure Interaction in Tunneling. *Proceedings of the 21st US Symposium on Rock Mechanics*, 10 pp.

Schwartz, Ch.W. and Einstein, H.H. (1980b). Improved Design of Tunnel Supports. Vol. 1. *Simplified Analysis for Ground Structure Interaction in Tunnels*. Series of Reports for USDOT.

Schwesig, M., Duddeck, H. (1985). *Beanspruchung des Tunnelausbaus infolge Quellverhaltens von Tonsteingebirge. Bericht Nr. 85-47*, Institut für Statik, Technische Universität Braunschweig, 57 pp.

Scott, J.J. (1976). Friction rock stabilizers – a new rock reinforcement method. *Monograph on Rock Mechanics Applications in Mining*. W.S. Brown, S.J. Green, and W.A. Hustrulid, Editors. Society of Mining Engineers, American Institute of Mining Metallurgy and Petroleum Engineers, New York, pp. 242–249.

Scott, J.J. (1983). Friction rock stabilizer impact upon anchor design and ground control practices. *Rock Bolting. Theory and Application in Mining and Underground Construction. Proceedings of the International Symposium on Rock Bolting*. O. Stephansson, Editor. A.A. Balkema, Rotterdam, The Netherlands, pp. 407–418.

Seeber, G. (1975). Neue Entwicklungen für Druckstollen und Druckschächte. *Osterreichische Ingenieur Zeitschrift*, Wien-New York, 5.

Seeber, G. (1984). Recent Developments in the Design and Construction of Power Conduits for Storage Power Stations. *Idraulica del Territorio Montano*, Bressanone, Italy, pp. 177–204.

Seeber, G. (1985). Power Conduits for High-head Plants. *International Water Power and Dam Construction*, Vol. 37, No. 7, pp. 50–54.

Seeber, G. (1999). *Druckstollen und Druckschächte: Bemessung-Konstruktion-Ausführung*. Enke im Thieme Verlag, Stutgart.

Seed, H.B., Wong, R.T., Idriss, I.M., and Tokimatsu, K. (1986). Moduli and Damping Factors for Dynamic Analyses of Cohesionless Soils. *Journal of the Geotechnical Engineering Division*, ASCE. Vol. 112, No. 11, pp. 1016–1032.

Selby, A.R. (1999). Tunnelling in Soils – Ground Movements, and Damage to Buildings in Workington, UK. *Geotechnical and Geological Engineering*, Vol. 17, pp. 351–371.

Selmer-Olsen, R. (1985). Experience Gained from Unlined High Pressure Tunnels and Shafts in Hydroelectric Power Stations in Norway. *Norwegian Hydropower Tunnelling*. Publication No. 3, Norwegian Soil and Rock Engineering Association, Tapir Publishers, University of Trondheim, Trondheim, Norway, pp. 31–40.

Semple, R.M., Hendron, A.J., and Mesri, G. (1973). *The Effect of Time-Dependent Properties of Altered Rock on Tunnel Support Requirements*. Report No. FRAORDD-75-30. University of Illinois at Urbana-Champaign, Ill. 217 pp.

Serafeimidis, K. and Anagnostou, G. (2012). On the Kinetics of the Chemical Reaction Underlying Swelling of Anhydritic Rocks. *Proceedings of the 2012 EUROCK Conference*, Stockholm, Sweden, 14 pp.

Serafeimidis, K. and Anagnostou, G. (2013). On the Time Development of Sulphate Hydrate on Anhydritic Swelling Rocks. *Rock Mechanics and Rock Engineering*, Vol. 26, No. 3, pp. 619–634.

Serafeimidis, K. and Anagnostou, G. (2014a). The Solubilities and Thermodynamic Equilibrium of Anhydrite and Gypsum. *Rock Mechanics and Rock Engineering*. doi: 10.1007/s00603-014-0557-1.

Serafeimidis, K. and Anagnostou, G. (2014b). On the Crystallization Pressure of Gypsum. *Enviornmental Earth Sciences*, Vol. 72, No. 12, pp. 4985–4994.

Serafim, J.L. and Pereira, J.P. (1983). Consideration of the Geomechanical Classification of Bieniawski. *Proceedings International Symposium on Engineering Geology and Underground Construction*, Lisbon, Portugal. Sociedade Portuguesa de Geotecnia, pp. 33–44.

Serrano, A. and Olalla, C. (1999). Tensile Resistance of Rock Anchors. *International Journal of Rock Mechanics and Mining Sciences*, Vol. 36, pp. 449–474.

Sharan, S.K. (2003). Elastic-brittle-plastic Analysis of Circular Openings in Hoek-Brown Media. *International Journal of Rock Mechanics and Mining Sciences*, Vol. 40, pp. 817–824.

Sharan, S.K. (2005). Exact and Approximate Solutions for Displacements around Circular Openings in Elastic-brittle-plastic Hoek-Brown Rock. *International Journal of Rock Mechanics and Mining Sciences*, Vol. 42, pp. 542–549.

Sharma, K.G. and Pande, G.N. (1988). Stability of Rock Masses Reinforced by Passive, Fully-grouted Rock Bolts. *International Journal of Rock Mechanics, Mining Sciences and Geomechanics Abstracts*, Vol. 25, No. 5, pp. 273–285.

Sharma, S. and Judd, W.R. (1991). Underground Opening Damage from Earthquakes. *Engineering Geology*, Vol. 30, pp. 263–276.

Shawky, A. and Maekawa, K. (1996). Collapse Mechanism of Underground RC Structures during Hanshin Great Earthquake. *Proceedings of the First International Conference on Concrete Structures*, Cairo, Egypt, pp. 1–17.

Shen, B. and Barton, N. (1997). The Disturbed Zone Around Tunnels in Jointed Rock Masses. *International Journal of Rock Mechanics and Mining Sciences*, Vol. 34, No. 1, pp. 117–125.

Shen, B. and Stephansson, O. (1994). Modification of the G-criterion for Crack Propagation Subjected to Compression. *Engineering Fracture Mechanics*, Vol. 47, No. 2, pp. 177–189.

Shen, B., Stephansson, O., Einstein, H.H., and Ghahreman, B. (1995). Coalescence of Fracture Under Shear Stresses in Experiments. *Journal of Geophysical Research*, Vol. 100, No. (B4), pp. 5,975–5,990.

Shi, G.H. (1992a). Discontinuous Deformation Analysis: A New Numerical Model for the Statics and Dynamics of Deformable Block Structures. *Engineering Computations*, Vol. 9, pp. 157–168.

Shi, G.H. (1992b). Modeling Rock Joints and Blocks by Manifold Method. *Proceedings of the 32nd U.S. Symposium on Rock Mechanics*, Santa Fe, New Mexico, pp. 639–648.

Shi, G.H. (1993). *Block System Modeling by Discontinuous Deformation Analysis*. Topics in Engineering, Vol. 11, C.A. Brebbia and J.J. Connor, Editors. Computational Mechanics Publications, Boston, USA.

Shi, G.H. (1996). Manifold method of material analysis. *Proceedings of IFDDA 1996*, Berkeley, CA, pp. 52–204.

Shi, G.H. (1997). Numerical Manifold Method. *Proceedings of the 2nd International Conference on Analysis of Discontinuous Deformation*, Kyoto, Japan, pp. 1–35.

Shi, G.H. (2001). Three dimensional discontinuous deformation analysis. *Rock Mechanics in the National Interest*, *Proceedings of the 38th U.S. Rock Mechanics Symposium*, D. Elsworth, J.P. Tinucci, and K.A. Heasley, Editors. American Rock Mechanics Association, Balkema: Rotterdam, Washington DC, pp. 1421–1428.

Shi, G.H. and Goodman, R.E. (1984). Discontinuous Deformation Analysis. *Proceedings of the 25th U.S. Symposium on Rock Mechanics*, pp. 269–277.

Shi, G.H. and Goodman, R.E. (1985). Two Dimensional Discontinuous Deformation Analysis. *International Journal for Numerical and Analytical Methods in Geomechanics*, Vol. 9, pp. 541–556.

Shi, J., Ortigao, J.A.R., and Bai, J. (1998). Modular Neural Networks for Predicting Settlements during Tunneling. *ASCE Journal of Geotechnical and Geoenvironmental Engineering*, Vol. 124, No. 5, pp. 389–395.

Shinde, P.P. and Shah, S. (2018). A Review of Machine Learning and Deep Learning Applications. *2018 Fourth International Conference on Computing Communication Control and Automation (ICCUBEA)*, 16–18 August 2018. doi: 10.1109/ICCUBEA.2018.8697857.

Shou, K.J. (2000). A Three-Dimensional Hybrid Boundary Element Method for Non-Linear Analysis of a Weak Plane Near and Underground Excavation. *Tunnelling and Underground Space Technology*, Vol. 15, No. 2, pp. 215–226.

Shou, K.J., Siebrits, E., and Crouch, S.L. (1997). A Higher Order Displacement Discontinuity Method for Three-dimensional Elastostatic Problems. *International Journal of Rock Mechanics and Mining Sciences*, Vol. 34, No. 2, pp. 317–322.

Simanjuntak, Y. (2015). *Prestresses Concrete-Lined Pressure Tunnels: Towards Improved Safety and Economical Design*. Ph.D. Thesis. CRC Press/Balkema, The Netherlands.

Simser, B., Andrieux, P., and Gaudreau, D. (2002a). Rockburst support at Noranda's Brunswick Mine, Bathurst, New Brunswick. *Proceedings of the North American Rock Mechanics Symposium: NARMS-TAC 2002*, pp. 805–813.

Simser, B., Joughin, W.C., and Ortlepp, W.D. (2002b). The Performance of Brunswick Mine's Rockburst Support System during a Severe Seismic Episode. *The Journal of the South African Institute of Mining and Metallurgy*, May/June 2002, Vol. 102, No. 4, pp. 217–223.

Simser, B.P. (2006). Mine Wide Seismic Monitoring at Falconbridge's Craig Mine. *41st U.S. Rock Mechanics Symposium*, Golden, CO. Paper 06-1024, 13 pages.

Simser, B.P. (2019). Rockburst Management in Canadian Hard Rock Mines. *Journal of Rock Mechanics and Geotechnical Engineering*, Vol. 11, No. 5, pp. 1036–1043.

Singh, B. (2000). *Proceedings Norwegian Method of Tunneling Workshop*. New Delhi.

Singh, B., Jethwa, J.L., and Dube, A.K. (1992). Correlation Between Observed Support Pressure and Rock Mass Quality. *Tunneling and Underground Space Technology*, Vol. 7, No. 1, pp. 59–74.

Singh, B., Viladkar, M.N., Samadhiya, N.K., and Sandeep (1995). A Semi-empirical Method for the Design of Support Systems in Underground Openings. *Tunnelling and Underground Space Technology*, Vol. 10, No. 3, pp. 375–383.

Singh, S.P. (1988). Technical Note: Burst Energy Release Index. *Rock Mechanic and Rock Engineering*, Vol. 21, pp. 149–155.

Singh, T.N. and Singh, V. (2005). An Intelligent Approach to Prediction and Control Ground Vibration in Mines. *Geotechnical and Geological Engineering*, Vol. 23, pp. 249–262.

Singh, T.N., Gupta, A.R., and Sain, R. (2006). A Comparative Analysis of Cognitive Systems for the Prediction of Drillability of Rocks and Wear Factor. *Geotechnical and Geological Engineering*, Vol. 24, pp. 299–312.

Singh, V.K., Singh, D., and Singh, T.N. (2001). Prediction of Strength Properties of Some Schistose Rocks from Petrographic Properties Using Artificial Neural Networks. *International Journal of Rock Mechanics and Mining Sciences*, Vol. 38, pp. 269–284.

Sinha, R.S. (1989). *Underground Structures: Design and Instrumentation*. R.S. Sinha, Editor. Elsevier, New York, NY.

Sinha, R.S. and Schoeman, K.D. (1983). Rock tunnels and rock reinforcement. *Rock Bolting. Theory and Application in Mining and Underground Construction. Proceedings of the International Symposium on Rock Bolting*. O. Stephansson, Editor. A.A. Balkema, Rotterdam, The Netherlands, pp. 333–344.

Sitar, N., MacLaughlin, M.M., and Doolin, D.M. (2005). Influence of Kinematics on Landslide Mobility and Failure Mode. *Journal of Geotechnical and Geoenvironmental Engineering*, Vol. 131, No. 6, pp. 716–728.

Skempton, A.W. (1954). The Pore Pressure Coefficients A and B. *Géotechnique*, Vol. 4, No. 4, pp. 143–147.

Skempton, A.W. and MacDonald, D.H. (1956). The Allowable Settlement of Buildings. *Proceedings of the Institution of Civil Engineers, Structures and Buildings*, Vol. 5, pp. 727–784.

Sofianos, A.I. and Nomikos, P.P. (2006). Equivalent Mohr-Coulomb and Generalized Hoek-Brown Strength Parameters for Supported Axisymmetric Tunnels in Plastic or Brittle Rock. *International Journal of Rock Mechanics and Mining Sciences*, Vol. 43, pp. 683–704.

Sofianos, A.I., Nomikos, P., and Tsoutrelis (1999). Stability of Symmetric Wedge Formed in the Roof of a Circular Tunnel: Nonhydrostatic Natural Stress Field. *International Journal of Rock Mechanics and Mining Sciences*, Vol. 36, pp. 687–691.

Sokolnikoff, I.S. (1956). *Mathematical Theory of Elasticity*. McGraw-Hill, New York.

Son, M. and Cording, E.J. (2005). Estimation of Building Damage Due to Excavation-Induced Ground Movements. *ASCE Journal of Geotechnical and Geoenvironmental Engineering*, Vol. 131, No. 2, pp. 162–177.

Son, M. and Cording, E.J. (2007). Evaluation of Building Stiffness for Building Response Analysis to Excavation-Induced Ground Movements. *ASCE Journal of Geotechnical and Geoenvironmental Engineering*, Vol. 133, No. 8, pp. 995–1002.

Son, M. and Cording, E.J. (2011). Responses of Buildings with Different Structural Types to Excavation-Induced Ground Settlements. *ASCE Journal of Geotechnical and Geoenvironmental Engineering*, Vol. 137, No. 4, pp. 323–333.

Sonmez, H., Gokceoglu, C., Nefeslioglu, H.A., and Kayabasi, A. (2006). Estimation of Rock Modulus: For Intact Rocks with an Artificial Neural Network and for Rock Masses with a New Empirical Equation. *International Journal of Rock Mechanics and Mining Sciences*, Vol. 43, pp. 224–235.

Soutas-Little, R.W. (1973). *Elasticity*. Dover Publications, Mineola, NY.

Southwell, R.V. (1946). *Relaxation Methods in Theoretical Physics*. Clarendon Press, Oxford, England.

Spang, K. and Egger, P. (1990). Action of Fully-Grouted Bolts in Jointed Rock and Factors of Influence. *Rock Mechanics and Rock Engineering*, Vol. 23, pp. 201–229.

Spaun, G. (1974). Tunnelbaugeologie - Bauvertrag - Bauabwicklung. Auswirkung geologischer Faktoren auf Bauabwicklung und Vertrag: Proceedings 22nd Salzburg Geomechanik kolloquium [monograph]. *Strassenforschung*, Vol. 18, pp. 7–14.

Spearing, A.J.S. and Naismith, W.A. (1999). The rationale, design and implementation of steel fibre reinforced wet shotcrete, in highly stresses hard rock tunnels, where rapid development advance is needed, in South Africa. *Rock Support and Reinforcement Practice in Mining, Proceedings of the International Symposium on Ground Support*. E. Villaescusa, C.R. Windsor, and A.G. Thompson, Editors. A.A. Balkema, Rotterdam, pp. 193–198.

Srinivasan, C., Arora, S.K., and Benady, S. (1999). Precursory Monitoring of Impending Rockbursts in Kolar Gold Mines from Microseismic Emissions at Deeper Levels. *International Journal of Rock Mechanics and Mining Sciences*, Vol. 36, pp. 941–948.

Srinivasan, C., Arora, S.K., and Yaji, R.K. (1997). Use of Mining and Seismological Parameters as Premonitors of Rockbursts. *International Journal of Rock Mechanics and Mining Sciences*, Vol. 34, No. 6, pp. 1001–1007.

St John, C.M. and Zahrah, T.F. (1985). Applicable Analytical Tools for a Seismic Design of Underground Excavations. *Proceedings of the Rapid Excavation and Tunneling Conference (RETC)*. C.D. Mann and M.N. Kelley, Editors. Society of Mining Engineers of the American Institute of Mining, Metallurgical, and Petroleum Engineers, New York, NY, pp. 84–103.

St John, C.M. and Zahrah, T.F. (1987). Aseismic Design of Underground Structures. *Tunnelling and Underground Space Technology*, Vol. 2, No. 2, pp. 165–197.

Stacey, T.R. and Ortlepp, W.D. (1994). Rockburst Mechanisms and Tunnel Support in Rockburst Conditions. *Proceedings of the International Conference Geomechanics 93*, Z. Rakowski, Editor. A.A. Balkema, Rotterdam, The Netherlands, pp. 39–46.

Statens Järnvägars Geotekniska Kommission (1922). *Slutbetänkande avgiet den Maj 31*, 1922, Stockholm.

Steiner, W. (1979). *Empirical Methods in Rock Tunneling*. Ph.D. Thesis, Massachusetts Institute of Technology, 553 pp.

Steiner, W. (2020). Sulphate Bearing Rocks in Tunnels (2020) – Lessons from Field Observations and In-situ Swelling Pressures. *Geomechanik und Tunnelbau*, Vol. 13, No. 3, pp. 286–301.

Steiner, W. and Einstein, H.H. (1980). *Improved Design of Tunnel Supports. Vol. 5 – Empirical Methods in Rock Tunneling – Review and Recommendations*. Final Report to U.S. DOT, Washington, DC, 541 pp.

Steiner, W. and Metzger, R. (1988). *IG Wisenberg-Tunnel, Bahn 2000-Wisenbergtunnel. Erfahrungen aus Tunneln in quellendem Gestein*.

Steiner, W., Einstein, H.H., and Azzouz, A.S. (1979). *Tunneling Practices in Austria and Germany*. Volume 4 of Report prepared for U.S. DOT, Contract No. Dot-TSC-1489. Also MIT Report No. 79-30, 475 pp.

Stephens, R.E. and Banks, D.C. (1989). Moduli for deformation studies of the foundation and abutments of the Portugues Dam - Puerto Rico. *Rock Mechanics as a Guide for Efficient Utilization of Natural Resources, Proceedings of the 30th US Symposium on Rock Mechanics*. A.W. Khair, Editor. A.A. Balkema, Morgantown, USA, Rotterdam, pp. 31–38.

Sterpi, D. and Cividini, A. (2004). A Physical and Numerical Investigation of the Stability of Shallow Tunnels in Strain Softening Media. *Rock Mechanics and Rock Engineering*, Vol. 37, No. 4, pp. 277–298.

Stillborg, B. (1994). *Professional Users Handbook for Rock Bolting: Series on Rock and Soil Mechanics*. Trans Tech Publications, Clausthal-Zellerfeld, Germany.

Stille, H. (1983). Theoretical aspects on the difference between prestressed anchor bolt and grouted bolt in squeezing rock. *Theory and Application in Mining and Underground Construction. Proceedings of the International Symposium on Rock Bolting*. O. Stephansson, Editor. A.A. Balkema, Rotterdam, The Netherlands, pp. 65–73.

Stille, H., Holmberg, M., and Nord, G. (1989). Support of Weak Rock with Grouted Bolts and Shotcrete. *International Journal for Rock Mechanics, Mining Sciences and Geomechanics Abstracts*, Vol. 26, No. 1, pp. 99–113.

Stini, J. (1950). *Tunnelbaugeologie*. Springer, Vienna, 366 pp.

Stini, J. (1956). Wassersprengung und Sprengwasser (Blasting Effects of Water). *Geologie u. Bauwesen*, Vol. 22, pp. 141–169.

Stone, K.J.L. (1988). *Modelling of Rupture Development in Soils*. Ph.D. Thesis, Wolfson College, Cambridge University, Cambridge, UK.

Stone, K.J.L. and Newsom, T.A. (2002). Arching Effects in Soil-Structure Interaction. *Proceedings of the International Conference on Physical Modelling in Geotechnics*, ICPMG '02, St. John's, Newfoundland. P. Guo, R. Phillips, R. Popescu, Editors. CRC Press, pp. 935–939.

Strack, O.E. and Verruijt, A. (2002). A Complex Variable Solution for a Deforming Buoyant Tunnel in a Heavy Elastic Half-plane. *International Journal for Numerical and Analytical Methods in Geomechanics*, Vol. 26, pp. 1235–1252.

Su, H.D., Xie, X.L., and Liang, Q.Y. (2003). Automatic Programming for High-order Numerical Method. *Development and Application of Discontinuum Modelling for Rock Engineering. Proceedings of the 6th International Conference on Analysis of Discontinuous Deformation*, M. Lu, Editor. A.A. Balkema Publishers, The Netherlands, pp. 153–157.

Su, S. and Stephansson, O. (1999). Effect of a Fault on in situ Stresses Studied by the Distinct Element Method. *International Journal of Rock Mechanics and Mining Sciences*, Vol. 36, pp. 1051–1056.

Sulem, J., Panet, M., and Guenot, A. (1987). An Analytical Solution for Time-Dependent Displacements in a Circular Tunnel. *International Journal of Rock Mechanics and Mining Sciences & Geomechanics Abstracts*, Vol. 24, No. 3, pp. 155–164.

Sun Jun, T., Zhang, L., and Li, C. (1984). The Coupled Creep Effect of Pressure Tunnels Interacted with its Osmotic Swelling Viscous Elastic-Plastic Surrounding Rocks. Proceedings of Tunnelling and Underground Works Advances in Tunnelling Technology Colloquium, Beijing, China.

Suwansawat, S. and Einstein, H.H. (2006). Artificial Neural Network for Predicting the Maximum Surface Settlement Caused by EPB Shield Tunneling. *Tunnelling and Underground Space Technology*, Vol. 21, No. 2, pp. 133–150.

Suwansawat, S. and Einstein, H.H. (2007). Describing Settlement Troughs over Twin Tunnels Using a Superposition Technique. *Journal of Geotechnical and Geoenvironmental Engineering*, Vol. 133, No. 4, pp. 445–468.

Sweby, G.J., Trifu, C.-I., Goodchild, D.J., and Morris, L.D. (2006). High-Resolution Monitoring at Mt. Keith Open Pit Mine. *41st U.S. Rock Mechanics Symposium*, Golden, CO. Paper 06-1159, 6 pages.

Swoboda, G. (1990). Numerical modelling of tunnels. *Numerical Methods and Constitutive Modelling in Geomechanics*. C.S. Desai and G. Gioda, Editors. Springer-Verlag, Wien, Austria, pp. 277–318.

Swoboda, G. and Abu-Krisha, A. (1999). Three-Dimensional Numerical Modelling for TBM Tunnelling in Consolidated Clay. *Tunnelling and Underground Space Technology*, Vol. 14, No. 3, pp. 327–333.

Takahashi, H. (1966). Tunnels. *Journal of the Japanese Society of Civil Engineers*, Vol. 51, No. 12, pp. 87–93.

Tanaka, T. and Sakai, T. (1993). Progressive Failure and Scale Effect of Trap-door Problem with Granular Materials. *Soils and Foundations*, Vol. 33, No. 1, pp. 11–22.

Tang, B., Mitri, H.S., and Marwan, J. (2000). Distress blasting on a mining face in high horizontal stress environment. *Proceedings of the 4th North American Rock Mechanics Symposium, NARMS 2000*, J. Girard, M. Liebman, C. Breeds, and T. Doe, Editors. A.A. Balkema, Rotterdam, The Netherlands, pp. 335–341.

Tannant, D.D. and Wang, C. (2002). Thin Rock Support Liners Modeled with Particle Flow Code. *Discrete Element Methods: Numerical Modeling of Discontinua*. Geotechnical Special Publication No. 117, ASCE, pp. 346–352.

Tannant, D.D. and Wang, C. (2004). Thin Tunnel Liners Modeled with Particle Flow Code. *Engineering Computations*, Vol. 21, No. 2/3/4, pp. 318–342

Tannant, D.D., Brummer, R.K., and Yi, X. (1995). Rockbolt Behaviour Under Dynamic Loading: Field Tests and Modelling. *International Journal of Rock Mechanics, Mining Sciences and Geomechanics Abstracts*, Vol. 32, No. 6, pp. 537–550.

Taylor, L.M. and Preece, D.S. (1992). Simulation of Blasting Induced Rock Motion using Spherical Element Models. *Engineering Computations*, Vol. 9, pp. 243–252.

Telles, J.C.F. and Brebbia, C.A. (1981a). Boundary Elements: New Developments in Elastoplastic Analysis. *Applied Mathematical Modeling*, Vol. 5, pp. 376–382.

Telles, J.C.F. and Brebbia, C.A. (1981b). Elasto-plastic boundary element analysis. *Nonlinear Finite Element Analysis*. W. Wunderlich, E. Stein, and K.J. Bathe, Editors. Springer-Verlag, Berlin, pp. 403–434.

Tenschert, E., Vigl, A. and Goricki, A (2003). Guideline for Geomechanical Design of Underground Structure with Continuous Excavation. *Felsbau*, Vol. 21, No. 4, pp. 20–25.

Terrametrics (1965a). *Final Report, Rock Mechanics Instrumentation, Straight Creek Tunnel Pilot Bore for Colorado Department of Highways*. Golden, Colorado.

Terrametrics (1965b). *Final Report. Rock Mechanics Instrumentation. Green River Marginal Route for Wyoming Department of Highways*. Golden, Colorado.

Terzaghi, K. (1923). Die Berechnung der Durchlässigkeitsziffer des Tones aus dem Verlauf der Hydrodynamischen Spannungerscheinungen. *Sitzungsberichte, Akademie der Wissenschaften in Wien*, Vol. 132, No. ¾, pp. 125–138.

Terzaghi, K. (1936). Stress Distribution in Dry and in Saturated Sand above a Yielding Trapdoor. *Proceedings 1st International Conference on Soil Mechanics and Foundation Engineering*. Graduate School of Engineering, Harvard University, Harvard University, MA, Vol. 1, pp. 307–311.

Terzaghi, K. (1943). *Theoretical Soil Mechanics*. John Wiley and Sons, New York.

Terzaghi, K. (1946). Rock defects and loads on tunnel support. *Rock Tunneling with Steel Supports*. R.V. Proctor and T. White, Editors. The Commercial Shearing and Stamping Co., Youngstown, Ohio pp. 15–99.

Terzaghi, K. (1950). Geologic Aspects of Soft Ground Tunneling. Chapter 11 in *Applied Sedimentation*, P.D. Trask, Editor. John Wiley and Sons, New York, NY, US.

Terzaghi, K. (1961). Past and Future of Applied Soil Mechanics. *Journal of the Boston Society of Civil Engineers*, Vol. 68, pp. 110–139.

Terzaghi, K. (1962). Stability of Steep Slopes on Hard Unweathered Rock. *Géotechnique*, Vol. 12, No. 4, pp. 251–270.

The British Tunneling Society, BTS, and The Institution of Civil Engineers, ICE (2004). *Tunnel Lining Design Guide*. Thomas Telford, London, England.

Theocaris, P.S. (1991). Peculiarities of the Artificial Crack. *Engineering Fracture Mechanics*, Vol. 38, No. 1, pp. 37–54.

Theocaris, P.S. and Petrou, L. (1989). From the Rectangular Hole to the Ideal Crack. *International Journal of Solids and Structures*, Vol. 25, No. 3, pp. 213–233.

Thompson, A.G. (1992). Tensioning reinforcing cables. *Rock Support in Mining and underground construction. Proceedings of the International Symposium on Rock Support*. P.K. Kaiser and D.R. McCreath, Editors. A.A. Balkema, Rotterdam, The Netherlands, pp. 285–291.

Timoshenko, S.P. and Goodier, J.N. (1970). *Theory of Elasticity*. McGraw-Hill, New York, NY.

Tonon, F. and Amadei, B. (2002). Effect of Elastic Anisotropy of Tunnel Wall Displacement Behind a Tunnel Face. *Rock Mechanics and Rock Engineering*, Vol. 35, No. 3, pp. 141–160.

Tsang, C.-F., Jing, L., Stephansson, O., and Kautsky, F. (2005). The DECOVALEX III Project: A Summary of Activities and Lessons Learned. *International Journal of Rock Mechanics and Mining Sciences*, Vol. 42, pp. 593–610.

Tsesarsky, M. and Hatzor, Y.H. (2006). Tunnel Roof Deflection in Blocky Rock Masses as a Function of Joint Spacing and Friction – A Parametric Study Using Discontinuous Deformation Analysis (DDA). *Tunnelling and Underground Space Technology*, Vol. 21, pp. 29–45.

Tsesarsky, M., Hatzor, Y.H., and Sitar, N. (2005). Dynamic Displacement of a Block on an Inclined Plane: Analytical, Experimental and DDA Results. *Rock Mechanics and Rock Engineering*, Vol. 38, No. 2, pp. 153–167.

Tsinidis, G., de Silva, F., Anastasopoulos, I., Bilotta, E., Bobet, A., Hashash, Y., He, Ch., Kampas, G., Knappett, J., Madabhushi, G., Nikitas, N., Pitilakis, K., Silvestri, F., Viggiani, G., and Fuentes, R. (2020). Seismic Behaviour of Tunnels: From Experiments to Analysis. *Tunnelling and Underground Space Technology*, Vol. 99. doi: 10.1016/j.tust.2020.103334.

Tunnelling Technology: an appraisal of the state of the art for application to transit systems (1976). Golder Associates, James F. MacLaren Limited. Published by Ontario Ministry of Transportation and Communications, Toronto, Canada.

Turner, M.J., Clough, R.W., Martin, H.C., and Topp, L.J. (1956). Stiffness and Deflection Analysis of Complex Structures. *Journal of the Aeronautical Sciences*, Vol. 23, pp. 805–823.

U.S. Army Core of Engineers (USACE) (1997). *Engineering and Design – Tunnels and Shafts in Rock*. Publication Number: EM 1110-2-2901. Department of the Army, USACE, Washington, DC.

U.S. Bureau of Reclamation (USBR) (2014). Water Conveyance Facilities, Fish Facilities, and Roads and Bridges. Chapter 4: Tunnels, Shafts, and Caverns. Design Standards No. 3, DS-3(4)-3: Phase 4 (Final) April 2014. U.S. Department of the Interior, USBR, Washington, DC.

Unal, E. (1983). *Design Guidelines and Roof Control Standards for Coal Mine Roofs*. Ph.D. Dissertation, The Pennsylvania State University, 355 pp.

Unlu, T. and Gercek, H. (2003). Effect of Poisson's Ratio on the Normalized Radial Displacements Occurring around the Face of a Circular Tunnel. *Tunnelling and Underground Space Technology*, Vol. 18, pp. 547–553.

Ushijima, R.S. and Einstein, H.H. (1985). Application of a Three-Dimensional Coupled Finite Element-Boundary Element Method. *Proceedings of Rock Masses: Modeling of Underground Openings/ Probability of Slope Failure/Fracture of Intact Rock*, C.H. Dowding, Editor. ASCE, pp. 21–37.

Van der Berg, J.P., Clayton, C.R.I., and Powell, D.B. (2003). Displacements Ahead of an Advancing NATM Tunnel in the London Clay. *Géotechnique*, Vol. 53, No. 9, pp. 767–784.

Vardoulakis, I., Gaf, B., and Gudehus, G. (1981). Trap-Door Problem with Dry Sand: A Statical Approach Based Upon Model Test Kinematics. *International Journal for Numerical and Analytical Methods in Geomechnics*, Vol. 5, pp. 57–78.

Vásárhelyi, B. and Bobet, A. (2000). Modeling of Crack Coalescence in Uniaxial Compression. *Rock Mechanics and Rock Engineering*, Vol. 33, No. 2, pp. 119–139.

Venkateswarteu, V. (1986). *Geomechanics Classification of Coal Measure Rocks vis-à-vis Rock Support*. Ph.D. Thesis, Indian School of Mines, Dhanbad, 251 pp.

Venturini, W.S. (1983). Boundary element method in geomechanics. *Lecture Notes in Engineering, 4*. C.A. Brebbia and S.A. Orszag, Editors. Springer-Verlag, Berlin, 246 pp.

Verman, M., Singh, B., Jethwa, J.L., and Viladkar, M.N. (1995). Determination of Support Reaction Curve for Steel-Supported Tunnels. *Tunnelling and Underground Space Technology*, Vol. 10, No. 2, pp. 217–224.

Vermeer, P.A., Ruse, N., and Marcher, T. (2002). Tunnel Heading Stability in Drained Ground. *Felsbau*, Vol. 20, No. 6, pp. 8–18.

Verruijt, A. (1997). A Complex Variable Solution for a Deforming Circular Tunnel in an Elastic Half-plane. *International Journal for Numerical and Analytical Methods in Geomechanics*, Vol. 21, pp. 77–89.

Verruijt, A. (1998). Deformations of an Elastic Half Plane with a Circular Cavity. *International Journal of Solids and Structures*, Vol. 35, No. 21, pp. 2795–2804.

Verruijt, A. and Booker, J.R. (1996). Surface Settlements due to Deformation of a Tunnel in an Elastic Half Plane. *Géotechnique*, Vol. 46, pp. 753–756.

Vijayakumar, S., Yacoub, T.E., and Curran, J.H. (2000). A Node-Centric Indirect Boundary Element Method: Three-Dimensional Displacement Discontinuities. *Computers and Structures*, Vol. 74, pp. 687–703.

Vitali, O., Bobet, A., and Celestino, T. (2018a). 3D Finite Element Modeling Optimization for Deep Tunnels with Material Nonlinearity. *Underground Space*, Vol. 3, No. 2, pp. 125–139.

Vitali, O., Celestino, T., and Bobet, A. (2018b). Analytical Solution for Tunnels not Aligned with Geostatic Principal Stress Directions. *Tunnelling and Underground Space Technology*, Vol. 82, pp. 394–405.

Vitali, O., Celestino, T., and Bobet, A. (2019). Shallow Tunnels Misaligned with Geostatic Principal Stress Directions: Analytical Solution and 3D Face Effects. *Tunnelling and Underground Space Technology*, Vol. 89, pp. 268–283. doi: 10.1016/j.tust.2019.04.006.

Vitali, O., Celestino, T., and Bobet, A. (2020). Analytical Solution for a Deep Circular Tunnel in Anisotropic Ground and Anisotropic Geostatic Stresses. *Rock Mechanics and Rock Engineering*, Vol. 53, pp. 3859–3884. doi: 10.1007/s00603-020-02157-5.

Vitali, O., Celestino, T., and Bobet, A. (2021). Construction Strategies for a NATM Tunnel in Sao Paulo, Brazil, in Residual Soil. *Underground Space*, Vol. 131, No. 2, pp. 193–201. doi: 10.1016/j.undsp.2021.04.002.

Vorster, T.E.B., Klar, A., Soga, K., and Mair, R.J. (2005). Estimating the Effects of Tunneling on Existing Pipelines. *ASCE Journal of Geotechnical and Geoenvironmental Engineering*, Vol. 131, No. 11, pp. 1399–1410.

VSL Systems Limited (1982). *Slab Post-Tensioning*. Switzerland.

Vucetic, M. and Dobry, R. (1991). Effect of Soil Plasticity on Cyclic Response. *Journal of the Geotechnical Engineering Division*, ASCE. Vol. 117, No. 1, pp. 89–107.

Wagner, H. (1984). Support requirements for Rockburst Conditions. *Proceedings of the 1st International Congress on Rockbursts and Seismicity in Mines, The South African Institute of Mining and Metallurgy Symposium Series, No. 6*. N.C. Gay and E.H. Wainwright, Editors. The South African Institute of Mining and Metallurgy, Johannesburg, South Africa, pp. 209–218.

Wang, B. and Garga, V.K. (1992). A numerical model for rock bolts. *Rock Support in Mining and Underground Construction. Proceedings of the International Symposium on Rock Support*. P.K. Kaiser and D.R. McCreath, Editors. A.A. Balkema, Rotterdam, The Netherlands, pp. 57–65.

Wang, C.-T. (1953). *Applied Elasticity*. McGraw-Hill Book Company, Inc., New York, NY.

Wang, C., Tannant, D.D., and Lilly, P.A. (2003a). Numerical Analysis of the Stability of Heavily Jointed Rock Slopes Using PFC2D. *International Journal of Rock Mechanics and Mining Sciences*, Vol. 40, pp. 415–424.

Wang, J.N. (1993). *Seismic Design of Tunnels, A State-of-the-Art Approach*. Monograph 7, Parsons Brickerhoff Quade & Douglas, Inc., New York.

Wang, J.-A. and Park, H.D. (2001). Comprehensive Prediction of Rockburst Based on Analysis of Strain Energy in Rocks. *Tunnelling and Underground Space Technology*, Vol. 16, pp. 49–57.

Wang, S.L., Feng, X.T., and Zheng, H. (2003b). Modelling Strong Discontinuities by Manifold Method. *Development and Application of Discontinuum Modelling for Rock Engineering. Proceedings of the 6th International Conference on Analysis of Discontinuous Deformation*, M. Lu, Editor. A.A. Balkema Publishers, The Netherlands, pp. 141–144.

Wang, W.L., Wang, T.T., Su, J.J., Lin, C.H., Seng, C.R., and Huang, T.H. (2001). Assessment of Damage in Mountain Tunnels due to the Taiwan Chi-Chi Earthquake. *Tunneling and Underground Space Technology*, Vol. 16, No. 3, pp. 133–150.

Wang, Y. (1996). Ground Response of Circular Tunnel in Poorly Consolidated Rock. *Journal of Geotechnical Engineering*, Vol. 122, No. 9, pp. 703–708.

Wanne, T. (2002). PFC^{3D} simulation procedure for compressive strength testing of anisotropic hard rock. *Numerical Modeling in Micromechanics via Particle Methods*. H. Konietzky, Editor. Balkema, Netherlands, pp. 241–249.

Wannenmacher, H., Heizmann, A., and Sabew, S. (2015). Operational analyses of pressure tunnel and shaft grouting operations. *EUROCK 2015 and 64th Geomechanics Colloquium*, Schubert and Kluckner, Editors. Salzburg, Austria. 7 pp.

Wanninger (-Huber), T. (2020). Experimental Investigations for the Modeling of Anhydritic Swelling Claystones. *Veröffentlichungen des Instituts für Geotechnik (IGT) der ETH Zürich*, Vol. 255, 321 pp.

Whyatt, J.K., Williams, T.J., and White, B.G. (2000). Ground conditions and the May 13, 1994 rock burst, Coeur d'Alene Mining District, northern Idaho. *Proceedings of the 4th North American Rock Mechanics Symposium, NARMS 2000*, J. Girard, M. Liebman, C. Breeds, and T. Doe, Editors. A.A. Balkema, Rotterdam, The Netherlands, pp. 313–318.

Wickham, G.E. and Tiedemann, H.R. (1972). *Research in Ground Support and its Evaluation for Coordination with Systems Analysis in Rapid Excavation*. Contract Report H0210038, U.S. Bureau of Mines (USBM), ARPA Program, NTIS Publication No. AD743100, 178 pp.

Wickham, G.E., Tiedemann, H.R., and Skinner, E.G. (1972). Support Determinations Based on Geologic Predictions. *Proceedings 1st RETC*, Chicago, IL. AIME, New York, pp. 43–64.

Wickham, G.E., Tiedemann, H.R., and Skinner, E.G. (1974a). *Ground Support Prediction Model – RSR Concept*. Contract Report H0220075, USBM, ARPA Program, NTIS Publication No. AD773018, 271 pp.

Wickham, G.E., Tiedemann, H.R., and Skinner, E.G. (1974b). Ground Support Prediction Model – RSR Concept. *Proceedings 2nd RETC*, San Francisco, CA. H.C. Pattison and E. D'Appolonia, Editors. AIME, New York, pp. 691–707.

Wiles, T.D. and Curran, J.H. (1982). General 3-d Displacement Discontinuity Method. *Proceedings of the 4th International Conference on Numerical Methods in Geomechanics, Edmonton, Canada.* Z. Eisenstein, Editor. Balkema, Rotterdam, pp. 103–111.

Williams, J.R. and Pentland, A.P. (1992). Superquadratics and Modal Dynamics for Discrete Elements in Interactive Design. *Engineering Computations*, Vol. 9, pp. 115–127.

Windsor, C.R. (1990). *Ferruled Strand.* Unpublished Memorandum. CSIRO, Perth, Australia.

Windsor, C.R. (1992). Cable bolting for underground and surface excavations. *Rock Support in Mining and Underground Construction. Proceedings of the International Symposium on Rock Support.* P.K. Kaiser and D.R. McCreath, Editors. A.A. Balkema, Rotterdam, The Netherlands, pp. 349–376.

Windsor, C.R. (1997). Rock Reinforcement Systems. *International Journal for Rock Mechanics and Mining Sciences*, Vol. 34, No. 6, pp. 919–951.

Windsor, C.R. (2001). Cable bolting. *Underground Mining Methods: Engineering Fundamentals and International Case Studies.* W.A. Hustrulid and R.L. Bullock, Editors. Society for Mining, Metallurgy and Exploration, Littleton, CO, pp. 555–561.

Windsor, C.R. and Thompson, A.G. (1993). Rock reinforcement – technology, testing, design and evaluation. *Comprehensive Rock Engineering. Principles, Practice & Projects.* J.A. Hudson, Editor. Pergamon Press, Oxford, UK, Vol. 4, pp. 451–484.

Windsor, C.R. and Thompson, A.G. (1999). The design of shotcrete linings for excavations created by drill and blast methods. *Rock Support and Reinforcement Practice in Mining, Proceedings of the International Symposium on Ground Support.* E. Villaescusa, C.R. Windsor, and A.G. Thompson, Editors. A.A. Balkema, Rotterdam, pp. 231–242.

Wittke, W. (1990). *Rock Mechanics - Theory and Application with Case Histories.* Springer, Berlin, Heidelberg, New York, Tokyo, 1075 pp.

Wittke, W. and Rissler, P. (1976). Dimensioning of the Lining of Underground Openings in Swelling Rock Applying the Finite Element Method. *Publication of the Institute for Foundation Engineering, Soil Mechanics, Rock Mechanics and Water Ways Construction*, RWTH (University), Aachen, Vol. 2, pp. 7–48.

Wittke, W., Wittke, M., Erichsen, C., Wittke-Schmitt, B., Wittke-Gattermann, P., and Schmitt, D. (2017). AJRM as Basis for Design and Construction of More Than 70 km of Tunnels of the Railway Project Stuttgart-Ulm. *Geomechanics and Tunnelling*, Vol. 10, No. 2, pp. 204–211.

Wondrad, M. and Chen, D. (2006). Application of mine seismic monitoring technology in mitigating geotechnical risks at Barrick's Darlot Gold Mine. *41st U.S. Rock Mechanics Symposium*, Golden, CO, 11 pages.

Wood, D.F. (1992). Specification and application of fibre reinforced shotcrete. *Rock Support in Mining and Underground Construction, Proceedings of the International Symposium on Rock Support.* P.K. Kaiser and D.R. McCreath, Editors. A.A. Balkema, Rotterdam, pp. 149–156.

Wu, J.H. (2007). Applying Discontinuous Deformation Analysis to Assess the Constrained Area of the Unstable Chiu-fen-erh-shan Landslide Slope. *International Journal for Numerical and Analytical Methods in Geomechanics*, Vol. 31, No. 5, pp. 649–666.

Wu, J.H., Ohnishi, Y., and Nishiyama, S. (2004). Simulation of the Mechanical Behavior of Inclined Jointed Rock Masses during Tunnel Construction Using Discontinuous Deformation Analysis (DDA). *International Journal on Rock Mechanics and Mining Sciences*, Vol. 41, pp. 731–743.

Wyllie, D.C. (1999). *Foundations on Rock.* E & FN Spon, New York, NY.

Wyllie, D.C. and Mah, C.W. (2004). *Rock Slope Engineering. Civil and Mining.* Spon Press, New York, NY.

Xie, H., Chen, Z., and Wang, J. (1999). Three-Dimensional Numerical Analysis of Deformation and Failure during Top Coal Caving. *International Journal on Rock Mechanics and Mining Sciences*, Vol. 36, pp. 651–658.

Xu, Z. (1992). *Applied Elasticity.* John Wiley and Sons, New York, NY.

Yang, Y. and Rosenbaum, M.S. (2002). The Artificial Neural Network as a Tool for Assessing Geotechnical Properties. *Geotechnical and Geological Engineering*, Vol. 20, pp. 149–168.

Yang, Y. and Zhang, Q. (1997). A Hierarchical Analysis for Rock Engineering Using Artificial Neural Networks. *Rock Mechanics and Rock Engineering*, Vol. 20, No. 4, pp. 207–222.

Yasitli, N.E. and Unver, B. (2005). 3D Numerical Modeling of Longwall Mining with Top-coal Caving. *International Journal of Rock Mechanics and Mining Sciences*, Vol. 42, pp. 219–235.

Yazici, S. and Kaiser, P.K. (1992). Bond Strength of Grouted Cable Bolts. *International Journal of Rock Mechanics, Mining Sciences and Geomechanics Abstracts*, Vol. 29, No. 3, pp. 279–292.

Yeung, M.R., Jiang, Q.H., and Sun, N. (2003). Validation of Block Theory and Three-Dimensional Discontinuous Deformation Analysis as Wedge Stability Analysis Methods. *International Journal of Rock Mechanics and Mining Sciences*, Vol. 40, pp. 265–275.

Yoo, C. (2005). Interaction between Tunneling and Groundwater – Numerical Investigation Using Three Dimensional Stress-Pore Pressure Coupled Analysis. *Journal of Geotechnical and Geoenvironmental Engineering*, Vol. 131, No. 2, pp. 240–250.

Yoo, C. and Shin, H.-K. (2003). Deformation Behaviour of Tunnel Face Reinforced with Longitudinal Pipes – Laboratory and Numerical Investigation. *Tunnelling and Underground Space Technology*, Vol. 18, pp. 303–319.

Young-Su, K. and Byung-Tak, K. (2006). Use of Artificial Neural Networks in the Prediction of Liquefaction Resistance of Sands. *Journal of Geotechnical and Geoenvironmental Engineering*, Vol. 132, No. 11, pp. 1502–1504.

Yu, H.-S. (2000). *Cavity Expansion Methods in Geomechanics*. Kluwer Academic Publishers Group, The Netherlands.

Yu, H., Chen, J., Bobet, A., and Yuan, Y. (2016). Damage Observation and Assessment of the Longxi Tunnel during the Wenchuan Earthquake. *Tunnelling and Underground Space Technology*, Vol. 54, pp. 102–116.

Yu, H., Yuan, Y., and Bobet, A. (2017). Seismic Analysis of Long Tunnels: A Review of Simplified and Unified Methods. *Underground Space*, Vol. 2, No. 2, pp. 73–87.

Yu, T.Z. and Xian, C.J. (1983). Behaviour of rock bolting as tunneling support. *Rock Bolting. Theory and Application in Mining and Underground Construction. Proceedings of the International Symposium on Rock Bolting*. O. Stephansson, Editor. A.A. Balkema, Rotterdam, The Netherlands, pp. 87–92.

Yuan, Y., Jiang, X., and Lee, C.F. (2000). Tunnel Waterproofing Practices in China. *Tunnelling and Underground Space Technology*, Vol. 15, No. 2, pp. 227–233.

Zeller, S.S. and Pollard, D.D. (1992). Boundary Conditions for Rock Fracture Analysis Using the Boundary Element Method. *Journal of Geophysical Research*, 97, No. B2, 1991–1997.

Zhang, G.X., Zhu, B.F., and Lu, Z.C. (2003). Cracking Simulation of the Wuqiangxi Ship Lock by Manifold Method. *Development and Application of Discontinuum Modelling for Rock Engineering. Proceedings of the 6th International Conference on Analysis of Discontinuous Deformation*, M. Lu, Editor. A.A. Balkema Publishers, The Netherlands, pp. 133–139.

Zhang, H.W. and Zhou, L. (2006). Numerical Manifold Method for Dynamic Nonlinear Analysis of Saturated Porous Media. *International Journal for Numerical and Analytical Methods in Geomechanics*, Vol. 30, pp. 927–951.

Zhang, L. and Einstein, H.H. (2004). Using RQD to Estimate the Deformation Modulus of Rock Masses. *International Journal of Rock Mechanics and Mining Sciences*, Vol. 41, No. 2, pp. 337–341.

Zhang, P., Wu, H.-N., Chen, R.-P., Dai, T., Meng, F.-Y., and Wang, H.-B. (2020). A Critical Evaluation of Machine Learning and Deep Learning in Shield-ground Interaction Prediction. *Tunnelling and Underground Space Technology*, Vol. 106. doi: 10.1016/j.tust.2020.103593.

Zhang, X., Lu, M., and Wegner, J.L. (2000). A 2-D Meshless Model for Jointed Rock Structures. *International Journal for Numerical Methods in Engineering*, Vol. 47, pp. 1649–1661.

Zhang, Z. and Kieffer, S. (2020). *Fundamentals of Shield Tunnelling: Design and Construction*. CRC Press, Great Britain.

Zhou, W.Y., Lin, P., Yang, R.Q., and Yang, Q. (2003). A Comparison of Dam Fracture Studies between Physical Model Tests and Numerical Analysis. *Development and Application of Discontinuum Modelling for Rock Engineering. Proceedings of the 6th International Conference on Analysis of Discontinuous Deformation*, M. Lu, Editor. A.A. Balkema Publishers, The Netherlands, pp. 129–132.

Zhu, W., Li, S., Li, S., Chen, W., and Lee, C.F. (2003). Systematic Numerical Simulation of Rock Tunnel Stability Considering Different Rock Conditions and Construction Effects. *Tunnelling and Underground Space Technology*, Vol. 18, pp. 531–536.

Zipf, K. (2007). Numerical Modeling Procedures for Practical Coal Mine Design. *Proceedings of the International Workshop on Rock Mass Classification in Underground Mining*. C. Mark, R. Pakalnis, and R.J. Tuchman, Editors. DHHS (NIOSH) Publication No. 2007-128, pp. 158–167.

Author Index

Pages in *italics* refer to figures and pages in **bold** refer to tables.

For Product Safety Concerns and Information please contact our EU
representative GPSR@taylorandfrancis.com
Taylor & Francis Verlag GmbH, Kaufingerstraße 24, 80331 München, Germany

www.ingramcontent.com/pod-product-compliance
Lightning Source LLC
LaVergne TN
LVHW081312110826
845149LV00006B/1492

* 9 7 8 1 0 3 2 3 5 8 4 5 1 *